Advances in Experimental Medicine and Biology
Volume 730

For further volumes:
http://www.springer.com/series/5584

Arthur N. Popper · Anthony Hawkins
Editors

The Effects of Noise on Aquatic Life

Springer

Editors
Arthur N. Popper
Department of Biology
University of Maryland
College Park, MD, USA
apopper@umd.edu

Anthony Hawkins
Loughine Limited
Kincraig, Blairs Aberdeen, UK
a.hawkins@btconnect.com

ISBN 978-1-4419-7310-8 e-ISBN 978-1-4419-7311-5
DOI 10.1007/978-1-4419-7311-5
Springer New York Dordrecht Heidelberg London

Library of Congress Control Number: 2011938466

Printed on acid-free paper

Springer is part of Springer Science+Business Media (www.springer.com)

This book is dedicated to our wives
Helen Popper and Susan Hawkins

We are grateful to them for putting up with us while we planned and organized the conference. Without their support, as well as their critical input in planning, the meeting could not have happened.

Tony Hawkins and Art Popper

Preface

These proceedings are the extended abstracts of the papers presented at the 2010 Second International Meeting on the Effects of Noise on Aquatic Life that took place in August in Cork, Ireland. The meeting brought together 248 scientists, regulators, and representatives from industry and environmental groups, representing 21 countries from all continents, to hear papers and discuss a broad range of topics focused on underwater sound and its effects on organisms living in the aquatic environment. This meeting followed from the immensely successful first conference that took place in 2007 in Nyborg, Denmark. The Cork meeting was, if anything, more successful than the first meeting in bringing people with different interests and experiences together and in allowing them to get to know one another, learn about new findings, and interact very successfully.

The basis for the first two meetings, and the third which will be held in Europe in August 2013, arises from concern that has been growing since at least the early 1990s. Humans are adding substantial noise to the aquatic environment, and this noise might have an impact on the quality of life, and even the survival, of aquatic organisms. While the original concern focused on marine mammals, this has changed in more recent years to the point where equal emphasis is now being placed upon fish and, most recently, on invertebrates. Indeed, while fish and marine mammals were well represented in Nyborg, with no representation of invertebrates, we are most pleased that a number of outstanding papers were presented in Cork on invertebrates.

The Cork conference was packed with papers; so many that we extended the idea, first tested at Nyborg, of having groups of rapid-fire presentations in the evening. These were as successful in arousing interest and provoking comment as the longer presentations given at the conference. We can have no doubt that the subject of underwater noise and its impact has come of age and that a community of people with strong interest in this topic has now been formed. We are planning to hold this community together in the period leading up to our next conference.

Arthur N. Popper, College Park, MD
Anthony Hawkins, Aberdeen, Scotland

Acknowledgments

The editors and the conference attendees are immensely grateful to the following agencies and organizations for providing the financial assistance that enabled us to support attendance of a large number of delegates at the Conference and to publish this volume.

Funding Agencies

Bureau of Ocean Energy Management, Regulation and Enforcement (USA)
Fisheries and Oceans (Arctic and Atlantic Divisions, Canada)
Marine Mammal Commission (USA)
National Oceanographic and Atmospheric Administration (USA)
National Science Foundation (USA) Grant No. 0952925
Office of Naval Research – Global (UK)
Office of Naval Research (USA)

Companies

ExxonMobil (USA)
Greenridge Associates (USA)
JASCO (Canada)
Loughine Ltd (UK)
Oil and Gas Producers Association (OGP)
SIMRAD (Norway)

Organizations

Acoustical Society of America
Fáilte Ireland
College of Chemical and Life Sciences & The Center for Comparative and Evolutionary Biology of Hearing, University of Maryland (USA)

The editors/organizers would also like to express their gratitude to a large number of people who have provided help in planning and executing the meeting.

We start with our wives, Susan Hawkins and Helen Popper, for substantial and invaluable guidance in selecting the meeting site and planning the non-science activities for the meeting. We also thank Helen Popper for the many hours spent in editing the manuscripts in preparation for publication.

We are grateful to Erica Casper for tireless efforts in keeping track of delegates, manuscripts, and other items.

The logistics and organization of the meeting in Cork, and the running of the meeting, could not have been carried out without the exceptional guidance, wisdom, and help of the staff of Venues Ireland. In particular, we acknowledge the enormous contributions of Terri Cullinane, Triona Dillon, and Fiona McGarry. We do not know what we would have done without their help.

Charlene Hurley, the Conference and Events Executive of the Clarion Hotel Cork, rendered invaluable assistance in arranging facilities for the conference at the hotel.

We thank Sara Therrien (University of Maryland) and Jenni Stanley (University of Auckland) for organization of the student participants helping with the meeting, and the many students in attendance who agreed to help during various parts of the meeting.

We also thank Patty Selby of the Department of Biology at the University of Maryland for her meticulous work to ensure that our funding was managed properly and that we were in the position to reimburse delegates getting support in the most efficient manner. Mr. James Parker, also of the Department of Biology, has been of major assistance in arranging the dispersal of funds.

We are grateful to Ann Avouris, our editor at Springer Science + Business Media, for showing great flexibility and resourcefulness as we worked to develop ways to rapidly produce a very high quality book as the work product of this meeting.

Finally, we want to express greatest gratitude and thanks to Roger Gentry, Marta Picciulin, Mark Tasker, and Doug Wartzok for serving on our steering committee and for providing invaluable insights and guidance as the meeting evolved. We also are most grateful to Elizabeth Burkhard, Peter Cott, Tom Carlson, Douglas Cato, John Dalen, Bob Gisiner, Bill Lang, Jennifer Miksis-Olds, Brandon Southall, Amy Scholik-Schlomer, Alexander Supin, Magnus Wahlberg, and Lidia (Eva) Wysocki for serving on our advisory committee and for being willing to provide their guidance during the development of the program.

Some Views of the Meeting

Susan Hawkins, Cork Lord Mayor Michael O'Connell, Tony Hawkins, and Colleen Le Prell

Michele Halvorsen, Thor Halvorsen
(the youngest attendee)

Some of the members of the organizing committee: Pete Cott, William Lang, Doug Wartzog, Roger Gentry, Mark Tasker, Tom Carlson, Jennifer Miksis-Olds, Arthur Popper, John Dalen, Tony Hawkins

Organizers Tony Hawkins and Arthur Popper along with Triona Dillon from Venues Ireland

Bruce Mate, Bill Ellison, Roger Gentry, Colleen LePrell

Meeting site at the Clarion Hotel, Cork

(forefront) Brandon Casper, Mark Tasker, Paul Nachtigall, Colleen LePrell

Carl Schilt, Tony Hawkins

Bob Dooling, Art Popper, Pete Rogers, Tony Hawkins, Dick Fay

River Lee in Cork

Brandon and Erica Casper, Mike and Tarralei Smith enjoying closing "shindig"

Jameson Brewery, site of the closing "shindig"

Contents

Part VI Population Effects

Part VII Anthropogenic Sound Sources and Their Measurement

Part I
Introduction

Noise and Marine Life: Progress From Nyborg to Cork in Science and Technology to Inform Decision Making

Brandon L. Southall

1 Introduction

The 1st International Conference on *The Effects of Noise on Aquatic Life* in Nyborg, Denmark, in August 2007 was a watershed event. There have been numerous sessions of scientific and professional conferences devoted to this topic, but the Nyborg meeting was unique in being specifically focused on environmental noise impacts while considering all aquatic species. It was also remarkable in having a truly international scope and participation and, perhaps most notably, in serving as an interdisciplinary nexus for scientists, regulators, policy makers, industry representatives, and resource sponsors. The subsequent period leading to this follow-on meeting in Cork, Ireland, in August 2010 saw remarkable advances in understanding the effects of human noise on aquatic life and how data are used to predict and minimize the environmental impacts.

With the rapid industrialization of many marine areas and increasing recognition of how we are impacting our ocean, there is an ever greater need for such interaction. The 2nd International Conference on *The Effects of Noise on Aquatic Life* and the new collaborations, revelations, and innovations it brings serves as a reminder of the opportunities and obligation for each of us to continue to make real and collective progress at an important for this international issue.

Globally, we are seeing 1) an emerging appreciation of how shipping noise and other industrial sounds affect the acoustic environment of biologically important areas and how this may change with shifting climate and ocean chemistry; 2) the expansion of conventional and alternative energy technologies into new environments with a host of acoustic sources and various potential impacts on aquatic life; 3) an increasing understanding of how intense sound sources (e.g., seismic air gun surveys and active military sonar systems) can affect animals; and 4) an overarching recognition that sublethal impacts (e.g., communication masking and significant behavioral responses) from chronic acoustic sources are likely the most important considerations for populations of animals, particularly as they interact with other stressors such as overfishing, habitat loss, entanglement, and pollution (see Arctic Council 2009; Bailey et al. 2010; Clark et al. 2009; European Wind Energy Association 2008; Nowacek et al. 2007; Reeves et al. 2005; Southall et al. 2007, 2008, 2009; US Department of Energy 2009; US National Marine Fisheries Service 2010; Van Parijs et al. 2009; Wright 2009; Wright and Highfill 2007).

B.L. Southall (✉)
Southall Environmental Associates, Inc., Santa Cruz, CA 95060, and Long Marine Laboratory, University of California, Santa Cruz, Santa Cruz, CA 95064, USA
e-mail: Brandon.Southall@sea-inc.net

A.N. Popper and A. Hawkins (eds.), *The Effects of Noise on Aquatic Life*, Advances in Experimental Medicine and Biology 730, DOI 10.1007/978-1-4419-7311-5_1,

Consequently, there is an increasing need for integrated and practically relevant research and data synthesis. We also require the consistent and transparent, yet sufficiently adaptable, application of current scientific knowledge in conservation management decisions. The research recommendations from National Research Council (2000, 2003, 2005), International Council for the Exploration of the Sea Ad Hoc Group on the Impact of Sonar on Ceatceans and Fish (2005), and Southall et al. (2007) remain largely relevant. However, several recent syntheses of more applied data needs have been presented by the European Science Foundation (Boyd et al. 2008) and by a US federal interagency task force on marine noise impacts (Southall et al. 2009). Many of the issues raised in these assessments are in fact being addressed in some of the areas described below, but a vast amount of work remains to be done.

2 Effects of Noise on Aquatic Life: Advances in Science and Technology

This brief manuscript does not attempt to cover specific progress in each area of science and technology regarding the effects of noise on aquatic life; an adequate treatment would require review of essentially all of the science presented in Cork and elsewhere. Rather, general trends and developments are considered briefly in several key areas here, with reference to other papers in this volume (see also papers associated with Southall and Nowacek 2009).

2.1 Acoustic Monitoring and Detection

One of the most significant areas of recent progress is in the use of evolving technologies to detect vocalizing aquatic animals and characterize natural and human contributions to the ambient acoustic environment on meaningful spatial scales (e.g., Hatch et al. 2008; McDonald et al. 2006; Van Parijs et al. 2009) as well as how they may impact vocal communication (e.g., Clark et al. 2009). Recent developments in the use of passive and active acoustic monitoring technologies around offshore industrial applications were reviewed in an interactive forum convened in November 2009 by the US Bureau of Ocean Energy Management. Important new directions in this expanding area of science and technology are discussed throughout this volume.

2.2 Measurements of Behavior and Responses to Sound

Some concerns remain regarding the effects of noise on auditory and other physiological systems, and there has been progress in these areas (discussed below). However, there is increasing recognition that issues related to behavioral responses and interference with communication are likely much more common and potentially the most important effects of noise for populations of aquatic animals.

Consequently, we have seen a rapid increase in both opportunistic measurements of behavior around realistic operations (e.g., DiMarzio et al. 2008) and field and laboratory controlled-exposure studies to measure responses (see Tyack 2009). Both opportunistic and experimental methods have advantages and disadvantages. An integrated approach, increasingly involving full-scale rather than scaled sound sources, is emerging as the most comprehensive and rigorous approach. Many of these methods have been aided by accelerating progress in the application of tagging and tracking technologies (e.g., Bailey et al. 2010; Johnson et al. 2009; Mate et al. 2007). This is an expansive area of current study in many countries.

2.3 Hearing and Effects of Noise on Hearing and Physiology in Aquatic Animals

We can simultaneously marvel at the remarkable progress in measuring basic hearing capabilities of aquatic animals using both behavioral and physiological tools and be overwhelmed by how many of the extant species (and entire taxa) remain completely unknown. Significant developments include recent research on hearing and noise impacts in nonmammalian vertebrates (e.g., Popper et al. 2007) as well as in some invertebrates (see Mooney et al., Chapter 28), advances in the use of electrophysiological techniques for measuring the effects of noise on hearing (see papers in André and Nachtigall 2007), and measurements related to potential physiological effects of noise exposure (e.g., Houser et al. 2009). This also is an active and key area of current science and technology as shown in many papers in this volume.

2.4 Reducing Acoustic Footprints: Advances in Quieting Technologies

Concomitant with advances in understanding hearing and the effects of noise on aquatic life and the use of sound to characterize marine environments has been the recognition that simply minimizing the noise associated with certain human activities is often possible, logical, and even beneficial (see Spence et al. 2007; Weilgart 2010).

We are beginning to see international progress in quieting technologies within some industries such as commercial shipping. We have seen a fairly rapid progression of this global issue from the initial workshops on this issue (see Southall and Scholik-Schlomer 2008; Wright 2008) and resulting follow-on efforts currently underway within the International Maritime Organization to engage the international shipping industry in implementing vessel quieting technologies (see United States 2008, 2009).

3 Advances in Assessing Acoustic Impacts

The Southall et al. (2007) noise-exposure criteria were a significant development in predicting how noise affects marine mammals (the taxonomic scope was more restricted than for the Nyborg/Cork meetings), but they were admittedly limited in many ways. They called specifically for targeted research in key areas and an increased sophistication in management decisions regarding noise impact. Significant progress has been made in both the former (see above) and latter areas.

Notably, building on the population consequences of acoustic disturbance (PCAD) conceptual framework (National Research Council 2005), an expert working group is applying empirical data on behavior and energetics in selected taxa to derive more biologically meaningful ways of assessing potential impacts on populations (see Costa, Chapter 96). Additionally, the integrated and potentially interacting impacts of chronic and/or multiple acoustic events and other stressors on aquatic species are now being systematically considered (e.g., Wright 2009) and will benefit from progress made in terrestrial systems (see Barber et al. 2010; Leu et al. 2008). Finally, we are seeing evolutions of thinking toward more sophisticated ways of assessing behavioral responses involving the integration of multiple, covarying contextual and acoustic variables associated with sound exposure (e.g., Southall et al. 2010).

4 Policy and Management Applications of Science on Noise Impacts: Progress and Challenges

As science continues to reveal the complexity and context specificity of noise impacts in aquatic species, policy makers and managers are faced with increasingly daunting challenges in establishing consistent and transparent regulations. The criteria recommendations of Southall et al. (2007) are beginning to be considered and, in some ways, applied in predictions of acoustic impacts (e.g., Tasker, Chapter 28; US Department of the Navy 2008) but perhaps more slowly and less comprehensively than some expected. As noted by the authors, however, the application of those recommendations is complicated by the myriad of legal, political, economic, and other practical realities in the decision-making processes across many jurisdictions. However, it is also worth reiterating that existing data on behavioral responses for many species do not converge on a simplistic received sound exposure level(s) as a reliable predictor of negative effects.

Thus, regulatory agencies are faced with the realization that science-based assessments of impact will require a more sophisticated approach than the simplistic descriptions and predictors that have largely being used (see Bejder et al. 2009). Decision makers must also increasingly consider noise in the context of marine spatial planning and regional management frameworks (see Hatch and Fristrup 2009) as well as the cumulative and interacting effects of multiple stressors. Although these challenges are seemingly overwhelming for managers striving for practical and science-based approaches to conservation management in the face of biological and legislative complexity, significant remaining uncertainty, and ever-present risk of litigation, we are seeing some progress in policies and decision making as well. Examples include the derivation of criteria for noise impacts within the European Union Marine Strategy Framework Directive (Tasker, Chapter 28) and acoustic exposure guidelines within the US regulatory process (see Johnson, Chapter 127; Scholik-Schlomer, Chapter 126).

5 Future Directions

The progress made in our ability to measure, understand, and apply data on noise impacts on aquatic life from Nyborg to Cork has been truly remarkable. However, these advances come as we are increasingly aware of a host of overwhelming environmental challenges. These global issues, including overfishing, habitat loss due to the industrialization of coastal ecosystems, climate change, ocean acidification, likely render noise impacts a second-order human impact on aquatic species. Nevertheless, it remains an important consideration, perhaps most importantly as it interacts with these other stressors placed on marine and freshwater ecosystems. Furthermore, some of the developments in acoustic tools and technologies are proving integral in measuring and understanding the impacts of other human impacts on protected species. Our ability to respond and adapt to current and future challenges regarding noise and other impacts depends on sustained progress in at least the following topical areas:

1. Integrating passive acoustic technologies into ocean-observing systems (Dunshaw et al. 2009; Southall et al. 2009) and regional management frameworks (Hatch and Fristrup 2009);
2. Sustained research and development in quantifying impacts of acoustic masking in biologically meaningful and realistic scenarios (see Clark et al. 2009; Reichmuth, Chapter 4);
3. Developing context-specific means of predicting individual impacts and empirically based means of quantifying population consequences of acoustic disturbance (see Costa, Chapter 96);
4. Developing and implementing consistent and transparent, yet adaptable, regulatory policies regarding behavioral impacts based on insights into biological significance of noise impacts;

5. Continuing to support interdisciplinary, integrated research programs where multimodal time series data streams are integrated to assess cumulative and ecosystem level impacts; and
6. Ensuring public awareness and understanding of the biology of acoustic impacts on aquatic species using visualizations of acoustic concepts and on-line and multimedia resources (see Vigness-Raposa et al., Chapter 30) and/or nontraditional media (e.g., social networking resources).

To some it may seem as if we have been working on understanding noise impacts on aquatic life for a very long time. To others, it may seem as if we are just beginning our collective journey. In some ways, both are true, but we as a community should recognize that, like all environmental issues, the noise issue has a time horizon where research funding and public interest will peak. While decades more work are clearly needed and likely will occur in some, particularly applied, areas, it is likely that we are at or near that apex. There are daunting opportunities and obligations for our field associated with that conclusion.

As we move forward from the new conclusions and collaborations made in Cork, how will we respond to these challenges? Are we making the best use of limited resources to address the most pertinent challenges facing us now while ensuring that we do not lose focus on understanding basic principles needed to address issues yet to come? How will we continue to advance our collective fields and be judged by future generations at this pivotal period?

References

André M, Nachtigall PE (2007) Electrophysiological measurements of hearing in marine mammals. Aquat Mamm 33:1–6.

Arctic Council (2009) Arctic marine shipping assessment 2009 report. Arctic Council, Akureyri, Iceland.

Bailey H, Mate BR, Palacios DM, Irvine L, Bograd SJ, Costa DP (2010) Behavioural estimation of blue whale movements in the Northeast Pacific from state-space model analysis of satellite tracks. Endang Species Res 10:93–106.

Barber JR, Crooks KR, Fristrup KM (2010) The costs of chronic noise exposure for terrestrial organisms. Trends Ecol Evol 25:180–189.

Bejder L, Samuels A, Whitehead H, Finn H, Allen S (2009) Impact assessment research: Use and misuse of habituation, sensitisation and tolerance in describing wildlife responses to anthropogenic stimuli. Mar Ecol Prog Ser 395:177–185.

Boyd IL, Brownell R, Cato D, Clark C, Costa D, Evans P, Gedamke J, Gentry R, Gisiner R, Gordon J, Jepson P, Miller P, Rendell L, Tasker M, Tyack P, Vos E, Whitehead H, Wartzok D, Zimmer W (2008) The effects of anthropogenic sound on marine mammals: A draft research strategy. Marine Board, European Science Foundation, Oxford, UK.

Clark CW, Ellison WT, Southall BL, Hatch L, Van Parijs SM, Frankel A, Ponirakis D (2009) Acoustic masking in marine ecosystems: intuitions, analysis, and implication. Mar Ecol Prog Ser 395:201–222.

DiMarzio N, Moretti D, Ward J, Morrissey R, Jarvis S, Izzi A (2008) Passive acoustic measurement of dive vocal behavior and group size of Blainville's beaked whale (*Mesoplodon densirostris*) in the Tongue of the Ocean (TOTO). Canad Acoust 36:166–173.

Dunshaw BW, Au WWL, Beszczynska-Möller A, Brainard R, Cornuelle BD, Duda T, Forbes MDA, Freitag L, Gascard J-C, Gavrilov A, Gould J, Howe B, Jayne SR, Johannessen OM, Lynch JF, Martin D, Menemenlis D, Mikhalevsky P, Miller JH, Moore SE, Munk WH, Nystuen J, Odom RI, Orcutt J, Rossby T, Sagen H, Sandven S, Simmen J, Skarsoulis E, Southall BL, Stafford K, Stephen R, Vigness-Raposa KJ, Vinogradov S, Wong KB, Worcester PF, Wunsch C. (2009) A global ocean acoustic observing network. Proceedings of the OceanObs '09 Symposium, 21–25 September 2009, Venice, Italy.

European Wind Energy Association (2008) Delivering offshore wind power in Europe: Policy recommendations for large-scale deployment of offshore wind power in Europe by 2020. European Wind Energy Association, Brussels, Belgium.

Hatch L, Clark C, Merrick R, Van Parijs S, Ponirakis D, Schwehr K, Thompson M, Wiley D (2008) Characterizing the relative contributions of large vessels to total ocean noise fields: A case study using the Gerry E. Studds Stellwagen Bank National Marine Sanctuary. Environ Manage 42:735–752.

Hatch LT, Fristrup KM (2009) No barrier at the boundaries: implementing regional frameworks for noise management in protected natural areas. Mar Ecol Prog Ser 395:223–244.
Houser DS, Dankiewicz-Talmadge LA, Stockard TK, Ponganis PJ (2009) Investigation of the potential for vascular bubble formation in a repetitively diving dolphin. J Exp Biol 213:52–62.
International Council for the Exploration of the Sea Ad Hoc Group on the Impact of Sonar on Cetaceans and Fish (2005) Ad-hoc group on the impact of sonar on cetaceans. International Council for the Exploration of the Seas, Copenhagen, Denmark.
Johnson M, Aguilar de Soto N, Madsen PT (2009) Studying the behaviour and sensory ecology of marine mammals using acoustic recording tags: A review. Mar Ecol Prog Ser 395:55–73.
Leu M, Hanser SE, Knick ST (2008) The human footprint in the west: A large-scale analysis of anthropogenic impacts. Ecol Appl 18:1119–1139.
Mate B, Mesecar R, Lagerquist B (2007) The evolution of satellite-monitored radio tags for large whales: One laboratory's experience. Deep-Sea Res Part II Top Stud Oceanogr 54:224–247.
McDonald MA, Hildebrand JA, Wiggins SM (2006) Increases in deep ocean ambient noise in the Northeast Pacific west of San Nicolas Island, California. J Acoust Soc Am 120:711–718.
National Research Council (2000) Marine mammals and low-frequency sound. National Academy Press, Washington, DC.
National Research Council (2003) Ocean noise and marine mammals. National Academy Press, Washington, DC.
National Research Council (2005) Marine mammal populations and ocean noise: Determining when noise causes biologically significant events. National Academy Press, Washington, DC.
Nowacek DP, Thorne LH, Johnston DW, Tyack PL (2007) Responses of cetaceans to anthropogenic noise. Mamm Rev 37:81–115.
Popper AN, Halvorsen MB, Kane A, Miller DL, Smith ME, Song J, Stein P, Wysocki LE (2007) The effects of high-intensity, low-frequency active sonar on rainbow trout. J Acoust Soc Am 122:623–635.
Reeves RR, Brownell RL Jr, Burdin A, Cooke JG, Darling JD, Donovan GP, Gulland FMD, Moore SE, Nowacek DP, Ragen TJ, Steiner RG, VanBlaricom GR, Vedenev A, Yablokov AV (2005) Report of the Independent Scientific Review Panel on the impacts of Sakhalin II phase 2 on western North Pacific gray whales and related biodiversity. International Union for the Conservation of Nature, Gland, Switzerland.
Southall B, Berkson J, Bowen D, Brake R, Eckman J, Field J, Gisiner R, Gregerson S, Lang W, Lewandoski J, Wilson J, Winokur R (2009) Addressing the effects of human-generated sound on marine life: An integrated research plan for US federal agencies. Interagency Task Force on Anthropogenic Sound and the Marine Environment of the Joint Subcommittee on Ocean Science and Technology, Washington, DC.
Southall BL, Bowles AE, Ellison WT, Finneran JJ, Gentry RL, Greene CR Jr, Kastak D, Ketten DR, Miller JH, Nachtigall PE, Richardson WJ, Thomas JA, Tyack PL (2007) Marine mammal noise exposure criteria: Initial scientific recommendations. Aquat Mamm 33:411–522.
Southall BL, Frankel AS, Ellison WT (2010) Predicting acoustic impact: Considering individuals versus populations. J Acoust Soc Am 127:1802.
Southall BL, Nowacek DP (2009) Acoustics in marine ecology: innovation in technology expands the use of sound in ocean science. Mar Ecol Prog Ser 395:1–3.
Southall BL, Scholik-Schlomer A (2008) Potential application of vessel-quieting technology on large commercial vessels. Final Report of the National Oceanic and Atmospheric Administration International Conference, 1-2 May 2007, Silver Spring, MD.
Spence J, Fischer R, Bahtiarian M, Boroditsky L, Jones N, Dempsey R (2007) Review of existing and future potential treatments for reducing underwater sound from oil and gas industry activities. NCE Report 07-001, prepared by Noise Control Engineering for Joint Industry Programme on E&P Sound and Marine Life.
Tyack P (2009) Acoustic playback experiments to study behavioral responses of free-ranging marine animals to anthropogenic sound. Mar Ecol Prog Ser 395:187–200.
United States (2008) Minimizing the introduction of incidental noise from commercial shipping operations into the marine environment to reduce potential adverse impacts on marine life. US Submission to the 58th Session of the Marine Environment Protection Committee, International Maritime Organization [MEPC 58/19].
United States (2009) Noise from commercial shipping operations and its impacts on marine life. US Submission to the 59th Session of the Marine Environment Protection Committee, International Maritime Organization [MEPC 59/19].
US Department of Energy (2009) Report to Congress on the potential environmental effects of marine and hydrokinetic energy technologies. Prepared by the Department of Energy Wind and Hydropower Technologies Program in Response to the Energy Independence and Security Act of 2007, Section 633(B).
US Department of the Navy (2008) Southern California range complex final environmental impact statement/overseas environmental impact statement. Department of the Navy, San Diego, CA.
US National Marine Fisheries Service (2010) Notice of intent to prepare an environmental impact statement on the effects of oil and gas activities in the Arctic Ocean. US Federal Register 75 (25):6175– 6177, RIN 0648–XU06,

National Marine Fisheries Service, National Oceanic and Atmospheric Administration, US Department of Commerce, Silver Spring, MD. Available via http://www.nmfs.noaa.gov/pr/permits/eis/arctic.htm.

Van Parijs SM, Clark CW, Sousa-Lima RS, Parks SE, Rankin S, Risch D, Van Opzeeland IC (2009) Management and research applications of real-time and archival passive acoustic sensors over varying temporal and spatial scales. Mar Ecol Prog Ser 395:21–36.

Weilgart LS (ed) (2010) Report of the workshop on alternative technologies to seismic airgun surveys for oil and gas exploration and their potential for reducing impacts on marine mammals. Okeanos-Foundation for the Sea, 31 August–1 September 2009, Monterey, CA. Available via http://www.sound-in-the-sea.org/download/AirgunAlt2010_en.pdf.

Wright AJ (ed) (2008) International workshop on shipping noise and marine mammals. Okeanos-Foundation for the Sea, 21–24 April 2008, Hamburg, Germany, Available via http://www.sound-in-the-sea.org/download/ship2008_en.pdf.

Wright AJ (ed) (2009) Report of the workshop on assessing the cumulative impacts of underwater noise with other anthropogenic stressors on marine mammals: From ideas to action. Okeanos-Foundation for the Sea, 26–29 August 2009, Monterey, CA. Available via http://www.sound-in-the-sea.org/download/CIA2009_en.pdf.

Wright AJ, Highfill L (2007) Considerations of the effects of noise on marine mammals and other animals. Int J Comp Psychol 20:89–316.

Listening Backward: Early Days of Marine Bioacoustics

William N. Tavolga

In 1953, the eminent undersea explorer Jacques Cousteau published *The Silent World*. This small volume was to become the generator of a huge upwelling of research on life in the ocean. The title, however, was a misnomer. The book came out years after the end of World War II (WWII) and the concomitant release of data and equipment from the Navy on the plethora of sounds produced by marine organisms. Probably the reason Cousteau could not hear the sounds was because the human ear is just not built for underwater hearing. Besides, the small amount of animal sound that got to his ears was masked by the bubbling of his breathing apparatus. In fact, information on sounds of fishes and cetaceans was available long before World War I. Moby Dick-era whalers listened for the songs of humpback whales through the hulls of their wooden ships. Historical details can be found elsewhere (Tavolga 1971a).

Antisubmarine warfare (ASW) was a major effort during WWII and continues to the present day. The techniques have been based primarily on acoustics, and the presence of what were called "biologicals" was often distracting and confusing. As wartime instrumentation became declassified and available, it was discovered that the seas are truly noisy places, with a cacophony of snapping shrimp, spawning fishes, and echolocating dolphins.

A landmark piece of research in the 1950s was the work of Dr. Marie Poland Fish (Fish 1954; Fish et al. 1952). She reported that a large number of fish species were capable of producing sounds. She and her co-workers literally "auditioned" individual fish in aquarium tanks by stimulating them with the equivalent of an electrical cattle prod. Unfortunately, it turned out that many of the sounds emitted were from violent muscle contractions, not sounds that the animals produced during normal behavior. Nevertheless, the basis for further study in marine bioacoustics was established.

The precise relationship of sounds to behavior is not always easy to determine. In some cases, the choruses of spawning fish (mainly sciaenids) can be heard above water, and such events have been known to fishermen since time immemorial. In most cases, however, the behavioral significance of fish sounds has required underwater listening equipment and careful observation. A combined video and audio system was installed off Bimini, Bahamas, in the 1960s by the Rosensteil Laboratory of the University of Miami, Coral Gables, FL. The combination of visual and acoustic observations contributed a large body of data on sonic fishes (Cummings et al. 1964; Steinberg and Koczy 1964). In my own studies on the frill-finned goby (*Bathygobius soporator*), I found that

W.N. Tavolga (✉)
Mote Marine Laboratory, Sarasota, FL 34236, USA
e-mail: tavolga@mote.org

A.N. Popper and A. Hawkins (eds.), *The Effects of Noise on Aquatic Life*,
Advances in Experimental Medicine and Biology 730, DOI 10.1007/978-1-4419-7311-5_2,

successful courtship and spawning involved a complex of acoustic, visual, and chemical stimuli (Tavolga 1956, 1958).

Of all marine animals, dolphins were (and still remain) most prominent in bioacoustic research. When the first captive and trained dolphins were exhibited in the 1940s at Marineland in Florida, all were immediately intrigued by the huge variety of sounds they produced, from high-pitched whistles to low buzzes. One well-known physiologist even proposed that dolphins could imitate human speech and had a vocal language of their own (Lilly 1961). The careful research by Schevill (1964) and Norris (1964) eventually separated the communication whistles from the trains of clicks that were used in echolocation.

The echolocating pulses of dolphins were found to be highly directional, and this was discovered to be a function of the oil-filled bulbous forehead, the so-called "melon" (Norris and Evans 1967; Norris et al. 1961). Norris's research, however, was preceded by a series of unpublished tests using dolphin heads and primitive audio equipment. The credit for these tests and the original idea for echolocation and melon function should go to the late F. G. Wood, Jr., then of Marineland in Florida. Indeed, the directional character of echolocation in dolphins is responsible for the high resolving power and range of this remarkable facility.

Echolocation was first discovered in bats and subsequently in several other mammalian species, including delphinid cetaceans. Short pulse length and high (ultrasonic) frequency enables the finest resolution and identification of targets. However, low-frequency sounds can serve an echolocation function also. Evidence was presented to demonstrate that the low-pitched sound bursts in the sea catfish (*Arius felis*) could serve both a social function in schooling and a coarse version of echolocation (Tavolga 1971b, 1976).

An interesting case was that of the curious very low pitched sounds detected by ASW stations. These were powerful pulses with a dominant frequency of ~20 Hz. The sound sources could be tracked and were immediately classified as top secret. The military minds were quick to blame an adversary, possibly a Russian submarine with a secret device. These 20-Hz sounds had been recorded and publically puzzled over off the Canadian east coast for many years but became designated as secret south of the 54°40′ parallel. Although known since the 1940s, these sounds were declassified during the Marine Bioacoustics Conference in Bimini in 1963, and their source was identified as a large cetacean, probably the fin whale (*Balaenoptera physalus*) (Patterson and Hamilton 1964; Schevill et al. 1964; Walker 1964).

In proposing to investigate the function of acoustics in the behavior of fishes, obvious experimental procedures involve the playback of sounds, natural and artificial. How loud should the stimulus sounds be? What about pitch and timbre? The question of how well can fish hear has to be addressed first. Lacking an external ear, it was assumed that fish were essentially deaf, but sporadic reports began to appear in the early 1900s. A dramatic and convincing demonstration was when the eminent Karl von Frisch trained catfish in his backyard pond to come when he whistled (von Frisch 1923). These early studies on fish hearing have been adequately reviewed elsewhere (Kleerekoper and Chagnon 1954; Moulton 1963; Tavolga and Wodinsky 1963). However, few of these reports contained any quantitative data. An attempt to put absolute values on hearing thresholds in fishes was a report using avoidance conditioning (Tavolga and Wodinsky 1963), and subsequent attempts to demonstrate masking and frequency discrimination (Tavolga 1967; Tavolga and Jacobs 1968). Current techniques using electrophysiological auditory brain stem response (ABR) have confirmed much of the data derived from behavioral conditioning (Fay 1988). In this way, the first evidence for ultrasound detection in fish was established (Mann et al. 1997). We are now capable of testing fragile species such as herring and anchovies. Indeed, we were able to demonstrate that a clupeid fish, the menhaden (*Brevoortia*), was capable of detecting sound frequencies from 40,000 to at least 80,000 Hz (Mann et al. 2001).

In goldfish (*Carassius auratus*) hearing, thresholds obtained by behavioral techniques and electrophysiological measurements were not significantly different (Fay 1988). This may not always be the case. Field observations on the bonefish (*Albula vulpes*) show the fish to be very sensitive and responsive to noises, yet laboratory tests of their hearing reveal thresholds not significantly different from those of other species in the same areas. In other words, some fish "listen" better but may not actually have better hearing (Tavolga 1974).

A few words here about a good friend and colleague, the late Arthur Myrberg. Art was a true ethologist and was a student with Konrad Lorenz. He and I spend many hours, far into the night, over a bottle of wine, arguing about "releasers," instinct, and innate behavior. He was a keen observer and a fine scientist, and his studies on behavior of reef fishes are true classics (Myrberg 1972). He bemoaned the fact that so little research was being done on acoustic behavior in fish. With all the advances we have made in the areas of acoustic reception in fishes over the past 10–15 years, our behavioral advances have lagged far behind. Armed with our understanding of audition, we should now be able to pursue the operation of acoustics at the higher levels of organization, e.g., behavior, ethology, and sociobiology. We now have the tools of psychophysics, and we have a substantial database on the acoustic sensory system in fishes and marine mammals. We should use this information to learn more about how these animals use this acoustic sense to survive. Here is an example. The ubiquitous freshwater catfish (*Ameiurus*) and carp (*Carassius*; the ancestor of the domesticated goldfish) possess a highly specialized hearing system of Weberian ossicles, yet produce no sounds of their own and appear to live in a quiet neighborhood. What is that fish listening to so intently? What does it need to hear to survive? In other words, how do fish use their bioacoustic properties to make a living?

References

Cousteau JY, Dumas F (1953) The silent world. Harper and Brothers Publishers, New York.

Cummings WC, Brady BD, Herrenkind WF (1964) The occurrence of underwater sounds of biological origin off the west coast of Bimini, Bahamas. In: Tavolga WN (ed) Marine bioacoustics. Pergamon Press, Oxford, UK, pp 27–43.

Fay RR (1988) Hearing in vertebrates. Hill-Fay Associates, Winnetka, IL.

Fish MP (1954) The character and significance of sound production among fishes of the western North Atlantic. Bull Bingham Oceanogr Collect 14:1–109.

Fish MP, Kelsey AS, Mowbray WM (1952) Studies on the production of underwater sound by North Atlantic coastal fishes. J Mar Res 11:180–193.

Kleerekoper H, Chagnon EC (1954) Hearing in fish, with special reference to *Semotilus atromaculatus atromaculatus* (Mitchill). J Fish Res Board Can 11:130–152.

Lilly JC (1961) Man and dolphin. Doubleday and Co., New York.

Mann DA, Lu Z, Popper AN (1997) A clupeid fish can detect ultrasound. Nature 389:341.

Mann DA, Tavolga WN, Higgs DM, Souza M, Popper AN (2001) Ultrasound detection by clupeiform fishes. J Acoust Soc Am 109:3048–3054.

Moulton JM (1963) Acoustic behaviour of fishes. In: Busnel RG (ed) Acoustic behaviour of animals. Elsevier, Amsterdam, pp 655–887.

Myrberg AA Jr (1972) Ethology of the bicolor damselfish, *Eupomacentrus partitus*. Anim Behav Monogr 5:199–283.

Norris KS (1964) Some problems of echolocation in cetaceans. In: Tavolga WN (ed) Marine bioacoustics. Pergamon Press, Oxford, UK, pp 317–336.

Norris KS, Evans WE (1967) Directionality of echolocation clicks in the rough-tooth porpoise, *Steno bredanensis* (Lesson). In: Tavolga WN (ed) Marine bio-acoustics, vol 2. Pergamon Press, Oxford, UK, pp 305–316.

Norris KS, Prescott JH, Evans WE (1961) An experimental demonstration of echolocation behavior in the porpoise, *Tursiops truncatus* (Montagu). Biol Bull 120:163–176.

Patterson B, Hamilton GR (1964) Repetitive 20 cycle per second biological hydroacoustic signals at Bermuda. In: Tavolga WN (ed) Marine bioacoustics. Pergamon Press, Oxford, UK, pp 125–146.

Schevill WE (1964) Underwater sounds of cetaceans. In: Tavolga WN (ed) Marine bioacoustics. Pergamon Press, Oxford, UK, pp 307–316.

Schevill WE, Watkins, WA, Backus RH (1964) The 20-cycle signals and *Balaenoptera* (fin whales). In: Tavolga WN (ed) Marine bioacoustics. Pergamon Press, Oxford, UK, pp 147–154.
Steinberg JC, Koczy FF (1964) An acoustic-video system for marine biological research. Objectives and requirements. In: Tavolga WN (ed) Marine bioacoustics. Pergamon Press, Oxford, UK, pp 1–10.
Tavolga WN (1956) Visual, chemical and sound stimuli as cues in the sex discriminatory behavior of the gobiid fish *Bathygobius soporator*. Zoologica 41:49–84.
Tavolga WN (1958) The significance of underwater sounds produced by males of the gobiid fish, *Bathygobius soporator*. Physiol Zool 31:259–271.
Tavolga WN (1967) Masked auditory thresholds in teleost fishes. In: Tavolga WN (ed) Marine bio-acoustics, vol 2. Pergamon Press, Oxford, UK, pp 233–245.
Tavolga WN (1971a) Sound production and detection. In: Hoar WS, Randall DJ (eds) Fish physiology, vol 5. Academic Press, New York, pp 135–205.
Tavolga WN (1971b) Acoustic orientation in the sea catfish, *Galeichthys felis*. Ann NY Acad Sci 188:80–97.
Tavolga WN (1974) Sensory parameters in communication among coral reef fishes. Mt Sinai J Med 41:324–340.
Tavolga WN (1976) Acoustic obstacle detection in the sea catfish (*Arius felis*). In: Schuijf A, Hawkins AD (eds) Sound reception in fish. Elsevier, Amsterdam, pp 185–204.
Tavolga WN, Jacobs DW (1968) Acoustic frequency discrimination in the goldfish. Anim Behav 16:67–71.
Tavolga WN, Wodinsky J (1963) Auditory capacities in fishes. Pure tone thresholds in nine species of marine teleosts. Bull Am Mus Nat Hist 126:179–239.
von Frisch K (1923) Ein Zwergwels der kommt, wenn man ihm pfeift. Biol Zentr 43:439–446.
Walker RA (1964) Some widespread high-level underwater noise pulses of apparent biological origin off Cape Cod. In: Tavolga WN (ed) Marine bioacoustics. Pergamon Press, Oxford, UK, pp 121–124.

Part II
Sound Detection by Aquatic Animals

Effects of Underwater Noise on Marine Mammals

Christine Erbe

1 Introduction

Public concern about the effects of underwater noise on marine mammals has steadily increased over the past few decades. Research programs have been developed around the globe to investigate noise impacts. Government departments in many countries regulate underwater noise emission. Industries, in particular the oil and gas industry, undertake environmental impact assessments of underwater noise expected from planned marine activities and submit these to regulatory agencies as part of a permit application process. Lawsuits have been brought against the Navy in an attempt to protect marine mammals from sonar testing. The number and diversity of stakeholders in the management of noise and marine animals is great. *Marine Mammals and Noise* (Richardson et al. 1995) was the first book to review and synthesize research on the noise effects on marine mammals. In the 15 years since then, a handful of review projects have been undertaken, with focus on specific aspects (e.g., Committee on Characterizing Biologically Significant Marine Mammal Behavior 2005; Committee on Potential Impacts of Ambient Noise in the Ocean on Marine Mammals 2003; National Research Council 2000; Nowacek et al. 2007; Southall et al. 2007).

Sources of anthropogenic noise include transportation, mineral and hydrocarbon exploration and production, and construction, sample spectra of which are shown in Figure 1, measured by the author or JASCO at some range and back-propagated to 1 m (Erbe 2002, 2009, 2010; Erbe and Farmer 2000), except for the mean large-vessel spectrum (Ross 1976; Scrimger and Heitmeyer 1991).

2. Potential Effects of Noise

Noise can affect marine mammals in many ways. At low levels, it might be merely detectable. At somewhat higher levels, it might interfere with animal communication and hinder acoustic signal detection. Noise can alter animal behavior. It can affect the auditory system and induce a shift in hearing threshold. Other systems potentially affected by noise include the vestibular, reproductive, and nervous systems. Noise might cause concussive effects, physical damage to tissues and organs

C. Erbe (✉)
Centre for Marine Science & Technology, Curtin University, Bentley, WA 6102, Australia;
formerly JASCO Applied Sciences, Australia
e-mail: C.Erbe@curtin.edu.au

A.N. Popper and A. Hawkins (eds.), *The Effects of Noise on Aquatic Life*,
Advances in Experimental Medicine and Biology 730, DOI 10.1007/978-1-4419-7311-5_3,

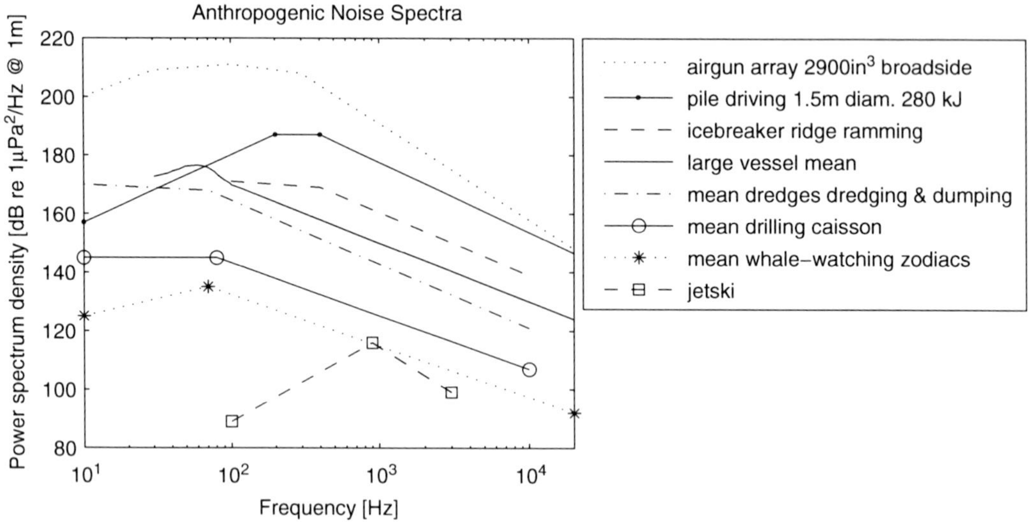

Fig. 1 Source spectra of selected anthropogenic sources

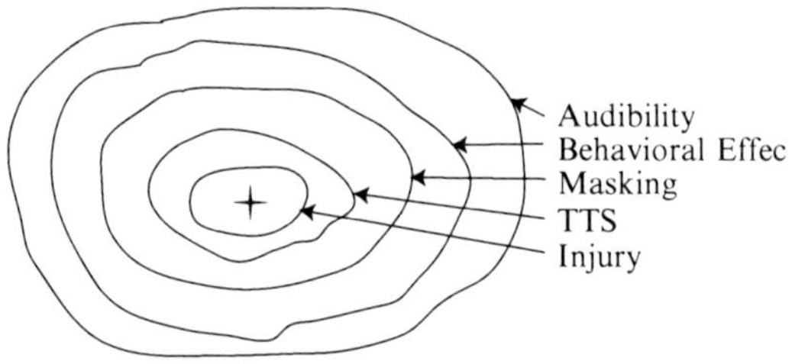

Fig. 2 Relative extent of different zones of impact around a noise source

(in particular gas filled), and cavitation (bubble formation). Stress is a physiological response to a stressor such as noise, aimed at surviving the immediate threat. Prolonged stress can cause serious health problems. The effects of noise and the ranges over which they happen depend on the acoustic characteristics of the source (e.g., noise level, duration, duty cycle, rise time, spectrum), the medium (hydro- and geoacoustic parameters of the environment, bathymetry), and the receiver (e.g., age, size, behavioral state, auditory capabilities). Figure 2 gives a bird's-eye view of the potential zones around a source over which some of these effects might happen.

2.1 Audibility

As sound spreads through the ocean, its acoustic energy decreases due to propagation losses. Audibility of a sound is limited by the sound dropping below either ambient noise levels or the animal's detection threshold. Audiograms, hearing thresholds as a function of frequency, have been measured for only about 20 marine mammal species and in only few individuals. The threshold is a statistical quantity, e.g., depending on the audiometric paradigm, the level at which the signal was heard 50% of the time. Figure 3 shows the lowest hearing thresholds measured for a

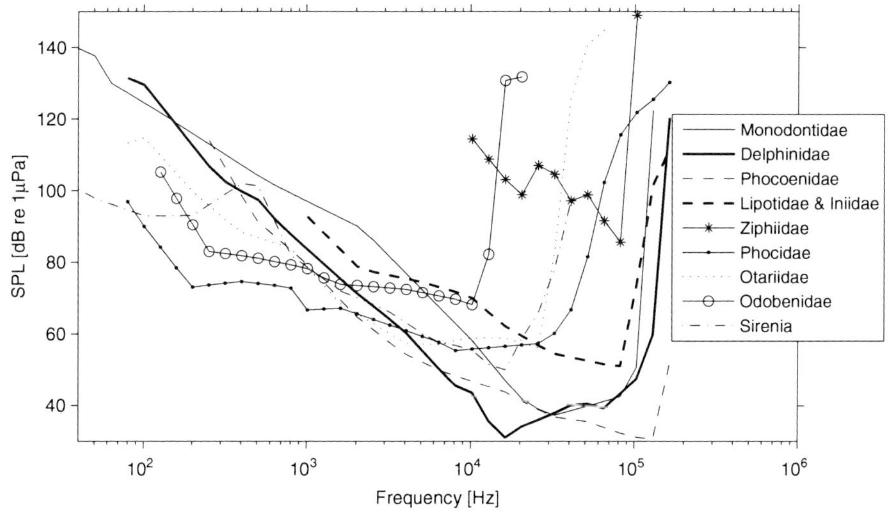

Fig. 3 Audiograms of marine mammal families. Modified from Erbe (2010)

number of families. Underwater audiograms have not yet been measured for *Ursus maritimus* (polar bear), *Mustelidae* (sea otters), *Physeteridae* (sperm whales), and *Balaenidae* (baleen whales). Indirect information on hearing stems from observed responses to sound and from anatomical studies. Furthermore, animals are expected to be very sensitive at the frequencies of their own calls.

2.2 *Behavioral Responses*

The zone of responsiveness is expected to be smaller than the zone of audibility because an animal will not likely respond to a sound that is barely detectable. However, long ranges of behavioral responses (up to 70 km) have been observed (Cosens and Dueck 1988; Finley et al. 1990) that were close to the maximum ranges of audibility (Erbe and Farmer 2000). Measured indicators include changes in swim direction and speed, dive duration, surfacing duration and interval, and respiration and changes in contextual behavior and acoustic behavior. Prior exposure (habituation vs. sensitization), age, gender, health, current behavioral state, and other factors affect the likelihood and severity of response. A dose-response curve (risk function) was used by the US Department of the Navy (2009) to predict the percentage of a population that might respond. Southall et al. (2007) ranked behavioral responses reported in the literature on a severity scale from zero to nine, compiled tables of the number of individuals or groups that reacted as a function of severity score and received root mean square (RMS) sound pressure levels (SPLs) because this is the most commonly reported metric. However, it might not be the one that correlates best with behavior. Behavioral analyses should be multivariate, considering the full range of metrics appropriate for the sound source (e.g., SPL_{RMS}, SPL_{peak}, SEL, and signal-to-noise ratio) and the full range of behavioral and contextual variables.

2.3 *Masking*

Noise can mask signals such as communication sounds, echolocation, predator and prey sounds, and environmental sounds. Figure 4 shows the bandwidths of sounds emitted by marine mammals. Masking depends on the spectral and temporal characteristics of signal and noise. At a low signal-to-noise ratio

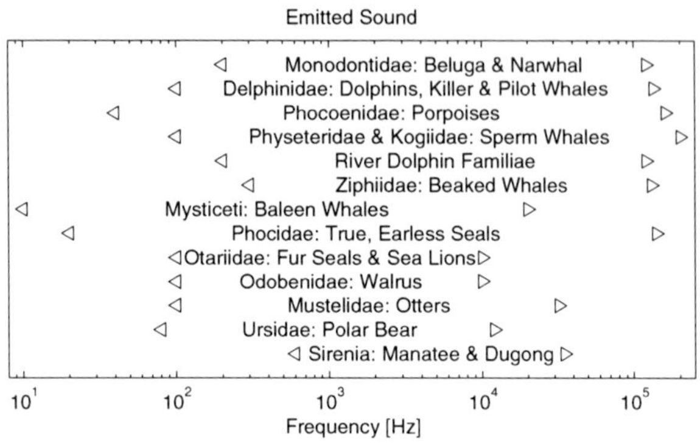

Fig. 4 Bandwidth of sounds emitted by marine mammals

(SNR), a signal might just be audible. A higher SNR is needed for signal recognition and discrimination and an even higher SNR for comfortable communication. The potential for masking is reduced by good frequency discrimination, temporal discrimination, and directional hearing abilities of the animal. Masking can be further reduced in some species if the noise is amplitude modulated over a number of frequency bands (comodulation masking release), if the noise has gaps or the signal is repetitive (multiple looks model), and by antimasking strategies such as deliberate increases in call level and repetition or frequency shifting (Erbe 2008). Models for the masking of complex calls by anthropogenic noise were developed by Erbe (2000) and Erbe et al. (1999) based on behavioral experiments (Erbe and Farmer 1998).

2.4 *Auditory Threshold Shift*

Noise exposure can result in a loss of hearing sensitivity, termed threshold shift. If hearing returns to normal after some quiet time, the effect is a temporary threshold shift (TTS); otherwise, it is a permanent threshold shift (PTS). TTS is considered auditory fatigue, whereas PTS is considered injury. TTS, but not PTS, has been measured experimentally in a few species of odontocetes and pinnipeds. Southall et al. (2007) derived initial noise-exposure criteria for marine mammals aimed at preventing injury. Data for TTS onset in marine mammals were combined with data for TTS growth as a function of noise level, and a 40-dB TTS was chosen as the onset of auditory injury (PTS). Marine mammal species were grouped into five functional hearing groups: low-, mid- and high-frequency cetaceans and pinnipeds in air and underwater. Spectral weighting functions (M-weighting) for the five functional hearing groups were applied to the noise in order to emphasize the frequency bands where acoustic exposures to high levels might cause auditory damage. Noise sources were grouped into single pulses, multiple pulses, and nonpulses based on the number of emissions per 24 h and on the level difference if measured with impulse time constants compared with continuous time constants. Thresholds in terms of peak SPL and sound exposure level (SEL) were derived; the one to be reached first was recommended for mitigation. Since then, TTS onset in a high-frequency cetacean has been shown at ~20 dB lower levels (Lucke et al. 2009).

2.5 *Nonauditory Physiological Effects*

Noise may impact nonauditory organs and systems, but data for marine mammals do not exist. Given that no damage to tissues and organs was observed in marine mammals during TTS experiments, levels will likely be higher. Stress is a physiological response that involves the release of the hormone adrenalin, which increases heart rate, gas exchange, acuity, and blood flow to the brain and muscles for a fight-or-flight response (Wright et al. 2009). Stress responses are intended to improve survival in the face of an immediate threat; however, repetitive or prolonged stress can negatively affect health in the long run. Chronic stress in humans can cause coronary disease, immune problems, anxiety, depression, cognitive and learning difficulties, and infertility. The onset of stress might correspond to fairly low noise levels that induce a behavioral disturbance or masking. Stress might be a direct result of noise, e.g., if an unknown noise is detected, or an indirect result of noise causing, e.g., masking.

3 Discussion

Many of the discussed effects can be related; a temporary shift in hearing threshold will affect the audibility of signals (e.g., of conspecific calls) and thus alter or prevent the "normal" behavioral response to such signals. Or noise received by a diving animal might induce stress leading to a so-called fight-or-flight response involving rapid surfacing that can cause decompression sickness and injury and ultimately death. There is no information on chronic effects of noise on marine mammals. Although it is feasible to model cumulative sound exposure over multiple sources, long durations, and large areas (Erbe and King 2009), the manner in which repeated exposure gets accumulated by the animals and the effects of cumulative exposure are unknown. Regulation and mitigation mostly address acute exposure from a single operation or event and direct damage. The biological significance of acoustic impacts is poorly understood. If critical behavior such as mating or nursing is repeatedly disrupted or if raised background noise causes chronic stress, it seems plausible that survival of the population might be affected. However, temporary and localized impacts are likely less significant. The population consequences of acoustic disturbance (PCAD) model (Committee on Characterizing Biologically Significant Marine Mammal Behavior 2005) provides a conceptual framework for linking acoustic disturbance to population effects. The ranking of noise among environmental stressors on marine mammals and the interaction of stressors are not understood. Other "stressors" affecting marine mammals include harvesting, culling, bycatch, ship strikes, chemical pollution, habitat degradation, prey overfishing, and climate change. An animal stressed by pollution or prey depletion might find it "harder" to cope with noise, and vice versa, an animal suffering from repeated or severe noise exposure might not be able to effectively cope with additional nonacoustic stressors.

4 Conclusions

Summarizing and synthesizing the effects of noise on marine mammals in six pages is difficult. The topic has received perhaps exponential attention over the past few decades, with great research undertaken across the oceans, so giving adequate credit to which is impossible here. What is still lacking is consent on measurement and reporting metrics and standards. Noise impacts should be viewed in context with other environmental stressors. Regulation would ideally not focus on a single operation limited in space and time but would instead consider cumulative impacts experienced by animals over time and space.

References

Committee on Characterizing Biologically Significant Marine Mammal Behavior (2005) Marine mammal populations and ocean noise: Determining when noise causes biologically significant effects. National Academies Press, Washington, DC.

Committee on Potential Impacts of Ambient Noise in the Ocean on Marine Mammals (2003) Ocean noise and marine mammals. National Academies Press, Washington, DC.
Cosens SE, Dueck LP (1988) Responses of migrating narwhal and beluga to icebreaker traffic at the Admiralty Inlet ice-edge, N.W.T. in 1986. In: Sackinger WM, Jeffries MO (eds) Port and ocean engineering under arctic conditions. Geophysical Institute, University of Alaska, Fairbanks, pp 39–54.
Erbe C (2000) Detection of whale calls in noise: Performance comparison between a beluga whale, human listeners and a neural network. J Acoust Soc Am 108:297–303.
Erbe C (2002) Underwater noise of whale-watching boats and potential effects on killer whales (*Orcinus orca*), based on an acoustic impact model. Mar Mamm Sci 18:394–418.
Erbe C (2008) Critical ratios of beluga whales (*Delphinapterus leucas*) and masked signal duration. J Acoust Soc Am 124:2216–2223.
Erbe C (2009) Underwater noise from pile driving in Moreton Bay, QLD. Acoust Aust 37:87–92.
Erbe C (2010) Underwater acoustics: Noise and the effects on marine mammals, 3rd edn. Pocketbook, printed by JASCO Applied Sciences, Brisbane, QLD, Australia.
Erbe C, Farmer DM (1998) Masked hearing thresholds of a beluga whale (*Delphinapterus leucas*) in icebreaker noise. Deep Sea Res II Top Stud Oceanogr 45:1373–1388.
Erbe C, Farmer DM (2000) Zones of impact around icebreakers affecting beluga whales in the Beaufort Sea. J Acoust Soc Am 108:1332–1340.
Erbe C, King AR (2009) Modeling cumulative sound exposure around marine seismic surveys. J Acoust Soc Am 125:2443–2451.
Erbe C, King AR, Yedlin M, Farmer DM (1999) Computer models for masked hearing experiments with beluga whales (*Delphinapterus leucas*). J Acoust Soc Am 105:2967–2978.
Finley KJ, Miller GW, Davis RA, Greene CR (1990) Reactions of belugas (*Delphinapterus leucas*) and narwhals (*Monodon monoceros*) to ice-breaking ships in the Canadian High Arctic. Can Bull Fish Aquat Sci 224:97–117.
Lucke K, Siebert U, Lepper PA, Blanchet MA (2009) Temporary shift in masked hearing thresholds in a harbor porpoise (*Phocoena phocoena*) after exposure to seismic airgun stimuli. J Acoust Soc Am 125:4060–4070.
National Research Council (2000) Marine mammals and low-frequency sound. National Academies Press, Washington, DC.
Nowacek DP, Thorne LH, Johnston DW, Tyack PL (2007) Responses of cetaceans to anthropogenic noise. Mamm Rev 37:81–115.
Richardson WJ, Greene CR Jr, Malme CI, Thomson DH (1995) Marine mammals and noise. Academic Press, San Diego, CA.
Ross D (1976) Mechanics of underwater noise. Pergamon Press, New York.
Scrimger P, Heitmeyer RM (1991) Acoustic source-level measurements for a variety of merchant ships. J Acoust Soc Am 89:691–699.
Southall BL, Bowles AE, Ellison WT, Finneran JJ, Gentry RL, Greene CR, Kastak D, Ketten DR, Miller JH, Nachtigall PE, Richardson WJ, Thomas JA, Tyack PL (2007) Marine mammal noise exposure criteria: Initial scientific recommendations. Aquat Mamm 33:412–522.
US Department of the Navy (2009) Atlantic fleet active sonar training environmental impact statement/overseas environmental impact statement. Available via http://afasteis.gcsaic.com/docs.aspx. Accessed 5 April 2010.
Wright AJ, Deak T, Parsons ECM (2009) Size matters: Management of stress responses and chronic stress in beaked whales and other marine mammals may require larger exclusion zones. Mar Pollut Bull 31 December 2009. Available via http://dx.doi.org/10.1016/j.marpolbul.2009.11.024:1879-3363.

Psychophysical Studies of Auditory Masking in Marine Mammals: Key Concepts and New Directions

Colleen Reichmuth

1 Introduction

In recent years, growing awareness of the potentially harmful effects of human-generated noise has led to concern over whether and how increasing ocean noise levels may adversely affect marine mammals by interfering with their detection of biologically important signals. The consequences of adding noise to an animal's environment may be studied, in part, by identifying and understanding the ways in which noise alters normal or baseline hearing capabilities. The process of auditory masking occurs when the perception of a given signal is negatively influenced by the presence of another sound. Masking effects may be manifested through spectral or temporal interference from noise that decreases the audibility of the signal relative to conditions when the masking noise is not present. In addition to auditory-masking effects, it should be noted that other auditory effects as well as nonauditory behavioral, physiological, or anatomical changes may also occur as a result of noise exposure. Furthermore, marine mammals are not the only aquatic animals that may be vulnerable to these effects. However, the scope of this paper is limited to consideration of the masking effects of noise on the auditory perception of marine mammals. The specific aims are to 1) briefly review the key concepts and methods drawn from a psychophysical approach to the study of auditory masking, 2) examine how masking studies have been applied thus far to improve understanding of noise effects on marine mammals, and 3) consider how future laboratory studies with marine mammals may incorporate progressively more complex and realistic listening scenarios into psychophysical testing programs.

2 Psychophysical Approach to Auditory Masking

Neural processing of auditory information in animals occurs at many levels, from the sensory receptors that receive sound cues from the external environment through progressively higher centers of the brain. Because auditory masking is a perceptual phenomenon, it is clear that studies of masking must take into account the whole animal and not merely the physical environment and the primary

C. Reichmuth (✉)
Long Marine Laboratory, Institute of Marine Sciences,
University of California Santa Cruz, Santa Cruz, CA 95060, USA
e-mail: coll@ucsc.edu

A.N. Popper and A. Hawkins (eds.), *The Effects of Noise on Aquatic Life*,
Advances in Experimental Medicine and Biology 730, DOI 10.1007/978-1-4419-7311-5_4,

receptor system. Psychophysics is the field of experimental psychology that uses precise behavioral methods to determine the relationship between the physical environment and an individual's subjective experience of that environment (Fechner 1860). Consequently, psychoacoustic approaches, which describe the relationship between the lowest possible level of audition (the sound stimulus) and the highest possible level (the sensation of that stimulus), provide the most comprehensive and effective perspectives on auditory processes such as masking (Fastl and Zwicker 2007).

Psychoacoustic parameters are not measured but rather are approximated based on the subjective impression of an individual averaged over many stimulus presentations. For example, hearing thresholds for sounds of a given type, frequency, and duration are typically determined as the lowest sound pressure level (SPL) that is detected by a subject in the absence of interfering noise over a specified percentage of experimental trials. To measure such probabilistic psychoacoustic thresholds, a trained listener must produce observable and reliable reporting responses when presented with a series of experimental cues because direct observation of perceptual events is possible only by introspection.

Within psychophysical testing paradigms, there are a few basic steps that are common to most studies of auditory masking. First, the hearing threshold for a given signal is determined through behavioral testing of a subject that reports detection of a signal presented at various SPLs. Next, a potential masker such as broadband noise or band-limited noise is added to the testing scenario. Finally, the hearing threshold of the signal is remeasured using the same methods, this time in the presence of the masking noise (see Gelfand 2001). The difference between the initial and final hearing thresholds reveals how much of an effect the masker had on the audibility of the signal.

Two of the most important metrics of auditory masking that can be derived from psychophysical experiments are the critical bandwidths and critical ratios. Both of these terms help to describe how the presence of noise influences hearing. A critical band describes the frequency region over which masking noise may interfere with detection of a given signal. If a signal and a masker are presented simultaneously, then only the masker frequencies falling within the critical bandwidth of the signal contribute to masking of the signal. Outside this critical-frequency band, the presence of noise does not alter the audibility of the signal. Therefore, understanding critical bandwidths as a function of signal frequency is essential for predicting the masking effects of noise. Generally, the frequency span of the critical band is proportional to the center frequency of the signal, so critical bandwidths increase in span with increasing signal frequency. Additionally, the shape of the auditory filter becomes asymmetrical with increasing masker level, generating an upward frequency spread of masking effects with increasing noise (Yost 2000). Critical ratios are related to critical bandwidths. A critical ratio is the minimum difference in decibels between the SPL of a just-audible pure-tone signal and the spectrum level of background white noise (the power contained in each 1-Hz band of noise) when the frequency span of the noise matches or exceeds the critical band. For example, a 5-kHz tone with a level of 80-dB SPL might be just audible when the spectrum level of the background noise is 60 dB, showing a critical ratio, or difference, of 20 dB; if the spectrum level of the noise should increase by 10 dB to 70 dB, then the level of the tone would need to increase to 90-dB SPL, still exceeding the noise by the critical ratio of 20 dB, to be audible to the same listener. Like critical bandwidths, critical ratios show a dependence on frequency, with critical ratios tending to increase with increasing signal frequencies except at relatively low frequencies (Fletcher 1940). Because these ratios hold across a range of signal and noise levels, they can facilitate the prediction of masking effects when the critical bandwidth and the characteristics of the masking noise are known.

Psychoacoustic experiments that systematically explore various masking phenomena, including critical bands and critical ratios, serve to reveal important details about auditory processing and the underlying physical and neural mechanisms of auditory systems. Comparative investigations of masking and auditory processing are essential because they highlight the generality of some effects and identify and quantify the relevant parameters that differ among species and taxonomic groups.

Both kinds of data are needed to establish a useful knowledge base from which informed assessments about potential noise effects on hearing can be made.

3 Psychophysical Studies of Auditory Masking in Marine Mammals

There have been many psychophysical studies of auditory masking in marine mammals since Johnson (1968) reported the first critical ratios for *Tursiops truncatus* (bottlenose dolphin). Most of these have been reviewed by Richardson et al. (1995), with additional recent studies of odontocete cetaceans (Branstetter and Finneran 2008; Erbe 2000; Erbe and Farmer 1998; Kastelein and Wensveen 2008; Kastelein et al. 2009; Lemonds 1999), sirenians (Gerstein 1999), and pinnipeds (Holt and Schusterman 2007; Southall et al. 2000, 2003; Turnbull 1994). The majority of these studies utilized pure-tone or narrowband signals and uniform masking noise to characterize masking effects on underwater hearing. As a result, the metrics of the critical bandwidth and critical ratio are reasonably well understood for representative marine mammal species for which absolute auditory sensitivity has been measured.

The available data on critical ratios in marine mammals provide clear and useful indicators of how these animals hear in noise. As expected, critical ratios in all marine mammals tested thus far increase with increasing frequency, except at very low frequencies. There are no significant deviations to this trend, suggesting that marine mammals are generalists with respect to frequency resolution. The critical ratios of amphibious marine mammals, including seals and sea lions, are the same in air as they are underwater despite differences in absolute hearing sensitivity between the two media. Notably, marine mammals (with the exception of the sirenian *Trichechus manatus latirostis*) tend be better than most terrestrial mammals at detecting signals in noise. This may be due, in part, to their reliance on acoustic detection and frequency resolution in naturally noisy environments where the use of other sensory modalities is constrained.

Marine mammals pose an interesting case among other mammals with respect to critical bandwidths. The underwater hearing capabilities of dolphins and seals extend across an extremely wide range of frequencies, from below 100 Hz to the high ultrasonics. Critical bandwidths may be measured directly in psychophysical experiments using bands of noise that are systematically varied in width around test frequencies, but more commonly, they are estimated from critical ratios that may be obtained across the entire frequency range of hearing. One-third octave bands of noise are typically considered reasonable minimum spans of the frequencies that contribute to masking in terrestrial mammals; however, estimates from critical ratio measurements suggest that the critical bandwidths of some marine mammals may become wider at low and very high frequencies, whereas at intermediate frequencies, they are likely to be significantly narrower (Richardson et al. 1995). Accurate critical bandwidth data are required to support predictions of if and how noise sources may interfere with an animal's ability to detect relevant signals and to estimate the distances at which such interference may occur. For example, the widening of critical bands beyond one-third octave at relatively low frequencies means that noise across a wider frequency range would contribute to the masking of low-frequency signals, whereas critical bands smaller than one-third octave would result in much narrower bandwidths of noise contributing to masking. Comparisons of empirically measured critical bandwidths and estimates derived from critical ratios show that directly measured critical bandwidths are required to accurately model potential masking effects (see Yost and Schofner 2009). At present, such data are available for only a few marine mammal species at select frequencies.

These findings with marine mammals show how psychoacoustic studies using simple stimuli can improve understanding of how auditory systems operate to extract signals from interfering noise. The results from these sorts of listening experiments are bolstered by complementary neurophysiological studies of masking that provide additional insight into the general and species-typical characteristics of hearing, especially at the level of the peripheral auditory nervous system. However, although the metrics of critical bandwidths and critical ratios have been directly applied to model the auditory effects of anthropogenic noise on marine mammals living in natural environments, they are not sufficient to describe how noise interferes with the perception of sounds as the acoustic environment becomes progressively more realistic. New research is needed to improve understanding of how marine mammals cope with background noise during functional hearing.

4 Auditory Masking in Progressively Realistic Hearing Scenarios

In natural environments, the interactions between signals and noise are complex and stimuli vary widely with respect to their temporal, spectral, and spatial characteristics. Marine mammals have adaptations on both the signal-production side and the signal-reception side to optimize their use of sound and to limit their susceptibility to auditory masking, and these include some higher order aspects of hearing that can only be accessed using psychoacoustic methods. Quantitative psychoacoustic approaches have been developed and refined to better understand acoustic communication in noise, including through studies of speech perception and birdsong. Many of these approaches have or can be expanded to marine mammal research in ways that will dramatically improve understanding of potential noise effects on hearing. Some concepts worthy of further development or consideration include the following:

1. *Release from masking.* There are a variety of conditions where the detection of signals in noise can be improved by auditory or behavioral "demasking" processes. Spatial release from masking (SRM) occurs when the masking effects of colocated signals and maskers are reduced because the signals and maskers are spatially segregated and directional hearing is sufficient to support enhanced detection. The comodulation masking release (CMR) occurs when the energy in masking noise is coherently modulated in time across frequency regions rather than randomly modulated, as often found in real noise environments (see Branstetter et al., Chapter 5). The lower critical ratios obtained in contexts such as these can result in larger detection ranges for sounds than would otherwise be predicted from simple models of auditory masking.
2. *Complexity of signals.* Complex signals, such as those often used as communication signals, are often easier to detect in noise than are tonal or narrowband signals. Sounds with pulsed characteristics, harmonic elements, frequency modulation, or amplitude modulation may require lower signal-to-noise ratios for detection due to auditory processes such as loudness summation across critical bands. To date, a few psychoacoustic studies have explored the masking of complex signals in marine mammals, and there is one study of hearing in *Delphinapterus leucas* (beluga whale) with both natural signals and realistic noise (Erbe and Farmer 1998). Masked hearing thresholds obtained for complex signals in noise can be productively compared with those predicted by pure-tone critical ratio and broadband critical bandwidth data to refine models of auditory masking in marine mammals.
3. *Masking and communication.* Discussion of masking in marine mammals focuses almost exclusively on detection thresholds and the corresponding potential communication distances. It is clear from studies of speech and birdsong, however, that progressively increasing signal-to-noise levels are required to move from the mere detection of sounds in noise to discrimination, recognition, and, ultimately, effective communication (see Lohr et al. 2003; Dooling and Therrien, Chapter 17). This issue truly gets at the perception of acoustic signals and brings higher level processes such as learning into the forefront. Marine mammals are especially good candidates for

explorations of informational masking due to their trainability and highly developed cognitive capabilities; the experiments are challenging for many reasons but will yield invaluable insight into how noise constrains functional hearing in realistic listening scenarios.

5 Summary

Current models of auditory masking in marine mammals oversimplify hearing in realistic environments. Systematic and progressive experiments using psychoacoustic methods will help us move "out of the ideal and into the real" to gain a more complete view of potential auditory-masking effects in these animals.

References

Branstetter BK, Finneran JJ (2008) Comodulation masking release in bottlenose dolphins (*Tursiops truncatus*). J Acoust Soc Am 124:25–33.

Erbe C (2000) Detection of whale calls in noise: Performance comparison between a beluga whale, human listeners, and a neural network. J Acoust Soc Am 108:297–303.

Erbe C, Farmer DM (1998) Masked hearing thresholds of a beluga whale (*Delphinapterus leucas*) in icebreaker noise. Deep-Sea Res II Top Stud Oceanogr 45:1373–1388.

Fastl H, Zwicker E (2007) Psychoacoustics: Facts and models. Springer, Berlin, Germany.

Fechner GT (1860) Elemente der psychophysik (Elements of psychophysics). Holt Rinehart & Winston, New York.

Fletcher H (1940) Auditory patterns. Rev Mod Phys 12:47–65.

Gelfand SA (2001) Essentials of audiology, 2nd edn. Thieme, New York.

Gerstein ER (1999) Psychoacoustic evaluations of the West Indian manatee (*Trichechus manatus latirostis*). Unpublished doctoral dissertation, Florida Atlantic University, Boca Raton, FL.

Holt MM, Schusterman RJ (2007) Spatial release from masking of aerial tones in pinnipeds. J Acoust Soc Am 121:1219–1225.

Johnson CS (1968) Masked tonal thresholds in the bottlenosed porpoise. J Acoust Soc Am 44:965–967.

Kastelein RA, Wensween, PJ (2008) Effect of two levels of masking noise on the hearing threshold of a harbor porpoise (*Phocoena phocoena*) for a 4.0 kHz signal. Aquat Mamm 34:420–425.

Kastelein RA, Wensween PJ, Hoek L, Au WWL, Terhune JM, de Jong CAF (2009) Critical ratios in harbor porpoises (*Phocoena phocoena*) for tonal signals between 0.315 and 150 kHz in random Gaussian white noise. J Acoust Soc Am 126:1588–1597.

Lemonds DW (1999) Auditory filter shapes in an Atlantic bottlenose dolphin *Tursiops truncatus*. Unpublished doctoral dissertation, University of Hawai'i, Honolulu, HI.

Lohr B, Wright TF, Dooling RJ (2003) Detection and discrimination of natural calls in masking noise by birds: Estimating the active space of a signal. Anim Behav 65:763–777.

Richardson WJ, Greene CR Jr, Malme CI, Thomson DH (1995) Marine mammals and noise. Academic Press, San Diego, CA.

Southall BL, Schusterman RJ, Kastak D (2000) Masking in three pinnipeds: Underwater, low frequency critical ratios. J Acoust Soc Am 114:1660–1666.

Southall BL, Schusterman RJ, Kastak D (2003) Auditory masking in three pinnipeds: Aerial critical ratios and direct critical bandwidth measurements. J Acoust Soc Am 114:1660–1666.

Turnbull SD (1994) Changes in masked thresholds of a harbor seal (*Phoca vitulina*) associated with angular separation of signal and noise sources. Can J Zool 72:1863–1866.

Yost WA (2000) Fundamentals of hearing: An introduction. Academic Press, San Diego, CA.

Yost WA, Schofner WP (2009) Critical bands and critical ratios in animal psychoacoustics: An example using chinchilla data. J Acoust Soc Am 125:315–323.

On the Relationship Between Environmental Noise, Critical Ratios, and Comodulation Masking Release in the Bottlenose Dolphin (*Tursiops truncatus*)

Brian K. Branstetter, Jennifer S. Trickey, and James J. Finneran

1 Introduction

When one sound interferes with the ability to detect another sound, masking occurs. The potential negative impacts of masking include the inability to effectively communicate, navigate, and forage. Due to the difficulty in acquiring, maintaining, and training marine mammals to participate in psychophysical testing, only a limited number of masking studies have been conducted. Most of these are critical-ratio and critical-band measurements. A primary finding of these studies is that masking typically occurs only when the frequency of the signal and masker are similar (Fletcher 1940). For example, critical-band and notched-noise experiments using white-noise maskers (i.e., the instantaneous amplitude is sampled from a Gaussian distribution and the frequency spectrum is flat) suggest that only noise frequencies within a narrow band centered on a tonal signal contribute to the masking of the signal (Au and Moore 1990; Finneran et al. 2002; Southall et al. 2003).

2 Power Spectrum Model of Masking

Findings from critical-ratio and critical-band studies have been used to develop what is often referred to as the power spectrum model (PSM) of auditory masking. The axioms of the PSM can be summarized as follows:

1. The auditory periphery behaves as a series of overlapping band-pass filters that are often referred to as auditory filters.
2. Only the spectral components of the noise within an auditory filter centered on a signal contribute to the masking of the signal.

B.K. Branstetter (✉)
National Marine Mammal Foundation, San Diego, CA 92106, USA
e-mail: brian.branstetter@nmmpfoundation.org

J.S. Trickey
G2 Software Systems Inc., San Diego, CA 92110, USA
e-mail: trickey@g2ss.com

J.J. Finneran
US Navy Marine Mammal Program, Space and Naval Warfare Systems Center Pacific, San Diego, CA 92152, USA
e-mail: james.finneran@navy.mil

A.N. Popper and A. Hawkins (eds.), *The Effects of Noise on Aquatic Life*,
Advances in Experimental Medicine and Biology 730, DOI 10.1007/978-1-4419-7311-5_5,

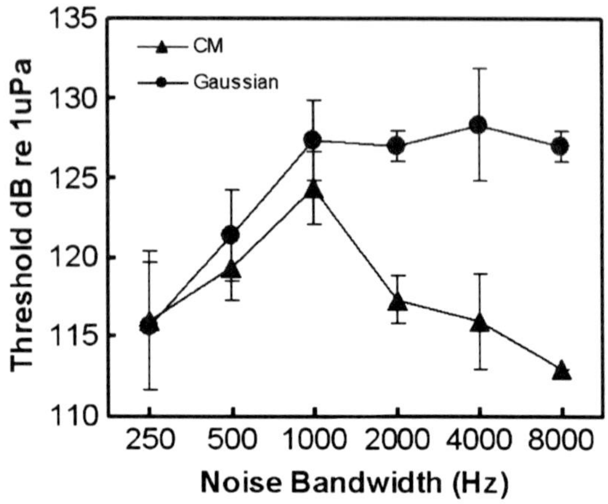

Fig. 1 Thresholds for a 10-kHz tone embedded in comodulated noise (CM) and Gaussian noise (Gaussian) of variable bandwidth. There is a noticeable release from masking in the CM condition for bandwidths that exceed 1,000 Hz. Adapted from Branstetter and Finneran (2008)

3. Signal detection is accomplished by an energy detector at the output of the auditory filter centered on the signal. The signal-plus-noise interval will have more energy than the noise-alone interval.
4. The level of the signal at threshold will be proportional to the noise level that passes through the signal auditory filter. Noise is represented by its long-term spectra.

Recent data from *Tursiops truncatus* (Branstetter and Finneran 2008) and *Delphinapterus leucas* (Erbe 2008) suggest that the assumptions of the PSM fail to describe patterns of masking with a variety of noise types that are not strictly Gaussian. For example, spectral components of comodulated noise wider than an auditory-filter bandwidth significantly reduced threshold levels (Fig. 1). This result is known as comodulation masking release and is difficult for the PSM to explain. In the present study, we test if the assumptions of the PSM generalize to different noise types including environmental noise.

3 Masking With Gaussian, Comodulated, and Environmental Noise

In experiment I, we measured masked hearing thresholds in *Tursiops truncatus* with three different masker types: 1) Gaussian noise, 2) comodulated noise, and 3) ambient noise recorded from San Diego Bay, CA (bay noise). All noise types were of equal bandwidth and equal spectral-density levels. To a *Homo sapiens* (human) listener, bay noise was dominated by *Synalpheus parneomeris* (snapping shrimp). Like comodulated noise, bay noise contained temporal fluctuations that were correlated across frequency regions. The PSM predicted identical masked thresholds for each noise condition. However, masked thresholds were significantly lower for both comodulated and bay noise conditions.

In experiment II, a band-widening study (critical-band paradigm) demonstrated different patterns of masking in *Tursiops truncatus* for Gaussian noise and bay noise of equal spectral-density levels. Gaussian noise resulted in the standard critical-band pattern of masking. Thresholds increased proportionally with bandwidth but became relatively constant beyond a critical bandwidth. For bay noise, thresholds were relatively flat for noise bandwidths up to 500 Hz (half a critical band) and thereafter exhibited a decreasing trend as bandwidths increased. The pattern of masking for bay noise is more similar to the comodulation masking release than Gaussian noise, but more investigation is required to determine the mechanisms behind the release from masking with bay noise.

4 Conclusions

The PSM does not fully describe or predict auditory masking with the comodulated and bay noise types in the present study. A simple energy detector after band-pass filtering has difficulty describing these data. Furthermore, the *Tursiops truncatus* auditory system clearly makes use of auditory information beyond a critical band when detecting a tonal signal embedded in noise. The temporal structure of environmental noise is more complex than Gaussian noise, and the auditory system appears to use this broadband complexity to aid in segregating the signal from the background noise. Metrics such as critical ratios and one-third octave noise measurement that rely on assumptions of the PSM should be used with caution when extrapolating the effects of masking if the noise is non-Gaussian.

5 References

Au WWL, Moore PWB (1990) Critical ratio and critical bandwidth for the Atlantic bottlenose dolphin. J Acoust Soc Am 88:1635–1638.

Branstetter BK, Finneran JJ (2008) Comodulation masking release in bottlenose dolphins (*Tursiops truncatus*). J Acoust Soc Am 124:625–633.

Erbe C (2008) Critical ratios of beluga whales (*Delphinapterus leucas*) and masked signal duration. J Acoust Soc Am 124:2216–2223.

Finneran JJ, Schlundt CE, Carder DA, Ridgway SH (2002) Auditory filter shapes for the bottlenose dolphin (*Tursiops truncatus*) and the white whale (*Delphinapterus leucas*) derived with notched noise. J Acoust Soc Am 112:322–328.

Fletcher H (1940) Auditory patterns. Rev Mod Phys 12:47–65.

Southall BL, Schusterman RJ, Kastak D (2003) Auditory masking in three pinnipeds: Aerial critical ratios and direct critical bandwidth measurements. J Acoust Soc Am 114:1660–1666.

Direct Measurements of Subjective Loudness in a Bottlenose Dolphin

Carolyn E. Schlundt and James J. Finneran

1 Introduction

For humans and terrestrial mammals, the variation in susceptibility to noise as a function of frequency is handled by "weighting" sound exposures to emphasize frequencies where auditory sensitivity is highest and lessen the importance of frequencies outside the audible range. This technique allows the use of single, weighted numeric values for impact or damage-risk criteria regardless of the sound frequency. Human weighting schemes were derived from measurements of equal-loudness curves obtained from subjective experiments where a listener compares the loudness of sounds at different frequencies. Previous terrestrial mammal data have shown that response latencies measured in the context of a simple acoustic-detection task may be used to construct equal-latency contours that are analogous to equal-loudness contours, albeit measured indirectly (Pfingst et al. 1975; Stebbins 1966). Until now, there were no empirical measures of equal-loudness curves or auditory weighting functions in marine mammals. This data gap became especially apparent following certain marine mammal experiments of temporary threshold shift (TTS). Limited data at 75 kHz (Schlundt et al. 2000) and more recent TTS data at frequencies up to 28 kHz (Finneran and Schlundt 2010; Finneran et al. 2007) have been compared with results of midfrequency data at 3 kHz (Finneran et al. 2010) and reveal substantial differences between onset TTS levels. Specifically, TTS will occur after lower exposure levels for these higher frequencies. Data at higher frequencies should be used to create more accurate frequency-dependent estimates for onset TTS (i.e., TTS weighting functions). Similarly, equal-loudness data would show the relationship between the frequency of sound and the subjective loudness of the sound. The objective of this effort was to develop auditory weighting functions for *Tursiops truncatus*. The weighting functions would be defined by measuring subjective loudness as a function of the sound frequency. Loudness contours may be more appropriate for assessing behavioral effects of sound, assuming behavioral reactions are more strongly related to loudness than to sound pressure level (SPL).

C.E. Schlundt (✉)
ITT Corporation, San Diego, CA 92154, USA
e-mail: carolyn.melka@itt.com

J.J. Finneran
US Navy Marine Mammal Program, Space and Naval Warfare Systems Center Pacific, San Diego, CA 92152, USA
e-mail: james.finneran@navy.mil

A.N. Popper and A. Hawkins (eds.), *The Effects of Noise on Aquatic Life*,
Advances in Experimental Medicine and Biology 730, DOI 10.1007/978-1-4419-7311-5_6,

2 Methods

The subject was a male *Tursiops truncatus* (26–27 yr, ~200 kg) whose hearing ability was within the "normal" range for *Tursiops truncatus* (Houser and Finneran 2006). The subject lived in floating netted enclosures (9 × 9 to 12 × 24 m) located in San Diego Bay, CA. The subject voluntarily beached for transport to an indoor, aboveground, vinyl-walled, seawater-filled, 4- × 6- × 1.5-m pool for testing each day and returned to its enclosure in the bay afterward. Sound fields in this pool were characterized by Finneran et al. (2007). All tests were approved by the Space and Naval Warfare Systems Center Pacific (SSC Pacific) Institutional Animal Care and Use Committee, San Diego, CA, and followed all applicable Department of Defense guidelines.

The pool contained an underwater test station composed of flooded polyvinylchloride tubing. The station contained a neoprene-covered "biteplate" on which the subject positioned itself as well as an underwater sound projector and calibration hydrophones. The biteplate was located middepth in the pool. The sound projector was located at a distance of ~1 m from the subject's ears when on the biteplate.

Testing was controlled using a personal computer and custom software to generate and calibrate sound stimuli (Finneran 2003). Additional custom software was used to control the order and method in which trials were presented and to record auditory stimuli and acoustic responses. Tones were projected in the direct field by an underwater sound projector (ITC 1032). Tones were frequency modulated (FM sine, 10% bandwidth) and 500 ms in duration, including 20-ms rise/fall times. FM tones were used to mitigate large sound pressure variations observed with pure-tone stimuli in the relatively small test pool (Finneran and Schlundt 2007). Tones were calibrated before and after each session.

Equal-loudness tests used a two-alternative, forced-choice, vocal-response paradigm where the subject was presented two sequential tones separated by a 500-ms gap. The subject was trained to whistle if the first tone was louder than the second and to produce a burst pulse or "buzz" response if the second tone was louder than the first. The majority of trials (~70%) featured "baseline" trials, stimulus pairs for which the loudness relationship between the two tone pairs was known. These included either two tones at the same frequency or those at frequencies one half-octave apart but with different SPLs. Baseline frequencies ($n = 13$) ranged from 1.8 to 113.1 kHz. The baseline trials allowed the animal's performance to be tracked within each session. The remaining trials were "probe" trials, consisting of a 10-kHz standard tone (A) with a fixed SPL (either 90, 105, or 115 dB re 1 μPa) and a comparison tone (B) whose frequency was fixed within a session but whose SPL varied. These are the trials of interest and the answers to these comparisons are unknown. Baseline and comparison frequencies ranged from 1.8 to 113.1 kHz ($n = 13$).

All baseline, standard, and comparison tones were at least 10 dB above the subject's threshold to ensure that both tones in a trial could be heard. SPLs between baseline tones ranged from 10 to 30 dB (minimum 15 dB for half-octave comparisons). Sessions were organized such that each baseline frequency and decibel difference between tones were represented equally and randomly throughout a session. There were an equal number of baseline trials to which the correct response should elicit the whistle or buzz response in order to avoid any response bias, and these were presented using a Gellerman (1933) distribution. There were two probe trials in every block of seven trials. Each standard-comparison pair (i.e., probe trial) was presented to the subject twice within the session in alternating order (i.e., A-B and B-A).

Tones of varying SPLs at 13 frequencies from 1.8 to 113.1 kHz were compared with a 10-kHz standard tone with a fixed SPL (either 90, 105, or 115 dB re 1 μPa) in *Tursiops truncatus*. The subject's responses to the probe trials were analyzed using logistic regression to derive curves relating the probability of the comparison tone being perceived louder for each comparison tone frequency. The 50% point then represented the SPL at which the comparison tone was equally as loud as the standard. This method is similar to methods used to derive equal-loudness curves in humans (Fletcher and Munson 1933; Robinson and Dadson 1956).

3 Results

Tursiops truncatus participated in more than 150 equal-loudness sessions and made direct decisions of subjective loudness in more than 16,000 trials. In general, the three equal-loudness contours closely followed the shape of the subject's audiogram and showed increased variability at the lowest and highest comparison frequencies. Included among the probe trials was a comparison frequency of 10 kHz (i.e., the same frequency as the standard tone). Analysis of this comparison was of particular interest because the SPL represented by the 50% point, where the comparison tone was equally as loud as the standard, was known. The results were in close agreement with the expected value, namely, 92, 105, and 120 dB for 10-kHz standard tones of 90, 105, and 115 dB re 1 μPa, respectively, lending confidence to the subject's responses to the probe trials at other comparison frequencies. Furthermore, performance on baseline trials was consistently well above the acceptable performance criterion of 80%. There was no response bias for response type or frequency regardless of whether baseline tone comparisons were within the same frequency or one half-octave apart. Performance improved as the difference between tones increased from 10 to 30 dB.

4 Conclusions

These data represent the first direct measurements of equal-loudness curves in any animal and show the relationship between the frequency and subjective loudness. The shape of the equal-loudness contours can be used to create weighting functions to properly emphasize frequencies at which auditory sensitivity is highest and lessen the importance of other frequencies, similar to human A- and C-weighting networks. Loudness contours may be more appropriate for assessing behavioral effects of sound, assuming behavioral reactions are more strongly related to loudness than to SPL.

References

Finneran JJ (2003) An integrated computer-controlled system for marine mammal auditory testing. Technical Report, Space and Naval Warfare Systems Center Pacific, San Diego, CA.

Finneran JJ, Carder DA, Schlundt CE, Dear R (2010) Growth and recovery of temporary threshold shift at 3 kHz in bottlenose dolphins: Experimental data and mathematical models. J Acoust Soc Am 127:3256–3266.

Finneran JJ, Schlundt CE (2007) Underwater sound pressure variation and bottlenose dolphin (*Tursiops truncatus*) hearing thresholds in a small pool. J Acoust Soc Am 122:606–614.

Finneran JJ, Schlundt CE (2010) Frequency-dependent and longitudinal changes in noise-induced hearing loss in a bottlenose dolphin (*Tursiops truncatus*). J Acoust Soc Am 128:567–570.

Finneran JJ, Schlundt CE, Branstetter B, Dear RL (2007) Assessing temporary threshold shift in a bottlenose dolphin (*Tursiops truncatus*) using multiple simultaneous auditory evoked potentials. J Acoust Soc Am 122:1249–1264.

Fletcher H, Munson WA (1933) Loudness, its definition, measurement and calculation. J Acoust Soc Am 5:82–108.

Gellerman LW (1933) Chance orders of alternating stimuli in visual discrimination experiments. J Gen Psychol 42:206–208.

Houser DS, Finneran JJ (2006) Variation in the hearing sensitivity of a dolphin population obtained through the use of evoked potential audiometry. J Acoust Soc Am 120:4090–4099.

Pfingst BE, Hienz R, Kimm J, Miller J (1975) Reaction-time procedure for measurement of hearing. I. Suprathreshold functions. J Acoust Soc Am 57:421–430.

Robinson DW, Dadson RS (1956) A re-determination of the equal-loudness relations for pure tones. Brit J Appl Phys 7:166–181.

Schlundt CE, Finneran JJ, Carder DA, Ridgway SH (2000) Temporary shift in masked hearing threshold of bottlenose dolphins, *Tursiops truncatus*, and white whales, *Delphinapterus leucas*, after exposure to intense tones. J Acoust Soc Am 107:3496–3508.

Stebbins WC (1966) Auditory reaction time and the derivation of equal loudness contours for the monkey. J Exp Anal Behav 9:135–142.

High Auditory Time Resolution in Bottlenose Dolphins Is Effective Protection Against Reverberation

Gennadi Zaslavski

1 Introduction

Bottlenose dolphins (*Tursiops truncates*) use very short clicks for echolocation, with theoretical time resolution as high as 15–20 µs (Au 1993). High time resolution of *Tursiops truncates* sonar clicks can provide extremely good protection against reverberation. However, the auditory time resolution of *Tursiops truncates* is widely believed to be ~300 µs (e.g., Au 1993; Supin et al. 2001) despite abundant behavioral data indicating that the auditory time resolution is as high as the theoretical time resolution of the echolocation clicks (e.g., Zaslavski 2003, 2006, 2008).

Because reverberation normally has the same frequency content as an outgoing sonar signal, a dolphin can hardly use frequency filtering for target detection or discrimination. The range resolution and directivity pattern of an underwater sonar system enable the sonar capability to operate in reverberant conditions. An Atlantic *Tursiops truncates* was able to detect a target lying on a sandy bottom at a 70-m range (Murchison 1980). Au and Turl (1983) and Turl et al. (1991) investigated the capability of echolocating *Tursiops truncates* and *Delphinapterus leucas* (beluga whale) to detect targets situated in front of a clutter screen. The experimental conditions corresponded to so-called backward masking that is customarily used for the auditory time resolution assessments. They found that *Tursiops truncates* and *Delphinapterus leucas* performances gradually degraded with the target-screen separation decrease that was typical for the backward masking. The degradation was faster for *Tursiops truncates* than for *Delphinapterus leucas*. Nevertheless, for an echo-to-reverberation ratio of ~0 dB (−3 to +3), the threshold (70–75% correct response) separation for both animals was within 2.5–5 cm, which corresponded to a 35- to 50-µs delay between the target echo and reverberation. These numbers could well stand for the range and time resolution of *Tursiops truncates* and *Delphinapterus leucas* sonar. However, the authors chose to characterize the sonar performance in reverberation by a threshold echo-to-reverberation ratio at a zero target-screen separation. At a coplanar position of the target and clutter screen, the threshold target echo-to-reverberation ratio for *Delphinapterus leucas* was 3–5 dB lower than the one found for *Tursiops truncates*.

The estimate of Atlantic *Tursiops truncates* sonar time resolution that could be made using the results of the experiment by Au and Turl (1983) is obviously better than the commonly accepted

G. Zaslavski (✉)
University Authority for Applied Research, RAMOT, Tel Aviv 61392, Israel
e-mail: gennadi.zaslavski@gmail.com

A.N. Popper and A. Hawkins (eds.), *The Effects of Noise on Aquatic Life*,
Advances in Experimental Medicine and Biology 730, DOI 10.1007/978-1-4419-7311-5_7,

300 μs (Au 1993). The range and time resolution of Black Sea *Tursiops truncates* sonar measured in similar reverberant conditions were found to be even better than estimated for both Atlantic *Tursiops truncates* and *Delphinapterus leucas*. Zaslavski and Novikov (1983) investigated the ability of Black Sea *Tursiops truncates* and *Phocoena phocoena* (harbor porpoise) to detect and discriminate targets situated in front of single solid-steel balls 7.6 cm in diameter. Two 7.6-cm masker balls were suspended in water at a 1-m depth on the left and right sides of a separation net. The dolphins were required to determine in front of which masker ball was a target and approach it. In case of target discrimination, comparison targets were placed in front of each masker ball.

Tursiops truncates was able to detect a 1-cm steel ball situated just 2.7 cm (between front surfaces of the balls) in front of the 7.6-cm masker ball. The threshold delay of ~35 μs (at 75% correct response level) was found to be as small as the one that could be estimated for *Delphinapterus leucas* (Turl et al. 1991) but at a much smaller echo-to-reverberation ratio (around −18 dB compared with 0 dB for *Delphinapterus leucas*). At the threshold target-masker ball separation of ~3.0 cm, *Tursiops truncates* was able to discriminate between a 3-cm steel and a brass solid sphere.

A slightly larger threshold echo-reverberation separation of ~6.5 cm was found in *Phocoena phocoena* for 1-cm steel ball detection as well as for discrimination between the 3-cm steel and brass solid spheres. The differences could be expected because *Phocoena phocoena* echolocation clicks are longer than *Tursiops truncates* echolocation clicks. It should be noted that despite the experiments being conducted in closed pool, a small target suspended on a long line could swing by at ~1.5 cm, giving a chance for the animals to detect the target at slightly larger separations than the recorded threshold. Still, the auditory time resolution corresponding to the threshold separation, including correction for the target swinging, proved to be almost as good as the theoretical time resolution of *Tursiops truncates* and *Phocoena phocoena* echolocation clicks of ~20 and 50 μs, respectively. Here I discuss the detection and discrimination of targets positioned in front of clutter screens by the same Black Sea *Tursiops truncates* tested in the experiment of Zaslavski and Novikov (1983).

2 Methods

The clutter screen consisted of ~80 asymmetrical round stones (sea pebbles) 5–7 cm arranged vertically in a rectangular area 60 × 50 cm. Two clutter screens were suspended at the left and right sides of a separation net at 1-m depth (center of the screen) 7 m from the animal's start position and at 30° azimuth separation. Solid aluminum and brass spheres 5 cm in diameter were used as targets. The dolphin's performance was investigated as a function of separation (range difference) between a target and the clutter screen. The dolphin was required to determine at which side of the separation net was a target and approach it. The position of a target (in front of the left or right clutter screen; Fig. 1b) in any particular trail was randomized. A transducer with transmit-and-receive beams similar to those of *Tursiops truncates* sonar recorded echoes from a target situated in front of the clutter screen (Fig. 1c). The transducer transmit click was similar to the *Tursiops truncates* echolocation click, with a peak frequency of ~120 kHz.

3 Results

The threshold target-clutter screen separation was found to be ~4 and 5 cm for 5-cm aluminum and brass sphere detection, respectively. At a 2.5-cm separation, the targets were half hidden between the clutter screen stones. Discrimination between the aluminum and brass spheres was above the 75% correct response at the target-clutter screen separation, larger than that at ~7 cm.

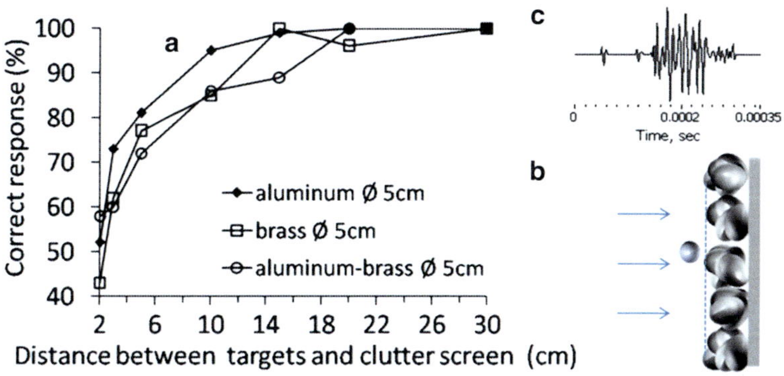

Fig. 1 (**a**) *Tursiops truncates* performance as a function of target-clutter screen separation. (**b**) A target positioned in front of the clutter screen (schematic). (**c**) The echo from a 5-cm brass solid sphere (two small highlights) and the clutter screen

These threshold values should be increased by 1–1.5 cm because of slight target swinging in front of the clutter screen. The threshold target-clutter screen separation of 4–5 cm can be converted to a threshold delay between the target echo and reverberation for 55–65 μs.

Although these threshold values appear to be slightly larger than those found for Atlantic *Tursiops truncates*, an echo-to-reverberation ratio for either target in our experiments was much smaller (−12 to −14 dB for peak-to-peak amplitudes and around −20 dB for energy ratio).

The backscatter record shown in Figure 1c was made for the clutter screen placed perpendicular to the transducer acoustic axis (90° grazing angle). Because the dolphin was not fixed at the start position, it could ensonify targets and clutter screens at 75–90° grazing angles, receiving slightly different echo-to reverberation ratios.

Even better results were obtained in earlier behavioral experiments with this dolphin on passive detection of a click (simulating a *Tursiops truncates* echolocation click) in a condition of backward masking by a noise pulse. For a 100-μs noise masker, a −20 dB click-to-masker (peak-to-peak) threshold ratio was reached at the masker delay of just 50 μs. The threshold click-to-masker ratio was as low as around −30 dB when the masker duration was decreased to 20–30 μs (Zaslavski 2003).

Although Black Sea *Tursiops truncates* appeared to perform in reverberant conditions better than Atlantic *Tursiops truncates* and *Delphinapterus leucas*, individual animal differences could easily account for a difference in the thresholds. A 6-dB or even larger difference in threshold values measured in the same target or signal detection and discrimination tasks between two *Tursiops truncates* was quite ordinary in many of our behavioral experiments. In any case, all dolphins tested in reverberant conditions indicated an auditory time resolution better than 50–60 μs. The results strongly support the notion that *Tursiops truncates* auditory time resolution is as high as the theoretical time resolution of the echolocation clicks.

References

Au WWL (1993) The sonar of dolphins. Springer-Verlag, New York.

Au WWL, Turl CW (1983) Target detection in reverberation by an echolocating Atlantic bottlenose dolphin (*Tursiops truncates*). J Acoust Soc Am 73:1676–1681.

Murchison AE (1980) Detection range and range resolution of porpoise. In: Busnel RG, Fish JF (eds) Animal sonar systems. Plenum Press, New York, pp 43–70.

Supin AY, Popov VV, Mass AM (2001) The sensory physiology of aquatic mammals. Kluwer Academic Publishers, Boston, MA.

Turl CW, Skaar DJ, Au WWL (1991) The echolocation ability of the beluga (*Delphinapterus leucas*) to detect target in clutter. J Acoust Soc Am 89:896–901.
Zaslavski GL (2003) Target echo processing in the dolphin's auditory system. Proceedings of the Oceans 2003 MTS/IEEE Conference, vol 2, 22–26 September 2003, San Diego, CA, pp 620–628.
Zaslavski GL (2006) Differences between the auditory system of humans and bottlenose dolphins. Advances in Bioacoustics 2, Dissertationes Classis IV: Historia Naturalis, Slovenian Academy of Sciences and Arts, (Ljubljana), XLVII-3:51–74.
Zaslavski GL (2008) The auditory time resolution in bottlenose dolphins: Behavioral experiments versus auditory evoked potential methods. Proceedings of the 9th European Conference on Underwater Acoustics (ECUA) 2008/Acoustics08 Paris Conference, 30 June to 4 July, 2008, Paris, France, pp 571–576.
Zaslavski GL, Novikov BA (1983) Range resolution of the bottlenose dolphin sonar. Proceedings of the 10th All-Union Acoustic Conference, Moscow, USSR, pp 5–7.

Frequency Selectivity in the Bottlenose Dolphin Auditory System

Gennadi Zaslavski

1 Critical Ratio and Critical Bandwidth

The auditory system is often modeled by a bank of band-pass filters. There are two different estimates of the bandwidth of the auditory filters, called the critical ratio (CR) and the critical bandwidth (CB), based on measuring detection thresholds of a pure tone in continuous broadband noise (Fletcher 1940). In humans, the CRs are ~2.5 times smaller than the CBs. In the bottlenose dolphin (*Tursiops truncates*), the relationship between the CR and CB is different than that in humans. Au and Moore (1990), using a behavioral-response paradigm and variable bandwidth noise, found that in *Tursiops truncates*, the CB was 11 times wider than the CR at 30 kHz, 8.2 times wider at 60 kHz, and 2.2 times wider at 120 kHz. The CBs at these frequencies were found to be 17, 25, and 45 kHz, respectively. Even the CR measured at 100 and 120 kHz was as large as that around 20 kHz. Lemonds et al. (2000), in behavioral experiments, estimated the auditory filter bandwidth in *Tursiops truncates* using notched noise. Equivalent rectangular bandwidth (ERB) of the suggested auditory roex(p,r) filters was found to range from 4.4 kHz at 40 kHz to 16 kHz at 100 kHz.

The behaviorally estimated ERBs and CBs (Au and Moore 1990) are substantially larger than ~3% (of the filter center frequency) ERBs reported by Popov et al. (1997) based on measurements of the *Tursiops truncates* auditory brain stem responses. Assuming that both the behavioral and electrophysiological data are correct (within some experimental and individual differences between animals), *Tursiops truncates* appears to be capable of voluntarily changing the auditory filter bandwidth at the same frequency, e.g., 100 kHz, from as small as 3 kHz to as large as 45 kHz.

2 Detection and Discrimination of Brief Signals

Although the use of a narrowband auditory filter is obviously advantageous for detection of a pure tone in noise, detection and discrimination of brief broadband signals similar to *Tursiops truncates* echolocation clicks may require broadband reception. There are numerous behavioral data on detection and discrimination of brief signals in noise. In many cases, *Tursiops truncates* was able to

G. Zaslavski (✉)
University Authority for Applied Research (RAMOT), Tel Aviv University,
Tel Aviv 61392, Israel
e-mail: gennadi.zaslavski@gmail.com

A.N. Popper and A. Hawkins (eds.), *The Effects of Noise on Aquatic Life*,
Advances in Experimental Medicine and Biology 730, DOI 10.1007/978-1-4419-7311-5_8,

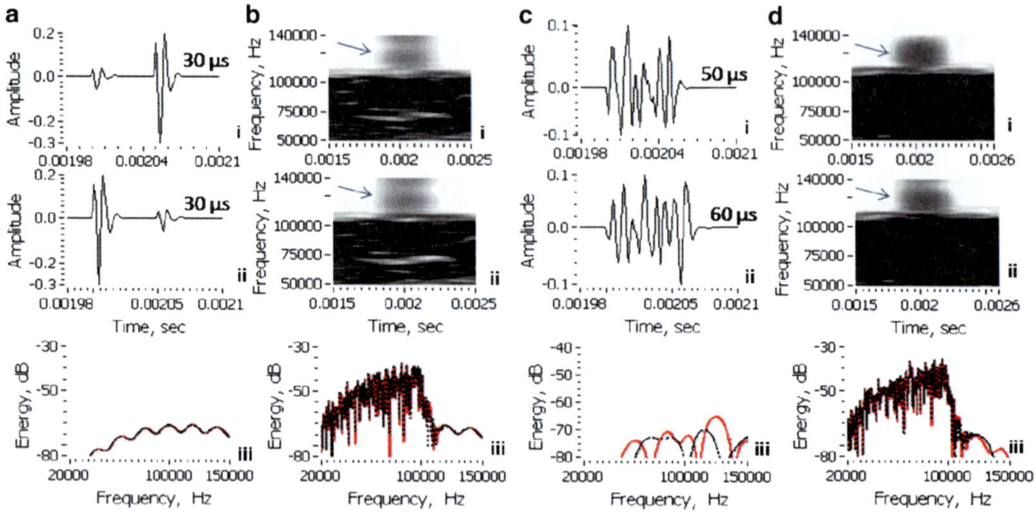

Fig. 1 (**a**) Time-reversed double clicks (i and ii) with identical energy spectra (iii). (**b**) Short-time Fourier transform (STFT) spectrogram (i and ii) and energy spectra (iii) of the double clicks mixed with broadband (10–100 kHz) continuous noise. (**c**) Noise pulses of 50 (i) and 60 (ii) μs and their energy spectra (iii). (**d**) STFT spectrograms of the noise pulses (i and ii) mixed with broadband (10–100 kHz) continuous noise and their energy spectra (iii). The spectrograms were generated using a 700-μs Hanning analysis window. Signal-to-noise ratio was 3–6 dB above the threshold. Arrows indicate the portion of the short-time spectra of the double clicks (**b**) and the noise pulses (**d**) unmasked by the noise

detect or discriminate signals even when most of the hearing frequency range was masked with broadband continuous noise (Zaslavski 2001, 2003, 2007a).

A frequency span of 20–25 kHz around 110 kHz to the high-frequency limit of *Tursiops truncates* hearing of 135–140 kHz (Fig. 1) was normally wide enough for the dolphins to detect or discriminate between most of the tested brief signals in our behavioral experiments. In reference to the behavioral data (Au and Moore 1990), this frequency span could be associated with a single auditory filter. On the other hand, the 20- to 25-kHz frequency range is wide enough to accommodate 6 to 8 auditory filters having a bandwidth of ~3 kHz (Popov et al. 1997). In some cases, however, it is possible to tell whether the frequency span used by the dolphin is a single broadband auditory filter or a set of narrowband filters.

To discriminate the signals given in Figure 1, a single broadband auditory filter appears to be much more appropriate than a set of narrowband auditory filters. Because the signals have identical (Fig. 1a) or random (Fig. 1c) energy spectra, discrimination most likely was based on the difference in the time waveform of the auditory filter reaction. The wider the auditory filter, the better the difference in duration of the input noise pulses is preserved at the auditory filter output (Zaslavski 2007a; Zaslavski and Ryabov 1991).

Tursiops truncates appeared to also use a single auditory filter to discriminate the double clicks having identical energy spectra (Fig. 1a). The discrimination was apparently based on time domain differences in the auditory filter response to the double clicks (Zaslavski 2001, 2003, 2006). To preserve the time domain features at the auditory filter output, the bandwidth of the filter should be large enough. I found that the smaller the interclick interval, the wider the frequency span required for the dolphin to discriminate the double clicks (Zaslavski 2001).

The frequency span required for *Tursiops truncates* to discriminate between brief clicks could be as wide as 60–70 kHz, from 40–50 kHz to ~120 kHz, which is almost as large as the frequency range of *Tursiops truncates* echolocation hearing (Zaslavski 2007b). In this experiment, the dolphin seemed to discriminate the time domain waveforms between two clicks separated by an interclick interval less than 100 μs. On the other hand, in behavioral experiments, *Tursiops truncates* was required to discriminate between brief signals with fixed differences in frequency spectra and the bandwidth of the auditory filter at high frequencies was found to be as small as 5–6 kHz (at 3-dB level). A possible relationship between the auditory filter bandwidth and discrimination task needs to be discussed.

References

Au WWL, Moore PWB (1990) Critical ratio and critical bandwidth for the Atlantic bottlenose dolphin. J Acoust Soc Am 88:1635–1638.
Fletcher H (1940) Auditory patterns. Rev Mod Phys 12:47–61.
Lemonds DW, Au WWL, Nachtigall PE, Roitblat HL, Vlachos SA (2000) High-frequency filter shapes in an Atlantic bottlenose dolphin. J Acoust Soc Am 108:2614.
Popov VV, Supin AY, Klishin VO (1997) Frequency tuning of the dolphin's hearing as revealed by auditory brainstem response with notch-noise masking. J Acoust Soc Am 102:3795–3801.
Zaslavski GL (2001) Temporal order discrimination in the dolphin. Proceedings of the Second International Symposium on Underwater Biosonar and Bioacoustic Systems, Loughborough University, 23–24 July 2001, Loughborough, Leicestershire, UK, pp 143–152.
Zaslavski GL (2003) Discrimination of the signals with identical energy spectra by the bottlenose dolphin. Proceedings of the Oceans 2003 Conference, San Diego, CA, pp 633–641.
Zaslavski GL (2006) Differences between the auditory system of humans and bottlenose dolphins. Advances in Bioacoustics 2, Razprave IV. Razreda sazu XLVII-3, Ljubljana, Slovenia, pp 51–74.
Zaslavski GL (2007a) Discrimination of time intervals by the Black Sea bottlenose dolphin. Proceedings of the Fourth International Conference on Bio-Acoustics, 10–12 April 2007, Loughborough University, Loughborough, Leicestershire, UK, pp 277–281.
Zaslavski GL (2007b) 100-microseconds time interval of the bottlenose dolphin's auditory system. Proceedings of the Fourth International Conference on Bio-Acoustics, 10–12 April 2007, Loughborough University, Loughborough, Leicestershire, UK, pp 21–29.
Zaslavski GL, Ryabov VA (1991) Time interval discrimination by a bottlenose dolphin. Proceedings of the 9th All-Union Acoustic Conference, Moscow, Russia, pp 33–35.

Auditory Brain Stem Responses Associated with Echolocation in an Atlantic Bottlenose Dolphin (*Tursiops truncatus*)

Songhai Li, Paul E. Nachtigall, and Marlee Breese

1 Introduction

Dolphins and toothed whales (odontocetes) possess highly developed sound production systems and hearing capabilities (Au 1993; Au et al. 2000). Because sound is transmitted much more efficiently than light and other possible stimuli through water, hearing plays a fundamental role as a primary sensory modality in dolphins and toothed whales and functions to aid in navigation, orientation, foraging, and communication (Au 1993; Nachtigall and Moore 1988; Richardson et al. 1995). Any sound in the water is detectable when the received level of the sound exceeds a certain hearing detection threshold of the animal. The efficiency of underwater sound propagation allows underwater noise created by ships and other human activities to be detected by aquatic animals far away from the source. Dolphins and toothed whales may be listening to many sounds from natural and human-made sources in addition to the sounds from themselves. Both natural and human-made sounds could have deleterious effects on the animals through interference with the animals' ability to detect signals from conspecifics and echoes of echolocation clicks.

Concerns have arisen about the effect of human-made noise on the dolphins and toothed whales. There has been a lot of research emphasizing the hearing of external signals or the effects of human-made noise on hearing external signals with some odontocete species; however, there are still many unstudied aspects of odontocete hearing and related noise effects such as how loudly the animals hear their own echolocation clicks and corresponding echoes, how human-made noise affects their ability to hear the echoes of echolocation clicks, and whether the animals can actively control their hearing under certain noise situations or not. To answer these questions, above all we need to investigate the basic mechanisms that allow the dolphins and toothed whales to hear their own signals and corresponding echoes. Although an attempt was made to record auditory evoked potentials (AEPs) to voluntary click emissions from a dolphin by Bullock and Ridgway (1972) several decades ago, a quantitative investigation on the ability of odontocetes to hear their own echolocation clicks and the corresponding echoes has only recently been reported in a single false killer whale (*Pseudorca crassidens*) (Nachtigall and Supin 2008; Supin et al. 2003, 2004). These studies showed that a noninvasive evoked-potential method may be a very effective way to investigate the basic mechanisms by which echolocating dolphins and toothed whales hear their own signals and the

S. Li (✉) • P.E. Nachtigall • M. Breese
Marine Mammal Research Program, Hawai'i Institute of Marine Biology,
University of Hawai'i, Kane'ohe, HI 96744, USA
e-mail: songhai@hawaii.edu; nachtiga@hawaii.edu

A.N. Popper and A. Hawkins (eds.), *The Effects of Noise on Aquatic Life*,
Advances in Experimental Medicine and Biology 730, DOI 10.1007/978-1-4419-7311-5_9,

corresponding echoes. In this study, we similarly investigated the auditory brain stem responses (ABRs) of an Atlantic bottlenose dolphin (*Tursiops truncatus*) to outgoing echolocation clicks and the corresponding echoes during echolocation by using the evoked-potential method.

2 Subjects and Methods

The experiments were carried out in the facilities of the Marine Mammal Research Program, Hawai'i Institute of Marine Biology, Kane'ohe, HI. The subject was an adult female *Tursiops truncatus*, named BJ, who was 25 years of age at the time of the experiment. She was housed in a wire-net enclosure in Kane'ohe Bay, HI. The animal was trained to wear soft latex suction cups containing human EEG electrodes to pick up the evoked potentials, to ensonify and recognize targets by echolocation, and to report the target presence or absence using a go/no-go reporting paradigm.

The animal was trained to detect the presence or absence of a target. Three targets were used in this study. They were hollow aluminum cylinders with an outer diameter of 38 mm (1.5 in.), an inner diameter of 25.4 mm (1 in.), and vertical axes 180, 90, and 46 mm long. The target strengths were −22, −28, and −34 dB, respectively, as measured by a short pulse produced by excitation of a 60-mm spherical piezoceramic transducer with 10-μs rectangular pulses (Supin et al. 2004). The targets were hung from a thin monofilament line at a distance ~2 m from the blowhole of the animal. ABR collection was triggered by echolocation clicks produced by the animal in both target-present and target-absent scenarios. The echolocation clicks were picked up by a RESON TC4013 hydrophone in front of the animal's head at a distance ~1.5 m from the blowhole of the animal. To extract low-amplitude ABRs from background noise, an off-line averaging procedure was used.

3 Results and Discussion

After averaging more than a thousand individual records, clearly visible response waves (each shorter than 1 ms) emerged. When a target was presented, all records contained two typical two-wave ABR sets, with 2.56–2.64 ms apart from each other. The ABR sets consisted of two typical alternate positive and negative waves similar to those described previously in a number of dolphin species (Supin et al. 2001). The latency of the first positive peak of the first ABR set from the start of the record is 2.1–2.2 ms; the next negative peak is at 2.4–2.5 ms; the third one is a positive peak at 2.8–2.9 ms; and the last one is negative peak at 3.3–3.4 ms. They are all independent of targets. Considering a distance difference of ~1.8 m between the region of the nasal sacs (an expected region of sound generation in dolphins) to the hydrophone and to the animal's ears, the response latency after a click emission would be ~1.2 ms longer than the latency from the start of the record. When the targets were absent, the records contained only 1 two-wave ABR set emerging in a consistent latency with that of the first ABR set of the target-present records. The two ABR sets of target-present records with 2.56–2.64 ms apart, corresponding to an ~4-m back and forth distance, were interpreted as responses to both outgoing echolocation clicks and echoes. The one ABR set of target-absent records was interpreted as a response to outgoing echolocation clicks without echoes. Averaging all the evoked responses triggered by the signals received from the hydrophone, the peak-to-peak amplitude of the click-related ABR was 0.75–0.95 μV for both target-present and target-absent scenarios. It should be noted, however, that there were probably some number of "false" evoked responses that were triggered by external noise, not echolocation clicks from the animal. Being averaged together with click-triggered evoked responses, the false evoked responses may result in a lower amplitude of the records. When sorted by received levels of echolocation click, the peak-to-peak amplitude of click-related ABRs increased proportionally with the received levels of the clicks for both target present and target absent, and the ABR sensitivities were comparable between both scenarios. Averaging all the evoked responses triggered by signals received from the hydrophone in the target-present scenario, the peak-to-peak amplitude of the echo-related ABRs was 0.24, 0.26, and 0.08 μV, which was 3.2, 3.6, and 9.4 times weaker than that of the click-related ABRs for the target strengths of −22, −28, and −34 dB, respectively. The echo-related ABRs are obviously lower than the click-related ABRs and apparently dependent on the target strength.

References

Au WWL (1993) The sonar of dolphins. Springer-Verlag, New York.

Au WWL, Popper AN, Fay RR (eds) (2000) Hearing by whales and dolphins. Springer-Verlag, New York.

Bullock TH, Ridgway SH (1972) Evoked potentials in the central auditory system of alert porpoises to their own and artificial sounds. J Neurobiol 3:79–99.

Nachtigall PE, Moore PWB (1988) Animal sonar: Processes and performance. Plenum, New York.

Nachtigall PE, Supin AY (2008) A false killer whale adjusts its hearing when it echolocates. J Exp Biol 211:1714–1718.

Richardson WJ, Greene CR Jr, Malme CI, Thomson DH (1995) Marine mammals and noise. Academic Press, San Diego, CA.

Supin AY, Nachtigall PE, Au WWL, Breese M (2004) The interaction of outgoing echolocation pulses and echoes in the false killer whale's auditory system: Evoked-potential study. J Acoust Soc Am 115:3218–3225.

Supin AY, Nachtigall PE, Pawloski J, Au WWL (2003) Evoked potential recording during echolocation in a false killer whale *Pseudorca crassidens*. J Acoust Soc Am 113:2408–2411.

Supin AY, Popov VV, Mass AM (2001) The sensory physiology of aquatic mammals. Kluwer Academic Publishers, Boston, MA.

Hearing, Noise, and Echolocating Odontocetes

Paul E. Nachtigall, Alexander Y. Supin, and Marlee Breese

1 Introduction

Anthropogenic noise may interfere with active echolocation, which is the primary foraging tool for odontocete cetaceans. Our work on temporary threshold shifts (Mooney et al. 2009a; Nachtigall et al. 2003, 2004) has shown that either lower levels of anthropogenic noise presented for long time periods or intense sonar pings for short time periods (Mooney et al. 2009b) can produce a temporary reduction in hearing sensitivity and temporary threshold shifts. Intense sounds intended to disrupt echolocation can also reduce echolocation performance (Mooney et al. 2009c). Our measurements of hearing during echolocation have shown that self-generated intense outgoing signals are managed by the whale's auditory system using a number of mechanisms (Nachtigall and Supin 2008; Supin et al. 2007, 2008, 2009) based on measurements of the auditory evoked potentials (AEPs) associated with what the animal hears of its own echolocation signals and echoes while echolocating (Supin et al. 2003, 2004).

2 Temporary Threshold Shifts

Dolphins have been shown to have robust hearing systems in response to noise exposure. Our first work (Nachtigall et al. 2003) using a behavioral task that took nearly 20 min to obtain thresholds after intense sound exposure demonstrated that noise levels of a fatiguing noise between 4 and 11 kHz, essentially equivalent to a noise pressure level of 179 dB re 1 μPa for 50 min, produced a dolphin temporary threshold shift (TTS) on average of 11 dB. Not being satisfied with the 20-min wait to test because recovery was likely occurring within that time, we began using AEP hearing measurements to quickly examine thresholds after exposure (Nachtigall et al. 2004). Noise exposure was reduced to 160 dB re 1 μPa and thresholds were tested within 5 min postexposure. After 30 min

P.E. Nachtigall (✉) • M. Breese
Marine Mammal Research Program, Hawai'i Institute of Marine Biology,
University of Hawai'i, Kane'ohe, HI 96744, USA
e-mail: nachtiga@hawaii.edu

A.Y. Supin
Institute of Ecology and Evolution, Russian Academy of Sciences,
119071 Moscow, Russia

A.N. Popper and A. Hawkins (eds.), *The Effects of Noise on Aquatic Life*,
Advances in Experimental Medicine and Biology 730, DOI 10.1007/978-1-4419-7311-5_10,

of exposure, in a hoop station, the dolphin showed a 5-dB shift, with total recovery to preexposure levels after 100 min. The recovery rate was shown to be ~1.5 dB per doubling of time.

Many sound exposures do not occur as continuous tones. Sonar pings are single energy bursts. A logical question arose as to whether equal amounts of TTS could be produced with equal amounts of energy (a short loud burst compared with a longer, lower, more sustained loud exposure of equal energy). This equal-energy hypothesis was examined (Mooney et al. 2009a) using the same basic experimental procedures and dolphin discussed above. The equal-energy hypothesis did not hold. It took more energy in a short amount of time to produce a TTS than energy spread out across a longer time period. This was further examined empirically using recorded navy sonar signals (Mooney et al. 2009b) to produce TTS. The data supported an increasing energy model to predict TTS and that a dolphin exposed to a 53-C sonar operating at 235 dB re 1 μPa would have to remain ~40 m from the ship's sonar source for 2–2.5 min to experience a TTS.

3 Echolocation Disruption With Noise

False killer whales are currently being hooked on tuna longlines, and acoustic devices have been developed to disrupt echolocation with noise. The Long-line Saver pinger was examined (Mooney et al. 2009c) to see whether or not it disrupted a false killer whale's ability to echolocate a small cylinder located 8 m away. When the 182 dB re 1 μPa device was initially used, performance was disrupted. Subsequent performance of the task with the pinger still present recovered, and reduction of the intensity of the signal resulted in total echolocation recovery. This short experiment appears to indicate that an experienced animal may quickly adapt its echolocation to overcome intense noise interference. Obviously, much work is required in this area given the importance of echolocation as a foraging tool.

4 Measuring Hearing During Echolocation

Odontocetes produce loud echolocation clicks, and their auditory systems must immediately process quiet returning echoes. Nachtigall and Supin (2008) and Supin et al. (2003, 2004, 2007, 2008, 2009) have been examining the hearing of the false killer whale during echolocation. The whale is trained to echolocate and report the presence or absence of targets presented at various distances. The animals swim into a hoop in a fixed position while wearing rubber suction-cup skin-surface electrodes. AEPs are triggered from the individual outgoing echolocation clicks, and hearing of both the outgoing click and the returning echoes are inferred to the amplitudes of the measured levels of the evoked potentials.

Our measurements of AEP hearing during echolocation have shown that self-generated intense outgoing signals are managed by the whale's auditory system in a variety of ways. First, the whale, in some way, "muffles" what is heard of the outgoing signal. Even though it is produced inside the head within 20 cm of the ear, the animal hears the outgoing signal 40 dB down from what it hears as an equivalent signal presented 1 m in front of it in the free field. Second, the whale apparently uses forward masking of the signal as a process to manage what it hears of echoes in an automatic gain control scenario. This was examined by playing two clicks to the whale and varying both amplitude and the time between clicks (Supin et al. 2007). Hearing recovers from the forward masking of the outgoing click as the echoes return. Within a short range, the greater the time between outgoing click and echo return, the more sensitive the ear becomes. The third mechanism used by the whale for the maximization of hearing quiet echoes is an active control of the hearing process itself, a true automatic gain control. This has been demonstrated in two ways. 1) As the whale echolocates, it generally hears its outgoing clicks at a stable level. If very small targets are presented, the animal will increase its sensitivity to hear both the echoes and the outgoing clicks at a more sensitive level. 2) If a comparison is made between hearing target-present outgoing clicks and target-absent outgoing clicks, there is a 20 dB re 1 μPa difference between the hearing levels of outgoing signals in the two conditions. It appears as though the animal is 20 dB more sensitive when searching for targets (Supin et al. 2008).

References

Mooney TA, Nachtigall PE, Breese M, Vlachos S, Au WWL (2009a) Predicting temporary threshold shifts in a bottlenose dolphin (*Tursiops truncatus*): The effects of noise level and duration. J Acoust Soc Am 125:1816–1826.

Mooney TA, Nachtigall PE, Vlachos S (2009b) Sonar induced temporary hearing loss in dolphins. Biol Lett 5:565–567.

Mooney TA, Pacini A, Nachtigall PE (2009c) False killer whale (*Pseudorca crassidens*) echolocation and acoustic disruption: Implications for long-line bycatch and depredation. Can J Zool, 87:726–733.

Nachtigall PE, Pawlowski JL, Au WWL (2003) Temporary threshold shifts and recovery following noise exposure in the Atlantic bottlenosed dolphin (*Tursiops truncatus*). J Acoust Soc Am 113:3425–3429.

Nachtigall PE, Supin AY (2008) A false killer whale adjusts its hearing when it echolocates. J Exp Biol 211:1714–1718.

Nachtigall PE, Supin AY, Pawloski JL, Au WWL (2004) Temporary threshold shifts after noise exposure in a bottlenose dolphin (*Tursiops truncatus*) measured using evoked auditory potentials. Mar Mamm Sci 20:673–687.

Supin AY, Nachtigall PE, Au WWL, Breese M (2004) The interaction of outgoing echolocation pulses and echoes in the false killer whale's auditory system: Evoked-potential study. J Acoust Soc Am 115:3218–3225.

Supin AY, Nachtigall PE, Breese M (2007) Evoked-potential recovery during double click stimulation in a whale: A possibility of biosonar automatic gain control. J Acoust Soc Am 121:618–625.

Supin AY, Nachtigall PE, Breese M (2008) Forward masking as a mechanism of automatic gain control in whale biosonar: A psychophysical study. J Acoust Soc Am 124:648–656.

Supin AY, Nachtigall PE, Breese M (2009) Forward-masking based gain control in odontocete biosonar: An evoked-potential study. J Acoust Soc Am 125:2432–2442.

Supin AY, Nachtigall PE, Pawloski J, Au WWL (2003) Evoked potential recording during echolocation in a false killer whale *Pseudorca crassidens* (L). J Acoust Soc Am 113:2408–2411.

Evoked Potential Audiometry in Aquatic Mammals

Alexander Y. Supin

1 Introduction

Investigation of the influence of noise on the hearing of aquatic mammals requires appropriate methods of audiometry to assess any temporal or permanent hearing sensitivity shift. The auditory evoked potential (AEP) method is widely used for this purpose. It does not require long training of the subject, is not time consuming, and therefore may be used in short-term captivity conditions (on the catch-and-release basis) and in wild conditions (in stranded animals). The method is the most productive in odontocetes because of the unique features of their auditory system (hypertrophy of the auditory nerve centers, high AEP amplitude, and capability to produce AEP at high rates of stimulus presentation [Supin et al. 2001]). However, the efficiency of AEP audiometry depends on the version of the method used. Here, the influences of probe stimulus type, data collection, and processing manner are considered.

2 Use of Single or Rhythmic Probe Signals

Originally, single-tone pips or band-filtered clicks were used as probe stimuli to find hearing thresholds (Popov and Supin 1990). This version of the probe is very effective in producing AEPs, which makes it usable for a wide variety of investigations. However, with the use of this probe, one comes up against a problem of comparing AEPs and psychophysical threshold estimates. In psychophysical measurements, the duration of the probe may be unlimited, thus exploiting the entire possible temporal summation of the probe energy in the auditory system. On the contrary, only short sound transients effectively produce AEPs, so AEP thresholds may be higher because of incomplete temporal summation. Therefore, correction for incomplete temporal summation must be made for comparison with psychophysical data.

In odontocetes, AEP audiometry may use rhythmically modulated sound stimuli instead of single clicks or tone pips. It is possible because of the unique properties of the odontocete auditory system that have a very high temporal resolution and are capable of responding to sound modulation with

A.Y. Supin (✉)
Severtsov Institute of Ecology and Evolution, Russian Academy of Sciences,
Moscow 119071, Russian Federation
e-mail: alex_supin@mail.ru

A.N. Popper and A. Hawkins (eds.), *The Effects of Noise on Aquatic Life*,
Advances in Experimental Medicine and Biology 730, DOI 10.1007/978-1-4419-7311-5_11,

a rate of more than 1,000 Hz (Dolphin 1995; Popov and Supin 1998; Supin and Popov 1995). The response to such stimuli (the envelope-following response [EFR]) is a high-rate rhythmic sequence of AEPs. Contrary to single stimuli, multicycle rhythmic responses may be more confidently extracted from background noise by Fourier analysis. Apart from that, rather long amplitude-modulated stimuli can be characterized by mean root mean square (RMS) level, which makes possible a direct comparison with psychophysical data.

However, among all aquatic mammals, the use of high-rate rhythmic stimuli for AEP audiometry is possible only for odontocetes. In other aquatic and semiaquatic mammals (pinnipeds and sirenians), the auditory system does not feature the ability to produce well-developed AEPs at stimulus rate of hundreds of hertz. Only a single-click/-pip paradigm of stimulation is applicable in those cases.

3 Use of Steady-State and Short-Burst Rhythmic Probe Stimuli

There are two main manners in presenting rhythmic stimuli: steady state and short burst. With the steady-state manner, a continuous rhythmic signal is presented throughout data collection; to extract the response from noise, fractions of the continuous record containing a few response cycles are averaged and subjected to Fourier transform to evaluate the magnitude of the response at the modulation frequency (Dolphin 1995). With the short-burst presentation manner, signals are presented as bursts lasting 10–30 ms, separated by several times longer silent intervals (Popov and Supin 1998; Supin and Popov 1995).

An advantage of the last manner of presentation is that it reveals the dynamics of the response. A lag of the response relative to the stimulus burst is an ideal control, ensuring that the response is of physiological origin and not a physical artifact. The steady-state presentation does not offer this opportunity; several early investigations using steady-state probes featured obvious artifacts (e.g., Dolphin 1995).

4 Sinusoidal Amplitude-Modulated Versus Pip-Train Rhythmic Stimuli

Among several version of rhythmic probes, sinusoidal amplitude-modulated (SAM) stimuli are widely used for AEP audiometry in odontocetes. Apart from many advantages (narrow frequency spectrum, possibility to characterize the level by long-term RMS value), SAM stimuli feature a disadvantage: at near-threshold intensities, EFR produced by such stimuli are low amplitude. The low efficiency of SAM stimuli results from their two peculiarities: 1) rather slow rise-fall of SAM waves; quick sound transients are most effective to provoke AEP and 2) narrow frequency band; AEP amplitude depends to a large extent on the stimulus frequency bandwidth rather than on level (Popov and Supin 2001). An obvious solution to the problem is to use signals as rhythmic pip trains, with each pip shorter than the modulation cycle. Shortening the pip duration markedly enhances the EFR. EFR records were obtained using pip-train stimuli with a pip rate of 1 kHz and a pip duration of 1 ms (Fig. 1a,b) and 0.25 ms (Fig. 1c,d). Comparison shows much higher EFR amplitude in the near-threshold intensity range when short-pip stimuli were used. The shortest pip duration for audiometric use is limited by the required precision of the audiogram on the frequency scale. For many cases, an audiogram obtained with one-fourth to one-half octave steps is acceptable. In these cases, a pip may contain no more than 8–16 carrier cycles, e.g., an 8-cycle pip has a spectrum bandwidth (at the half-level) from −1/4 to +1/16 octave.

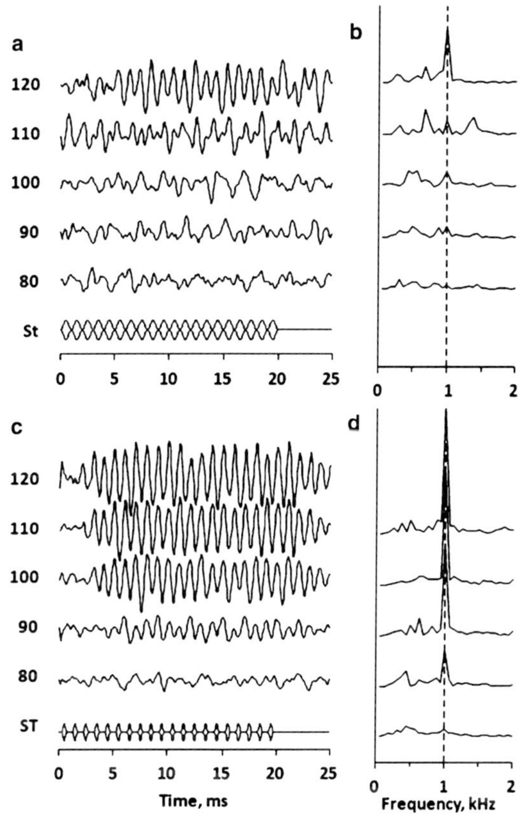

Fig. 1 (**a**) Envelope-following response (EFR) in a bottlenose dolphin *Tursiops truncatus* to a sinusoidal amplitude-modulated (SAM) 64-kHz stimulus; levels (dB re 1 μPa) are indicated near the records. Note definite response at high-stimulus level (120 dB) and nondefinite responses at lower levels. (**b**) Frequency spectra of records presented in A reveal responses (1-kHz spectrum peak) at levels of 100–110 dB; however, difference between responses to low-level stimuli (80–90 dB) is poorly detectable. (**c**) and (**d**) Same as A and B, respectively, but with short pip-train stimulus. Responses are definite at levels down to 90 dB; the difference between suprathreshold responses (90 dB) and subthreshold responses (80 dB) is obvious. Lag of EFR relative to the stimulus is well visible. ST/St, stimulus envelope

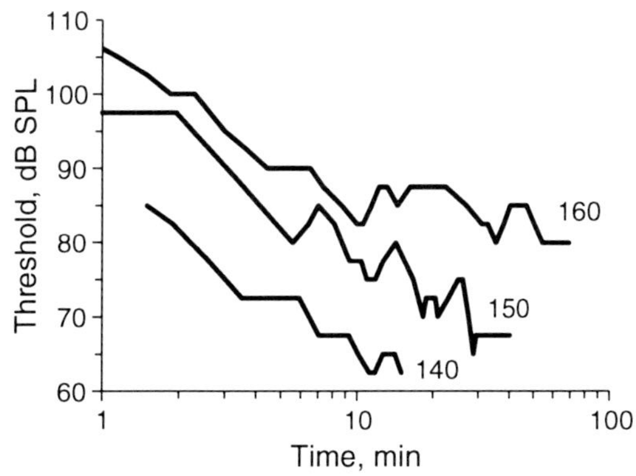

Fig. 2 Short-term threshold recovery in a finless porpoise, *Neophocaena focaenoides*, after noise exposure. Test frequency was 45 kHz; fatiguing noise was 32 kHz ± 0.25 octave, 3 min, 140–160 dB sound pressure level (SPL), as indicated

Apart from better confidence of threshold determination, the use of effective short-pip rhythmic stimuli allows a quicker threshold determination. Due to the high effectiveness of the stimulus, each EFR record can be obtained in 15–20 s (300–500 presentations at a rate of 20/s), thus finding a threshold about every minute. This makes tracing the rapid threshold variation possible, in particular, an investigation into the short-term recovery after temporary threshold shifts (TTSs), which is hardly possible with the use of other techniques (Fig. 2).

5 Conclusions

The evoked potential audiometry, although widely used in aquatic mammals, may need further elaboration and standardization with respect to the used stimuli, data collection, and processing. It may make audiometric measurements more easy and precise, thus enlarging the area of the use of the evoked potential audiometry.

References

Dolphin WF (1995) Steady-state auditory-evoked potentials in three cetacean species elicited using amplitude-modulated stimuli. In: Kastelein RA, Thomas JA, Nachtigall PE (eds) Sensory systems of aquatic mammals. De Spil, Woerden, The Netherlands, pp 25–47.

Popov VV, Supin AY (1990) Auditory brain stem responses in characterization of dolphin hearing. J Comp Physiol A 166:385–393.

Popov VV, Supin AY (1998) Auditory evoked responses to rhythmic sound pulses in dolphins. J Comp Physiol A 183:519–524.

Popov VV, Supin AY (2001) Contribution of various frequency bands to ABR in dolphins. Hear Res 151:250–260.

Supin AY, Popov VV (1995) Envelope-following response and modulation transfer function in the dolphin's auditory system. Hear Res 92:38–46.

Supin AY, Popov VV, Mass AM (2001)The sensory physiology of aquatic mammals. Kluwer Academic Publishers, Boston, MA.

Prediction of a Mysticete Audiogram via Finite Element Analysis of the Middle Ear

Andrew Tubelli, Aleks Zosuls, Darlene Ketten, and David C. Mountain

1 Introduction

The impact of anthropogenic sound on marine mammals is difficult to assess, especially for species without available audiograms. There are currently no audiograms for any species of mysticete because of their size and, in many cases, their endangerment status. Consequently, insight into the hearing range of all mysticete species comes from indirect sources such as vocalization recordings. In contrast to mysticetes, several odontocete species have published audiograms.

Both the middle ear and the cochlea play an important role in shaping the audiogram of any mammal. The transfer function of the middle ear shapes the low-frequency portions of an audiogram, whereas the high-frequency portion of the audiogram is shaped by the frequency place map of the cochlea (Rosowski 1994).

Biophysical models of the cetacean middle ear can be developed using finite element (FE) techniques. FE modeling has been successfully used to provide an understanding of how several terrestrial mammalian middle ear systems work. The advantage to using FE models is that they directly incorporate the geometry and material properties of the structures of interest. For this study, the middle ear of a mysticete species, *Balaenoptera acutorostrata* (minke whale), was modeled using FE methods. The same methods were used to develop a model for the *Tursiops truncatus* (bottlenose dolphin) middle ear, a control species that has a behaviorally derived audiogram to verify the modeling approach.

2 Model

A formalin-fixed *Balaenoptera acutorostrata* ear and a thawed *Tursiops truncatus* ear were scanned using computed tomography. The resulting stacks of images were segmented for structures of interest (malleus, incus, stapes, tympanic bone, incudostapedial joint, incudomalleolar joint, and annular ligament) using Amira (Mercury Computer Systems, Chelmsford, MA). The three-dimensional volume generated from the segmented slices was subsequently used for FE analysis (COMSOL, Stockholm, Sweden).

A. Tubelli (✉) • A. Zosuls • D.C. Mountain
Hearing Research Center, Boston University, Boston, MA 02215, USA
e-mail: atubelli@bu.edu

D. Ketten
Woods Hole Oceanographic Institution, Woods Hole, MA 02543, USA

Harvard Medical School, Boston, MA 02115, USA

A.N. Popper and A. Hawkins (eds.), *The Effects of Noise on Aquatic Life*,
Advances in Experimental Medicine and Biology 730, DOI 10.1007/978-1-4419-7311-5_12,

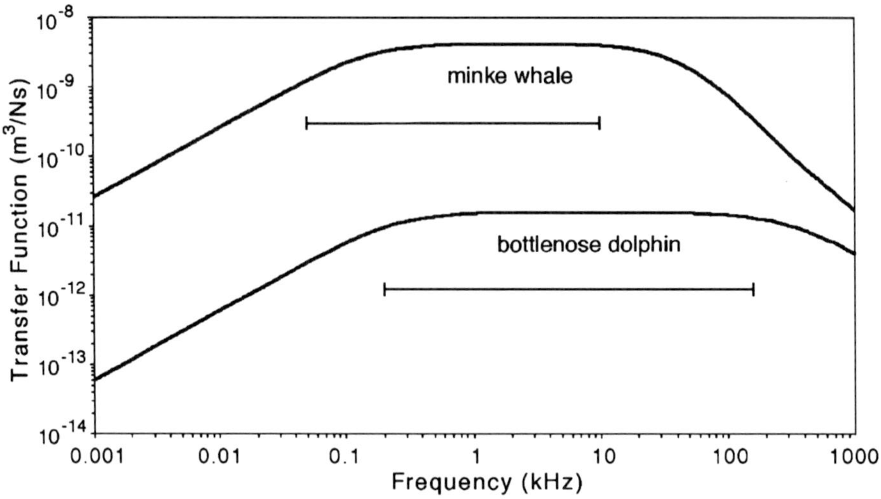

Fig. 1 Transfer functions for *Balaenoptera acutorostrata* and *Tursiops truncatus*. The solid lines under each curve represent the respective ranges of vocalization recordings (Gedamke et al. 2001; Ketten 1997)

Material properties were derived from a combination of literature and laboratory measurements and included density, Young's moduli, spring constants, cochlear damping, Rayleigh damping, and Poisson's ratio (Gan et al. 2004; Koike et al. 2002; Nummela et al. 1999).

An input force was applied to the malleus at the attachment point of the tympanic ligament. The model was fixed along the edge of the tympanic bone and the annular ligament to simulate the differential motion between the tympanic bone and the periotic bone. The soft tissues were modeled as springs and the cochlear input impedance was assumed to be resistive.

The models were calibrated using direct experimental measurements of middle ear stiffness (Miller et al. 2006).

3 Results

The model predicts that the passband for the middle ear (i.e., between the −3-dB cutoff frequencies) for *Balaenoptera acutorostrata* is between 100 Hz and 30 kHz (Fig. 1). Vocalizations for *Balaenoptera acutorostrata* occur at frequencies between 50 Hz and 9.4 kHz (Gedamke et al. 2001). In contrast, the FE model predicts the middle ear passband for *Tursiops truncatus* to be between 200 Hz and 300 kHz. The experimental audiogram for *Tursiops truncatus* (Johnson 1967) shows the range of best sensitivity of hearing to be between ~6 kHz and 140 kHz. Vocalizations occur at frequencies between 0.2 kHz and 150 kHz (Ketten 1997).

References

Gan RZ, Feng B, Sun Q (2004) Three-dimensional finite element modeling of human ear for sound transmission. Ann Biomed Eng 32:847–859.

Gedamke J, Costa DP, Dunstan A (2001) Localization and visual verification of a complex minke whale vocalization. J Acoust Soc Am 109:3038–3047.

Johnson CS (1967) Sound detection thresholds in marine mammals. In: Tavolga WN (ed) Marine bio-acoustics II. Pergamon Press, Oxford, UK, pp 247–260.

Ketten DR (1997) Structure and function in whale ears. Bioacoustics 8:103–136.

Koike T, Wada H, Kobayashi T (2002) Modeling of the human middle ear using the finite-element method. J Acoust Soc Am 111:1306–1317.

Miller BS, Zosuls AL, Ketten DR, Mountain DC (2006) Middle-ear stiffness of the bottlenose dolphin *Tursiops truncatus*. IEEE J Oceanic Eng 31:87–94.

Nummela S, Wägar T, Hemilä S, Reuter T (1999) Scaling of the cetacean middle ear. Hearing Res 133:71–81.

Rosowski JJ (1994) Outer and middle ears. In: Fay RR, Popper AN (eds) Comparative hearing: Mammals. Springer-Verlag, New York, pp 172–247.

Reverse Engineering the Cetacean Ear to Extract Audiograms

Aleks Zosuls, Seth O. Newburg, Darlene R. Ketten, and David C. Mountain

1 Introduction

The cochlear frequency-place map is believed to be an important determinant of the frequencies that a species can hear as well as the bandwidth of cochlear filters. Both features impact an animal's ability to detect biologically significant sounds in noise. The cochlear frequency-place map is created in part by a stiffness gradient in the basilar membrane (BM) in which stiff regions respond best to high frequencies and more compliant regions respond best to low frequencies.

The goal of this research is to build cochlear models that predict audiograms of species for which it is impractical to obtain an audiogram through behavioral testing (e.g., large marine mammals). In this study, we measured BM stiffness in *Tursiops truncatus*, *Meriones unguiculatus*, *Phocoena phocoena*, and *Delphinus delphis*, all species with known audiograms. The results will be used to calibrate cochlear models for estimating the audiograms of species that cannot be measured behaviorally.

2 Methods

A custom piezoelectric force probe was constructed based on Olson and Mountain (1991) and Naidu and Mountain (1998). The probe consists of two displacement transducers and a force sensor in series, terminating at a sharp tip placed in contact with the underside of the BM. The first displacement transducer was mounted to a micromanipulator and was used to apply static displacements to the probe, displacing the probe tip toward the membrane in 1-mm steps. The second displacement transducer applied a 50-nm peak-to-peak 80-Hz sinusoidal signal to the probe tip.

A. Zosuls (✉) • D.C. Mountain
Boston University, Boston, MA 02215, USA
e-mail: azosuls@bu.edu; dcm@bu.edu

S.O. Newburg
Massachusetts Institute of Technology, Cambridge, MA 02142, USA
e-mail: setho@mit.edu

D.R. Ketten
Woods Hole Oceanographic Institution, Woods Hole, MA 02543, USA
e-mail: dketten@whoi.edu

A.N. Popper and A. Hawkins (eds.), *The Effects of Noise on Aquatic Life*,
Advances in Experimental Medicine and Biology 730, DOI 10.1007/978-1-4419-7311-5_13,

The force sensor was a piezoelectric bimorph with the glass probe tip bonded to its center. As the second displacement sensor applied the sinusoidal stimulus, the force sensor measured the return force of the membrane. A computer with Tucker-Davis Technologies and National Instruments data-acquisition hardware running custom MATLAB scripts was used to control the experiments.

Inner ear preparations varied by species. In *Meriones*, the animals were deeply anesthetized, then decapitated according to Institutional Animal Care and Use Committee-approved protocols. The bulla was removed and placed in oxygenated L-15 culture medium (Sigma-Aldrich). The scala tympani was opened, exposing the underside of the BM, and mounted on a holder with cyanoacrylate glue (Great Planes). The force probe was positioned orthogonal to the BM using a surgical microscope. A radial profile of positions was obtained by scanning from the spiral lamina to the spiral ligament. Longitudinal location was recorded by digital images. In *Meriones*, only one longitundinal location was taken per preparation to ensure the most physiologically relevant data.

In *Tursiops*, *Phocoena*, and *Delpinius*, a different approach was required because the ears were harvested postmortem. Legal restrictions prevent euthanasia perfusion for research; therefore, fresh samples were obtained opportunistically. Fixatives can also change mechanical tissue properties. After an animal was pronounced dead naturally or euthanized for medical reasons, its ears were extracted at the site of stranding or at the Marine Mammal Facility, Woods Hole Oceanographic Institution, Woods Hole, MA, scanned in a CT unit, and transported immediately to Boston University, Boston, MA, for measurement. In many cases, the experiments were performed 8–24 h postmortem.

The bullar complex of marine mammals is composed of dense, fully ossified bone, second only to teeth in density and hardness. A Dremel Moto tool and a dental drill equipped with carbide burrs were used to open the scala tympani to expose the underside of the BM. During this process, the ear was bathed in normal saline solution to cool and maintain moisture. Bone dust was removed by vacuum to prevent contamination of the BM. The periotic bone was ground very near to the canals. The remaining bone was carefully chipped with a scalpel to minimize spiral lamina, spiral ligament, and BM damage. The ear was then mounted on a large ear bar with cyanoacrylate glue. The bar was positioned under the probe, and a radial profile was collected. Longitudinal location was documented with photographs. The ear was removed from the probe system, and a new longitudinal access location was opened in the bone. During machining, existing holes were sealed with bone wax to prevent contamination with bone dust. The process was repeated for multiple locations base to apex until the preparation deteriorated or the cochlea collapsed.

3 Results

In all ears measured, the stiffness values decreased from base to apex (Fig. 1). Higher frequency species had the highest basal turn stiffness (Fig. 1). These results are consistent with other measurements made on BM stiffness.

Probe noise floor limited the ability to make reliable measurements in the very low stiffness apical regions. New strategies are being developed to measure these regions in a timely matter. They currently require many averages and finer probe advancing steps, which is problematic with the rapid deterioration of the tissue.

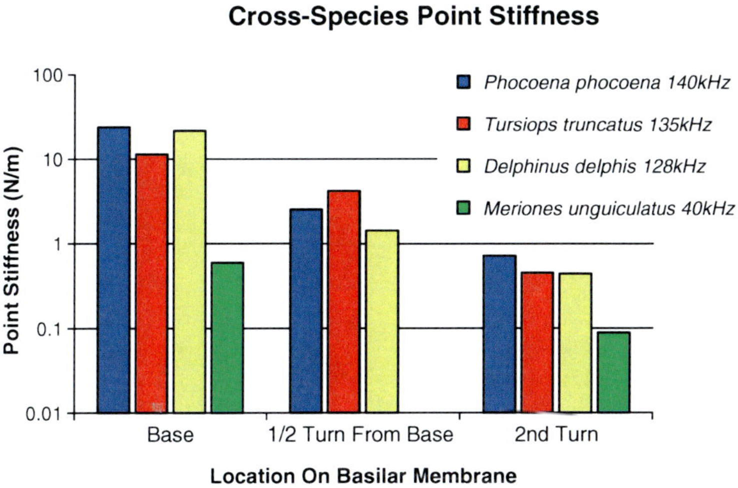

Fig. 1 Point stiffness for multiple species at three longitudinal locations along the length of the basilar membrane. Radial data from each longitudinal position were averaged. Frequencies refer to high-frequency cutoffs in measured audiograms of *Phocoena* (Kastelein et al. 2002), *Delphinus* (Popov and Klishin (1998), *Tursiops* (Ljungblad et al. 1982), and *Meriones* (Ryan 1976)

References

Kastelein RA, Bunskoek P, Hagedoorn M, Au WWL, de Haan D (2002) Audiogram of a harbor porpoise (*Phocoena phocoena*) measured with narrow-band frequency-modulated signals. J Acoust Soc Am 112:334–344.

Ljungblad DK, Scoggins PD, Gilmartin WG (1982) Auditory thresholds of a captive Eastern Pacific bottle-nosed dolphin *Tursiops* spp. J Acoust Soc Am 72:1726–1729.

Naidu RC, Mountain DC (1998) Measurements of the stiffness map challenge a basic tenet of cochlear theories. Hear Res 124:124–131.

Olson ES, Mountain DC (1991) In vivo measurement of basilar membrane stiffness. J Acoust Soc Am 89:1262–1275.

Popov VV, Klishin VO (1998) EEG study of hearing in the common dolphin, *Delphinus delphis*. Aquat Mamm 24:13–20.

Ryan A (1976) Hearing sensitivity of the Mongolian gerbil, *Meriones unguiculatis*. J Acoust Soc Am 59:1222–1226.

Validation of a Vibroacoustic Finite Element Model Using Bottlenose Dolphin Experiments

Petr Krysl, Vanessa Trijoulet, and Ted W. Cranford

1 Introduction

Our understanding of potential impacts of anthropogenic sounds needs to grow so that environmental consequences of ocean noise levels may be evaluated. Experimentally exposing animals is expensive and often impractical or unethical, and thus simulation is among the most promising approaches, especially finite element modeling (FEM). FEM techniques have been successfully used in the area of bioacoustics of marine animals, e.g., the sonar anatomy and acoustic pathways for Cuvier's beaked whale (Cranford et al. 2008). This animal is at the forefront of concerns about the potential impacts from Navy sonar. An obvious problem in working with this species is that we know so little about it. Neither do we have sufficient experimental observations against which to assess the validity of the simulation results. The work described here remedies that situation by validating our simulation framework using some of the experimental results available for the bottlenose dolphin.

In the current paradigm of computer-based prediction, the role of physical experiments is to provide data that can be compared with simulation results to validate the computer code for an entire class of predictions. In the context of model validation, confidence is established by comparing a consistent body of evidence from physical testing with the predictions of FEM models. Our vibroacoustic toolkit is based on a model that has a well-established domain of validity, small-deformation, small-strain, viscoelastic solids, possibly in combination with compressible fluids (Krysl et al. 2008). However, vibroacoustic problems in biological systems possess attributes that push the boundaries of the simple domain for which our toolkit has been validated. The geometries are to a considerable degree approximate. Tissue interfaces can be diffuse or graded, and structures can be difficult to resolve adequately. The material response is often complex, especially with respect to incompressibility and a time-dependent or dissipative response. The coupling with fluids is both internal (nearly incompressible fluid-saturated tissues, air cavities) and external (ambient seawater).

We performed validation of our vibroacoustic model for a representative problem in bioacoustics, the formation and geometry of the echolocation beam for the bottlenose dolphin (*Tursiops truncatus*).

P. Krysl (✉) • V. Trijoulet
University of California, San Diego, La Jolla, CA 92093, USA
e-mail: pkrysl@ucsd.edu

T.W. Cranford
San Diego State University, San Diego, CA 92182, USA
e-mail: tcranfor@mail.sdsu.edu

A.N. Popper and A. Hawkins (eds.), *The Effects of Noise on Aquatic Life*,
Advances in Experimental Medicine and Biology 730, DOI 10.1007/978-1-4419-7311-5_14,

In particular, we 1) predicted the horizontal and vertical beam patterns in terms of the relative sound pressure level at some distance from the rostrum of animal and 2) identified some of the so-called modeling errors (assumptions or simplifications in the formulation of the conceptual model that cause differences between computational and experimental data) and assessed their effects on the formation of the echolocation beam.

2 Methods

Two bottlenose dolphin CT scan datasets were available. The first was a postmortem CT scan (specimen D1; resolution 0.976 × 0.976 × 0.976 mm). The second model was from a live bottlenose dolphin CT scan (specimen D2; resolution 0.941 × 0.941 × 0.941 mm). Note that neither animal was in "echolocation mode" in the scanner, and we should not necessarily expect the configuration of the anatomy in either situation to be close to that used for biosonar.

The so-called phonic lips are hypothesized to be the biosonar sound source in the bottlenose dolphin (Cranford 2000). The phonic lips consist of constrictions in the spiracular nasal passage. The walls of the lips contain pairs of fat bodies ensheathed in connective tissue (bursae). During sound generation, air is pushed through the phonic lips, setting them into vibration. As the opposing walls or lips vibrate, they impact one another and generate short pulses of sound. In this work, we produce sound in a simplified but related manner: tissue at the two locations of the bursae is given an initial velocity that starts the tissue "blobs" in opposing directions so that they collide and thereby produce a pressure wave, sound. The location of the phonic lips could play a role in forming the echolocation beam. The musculature associated with the dolphin's melon suggests that it may be actively distorted or shaped by the animal. Evidence for "beam steering" in odontocetes does exist (e.g., Moore et al. 2008). Thus we also consider the location of the phonic lips among the sources of modeling error.

We have studied several models. Spherical bursae (SBx) or block bursae (BBx) were used for the sound source. The SBx approximates a point source while the "BBx" approximates the structures as they are found in actual specimens (Cranford et al. 1996). The suffix "x" indicates the various locations of the sound source (eight different locations). In addition to the original location in the CT scans, alternative locations for the sound source were simulated by shifting them vertically or horizontally ~6 mm in the sagittal plane.

1. Model D1-SO-SBx. Only the skull surrounded entirely by seawater.
2. Model D1-T-SBx. Soft tissues (connective tissue, muscle, melon blubber, melon proper) and bone were modeled. Air cavities were considered by the introduction of pressure relief surfaces.
3. Model D1-TSM-SBx. Same as the previous model with additional refinement of the acoustic fat properties (Norris and Harvey 1974).
4. Model D1-TSM-A-SBx. Improved resolution of the air spaces; otherwise identical to model 3.
5. Model D1-TSM-A-BBx. Identical to model 4 but for BBx sound source.
6. Model D2-TSM-SBx. The live dolphin specimen D2. Otherwise identical to model 3.
7. Model D2-TSM-BBx. The BBx sound source is used. Otherwise identical to model 6.

3 Results and Discussion

Figure 1 illustrates the results obtained for one particular model, D2-TSM-SB1. The sound pressure level (SPL) with respect to the maximum over the entire beam is shown on a sphere centered at the sound source. The horizontal and vertical beam pattern is shown on the right in comparison with

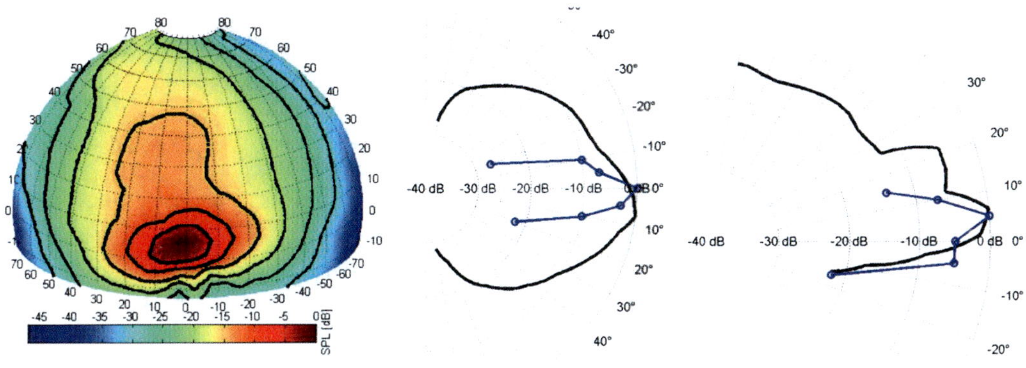

Fig. 1 Model D2-TSM-SB1. The simulated echo location beam is plotted on an imaginary screen in front of the animal, Sound pressure level (SPL) is coded according to the color bar. The horizontal and vertical sections are the black lines on the right. Blue lines with circular markers are from Au et al. (1986)

Table 1 Directivity index for the various models and sound source locations

	Directivity Index, dB		
Model	Mean	Minimum	Maximum
D1-SO-SBx	13.5	10.5	15.5
D1-T-SBx	16.5	14.8	18.0
D1-TSM-SBx	19.4	17.8	20.7
D1-TSM-BBx	20.9	20.6	21.1
D1-TSM-A-SBx	19.6	18.6	20.1
D1-TSM-A-BBx	21.8	21.7	21.9
D2-TSM-SBx	18.7	16.4	19.6
D2-TSM-BBx	20.3	19.2	21.4

Au et al. (1986) computed the directivity index from their experimental data as 26.5 dB. SBx, spherical bursae; BBx, block bursae.

the experimental results of Au et al. (1986). Table 1 compares the directivity index (DI) of the transmitting beam for the models considered.

Our results support the hypothesis that the primary structural element in the sound generation system is the odontocete skull (e.g., Evans et al. 1964). Furthermore, all model improvements (adding soft tissue, realistic melon tissue gradient, adding nasal air spaces, more realistic model of sound source) contribute to the formation of the beam. Our results also present the first evidence that small changes in the relative position of bursae can produce changes in beam direction. The differences in the directivity index between the postmortem D1 and live D2 dolphins are on the order of only ~5%. The predicted DI differs from the 26.5 dB computed from the experimental DI (Au et al. 1986) by ~20% for the most sophisticated models for both specimens. Strictly speaking, this difference should not be considered an error because neither the subject of the physical experiment nor the specimens used in the computational studies are known to be statistically representative of the species. The unraveling of the observed differences and further improvements to the conceptual models are the subject of ongoing studies.

Acknowledgments We acknowledge the support of Dr. Frank Stone and Dr. Ernie Young, Chief of Naval Operations Environmental Readiness Division, Washington, DC; Dr. Curtis Collins, Naval Postgraduate School, Monterey, CA; and Dr. Michael Weise and Dr. James Eckman, Office of Naval Research, Arlington, VA. The live dolphin CT scan dataset was provided by the US Navy Marine Mammal Program, San Diego, CA.We also thank John Hildebrand, Scripps Institution of Oceanography, La Jolla, CA, for his contribution to this work and Dr. Robert Gisiner, currently with the Marine Mammal Commission, Bethesda, MD.

References

Au WWL, Moore PWB, Pawloski D (1986) Echolocation transmitting beam of the Atlantic bottlenose dolphin. J Acoust Soc Am 80:688–691.

Cranford TW (2000) In search of impulse sound sources in odontocetes. In: Au WWL, Popper AN, Fay RR (eds) Hearing by whales and dolphins. Springer, New York, pp 109–156.

Cranford TW, Amundin M, Norris KS (1996) Functional morphology and homology in the odontocete nasal complex: Implications for sound generation. J Morphol 228:223–285.

Cranford TW, Krysl P, Hildebrand JA (2008) Acoustic pathways revealed: Simulated sound transmission and reception in Cuvier's beaked whale (*Ziphius cavirostris*). Bioinsp Biomim 3:1–10.

Evans WE, Sutherland WW, Beil RG (1964) The directional characteristics of delphinid sounds. In: Tavolga WN (ed) Marine Bio-Acoustics, vol 2. Pergamon Press, New York, pp 353–372.

Krysl P, Cranford TW, Hildebrand JA (2008) Lagrangian finite element treatment of transient vibration/acoustics of biosolids immersed in fluids. Int J Numer Meth Eng 74:754–775.

Moore PW, Dankiewicz LA, Houser DS (2008) Beamwidth control and angular target detection in an echolocating bottlenose dolphin (*Tursiops truncatus*). J Acoust Soc Am 124:3324–3332.

Norris KS, Harvey GW (1974) Sound transmission in the porpoise head. J Acoust Soc Am 56:659–664.

Acoustic Function in the Peripheral Auditory System of Cuvier's Beaked Whale (*Ziphius cavirostris*)

Ted W. Cranford and Petr Krysl

1 Introduction

The potential for sound to impact living marine resources has become an important topic in the last decade, initiated by stranded whales associated with Navy sonar operation, heightened public concern, and at least one court case reaching all the way to the United States Supreme Court. Recently, evidence has surfaced that suggests that human-generated sound can also have detrimental effects on fish hearing, reproductive habits, and stress levels (Popper and Hastings 2009).

We have developed a suite of techniques that, when combined, allow "finite element modeling" (FEM) of the vibroacoustic environment inside a whale's head. This model allows us to simulate what happens when the anatomy of the whale interacts with sound. It is now possible to decipher the physics and physiology of sound production and sound reception in a model of the head of an adult male *Ziphius cavirostris* (Cuvier's beaked whale).

2 Building an FEM Model

The first step in assembling a computational model was to devise a method for collecting accurate representations of the in situ anatomic structure, the "anatomic geometry," from large whales. We have scanned several dead whales (or parts of them) using an industrial X-ray CT scanner normally utilized to inspect solid-fuel rockets for flaws.

After scanning, we take the specimen apart systematically and measure the elastic properties of the various tissues (Soldevilla et al. 2005). These values for tissue elasticity and the tissue density values given by the CT scans are the two primary building blocks for the computer model. The toolbox that takes the building blocks and constructs the model using computer software, the Vibroacoustic Toolkit, was custom-built for this purpose.

T.W. Cranford (✉)
San Diego State University, San Diego, CA 92182, USA
e-mail: tcranfor@mail.sdsu.edu

P. Krysl
University of California, San Diego, La Jolla, CA 92093, USA
e-mail: pkrysl@ucsd.edu

A.N. Popper and A. Hawkins (eds.), *The Effects of Noise on Aquatic Life*,
Advances in Experimental Medicine and Biology 730, DOI 10.1007/978-1-4419-7311-5_15,

3 Simulating Sound Reception in Cuvier's Beaked Whale

After scanning the head of an adult male *Ziphius cavirostris*, we published the first quantitative description of biosonar anatomy for any toothed whale (Cranford et al. 2008b). Numerical analysis using FEM led to another published paper that reported the discovery of a new channel for sound reaching the ears through a gular pathway in *Ziphius cavirostris* (Cranford et al. 2008a).

The computer simulations show that most sounds arriving from directly in front of the animal enter the head from underneath the tongue region, pass through the throat and an opening in the posterior part of hollow lower jaws, and propagate along the mandibular fat bodies to the bony ear complex (tympanoperiotic complex [TPC]).

For this sound reception pathway to function, the bony wall on the medial (inner) side of the lower jaws must be absent; this "door" must be open. As it turns out, all living toothed whales have this open door. And some of the earliest fossils also show the same excavated jaw structure with the open door. This suggests that this sound reception pathway developed early in the evolution of ancient whales.

The computer model allows us to visualize the pathway or "river" of sound as it "flows" from the front of the head to the ear complex (Fig. 1). We call this the "gular" pathway because it passes through the throat region of the animal.

The computer model allows us to predict which sounds reach the TPC and how the TPC will vibrate in response to sound. An example of this vibrational analysis is shown in Figure 2. The vibrational pattern is the result of the stiffness and mass configurations that interact to produce the collective motion of the entire TPC based on elementary physics. This analysis also indicates that the middle ear ossicles are an integral part of the vibrational characteristics of the TPC and transmit unique motions to the oval window for every natural mode of vibration (or resonant frequency) calculated (Cranford et al. 2010).

The gular pathway transmits sound to the TPC. Figure 3 shows maps of sound pressure on the TPC from sounds that arrive head on (from directly in front of the animal). Analysis suggests that

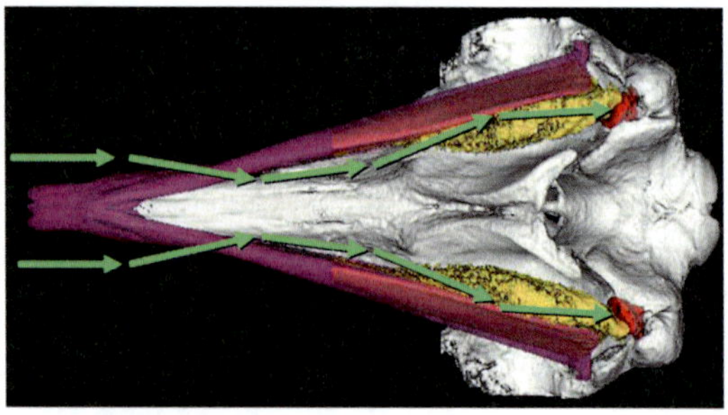

Fig. 1 Diagram of the generalized pathway for sound entering the head and "flowing" to the ears in *Ziphius cavirostris*

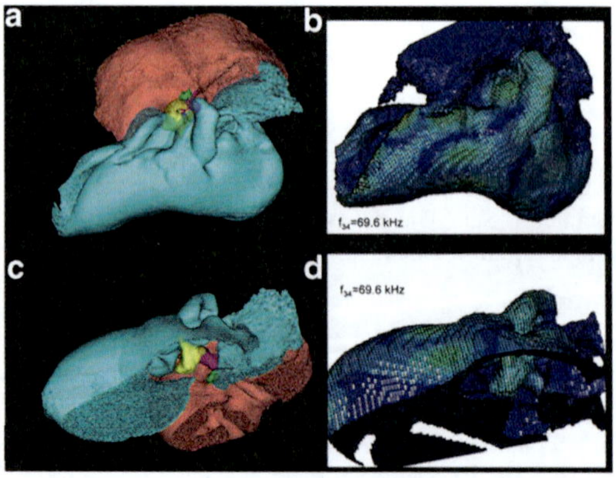

Fig. 2 Vibrational pattern of a bony ear complex (tympanoperiotic complex [TPC]) from *Tursiops truncatus* (bottlenose dolphin). (**a**) Left lateral view of TPC from CT scans. (**c**) Medial view of upside-down TPC cut to show middle ear. Yellow, malleus; magenta, incus; green, stapes). (**b**) and (**d**) Vibration pattern at 69.6 kHz where cold colors indicate little or no motion and warm colors indicate various amplitudes of displacement

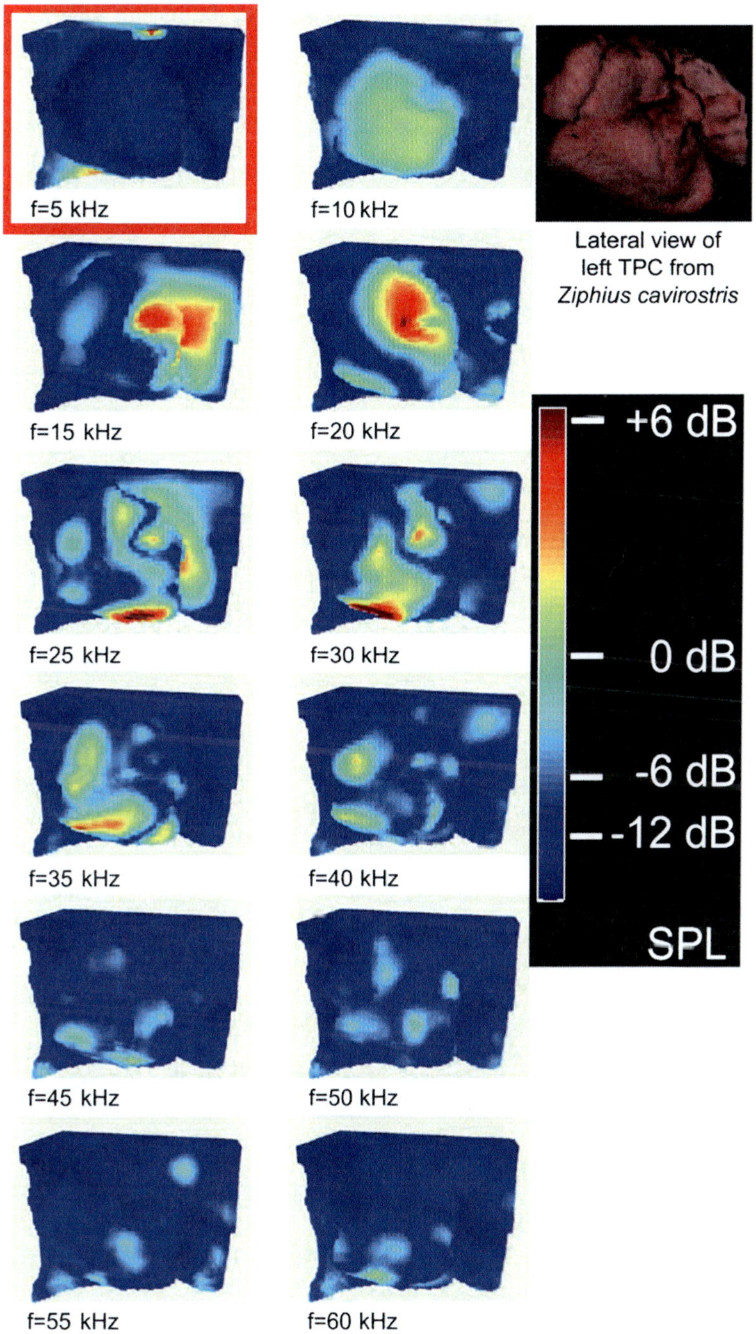

Fig. 3 Received sound pressure over the surface of the TPC with respect to the sound pressure incident on the head from directly in from of this *Ziphius cavirostris*. Each panel shows a side view of the TPC, over which the distribution of the color-coded sound pressure is depicted on the surface of the TPC. Green indicates the pressure that is equal to the sound pressure incident on the head (0 dB). Blue indicates that the sound pressure is –12 dB below the incident pressure or 4 times less than the incident pressure. Red indicates that the sound pressure is +6 dB or twice the incident pressure. Interestingly, the panel in the red box indicates that this frequency (5 kHz), within the range produced by midfrequency active sonar, is largely filtered out before it reaches the ear complex. Rows 2–4 (15–35 kHz) show that those frequencies are particularly effective at driving the vibrations of the TPC. SPL, sound pressure level

Navy sonar sounds reach the ears with reduced amplitude, but the biosonar frequencies (Johnson et al. 2004) used by *Ziphius cavirostris* to catch prey are amplified. This is evidence that there is selective amplification of the particular frequencies associated with biosonar in *Ziphius cavirostris*, which is apparently given by the anatomic components of the peripheral auditory system.

4 Conclusions

Computer models are powerful tools for investigation and discovery of bioacoustic physiology. This is particularly valuable because it provides a means to simulate across a broad range of scales and taxonomic groups, from whales to fish (see Krysl et al., Chapter 14; Schilt et al., Chapter 23). Virtual experiments can also assess the potential for and mechanism(s) of physical injury. Our techniques also provide a means for evaluating and directing mitigation efforts.

Acknowledgments We acknowledge the continued support of Dr. Frank Stone and Dr. Ernie Young, Chief of Naval Operations, Environmental Readiness Division (CNO 45), Washington, DC; Dr. Curtis Collins, Naval Postgraduate School, Monterey, CA; and Dr. Michael Weise and Dr. James Eckman, Office of Naval Research, Arlington, Va. In addition, we thank John Hildebrand, Scripps Institution of Oceanography, La Jolla, CA, for his contribution to this work. Key early development support came from Dr. Robert Gisiner, currently with the Marine Mammal Commission, Bethesda, MD.

References

Cranford TW, Krysl P, Amundin M (2010) A new acoustic portal into the odontocete ear and vibrational analysis of the tympanoperiotic complex. PLoS ONE 5:e11927. Doi:10.1371/journal.pone.0011927.

Cranford TW, Krysl P, Hildebrand JA (2008a) Acoustic pathways revealed: Simulated sound transmission and reception in Cuvier's beaked whale (*Ziphius cavirostris*). Bioinsp Biomim 3:1–10.

Cranford TW, McKenna MF, Soldevilla MS, Wiggins SM, Goldbogen JA, Shadwick RE, Krysl P, St. Leger JA, Hildebrand JA (2008b) Anatomic geometry of sound transmission and reception in Cuvier's beaked whale (*Ziphius cavirostris*). Anat Rec 291:353–378.

Johnson M, Madsen PT, Zimmer WMX, Aguilar de Soto N, Tyack PL (2004) Beaked whales echolocate on prey. Proc R Soc Lond Ser B Biol Sci Suppl 271:S383–S386.

Popper AN, Hastings MC (2009) The effects of human-generated sound on fish. Integr Zool 4:43–52.

Soldevilla MS, McKenna MF, Wiggins SM, Shadwick RE, Cranford TW, Hildebrand JA (2005) Cuvier's beaked whale (*Ziphius cavirostris*) head tissues: Physical properties and CT imaging. J Exp Biol 208:2319–2332.

Auditory Evoked Potential Measurement of Hearing Sensitivity in Pinnipeds

Jason Mulsow, Colleen Reichmuth, Dorian Houser, and James J. Finneran

1 Background

The majority of research on marine mammal hearing sensitivity has focused on the odontocete cetaceans (dolphins and porpoises) who possess morphological and neural adaptations that support sensitive passive hearing and a refined echolocation system. Fewer studies have examined the hearing sensitivity of the amphibious pinnipeds (sea lions, seals, and the walrus) who do not possess a sophisticated echolocation system. Passive hearing in pinnipeds is, nonetheless, important in behaviors related to reproduction, foraging, and predator avoidance. Many of these studies have used behavioral psychophysical methods to directly measure an animal's perceptual experience. Although the high-quality data provided by psychophysical methods are the most accurate description of hearing sensitivity, the methods are limited because they require trained subjects tested in captive environments. As a result, psychophysically measured profiles describing hearing sensitivity as a function of frequency (audiograms) are available for a small proportion of pinniped and odontocete species, and each examined species is usually represented by only a few individuals.

Auditory evoked potential (AEP) audiometric methods offer the most promising supplement to psychophysical procedures for measuring hearing sensitivity in a larger number of marine mammals. AEP procedures use electrodes to measure the voltages generated by the auditory system in response to acoustic stimuli and do not require the active participation of a subject. Auditory threshold can be assessed by determining the acoustic stimulus levels at which an electrophysiological response disappears. The relatively large voltages generated by hypertrophic structures in the odontocete auditory nervous system, in addition to research interest in echolocation, have facilitated the

J. Mulsow (✉) • J.J. Finneran
US Navy Marine Mammal Program, Space and Naval Warfare Systems Center Pacific,
San Diego, CA 92152, USA
e-mail: jason.mulsow@gmail.com; james.finneran@navy.mil

C. Reichmuth
Long Marine Laboratory, Institute of Marine Sciences, University of California,
Santa Cruz, Santa Cruz, CA 95060, USA
e-mail: coll@ucsc.edu

D. Houser
National Marine Mammal Foundation, San Diego, CA 92106, USA
e-mail: dorian.houser@nmmfoundation.org

A.N. Popper and A. Hawkins (eds.), *The Effects of Noise on Aquatic Life*,
Advances in Experimental Medicine and Biology 730, DOI 10.1007/978-1-4419-7311-5_16,

refinement of AEP audiometric methods with odontocetes (see Supin et al. 2001). AEP audiometric methods for assessing hearing sensitivity in pinnipeds are comparatively underdeveloped despite a similar need for data outside of existing psychophysical audiograms.

2 Amphibious Hearing Sensitivity in Pinnipeds

Pinnipeds, with a few exceptions such as *Mirounga angustirostris* (northern elephant seal), are sensitive to aerial sound over a wide range of frequencies. Aerial audiograms for both otariids (sea lions and fur seals) and phocids (true seals) usually have a characteristic mammalian "U-shape," with a shallow roll-off in sensitivity at low frequencies below 1 kHz and a sharp high-frequency hearing limit that occurs in the ultrasonic range between 20 and 40 kHz. Thresholds are commonly less than 10 dB re 20 μPa in the region of best sensitivity for many species. Underwater audiograms for pinnipeds, like their aerial audiograms, typically display a U-shape with a shallow low-frequency roll-off and a sharp high-frequency hearing limit. Although their lowest absolute detection thresholds are higher than those reported for odontocetes, thresholds for pinnipeds in the range of best hearing are generally between 50 and 70 dB re 1 μPa. One of the most striking features of the underwater hearing of pinnipeds is the difference between the underwater audiograms of otariid and phocid species: whereas the otariid underwater audiogram is essentially similar to the aerial audiogram in terms of frequency range of hearing, the underwater high-frequency limit for many phocid species is in the region of 70 to 100 kHz, markedly higher than the aerial high-frequency limit (see Hemilä et al. 2006).

3 Development of Noninvasive Audiometric Methods

The first evoked potential measurements of hearing sensitivity with pinnipeds were conducted using intracranial electrodes, allowing for the reduction of extraneous electrical noise levels relative to AEP voltages (Bullock et al. 1971; Ridgway and Joyce 1975). These studies demonstrated that the electrophysiological responses evoked by frequency-specific acoustic stimuli could be used to estimate the audiogram of pinnipeds.

Recent AEP studies with pinnipeds have recorded AEPs using small needle electrodes that are placed superficially under a subject's skin. The ratio of AEP signal to extraneous noise that is provided by these surface electrodes is significantly lower that those obtained with intracranial electrodes. Despite this, studies with a phocid, *Phoca vitulina* (harbor seal; Wolski et al. 2003), and an otariid, *Eumetopias jubatus* (Steller sea lion; Mulsow and Reichmuth 2010)], have demonstrated that these noninvasive recordings can provide a rapid estimate of a subject's psychophysical aerial audiogram. Most notably, relative sensitivity and the high-frequency hearing limit of the audiogram are accurately reproduced using AEP methods. The subjects of both studies were chemically sedated or anesthetized for the duration of data collection. The AEPs of interest were not markedly affected by the chemical agents and electrical artifacts related to subject movement during testing were largely eliminated, resulting in signal-to-noise conditions favorable for detecting the presence of an electrophysiological response. The stimuli used to elicit responses in *Phoca vitulina* were tone bursts, and the experimenters visually determined the presence or absence of an AEP in the electrophysiological record. For testing with *Eumetopias jubatus*, sinusoidally amplitude-modulated (SAM) tones were used to elicit rhythmic AEPs that were phase-locked to the rate of amplitude modulation imposed on the stimulus. This phase locking allows for frequency-domain analysis of the AEP as opposed to traditional time-domain analysis. After Fourier analysis of the time-domain

AEP waveform, the response is detectable as a spectral peak at the stimulus amplitude-modulation rate. The low-level AEPs (on the order of tens of nanovolts) that persist at near-threshold stimulus levels can then be objectively detected using frequency-domain signal-to-noise statistics.

Although AEP procedures with pinnipeds do not yet possess the same level of refinement as those with odontocetes, procedures using SAM tone stimuli with anesthetized subjects can provide an estimate of a subject's aerial audiogram in less than an hour. The AEP audiograms obtained for numerous untrained *Eumetopias jubatus* and *Zalophus californianus* (California sea lion) individuals have been similar to previously reported psychophysical audiograms in terms of relative sensitivity and high-frequency hearing limit. These results suggest that psychophysical audiograms obtained for a few individuals are representative of larger populations and reinforce the idea that the otariids form a functional hearing group. Profoundly elevated thresholds have been detected in a few subjects of both species, demonstrating that AEP methods can provide a tool for the rapid detection of hearing loss.

4 Limitations of AEP Measurements

Thresholds obtained with AEP methods are almost uniformly elevated relative to psychophysical thresholds. This relative elevation is normally largest at the low-frequency end of the audiogram, the frequency range in which anthropogenic noise predominantly occurs. There is a high degree of intersubject variability in AEP thresholds; standard deviations of 10–20 dB at each frequency are common. This is quite large compared with the level of intersubject variability in existing psychophysical data, and it is not yet known whether this is inherent to the AEP methods or due to actual variability in the hearing sensitivities of the individuals that have been tested thus far using these methods.

All noninvasive AEP studies with pinnipeds have been limited to measuring the aerial hearing sensitivity of subjects. Although concerns regarding the negative effects of aerial anthropogenic noise have to some extent motivated AEP studies with pinnipeds, primary interests lie in understanding the potential effects of underwater noise. The use of chemical sedation or anesthesia during pinniped AEP studies unfortunately makes the submersion of a subject a difficult prospect. Some conclusions regarding underwater sensitivity can, however, be based aerial audiograms obtained during AEP procedures (see below). Direct measurements of underwater hearing will likely continue to be one of the main challenges facing AEP hearing sensitivity measurements with pinnipeds.

5 Future Research Directions

AEP audiometric techniques for pinnipeds currently possess a level of refinement that is likely sufficient for testing with any of the otariids, a family for which audiometric data is available for only 3 of 14 species. The signal-to-noise ratios in AEP recordings with phocids are, however, generally much smaller than those for otariids. Modifications of the objective audiometric techniques that have proven useful for otariids should be a focus of future research aimed at measuring hearing sensitivity in phocids.

Some of the results of aerial AEP procedures can probably be extrapolated to describe certain features of underwater hearing. For example, both the aerial and underwater high-frequency hearing limits of otariids are limited by cochlear sensitivity and are therefore very similar (Hemilä et al. 2006). The aerial high-frequency hearing limit is accurately represented using AEP methods, and it most likely can provide a rapid estimate of underwater frequency range of hearing. Other auditory processes that primarily involve the cochlea (e.g., energetic masking, temporary threshold shift) are likely to have similarities in air and underwater. Future studies may therefore be able to simultaneously predict the effects of aerial and underwater noise on the pinniped auditory system using a large sample size of untrained subjects.

6 Conclusions

Although AEP audiometric methods for pinnipeds still do not possess the level of sophistication of those for odontocetes, they are currently providing a useful supplement to traditional behavioral psychophysics for measuring hearing sensitivity. Within-subject comparisons of electrophysiological and psychophysical methods have demonstrated that AEP thresholds obtained from sedated or anesthetized pinnipeds provide a rapid estimation of a subject's aerial audiogram. The features of relative sensitivity and high-frequency hearing limit have been similar in psychophysical and AEP audiograms for *Eumetopias jubatus* and *Zalophus californianus*, suggesting that the high-quality psychophysical data from a few individuals are representative of larger populations of animals. Drawbacks of AEP audiometric methods include a tendency for a high degree of intersubject variability in frequency-specific thresholds and a near-uniform elevation of thresholds relative to psychophysical data.

Future studies of pinniped hearing should focus on the further development of AEP audiometric methods, especially those that will help to resolve the challenge of testing underwater hearing sensitivity and promote testing with phocid species. In the absence of AEP methods that can examine underwater hearing, some features of aerial hearing, such as the high-frequency hearing limit of otariid subjects, can likely be used to predict underwater hearing capabilities. The generalization of results from aerial AEP procedures to underwater function may also be a promising means of investigating noise effects such as masking and temporary threshold shift.

References

Bullock TH, Ridgway SH, Nobuo S (1971) Acoustically evoked potentials in midbrain auditory structures in sea lions (pinnipedia). Z Vergl Physiol 74:372–387.

Hemilä S, Nummela S, Berta A, Reuter T (2006) High-frequency hearing in phocid and otariid pinnipeds: An interpretation based on inertial and cochlear constraints. J Acoust Soc Am 120:3463–3466.

Mulsow J, Reichmuth C (2010) Psychophysical and electrophysiological aerial audiograms of a Steller sea lion (*Eumetopias jubatus*). J Acoust Soc Am 127:2692–2701.

Ridgway SH, Joyce PL (1975) Studies on seal brain by radiotelemetry. Rapp P-V Reun Cons Int Explor Mer 169:81–91.

Supin AY, Popov VV, Mass AM (2001) The sensory physiology of aquatic mammals. Kluwer Academic Publishers, Boston, MA.

Wolski LF, Anderson RC, Bowles AE, Yochem PK (2003) Measuring hearing in the harbor seal (*Phoca vitulina*): Comparison of behavioral and auditory brainstem response techniques. J Acoust Soc Am 113:629–637.

Hearing in Birds: What Changes From Air to Water

Robert J. Dooling and Sara C. Therrien

1 Introduction

There are ~10,000 species of birds in the world. Of these, there are ~820 diverse species that live on or near water, many of whom dive when foraging for food. At one end of the diving continuum, diving ducks of the family Anatidae, which contains 158 species of ducks, geese, and swans, can dive to depths of tens of meters for seconds up to a minute or two. At the other end of the continuum is the family Spheniscidae, which includes 17 species of penguins. The most extreme diver among all birds, the Emperor penguin, can dive to depths of 500 m and remain submerged for over 20 min (Meir et al. 2008). As our oceans and waterways become increasingly noisy, it is important to consider the potential impact of this noise on these birds. Although vision is surely important to birds, many of these diving bird species dive to depths where there is little light available, often dive at night, and forage cooperatively (Croxall et al. 1988; Hunt et al. 1993; Kooyman et al. 1992; Speckman et al. 2003). These behaviors probably depend on a sense other than vision. The use of sound for communication, navigation, and foraging in many marine mammals and fish makes hearing a prime candidate. Indeed, hearing is quite important for many diving bird species in the air. Both Emperor and King penguins identify their partners through individually distinctive vocalizations among several thousands of unrelated birds in the tightly packed, noisy colony (Aubin et al. 2000; Jouventin 1982). However, whether diving birds use their hearing in support of similar behaviors underwater is unclear.

In terms of both hearing and the effects of anthropogenic noise, birds present a unique and interesting situation. On the one hand, for landbirds, there is a considerable amount of empirical data from both the laboratory and the field on hearing and acoustic communication and on the effects of anthropogenic noise on both auditory function and behavior. On the other hand, the state of our knowledge about hearing in diving birds or the effects of anthropogenic noise is virtually nonexistent. In this paper, we summarize what is known about hearing and the effects of anthropogenic noise on landbirds, consider whether similar effects might occur in diving birds, and suggest approaches to understanding both underwater hearing in birds and the effects of anthropogenic noise.

R.J. Dooling (✉) • S.C. Therrien
University of Maryland, College Park, College Park, MD 20742, USA
e-mail: dooling@psyc.umd.edu

A.N. Popper and A. Hawkins (eds.), *The Effects of Noise on Aquatic Life*,
Advances in Experimental Medicine and Biology 730, DOI 10.1007/978-1-4419-7311-5_17,

2 Hearing in Birds

Audiograms are available for over 50 species of birds, showing that birds on average hear best between 2 and 5 kHz with absolute thresholds approaching 0 dB sound pressure level (SPL) in air. The average avian audiogram shows a loss of sensitivity below 1 kHz of ~20 dB/octave and a loss of sensitivity at high frequencies above 4 kHz of ~60 dB/octave (Dooling et al. 2000). The inner ears of archosaurs (birds and crocodiles) show a high degree of structural similarity, strong correlations with body mass, and strong correlations between the length of the auditory sensory epithelium and behavioral characteristics of hearing such as the frequency and the high-frequency limit of hearing (Gleich et al. 2005). From such data, we would expect that the hearing of diving birds in air would follow the general hearing pattern for birds. Indeed, a behavioral audiogram for the duck (Trainer 1946) and a cochlear microphonic audiogram for the blackfooted penguin (Wever et al. 1969) are completely in line with these expectations.

2.1 *Effects of Anthropgenic Noise on Hearing in Birds*

The effects of anthropogenic noise (in air) on landbirds include auditory system damage and permanent threshold shift (PTS) from acoustic overexposure, temporary threshold shift (TTS) from acoustic overexposure, masking of important biological sounds, and other physiological and behavioral responses. In all but the last case, these auditory effects depend strongly on the level of noise exposure, which is highly correlated with the proximity of the bird to the noise source. These relationships are schematically represented in Figure 1 showing that different but overlapping effects of

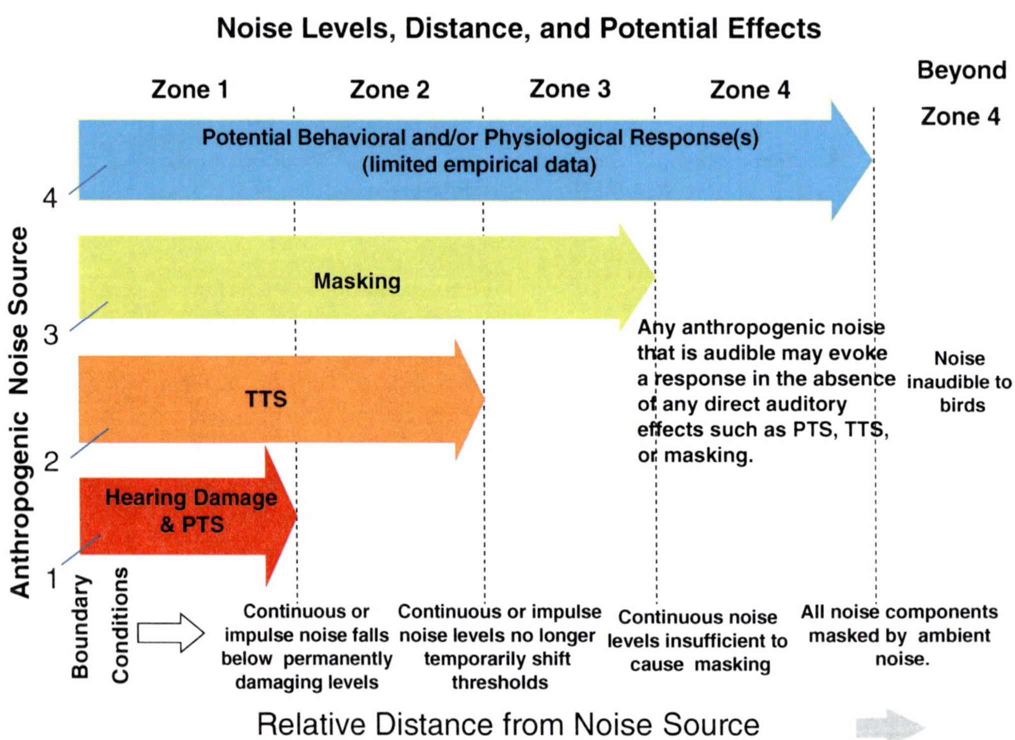

Fig. 1 Relationship between four potential overlapping effects of anthropogenic noise with respect to distance from the noise source. Adapted from Dooling and Popper (2007)

anthropogenic noise are highly dependent on distance from the source (bottom). The wealth of laboratory data on noise exposure in landbirds provides a clear and quantitative picture of noise levels leading to the various adverse effects.

For birds in air, continuous noise exposure at levels above 110 dB(A) SPL or blast noise over 140 dB SPL can result in physical damage of the auditory system and PTS. Birds are generally more resistant to auditory system damage and PTS from noise exposure than mammals. Continuous noise exposure at levels above 90–95 dB SPL, although it does not cause permanent damage, can cause a temporary elevation of hearing thresholds, mask important communication signals, and possibly result in other effects. In addition to direct effects on the auditory system, noise levels that are at or above the natural ambient noise level can increase masking of acoustic communication signals or other biologically important signals. Finally, anthropogenic noises at levels too low to cause masking may, if audible, still result in other behavioral and/or physiological effects that are detrimental.

3 Considerations in Moving From Air to Water

In the complete absence of any data on underwater hearing in birds, the only recourse is to extrapolate from what is known about hearing in landbirds and other land vertebrates and from what is known about hearing in other underwater vertebrates such as fish and marine mammals. To be sure, this is a highly speculative endeavor but a useful first step in developing questions and designing approaches to learn about underwater hearing in this group of vertebrates.

3.1 Adaptations in Diving Birds for Underwater Hearing?

Perhaps there are no adaptations in diving birds for hearing underwater. Absent any data on birds, measures of human hearing underwater are instructive. Data show that thresholds are elevated over 20 dB at low frequencies to as much as 70 dB at higher frequencies of 8 kHz and beyond (Brandt and Hollien 1967; Parvin and Nedwell 1995). In effect, the frequency of best hearing in humans shifts from ~2 kHz in air to ~800 Hz in water (Parvin and Nedwell 1995). If similar patterns were observed in diving birds, it would suggest that hearing may not serve a useful function for birds underwater. It would also argue for increased protection in birds from intense noise exposure in water.

3.2 Outer and Middle Ear Adaptations for Diving and Their Effects on Underwater Hearing

What little is known from anatomical studies on the outer and middle ear of diving birds suggests that there are adaptations to protect the tympanum and middle ear from the large, rapid pressure changes that occur while diving. Feather covering in diving birds is evenly packed over the surface of the body, and muscles attached to the feather shaft are contracted to create a waterproof seal when diving (Dawson et al. 1991; Kooyman et al. 1976). These adaptations probably prevent water from entering the auditory meatus and may also affect hearing. Similar to the walrus and hooded seal, there are muscles and blood vessels surrounding the external ear canal that may cause the canal to contract and shut on submergence (Sadé et al. 2008). Some penguins also have a venous cavernous

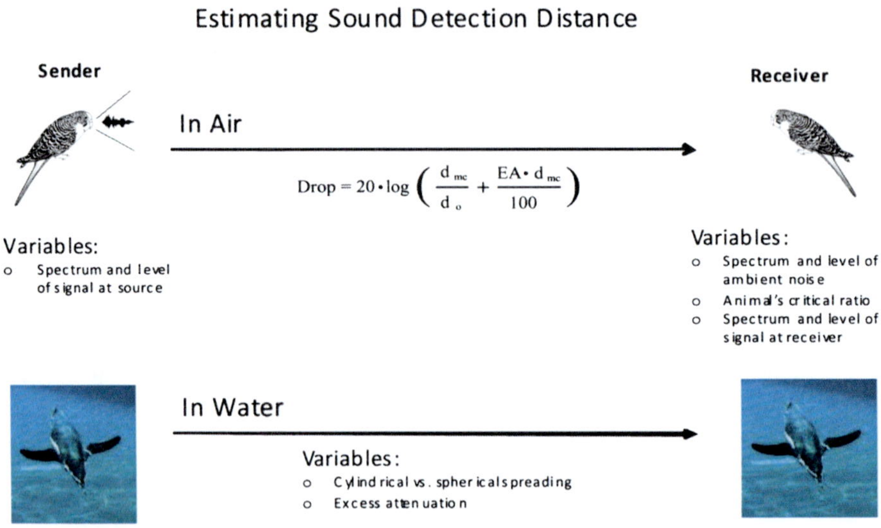

Fig. 2 Comparison of the variables used in estimating maximum sound communication distance between two birds in air and in water. The primary variables are transmission loss from spherical spreading (or possibly cylindrical in water) and excess attenuation

sinus beneath the middle ear mucosa (similar to many marine mammals), enabling these birds to equalize middle ear pressure with surrounding pressure during deep dives (Sadé et al. 2008). A study on small birds showed that even slight changes in middle ear pressure resulted in large inefficiencies in sound conduction and considerable protection from hair cell loss and permanent threshold shift due to acoustic overexposure (Ryals et al. 1999). Whether these same kinds of pressure changes occur in diving birds as part of the generalized reflex in birds is also unknown. If they do occur as part of a diving reflex, it would stand to reason that the only valid test of underwater hearing in these birds would have to be with awake behaving animals rather than anesthetized preparations.

3.3 Are the Effects of Masking the Same in Air and Water?

Masking is the interference with the detection of one sound by another. More precisely, masking refers to the increase in the threshold for detection or discrimination of sounds in the presence of another sound. In air, the masking effects from ambient noise (whether from natural or anthropogenic sources) at the receiver play a large role in determining the distance at which communication signals, other important biological sounds, or even other anthropogenic sounds can be heard by a bird. Whatever acoustic effects occur in moving from air to water, we can fairly safely assume that they affect both the signal and the noise in a similar manner. Thus the animal's masked thresholds are the proper data metric to use in determining the maximum distance at which a sound, whether a conspecific vocalization or an anthropogenic noise, can be heard. There are real questions about whether underwater hearing is important for diving birds for any purpose including foraging, acoustic communication, or predator avoidance. But if hearing is important underwater, the distance over which a sound may be heard in air will almost always limited by the animal's critical ratio rather than by its absolute auditory threshold.

Figure 2 schematically represents the variables involved in estimating sound detection distance. Whether in air or water, the spectrum and level of the source are required. In air, there is a 6-dB loss per doubling of distance due to spherical spreading and a linear loss due to excess attenuation. In water, the variables are similar, but the formulas are more complicated and depend on a number of factors including the depth of the water and the animal's position in the water column. In both cases, the bird can detect the signal after transmission loss as long as the signal is at least one critical ratio about the level of the background noise.

4 Summary and Conclusions

We reviewed what is known about hearing in birds, including the effects of anthropogenic noise, to speculate on the nature of underwater hearing in diving birds. Taking into consideration changes in human hearing underwater, the effects of changes in middle ear pressure on hearing in humans, and the protective effects against acoustic overexposure in birds from changes in middle ear pressure, we suggest that if similar patterns hold for diving birds, they may not hear well underwater. Moreover, the frequency of best hearing sensitivity may shift to frequencies below 2 and 4 kHz. Trapped air may allow the middle ear cavity of birds to function much as the swim bladder functions in fish.

Diving birds present important auditory and environmental issues. To resolve these issues properly requires comparative anatomical investigations of the middle and inner ears of diving birds. More importantly, behavioral measures of hearing in diving birds must be conducted both in air and in water. Finally, behavioral studies of these animals in their natural habitats are required to determine whether they use sound underwater for communication, foraging, predator avoidance, or other behaviors.

References

Aubin T, Jouventin P, Hildebrand C (2000) Penguins use the two-voice system to recognize each other. Proc R Soc Lond Ser B Biol Sci 267:1081–1087.

Brandt JF, Hollien H (1967) Underwater hearing thresholds in man. J Acoust Soc Am 42:1966–1971.

Croxall JP, Davis RW, O'Connell MJ (1988) Diving patterns in relation to diet of gentoo and macaroni penguins at South Georgia. Condor 90:157–167.

Dawson C, Vincent JFV, Jeronimidis G, Rice G, Forshaw P (1991) Heat transfer through penguin feathers. J Theor Biol 199: 291–295.

Dooling RJ, Lohr B Dent ML (2000) Hearing in birds and reptiles. In: Dooling RJ, Popper AN, Fay RR (eds) Comparative hearing: Birds and reptiles. Springer-Verlag, New York, pp 308–359.

Dooling RJ, Popper AN (2007) The effects of highway noise on birds. Report for The California Department of Transportation, Division of Environmental Analysis, Sacramento, CA. Available via http://www.dot.ca.gov/hq/env/bio/files/caltrans_birds_10-7-2007b.pdf.

Gleich O, Dooling RJ, Manley GA (2005) Audiogram, body mass and basilar papilla length: correlations in birds and predictions for extinct archosaurs. Naturwissenshaften 92:595–598.

Hunt GL, Harrison NM, Piatt JF (1993) Foraging ecology as related to the distribution of planktivorous auklets in the Bering Sea. In: Vermeer K, Briggs KT, Morgan KH, Siegel-Causey D (eds) The status, ecology and conservation of marine birds of the North Pacific. Can Wildl Serv Spec Publ Ottawa, pp 18–26.

Jouventin P (1982) Visual and vocal signals in penguins, their evolution and adaptive characters. Adv Ethol 24: 1–149.

Kooyman GL, Cherel Y, Le Maho Y, Croxall JP, Thorson PH, Ridoux V, Kooyman CA (1992) Diving behavior and energetics during foraging cycles in king penguins. Ecol Monogr 62):143–163.

Kooyman GL, Gentry RL, Bergman WP, Hammel HT (1976) Heat loss in penguins during immersion and compression. Comp Biochem Physiol A 54:75–80.

Meir JU, Stockard TK, Williams CL, Ponganis KV, Ponganis PJ (2008) Heart rate regulation and extreme bradycardia in diving emperor penguins. J Exp Biol 211:1169–1179.

Parvin SJ, Nedwell JR (1995) Underwater sound perception by divers and the development of an underwater noise weighting scale. J Soc Underwater Technol 21:12–19.

Ryals BM, Dooling RJ, Westbrook E, Dent ML, MacKenzie A, Larsen ON (1999) Avian species differences in susceptibility to noise exposure. Hear Res 131:71–88.

Sadé J, Handrich Y, Bernheim J, Cohen D (2008) Pressure equilibration in the penguin middle ear. Acta Oto-Laryngol 128:18–21.

Speckman SG, Piatt JF, Springer AM (2003) Deciphering the social structure of marbled murrelets from behavioral observations at sea. Waterbirds 26:266–274.

Trainer JE (1946) The auditory acuity of certain birds. Unpublished dissertation, Cornell University, Ithaca, NY.

Wever EG, Herman PN, Simmons JA, Hertzler DR. (1969) Hearing in the blackfooted penguin, *Spheniscus demerus*, as represented by the cochlear potentials. Proc Natl Acad Sci USA 63:676–680.

Amphibious Hearing in Sea Turtles

Wendy E. Dow Piniak, David A. Mann, Scott A. Eckert, and Craig A. Harms

1 Introduction

Despite increasing levels of anthropogenic noise throughout the oceans, we know very little about the hearing capabilities of sea turtles or how they might behaviorally and physiologically respond to potentially harmful sources of noise. Sea turtles are among the evolutionarily oldest and most endangered marine species. Seven species of sea turtle exist worldwide, including *Dermochelys coriacea* (leatherback), *Eretmochelys imbricata* (hawksbill), *Chelonia mydas* (green), *Caretta caretta* (loggerhead), *Lepidochelys kempi* (Kemp's ridley), *Lepidochelys olivacea* (olive ridley), and *Natator depressus* (flatback). With the exception of *Natator depressus* (for which we have insufficient data), all are classified as critically endangered or endangered by the International Union for the Conservation of Nature's Red List of Threatened Species (2010). Sea turtles are found in nearly all temperate and tropical marine environments and are highly migratory, traveling great distances between developmental, foraging, and nesting habitats. Given their endangered status, understanding the effects of noise on sea turtles is both timely and critically important.

W.E.D. Piniak (✉)
Duke University Marine Laboratory, Beaufort, NC 28516, USA
e-mail: wed3@duke.edu

D.A. Mann
College of Marine Science, University of South Florida, St. Petersburg, FL 33701, USA
e-mail: dmann@marine.usf.edu

S.A. Eckert
Principia College, Elsah, IL, 62028

Wider Caribbean Sea Turtle Conservation Network, Ballwin, MO 63011, USA
e-mail: seckert@widecast.org

C.A. Harms
Center for Marine Science and Technology, College of Veterinary Medicine,
North Carolina State University, Morehead City, NC 28557, USA
e-mail: craig_harms@ncsu.edu

A.N. Popper and A. Hawkins (eds.), *The Effects of Noise on Aquatic Life*,
Advances in Experimental Medicine and Biology 730, DOI 10.1007/978-1-4419-7311-5_18,

2 Sea Turtle Ear Morphology

Sea turtles lack an external pinna or ear canal. Like all terrestrial tortoises and turtles, their tympanum, located posterior to the midline of the skull and level with the corner of the mouth, is an extension of the facial tissue (Bartol and Musick 2003; Wever 1978). Unlike terrestrial turtles, marine turtles have a thick layer of subtympanal fat, coupling the thick surface scutes with the extracolumellar knob (Wever 1978).

The middle ear cavity is medial to the tympanum, filled with air, and connected to the throat via the Eustachian tube. The sea turtle ossicular mechanism consists of a mushroom-shaped cartilaginous extracolumella that lies beneath the tympanum, connected by ligaments to the columella. The columella is a long, thin, curved bone that expands into a cone-shaped stapes that, in turn, expands throughout the oval window (Wever 1978). Small fibrous strands, or stapedosaccular strands (unique to turtles), connect the stapes and oval window to the saccule. These strands are hypothesized to relay vibrational energy to the saccule (Lenhardt et al. 1985; Wever 1978; Wever and Vernon 1956). A fluid-filled cavity, the perocapsular recess, surrounds the otic capsule. This recess is separated from the perilymphatic duct by the pericapsular membrane and is often highly divided by intracapsular tissue. An opening at the posterior wall of the otic capsule, the foramen, completes a fluid circuit extending from the inner surface of the stapedial footplate to the pericapsular recess to the outer surface of the footplate (activated by the inward and outward movement of the stapes). This reentrant fluid circuit is responsible for the movement and pressure relief of cochlear fluids (Bartol and Musick 2003; Wever 1978). The presence of the circuit and the volume of fluid in the circuit may limit high-frequency hearingbecause the amount of sound pressure needed to move the columella increases with increasing frequency (Wever 1978; Wever and Vernon 1956).

The few studies that have examined the inner ear of turtles have found that the cochlear duct lies in the otic capsule and the basilar membrane lies on the medial wall supported by the limbic plate. The cochlea elongated and slightly curved. The basilar membrane contains both hair cells and support cells. Most ciliary tufts of hair cells are posterior or posterolateral oriented, but some hair cells are oriented in the opposite direction (Wever 1978). When these hair cells are stimulated, they open ion channels converting the mechanical movement of waves of sound to an electrochemical signal, which is received by the auditory nerve. Each hair cell has a characteristic frequency correlated with its position along the basilar membrane. Cells detecting low frequencies are located toward the apical end and cells detecting high frequencies are located toward the basal end (Crawford and Fettiplace 1980).

The functional morphology of the sea turtle ear remains poorly understood. Lenhardt et al. (1985) suggest that the ear is adapted for hearing via bone conduction in water and is a poor aerial receptor. In this case, the whole body serves as a receptor, with sound passing through bones and soft tissue to stimulate the inner ear (Lenhardt 1982). High frequencies are attenuated by bone, decreasing high-frequency hearing sensitivity. Computerized tomography has shown that sea turtles possess well-organized bundles of coherent fatty tissues connected to the middle ear. Densities of these fats are consistent with sound speeds in seawater, indicating the sea turtle ear may be well adapted for underwater sound conduction (Ketten 2008). Lenhardt and Hawkins (1983) found that both vibratory and acoustic stimuli are processed by the auditory system, so responses are likely due to a combination or summation of these signals.

3 Sea Turtle Auditory Capabilities

3.1 Aerial and Partially Submerged Hearing in Sea Turtles

Until the mid-1900s, turtles were generally believed to be deaf due to the lack of an outer ear and the lack of responsiveness to acoustic stimuli (Wever 1978). Ridgway et al. (1969) collected the first successful measurements of sea turtle hearing sensitivity by using both aerial and vibrational sound stimuli between 50 and 2,000 Hz to collect measurements of the cochlear response potential of 3 juvenile *Chelonia mydas*. Turtles responded to aerial stimuli between 100 and 1,000 Hz and vibrational stimuli between 100 and 700 Hz, with maximum sensitivity between 300 and 400 Hz for both stimuli, with a rapid decline in sensitivity in lower and higher frequencies. They found that 2,000 Hz was the upper limit for observation of cochlear potentials without injury and suggested that the practical hearing range did not exceed 1,000 Hz.

Recent measurements of sea turtle hearing sensitivity have been made by recording auditory evoked potential (AEP) responses to known sound stimuli. AEPs are produced by the synchronous discharge of neurons in the auditory pathway after stimulation by sound. This technique is a rapid, noninvasive method for measuring hearing in noncommunicative species. Bartol et al. (1999) measured electrophysiological responses to aerial sound stimuli in 35 juvenile *Caretta caretta* by collecting auditory brain stem responses (ABRs; which are AEPs measured within the first 8–10 ms of stimulation) derived from 2 types of vibrational stimuli: low-frequency clicks and tone bursts delivered directly to the tympanum. Bartol et al. measured a mean click threshold of −10.8 dB re 1 g root mean square (RMS) ± 2.3 dB SD and an effective hearing range from tone bursts of 250–750 Hz. The most sensitive threshold was the lowest frequency tested, 250 Hz with a mean threshold of −23.3 dB re 1 g RMS ± 2.3 dB SD (Bartol et al. 1999).

In an effort to develop physiologically based methodologies to reduce fisheries bycatch of sea turtles, Bartol and Ketten (2006) measured ABRs in two juvenile and six subadult *Chelonia mydas* and two juvenile *Lepidochelys kempi*. They measured ABRs in partially submerged turtles (ear submerged, with top of head and portions of carapace in air) using a three-electrode array and an aerial stimulus. Pacific subadult *Chelonia mydas* responded to stimuli between 100 and 500 Hz, with the highest sensitivity between 200 and 400 Hz and Atlantic juvenile *Chelonia mydas* responded to stimuli between 100 and 800 Hz, with the highest sensitivity between 600 and 700 Hz. *Lepidochelys kempi* responded to stimuli between 100 and 500 Hz, with maximum sensitivity between 100 and 200 Hz (Bartol and Ketten 2006).

3.2 Amphibious Hearing in Sea Turtles

Due to their aquatic lifestyle, an understanding of underwater hearing sensitivity is critical to evaluating the potential effects of underwater anthropogenic noise on sea turtles. We have developed protocols to measure both in-air and in-water hearing sensitivity in hatchling, juvenile, and adult sea turtles. To test these methodologies, we recorded in-air and in-water AEPs to click and tonal stimuli from 50 to 3,200 Hz in 5 juvenile *Chelonia mydas*. We presented stimuli with an underwater speaker calibrated with a hydrophone. Before testing, we isolated turtles from noise and vibrations and lightly restrained them to prevent excessive movement. For underwater measurements, the turtles were completely submerged at a depth of 10 cm (measured at the location of the ear). A Tucker-Davis Technologies AEP workstation with SigGen and BioSig software generated click and tonal

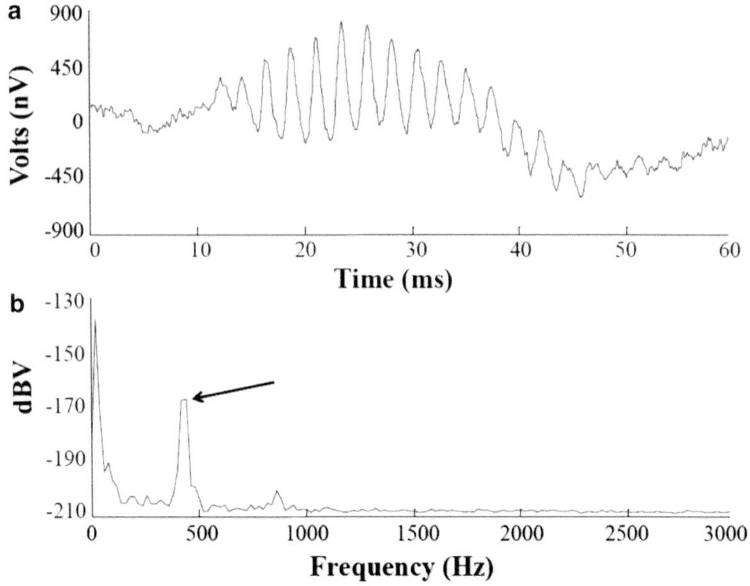

Fig. 1 (**a**) A 200-Hz auditory evoked potential (AEP) for a juvenile *Chelonia mydas*. (**b**) Fast Fourier transform of the 200-Hz AEP showing frequency-doubling response at 400 Hz (arrow)

stimuli and recorded AEP responses from subdermal electrodes. We developed underwater anesthesia protocols using medetomidine (50 μg/kg IV) and ketamine (5 mg/kg IV) and ventilating via a custom-designed double-cuffed extended endotracheal tube to provide a watertight seal at sufficient ear depth (Harms et al. 2009). The medetomidine was reversed with atipamezole (0.25 mg/kg). Anesthesia was helpful to eliminate myogenic artifacts in turtles that were not amenable to manual restraint, but this was not required for all turtles. Manual restraint was superior to anesthesia for turtles that did not resist restraint (better venous blood oxygenation, acceptable AEPs), but anesthesia was superior to manual restraint for turtles that did resist (marked lactic acidosis and AEPs not possible).

Sea turtle AEP signals exhibited a frequency-doubling signature similar that seen fish (Fig. 1). Juvenile *Chelonia mydas* responded to stimuli between 50 and 1,600 Hz in water and 50 and 800 Hz in air, with ranges of maximum sensitivity between 50 and 400 Hz in water and 300 and 400 Hz in air. In both water and air, sensitivity decreased sharply after 400 Hz.

Juvenile *Chelonia mydas* have a narrow range of hearing sensitivity in air and in water and are most sensitive to low frequencies. They hear well both in water and in air, particularly at frequencies below 1,000 Hz, but hearing sensitivities in the two media are different and in-water results show a broader and higher frequency range of sensitivity than reported by previous research in-air and at the water's surface.

4 Conclusions

Sea turtles likely use sound for navigation, locating prey, avoiding predators, and environmental awareness. Because sea turtles can be found in nearly all temperate, tropical, coastal, and offshore habitats, the potential overlap between sea turtle habitat and marine anthropogenic noise is vast. Sea turtle ear morphology lends itself to the reception of low-frequency sounds, and research shows they are most sensitive to frequencies below 1,000 Hz (Bartol et al. 1999; Bartol and Ketten 2006; Ridgway et al. 1969). Our research shows that sea turtles are particularly sensitive to frequencies below 1,000 Hz in air and water but are able to detect slightly higher frequencies in water, indicating that sea turtles are able to hear much of the pervasive low-frequency and high-intensity anthropogenic noise in the ocean, including sonar, shipping, and oil and gas exploration. More research is urgently needed to investigate the potential physiological and behavioral effects of anthropogenic noise on sea turtles.

References

Bartol S, Ketten DR (2006) Turtle and tuna hearing. In: Swimmer Y, Brill R (eds) Sea turtle and pelagic fish sensory biology: Developing techniques to reduce sea turtle bycatch in longline fisheries. Technical Memorandum NMFS-PIFSC-7, National Ocean and Atmospheric Administration (NOAA), US Department of Commerce, pp 98–105.

Bartol SM, Musick JA (2003) Sensory biology of sea turtles. In: Lutz PL, Musick JA, Wyneken J (eds) Biology of sea turtles, vol II. CRC Press, Boca Raton, FL, pp 79–102.

Bartol SM, Musick JA, Lenhardt M (1999) Auditory evoked potentials of the loggerhead sea turtle (*Caretta caretta*). Copeia 3:836–840.

Crawford AC, Fettiplace R (1980) The frequency selectivity of auditory nerve fibres and hair cells in the cochlear of the turtle. J Physiol 306:79–125.

Harms CA, Eckert SA, Jones TT, Dow Piniak WE, Mann DA (2009) A technique for underwater anesthesia compared with manual restraint of sea turtles undergoing auditory evoked potential measurements. J Herpetol Med Surg 19:8–12.

International Union for the Conservation of Nature (2010) 2010 IUCN red list of threatened species. Available via http://www.iucnredlist.org. Accessed 1 May 2010.

Ketten DR (2008) Underwater ears and the physiology of impacts: Comparative liability for hearing loss in sea turtles, birds, and mammals. Bioacoustics 17:312–315.

Lenhardt ML (1982) Bone conduction hearing in turtles. J Aud Res 22:153–160.

Lenhardt ML, Harkins SW (1983) Turtle shell as an auditory receptor. J Aud Res 23:251–260.

Lenhardt ML, Klinger RC, Musick JA (1985) Marine turtle middle-ear anatomy. J Aud Res 25:66–72.

Ridgway SH, Wever EG, McCormick JG, Palin J, Anderson JH (1969) Hearing in the giant sea turtle, *Chelonia mydas*. Proc Natl Acad Sci USA 64:884–890.

Wever EG (1978) The reptile ear: Its structure and function. Princeton University Press. Princeton, NJ.

Wever EG, Vernon JA (1956) Sound transmission in the turtle's ear. Proc Natl Acad Sci USA 42:292–299.

Hearing Capabilities of Loggerhead Sea Turtles (*Caretta caretta*) Throughout Ontogeny

Ashley L. Lavender, Soraya M. Bartol, and Ian K. Bartol

1 Introduction

Sea turtles are one group of endangered marine animals potentially impacted by anthropogenic sound production. Similar to other migratory marine species, sea turtles occupy different ecological niches throughout ontogeny, each characterized by unique acoustic conditions. Developing sea turtle hatchlings remain in pelagic habitats but then migrate to shallower neritic waters to forage as juveniles and reproduce as adults. Near-shore habitats are typically inundated with low-frequency human-generated sounds (Hawkins and Myrberg 1983) including recreational boating, commercial shipping, and high-intensity signals associated with seismic exploration (Greene and Moore 1995).

The effects of anthropogenic sound on sea turtles are not known because of limited research on the auditory systems of sea turtles throughout their life history stages. According to a few electrophysiological studies (Bartol et al. 1999; Ridgway et al. 1969), sea turtles appear to be low-frequency specialists (<1 kHz). However, these electrophysiological studies were conducted in air and have not been correlated with behavioral responses, an important step for comprehensive hearing assessment. Due to the global nature of auditory brain stem responses (ABRs), audiograms solely based on ABRs underestimate the auditory threshold at low frequencies (Kenyon et al. 1998). Moreover, these previous studies do not explore the hearing capabilities of sea turtles in multiple life history stages.

For this study, we are collecting both behavioral audiograms and ABRs from loggerhead sea turtles (*Caretta caretta*) in multiple developmental stages. The objectives of this project are 1) to compare sensitivity thresholds yielded by the two methods to determine how closely correlated ABR and behavioral audiogram data are, 2) to determine if sea turtle hearing capabilities change throughout ontogeny, and 3) to develop a comprehensive assessment of hearing in loggerheads that may be used for future assessment plans that address the potential impacts of anthropogenic sound exposure on sea turtles.

A.L. Lavender (✉) • I.K. Bartol
Old Dominion University, Norfolk, VA 23529, USA
e-mail: alave002@odu.edu; ibartol@odu.edu

S.M. Bartol
Virginia Wesleyan College, Norfolk, VA 23502, USA
e-mail: sbartol@vwc.edu

A.N. Popper and A. Hawkins (eds.), *The Effects of Noise on Aquatic Life*,
Advances in Experimental Medicine and Biology 730, DOI 10.1007/978-1-4419-7311-5_19,

2 Study Site

Experiments were conducted at the National Oceanic and Atmospheric Administration (NOAA) Fisheries Service Galveston Laboratory, Galveston, TX, which maintains ~400 captive-reared loggerheads (4–50+ cm straight carapace length [SCL]) from Florida nests for scientific studies. All turtles were held under several federal and state permits (US Fish and Wildlife Service [FWS] Permit TE676379-3; Florida Fish and Wildlife Conservation Commission [FWC] Permit TP 015; and Texas Parks and Wildlife Department [TPWD] Permit SPR-0390-038). ABR and behavioral trials were performed in two separate large (15,142-l) saltwater tanks of equal dimensions (3.7 m diameter; 1.5 m depth).

3 ABRs

ABRs are averaged electrophysiological recordings of simultaneous neural firing (or excitation) that are induced by an acoustic stimulus in the auditory center of the brain,. Historically, ABRs have been used as a method for testing audition and acoustic threshold on nonverbal, conscious animals (Bullock 1981; Corwin et al. 1982). For this project, we used Tucker-Davis Technologies (TDT) System 3 hardware and complementary software (SigGen and BioSig) together with an amplifier and J9 underwater transducer (speaker) to deliver underwater acoustic stimuli and acquire time-locked bioelectrical data. A hydrophone recorded sound pressure levels at the turtle during stimulus delivery.

We developed an effective protocol for restraining the turtle, which involves lowering it into the water using a series of pulleys and submerging the animal to a depth sufficiently deep to cover the tympanic scutes but shallow enough to facilitate voluntary breathing. Before the turtles were lowered into the tank, three subdermal electrodes (i.e., recording, reference, and ground) were inserted. The recording and reference electrodes were positioned dorsally along the frontoparietal scute and sealed with petroleum-based ointment, and the ground electrode was inserted into the inactive skin of the lateral neck and secured with a liquid bandage. Tone bursts of known frequencies (<1 kHz) were presented in descending order of intensity in 5-dB steps. Electrical and myogenic noise was removed by averaging ABR traces from stimuli presented in opposite polarities.

4 Behavioral Audiograms

Behavioral audiograms were recorded using a two-response, forced-choice approach (Blough and Blough 1977) where the turtles were required to vary behavior according to small acoustic stimuli differences, permitting a behavioral measure of acoustic sensitivity. Operant-conditioning methods have been successfully employed in studies of marine turtle chemoreception (Manton et al. 1972), learning (Mellgren and Mann 1996), and visual acuity (Bartol et al. 2003) at different ontogenetic stages.

A stimulus delivery and data-acquisition system was developed in-house specifically for this project using National Instruments hardware and LabVIEW software. TDT hardware and software were also utilized to generate acoustic stimuli, collect hydrophone data, and calculate sound pressure levels. An observing key was positioned in the middle of the tank in front of the J9 speaker. A white light was suspended above the observing key, which illuminated at trial onset, and a mounted video camera enabled real-time monitoring of trials by the researcher. Two response chutes were located equidistant from the observing key along the tank walls. Each response chute was constructed from PVC pipe and was designated as the "signal" or "no-signal" key. Individual turtles were subjected to a multistep conditioning procedure to establish associations between the observing key and trial onset (light) and between the presence/absence of the acoustic signal and the appropriate response chute. Out of the animal's view, squid were dispensed through the response chutes to reinforce correct responses. Once the turtles demonstrated proficiency in selecting the correct response chute in nonrandom signal presentations, the acoustic signals were then randomized. To qualify for trials, turtles must correctly respond to randomized sound/no-sound presentations with a 70–80% success rate. A training signal frequency of 300 Hz was used, with a sound pressure level of ~125 dB re 1 μPa.

5 Results and Discussion

Current ABR and behavioral audiogram analyses indicate that subadult sea turtles respond to sounds in the low-frequency range of 200–700 Hz. Furthermore, we have recorded a correlation between behavioral response times and sound intensity, with response times increasing near threshold. High levels of variability were recorded in behavioral trials, with some turtles demonstrating inconsistent positioning with respect to response chutes, nonrepeatable behavioral responses, inconsistent response times, and lack of motivation. For example, in 2009, 10 turtles belonging to 1 size class were trained extensively, but only 3 showed consistent repeatable responses and advanced to trials, with only 1 turtle performing at a consistently high level. The difficulties associated with acquiring reliable, robust behavioral data are not trivial and training alone requires a significant time investment by the researcher. We found that constant refinement of triggering keys, reward chutes, and speaker orientation was necessary to facilitate reliable behavioral data collection. Despite inherent challenges associated with data collection, behavioral data are necessary to evaluate the accuracy of ABR sensitivity thresholds and assess hearing in sea turtles throughout ontogeny. Additional ABR and behavioral work with early life history stages of sea turtles is ongoing. These data together with current subadult ABR/behavioral datasets promise to serve as a valuable foundation for hearing assessment in sea turtles throughout ontogeny and for validating prevailing electrophysiological methods.

Acknowledgments We gratefully acknowledge financial support from the Exploration and Production Sound and Marine Life Joint Industry Programme, International Association of Oil and Gas Producers, London, UK (JIP22 07–14). We also thank Ben Higgins and the helpful Protected Species Branch staff, National Oceanic and Atmospheric Administration Fisheries Service Galveston Laboratory, Galveston, TX.

References

Bartol SM, Mellgren RL, Musick JA (2003) Visual acuity of juvenile loggerhead sea turtles (*Caretta caretta*): A behavioral approach. Int J Comp Psychol 16:143–155.

Bartol SM, Musick JA, Lenhardt M (1999) Auditory evoked potentials of the loggerhead sea turtle (*Caretta caretta*). Copeia 3:836–840.

Blough D, Blough P (1977) Animal psychophysics. In: Honig WK, Staddon JER (eds) Handbook of operant behavior. Prentice-Hall, Englewood Cliffs, NJ, pp 514–539.

Bullock TH (1981) Neuroethology deserves more study of evoked responses. Neuroscience 6:1203–1215.

Corwin JT, Bullock TH, Schweitzer J (1982) Auditory brainstem response in five vertebrate classes. Electroenceph Clin Neurophysiol 54:629–641.

Greene CR Jr, Moore SE (1995) Man-made noise. In: Richardson WJ, Greene CR Jr, Malme CI, Thomson DH (eds) Marine mammals and noise. Academic Press, San Diego, CA, pp 101–158.

Hawkins AD, Myrberg AA Jr (1983) Hearing and sound communication under water. In: Lewis B (ed) Bioacoustics: A comparative approach. Academic Press, London, pp 347–404.
Kenyon TN, Ladich F, Yan HY (1998) A comparative study of hearing ability in fishes: The auditory brainstem response approach. J Comp Physiol A 182:307–318.
Manton ML, Karr A, Ehrenfeld DW (1972) An operant method for the study of chemoreception in the green turtle, *Chelonia mydas*. Brain Behav Evol 5:188–201.
Mellgren RL, Mann MA (1996) Comparative behavior of hatchling sea turtles. In: Keinath JA, Barnard DE, Musick JA, Bell BA (eds). NOAA Technical Memorandum NMFS-SEFSC-387, Proceedings of the Fifteenth Annual Symposium on Sea Turtle Biology and Conservation, 20–25 February 1995, Hilton Head, SC, pp 202–204.
Ridgway SH, Wever EG, McCormick JG, Palin J, Anderson JH (1969) Hearing in the giant sea turtle, Chelonia mydas. Proc Natl Acad Sci USA 64:884–890.

Are Sharks Even Bothered by a Noisy Environment?

Brandon M. Casper, Michele B. Halvorsen, and Arthur N. Popper

1 Introduction

There is growing concern that sounds produced by anthropogenic sources have the potential to impact bony fishes. However, there are no data as to whether elasmobranch fishes (sharks, rays, and skates) could be affected by exposure to anthropogenic sources.

According to the International Union for the Conservation of Nature (IUCN), ~60% of elasmobranch species are considered threatened with extinction due to overfishing and habitat degradation (Godin and Worm 2010). Elasmobranchs are important from an evolutionary perspective because they have evolved little over hundreds of millions of years and represent a unique opportunity to examine one of the more basal stages within the evolution of vertebrates. This paper considers the possibility that anthropogenic noise may have an effect on elasmobranch fishes. The analysis is based on the results from noise-exposure studies in teleosts as well as knowledge of elasmobranch anatomy and physiology. A review of how elasmobranchs detect sound and their hearing abilities is addressed, with the goal of touching on areas in need of further exploration.

2 Elasmobranch Sound Detection

Elasmobranchs detect sound using inner ear end organs (see Myrberg 2001 for review). It is likely that the saccule, a portion of the utricle, and the macula neglecta are the acoustically sensitive organs, whereas the lagena and the other portion of the utricle are utilized for detection of gravity and rotational stimuli. Unlike the hardened otoliths in teleosts, the sensory epithelia (maculae) of the saccule, utricle, and lagena in elasmobranchs are covered by otoconia, a gelatinous matrix of calcium carbonate granules (a pattern also found in primitive teleost fish and all terrestrial vertebrates). In contrast, the fourth end organ, the macula neglecta, is covered by a gelatinous cupula that is similar to the cupula found in the lateral line organs and ampullae of the semicircular canals. The macula neglecta is not unique to elasmobranchs, but these are the only vertebrates in which there is evidence that it serves a role in acoustic detection. The elasmobranch auditory system is also unique

B.M. Casper (✉) • M.B. Halvorsen • A.N. Popper
Department of Biology, University of Maryland, College Park, MD 20742, USA
e-mail: bcasper@umd.edu

A.N. Popper and A. Hawkins (eds.), *The Effects of Noise on Aquatic Life*,
Advances in Experimental Medicine and Biology 730, DOI 10.1007/978-1-4419-7311-5_20,

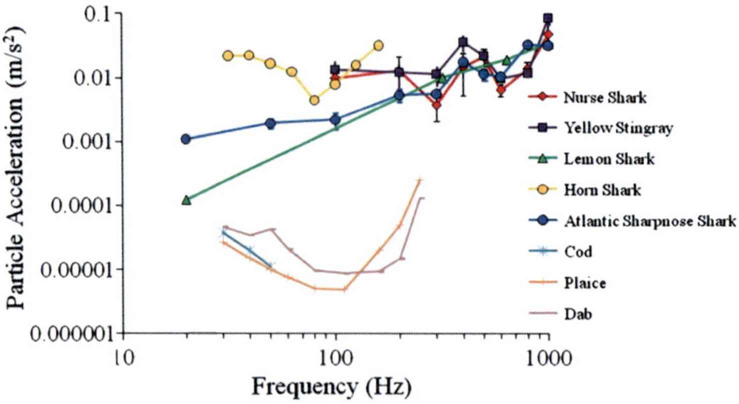

Fig. 1 Particle motion audiogram describing hearing thresholds of elasmobranch and teleost fishes. Modified from Casper and Mann (2009) and Fay (1988)

in having a direct connection from the saccular chamber to the surface of the head and the outside environment through the endolymphatic duct. However, no direct evidence has linked the duct to any specific role in the detection of sound.

There are two proposed pathways for sound to travel to the inner ear of elasmbranchs (Corwin 1981). First, the otoconial pathway involves the saccule and utricle end organs. The elasmobranch body is approximately equal in density to the surrounding water and is therefore acoustically transparent. As a consequence, sound waves travel through the fish until they come into contact with a structure of greater density such as the otoconia. These otoconia lag in movement relative to the surrounding tissues. This results in bending of the cilia of the sensory hair cells that underlie the otoconia, and this, in turn, generates a physiological response that results in sound detection.

The second, nonotoconial sound pathway involves the macula neglecta. This sensory structure is located in the dorsal portion of the ear in the posterior canal duct. Sounds travel from above the elasmobranch head and through an area of tissue located above the ear called the parietal fossa. In the ventral end of the parietal fossa is a small membrane, the fenestra ovalis, that leads to the posterior canal duct. Sound waves depressing this membrane produce a flow of fluid through the posterior canal duct, shifting the position of the cupula of the macula neglecta and stimulating the sensory hair cells.

Measures of hearing bandwidth and frequency (the audiogram) for particle motion have been obtained in five species of elasmobranch fishes using classical conditioning or auditory evoked potential methods (Fig. 1; see Casper and Mann 2009 for a review). Because elasmobranchs do not have a swim bladder or any other air-filled cavity, they are incapable of detecting sound pressure. Thus particle motion is presumably the only sound stimulus that can be detected. The hearing bandwidth for elasmobranchs is from ~20 Hz up to 1 kHz, with similar thresholds in all species above 100 Hz (Casper and Mann 2009). Below 100 Hz, however, the two more active swimming piscivorous species, *Rhizoprionodon terraenovae* (Atlantic sharpnose shark) and *Negaprion brevirostris* (lemon shark), have more sensitive hearing, suggesting that hearing could be more important for the detection of prey. The other three species, *Ginglymostoma cirratum* (nurse shark), *Heterodontus francisci* (horn shark), and *Urobatis jamaicensis* (yellow stingray), are demersal species and likely use other senses including the lateral line and electroreception to find buried prey. Thus, although it is clear that elasmobranchs can detect particle motion, they do not appear to be as sensitive as teleosts measured in comparable ways (Fig. 1). One explanation for this difference in hearing sensitivity could be due to the composition of the denser otoliths in teleosts compared with the otoconia in elasmobranchs. A denser otolith might be more sensitive to particle motion and therefore yield more sensitivity to the auditory system. However, knowledge of the hearing of elasmobranch fishes is based on data from only a few of the hundreds of species, and so one must be cautious in making generalizations about an entire subclass of fishes based on these data.

2.1 *Shark Behavioral Responses to Sound*

The US Navy became interested in sounds that might attract or repel sharks following repeated observations of the presence of sharks in areas where ships were sunk by torpedoes during World War II. Acoustic attraction studies revealed that coastal and oceanic sharks (18 species observed) would often approach underwater speakers broadcasting low-frequency, erratically pulsed sounds from as far away as several hundreds of meters (Myrberg 2001). A few studies also attempted to determine the features of sounds that might cause sharks to leave a location. They found that sudden onset, loud (20–30 dB above ambient noise levels) sounds played when a shark approached a location would result in startling the shark and it would turn away from the area. In most cases involving attraction and repelling, the sharks would habituate to the stimuli after a few trials. There have been no experiments exploring behavioral responses to sound in either skates or rays. There have also been no studies examining the effects of exposure to anthropogenic sound sources in any species of elasmobranch.

3 Sources of Anthropogenic Noise That Could Affect Elasmobranch Fishes

There are many human-based activities that produce anthropogenic noise, including sonar, aquatic construction, air guns, boat activity, and offshore wind farms, that could potentially threaten aquatic inhabitants. Based on the location of sources, rate of occurrence, frequency ranges, and damaging effects associated with exposure, several of these sounds could have negative effects on elasmobranch fishes.

3.1 *Aquatic-Based Construction*

Pile driving is used for construction, including installation and repair of bridges, docks, and other structures, in aquatic environments. There is documentation that elasmobranchs tend to aggregate around coastal and offshore man-made structures (Stanley and Wilson 1991). A major concern is that elasmobranchs congregating near such structures could be impacted by the intense sounds during pile driving. Sound levels can reach 237 dB re 1 μPa at frequencies within the range of hearing of elasmobranchs (100–1,000 Hz; Hildebrand 2009). Sounds at such high levels could yield hearing damage in the form of temporary threshold shift (TTS), resulting in a short-term decrease in auditory sensitivity. However, the more likely source of damage would be barotrauma as a result of the impulsive energy produced when the hammer hits the pile. Recent evidence (see Halvorsen et al., Chapter 52) suggests that some of the barotrauma damage found in teleosts when exposed to pile-driving stimuli is in the liver, kidneys, and intestines, all of which are very prominent in the elasmobranch body plan. Another consideration is for demersal elasmobranchs that are almost constantly in contact with the substrate. The intense vibrations within the sediment from pile driving could also be damaging, especially when considering the body shape of skates and rays. Many of the organs of these dorsoventrally flattened fishes are in close proximity to the ventral body surface, providing little protection from pile-driving vibrations.

3.2 *Offshore Wind Farms*

With the need for cost-effective forms of electricity, more countries are exploring the application of offshore wind farms. Wind farm installation generally involves pile-driving construction, with the associated noise issues as discussed in the previous section. Once completed and operating, the rotation of the turbines produces a constant low-frequency noise (~60–300 Hz) at sound levels of ~150 dB re 1 μPa (Hildebrand 2009). These levels are likely not loud enough to cause any hearing damage (TTS), but there could be the potential of masking of sounds that elasmobranchs might use to detect prey or avoid predators. Because the wind mills are anchored to the substrate, there is also the potential for vibrational stimuli traveling through the structure that could impact demersal elasmobranchs.

3.3 *Boat Noise*

The number of vessels in the worldwide shipping fleet has grown dramatically over the last 50–60 yr. Obviously, the size, speed, and other features of the ships can affect the type of noise produced as they travel through the water, but in general, the sounds produced can be quite dramatic. A typical shipping vessel can produce sounds of ~190 dB re 1 μPa at very low frequencies (40–100 Hz; Hildebrand 2009). At these sound levels, it is unlikely that hearing damage would occur in elasmobranchs, but the sounds would certainly be loud enough to mask detection of biologically relevant sounds. A few studies have examined the effects of shipping noise and other noise exposure on the production of stress hormones in teleost fishes. Extended exposure resulted in increased levels of cortisol, which can affect a variety of health parameters in fishes (Wysocki et al. 2006). No similar studies have yet been conducted on elasmobranch fishes, but there is the potential of similar effects to those encountered in teleosts.

4 Summary

Elasmobranch fishes have been around for hundreds of millions of years with very little evolutionary changes, yet our understanding of their hearing abilities is limited to only a few of the hundreds of extant species. Our general understanding suggests a relatively narrow hearing range with relatively poor sensitivity, particularly compared with many teleosts. This lack of knowledge makes it difficult to evaluate the potential effects that could be associated with exposure to anthropogenic noise. However, given the combination of the worldwide increase in anthropogenic aquatic noise as well as the drastic population decline in many species of elasmobranch fishes, it is imperative that noise-exposure studies be conducted to determine whether these fishes are being further threatened by our noise pollution.

References

Casper BM, Mann DA (2009) Field hearing measurements of the Atlantic sharpnose shark, *Rhizoprionodon terraenovae*. J Fish Biol 75:2768–2776.

Corwin JT (1981) Peripheral auditory physiology in the lemon shark: Evidence of the parallel otolithic and non-otolithic sound detection. J Comp Physiol 142:379–390.

Fay RR (1988) Hearing in vertebrates: A psychophysics databook. Hill-Fay Associates, Winnetka, IL.

Godin AC, Worm B (2010) Keeping the lead: How to strengthen shark conservation and management policies in Canada. Mar Policy 34:995–1001. Doi:10.1016/j.marpol.2010.02.006.

Hildebrand JA (2009) Anthropogenic and natural sources of ambient noise in the ocean. Mar Ecol Prog Ser 395:5–20.

Myrberg AA Jr (2001) The acoustical biology of elasmobranchs. Environ Biol Fish 60:31–45.

Stanley DR, Wilson CA (1991) Factors affecting the abundance of selected fishes near oil and gas platforms in the Northern Gulf of Mexico. Fish Bull 89:149–159.

Wysocki LE, Dittami JP, Ladich F (2006) Ship noise and cortisol secretion in European freshwater fishes. Biol Conserv 128:501–508.

Optimal Auditory Sensitivity Under Variable Background Noise Conditions: A Theoretical Model

Marco Lugli

1 Introduction

Sound signals relevant for mating and survival are very often masked by background noise, which makes their detection and recognition by organisms difficult (i.e., communication often takes place under partially masked conditions; e.g., Zwicker and Fastl 1990). Ambient noise (AN) varies in level and shape among different habitats, but remarkable variations in time and space also occur within the same habitat. Variable AN conditions mask hearing thresholds of the receiver in complex and unpredictable ways, thereby causing distortions in sound perception. For instance, sound and speech recognition in animal and human subjects quickly deteriorates with decreasing signal-to-noise ratio under nonstationary noise conditions. Furthermore, no sound with energy lower than the noise can be heard. These observations suggest that when communication takes place in a noisy environment, a highly sensitive system may confer no advantage to the receiver compared with a less sensitive one (Hawkins and Myrberg 1983). Fishes live in all types of underwater habitats differing widely for AN conditions, from quiet deep oceans and shallow ponds to noisy coastal waters and stony streams. Notably, they show an impressive variety of audiograms that differ in shape, level, and frequency range. Lugli et al. (2003) showed that the best hearing range and the dominant frequencies of sounds of the two freshwater gobies (*Padogobius bonelli* and *Gobius nigricans*) fit within a relatively quiet window in the low-frequency spectrum of the stream AN. Amoser and Ladich (2005) noted that teleosts with the best hearing (e.g., Cyprinids) live in quieter habitats than those with poor hearing abilities (e.g., Percids). These findings suggest that AN may be an important selective factor in the evolution of hearing sensitivity of a species. The way it would select for the level of hearing sensitivity of a species is unclear, however. Here I describe a simple fitness model for the detection and discrimination (and recognition) of sounds under variable AN conditions. I assume that noise masking significantly impairs all the above functions. The model predicts high sensitivity (i.e., low hearing thresholds) as the best strategy for species living in quiet habitats and low sensitivity (i.e., high hearing thresholds) as the best strategy for those living in noisy habitats, provided the cost of incorrect discrimination is not low.

M. Lugli (✉)
Università di Parma, I43121 Parma, Italy
e-mail: marco.lugli@unipr.it

A.N. Popper and A. Hawkins (eds.), *The Effects of Noise on Aquatic Life*,
Advances in Experimental Medicine and Biology 730, DOI 10.1007/978-1-4419-7311-5_21,

2 The Model

Let's consider two AN conditions: quiet and noisy. The quiet condition, characterized by a single, low-energy, flat noise spectrum, simulates the low AN levels of ponds, pools. and slow rivers inhabited by fishes with higher hearing sensitivity (e.g., Cyprinids).The noisy condition, characterized by high-energy spectra of different shapes, simulates the variety of masking-noise spectra found in small streams, creeks, and coastal waters inhabited by fishes with lower hearing sensitivity (e.g., Gobiids, Perciids, and Blennids). Consider a simple sound signal that can be detected at two intensity levels, low and high. Suppose that the high-amplitude signal is always detectable, whereas the low-amplitude signal is detectable only under the quiet condition. Suppose also that the detected signal must be discriminated against (e.g., a sound template) or recognized as such (except for sound amplitude) by the receiver. I assume that the variability of the masking pattern found in the noisy condition determines the incorrect discrimination and recognition of the high-amplitude signal. Assuming that the receiver always responds whenever a signal is detected, three outcomes and associated payoffs are possible: the signal is not detected, yielding the payoff E_0 (no detection); the signal is detected but incorrectly discriminated, yielding the payoff E_1 (incorrect discrimination and recognition; $E_1 > E_0$ or $E_1 \leq E_0$); or the signal is detected and correctly discriminated, yielding the highest payoff E_2 (correct discrimination and recognition). Given the above conditions, a receiver must choose between two hearing systems: a high-sensitive system ($\mathbf{W}_0$) and a low-sensitive system ($\mathbf{W}_1$). Auditory thresholds of $\mathbf{W}_0$ are masked only in the noisy condition, whereas those of $\mathbf{W}_1$ are never masked. Let Q be the probability for the receiver detecting the signal under the quiet condition (N_0) and SH be the probability of detecting a high-amplitude signal. The average fitness payoff for the receiver adopting the high-sensitive system, $E(\mathbf{W}_0)$, and that for the receiver adopting the low-sensitive system, $E(\mathbf{W}_1)$, can now be calculated, and the conditions favoring one system over the other can be determined.

3 Results and Conclusions

The model predicts that $\mathbf{W}_0$ is always favored (i.e., best strategy) in quiet environments, whereas $\mathbf{W}_1$ is favored in noisy environments provided that 1) most of detected signals are loud (close-range communication system) and 2) the payoff difference between E_2 and E_1 is large. Otherwise, $\mathbf{W}_0$ is favored. The results of the model emphasize the tradeoff between the detection and discrimination of sound signals when communication occurs in noisy habitats with variable masking-noise conditions. Furthermore, it addresses the importance of the AN as a possible main selective factor shaping the absolute hearing sensitivity of a species.

References

Amoser S, Ladich F (2005) Are hearing sensitivities of freshwater fishes adapted to the ambient noise of their habitats? J Exp Biol 208:3533–3542.

Hawkins AD, Myrberg AA Jr (1983) Hearing and sound communication under water. In: Lewis B (ed) Bioacoustics: A comparative approach. Academic Press, London, pp 347–405.

Lugli M, Yan HY, Fine ML (2003) Acoustic communication in two freshwater gobies: The relationship between ambient noise, hearing thresholds and sound spectrum. J Comp Physiol A 189:309–320.

Zwicker E, Fastl H (1990) Psychoacoustics: Facts and models. Springer-Verlag, Berlin.

A Critical Reevaluation of the Role of Acoustic Pressure in Source Localization by Fish

Peter H. Rogers, James S. Martin, and John R. Bogle

1 Introduction

The notion that the otolithic organs in the fish ear function as accelerometers has been accepted for over 50 years. By direct excitation, they are capable of sensing the acoustic particle acceleration of an incident acoustic wave. The particle acceleration is a vector field that, for monopole sources, is directed along the source bearing. There are, in effect, many tens of thousands of these accelerometers in each of the six otolithic organs. The spatial orientation of their response axes are ordered but complex. The compliant swim bladder functions as a secondary generator of acceleration at the otoliths, either by direct connection as in otophysans or by acoustic scattering. The swim bladder (or other gas cavities) thus makes it possible for some fish to sense acoustic pressure as well as particle velocity. This mode of excitation of the otolithic organs is referred to as the "indirect" path. The effect of a coupled swim bladder is to extend the hearing range and lower the threshold of hearing. The direct signal is related to a vector and contains information about the source direction. As both directional direct-path and nondirectional indirect signals impinge simultaneously on the otolithic organs, some means of segregating them is required for directionalization to occur. This segregation has been demonstrated by Buwalda (1981). Exactly how source bearing is extracted from the direct signals is unknown as is the mechanism for separating the direct and indirect signals, but theories abound (e.g., Buwalda 1981; Rogers et al 1988; Schuijf 1975). It is generally believed that the indirect (pressure-related) signal serves no role in directionalization other than serving to resolve the 180° ambiguity that occurs because particle acceleration alone only yields the line of bearing to the source (Schuijf and Buwalda 1975). The pressure component is otherwise believed to have no role or to be a negative factor in directionalization. This paper proposes that this viewpoint is not supported by the data and suggests how pressure detection might contribute to directionalization.

2 Thresholds

In the 1970s, a remarkable series of high-quality experiments investigated the hearing capabilities of the Atlantic cod (*Gadus morhua*; e.g., Buwalda and van der Steen 1979; Sand and Enger 1973; Schuijf 1975; Schuijf and Buwalda 1975). This study focuses on the results of papers by

P.H. Rogers (✉) • J.S. Martin • J.R. Bogle
Georgia Institute of Technology, Atlanta, GA 30332, USA
e-mail: peter.rogers@gatech.edu

A.N. Popper and A. Hawkins (eds.), *The Effects of Noise on Aquatic Life*,
Advances in Experimental Medicine and Biology 730, DOI 10.1007/978-1-4419-7311-5_22,

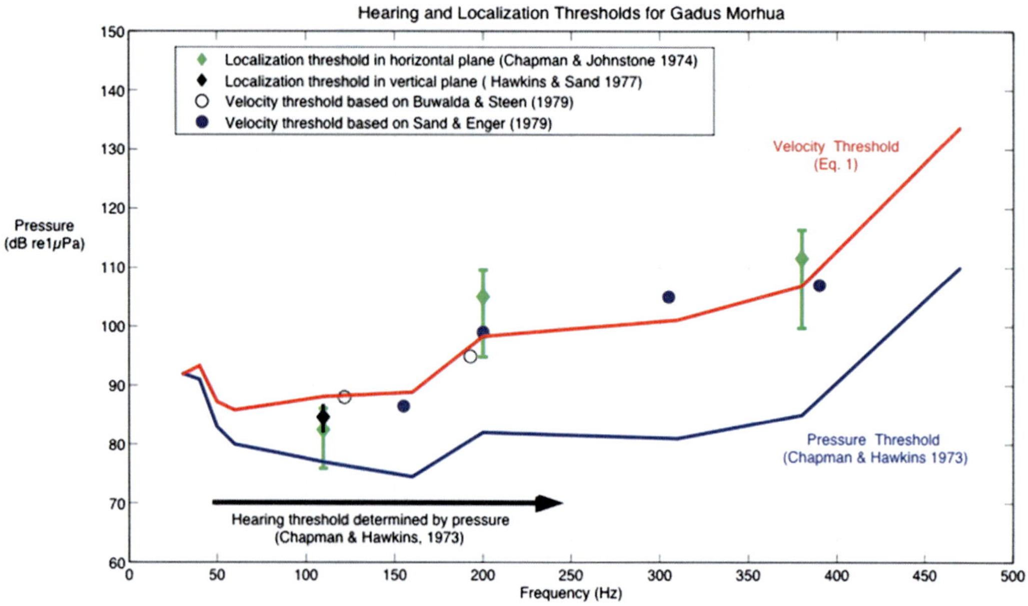

Fig. 1 Measured pressure audiogram and synthesized velocity audiograms compared with localization thresholds for the horizontal plane and the vertical plane

Chapman and Hawkins (1973), Chapman and Johnstone (1974), and Hawkins and Sand (1977). All were done in deep water under far-field conditions and utilized heart rate conditioning. They were generally done under low-noise conditions so that near-absolute thresholds for hearing and directionalization could be obtained. These conditions produced unambiguous results that are quite rare in the literature, before or since. A far-field audiogram for *Gadus* ($n = 43$; Fig. 1, blue curve) was obtained Chapman and Hawkins (1973). The data were acquired for a range of noise conditions including some at sea state zero. The means shown are believed to be just 2 dB above the absolute thresholds. They also demonstrated (by examining the range dependence of the thresholds) that for frequencies ≥ 50 Hz, the thresholds were determined by acoustic pressure, not particle velocity. No comparable threshold curve exists for the acoustic particle velocity, but it possible to construct one. The underlying transduction mechanism is the same for both pressure and particle velocity. The acoustic particle acceleration at the otolithic sensor is known for any incident plane wave pressure. Well below swim bladder resonance, the gain [$G(f)$], defined as the ratio of the acceleration at the ear due to scattering from the swim bladder to the directly induced acceleration at the ear, is proportional to the frequency (Sand and Hawkins 1973). This gain is equal to the ratio of the velocity threshold [$TV(f)$] to the pressure threshold [$TP(f)$] when both are expressed in the same units. Hence, if $TP(f)$ is the pressure threshold in decibels, then the velocity threshold must be given by

$$TV(f) = TP(f) + G(f) = TP(f) + A + 20\log_{10}(f/f_A) \tag{1}$$

Note that $G(f)$ is also the ratio of the response of the ear to a pressure-only signal of amplitude p to that of a velocity-only signal of amplitude $p/\rho c$ and also the ratio of the response of the ear to a plane wave signal with and without a swim bladder. The gain $G(f)$ was measured by Sand and Enger (1973) using swim bladder deflation and by Buwalda and van der Steen (1979) using multiple sources to produce pressure-only and velocity-only stimuli. Both experiments used saccular microphonics to measure the response of the ear.

We use Buwalda and van der Steen's (1979) result of 12 dB at 122 Hz to get $A = 12$ and $f_A = 122$ for the unknown coefficients in Equation 1. The resulting threshold for velocity detection (expressed in units of input pressure) is shown by the red curve in Figure 1. The circles shown in the figure are the velocity threshold values obtained from the pressure detection threshold using the individual gains measured by Sand and Enger (1973) and Buwalda and van der Steen (1979). They agree well with the red curve. Further evidence of the reasonableness of the red curve is derived by noting that it agrees with Chapman and Hawkins' (1973) observation that cod detect acoustic pressure for frequencies $\geq$ 50 Hz and particle velocity at lower frequencies.

3 Localization Thresholds

Figure 1 also shows localization thresholds obtained by Chapman and Johnstone (1974) for the horizontal plane and Hawkins and Sand (1977) for the vertical plane. In both cases, heart rate conditioning was used to determine whether the fish could discern when the origin of a sound is switched between two sources that differ only in their angular location. The fish were in the far field of the sources, and both the sources and fish were far from any reflecting surfaces. Noise levels were sufficiently low to permit determination of absolute thresholds. Chapman and Johnstone (1974) found thresholds at 105, 200, and 380 Hz, whereas all of Hawkins and Sand's (1977) results were at 105 Hz. In both cases, the thresholds were relatively constant at larger angular separations between the two sources but increased rapidly as an angular limit was approached. The thresholds shown are an average over the thresholds in this flattish region. The error bars and means ± σ.

4 Discussion

Because localization is a more difficult task than detection, one would expect that the velocity threshold for localization would exceed the corresponding threshold for detection by at least 6 dB (the typical just noticeable difference for the cod; Chapman and Johnstone 1974). From Figure 1, we see that this is the case at 200 and 380 Hz. The very large (24-dB) difference between the overall detection threshold and the localization threshold is often cited as "proof" that only velocity is involved in localization, as predicted by the prevailing theory. It is equally apparent for both the horizontal and vertical separation cases that at 105-Hz, localization is occurring at stimulus levels that are subliminal for particle velocity sensing. The thresholds are, in fact, just 5.5–7.7 dB above the pressure threshold. It is interesting to note how close to one another the vertical and horizontal thresholds are. It is evident that something quite different is occurring at the lower frequency. The results suggest that either there is directional information in the indirect signal at 105 Hz or that the presence of an indirect signal enhances the detectability of the direct signal. The possibility that directional information could be present in the indirect signal is based on the fact that a plane wave has a uniaxial and not an isotropic strain, that the swim bladder is elongated, and that the fish body has a finite shear modulus. This possibility is investigated with finite element and experimental models of the fish swim bladder system.

References

Buwalda RJA (1981) Segregation of directional and non-directional acoustic information in cod. In: Tavolga WN, Popper AN, Fay RR (eds) Hearing and sound communication in Fishes. Springer-Verlag, New York, pp 139–171.

Buwalda RJA, van der Steen J (1979) The sensitivity of the cod sacculus to directional and non-directional sound stimuli. Comp Biochem Physiol 64A:467–471.

Chapman CL, Hawkins AD (1973) A field study of hearing in the cod *Gadus mohua*. J Comp Physiol 85:147–167.

Chapman CJ, Johnstone ADF (1974) Some auditory discrimination experiments on marine fish. J Exp Biology 61:521–528.

Hawkins AD, Sand O (1977) Directional hearing in the median vertical plane by the cod. J Comp Physiol 122:1–8.

Rogers PH, Popper AN, Hastings MC, Saidel WM (1988) Processing of acoustic signals in the auditory system of bony fish. J Acoust Soc Am 83:338–349.

Sand O, Enger PS (1973) Evidence for an auditory function of the swimbladder in the cod. J Exp Biol 59:405–414.

Sand O, Hawkins AD (1973) Acoustic properties of the cod swimbladder. J Exp Biol 58:797–820.

Schuijf A (1975) Directional hearing of cod (*Gadus mohua*) under approximate free field conditions. J Comp Physiol 98:307–332.

Schuijf A, Buwalda RJA (1975) On the mechanism for directional hearing in cod (*Gadus mohua* L.). J Comp Physiol 98:333–343.

Vibration of the Otoliths in a Teleost

Carl R. Schilt, Ted W. Cranford, Petr Krysl, Robert E. Shadwick, and Anthony D. Hawkins

Fish populations comprise essential parts of marine and freshwater ecosystems as well as being the foundations of considerable human nutrition, industry, and economy. There is growing concern about the possible harmful effects from human-generated sound on fish. Sounds and other human-generated physical phenomena, such as explosions and shock waves, may have a variety of deleterious effects on aquatic animals, ranging from acute injury and death through masking of needful environmental sound stimuli to minor disturbances with little long-term effect on the animals.

1 Fish as Sound Receivers

Fish are sensitive to waterborne disturbances including bulk flows and the particle motions inherent in propagated sound waves. Bulk flows and particle motions attenuate abruptly with distance from the source of disturbance, especially for higher frequencies, and so limit absolute and bandwidth sensitivities in fish (Webb et al. 2008).

The bodies of many kinds of fish enclose gas-filled pockets that convert scalar pressure changes into vector motions accessible to the fish's mechanosensory system. Our finite element modeling (FEM) effort will eventually enable a time- and cost-effective means of answering questions about the effects of simulated sounds on animal models under selected conditions.

C.R. Schilt (✉)
Bigleaf Science Services, North Bonneville, WA 98639, USA
e-mail: bigleaf@gorge.net

T.W. Cranford
Quantitative Morphology Consulting, Inc., San Diego, CA 92123, USA
e-mail: Ted.W.Cranford@gmail.com

P. Krysl
University of California, San Diego, La Jolla, CA 92093, USA
e-mail: pkrysl@sdsu.edu

R.E. Shadwick
University of British Columbia, Vancouver, BC V6T 1Z4, Canada
e-mail: shadwick@zoology.ubc.ca

A.D. Hawkins
Loughine Limited, Kincraig, Blairs, Aberdeen AB12 5YT, UK
e-mail: a.hawkins@btconnect.com

A.N. Popper and A. Hawkins (eds.), *The Effects of Noise on Aquatic Life*,
Advances in Experimental Medicine and Biology 730, DOI 10.1007/978-1-4419-7311-5_23,

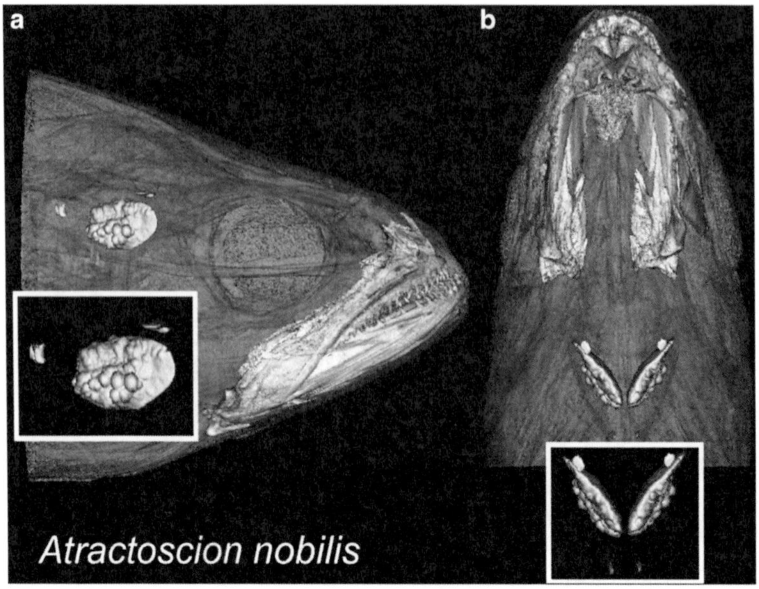

Fig. 1 Reconstruction of a young *Atractoscion nobilis* from micro-CT scan images (60-μm^3 voxels), with three pairs of otoliths in lateral view (**a**) and dorsal view (**b**). Insets: Magnified white otoliths. The cranial bones and soft tissues are semitransparent gray; the premaxilla and dentary are white. The bones, swim bladder, and soft tissues are not part of the current model

All vertebrates employ "hair cells" that mediate the neural response to differential motion (shear) from the cell bodies to the stiff ciliary bundles that grow from the hair cells. Dense, calcareous secretions called "otoliths" or ear stones (Popper et al. 2005) overlay dense beds of hair cells (auditory maculae), and due to the relatively large otolith masses, they exhibit inertial drag as the fish moves in the acoustic-particle motion field. The maculae and associated otoliths act as mass-loaded accelerometers that convey frequency, amplitude, and directional information to the fish's brain (Webb et al. 2008).

2 FEM and Virtual Bioacoustics Experiments

We are exploring a novel means of simulating experiments to advance understanding the interactions between selected sound stimuli and animal structures. Similar experiments have recently provided insights into the bioacoustics of a beaked whale, *Ziphius cavirostris* (Cranford et al. 2008a,b).

We used methods and tools developed by Cranford et al. (2008a,b) to build a FEM. We collected anatomic data from a small (21-cm total length) dead *Atractoscion nobilis* (Sciaenidae) from southern California by means of a micro-CT scanner (Fig. 1). This was the basis of a very simple FEM of the dynamic response of the fish otoliths to incident planar acoustic waves. The otoliths are modeled as embedded in a shear-soft fluidlike jelly. The current model does not include limits such as the nearby cranial bones nor influences such as the swim bladder. The model space was stimulated with two different sinusoidal signals (200 and 400 Hz) from several different directions vis-à-vis the fish.

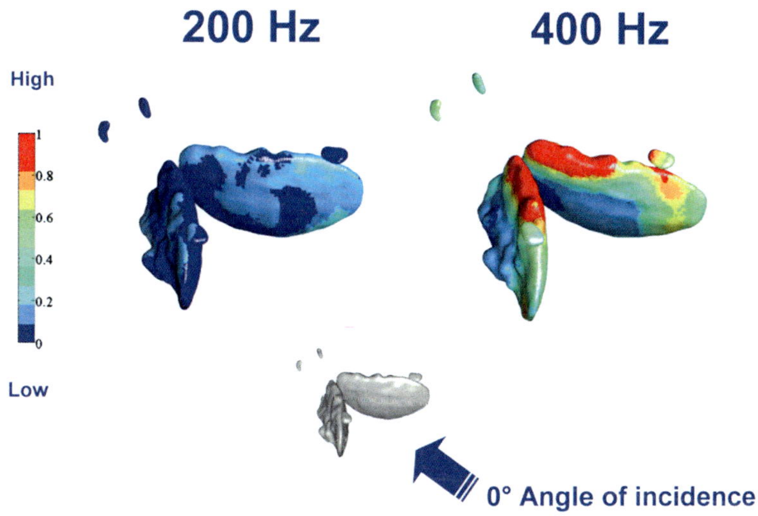

Fig. 2 A single instant in time for sinusoidal sounds that arrive from directly in front of the fish otoliths (angle of incidence = 0°) at each of two frequencies. Color variation indicates the magnitude of shear between the surfaces of the otoliths and the surrounding jellylike material

Figure 2 illustrates the shear forces that result from the relative motion between the otolith surfaces and the shear-soft jelly that currently surrounds them in the model space.

It is evident that the 400-Hz simulation produces greater shear values (due to larger displacements of the otoliths), particularly in the dorsoventral dimension, than does the 200-Hz signal from the same magnitude and direction. The FEM simulations apparently produce informative results, even at this early stage of model development. Future iterations will follow two contrasting trajectories. First, we will include more anatomic components within the model. Second, we anticipate simulating the responses of simple "spherical" otoliths without their distinctive shapes and sculpting. At a later stage in the development of this FEM effort, we will address the validation of our methods (see Krysl et al., Chapter 14).

Acknowledgments We appreciate the support of James Eckman and Michael Weise, Office of Naval Research, Arlington, VA; Mike Shane, Hubbs-Sea World Research Institute, San Diego, CA; Tom Nelson, University of California, San Diego (UCSD) Medical School, La Jolla, CA; and Jacqueline Corbeil and Robert Mattrey, Moores UCSD Cancer Center, La Jolla, CA.

References

Cranford TW, Krysl P, Hildebrand JA (2008a) Acoustic pathways revealed: Simulated sound transmission and reception in Cuvier's beaked whale (*Ziphius cavirostris*). Bioinsp Biomim 3:1–10. Doi: 10.1088/1748-3182/3/1/016001.

Cranford TW, McKennan MF, Soldevilla SS, Wiggins SM, Goldbogen JA, St Leger JA, Hildebrand JA (2008b) Anatomic geometry of sound transmission and reception in Cuvier's beaked whale (*Ziphius cavirostris*). Anat Rec 291:353–378. Doi: 10.1002/ar.20652.

Popper AN, Ramicharitar J, Campagna SF (2005) Why otoliths? Insights from fish inner ear physiology and fisheries biology. Mar Freshw Res 56:497–504.

Webb JF, Popper AN, Fay RR (eds) (2008) Fish bioacoustics. Springer, New York.

Hearing Sensitivity of the Painted Goby, *Pomatoschistus pictus*

Marta Bolgan, Silvia S. Pedroso, Raquel O. Vasconcelos, Joana M. Jordão, M. Clara P. Amorim, and Paulo J. Fonseca

1 Introduction

Pomatoschistus pictus is a coastal and estuarine species that inhabits shallow gravel and sand substrate areas of the eastern Atlantic Ocean and Mediterranean Sea (Miller 1986). Studies on *P. pictus* show that, like in other species of this genus (Malavasi et al. 2008), males produce sounds during courtship and agonistic contexts (Amorim and Neves 2007, 2008).

The amount of anthropogenic noise pollution has been increasing significantly in the last decades in coastal environments (Codarin et al. 2009; Ross 2005). A noisy coastal environment may strongly impact the ability of such a vocal species to communicate and, ultimately, reproduce. A first step in understanding the impact of anthropogenic noise is to describe this species' hearing sensitivity. However, nothing has been described in terms of the auditory abilities of this genus to date. This study presents the first data on the hearing sensitivity of *P. pictus*.

2 Materials and Methods

Six adult *P. pictus* were caught in April 2010 at Parede (38°41′ N, 009°21′ W), Portugal. The fish were maintained in aquaria at 18 ± 1°C.

Hearing thresholds were estimated using the auditory evoked potential (AEP) recording technique. Test subjects were mildly immobilized with 47.9 μg/g of pancuronium bromide diluted in teleost saline solution. The subjects were placed just below the water surface of a plastic tank (diameter 36 cm, water depth 13 cm), 7 cm above the center of the underwater speaker disk. Fish respiration was secured through a temperature-controlled (20 ± 1°C) gravity-fed seawater circulation system

M. Bolgan (✉)
Facoltà di Scienze Matematiche, Fisiche e Naturali, Dipartimento di Scienze della Vita-CSEE, University of Trieste, 34127 Trieste, Italy
e-mail: marta.bolgan@gmail.com

S.S. Pedroso • M.C.P. Amorim
Instituto Superior de Psicologia Aplicada, Unidade de Investigação em Eco-Etologia, 1149–041 Lisbon, Portugal

R.O. Vasconcelos • J.M. Jordão • P.J. Fonseca
Centro de Biologia Ambiental, Departamento de Biologia Animal, Faculdade de Ciências da Universidade de Lisboa, 1749–016 Lisbon, Portugal

A.N. Popper and A. Hawkins (eds.), *The Effects of Noise on Aquatic Life*, Advances in Experimental Medicine and Biology 730, DOI 10.1007/978-1-4419-7311-5_24,

using a micropipette tip inserted in the subject's mouth. The experimental tank was positioned on a vibration-insulated table surrounded by a Faraday cage.

The AEPs were recorded using platinum electrodes (diameter 1 mm). The recording electrode was placed above the brain stem and the reference electrode was close to the nares. Electrode leads were connected to a differential AC amplifier (CP 511, Grass Technologies). The AEP signal was monitored with an oscilloscope and digitized using an analog-to-digital (A/D) converter (Edirol UA25, Roland) connected to a personal computer running Adobe Audition 3.0 (Adobe Systems).

Sound stimuli were created with Adobe Audition 3.0 and consisted of tone pulses presented 1,000 times at opposite polarities. Hearing thresholds were estimated at 15 Hz with a repetition rate of 5 s^{-1}, 30 and 60 Hz with a repetition rate of 10 s^{-1}, and 100, 200, 300, 400, 500, 800, and 1,000 Hz with a repetition rate of 20 s^{-1}, randomly presented. Sound stimuli ranged from 2 (15–100 Hz) to 5 complete cycles. Stimuli, presented in 4-dB steps from 92 to 136 dB re 1 μPa, were fed to a home-made underwater speaker device and amplifier (P. J. Fonseca) using the laptop and an A/D converter (Edirol UA 25). Before each experiment, the sound stimuli were calibrated with a hydrophone (Brüel and Kjaer 8103) connected to a sound level meter (Brüel and Kjaer Mediator 2238) placed in the same position as the fish. AEPs were averaged to minimize stimulus artifacts using home-made software (P. J. Fonseca).

3 Preliminary Results

Some of the AEP waveforms obtained indicated a clear double-frequency effect, which was further reassurance of a biological response. The audiogram showed that *P. pictus* sensitivity is higher at low frequencies between 15 and 400 Hz, with the lowest hearing threshold of 105 dB re 1 μPa at 15 Hz (Fig. 1). This best hearing range matches the main sound energy of both courtship (thump and drum) and agonistic (drum) calls, i.e., between ~83 and 297 Hz (Amorim and Neves 2007, 2008).

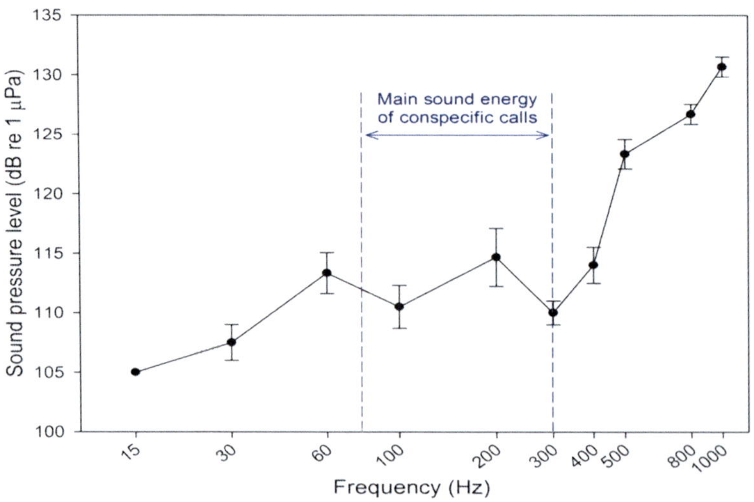

Fig. 1 Hearing thresholds of *Pomatoschistus pictus* showing the range of the main sound energy of agonistic and courtship calls. Values are averages ± SD

4 Discussion

The hearing sensitivity of *P. pictus* seems adapted to detect conspecific sounds, indicating that acoustic communication provides essential information during species-specific interactions. In addition, we suggest that the enhanced low-frequency sensitivity (below 60 Hz) could be the result of an evolutionary adaptation that, for a benthonic species, maximizes the ability to detect prey, predators, and mates.

Noise pollution is a threat to marine gobies (Codarin et al. 2009). Noise energy of man-made activity is mainly concentrated below 1 kHz (Nakahara 1999). Because *P. pictus* acoustic communication occurs within this frequency range, the concern is that anthropogenic noise might be strongly masking their hearing and hence their ability to communicate and to react to relevant acoustic stimuli. Future work is needed to test the masking effect of noise pollution on hearing in *P. pictus*.

References

Amorim MCP, Neves ASM (2007) Acoustic signaling during courtship in the painted goby, *Pomatoschistus pictus*. J Mar Biol Assoc UK 87:1017–1023.

Amorim MCP, Neves ASM (2008) Male painted gobies (*Pomatoschistus pictus*) vocalize to defend territories. Behaviour 145:1065–1083.

Codarin A, Wysocki LE, Ladich F, Picciulin M (2009) Effect of ambient and boat noise on hearing and communication in three fish species living in a protected area (Miramare, Italy). Mar Pollut Bull 58:1880–1087.

Malavasi S, Collatuzzo S, Torricelli P (2008) Interspecific variation of acoustic signal in Mediterranean gobies (Perciformes, Gobiidae): Comparative analysis an evolutionary outlook. Biol J Linn Soc 93:763–778.

Miller PJ (1986) Gobiidae. In: Whitehead PJP, Bauchot ML, Hureau JC, Nielsen J, Tortonese E (eds) Fishes of the north-eastern Atlantic and the Mediterranean, vol. 3. UNESCO, Paris, pp 1019–1085.

Nakahara F (1999) Influences of the underwater man-made noise on acoustic behavior of dolphin. Otsuchi Mar Sci 24:18–23.

Ross D (2005) Ship sources of ambient noise. IEEE J Ocean Eng 30:257–262.

Acoustic Communication in *Pomatoschistus* spp.: A Comparison Between Closely Related Species

Silvia S. Pedroso, Marta Bolgan, Joana M. Jordão, Paulo J. Fonseca, and M. Clara P. Amorim

1 Introduction

Many vertebrates, including teleosts, rely on acoustic signals for communication (Myrberg and Lugli 2006). Sound production has been documented in four species of sand gobies (*Pomatoschistus*). Males of the sand (*P. minutus*), marbled (*P. marmoratus*), and canestrini (*P. canestrinii*) gobies produce low-frequency pulsed sounds when females enter the nest (Lugli and Torricelli 1999). The male painted goby (*P. pictus*) also produces similar drumming sounds outside the nest during courtship (Amorim and Neves 2007). These drumming sounds present interspecific differences, suggesting that they could potentially be used in species-specific recognition (see Table 1).

Noise in the aquatic environment has significantly increased due to anthropogenic causes such as shipping (e.g., Smith et al. 2006). This increase in noise level may impair the detection of sound required for fish to communicate acoustically (Amoser et al. 2004). To better understand whether acoustic communication is impaired due to increased aquatic anthropogenic noise, studies regarding species vocal behavior and detailed analysis of sound parameters are required. Several studies have been carried out on acoustic communication in gobies (Myrberg and Lugli 2006); however, pre-spawning sounds have not been described for *P. pictus* and *P. microps*. We investigated the acoustic repertoire of *P. pictus* and, for the first time, the vocal behavior of *P. microps*.

S.S. Pedroso (✉) • M.C.P. Amorim
Instituto Superior de Psicologia Aplicada, Unidade de Investigação em Eco-Etologia, 1149–041 Lisbon, Portugal
e-mail: silvia.pedroso@yahoo.co.uk

M. Bolgan
Facoltà di Scienze Matematiche, Fisiche e Naturali, Dipartimento di Scienze della Vita-CSEE, University of Trieste, 34127 Trieste, Italy

J.M. Jordão • P.J. Fonseca
Centro de Biologia Ambiental, Departamento de Biologia Animal, Faculdade de Ciências da Universidade de Lisboa, 1749–016 Lisbon, Portugal

A.N. Popper and A. Hawkins (eds.), *The Effects of Noise on Aquatic Life*, Advances in Experimental Medicine and Biology 730, DOI 10.1007/978-1-4419-7311-5_25,

Table 1 Comparison of breeding sound characteristics between different sand gobies (*Pomatochistus* spp.)

Species	No. of Fish	Sound Duration, ms	No. of Pulses	Pulse Repetition Rate, Hz	Dominant Frequency, Hz	Reference
P. microps	2	504 ± 153 (334–630)	15 ± 4 (11–17)	31 ± 2 (29–33)	207 ± 103 (89–281)	Present study
P. pictus	20	663 ± 464 (211–2981)	29 ± 18 (9–114)	23 ± 4 (18–31)	298 ± 122 (156–431)	Amorim and Neves 2007
P. minutus	5	583 ± 68	15 ± 2	27 ± 2	99 ± 8	Malavasi et al. 2008
P. marmoratus	5	694 ± 314	17 ± 2	26 ± 2	125± 23	Malavasi et al. 2008

Values are means ± SD and range.

2 Materials and Methods

Fish were collected in Parede (38°41′ N, 009°21′ W) and Lagoa de Albufeira (38°30′ N, 009°10′ W), Portugal, during the breeding season (January to June 2010). Males and females were kept in separate stock aquaria (~18.4 l) with sand and shell shelters at 18 ± 1°C with a natural photoperiod. Experimental aquaria (~35 l) were divided with two acrylic partitions into three distinct compartments. Lateral compartments were provided with shelters (5.5 × 3 × 7 cm). A single male was placed in each lateral compartment while two gravid females occupied the middle compartment. Territorial males showed nuptial coloration and covered their nests with sand.

Sounds were recorded with three hydrophones, two High Tech 94 SSQ and a reference Brüel and Kjaer 8104, placed just above the male's nest, connected to a multitrack analog-to-digital (A/D) converter, and controlled by Adobe Audition 3.0 (Adobe Systems). At the start of the 20-min recordings, one partition was removed, allowing the gravid females to interact with the territorial male. Sounds were analyzed using Raven 1.2.1 for sound duration, number of pulses, pulse period, and dominant frequency.

3 Results

Males of *P. pictus* produced drumming sounds when the females entered the nest to spawn. These prespawning sounds are being analyzed.

With the same sampling effort, fewer males of *P. microps* produced prespawning sounds and vocal activity was never registered outside the nest. Table 1 summarizes the acoustic features of the breeding sounds produced by different species of sand gobies including *P. microps* (drumming shown in Fig. 1).

4 Discussion

We show that *P. microps* produce breeding sounds, although less frequently than *P. pictus*. This could indicate that *P. microps* uses other communication channels, e.g., chemical or visual, to communicate within its aquatic environment or that the studied *P. microps* males were not in a full-breeding motivation. Analysis of Table 1 suggests that *P. microps* produce shorter drumming sounds with fewer pulses in comparison to other *Pomatoschistus* species. Drumming sounds made by *P. pictus* have more pulses and faster pulse rates than the remaining congeners. Damselfish

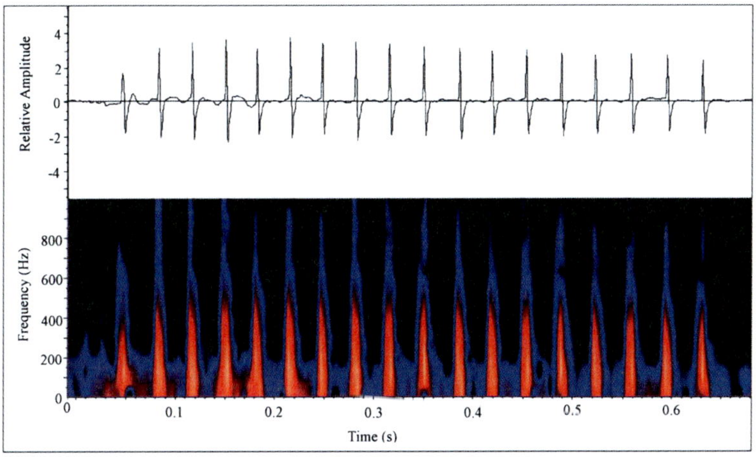

Fig. 1 Oscillogram and a spectrogram of a drumming sound produced by *Pomatoschistus microps*. The drumming sound has 18 pulses and a duration of 632 ms. Filter bandwidth 124 Hz, Hamming window

(*Stegastes*) are able to distinguish species-specific courtship sounds of two ecologically overlapping congeners, *S. planifrons* and *S. leucostictus*, that differ in the number of pulses and pulse intervals (Myrberg and Lugli 2006). Several *Pomatoschistus* species are also sympatric and their breeding seasons overlap. However, further investigation is required to establish if the suggested interspecific differences illustrated in Table 1 can promote species–specific recognition in *Pomatoschistus* as in *Stegastes* Future research will be carried out regarding intra- and interspecific recognition using playbacks. In addition, playbacks under noise conditions could be utilized to evaluate the noise effect on species-specific recognition.

References

Amorim MCP, Neves ASM (2007) Acoustic signaling during courtship in the painted goby, *Pomatochistus pictus*. J Mar Biol Assoc UK 87:1017–1023.

Amoser S, Wysocki LE, Ladich F (2004) Noise emission during the first powerboat race in an Alpine lake and potential impact on fish communities. J Acoust Soc Am 116:3789–3797.

Lugli M, Torricelli P (1999) Pre-spawning sound production in mediterranean sand-gobies. J Fish Biol 54:691–694.

Malavasi S, Collatuzzo S, Torricelli P (2008) Interspecific variation of acoustic signals in Mediterranean gobies (Perciformes, Gobiidae): Comparative analysis and evolutionary outlook. Biol J Linn Soc 93:763–778.

Myrberg AA Jr, Lugli M (2006) Diversity of sound production in fish. In: Ladich F, Collin SP, Moller P, Kapoor BG (eds) Communication in fishes, vol. 1, Science Publishers, Enfield, NH, pp 149–176.

Smith ME, Coffin AB, Miller DL, Popper AN (2006) Anatomical and functional recovery of the goldfish (*Carassius auratus*) ear following noise exposure. J Exp Biol 209: 4193–4202.

Importance of Hearing for Survival of *Danio rerio* (Zebrafish) in an Experimental Predator/Prey Environment

Rikke Agner Jørgensen, Christian Brandt, Magnus Wahlberg, and Jakob Christensen-Dalsgaard

1 Introduction

Throughout the past 10 years, there has been an increasing interest regarding the influence of man-made noise on life in the sea. Behavioral studies show that hearing in fish (and other animals) can be impaired, either temporarily or permanently, by exposure to intense sound (Smith et al. 2004). Also, physiological studies such as auditory brain stem response (ABR) measurements have shown that hearing thresholds shift when exposed to intense sounds (Kenyon et al. 1998). It is known that some fish use hearing for sound communication and auditory scene analyses (Popper et al. 2003), but little is known about how threshold shifts will affect their fitness through behavioral changes in, for example, predator/prey interactions.

With this pilot experiment, our objective is to test if temporary hearing threshold shift affects the ability of *Danio rerio* (zebrafish) to avoid predation. Previous experiments revealed that 40 h of exposure to Gaussian noise at a minimum of 130 dB re 1 μPa root mean square (RMS) produced a shift in threshold in this species. Thereafter, in an experimental predator/prey environment, we determined if such a temporary threshold shift had implications for the survival of *Danio rerio*. The intention was to provide information on the importance of hearing in this species as well as on environmental issues such as noise pollution.

2 Methods

Forty hours before ABR measurements, ten *Danio rerio* were moved to a separate ten-liter plastic tank. The fish were then exposed to Gaussian noise filtered by a first-order Biquad filter with a center frequency of 1,000 Hz and a bandwidth of 900 Hz. The sound intensity level was between 130 and 150 dB re 1 μPa RMS. The sound source, a subwoofer, was placed outside the experiment tank. The noise was generated by a digital sound processor (TDT RP2 Processor) and delivered

R.A. Jørgensen (✉) • C. Brandt • J. Christensen-Dalsgaard
Institute of Biology, University of Southern Denmark, DK-5230 Odense M, Denmark
e-mail: rijoe05@student.sdu.dk

M. Wahlberg
Institute of Biology, University of Southern Denmark, DK-5230 Odense M, Denmark

Fjord&Baelt and University of Southern Denmark, DK-5300 Kerteminde, Denmark

A.N. Popper and A. Hawkins (eds.), *The Effects of Noise on Aquatic Life*,
Advances in Experimental Medicine and Biology 730, DOI 10.1007/978-1-4419-7311-5_26,

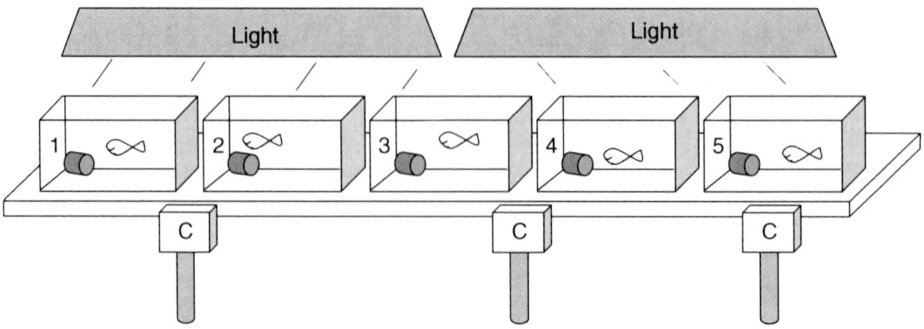

Fig. 1 The setup with the predators in aquaria. Control and sound-exposed groups of *Danio rerio* were introduced to each predator in groups of five. C: Video camera positions

through a power amplifier (XELEX DD-8 2 × 50 W). The system was controlled by a PC using software developed by Christian Brandt (QuickABR, University of Southern Denmark, Odense). After noise exposure, the fish were anesthetized with 3-aminobenzoic acid ethyl ester (MS-222) and the click ABR was measured by needle electrodes inserted subdermally, one over the ear, one over the brain stem, and one in the dorsal lateral musculature. Different intensity series of 100 clicks were played, and an average of the ABR to each series was registered. Changes in the averaged ABR for the sound-exposed group were correlated with the measurements from the control group to determine differences in thresholds. The threshold was defined as the lowest response above background noise.

Thereafter, groups of *Danio rerio* would be exposed to the same sound stimulation as mentioned above (Gaussian noise, bandwidth 550–1,450 Hz, noise level between 130 and 150 dB re 1 μPa RMS) before being introduced to predatory *Parachromis managuense* (jaguar cichlids) within 5–10 min after exposure. We had four rounds of experiments and each round had five groups of five *Danio rerio* (i.e., five separate predator tanks with five zebrafish in each). We used both control groups and groups of exposed *Danio rerio* in each round. In total, 10 groups, each with 5 *Danio rerio*, were sound exposed and an equal number participated as control animals. Video images of predator-exposed *Danio rerio* were analyzed.

With the video analyses, it was possible to determine the time between the release of the five *Danio rerio* in each group and their capture. The survival time for each *Danio rerio* introduced to predators was recorded for each group of five. Over a period of 7 wk, each of the 5 predators was introduced to 2 control groups and 2 sound-exposed groups of *Danio rerio*.

3 Results

Danio rerio exposed to Gaussian noise in the frequency band of 550–1,450 Hz showed a change in hearing sensitivity. The threshold measurements with click ABR revealed a change in sensitivity from 95 dB re 1 μPa RMS for the control group to 105 dB re 1 μPa RMS for the sound-exposed group.

Predation by the jaguar cichlids was very robust, and usually *Danio rerio* were taken within a few minutes of release in the experimental tanks. However, there was no significant difference in predation avoidance behavior between the control group and the sound-exposed group. All statistical tests were performed in SAS (Version 9.1 for Windows, SAS Institute, Inc., Cary, NC), with a significance level at a = 0.05.

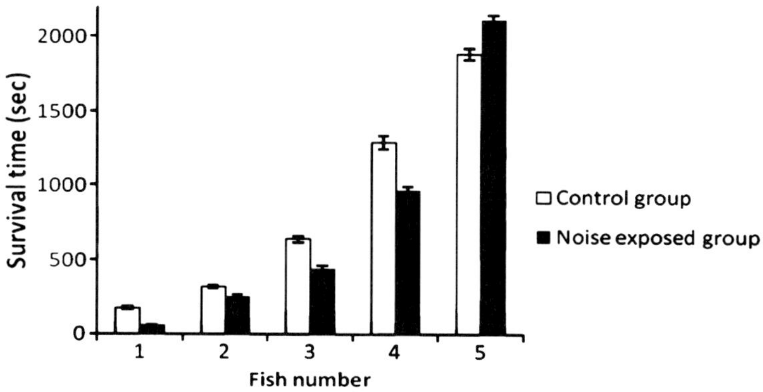

Fig. 2 Time (means ± SE) from introduction to capture of *Danio rerio* (in groups of five)

Figure 2 shows that there was a clear trend, on average, toward more severe predation of the sound-exposed group. Also, the time from release to capture tended to be higher and the capture rate lower for the control group, indicating a higher survival rate.

4 Discussion

Noise exposure creates a shift in threshold in the hearing sensitivity of *Danio rerio*. This shows that the method used here is applicable for experimental investigations of the effects of noise pollution.

The behavioral experiment showed no significant difference between the sound-exposed and control groups in their ability to survive in a predator/prey environment. With the current experimental setup, conclusions cannot be drawn as to whether sound exposure and the subsequent loss in hearing sensitivity decreases predation avoidance in *Danio rerio*. However, it is likely that the lack of significance is due to the small sample size in the experiments, and the results from this behavioral experiment can be used to calculate the correct statistical power. It is important to select species for study that are representative of other species to be able to see a model effect and understand the consequences. *Danio rerio* is a model species for the Ostariophysi suborder, the second largest suborder of fish, known among other things for their great hearing abilities. Thus our findings have the potential to give insight into hearing and predation avoidance in other ostariophysians, and we think that *Danio rerio* can be used as a model for the Ostariophysi suborder. Future studies are required to investigate the details of the potential predatory cue to which *Danio rerio* may react.

References

Kenyon TN, Ladich F, Yan HY (1998) A comparative study of hearing ability in fishes: The auditory brainstem response approach. J Comp Physiol A 182:307–318.

Popper AN, Fay RR, Platt C, Sand O (2003) Sound detection mechanisms and capabilities of teleost fishes. In: Collin SP, Marshall NJ (eds) Sensory processing in aquatic environments. Springer, New York.

Smith ME, Kane AS, Popper AN (2004) Acoustical stress and hearing sensitivity in fishes: Does the linear threshold shift hypothesis hold water? J Exp Biol 207:3591–3602.

Lateral Line Canal Morphology and Noise Reduction

Adrian T. Klein and Horst Bleckmann

1 Introduction

With the aid of the lateral line, fish perceive minute water motions and pressure gradients (Coombs 2002). The lateral line of fish is important for many behaviors including spatial orientation, prey detection, shoaling, and intraspecific communication (Bleckmann 1994). In addition, fishes use lateral line information for station holding and entraining (Liao 2007).

The smallest sensory unit of the lateral line is the neuromast. A single neuromast may contain up to 3,000 hair cells covered by a mucous cupula (Bleckmann 1994). Lateral line systems have a band-pass characteristic with mechanical properties primarily influenced by the stiffness of the ciliary bundles of the hair cells, the size and shape of the cupulae (van Netten 2006), and the dimensions of the canals. Lateral line canals enhance the ability to detect a vibrating sphere under flow conditions (Engelmann et al. 2002).

In their natural habitat, fish not only face animal-generated water motion but also water motion caused by inanimate sources. In general, the detection and discrimination of meaningful signals are crucial for the survival of animals; therefore, animals should be equipped with sensor systems that enable them to separate meaningful signals from noise.

2 Diversity of Lateral Line Systems

Fish live in various habitats that differ in their hydrodynamic noise (e.g., in lakes and ponds or in fast running rivers). For this reason, it is not surprising that the morphology of the peripheral lateral line is highly diverse (Coombs et al. 1988). For instance, neuromasts may occur freestanding on the skin (superficial neuromasts), in grooves, or in fluid-filled canals (canal neuromasts). The number of hair cells as well as the size and shape of the cupulae may differ in individual fish species. Head lateral line canals may be simple (Perciformes), groovelike (Holeocephali), or furcated (Clupeomorpha). Although most fish species have one trunk lateral line canal on each side of the body (e.g., *Carrasius auratus*), some fish (e.g., *Brachydanio rerio*) have either no trunk canal or up

A.T. Klein (✉) • H. Bleckmann
University of Bonn, 53113, Bonn, Germany
e-mail: adrian@uni-bonn.de; bleckmann@uni-bonn.de

A.N. Popper and A. Hawkins (eds.), *The Effects of Noise on Aquatic Life*,
Advances in Experimental Medicine and Biology 730, DOI 10.1007/978-1-4419-7311-5_27,

to four (e.g., *Xiphister*) on each body side (Webb 1989). The diameter of the canals, the size of the canal pores, and the interpore distances may also vary across species. Despite all this variability, the functional significance of the lateral line diversity is still not understood.

3 Artificial Lateral Line System

For many reasons, it is difficult to study form-function relationships in natural lateral lines. In most cases, lateral line systems are so diverse that even the influence of a single parameter (e.g., canal diameter or interpore distance) can hardly be uncovered. When studying form-function relationships, it is advantageous to use artificial lateral line canals over natural ones because any canal parameter (e.g., canal diameter or interpore distance) can easily be controlled and manipulated. For this reason, we used artificial lateral line canals equipped with highly sensitive artificial neuromasts to investigate lateral line filter properties.

4 Stimuli and Results

Most biotic and abiotic hydrodynamic stimuli can be described by a composition of monopoles, dipoles (e.g., a vibrating sphere), or their vortex patterns. The hydrodynamic field caused by monopoles and dipoles is well understood (Kalmijn 1988). More complex water motions, such as vortex streets, occur behind undulatory swimming fish and downstream of many objects exposed to running water (Liao 2007). The water motion in shore regions, in ruffles, and in fast running rivers may also be highly complex because of the synchronous presence of many superimposed stimuli. In contrast to the complexity of the natural hydrodynamic world, most experiments designed to uncover lateral line function use only simple hydrodynamic environments (still water, laminar flow, vortex streets) and stimuli (dipoles, moving objects) to study lateral line function. To uncover a possible form-function relationship of lateral line canals, we exposed artificial canals to dipole stimuli and to vortex streets in both laminar and turbulent flow. Our experiments clearly show that certain canal parameters greatly improve the signal-to-noise ratio.

Acknowledgments This work was supported by the Deutsche Forschungsgemeinschaft (DFG), Bonn, Germany, and the Bundesministerium für Bildung und Forschung (BMBF), Berlin, Germany.

References

Bleckmann H (1994) Reception of hydrodynamic stimuli in aquatic and semiaquatic animals. In: Rathmayer W (ed) Progress in zoology, vol 41. Gustav Fischer Verlag, Stuttgart, Germany, pp 1–115.

Coombs S (2002) Imaging of the hydrodynamic environment by the peripheral lateral line system. Bioacoustics 12:148–151.

Coombs S, Janssen J, Webb J (1988) Diversity of lateral line systems: Evolutionary and functional considerations. In: Atema J, Fay RR, Popper AN, Tavolga WN (eds) Sensory biology of aquatic animals. Springer-Verlag, New York, pp 553–594.

Engelmann J, Hanke W, Bleckmann H (2002) Lateral line reception in still- and running water. J Comp Physiol A Neuroethol Sens Neural Behav Physiol 188:513–526.

Kalmijn AJ (1988) Hydrodynamic and acoustic field detection. In: Atema J, Fay RR, Popper AN, Tavolga WN (eds) Sensory biology of aquatic animals. Springer-Verlag, New York, pp 83–130.

Liao JC (2007) A review of fish swimming mechanics and behaviour in altered flows. Philos Trans R Soc Lond B Biol Sci 362:1973–1993.
van Netten SM (2006) Hydrodynamic detection by cupulae in a lateral line canal: Functional relations between physics and physiology. Biol Cybern 94:67–85.
Webb JF (1989) Developmental constraints and evolution of the lateral line system in teleost fishes. The mechanosensory lateral line. In: Coombs S, Görner P, Münz H (eds) Neurobiology and evolution. Springer-Verlag, New York, pp 79–98.

Potential for Sound Sensitivity in Cephalopods

T. Aran Mooney, Roger Hanlon, Peter T. Madsen, Jakob Christensen-Dalsgaard, Darlene R. Ketten, and Paul E. Nachtigall

1 Introduction

Hearing is a primary sense in many marine animals, and we now have a reasonable understanding of what stimuli generate clear responses, the frequency range of sensitivity, expected threshold values, and mechanisms of sound detection for several species of marine mammals and fishes (Au et al. 2000; Fay 1988). For marine invertebrates, our knowledge of hearing capabilities is relatively poor and a definition or even certainty of sound detection is not agreed on (Webster et al. 1992) despite their magnitude of biomass and often central role in ocean ecosystems.

Cephalopods (squid, cuttlefish, octopods, and nautilus) are particularly interesting subjects for invertebrate sound detection investigations for several reasons. Ecologically, they occupy many of the same niches as sound-sensitive fish (Budelmann 1994) and may benefit from sound perception and use for the same reasons, such as to detect predators, navigate, or locate conspecifics.

T.A. Mooney (✉)
Biology Department, Woods Hole Oceanographic Institution, Woods Hole, MA 02543, USA

Marine Biological Laboratory, Woods Hole, MA 02543, USA
e-mail: amooney@whoi.edu

R. Hanlon
Marine Biological Laboratory, Woods Hole, MA 02543, USA
e-mail: rhanlon@mbl.edu

P.T. Madsen
Biology Department, Woods Hole Oceanographic Institution, Woods Hole, MA 02543, USA

Department of Biological Sciences, University of Aarhus 8000, Aarhus G, Denmark
e-mail: Peter.madsen@biology.au.dk

J. Christensen-Dalsgaard
Institute of Biology, University of Southern Denmark, 5230 Odense M, Denmark
e-mail: jcd@biology.sdu.dk

D.R. Ketten
Biology Department, Woods Hole Oceanographic Institution, Woods Hole, MA 02543, USA

Department of Otology and Laryngology, Harvard Medical School, Boston, MA 02115, USA
e-mail: dketten@whoi.edu

P.E. Nachtigall
Hawai'i Institute of Marine Biology, University of Hawai'i, Kaneohe, HI 96744, USA
e-mail: nachtiga@hawaii.edu

A.N. Popper and A. Hawkins (eds.), *The Effects of Noise on Aquatic Life*,
Advances in Experimental Medicine and Biology 730, DOI 10.1007/978-1-4419-7311-5_28,

Squid, e.g., are often the prey of loud, echolocating marine mammals (Clarke 1996) and may therefore be expected to have evolved hearing to avoid predators. Anatomically, squid have complex statocysts that are considered to serve primarily as vestibular and acceleration detectors (Nixon and Young 2003). However, statocysts may also be analogs for fish otolithic organs, detecting acoustic stimuli (Budelmann 1992). Previous studies have debated the subject of squid hearing, and recently, there has been a revival of research on the subject. Here, we briefly review what is known about squid sound detection, revisit hearing definitions, discuss potential squid susceptibility to anthropogenic noise, and suggest potential future research directions to examine squid acoustic sensitivity.

2 Sensitivity of Squid to Acoustic Stimuli

Early anecdotal reports suggested that cephalopods might detect sounds because squid were attracted to 600-Hz tones and cuttlefish (*Sepia officinalis*) elicited startle responses to 180-Hz stimuli (Dijkgraaf 1963; Maniwa 1976). Norris and Møhl (1983) later postulated that squid might be debilitated by the acoustic intensity of foraging odontocete (toothed whale and dolphin) echolocation clicks. This hypothesis led Moynihan (1985) to suggest that squid might, in turn, be deaf to acoustic stimuli to avoid this "stunning." However, anatomical evidence of squid statocysts indicates that the organ acts as an accelerometer (Budelmann 1976) potentially used for acoustic detection (Budelmann 1992). Behavioral conditioning experiments later confirmed that squid (*Loligo vulgaris*), octopus (*Octopus vulgaris*), and *S. officinalis* can detect acceleration stimuli from 1 to 100 Hz, presumably by using the statocyst organ as an accelerometer detecting the body movements of the squid in the sound field (Packard et al. 1990). This and a follow-up study (Kaifu et al. 2008) showed that cephalopods can detect the low-frequency particle-motion component of a sound field, but the question whether cephalopods are also sensitive to higher frequencies and sound pressures still remained. Recent laboratory experiments have demonstrated that squid do not exhibit anti-predator responses in the presence of odontocete echolocation clicks (Wilson et al. 2007), indicating that they cannot detect the ultrasonic pressure component of a sound field. However, recently, Hu et al. (2009) suggested that squid (*Sepiotheutis lessoniana*) can detect sound pressure stimuli using their statocyst organ. Unfortunately, these data had several methodological issues including no calibrations of particle motion and placing squid at the water's surface where discrepancies between sound pressure and particle motion are greatest. Thus cephalopods seem to be sensitive to the low-frequency particle-motion component of the sound field and not pressure. Yet it appears necessary to discuss whether such a sensitivity can be understood as hearing.

3 Defining "Hearing" For Squid and Other Marine Organisms

There is no inherent reason why the definition of "hearing" should be taxon specific. It arguably should be based on the ability to detect acoustic stimuli per se and on the ability to analyze acoustic properties of a stimulus. Thus this should not be a discussion about whether hearing fits squid perceptual mechanisms but whether squid perceptual abilities fit the broader scope of "hearing."

Yost (1994) defines hearing broadly as sensitivity to sounds in the environment and encompassing coding, processing, integration, and response of sound. An alternative definition suggests that hearing is the auditory detection of sound (Webster et al. 1992). Both definitions are similarly open-ended, probably intentionally so. Both also presume a definition of sound that is a well-defined hydrodynamic wave that propagates in an elastic medium by pressure and particle-motion oscillations (Kalmijn 1988). In the marine environment, an air cavity or at least some compressible tissue in the vicinity of the fenestrae is considered crucial to detect sound pressure (Fay and Popper 1974; Ketten 1992) because the air/tissue acts as a pressure-to-particle motion transducer. Particle motion alone can be detected by inertial delays in acceleration of higher density objects versus macular receptors. Taking either case to be valid, hearing may be considered the detection of acoustically generated sound pressure differentials or particle motion or both. Detection of the particle-motion component of a sound field has been demonstrated in many marine organisms including cartilaginous and teleost fish that are without specialized adaptations to detect or relay sound pressure reception to the inner ear (de Vries 1950).

Thus we argue that hearing can be validly considered as the detection of broad or isolated elements of sound-generated stimuli and that it may be achieved by a single or multifunctional organ adapted for that purpose. The ability to detect sound pressure and/or acoustic particle motion therefore is a sufficient qualifier. Given this premise, many marine organisms capable of detecting acoustically generated particle motion are not, as has often been presumed, unreceptive to sound or its impacts, thereby opening a new area of exploration for the physiology of hearing. We discourage the notion that hearing would be limited to only sound pressure detection; thus many marine organisms, although capable of detecting acoustically generated particle motion, would be considered deaf. It is important to note that according to our definition, a hearing organ may also respond to mechanical disturbances other than sound (such as vibrations, local hydrodynamic waves). We think that this is an important feature, especially in marine particle-motion sensors, and that it will be difficult to understand the evolution of hearing without acknowledging this dual function of the sound-receiving organs.

4 Squid Susceptibility to Aquatic Noise

Any impacts of aquatic noise on cephalopods have yet to be established and are poorly understood. Ambient and anthropogenic ocean noise are substantial at lower frequencies where squid are sensitive (Packard et al. 1990; Urick 1983), suggesting that they will be susceptible to masking or other physiological or behavioral impacts of anthropogenic noise (McCauley et al. 2000). Statocyst or lateral line hair cells could be impacted by sound energy (either long duration or brief, high-intensity noise). Such hair cell damage and related temporary hearing loss has been demonstrated in fishes (McCauley et al. 2003), and this has been suggested for squid (which do have a lateral line analogue; Budelmann 1994).

5 Future Research Directions

Ideally, detection capabilities would be measured both behaviorally and physiologically. Directed research should identify whether the statocyst acts as the organ of acoustic reception or whether the lateral line analogue may be involved. Finally, the biological relevance of the acoustic stimuli should be addressed. This includes whether squid may hear fish and odontocete predators, the neuroanatomy of auditory structures, and whether squid are susceptible to the impacts of anthropogenic noise.

Although the field of cephalopod audition has been debated in the past, almost all questions of acoustic detection remain to be addressed. This leaves room for significant and groundbreaking work on the sensory ecology of animals that are often considered keystone species in many ecological webs. Here we have attempted to introduce a preliminary road map for such progress and we expect the issue of squid sound detection to be resolved in due time.

References

Au WWL, Popper AN, Fay RR (eds) (2000) Hearing by whales and dolphins. Springer-Verlag, New York.

Budelmann BU (1976) Hair cell polarization in the gravity receptor systems of the statocysts of the cephalopods *Sepia officinalis* and *Loligo vulgaris*. Brain Res 164:1–5.

Budelmann BU (1992) Hearing in non-arthropod invertebrates. In: Webster DB, Fay RR, Popper AN (eds) The evolutionary biology of hearing. Springer-Verlag, New York, pp 141–155.

Budelmann BU (1994) Cephalopod sense organs, nerves and the brain: Adaptations for high performance and life style. Mar Behav Physiol 25:13–33. Doi:10.1080/10236249409378905.

Clarke MR (1996) Cephalopods as prey. III. Cetaceans. Philos Trans R Soc Lond Ser B Biol Sci 351:1053–1056. Doi:10.1098/rstb.1996.0093.

de Vries HL (1950) The mechanics of labyrith otoliths. Acta Oto-Laryngol 38:262–273.

Dijkgraaf S (1963) Verusche uber Schallwahrnehmung bei Tintenfischen. Naturwissenschaften 50:50. Doi:10.1007/BF00622830.

Fay RR (1988) Hearing in vertebrates: A psychophysics databook. Hill-Fay Associates, Winnetka, IL.

Fay RR, Popper AN (1974) Acoustic stimulation of the ear of the goldfish (*Carassius auratus*). J Exp Biol 61:243–260.

Hu M, Yan HY, Chung WS, Shiao JC, Hwang PP (2009) Acoustical evoked potentials in two cephalopods inferred using the auditory brainstem response (ABR) approach. Comp Biochem Physiol A 153:278–283. Doi:10.1016/j.cbpa.2009.02.040.

Kaifu K, Akamatsu T, Segawa S (2008) Underwater sound detection by cephalopod statocyst. Fish Sci 74:781–786. Doi:10.1111/j.1444-2906.2008.01589.x.

Kalmijn AD (1988) Acoustic and hydrodynamic field detection. In: Atema J, Fay RR, Popper AN, Tavolga WN (eds) Sensory biology of aquatic animals. Springer-Verlag, New York, pp 83–131.

Ketten DR (1992) The marine mammal ear: specializations for aquatic audition and echolocation. In: Webster DB, Fay RR, Popper AN (eds) The evolutionary biology of hearing. Springer-Verlag, New York, pp 717–750.

Maniwa Y (1976) Attraction of bony fish, squid and crab by sound. In: Schuijf A, Hawkins AD (eds) Sound reception in fish. Elsevier, Amsterdam, pp 271–283.

McCauley RD, Fewtrell J, Duncan AJ, Jenner C, Jenner M-N, Penrose JD, Prince RIT, Adihyta A, Murdoch J, McCabe K (2000) Marine seismic surveys: Analysis and propagation of air gun signals and effects of air-gun exposure on humpback whales, sea turtles, fishes and squid. Centre for Marine Society and Technology and the Australian Petroleum Production and Exploration Association, Curtin University, Perth, WA, Australia.

McCauley RD, Fewtrell J, Popper AN (2003) High intensity anthropogenic sound damages fish ears. J Acoust Soc Am 113:638–642. Doi:10.1121/1.1527962.

Moynihan M (1985) Why are cephalopods deaf? Am Nat 125:465–469. Doi:10.1086/284353.

Nixon M, Young JZ (2003) The brains and lives of cephalopods. Oxford University Press, New York.

Norris KS, Møhl B (1983) Can odontocetes debilitate prey with sound. Am Nat 122:85–104. Doi:10.1086/284120.

Packard A, Karlsen HE, Sand O (1990) Low frequency hearing in cephalopods. J Comp Physiol A 166:501–505. Doi:10.1007/BF00192020.

Urick RJ (1983) Principles of underwater sound. McGraw-Hill, New York.

Webster DB, Fay RR, Popper AN (eds) (1992) The evolutionary biology of hearing. Springer-Verlag, New York.

Wilson M, Hanlon RT, Tyack PL, Madsen PT (2007) Intense ultrasonic clicks from echolocating toothed whales do not elicit anti-predator responses or debilitate the squid *Loligo pealeii*. Biol Lett 3:225–227. Doi:10.1098/rsbl.2007.0005.

Yost WA (1994) Fundamentals of hearing: An introduction. Academic Press, New York.

Listening in Noise

Richard R. Fay

1 Introduction

For all animals that hear, the detection of a sound source depends on, among other things, the amount or level of the potentially interfering sounds (maskers) that exist simultaneously with the sound to be detected ("signal"). The potentially interfering sounds are generically referred to as "noise," whereas the sound of interest to be detected is generically referred to as the "signal." "Signals" and "noise" may be any sounds at all under the usual circumstances of a fish listening for critical environmental information. For example, the "signal" may be a noise waveform that conveys information that the fish needs to acquire to increase awareness of its surroundings so that it can behave appropriately with respect to its source. The interfering "noise" waveform may likewise be a wide band, narrow band, or a tonelike sound (e.g., a sonar sound) with an arbitrary spectrum.

Under many circumstances in usual environments, the background or ambient noise of natural origin (e.g., wind, rain, surf) may be interfering with signal detection (e.g., Chapman and Hawkins 1973). When this interference takes place, sounds to be detected must be at a higher sound pressure level to be heard than they would be in quiet background conditions. This suggests the definition of masking as the circumstance in which the presence of one sound (the masker) results in the threshold for the detection of another sound (the signal) to be raised above the thresholds obtained in the quiet. In effect, the animal becomes less sensitive (has a higher threshold or is, in effect, hearing impaired) with respect to the signal sounds. Masking effects depend on several factors including signal duration, the spectral features of the source, the location of the signal source relative to interfering sources, and the level and spectral features of interfering sounds (usually noise). This chapter focuses on the interfering effects of maskers on signal detection by fishes.

In the case of anthropogenic sounds, the background sound levels may be raised by anthropogenic sources (e.g., shipping noise) above what is considered normal or usual, resulting in more masking or signal threshold elevation than would be the case under normal ambient noise levels and an increased risk to animals having to detect the signals. The risk derives from at least two noise effects: the loss of sensitivity to all sounds that the fish can normally hear (e.g., environmental sounds necessary for soundscape orientation [Slabbekoorn and Bouton 2008] and communication sounds [Myrberg 1981]) and a contraction of the distance range at which all normally audible sources can be detected.

Anthropogenic noise may have an arbitrary temporal pattern and spectral shape, and the signals to be detected may be similarly diverse in center frequencies, spectra, and bandwidths. For most of

R.R. Fay (✉)
Marine Biological Laboratory, Woods Hole, MA, 02543, USA
e-mail: rfay@luc.edu

A.N. Popper and A. Hawkins (eds.), *The Effects of Noise on Aquatic Life*,
Advances in Experimental Medicine and Biology 730, DOI 10.1007/978-1-4419-7311-5_29,

these noises and signals, there are few or no data on the masking effects of specific or arbitrary maskers and signals, and we are therefore limited in our predictions of the risk to animals of various levels of anthropogenic noise.

However, there are systematic data on masking from laboratory studies using tones as signals and white flat spectrum noise as maskers. These are simplifying conditions that make the results of masking experiments generally interpretable and applicable across species. Most masking experiments proceed from the assumptions of the "power spectrum model" of sound detection and masking that is widely used in human and animal psychoacoustics (Fletcher 1940). Briefly, this model holds that when a tone signal is detected, it activates an optimal hypothetical detection channel or filter centered on the frequency of the tone. These channels are assumed to have a best frequency at the frequency of the tone to be detected and to have a certain finite bandwidth. These channels respond to the noise too, and the power (amount) of the noise admitted by the detection filter is determined by the filter's bandwidth; wider filters admit noise with more power and therefore produce greater masking of the signal.

When the signal power equals the noise power passing through the filter, the signal is assumed to be at masked threshold or "just masked" or "just detected." The noise power through the filter is calculated by summing up the noise components that pass through the filter. An indirect estimate of the bandwidth of the filter is calculated by determining the signal-to-noise ratio at the threshold of the masked tone. This is termed the critical masking ratio (CR) and is the level of the tone at masked threshold minus the level of the masking noise (specified as a spectrum level or level per hertz and assumed to have a uniform [flat] spectrum within the calculated bandwidth). For example, if the level of the tone at threshold is 60 dB re 1 μPa and the level per hertz of the masking noise is 40 dB re 1 μPa, the CR is 20 dB. Then the CR can be used to estimate the equivalent rectangular bandwidth (ERB) of the noise effective in masking the tone in the following manner: ERB (in Hz) = $10^{CR/10}$. In other words, the bandwidth of the noise passing through the hypothetical detection filter and responsible for masking the tone would be 100 Hz in this case. Using this model, only the noise components in a rectangular band centered on the signal frequency is assumed effective in masking the tone. These assumptions are likely wrong in some cases but can be considered to be good rules of thumb.

These CR values for pure tones masked by uniform spectrum noise have been measured for several fish species. These species are goldfish (*Carassius auratus*; Fay 1974), marine catfish (*Arius felis*; Tavolga 1982), African mouthbreeder (*Tilapia macrocephala*- Tavolga 1974), pinfish (*Lagodon rhomboids*; Tavolga 1974), Atlantic cod (*Gadus morhua*; Buerkle 1968; Chapman 1973; Chapman and Hawkins 1973; Chapman and Johnstone 1974; Hawkins and Sand 1977), haddock (*Melanogrammus aeglefinus*; Chapman 1973), and Atlantic salmon (*Salmo salar*; Hawkins and Johnstone 1978).

In general, CR values range from ~13 dB (20 Hz) at the low frequencies to over 25 dB (316 Hz) at higher frequencies. For the goldfish (Fay 1974), the function is approximately linear, with log frequency from 100 to 1,200 Hz. The data for the other species are more variable but, in aggregate, can be described as increasing with frequency at ~3 dB/octave, much like the goldfish data.

Independent measures of the effective bandwidths and shapes of the detection filters have also been made for several species using different masking techniques: goldfish, Atlantic cod, Atlantic salmon, elephant nose (*Gnathonemus petersii*), clown knifefish (*Notopterus chitala*), arawana (*Osteoglossum bicirrhosum*), and Hawaiian squirrelfish (*Adioryx zantheryhrus*; see Fay 1988 for references to these papers). In general, these filter shapes are approximately Gaussian shaped and help confirm that only the noise-frequency components surrounding the signal frequency are effective in causing masking.

The interpretation of the increasing CRs with signal frequency in Figure 1 is that the hypothetical detection filters widen in bandwidth at the higher frequencies. Thus each detection filter is generally narrower than the range of frequencies that a given species can hear. This means that for the detection of a given tone frequency, only a masker corresponding to some fraction of the audible frequency range activates the hypothetical detection filters and causes masking.

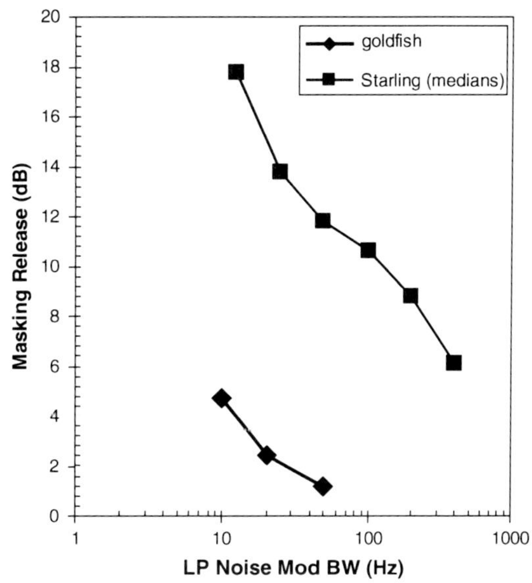

Fig. 1. The release from masking that occurs when the masking noise is randomly amplitude modulated at various low-pass corner frequencies for goldfish compared with the European starling (Klump and Langemann 1995)

The measurement of CRs has practical applications in estimating whether or not a given signal will be detected in a given noise environment. Given that the noise level is high enough to cause masking, a measurement of the CR predicts what the animal can and cannot detect. The audiogram in quiet cannot make this prediction except in the unusual case of "quiet" environmental conditions (no masking at all). In this sense, the CR measurement is most important in assessing the effects of elevated noise levels on hearing performance and risk to fishes.

2 Potential Problems With This Scenario

The power-spectrum model outlined above is incomplete. First, it is known that in humans and other terrestrial animals the temporally amplitude-modulated noise tends to lose some of its masking effects depending on the frequency and bandwidth of modulation. This is the basis for the masking effect known as comodulation masking release (CMR) (e.g., Klump and Langeman 1995). CMR phenomena indicate that the power-spectrum model of masking needs revision. Second, the mere detection of a signal may not provide enough information for the sound source to be segregated or discriminated from other sources (Lohr et al. 2003); signal levels must be somewhat higher than the masked detection threshold for the animal to gain biologically useful information about the source detected. So CRs by themselves don't tell the whole story of the risks associated with elevated noise levels.

3 An Experiment on These Issues

3.1 Effects of Amplitude-Modulated Noise Maskers

The effects of masking noise amplitude modulation and the signal-to-noise ratios required for signal detection were studied in 6-in. common goldfish (*Carassius auratus*) using classical respiratory conditioning (e.g., Fay 1995). Five groups of eight animals each were trained and tested. In experiment 1, four groups of animals detected a 6-s 400-Hz pure tone in the presence of wideband noise that was either unmodulated or modulated by different temporal patterns. The modulation was a random noise waveform that was low-pass filtered at 10, 20, and 50 Hz (corner frequency). Conditioned respiratory suppression was measured with electric shock as the unconditioned stimulus. Psychometric functions were determined for signal detection by fixing the tone signal level and manipulating masker level over a 40-dB range. Thresholds were defined as the sound-to-noise levels corresponding to 50% of maximum respiratory suppression.

Results showed that the maximum masking effect occurred in the presence of the unmodulated masker and that there was a slight release from this masking effect as the random modulation was created at lower frequencies (Fig. 1, diamonds). For comparison, data for the European starling (Klump and Langeman 1995) obtained in a similar experiment are plotted (squares). The masking release caused by modulation of the masker noise reaches a maximum of ~5 dB for a 10-Hz low-pass modulation and declines as modulation frequency is raised. Although this trend is similar for the starling (and for humans), the modulation masking release for the goldfish is particularly small.

Thus we conclude that the goldfish, as with all tetrapods investigated, shows some (but small) release from masking for amplitude-modulated noise maskers, with the release growing with reductions in the low-pass corner frequency of the noise masker.

3.2 The Signal-to-Noise Ratio Required for Source Determination

The masked signal level required for source determination or segregation was investigated in an additional group of eight goldfish by using classical respiratory conditioning in combination with a stimulus generalization paradigm (e.g., Fay 1995). Goldfish conditioned to a pure tone will generalize to other stimuli to the extent that the novel test stimuli have features in common or similar to the conditioning stimulus. It was reasoned that a 400-Hz tone just at masked threshold (just detected) would not appear to be very similar or identical to a clearly audible 400-Hz tone (i.e., would not be determined or well segregated from the noise) and that to be determined, its level would have to be higher than at the masked threshold. This experiment estimated how much higher in level it would have to be to be determined against a background of noise.

The psychometric function and threshold in unmodulated noise were compared to a comparable psychometric function obtained in a stimulus generalization paradigm. Fish were first conditioned (40 trials) to a clearly audible 400-Hz tone (~40 dB above the masked threshold) with a low-level background noise varying in amplitude from trial to trial. Following this conditioning, the fish received 40 additional test trials in which the same 40-dB sensation level tone was presented in a noise background that varied in level from trial to trial by up to 50 dB so that on some trials the tone was completely masked, whereas on others it was clearly audible. It was reasoned that if the tone was far enough above the masked threshold, it would be completely determined (e.g., its frequency would be completely known) and it would result in near 100% generalization (it would be determined to be near 100% similar to the clearly audible tone to which the fish was initially conditioned). On the other hand, at the masked threshold and at other signal-to-noise ratios near threshold, the tone would only be incompletely determined, and this would result in suboptimal generalization.

Figure 2 shows the psychometric function and the generalization function that resulted from these two conditions. The signal-to-noise ratio function for generalization requires about a 4-dB higher value than is required for signal detection. This difference was interpreted to be the signal level "gap" between what is necessary for detection and what is required for source determination or segregation. This gap is ~4 dB, meaning that the signal has to be ~4 dB above the masked threshold (the CR) to be clearly heard, determined, or segregated.

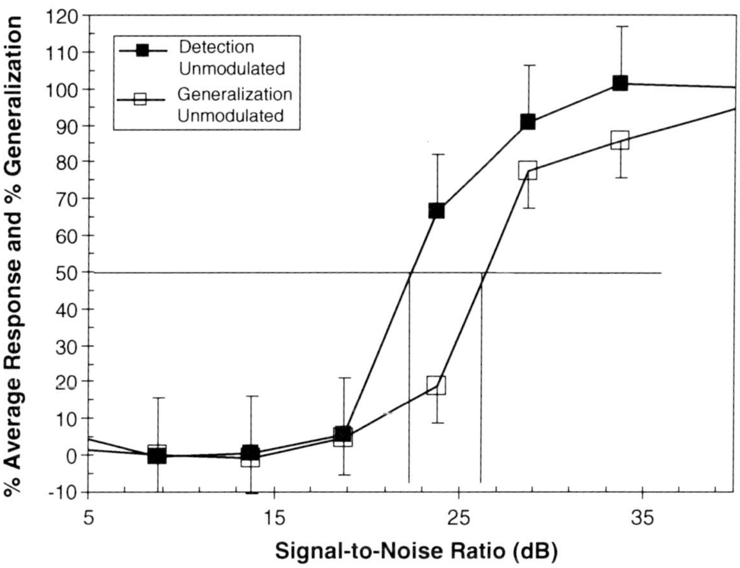

Fig. 2. The psychometric function at 400 Hz (filled squares) compared with the generalization function at 400 Hz (open squares)

4 Conclusions

All continuous masking noise has the effect of causing detection thresholds for all normally detectable signals to be raised above the quiet case, in effect causing a hearing impairment. This renders some low-level sounds undetectable and reduces the distance range over which all sources can be detected by fishes. This effect depends on the type of transmission loss with distance in a given environment. For goldfish, a partial release from this masking effect can occur when the noise is amplitude modulated at low rates (~5 dB). However, this masking effect increases by ~4 dB when the criterion for response is changed from mere detection to the ability to determine the details of the signal source (segregation). In general, ~4 dB should be added to the signal-to-noise ratio (CR) required for detection when estimating the interfering effects of noise on fishes. The masking caused by more charismatic but intermittent sound sources such as pile driving, seismic shooting, and sonar systems has not been studied and remains a topic for future research.

References

Buerkle U. (1968) Relation of pure tone thresholds to background noise levels in the Atlantic cod (*Gadus morhua*). J Fish Res Board Can 25:1155–1160.

Chapman CJ. (1973) Field studies of hearing in teleost fish. Helgol wiss Meeresunters 24:371–390.

Chapman CJ, Hawkins AD. (1973) A field study of hearing in the cod. J Comp Physiol 85:147–167.

Chapman CJ, Johnstone ADF. (1974) Some auditory discrimination experiments on marine fish. J Exp Biol 61:521–528.

Fay RR. (1974) Masking of tones by noise for the goldfish (*Carassius auratus*). J Comp Physiol Psychol 87:708–716.

Fay RR. (1988) Hearing in vertebrates: A psychophysics databook. Hill-Fay Associates, Winnetka, IL.

Fay RR. (1995) Psychoacoustical studies of the sense of hearing in the goldfish using conditioned respiratory suppression. In: Klump GM, Dooling RJ, Fay RR, Stebbins WC (eds) Methodology in animal psychophysics. Birkhauser Verlag, Berlin, pp 246–262.

Fletcher H. (1940) Auditory patterns. Rev Mod Phys 12:47–65.

Hawkins AD, Johnstone ADF. (1978) The hearing of the Atlantic salmon, *Salmo salar*. J Fish Biol 13:655–673.

Hawkins AD, Sand O. (1977) Directional hearing in the median vertical plane by the cod. J Comp Physiol 122:1–8.
Klump GM, Langemann U. (1995) Comodulation masking release in a songbird. Hear Res 87:157–164.
Lohr B, Wright TF, Dooling RJ. (2003) Detection and discrimination of natural calls in masking noise by birds: Estimating the active space of a signal. Anim Behav 65:763–777.
Myrberg A. (1981) Sound communication and interception by fishes. In: Webster D, Fay RR, Popper AN (eds) Hearing and sound communication by fishes. Springer-Verlag, New York, pp 395–426.
Slabbekoorn H, Bouton N. (2008) Soundscape orientation: A new field in need of sound investigation. Anim Behav 76:e5–e8.
Tavolga WN. (1974) Signal/noise ratio and the critical band in fishes. J Acoust Soc Am 55:1323–1333.
Tavolga WN. (1982) Auditory acuity in the sea catfish (*Arius felis*). J Exp Biol 96:367–376.

Discovery of Sound in the Sea: An Online Resource

Kathleen J. Vigness-Raposa, Gail Scowcroft, James H. Miller, and Darlene Ketten

1 Introduction

The scientific community and the general public have become increasingly aware of and concerned about underwater sound. In addition, abstract physical science concepts can be challenging to understand. As the interest in underwater sound and its potential effects on the marine environment increases, there is a growing need for resources on the fundamental concepts of underwater sound. The Discovery of Sound in the Sea (DOSITS) Web site (http://www.dosits.org) has been designed to provide comprehensive scientific information on underwater sound at a level appropriate for the general public and for educational and media professionals (Vigness-Raposa et al. 2008). The DOSITS Web site introduces the physical science of underwater sound and how people and animals use sound to accomplish various tasks in three science sections, three resources sections, and three galleries. DOSITS was recently redesigned to include an interactive front page, an interactive Audio Gallery, and a redesigned Scientist Gallery.

2 Web Site Overview

Underwater sound is a complex topic and there is a shortage of resources available at an introductory level. The DOSITS Web site introduces underwater sound in an interactive, online learning environment with material appropriate for ages from middle school through adult. However, even

K.J. Vigness-Raposa (✉)
Marine Acoustics, Inc., Middletown, RI 02842, USA
e-mail: kathleen.vigness@marineacoustics.com

G. Scowcroft
Graduate School of Oceanography, University of Rhode Island, Narragansett, RI 02882, USA
e-mail: gailscow@gso.uri.edu

J.H. Miller
Department of Ocean Engineering, University of Rhode Island, Narragansett, RI 02882, USA
e-mail: miller@egr.uri.edu

D. Ketten
Woods Hole Oceanographic Institution, Woods Hole, MA 02543, USA
e-mail: dketten@whoi.edu

A.N. Popper and A. Hawkins (eds.), *The Effects of Noise on Aquatic Life*,
Advances in Experimental Medicine and Biology 730, DOI 10.1007/978-1-4419-7311-5_30,

elementary students will find useful and understandable content, particularly in the galleries. Scientific information is presented in three major sections including the Science of Sound in the Sea, People and Sound in the Sea, and Animals and Sound in the Sea. Content targeting specific audiences is provided in resource sections for teachers, students, and the media. Three galleries highlight multimedia resources of sounds, technologies, and scientists.

2.1 Science Sections

Scientific content introducing the physical science concepts related to underwater sound and how people and animals use sound to accomplish various tasks is presented in three major science sections. The Science of Sound in the Sea section provides a thorough introduction to fundamental concepts of underwater sound, including topics such as sound movement, sound measurement, and the difference between sound in air and sound in water. Several advanced concepts are also presented, e.g., an introduction to decibels, cylindrical versus spherical spreading, and ocean noise variability, and noise budgets are provided.

The People and Sound in the Sea section includes information on the many everyday activities that depend on underwater sound. Examining the Earth's history, investigating marine animals, and studying weather are just a few examples of human activities that require the use of underwater sound. A new section on the history of underwater sound details the discovery of underwater acoustics and its development and uses since the 1800s.

The Animals and Sound in the Sea section includes information on the use, production, and reception of sound by marine life. New content on how sea turtles hear has recently been added, along with additional details on how hearing is measured. Animals and Sound in the Sea also includes an extensive discussion on the current state of knowledge of the effects of underwater sound on marine mammals and fishes. The process of risk analysis and the methods for measuring a marine animal's reaction to sound are discussed. In addition, the best available, peer-reviewed, published literature on the effects of underwater sound on the marine environment is summarized in comprehensible, efficient pages. These sections provide a thorough introduction to underwater acoustics and its many uses.

2.2 Resources Sections

The content of the Web site has been organized into resources sections that target specific audiences: teachers, students, and the media. The Teacher Resources include classroom activities developed by educators along with a list of links to helpful classroom resources. Structured tutorials provide linear teaching paths with a progressive development of knowledge on the science topics of underwater sound, the technologies used with underwater sound, and the effects of underwater sound on marine life. PowerPoint presentations of the Web site content are available to download for in-classroom use. The Student Resources also include the structured tutorials. The Media Resources include a backgrounder on how marine animals hear underwater, frequently asked questions (FAQ) that answer the most commonly asked questions about underwater sound, and a facts and myths quiz. The facts and myths quiz tests the reader's understanding of the complex topic of underwater sound, providing answers and links to pages on the DOSITS Web site with further information.

2.3 *Galleries*

The DOSITS Web site also includes three galleries: Audio Gallery, Technology Gallery, and Scientist Gallery. The Audio Gallery contains audio files in QuickTime and MP3 formats of more than 65 underwater sounds generated by marine mammals, marine invertebrates, fishes, human activities, and natural phenomena. The Technology Gallery contains images and descriptions of the scientific and commercial equipment that are used for everyday tasks such as observing ocean currents and temperature and locating objects underwater. The Scientist Gallery highlights five renowned scientists studying or using underwater acoustics in their research. A summary of their research, along with a biography and videotaped interviews, are provided for each scientist.

3 Redesign Highlights

The DOSITS Web site was launched in November 2002 when Internet speeds were a significant consideration and many schools had limited Internet access. However, as Web technology advanced, substantial improvements in speed and user ability opened the door for enhanced features on the DOSITS Web site. Working with a professional Web design team, the DOSITS Web site was relaunched in March 2010 with a fresh "look and feel" that maintains functionality and content. The redesign includes an interactive front page, an interactive Audio Gallery, and a redesigned Scientist Gallery.

The redesigned front page of the DOSITS Web site includes a Flash-based interactive that showcases five topic areas (Fig. 1). This allows readers to grasp the wide breadth of content covered on the Web site at an easy glance. New content added to the Web site can also be highlighted, pointing out recent developments that might otherwise not be readily discovered.

Fig. 1 Front page of the Discovery of Sound in the Sea (DOSITS) Web site (http://www.dosits.org)

Within the Audio Gallery, the interactive allows the reader to efficiently select and move between sound sources, providing content in a simple, straightforward manner. The Scientist Gallery has been redesigned to focus attention on videotaped interviews. Teachers continually request video content, but many were not aware of the scientist interviews included on the Web site. Individual questions can be selected, a transcript of the interview can be reviewed while listening to the interview, and both the interview and transcript are able to be downloaded.

4 Conclusions

The DOSITS Web site provides easy, efficient access to timely, peer-reviewed information on the science of underwater sound and the current state of knowledge on the effects of underwater sound on marine mammals and fishes. The redesigned format takes advantage of Web technology developments to create an interactive learning environment that showcases physical science concepts in the natural world to develop a deeper understanding of the science of underwater sound.

References

Vigness-Raposa KJ, Scowcroft G, Knowlton C, Worcester PF (2008) The *Discovery of Sound in the Sea* Web site: An educational resource. Bioacoustics 17:348–350.

Part III
Sound Production by Aquatic Animals

Whistles of Bottlenose Dolphins: Group Repertoires and Geographic Variations in Brazilian Waters

Lisa S. Hoffmann, Elton Ferlin, Pedro F. Fruet, Rodrigo C. Genovês, Fernanda P. Valdez, Juliana Di Tullio, Glauco Caon, and Thales R. Freitas

1 Introduction

1.1 Tursiops truncatus

Tursiops truncatus (bottlenose dolphin) is the best known of all marine mammals, being widely distributed in almost all the oceans in the world. In some regions, two distinctive forms, offshore and inshore, are recognized, with variations in color, size, and genotype of the animals (Hoelzel et al. 1998).

Although *Tursiops truncatus* has been recently classified as "least concern" in the last evaluation of the International Union for Conservation of Nature (IUCN; 2008), many threats are operating in some coastal populations due to the antropic activities. The geographic variation is still poorly known due to the difficulty in obtaining data from oceanic animals. The recent discovery of a group in the Saint Peter and Saint Paul Archipelago in the central part of the equatorial Atlantic Ocean offers a unique opportunity to study the species in the pelagic environment.

1.2 Vocalizations

Many gregarious species have group-specific signaling, and the sign specificity of individuals or populations can reflect the social system and behavioral context. *Tursiops truncatus* produces a wide repertoire of complex vocalizations, and the whistle is one of the categories that receives more attention due to the social context in which it occurs (Janik and Slater 1998).

L.S. Hoffmann (✉) • F.P. Valdez • G. Caon • T.R. Freitas
Universidade Federal do Rio Grande do Sul, Porto Alegre, RS, CEP 90050-170 Brazil
e-mail: liliansander@gmail.com

E. Ferlin
Serviço de Engenharia Biomédica, Hospital de Clínicas de Porto Alegre (HCPA),
Porto Alegre, RS, CEP 90035-903 Brazil

P.F. Fruet • R.C. Genovês • J.D. Tullio
Laboratório de Mamíferos Marinhos, Museu Oceanográfico do Rio Grande (MORG)/Fundação
Universidade Federal do Rio Grande (FURG), Rio Grande, RS, CEP 96200-970 Brazil

A.N. Popper and A. Hawkins (eds.), *The Effects of Noise on Aquatic Life*,
Advances in Experimental Medicine and Biology 730, DOI 10.1007/978-1-4419-7311-5_31,

The quantification of the structure and variability of the vocalizations is an important aspect of studies evaluating repertoire differences between populations (Rendell et al. 1999; Wang et al. 1995). This work describes the geographic variability of whistle repertoires in two distant groups of *Tursiops truncatus* in the southwestern Atlantic Ocean.

2 Materials and Methods

The acoustic data were collected from 2000 to 2009 from two populations: a coastal group inhabiting the Tramandaí channel (29°58′ S 50°07′ W), southern Brazil (Hoffmann 2004), and an oceanic group occurring in the surrounding waters of the Saint Peter and Saint Paul Archipelago, 1,010 km away from the northern coast of Brazil (0°56′ N 29°22′ W).

Vocalizations were recorded with C-53 and C-54XRS hydrophones (Cetacean Research Technology, Seattle, WA) using a DAT Sony TCD-D8 and a Fostex FR-2. The spectrograms were generated with Raven 1.1 (Cornell Laboratory of Ornithology, Ithaca, NY). The configuration was fast Fourier transform (FFT) 512 samples, Hanning window, and 50% superimposition. The following parameters were extracted: maximal, minimal, initial and final frequencies and duration and range of the frequency variation. To compare the data between the areas, the Mann-Whitney test was used for whistle duration and the independent sample *t*-test for all the other parameters.

3 Results and Discussion

In the repertoire analysis, 1,768 whistles were analyzed from Tramandaí and 720 from the archipelago group (Fig. 1). The differences between the areas were significant ($P < 0.0001$) for all whistle parameters except for final frequency ($P = 0.299$). The whistles from the archipelago presented bigger values for maximal frequency (maximum = 24 kHz; average = 15.033 ± 3.438 kHz), minimal frequency (maximum = 17.410 kHz; average = 6.401 ± 2.076 kHz), initial frequency (maximum = 24 kHz; average = 9.816 ± 5.134 kHz), duration (maximum = 2.727 s; average = 0.804 ± 0.404 s), and frequency variation (maximum = 18.701 kHz; average = 8.631 ± 3.011 kHz) compared with the values from Tramandaí (average = 10.378kHz, 4.729 kHz, 5.641 kHz, 0.392 s, and 5.647 kHz, respectively).

Different causal explanations for the geographic variations can be allocated, considering ecologic, genetic, and social factors. Usually, the biggest differences in whistle variations are found in populations of nonadjacent areas (Wang et al. 1995). In this work, a possible genetic isolation could be underlying the difference found because such a condition would mean the absence of constant contact with other groups and a consequent isolation of the acoustic repertoire of its members.

Whistle structure can also vary according to the environment where the dolphins seem to alter some parameters to adapt to specific environmental noise levels. The absence of interfering obstacles in open waters seems to favor the use of higher frequencies in pelagic species, given that such characteristics allow a better use of the binaural clues (Wang et al. 1995). The archipelago group lives in a region with low noise levels, great depth (to 1,400 m), and transparent waters (visibility can reach 30 m), unlike the coastal group that occupies a shallow channel (5 m) with low visibility and antropic activities.

In studies of the social organization, *Tursiops* populations around the world have been described as fission-fusion societies (Connor et al. 2000) and changes in their social organization seem to respond mainly to the area characteristics. In Tramandaí, the predominance of lone dolphins inside

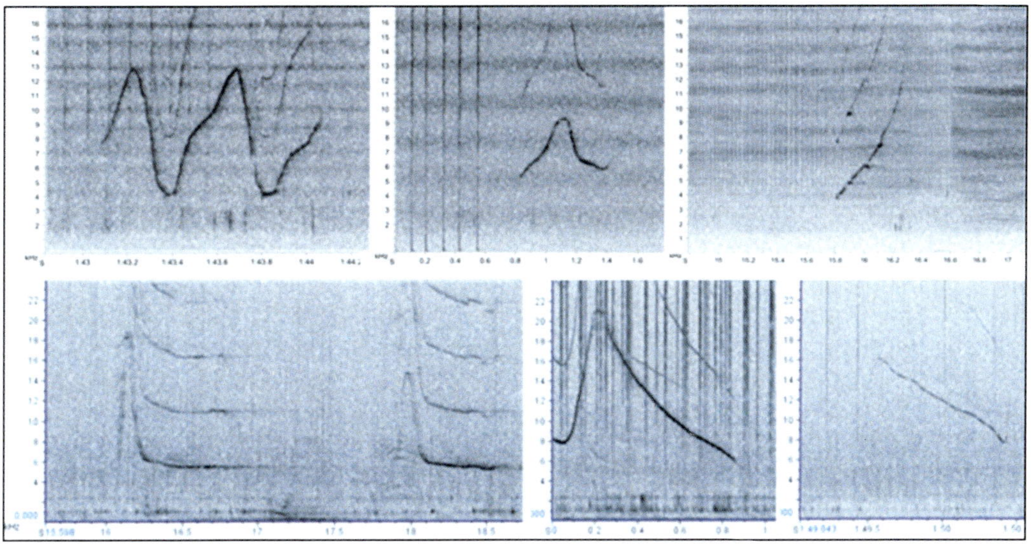

Fig. 1 Spectrograms showing some whistles recorded in Tramandaí (top) and in the Saint Peter and Saint Paul Archipelago (bottom). The frequency values are in kilohertz (*y*-axis), and time is in seconds (*x*-axis)

the channel was related to the facility of prey capture without the necessity of associations (Hoffmann 2004). In the archipelago, the depth, food offering, water currents, and presence of predators are completely diverse; thus it is expected that such differences would be reflected in their vocalizations, considering that their use is related to social organization and prey capture.

4 Conclusions

The differences in the whistle parameters between the areas seem to be related to differences in the environment and water characteristics, allowing the use of higher frequencies and longer vocalizations as well as whistles with a broader range of frequency variation. Future comparisons with other groups' repertoires can test the hypothesis presented and improve the current discussion. Acoustic tests to assess the sound propagation regarding the noise level and suspended sediments could also clear the exact influence of the environment in the whistle acoustic parameters.

References

Connor RC, Wells, RS, Mann J, Read AJ (2000) The bottlenose dolphin: Social relationships in a fission–fusion society. In: Mann J, Connor RC, Tyack PL, Whitehead H (eds) Cetacean societies: Field studies of dolphins and whales. University of Chicago Press, Chicago, IL, pp 91–126.

Hoelzel AR, Potter CW, Best PB (1998) Genetic differentiation between parapatric "nearshore" and "offshore" populations of the bottlenose dolphins. Proc R Soc Lond B Biol Sci 265:1177–1183.

Hoffmann LS (2004) Um estudo de longa duração de um grupo costeiro de golfinhos *Tursiops truncatus* no sul do Brasil: Aspectos de sua biologia e bioacústica. Dissertation, Universidade Federal do Rio Grande do Sul, Porto Alegre, Brazil. Available via http://hdl.handle.net/10183/16473.

International Union for Conservation of Nature (2008) Red data list 2008. International Union for Conservation of Nature, Gland, Switzerland. Available via http://www.redlist.org. Accessed 29 April 2010.

Janik VM, Slater PJB (1998) Context-specific use suggests that bottlenose dolphin signature whistles are cohesion calls. Anim Behav 56:829–838.

Rendell LE, Matthews JN, Gill A, Gordon JD, Macdonald DW (1999) Quantitative analysis of tonal calls from five odontocete species, examining interspecific and intraspecific variation. J Zool 249:403–410.

Wang W, Würsig B, Evans WE (1995) Whistles of bottlenose dolphins: Comparisons among populations. Aquat Mamm 21:65–77.

Detection and Classification of Vocalizations for the Study of Marine Mammal Distributions in the Chukchi Sea

David Hannay

1 Introduction

Automatic detection and classification of marine mammal vocalizations were performed on a large acoustic dataset collected almost continuously between July 2007 and October 2009 in the Alaskan Chukchi Sea. The purpose of this work was to determine spatial and temporal distributions of marine mammals over a wide area of the Chukchi Sea and to characterize ambient and anthropogenic noise. The acoustic data were obtained from multiple consecutive deployments of between 8 and 44 underwater acoustic recorders sampling at 16 kHz. Median filter and split-window normalizer detection processors were implemented to effectively detect vocalization events. The classification of calls by species was found to be more difficult due to a wide range of vocalization types produced by at least nine species; vocalizations were identified from bowheads, belugas, gray whales, fin whales, killer whales, walruses, bearded and ribbon seals, and arctic cod. Many of these species produced multiple call types and some call types evolved seasonally. Several classification approaches were implemented, and their performances were quantified by comparing classifier outputs with the results from manual classification analyses of a subset of the data. This presentation discusses the classification approaches implemented and the performance evaluations of the classifiers for selected species.

2 Chukchi Sea Acoustic-Monitoring Project

This acoustic-monitoring project was performed for Shell and ConocoPhillips Alaska, Inc., in the eastern Chukchi Sea to obtain information about marine mammal presence and ambient noise near areas of interest for oil and gas exploration. The programs have used JASCO Applied Science's (Victoria, BC) autonomous multichannel acoustic recorders (AMARs), Cornell University's (Ithaca, NY) marine autonomous recoring units (MARUs), and Multi-Electronique's (Rimouski, QC) autonomous underwater recorders for acoustic listening (AURALs). Although the multiple-season datasets have been analyzed using similar methods, we discuss here only the results from the 2008-2009 overwinter dataset. Figure 1 shows the recorder positions for this acoustic program.

D. Hannay (✉)
JASCO Applied Sciences, Victoria, BC, V8Z 7X8 Canada
e-mail: David.Hannay@jasco.com

A.N. Popper and A. Hawkins (eds.), *The Effects of Noise on Aquatic Life*,
Advances in Experimental Medicine and Biology 730, DOI 10.1007/978-1-4419-7311-5_32,

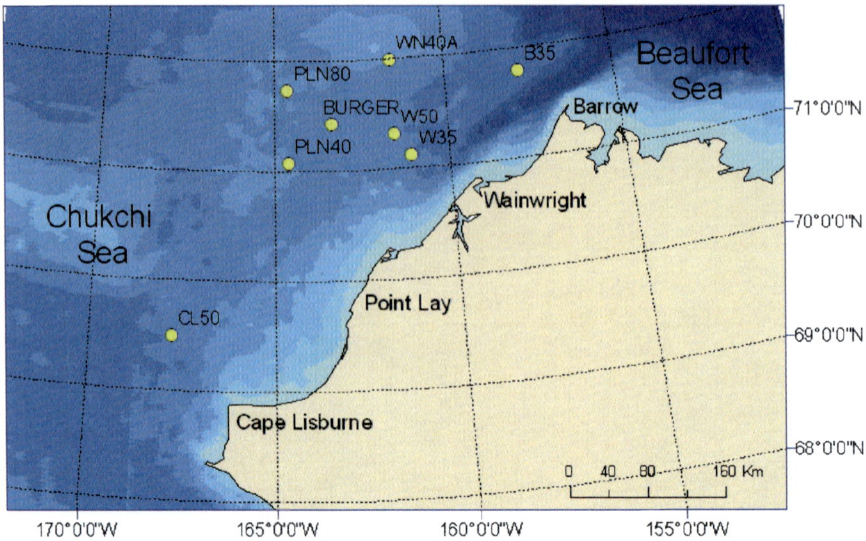

Fig. 1 Acoustic recorder deployment locations for 2008–2009 winter monitoring program

3 Data-Analysis Methods

The primary analysis goal was to identify calls from the species mentioned in the introduction to confirm, first, presence and, second, the rate of detected sound-generating activity as a function of time at each recorder location. Animal presence is confirmed by calling activity detections, but lack of detected calling activity does not preclude presence. Likewise, measurement of detection rates of marine mammal sounds does not necessarily reflect the numbers of animals present unless it is assumed that individual calling rates and call characteristics are constant.

Due to the large size of the acquired datasets, we used a hybrid manual and automatic analysis approach to detect, classify, and count marine mammal sounds. Manual analysis was performed on 5% of the data while automatic classification was applied to the entire datasets. The manual analysis examined a 2-min section of each 40-min recording file for the presence of calls by species. A more detailed manual analysis of the 2-min section from every 5th file, representing 1% of the data, identified the time of each call and logged its spectral range, spectral variation (e.g., frequency upsweep or downsweep), and duration and classified it by species. Seismic survey sounds and vessel sounds were also manually logged in this analysis. The information collected through manual analysis allowed us to quantify the performance of the automatic classifiers by comparing the results of the two methods for the same time periods as discussed in Section 3.1.

The automatic detection and classification methods included simple energy-in-frequency-band threshold detectors for belugas and seals, knock detectors for walruses, and more complex hybrid tonal detectors (Mellinger et al. 2009) combined with a random-forest decision tree using 48 characteristics for the bowhead classifier.

3.1 Automatic Classifier Performance Evaluation

We required performance parameters that would allow us to adjust the raw automatic classifier counts for each species to match the manual counts. The method of precision and recall was used. Decisions made by the automatic classifier can be classified as true positive (TP), false positive (FP), true negative (TN), and false negative (FN), where each of these values represents the counts of each category over some time period. A TP corresponds to the case where the call is correctly classified. A FN occurs when the call is classified incorrectly as noise. A FP occurs when noise is incorrectly classified as a call. A TN occurs when noise is correctly classified as noise. TP, FP, and FN were calculated for each classifier by comparing the automatic classification results for each call detection with the corresponding manual detection results.

The performance metric recall (R) represents the fraction of sounds correctly detected and classified. The metric precision (P) represents the fraction of sounds detected and classified that are attributed to the correct species. These metrics can be computed directly from TP, FP, and FN

R = TP/(TP + FN)
P = TP/(TP + FP)

R and P are used to adjust the automatic detection counts for given time periods. It is fairly easy to show that the correct value for adjusted number (N_{adj}) of detections can be calculated as N_{adj} = P*N/R, where N is the raw number of detections produced by the classifier.

4 Results

We used the approach described above to evaluate P and R for the detector and classifiers of the four species shown in Table 1.

The P and R values were applied to adjust the automatic classifier call count results for the four species in Table 1 for time periods within the deployment timeframes of the winter 2008–2009 monitoring season. An example of the adjusted call count rates for bowhead call detections from 15–30 November 2008 is shown in Figure 2.

Table 1 Precision and recall of the automatic detector and classifiers

Species	Recall, %	Precision, %	Detection Method
Bowhead	54.1	59.8	Tonal detector + classifier
Walrus	43.7	48.3	Median detector
Beluga	81.8	15.3	Median detector
Bearded seal	77.2	11.4	Median detector

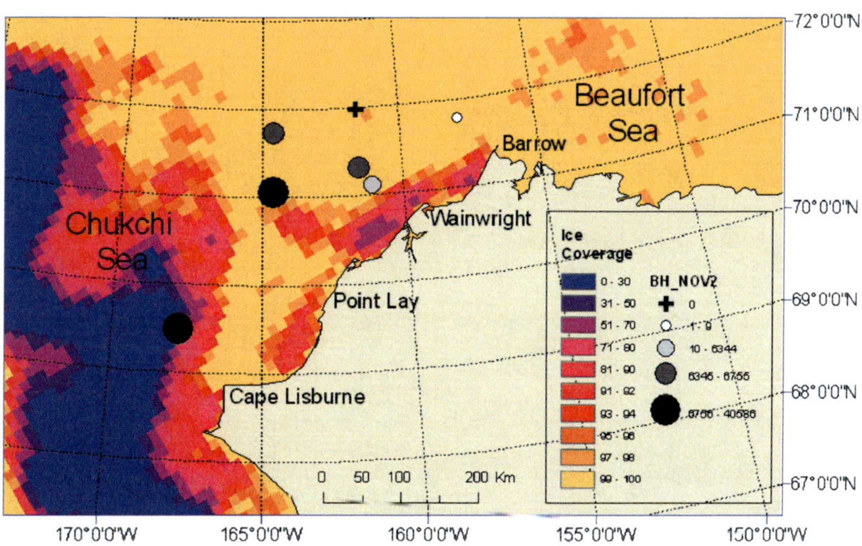

Fig. 2 Automatic call classification counts adjusted using precision and recall performance values from Table 1. BH, bowhead

References

Mellinger DK, Martin S, Morrissey R, DiMarzio N, Maretti D, Thomas L (2009) An algorithm for detection of whistles, moans, and other tonal sounds. Fourth Internal Workshop on Detection, Classification, and Localization of Marine Mammals Using Passive Acoustics, 10–13 September 2009, Pavia, Italy.

Acoustic Ecology and Behavior of Minke Whales in the Hawaiian and Marianas Islands: Localization, Abundance Estimation, and Characterization of Minke Whale "Boings"

Thomas Norris, Stephen Martin, Len Thomas, Tina Yack, Julie N. Oswald, Eva-Marie Nosal, and Vincent Janik

1 Introduction

Balaenoptera acutorostrata (minke whale) is a small and elusive baleen whale that is rarely sighted in the tropical waters of the North Pacific Ocean. During winter and spring, complex sounds called "boings" are frequently detected around the Hawaiian Islands and other Pacific island regions (Norris et al. 2009; Thompson and Friedl 1982). Although boings were described over 45 years ago (Wenz 1964), they were not attributed to minke whales until very recently (Rankin and Barlow 2005). Sightings of *Balaenoptera acutorostrata* are uncommon in tropical and subtropical waters; however, boings are frequently detected around the Hawaiian Islands using seafloor hydrophones and from hydrophone arrays towed from research vessels. Even today, very little is known about acoustic behaviors and ecology of *Balaenoptera acutorostrata.* The long-term objective of this research effort is to improve our understanding of the acoustic ecology and behavior of *Balaenoptera acutorostrata* in their breeding habitat.

A primary goal of this study is to compare characteristics of boings recorded in the Hawaiian Islands (central North Pacific) to other regions in the central and western North Pacific (e.g., the Northern Mariana Islands). These results will be used to elucidate stock identities and population characteristics for *Balaenoptera acutorostrata* in the Pacific Islands. Another goal is to estimate the local abundance of calling *Balaenoptera acutorostrata* for our main study sites off the Hawaiian Island of Kaua'i and around the Marianas Islands. Finally, we are collecting information

T. Norris (✉) • J.N. Oswald
Bio-Waves Inc., Encinitas, CA 92024, USA
e-mail: thomas.f.norris@bio-waves.net

S. Martin
SPAWAR Systems Center Pacific, San Diego, CA 92152, USA

L. Thomas • V. Janik
University of St. Andrews, St. Andrews, Fife, Scotland KY16 9LZ, UK

T. Yack
Bio-Waves Inc., Encinitas, CA 92024, USA

Southwest Fisheries Science Center, La Jolla, CA 92037, USA

E.-M. Nosal
School of Ocean and Earth Science and Technology (SOEST),
University of Hawai'i, Honolulu, HI 96822, USA

A.N. Popper and A. Hawkins (eds.), *The Effects of Noise on Aquatic Life*,
Advances in Experimental Medicine and Biology 730, DOI 10.1007/978-1-4419-7311-5_33,

that is being used to assess the calling rates of *Balaenoptera acutorostrata*. This information is necessary to validate cue-counting methods that are being developed to estimate densities of *Balaenoptera acutorostrata* exclusively from their calls (Marques et al. 2009; Martin et al. 2009; Thomas et al. 2008).

2 Methods

To accomplish these objectives, we used several types of passive acoustic methods to record and analyze data from vocalizing *Balaenoptera acutorostrata*. We used an acoustically quiet, 25-m motor-sailing vessel equipped with 2- to 6-element towed hydrophone arrays (effective bandwidth: ~100 Hz to 48 kHz) to conduct localization experiments in 2009 and an acoustic-visual line-transect survey in 2010. Bioacousticians monitored and processed acoustic data in real time throughout on-effort periods using various software including Ishmael, PAMGUARD, and WhaletrackII. In 2010, AN/SSQ-53F DIFAR sonobuoys were also used. Localizations of individual *Balaenoptera acutorostrata* were estimated using target-motion analysis techniques when possible.

Acoustic data were simultaneously recorded from the Barking Sands Underwater Range Expansion (BSURE) test site seafloor hydrophone array that is part of the Navy's Pacific Missile Range Facility (PMRF) off the west coast of Kaua'i. This Navy test range encompasses a large (>2,000 km^2), deep-water area northwest of the island of Kaua'i and includes 17 bottom-mounted hydrophones (effective bandwidth: ~100 Hz to 18 kHz) that were used for this project. PMRF seafloor array data were postprocessed using two localization methods: time-of-arrival (TOA) hyperbolic localization methods (manual and automated) and a propagation model-based time-of-arrival (PMTOA) localization method (automated). For the first method, manual techniques were initially used to identify and associate boings from the same calling animal on multiple PMRF seafloor hydrophones. Once associations and accurate TOAs were obtained, 2-dimensional (2-D) localizations were performed using standard hyperbolic techniques. Sound speed profiles (SSPs) were obtained from expendable bathythermographs (XBTs) deployed each day off the research vessel. For the second method (PMTOA), the upper 760 m of SSPs were averaged from several XBT deployments, whereas for depths below 760 m, SSPs were estimated from historical data.

In 2009, efforts were focused on obtaining localizations from the towed hydrophone array to compare with and validate those obtained from the BSURE seafloor array. In 2010, efforts were focused on conducting an acoustic-visual line-transect survey of the BSURE range for estimating abundance and comparison to the seafloor hydrophone array dataset.

Finally, two additional sources of data were included in this study: 1) data collected using a bottom-mounted hydrophone located at the Station ALOHA Cabled Observatory (ACO) were analyzed to examine seasonal and diurnal variability and 2) data from an acoustic-visual line-transect survey conducted in winter/spring 2006 for a large region surrounding the Mariana Islands that will be analyzed to derive abundance estimates and assess population structure from boing signal characteristics.

3 Results

We have completed two winter/spring field seasons (2009 and 2010) and are in the process of analyzing the acoustic data and developing automated analysis methods. In 2009, a vessel-based localization effort was conducted inside the BSURE area for 21 days between 15 March and 28 April and resulted in ~850 km of survey effort. In 2010, line-transect surveys were conducted for 2.5 mo

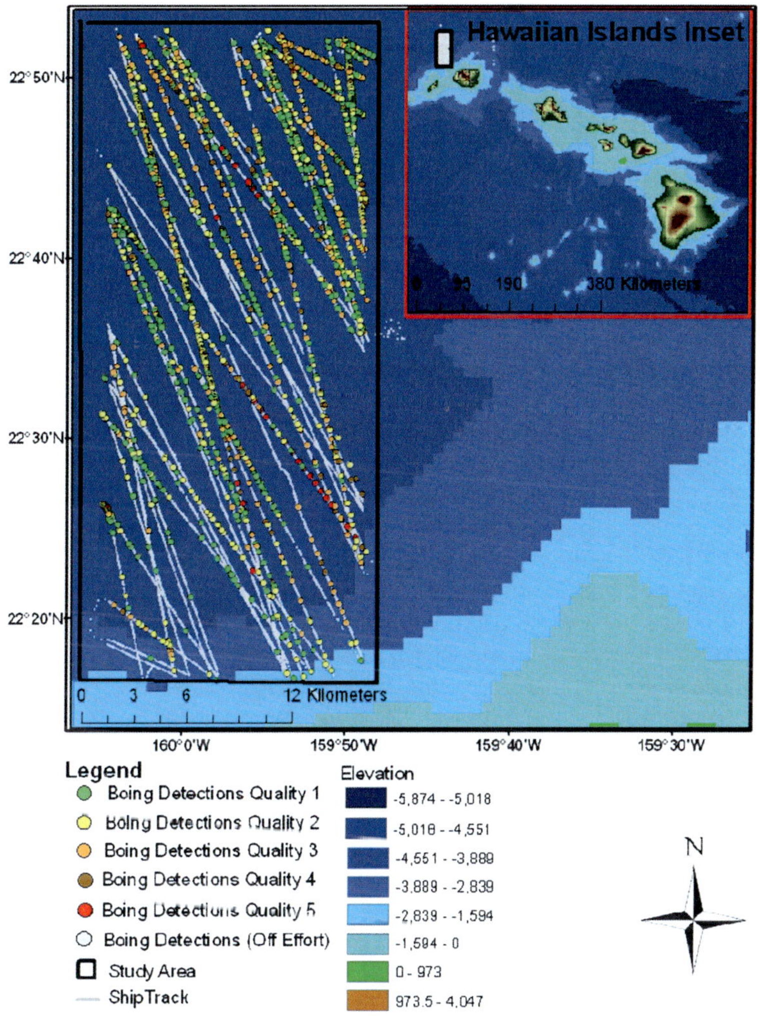

Fig. 1 Map of acoustic detections of minke whale boings made during 2010 vessel-based line-transect surveys of Kauai study area (~650 km^2). Locations of boings are plotted at location detected on ship track and do not necessarily indicate locations of calling animals

from 12 March to 11 April, resulting in 13 days and over 1,520 km of line-transect effort. Over 260 h of recordings were made for both field seasons.

In 2009, we obtained bearings for at least 777 boings that were manually detected using the towed array. Preliminary results from the just-completed 2010 field season indicate that at least 1,598 boings were detected manually, of which 1,198 bearings were obtained (Fig. 1). Over 50 localizations were made during the line-transect survey effort.

Automated detection and localization methods in PAMGUARD and Ishmael were attempted in real time during the 2010 surveys but were deemed too demanding on the available computer resources so they were discontinued after leg I. Therefore, the acoustic data are being postprocessed using automated and semiautomated techniques. These methods are expected to yield more detections and possibly localizations. The results of the automated detections and localizations will be reviewed and validated manually to ensure their reliability. Localizations obtained from the towed hydrophone array data will be compared with localizations collected from the BSURE seafloor hydrophone array. These automated methods will be used to assess localization estimates of animals and track movements and eventually to estimate the density and abundance of calling animals in the two main study areas using modified distance sampling analysis methods.

Assessment of localization accuracy is important to verify the assumptions of statistical methods being developed in a related effort called the density estimation of cetaceans from passive acoustic fixed sensors (DECAF) to estimate densities of calling animals from fixed hydrophones (Thomas et al. 2008). A case study was conducted from the BSURE seafloor hydrophone dataset containing over 6,000 boings automatically detected over a 6-h period in late April 2009. Comparisons of localizations from the two seafloor array methods described above indicated good agreement (mean difference = 142.7 m; range: 67-280 m).

Researchers on the survey vessel were able to acoustically detect, track, and sight the same individual that was being tracked from the seafloor array. The position of the sighted animal indicated relatively good accuracy (within a few hundred meters) of the positions obtained using the two seafloor localization techniques described earlier. Interestingly, the localizations determined from the towed array, although relatively precise, indicated biases based on the different algorithms used to plot the bearings from the towed hydrophone array to the calling animal. These errors were investigated further by manually plotting data. Results indicated that uncertainties in the true heading of the towed array can significantly affect localization error.

Based on the results of this case study, more fully automated techniques are being developed to facilitate the localization analysis. In addition, improvements were made to existing Matlab-based detectors used to detect boings for the automated localization algorithm. These automated methods were used to reduce processing time during the 2010 field effort.

One year of data collected at the ACO were analyzed (February 2007 to February 2008). These data showed that boings occurred seasonally from October to May, with little diurnal variation.

4 Discussion

Analyses of minke whale boings are underway to identify signal characteristics that might be useful for individual identification and as indicators of population structure. We have already found statistically significant differences in the pulse repetition rates of boings from Hawaiian waters compared with those recorded in the Mariana Islands in the western North Pacific. Interestingly, preliminary results from analysis of boings recorded on the seafloor hydrophone array (work conducted by S. Martin) indicate that there may be reliable signal characteristics that can be used for individual recognition. If so, these signal characteristics can be used to identify and track individuals using passive acoustic methods.

We are continuing to analyze data from both Hawai'i field seasons and the Marianas effort. Our immediate efforts are focused on obtaining density estimates for our two main study areas. We are also examining the effects of noise produced by our own vessel on the calling rates and acoustic behavior patterns of minke whales. It is important to assess any effects of vessel noise on calling rates in order to evaluate any biases in these data caused by our survey vessel and in future vessel-based surveys. The results of these investigations will provide important information about a species that is common in the subtropical waters of the North Pacific Ocean but about which there is little information regarding their ecology and behavior.

Acknowledgments This work was funded by the Office of Naval Research Marine Mammal Program. Station ALOHA Cabled Observatory access was provided by Dr. Duennebier.

References

Marques TA, Thomas L, Martin SW, Mellinger DK, Jarvis S, DiMarzio N, Morrisey RP (2009) Spatially explicit capture-recapture methods to estimate minke whale abundance from data collected at bottom mounted hydrophones. EURING 2009 Analytical Meeting and Workshop, 14–20 September 2009, Pescara, Italy.

Martin SW, Marques TA, Thomas L, Moretti D, Morrissey R, DiMarzio N, Jarvis S, Mellinger DK, Norris T (2009) Estimating minke whale boing density at PMRF, Hawaii. Symposium on Estimating Cetacean Density from Passive Acoustics, 16 July 2009, San Diego, CA. Available via http://www.creem.st-and.ac.uk/decaf/meetings/martin.pdf/view.

Norris T, Yack T, Oswald J, Martin S, Thomas L, Janik V (2009) Boing! Acoustic localization, characterization and comparison of minke whale songs from the Hawaiian Islands and other areas in the Northern Pacific. Fourth International Workshop on Detection, Classification and Localization of Marine Mammals Using Passive Acoustics, 10–13 September 2009, University of Pavia, Pavia, Italy,. Available via http://www.creem.st-and.ac.uk/decaf/meetings/Workshops-DCL-DE-Abstracts-Pavia-2009.pdf.

Rankin S, Barlow J (2005) Source of the North Pacific "boing" sound attributed to minke whales. J Acoust Soc Am 118:3346–3351.

Thomas L, Marques T, Borchers D, Harris C, Moretti D, Morrissey R, DiMarzio N, Ward J, Mellinger D, Martin S, Tyack P (2008) DECAF – density estimation for cetaceans from passive acoustic fixed sensors. National Oceanographic Partnership Program Annual Report, Office of Naval Research. Available via http://www.creem.st-and.ac.uk/decaf/outputs/NOPP%20PI%20Annual%20Report%202008.pdf.

Thompson P, Friedl W (1982) A long term study of low frequency sounds from several species of whales off Oahu, Hawaii. Cetology 45:1–19.

Wenz GM (1964) Curious noises and the sonic environment in the ocean. In: Tavolga WN (ed) Marine Bio-Acoustics, Pergamon Press, New York, pp 101–119.

What You See Is Not What You Hear: The Relationship Between Odontocete Echolocation Click Production and Hearing

Laura N. Kloepper, Paul E. Nachtigall, and Marlee Breese

1 Introduction

The improvement in tagging technology and passive listening devices has allowed researchers to measure the echolocation clicks of many species of free-ranging odontocetes. Although the data collected by these instruments provide valuable information on the clicks these animals produce, these tags cannot provide information on the hearing abilities of these species. A reasonable assumption is that animals produce sounds in the same frequency regions of hearing, but recent studies suggest this may not be the case. The development of a portable auditory evoked potential system has allowed for recent hearing measurements of stranded and rehabilitated animals. The white-beaked dolphin (*Lagenorhynchus albirostris*) has the most sensitive hearing between 45 and 128 kHz (Nachtigall et al. 2008), but free-ranging individuals produce echolocation clicks with considerable energy in frequencies up to 250 kHz (Rasmussen and Miller 2002). Risso's dolphin (*Grampus griseus*) has the most sensitive hearing between 22 and 90 kHz (Nachtigall et al. 2005), but free-ranging individuals produce echolocation clicks with considerable energy in frequencies up to 120 kHz (Madsen et al. 2004).

2 Relationship Between Click Production and Hearing

Thus far, the data with free-ranging odontocetes suggest that animals produce signals with higher frequencies than they can hear. The relationship between hearing and echolocation was further studied in a laboratory setting with the same subject, a female false killer whale (*Pseudorca crassidens*). The whale was housed in the floating pen structure at the Hawai'i Institute of Marine Biology, Kaneohe Bay, HI. Hearing measurements were obtained with auditory evoked potential techniques, a noninvasive method for rapidly collecting the hearing capabilities of trained and untrained animals. As an older animal, the best hearing sensitivity was found between 15 and 27 kHz, with a high-frequency cutoff above 34 kHz (Yuen et al. 2005). Clicks were collected with LABVIEW software with a star-shaped array of 16 hydrophones while the animal performed an echolocation task,

L.N. Kloepper (✉) • P.E. Nachtigall • M. Breese
Marine Mammal Research Program, Hawai'i Institute of Marine Biology,
University of Hawai'i, Kaneohe, HI 96744, USA
e-mail: kloepper@hawaii.edu

A.N. Popper and A. Hawkins (eds.), *The Effects of Noise on Aquatic Life*,
Advances in Experimental Medicine and Biology 730, DOI 10.1007/978-1-4419-7311-5_34,

discriminating between hollow aluminum cylinders with differing inner wall thicknesses. On-axis clicks were identified from the maximum peak-peak source level (dB re 1 μPa) and spectral characteristics were obtained for each on-axis click with Fourier analysis and custom analysis programs in MATLAB. The peak frequency of recorded signals averaged 35 kHz, which was outside the range of best hearing for the animal. Additionally, the majority of clicks contained components within 3 dB of the peak frequency that extended up to 70 kHz, which is well beyond the hearing capabilities for this animal. Thus, this animal consistently produced echolocation clicks with frequencies higher than she could hear.

The question of why an animal would produce clicks outside her range of hearing is one for which, at present, we can only speculate an answer. Because it is hypothesized that higher frequency echoes may provide better target resolution, animals may simply be producing clicks with the highest audible frequencies possible. Perhaps it is simply an artifact of the click production mechanism that remains fixed after the loss of high-frequency hearing. Alternatively, the higher frequencies in the clicks may serve some yet unknown function. Regardless, it is evident that many odontocetes produce signals with frequencies outside the range of hearing. This has led to the hypothesis that odontocetes only pay attention to a small band of their echolocation signals, a concept termed "functional bandwidth" (Ibsen et al. 2009). This functional bandwidth can be reduced with presbycusis; as an animal loses its ability to hear at higher frequencies as it ages, its ability to hear at higher frequencies is diminished. To fully understand the relationship between odontocete click production and hearing, it is imperative that both hearing and echolocation click production parameters are studied in concert.

References

Ibsen SD, Au WWL, Nachtigall PE, Breese M (2009) Functional bandwidth of an echolocating Atlantic bottlenose dolphin (*Tursiops truncatus*). J Acoust Soc Am 125:1214–1221.

Madsen PT, Kerr I, Payne R (2004) Echolocation clicks of two free-ranging, oceanic delphinids with different food preferences: False killer whales *Pseudorca crassidens* and Risso's dolphins *Grampus griseus*. J Exp Biol 207:1811–1823.

Nachtigall PE, Mooney TA, Taylor KA, Miller LA, Rasmussen MH, Akamatsu T, Jeilmann J, Linnenschmidt M, Vikingsson GA (2008) Shipboard measurements of the hearing of the white-beaked dolphin *Lagenorhynchus albirostris*. J Exp Biol 211:642–647.

Nachtigall PE, Yuen MML, Mooney TA, Taylor KA (2005) Hearing measurements from a stranded infant Risso's dolphin, *Grampus griseus*. J Exp Biol 208:4181–4188.

Rasmussen MH, Miller LA (2002) Whistles and clicks from white-beaked dolphins, *Lagenorhynchus albirostris* recorded in Faxafloi Bay. Aquat Mamm 28:78–89.

Yuen MML, Nachtigall PE, Breese M, Supin AY (2005) Behavioral and auditory evoked potential audiograms of a false killer whale (*Pseudorca crassidens*). J Acoust Soc Am 118:2688–2695.

Sound Production and Reception in Southern Sea Otters (*Enhydra lutris nereis*)

Asila Ghoul and Colleen Reichmuth

1 Introduction

The rapidly increasing pressure of human activity in coastal and pelagic marine environments has led to concern regarding the effects of anthropogenic noise on marine mammals. Recent studies dealing with acoustic communication and behavioral responses to sound along with investigations of auditory anatomy, acoustic sensitivity, and noise impacts have established a knowledge base that has proven vital to regulators charged with determining safe sound exposure limits for some cetaceans and pinnipeds (Southall et al. 2007). Comparable data are presently unavailable for sea otters (*Enhydra lutris*), which have been largely ignored in the context of this issue. To date, only one study has addressed the bioacoustics of *Enhydra lutris* from the standpoint of sound production and communication (McShane et al. 1995), and there have been no formal anatomical, physiological, or behavioral studies of sound reception in this species. Due to the lack of available data, decision makers must use less than optimal, indirect evidence to evaluate the potential effects of anthropogenic noise on *Enhydra lutris*. This information comprises mainly unpublished technical reports describing observed behavioral reactions to various noise sources; most of these studies are not systematic and none of them consider auditory sensitivity to airborne or waterborne sound sources (Richardson 1995). To begin addressing the critical data gaps that exist concerning the bioacoustics of *Enhydra lutris*, we used a twofold approach to systematically evaluate captive individuals of this species as both sound emitters and sound receivers. We opted to begin these studies in air because these amphibious marine mammals spend a majority of their time at the water's surface where they carry out many important life functions, including those related to communication.

A. Ghoul (✉)
Long Marine Laboratory, Department of Ocean Sciences, University of California
Santa Cruz, Santa Cruz, CA 95064, USA
e-mail: asila@ucsc.edu

C. Reichmuth
Long Marine Laboratory, Institute of Marine Sciences, University of California
Santa Cruz, Santa Cruz, CA 95060, USA
e-mail: coll@ucsc.edu

A.N. Popper and A. Hawkins (eds.), *The Effects of Noise on Aquatic Life*,
Advances in Experimental Medicine and Biology 730, DOI 10.1007/978-1-4419-7311-5_35,

2 Sound Production

Currently, the only available information regarding the significance of sound to *Enhydra lutris* comes from observations of vocal communication. Southern sea otters (*Enhydra lutris nereis*) produce a range of aerial vocalizations in different social contexts. However, sound production underwater has never been observed, and there is no information on how aerial calls may be transmitted beneath the water's surface. Although many of the vocalizations emitted by *Enhydra lutris nereis* are thought to be used for short-range communication at the surface, scream calls are relatively loud signals with potentially large effective communication distances. These harsh calls are most commonly produced by mothers and their young when separated from one another and can be heard by human listeners up to 1 km away (McShane et al. 1995). The ranges over which these signals may be detected by conspecifics is unknown, in part because the source levels of these calls have not been measured.

To obtain quantitative measures of the source level and full spectral content of scream vocalizations, we recorded these aerial calls from adult females and dependent pups at close range (1–3 m) in a captive setting. Vocalizations were analyzed to determine duration, frequency range, dominant frequency, and source sound pressure level (SPL), given here as root mean square decibels (dB_{RMS}) re 20 μPa. Durations varied from 0.5 to 2 s, and all screams were harmonic in structure and extremely broadband, with energy extending above 60 kHz. Dominant frequencies ranged from 6 to 8 kHz in screams from adult females and from 4 to 7 kHz in screams from dependent pups. Source level measurements from all age and sex classes showed a high degree of variability both within and between individuals and ranged from 50 to 113-dB SPL re 20 μPa.

The extremely wide bandwidth of these contact calls raises questions about what portion of the signals are available for functional communication in natural settings. Specifically, frequencies above 10 kHz represent only a small percentage of the total energy contained in these signals, and these high-frequency components are likely to undergo high transmission loss when emitted in air. To estimate potential communicative ranges, the source level and frequency range characteristics measured in this study must be considered in the context of vocal behavior and combined with direct information on auditory sensitivity and the frequency bandwidth of hearing in this species.

3 Sound Reception

In the absence of any quantitative measures of hearing capabilities in *Enhydra lutris*, we developed a simple behavioral approach to conservatively estimate the upper and lower frequency limits of the auditory system. This controlled exposure experiment (CEE) began as a part of a larger project aimed at investigating hearing sensitivity of *Enhydra lutris nereis* and was designed to provide estimates of the aerial frequency range of hearing from multiple captive subjects in a relatively short period of time.

The design of the CEE called for subjects to be exposed to 1-s pure-tone stimuli while resting calmly in an acoustically mapped enclosure. The subjects were not trained for active participation in this study. Exposures were only presented when the subject was in a relaxed behavioral state, with its head above water. Behavior during and immediately after each sound exposure was compared with baseline behavior to determine if a positive response was observed. Blank exposures, during which no sound was present, were included as control trials and responses were evaluated in the same manner. The auditory stimuli were presented at fixed source SPLs of 80–100 dB re 20 μPa, generating received levels exceeding 70 dB re 20 μPa. Sessions were remotely conducted by an experimenter who viewed the animals on high-definition video. Each session was recorded, and both sound exposures and blank exposures were identically visually marked using video editing software. Responses to both trial types were later scored by blind observers. Statistically reliable differences in exposure versus blank conditions were used to determine audible frequencies between 0.125 and 45.3 kHz.

These data revealed an upper frequency hearing limit extending to at least 32 kHz and a low-frequency limit below 0.125 kHz, results that are generally consistent with comparable data for other carnivores, including some terrestrial mustelids (Heffner and Heffner 1985). Reasonable estimates of frequency bandwidth of hearing will allow decision makers, for the first time, to identify or exclude potential sounds of concern when evaluating *Enhydra lutris nereis* in the context of environmental noise impacts. These data also improve understanding of vocal communication by suggesting that the portion of vocalizations below 32 kHz are most likely to contain biologically relevant information.

4 Summary

Because of their dependence on a highly restricted coastal habitat, *Enhydra lutris* is especially vulnerable to a variety of different environmental and anthropogenic threats. This species is presently listed as threatened and is protected throughout the northern and southern portions of its range. Resource managers are presently faced with uncertainty when responding to and prioritizing potential threats to these animals due to insufficient understanding of the factors that may disturb or disrupt normal behavior patterns both above and below the water's surface. The objective of these studies was to obtain direct measurements of the source characteristics of vocalizations and the limits of auditory reception in *Enhydra lutris*. These data are necessary to form a basic but essential understanding of bioacoustics in this species. To further develop this knowledge base, psychoacoustic profiles of aerial and underwater hearing sensitivity as a function of sound frequency are imperative to adequately consider sea otters alongside other marine mammals within the issue of anthropogenic noise impacts. These studies are presently ongoing in our laboratory. As these coastal-living carnivores have only recently transitioned to a marine lifestyle, an improved understanding of their acoustic communication and auditory adaptations will also provide insight into their evolutionary biology and behavioral ecology as well as the evolutionary pressures shaping underwater perception in marine mammals.

Acknowledgments This work was conducted with permission of the United States Fish and Wildlife Service (Permit MA186914) and the University of California Santa Cruz Institutional Animal Care and Use Committee and with support from the United States Bureau of Ocean Energy Management, Regulation and Enforcement, Washington, DC.

References

Heffner RS, Heffner HE (1985) Hearing in mammals: The least weasel. J Mamm 66:745–755.

McShane LJ, Estes JA, Riedman ML, Staedler MM (1995) Repertoire, structure and individual variation of vocalizations in the sea otter. J Mamm 76:414–427.

Richardson WJ (1995) Documented disturbance reactions. In: Richardson WJ, Greene CR Jr, Malme CI, Thomson DH (eds) Marine mammals and noise. Academic Press, San Diego, CA, pp 241–324.

Southall BL, Bowles AE, Ellison WT, Finneran JJ, Gentry RL, Greene CR Jr, Kastak D, Ketten DR, Miller JH, Nachtigall PE, Richardson WJ, Thomas JA, Tyack PL (2007) Marine mammal noise exposure criteria: Initial scientific recommendations. Aquat Mamm 33:412–522.

Nocturnal Acoustic Activity in the Shallow Waters of the WWF-Miramare Natural Marine Reserve (Trieste, Italy)

Antonio Codarin, Marta Picciulin, Linda Sebastianutto, Giuliana Calcagno, Maurizio Spoto, and Enrico A. Ferrero

1 Introduction

A growing body of scientific evidence shows that the increasing noise pollution in coastal environments due to human activities has detrimental effects on marine animals, including many endangered species. As a consequence, research studies and subsequent conservational actions are needed to mitigate the effects of noise pollution. One important step is to safeguard sensitive areas known as marine protected areas (MPAs) from anthropogenic noise; these ecologically rich areas, critical habitats for key species, are often located in highly populated coastal zones. This is the case of the WWF-Miramare Natural Marine Reserve, a UNESCO-Man and Biosphere Programme (MAB) Biosphere Reserve located in the Gulf of Trieste (Italy) and considered an important seasonal nursery area in the North Adriatic Sea (Guidetti et al. 2005). Although a recent study (Codarin et al. 2008), based on daytime acoustic monitoring, shows that the Miramare fish population is living in a heavily noisy underwater environment year-round, relatively little is known about the features and anthropogenic factors of nocturnal sea ambient noise (SAN) in the Reserve. This information is particularly relevant considering that intraspecific communication of many marine species (i.e., the brown meager, *Sciaena umbra*, during spawning season; Bonacito et al. 2001) occurs after sunset and during night. As a consequence, the present study aims 1) to quantify the nocturnal ambient noise levels at the Reserve, 2) to distinguish and quantify the biological versus anthropogenic components of the noise, and 3) to discuss their possible interaction.

A. Codarin (✉)
Regional Agency for Environmental Protection, 33170 Pordenone, Italy

M. Picciulin
CSEE-Department of Life Sciences, University of Trieste, 34127 Trieste, Italy

WWF-Natural Marine Reserve of Miramare, 34014 Trieste, Italy

L. Sebastianutto
Cognitive Neuroscience Sector, International School of Advanced Studies (SISSA), 34135 Trieste, Italy

G. Calcagno • E.A. Ferrero
CSEE-Department of Life Sciences, University of Trieste, 34127 Trieste, Italy

M. Spoto
WWF-Natural Marine Reserve of Miramare, 34014 Trieste, Italy

A.N. Popper and A. Hawkins (eds.), *The Effects of Noise on Aquatic Life*, Advances in Experimental Medicine and Biology 730, DOI 10.1007/978-1-4419-7311-5_36,

2 Materials and Methods

Ten separate surveys were carried out from dusk until dawn inside the core zone of the Reserve, close to an artificial submerged rocky reef with high animal density, from June to September 2009. A new prototype of an autonomous underwater recorder was used to record continuously 11 h/survey. The sonobuoy was provided by a preamplified Reson TC 4013 hydrophone (sensitivity −170 dB re 1 V/μPa; frequency range 1 Hz to 170 kHz) and a Gemini iKey Plus Recording Device, which created 10-min samples in WAV format at a rate of 44.1 kHz, 16 bit. During the monitoring sessions, the buoy was suspended in a midwater column from a PVC frame (10-m water depth). Water temperature was measured at the same depth with a multiparametric Ocean Seven 316 CTD-Idronaut probe and ranged from 15 to 20°C during the study period.

A subset of the collected acoustic data (for a total of 332 10-min samples or 55 h) has been analyzed and described minute by minute by hearing and by visual assessment using Avisoft SASLab Pro software. The source of each noise, where possible, was described. In addition, a subset of 264 10-min samples (43 h) was analyzed with SPECTRA RTA software calibrated with a signal of 100 mV root mean square (RMS) @ 1 kHz and hydrophone sensitivity, obtaining a one-third octave-band analysis and an equivalent continuous sound pressure level (SPL) value ($L_{\mathrm{Leq'10min}}$) for each acoustic sample. Considering that most of the fish vocalize within 2,500 Hz (Amorim 2006), noise levels in the 50- to 2,500-Hz frequency band were also calculated over each 10-min sample. Analyses of variance on $L_{\mathrm{Leq'10min}}$ and on 50- to 2,500-Hz SPL were performed to assess the differences among different nocturnal phases and recording day.

3 Results

The nocturnal noise at the Miramare MPA was mainly biological in origin. The sources were identified as snapping shrimp "crackle," an unknown biological sound source, and vocalizations of *S. umbra* (Sciaenidae). The latter produces three different pattern of acoustic emissions (Bonacito et al. 2001 2002) named "irregular sounds" (I), "regular sounds" (R; where the temporal delay between following emissions is highly constant), and "chorus" (C; characterized by overlapping and merging of sounds). Three phases have been distinguished: the first one (20.00-24.00) is characterized by high acoustic activity of *S. umbra* (43% C sounds, 40%, R sounds, 7% I sounds), a second phase (00.00-04.00) with a sharp decrease in *S. umbra* vocalizations (17% R sounds, 83% no sounds) and an increasing occurrence of the unknown sound, and a third phase (04.00-08.00) when *S. umbra* vocalizes again (44% R sounds, 3% I sounds, 53% no sounds). Boat noises have also been identified. They are produced by outboard and/or small inboard engines during the first and the third night phases (Fig. 1).

The average wideband $L_{\mathrm{Leq,10\,min}}$ and 50- to 2,500-Hz SPL of the nocturnal SAN are equal to 123.7 and 104.9 dB re 1 μPa, respectively. During the day, these values were 114 and 113.8 dB re 1 μPa SPL, respectively (Picciulin and Codarin, unpublished data). Nocturnal $L_{\mathrm{Leq,10\,min}}$ ranges from 115.98 to 133.26 dB re 1 μPa during the night and the 50- 2,500-Hz SPL ranged from 107.22 to 132.55 dB re 1 μPa. There was a significant effect of the recording day ($F_{(3,222)} = 209.25$; $P < 0.001$), the phase of the night ($F_{(2,222)} = 21.22$; $P < 0.001$), and their interaction ($F_{(6,222)} = 3.10$; $P = 0.006$ for $\mathrm{L}_{\mathrm{Leq,10\,min}}$). The same occurred for the 50- to 2,500-Hz SPL (recording day: $F_{(3,222)} = 43.63$; $P < 0.001$; phase of the night: $F_{(2,222)} = 44.37$; $P < 0.001$; and interaction: $F_{(6,222)} = 2.58$; $P = 0.02$).

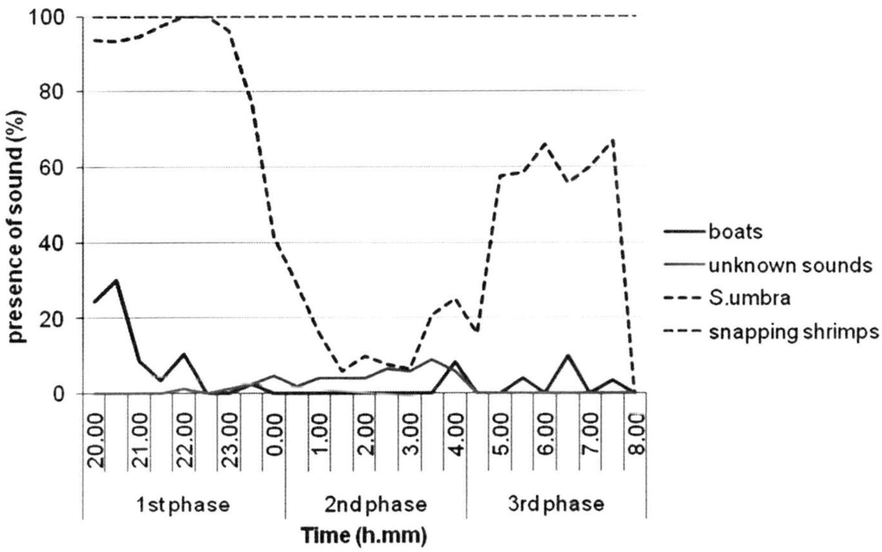

Fig. 1 Temporal distribution of nocturnal acoustic activity within the Miramare marine protected area and the identified sound sources

4 Discussion

The present study investigated the changes in nocturnal ambient noise at the Miramare Reserve during the summer. The results show some clear differences regarding the prevailing noise sources and the broadband noise levels during the night. The most consistent source is due to the activity of the snapping shrimp, which are considered an almost ubiquitous biological component of ambient noise in the frequency band of 2,000 Hz to 15 kHz (Radford et al. 2008). The vocalizations of the brown meager are the second largest component; most of these sounds are produced from late evening up to midnight. Unfortunately, from 20.00 to 24.00, we also found the highest frequency of occurrence of boat noises during the night. This is a cause of concern because boat noises recorded within the Reserve may diminish fish hearing ability, mask intraspecific relevant signals (Codarin et al. 2009), and cause behavioral changes in local soniferous fish species (Picciulin et al. 2010).

The wideband $L_{\mathrm{Leq,10\ min}}$ data integrated by the previous daytime SAN sampling (Codarin et al. 2008) showed that during the summer, day is the quietest time in the Reserve followed by night whereas dawn and dusk are the noisiest moments. Nevertheless, if we consider only the frequency range below 2,500 Hz, the opposite trend is emerging: day is the noisiest time, followed by dawn and dusk, whereas night is the quietest time despite the high biological acoustic activity. This is likely due to the daytime low-frequency noise energy input caused by port activity and shipping.

Codarin et al. (2009) showed that a daytime-recorded Miramare SANsample, with $L_{\mathrm{Leq,1min}}$ equal to 97 dB re 1 mPa, barely masked the hearing thresholds of *S. umbra*. Nevertheless, the average wideband $L_{\mathrm{Leq,10\ min}}$ and 50- to 2,500-Hz SPL of the SAN recorded at the submerged reef are ~13–15 dB higher during the day (Picciulin and Codarin, unpublished data) and 8–27 dB higher during the night than the above value. Considering that SPLs show a high variability in intensity and frequency composition during the 24 h and the summer season, further studies are needed to determine to what degree SAN influences acoustic communication and other important activities such as orientation (ambient sound is a source of orientation information for animals; Simpson et al. 2005) in local species.

References

Amorim MCP (2006) Diversity of sound production in fish. In: Ladich F, Collin SP, Moller P, Kapoor BG (eds) Communication in fishes. Science Publishers, Enfield, NH, pp 71–104.

Bonacito C, Costantini M, Casaretto L, Hawkins AD, Spoto M, Ferrero EA (2001) Acoustical and temporal features of sounds of *Sciaena umbra* (Sciaenidae). Proceedings of the XVIII International Bioacoustics Council Meeting, 3–6 September 2001, Cogne, Italy.

Bonacito C, Costantini M, Picciulin M, Ferrero EA, Hawkins AD (2002) Passive hydrophone census of *Sciaena umbra* (Sciaenidae) in the gulf of Trieste (Northern Adriatic Sea, Italy). Bioacoustics 12:292–293.

Codarin A, Spoto M, Picciulin M (2008) One-year characterization of sea ambient noise in a coastal marine protected area: A management tool for inshore marine protected areas. Bioacoustics 17:24–25.

Codarin A, Wysocki LE, Ladich F, Picciulin M (2009) Effects of ambient and boat noise on hearing and communication in three fish species living in a marine protected area (Miramare, Italy). Mar Pollut Bull 58:1880–1887.

Guidetti P, Verginella L, Viva C, Odorico R, Boero F (2005) Protection effects on fish assemblages, and comparison of two visual-census techniques in shallow artificial rocky habitats in the northern Adriatic Sea. J Mar Biol Assoc UK 85:247–255.

Picciulin M, Sebastianutto L, Codarin A, Farina A, Ferrero EA (2010) *In situ* behavioral responses to boat noise exposure of *Gobius cruentatus* (Gmelin, 1789; fam. Gobiidae) and *Chromis chromis* (Linnaeus, 1758; fam. Pomacentridae) living in a Marine Protected Area. J Exp Mar Biol Ecol 386:125–132.

Radford CA, Jeffs AG, Tindle CT, Montgomery JC (2008) Temporal patterns in ambient noise of biological origin from a shallow water temperate reef. Oecologia 156:921–929.

Simpson SD, Meekan MG, Montgomery JC, McCauley RD, Jeffs A (2005) Homeward sound. Science 308:221.

Acoustic Ecology of the California Mantis Shrimp *(Hemisquilla californiensis)*

Erica R. Staaterman, Christopher W. Clark, Austin J. Gallagher, Thomas Claverie, Maya S. deVries, and Sheila N. Patek

1 Introduction

Acoustic communication plays a major role in the behavioral ecology of various marine organisms (Busnel 1963), especially marine mammals and fish. However, little attention has been given to acoustic communication in marine crustaceans (Popper et al. 2001). Furthermore, the interplay between anthropogenic noise and the acoustic ecology of marine crustaceans remains virtually unexplored. In this study, we investigated the acoustic environment of a benthic stomatopod crustacean, the California mantis shrimp (*Hemisquilla californiensis*, Crustacea, Stomatopoda).

California mantis shrimp produce a "rumble" sound that has been anecdotally observed in the field (Haderlie et al. 1980) and first documented in the scientific literature in 2006 (Patek and Caldwell 2006). Patek and Caldwell's recordings were obtained in tanks and sounds were recorded when the animals were physically handled or approached by a stick. Fifty percent of the adult males produced rumbles, whereas none of the adult females produced sound. The rumbles were produced by vibrations of a pair of muscles that attach to the edge of the carapace. Rumbles lasted less than 2 s and the mean dominant frequency was 45 ± 10 (SD) Hz ($n = 53$ rumbles). The general function of this sound and whether or not females are capable of generating it remains unknown.

Although the laboratory-based recordings of rumbles provide a starting point for identifying the source of the sound, field recordings are essential for interpreting the rumble's function and role in the ecology of mantis shrimp. However, to our knowledge, no field recordings have been published for any stomatopod crustacean. The three primary goals of this study were to 1) characterize the sounds of *H. californiensis* in its natural habitat, 2) describe diel patterns of behavior and sound production, and 3) examine the presence of anthropogenic noise in the acoustic habitat of the California mantis shrimp. We employed several tools and techniques to accomplish these goals, including a coupled audio-video system and a passive acoustic device.

E.R. Staaterman (✉) • T. Claverie • S.N. Patek
Biology Department, University of Massachusetts Amherst, Amherst, MA 01003, USA
e-mail: e.staaterman@gmail.com

C.W. Clark
Bioacoustics Research Program, Cornell Laboratory of Ornithology, Ithaca, NY 14850, USA

A.J. Gallagher
Marine Science Center, Northeastern University, Nahant 01908, MA, USA

M.S. deVries
Department of Integrative Biology, University of California, Berkeley, Berkeley, CA 94720, USA

A.N. Popper and A. Hawkins (eds.), *The Effects of Noise on Aquatic Life*,
Advances in Experimental Medicine and Biology 730, DOI 10.1007/978-1-4419-7311-5_37,

2 Recording Methods and Results

Recordings were obtained in the naturally occurring communities of *H. californiensis* off the coast of Santa Catalina Island, CA, in March 2009. A coupled audio-video system was placed in front of several animals' burrows by a SCUBA diver, and the resulting footage was later imported into digital audio files for analysis. These recordings (48 kHz, 16-bit sampling rate) were used to describe general characteristics of the rumbles (Fig. 1). The average rumble had a dominant frequency of 167 ± 40.9 (SD) Hz and lasted 0.2 ± 0.08 (SD) s (n = 3,858 rumbles from an undetermined number of individuals). We found that the rumbles were typically produced in groups of twos, threes, and fours, which we now refer to as "rumble bouts." The leading rumble of each bout was louder by an average of 4.8 ± 0.185 dB and longer in duration by an average of 0.093 ± 0.005 s than the second rumble in the bout (n = 304 rumbles from ~17 individuals). Despite these general similarities, recordings from various individuals' burrows yielded rumbles that differed in dominant frequency and temporal patterning.

An autonomous recording unit (see Clark and Clapham 2004) was deployed ~9 km from the first site in a different mantis shrimp population and was run continuously (32 kHz, 16-bit sampling rate) for an 8-day period. We scanned this recording's spectrogram both visually and aurally and found distinct trends across the 8 days. During crepuscular periods, loud rhythmic rumbles were audible. At night, the mantis shrimp were acoustically active, but their rumbles were quieter and lower in frequency than during the day. We observed few rumbles during hours of peak sunlight.

We found that at the site of the 8-day recorder, boat activity was substantially higher during the day and during weekdays than during nights and weekends (Fig. 2). When averaging the acoustic energy across 1-h periods, we saw that midday energy values in the 100- to 500-Hz range, e.g., were, on average, 15.6 ± 0.7 dB louder than midnight energy values (Fig. 3; matched-pairs t-test; n = 8; t-ratio = 23.08; $P < 0.0001$).

Rumbles produced amid boat noise were compared with rumbles produced during normal ambient noise conditions. When the rumbles were not completely obscured by boat noise, rumble dominant frequencies decreased when boats were present (difference = 37 ± 1.3 [SD] Hz, t-test; degrees of freedom [DF] = 3893.01; $P < 0.0001$).

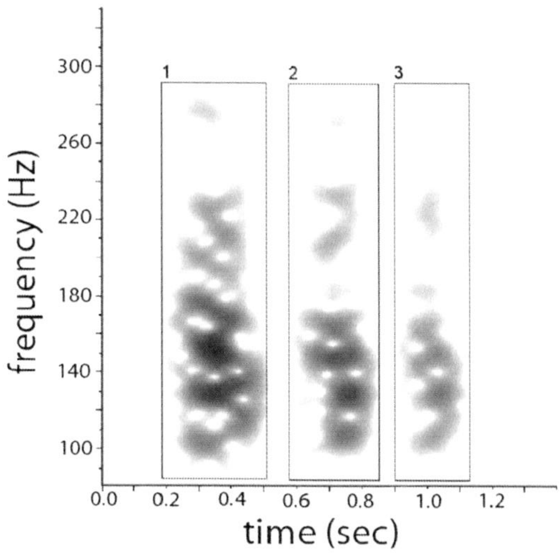

Fig. 1 An example of a rumble bout consisting of 3 rumbles. Spectrogram settings: discrete Fourier transform (DFT) = 8192; Hann window: 0.0512 s; 3-dB filter bandwidth at 5.62-Hz resolution

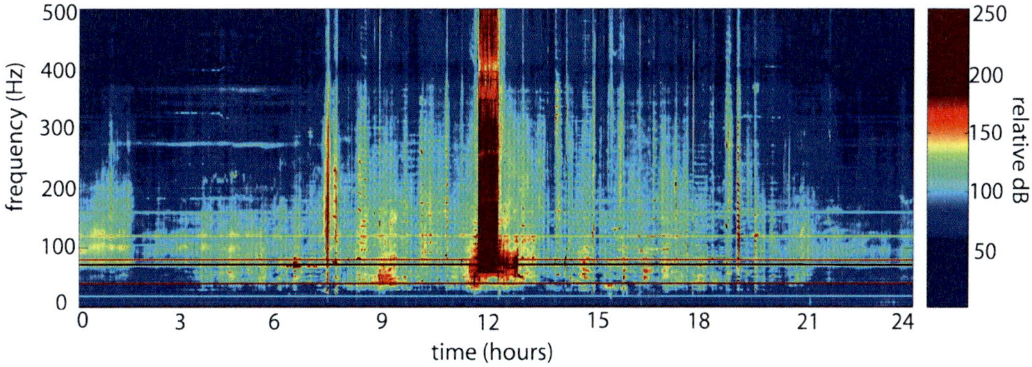

Fig. 2 A typical 24-h spectrogram from the autonomous recording unit showed substantial variation in noise levels across each day. The continuous lines at 50, 75, 125, and 160 Hz are an artifact of the recorder's hard drive. The intense (red) bands of broadband energy, especially between ~0700–1900, are a result of vessel noise. Spectrogram settings: DFT = 819; Hann window = 0.435 s

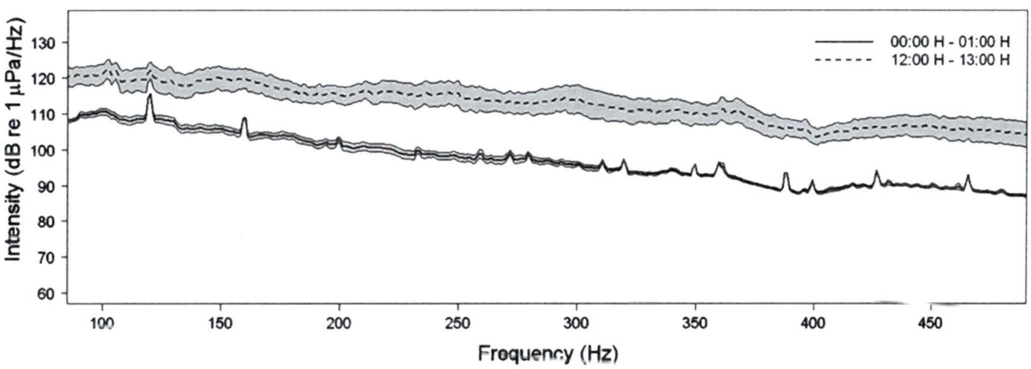

Fig. 3 Power spectra for the 8-day period comparing the distribution of acoustic energy at midnight (0000–0100) and midday (1200–1300). Shaded regions are SE. Peaks at 125 and 160 Hz are artifacts of the recorder's hard drive. Daytime periods were significantly louder than nighttime periods due to vessel traffic

3 Conclusions

We found an active acoustic scene in the benthos off the coast of California, an area that was previously unexplored acoustically. Furthermore, we demonstrated that the sounds produced by *H. californiensis* are highly variable; different individuals produce rumbles that differ in dominant frequency and number of rumbles per bout. Our recordings took place during the early part of the mating season when males are highly competitive for burrow space and actively attempt to recruit females into their burrows to mate (Basch and Engle 1993; J. Engle, personal communication). It is possible that the rumble plays a role in establishing territories and/or attracting potential mates.

We also observed that *H. californiensis* species spends a large portion of its day producing sound, which highlights the potentially important contribution of the rumble to this species' behavioral ecology. One interesting finding was that during crepuscular periods, when *H. californiensis* is typically found guarding its burrow entrance (Basch and Engle 1993), rumbles were loud and produced in rhythmic sequences. During times when the burrow is typically closed for protection from predators

(Basch and Engle 1993), we observed very few rumbles or rumbles that were lower in frequency and relatively less intense. This may indicate that sound production continues even when the burrow is closed. This initial exploratory study reveals a system that is rich with future questions and discovery, including the central question: what is the function of the rumble?

The waters near Santa Catalina Island are frequented by small boats and large shipping vessels, which collectively produce a tremendous amount of acoustic energy in the communication band of *H. californiensis* (i.e., 100–500 Hz). During periods of intense vessel noise, we could not resolve whether mantis shrimp ceased rumbling or whether complete acoustic masking was taking place. However, given the fact that vessel noise was detectable during a large portion of the day, we suspect that acoustic masking is a frequent phenomenon in this habitat. Given the deleterious effects of acoustic masking in other taxa (e.g., Clark et al. 2009; Nowacek et al. 2007; Popper and Hastings 2009), this omnipresent and acoustically overlapping vessel noise may substantially impact the acoustic ecology of the California mantis shrimp. Marine invertebrates should be included in future studies and consideration of the effects of anthropogenic noise on aquatic animals.

References

Basch LV, Engle JM (1993) Biogeography of *Hemisquilla ensigera californiensis* (Crustacea: Stomatopoda) with emphasis on Southern California bight populations. In Hochberg FG (ed) Third California Islands Symposium, Santa Barbara Museum of Natural History, Santa Barbara, CA, pp 211–220.

Busnel RG (1963) Acoustic behavior of animals. Elsevier Publishing Company, New York, NY.

Clark CW, Clapham PJ (2004) Acoustic monitoring on a humpback whale (*Megaptera novaeangliae*) feeding ground shows continual singing into late spring. Proc Biol Sci 271:1051–1057.

Clark CW, Ellison WT, Southall BL, Hatch L, Van Parijs SM, Frankel A, Ponirakis D (2009) Acoustic masking in marine ecosystems: Intuitions, analysis, and implication. Mar Ecol Prog Ser 395:201–222.

Haderlie EC, Abbott DP, Caldwell RL (1980) Three other crustaceans: a copepod, a leptostracan, and a stomatopod. In: Morriss RH, Abbott DP, Haderlie EC (eds) Intertidal invertebrates of California. Stanford University Press, Palo Alto, CA, pp 631–635.

Nowacek DP, Thorne LH, Johnston DW, Tyack PL (2007) Responses of cetaceans to anthropogenic noise. Mamm Rev 37:81–115.

Patek SN, Caldwell RL (2006) The stomatopod rumble: Sound production in *Hemisquilla californiensis*. Mar Freshw Behav Physiol 39:99–111.

Popper AN, Hastings MC (2009) The effects of human-generated sound on fish. Integr Zool 4:43–52.

Popper AN, Salmon M, Horch KW (2001) Acoustic detection and communication by decapod crustaceans. J Comp Physiol A 87:83–89.

Influence of Turbidity on the Incidence of Sound Production in Atlantic Croaker (*Micropogonias undulatus*) in Pamlico Sound, North Carolina

Cecilia S. Krahforst, John P. Walsh, Mark W. Sprague, Devon O. Eulie, D. Reide Corbett, and Joseph J. Luczkovich

1 Introduction

Increased sound production by fishes, which is used for communication during mating, in territorial defense, and possibly in echolocation, has been associated with decreased light and increased temperature and salinity (Luczkovich et al. 2008; Mok and Gilmore 1983). There has not been an attempt to associate changes in sound production with other environmental factors such as turbidity. Sediment deposition and resuspension commonly occur in estuaries due to changes in current velocity and direction, water runoff, and wave height. These factors can lead to shearing on the bed surface and thus an overall increase in water column turbidity (Whitehouse et al. 2000). It has been hypothesized that increased water column turbidity will lead to increased sound production in fishes because visual cues will be impaired. The goal of this research is to associate the incidence of sound production by *Micropogonias undulatus* (Atlantic croaker) to variations in estuarine temperature, salinity, dissolved oxygen, and particularly turbidity.

C.S. Krahforst (✉)
Department of Biology, East Carolina University, Greenville, NC 27858, USA

Coastal Resource Management PhD Program, East Carolina University, Greenville, NC 27858, USA
e-mail: csk1021@ecu.edu

J.P. Walsh
Department of Geological Sciences, East Carolina University, Greenville, NC 27858, USA

Institute for Coastal Science and Policy, East Carolina University, Greenville, NC 27858, USA

Renaissance Computing Institute, East Carolina University, Greenville, NC 27858, USA

M.W. Sprague
Department of Physics, East Carolina University, Greenville, NC 27858, USA

D.O. Eulie
Coastal Resource Management PhD Program, East Carolina University, Greenville, NC 27858, USA

Department of Geological Sciences, East Carolina University, Greenville, NC 27858, USA

D.R. Corbett
Department of Geological Sciences, East Carolina University, Greenville, NC 27858, USA

Institute for Coastal Science and Policy, East Carolina University, Greenville, NC 27858, USA

J.J. Luczkovich
Department of Biology, East Carolina University, Greenville, NC 27858, USA

Institute for Coastal Science and Policy, East Carolina University, Greenville, NC 27858, USA

A.N. Popper and A. Hawkins (eds.), *The Effects of Noise on Aquatic Life*,
Advances in Experimental Medicine and Biology 730, DOI 10.1007/978-1-4419-7311-5_38,

2 Materials and Methods

Water quality and sound production by *M. undulatus* were monitored at one site (PCS) in Pamlico River, NC, from March through December 2008. A long-term acoustic recording system (LARS), which recorded 10-s wave files every 10–15 min at frequencies < 10 kHz, was used to monitor fish calls. Environmental conditions were logged one to three times per hour with a Hydrolab DS5X (to monitor temperature, salinity, and dissolved oxygen) and an optical backscatter sensor (D&A OBS-3, 8-Hz sampling rate; to monitor turbidity). All LARS files were analyzed with extensible bioacoustics tools (http://www.XBAT.org) developed for MATLAB (version 7.0.1). A spectrogram detector with a minimum correlation rate of 54%, using known *M. undulatus* sounds, identified sound recordings of this species throughout the recording period. Any sounds with a correlation below 60% were analyzed by hand. Sounds that were not made by *M. undulatus* were removed from further analyses. *Micropogonias undulatus* sounds were analyzed using the number of croaks per hour and normalized by the number of recorded seconds in a given sampling hour. Sounds were averaged over a day to obtain the mean number of sounds per recorded hour per day. Correlations with water quality parameters were produced.

3 Results

At PCS, *M. undulatus* sound production was greatest in October and was highly correlated to water quality. Sound production was positively correlated to temperature (0.72; $P < 0.001$) and turbidity (0.61; $P < 0.001$), whereas salinity was negatively correlated with *M. undulatus* sound production (−0.65; $P < 0.001$) (Fig. 1). Dissolved oxygen levels did not significantly influence sound production (0.30; $P = 0.114$) at PCS.

4 Discussion

Sound production by *M. undulatus* is related to a variation in water quality parameters. Luczkovich et al. (2008) found that salinity, temperature, depth, and location in the estuary are contributing factors for four other sciaenid sound-producing fishes; however, sound production intensity, as measured by a qualitative index, was dependent on species and time of year. Luczkovich et al. (2008) did not look at *M. undulatus* sound production nor did they assess turbidity levels. In this study, we found that the most influential factors in *M. undulatus* sound production were temperature, turbidity, and salinity. Temperature and turbidity were positively correlated with sound production, whereas salinity was negatively correlated with *M. undulatus* sound production. The relationship with temperature and sound production was expected because other sciaenids have been shown to become reproductively active as the temperature increases and dissolved oxygen remains above a threshold (Luczkovich et al. 2008). However, the increased sound production in high turbidity has never been documented and suggests that sound cues may become more useful to Atlantic croaker when light levels are diminished by sediment resuspension events (e.g., storms and waves). Alternatively, Atlantic croaker activity levels may increase for other reasons and cause resuspension of sediments due to their feeding activities. Because temperature and salinity changes, low dissolved oxygen (hypoxia), and high-turbidity events may occur only for short periods of time during the passage of weather systems and stratification of the water column, a continuous recording system of fish sounds and environmental parameters was required in this study. Hurricanes cause massive sediment resuspension (Goni et al. 2007), yet the passage of hurricanes has not been found to

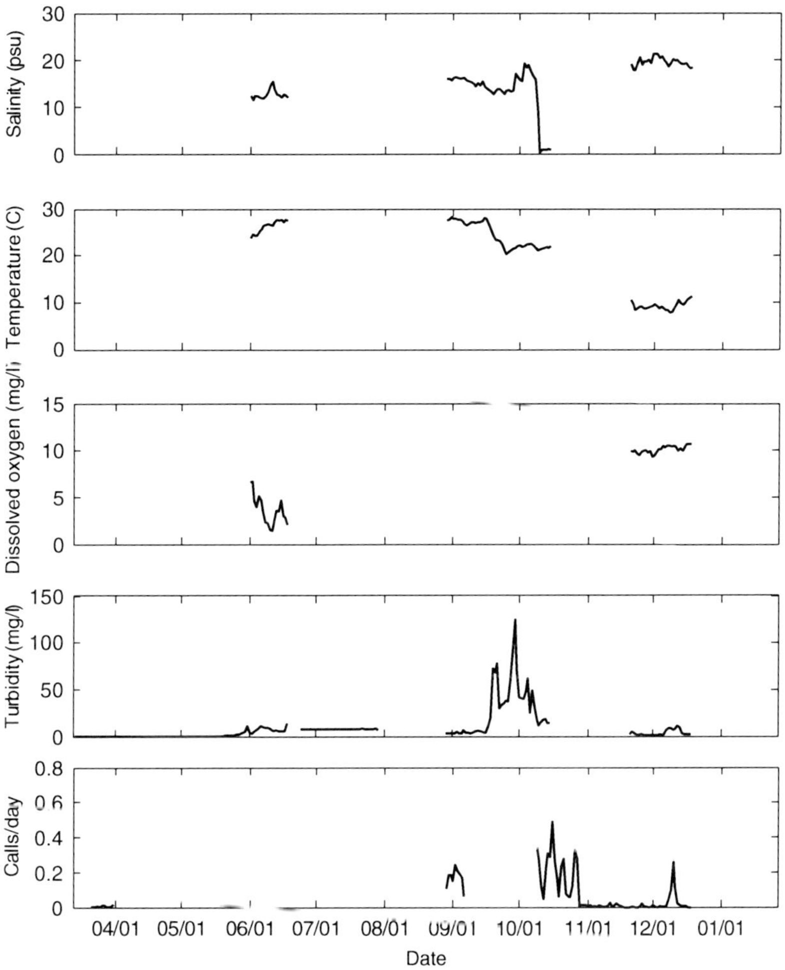

Fig. 1 Salinity (practical salinity units [psu]), temperature, dissolved oxygen, turbidity, and sound production by the Atlantic croaker (mean number of calls per recorded hour per day) for the PCS site in Pamlico River, NC, from March through December 2008. Breaks indicate periods of instrumentation failure

influence the sound production of spawning sand sea trout (Locascio and Mann 2005). This suggests that not all scaienids react similarly to changes in their environment. It is evident that *M. undulatus* sound production is tied to turbid events within the estuary.

References

Goni M, Alleau Y, Corbett R, Walsh JP, Mallinson D, Allison MA, Gordon E, Petsch S, Dellapenna TM (2007) The effects of hurricanes Katrina and Rita on the seabed of the Louisiana shelf. Sediment Rec 5:4–9.

Locascio JV, Mann DA (2005) Effects of Hurricane Charley on fish chorusing. Biol Lett 1:362–365.

Luczkovich JJ, Pullinger RC, Johnson SE, Sprague MW (2008) Identifying the critical spawning habitats of sciaenids using passive acoustics. Trans Am Fish Soc 137:576–605.

Mok HK, Gilmore RG (1983) Analysis of sound production in estuarine aggregations of *Pogonias cromis, Bairdiella chrysoura*, and *Cynoscion nebulosus* (Sciaenidae). Bull Inst Zool Acad Sin (Taipei) 22:157–186.

Whitehouse RJS, Soulsby R, Roberts W, Mitchener HJ (2000) Dynamics of estuarine muds. Thomas Telford Publishing, London.

Propagation of Lusitanian Toadfish Sounds in Estuarine Shallow Waters

Andreia Ramos, M. Clara P. Amorim, and Paulo J. Fonseca

1 Introduction

Many aquatic organisms, including teleost fishes, rely on acoustic signals for communication (Bass and Clark 2003). Open-water animals can communicate with low-frequency sounds over kilometers due to reduced thermodynamic absorption (down to 0.001 dB/km) and increased sound velocity (5 times higher) compared with terrestrial habitats (Rogers and Cox 1988). Many coastal teleost fish, however, face increased sound propagation constraints due to strong attenuation of their low-frequency communication sounds in shallow water (Bass and Clark 2003). In addition, interaction of the sound waves at boundary layers affects propagation, and, typically, a signal produced near the bottom or at the surface will not propagate as far as one produced in midwater (Fine and Lenhardt 1983). Indeed, bottom type and water depth limit which frequencies propagate, and low frequencies may be strongly affected in shallow water if they are below the cutoff frequency (Rogers and Cox 1988). For example, with a sandy bottom, a sound with a frequency of 100 Hz will only efficiently propagate in a water depth above 10 m, the theoretical cutoff frequency (Rogers and Cox 1988). Consequently, only when the sound frequency is above the cutoff frequency is its spreading loss the major source of fish sound attenuation (Mann 2006). Thus to understand the acoustic communication system of social fish, it is important to determine how sounds propagate in their natural habitat because it will constrain the communication range and the maintenance of signal quality as it spreads away from the source.

A. Ramos (✉) • P.J. Fonseca
Centro de Biologia Ambiental, Departamento de Biologia Animal, Faculdade de Ciências da Universidade de Lisboa, 1749-016 Lisbon, Portugal
e-mail: Andreia_ramos@hotmail.com

M.C.P. Amorim
Instituto Superior de Psicologia Aplicada, Unidade de Investigação em Eco-Etologia, 1149-041 Lisbon, Portugal

A.N. Popper and A. Hawkins (eds.), *The Effects of Noise on Aquatic Life*, Advances in Experimental Medicine and Biology 730, DOI 10.1007/978-1-4419-7311-5_39,

2 Study Species

Male Lusitanian toadfish, *Halobatrachus didactylus* (Batrachoididae), produce long and tonal advertising calls (boat whistles) in shallow waters to attract females at distance during the breeding season (Amorim et al. 2006). The sounds are produced by vibration of the swim bladder that is produced by a pair of embedded sonic muscles (Amorim and Vasconcelos 2008) that determine the fundamental frequency of boat whistles at ~50–60 Hz. Sound energy is restricted to low frequencies, with the dominant frequency usually corresponding either to the first or to the fundamental harmonic (Amorim and Vasconcelos 2008).

Previous studies suggested that the sound features of this call, such as the fundamental frequency and harmonic structure, are distorted with distance.

3 Our Study

We tested the propagation of boat whistles and low-frequency sound sweeps at different water depths by playing back these sounds in a natural toadfish breeding area in the Tagus estuary. Moreover, we broadcast the boat whistles of different males and measured the propagation loss at different water depths to estimate the distance over which a female may distinguish an individual male's call.

Finally, by comparing the properties of the sounds recorded at different distances and water depths with the measured hearing sensitivity of the toadfish, an estimation of the possible acoustic communication range is obtained.

4 Preliminary Results

Our preliminary results showed that attenuation of boat whistles, especially the one with a 60-Hz dominant frequency, increased considerably with the decrement of water depth, in particular below 2.5 m (Fig. 1). In higher water depths, the acoustic signals propagated better up to a 30-m distance (to the sound speaker), whereas in more shallow waters, played-back sounds were no longer recognizable much beyond 8 m. Previous observations indicated that the male toadfish calling rate decreases in lower tides, which might be an adaptive behavior due to this strong sound attenuation. Ongoing research includes a more detailed analysis of propagation loss of toadfish communication signals in lower water levels.

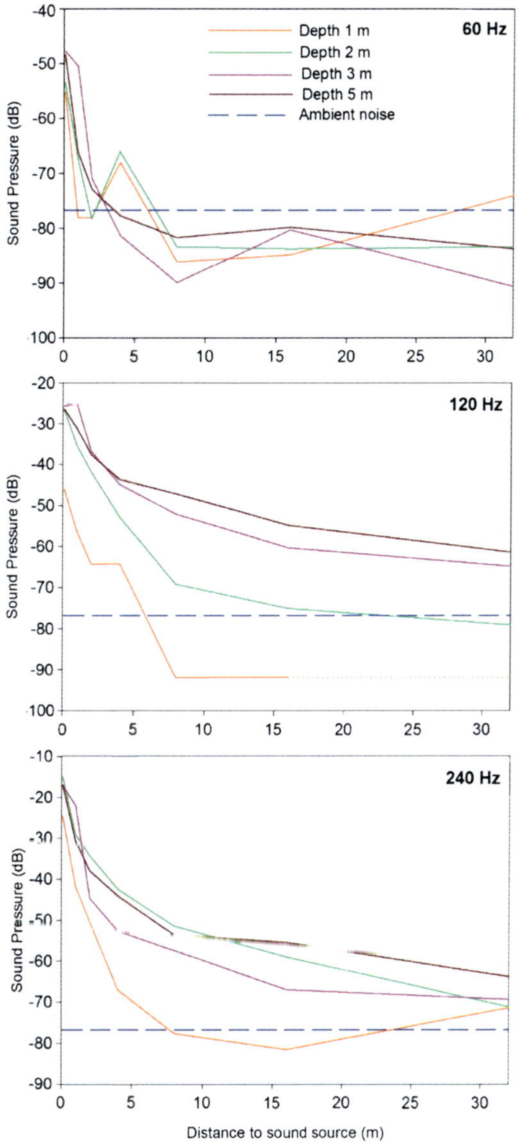

Fig. 1 Propagation loss registered for boat whistles with different dominant frequencies (60, 120, and 240 Hz). Sound pressure levels (relative amplitudes) were measured at different distances to the underwater speaker at different depths

References

Amorim MCP, Vasconcelos RO (2008). Variability in the mating calls of the Lusitanian toadfish *Halobatrachus didactylus*: Potential cues for individual recognition. J Fish Biol 73:1267–1283.

Amorim MCP, Vasconcelos RO, Marques JF, Almada F (2006). Seasonal variation of sound production in the Lusitanian toadfish, *Halobatrachus didactylus*. J Fish Biol 69:1892–1899.

Bass AH, Clark CW (2003) The physical acoustics of underwater sound communication. In: Simmons A, Fay RR, Popper AN (eds) Acoustic communication. Springer-Verlag, New York, pp 15–64.

Fine ML, Lenhardt ML (1983) Shallow-water propagation of the toadfish mating call. Comp Biochem Physiol 76:225–231.

Mann DA (2006) Propagation of fish sounds. In: Ladich F, Collin SP, Moller P, Kapoor BG (eds) Communication in fishes, vol 1. Science Publishers, Enfield, NH, pp 107–120.

Rogers PH, Cox H (1988) Underwater sound as a biological stimulus. In: Atema J, Fay RR, Popper AN, Tavolga WN (eds) Sensory biology of aquatic animals. Springer-Verlag, New York, pp 131–149.

Sound Production in Some Physostomous Fish Species and Effects of Biological Sounds on Fish

Michail Y. Kuznetsov and Yury A. Kuznetsov

1 Introduction

One of the basic channels of communication and orientation of fish in an environment is the sound channel. A highly developed system of acoustic reception by fishes and the sound production and acoustical signaling abilities of many sea hydrobionts represent this channel as the most perspective for the control of behavior of fishes from positions, distant opportunities, and multipurpose of influence. Other channels of communication and the corresponding physical fields (light and electric) have much smaller distant opportunities and serve in water, basically, for near orientation of hydrobionts.

There are two types of impact of a sound field on fishes: energetic and informational (Protasov 1978). The sound field causes painful or other unpleasant sensations in fishes at energetic impact and they aspire to go away from a dangerous zone. The reaction is carried out under the pattern of the existing stereotype of the response determined by natural programs of behavior according to the biological contents and structure of a signal at information impact.

2 Sound Production in Fish

The acoustic behavior of the five species of Far East physostomous fishes *Oncorhynchus keta* (chum salmon), *Oncorhynchus gorbuscha* (pink salmon), *Clupea harengus pallasii* (pacific herring), *Sardinops sagax melanosticta* (Japanese sardine), and *Engraulis japonicus* (Japanese anchovy) has been investigated.

M.Y. Kuznetsov (✉)
Pacific Scientific Research Fisheries Center, Vladivostok, 690950 Russia
e-mail: kuznetsovm@tinro.ru

Y.A. Kuznetsov
Far East State Technical Institute of Fishery and Fish Industry, Vladivostok, 690950 Russia
e-mail: imf.dalrybvtuz@mail.ru

A.N. Popper and A. Hawkins (eds.), *The Effects of Noise on Aquatic Life*,
Advances in Experimental Medicine and Biology 730, DOI 10.1007/978-1-4419-7311-5_40,

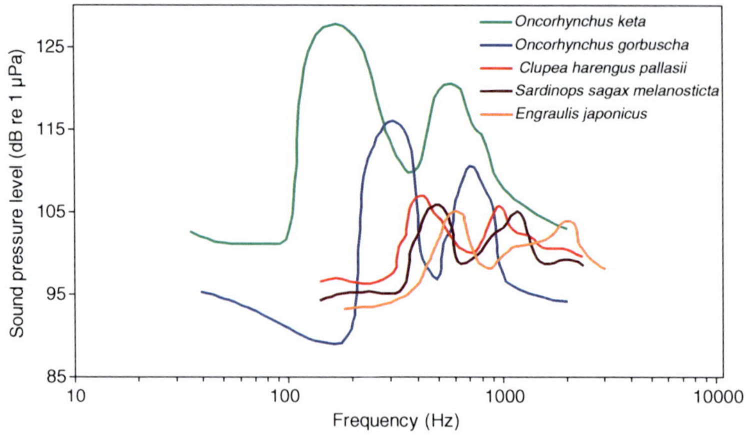

Fig. 1 Generalized spectral-energy characteristics of signals of fishes. Modified from Kuznetsov (2009)

2.1 Daily Rhythm of Acoustic Activity of Fish

All these species have high acoustic activity. The daily distribution of sounds shows certain cyclicity. The maximum number of sounds by salmon was recorded at twilight. There were two peaks of acoustic activity of salmon: in the evening from 10 p.m. to 2 a.m., with some alleviation at midnight, and at daybreak from 6 to 10 a.m. *Clupea harengus pallasii*, *Sardinops sagax melanosticta*, and *Engraulis japonicus* have maximal acoustic activity also at night. In contrast to Salmonidae, they have a monophasic type of diurnal distribution of sounds. The quantity of emitted sounds increases at twilight. Until midnight, the acoustic activity reaches a maximum and then it slowly decreases, with a slight increase in the morning. The amplification of function of an acoustic channel in communication and orientation between individuals in a school in dark time is assumed because of difficult visual contacts.

2.2 Sounds of Fish

The generalized spectral-energy characteristics of biosignal processing are shown in Figure 1. The spectrum of most frequent meeting sounds of physostomous fishes is concentrated mainly in two certain frequency subranges, different for each species of fish. Spectral-energy distribution and the level of the signals depend on the size (species) of fish: the smaller the object, the higher the frequency and the lower the amplitude of a signal. Characteristic attributes of sounds by fishes are the pulse structure of signals with a smooth increase and an exponential decrease in the amplitude in pulses, the deviation in frequency of a spectral maximum in certain zones of a spectrum from the beginning to the end of a pulse, the high variability in duration (0.25-3.0 s), and the amplitude modulation of the signals.

2.3 Sound Production Mechanism

We have confirmed that the sound-producing organs of physostomous fishes are the swim bladder and pneumatic duct with the muscular sphincter (Kuznetsov 2009). Specific distribution of the spectral energy of signals in two frequency subranges (Fig. 1) is related to the simultaneous participation of these organs in the sound production by fish. Experiments with swim bladders of various physostomous fishes have shown that the increase in the sound pressure level in the first frequency subrange is produced by effective pulsations of the swim bladder at the frequency corresponding to its resonance size and to the pressure of air in it. A rise in the level of signals in the second frequency subrange of the spectrum is created by vibrations of the pneumatic duct walls. Deviation in the spectral maximum within the subranges of the spectrum, typical of most signals from fish, depends on flow rate in the pneumatic duct and air pressure in the swim bladder (and thus its volume), regulated in a random way by the force of compression (relaxation) of the swim bladder walls and the locking muscles of the sphincter during radiation.

3 Effects of Biological Sounds on Fish

The second purpose of this work is an estimation of the reactions of fish to acoustic signals of transducers, imitating sounds of fish. The radiation of biosignals causes the reinforcement of moving and acoustic activity of feeding *Clupea harengus pallasii* and *Sardinops sagax melanosticta* in a cage at distances to 100 m (confidence level > 0.05) and tracks fish moving to the source of the sound. The average location of the fish characterizes the preferred value of their stay near the transducer zones of the cage.

Studies of the impact of a fish sound simulator on the behavior of prespawning *Oncorhynchus keta* in natural conditions of the habitat on the catch of the checking nets were performed. The nets were installed 90° from the main paths of fish migration in the bay (Fig. 2). The estimates of the

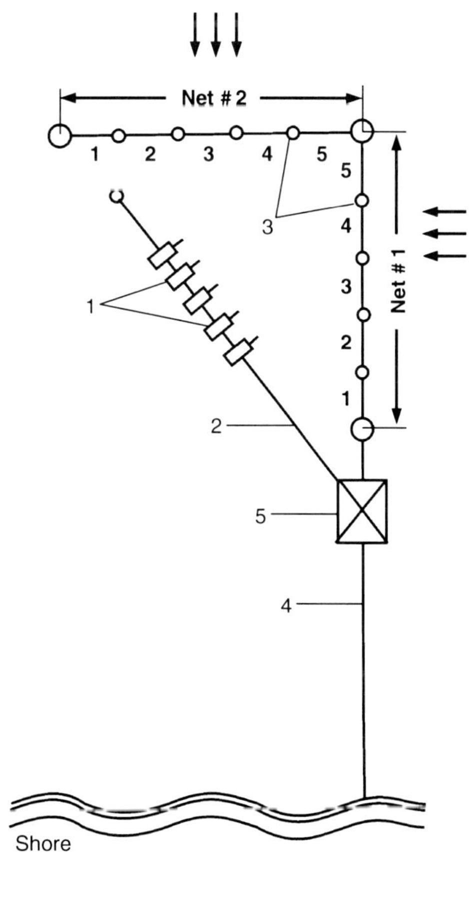

Fig. 2 Schematic of the experimental installation on the range. 1, Acoustic transmitters; 2, hose; 3, control nets with buoys; 4, cable; 5, air power supply

Table 1 Distribution of catches in control nets

					Stimulus		Background	
	N_S	N_B	N_S/N_B	P	$N_{4\text{-}5}/N_{1\text{-}3}$	P	$N_{1\text{-}3}/N_{4\text{-}5}$	P
Net #1	20	15	1.33	>0.05	4.0	0.01	1.2	0.02
Net #2	22	3	7.33	0.001	2.7	0.05	1.8	0.05
Total	42	18	2.33	0.01	3.2	0.01	1.5	0.05

N_S, number of fish at acoustic stimulus; N_B, number of fish under background conditions; $N_{4\text{-}5}$ and $N_{1\text{-}3}$, number of fish in nets in zones 4-5 and 1-3, respectively

net catches under background conditions and under impact of the acoustic stimulus are presented in Table 1. The catch ratio in net #2 proves the attracting influence of signal simulators on passing fish ($P < 0.001$). Fish approach from the sea side and, accordingly, catches in net #2 were only accidental under background conditions. The catches sharply increased under the impact of the acoustic stimulus in zones 4–5 of the nets (in the area of the directed radiation) compared with those in zones 1–3 (Table 1). The comparative estimate of the moving reaction of fish in the open basin shows the signal source localization capability of the fish and the presence of a directed reaction accompanying the attraction of fish from the main path of migration to the source of the sound.

References

Kuznetsov MY (2009) Traits of acoustic signalization and generation of sounds by some schooling physostomous fish. J Acoust Phys 55:866–875.
Protasov VR (1978) Behaviour of fishes. Food Industry Press, Moscow (in Russian).

Is Biological Sound Production Important in the Deep Sea?

Rodney A. Rountree, Francis Juanes, Clifford A. Goudey, and Kenneth E. Ekstrom

1 Introduction

Over 50 years ago, N. B. Marshall of the British Museum of Natural History hypothesized, based on anatomical studies, that biological sound production by fishes is widespread in the deep sea (Marshall 1954, 1967). Since that time, only a handful of studies have reported observations of possible sound production by deep-sea fishes based on acoustic recordings (Cato 1978; Griffin 1955; Kelly et al. 1985; Mann and Jarvis 2004; McDonald et al. 2006), but no direct attempts to use acoustics to address Marshall's hypothesis have yet been published. The failure of scientists to investigate the ecological importance of deep-sea sound production has resulted partially from a lack of adequate low-cost technology (Rountree 2008; Rountree et al. 2006). As the first step to begin to address Marshall's hypothesis, we developed a simple deep-water autonomous underwater listening system (DAULS) for use by ships of opportunity, funded by an MIT Sea Grant College Program seed grant and in collaboration with commercial fishermen.

2 Methods

The housing for the DAULS was adapted from a deep-water trawl float (*Panther Plast A/S*, Vordingborg, Denmark) with an 1,800-m working depth that was split in half and fitted with an alignment groove and O-ring. The DAULS housing contained a Nomad Jukebox (Creative Labs, Inc., Milpitas, CA) digital recorder programmed to record continuously at an 11-kHz sampling rate

R.A. Rountree (✉)
Marine Ecology and Technology Applications, Inc, East Falmouth, MA 02536, USA
e-mail: rrountree@fishecology.org

F. Juanes
Department of Environmental Conservation, University of Massachusetts Amherst, Amherst, MA 01003, USA
email: juanes@nrc.umass.edu

C.A. Goudey
C.A. Goudey and Associates, Newburyport, MA 01950, USA
e-mail: cliffgoudey@gmail.com

K.E. Ekstrom
Laboratory for Acoustic Analysis, Cambridge, MA 02142, USA
e-mail: ken.ekstrom.email@gmail.com

A.N. Popper and A. Hawkins (eds.), *The Effects of Noise on Aquatic Life*,
Advances in Experimental Medicine and Biology 730, DOI 10.1007/978-1-4419-7311-5_41,

for up to 60 h. The DAULS included two HTI-96-MIN hydrophones (High Tech Industries, Gulfport, MS; sensitivity −165 dB re 1 V/μPa; frequency response 2 Hz to 30 kHz) mounted 1 m apart. Fishermen deployed the DAULS inside a commercial deep-sea crab trap within a string of 100 traps during normal fishing operations along the shelf break and retrieved it when the traps were hauled after ~24 h. The spectral density of selected frequencies was determined at 15-min intervals over the deployment period using the method described by Locascio and Mann (2008). This method is particularly useful for detecting chorus sounds. In addition, the entire recording was monitored by an observer to detect potential fish and other biological sounds.

3 Results and Discussion

We successfully obtained a 24-h recording on 18 October 2005 from the seafloor in 682 m of water in Welkers Canyon located south of Georges Bank (latitude N 40.0883, longitude W 68.53167). The recording contained a wealth of biological sounds including sounds from fin whales (*Balaenoptera physalus*), humpback whales (*Megaptera novaeangliae*), pilot whales (*Glopicephala* spp.), and dolphins (Delphinidae) as well as frequent examples of at least 12 unique unidentified sounds that we attribute to either undescribed cetacean or fish sounds. An examination of spectral density peaks within selected frequency bands did not detect chorus activity (Fig. 1, top). Peaks in low- to

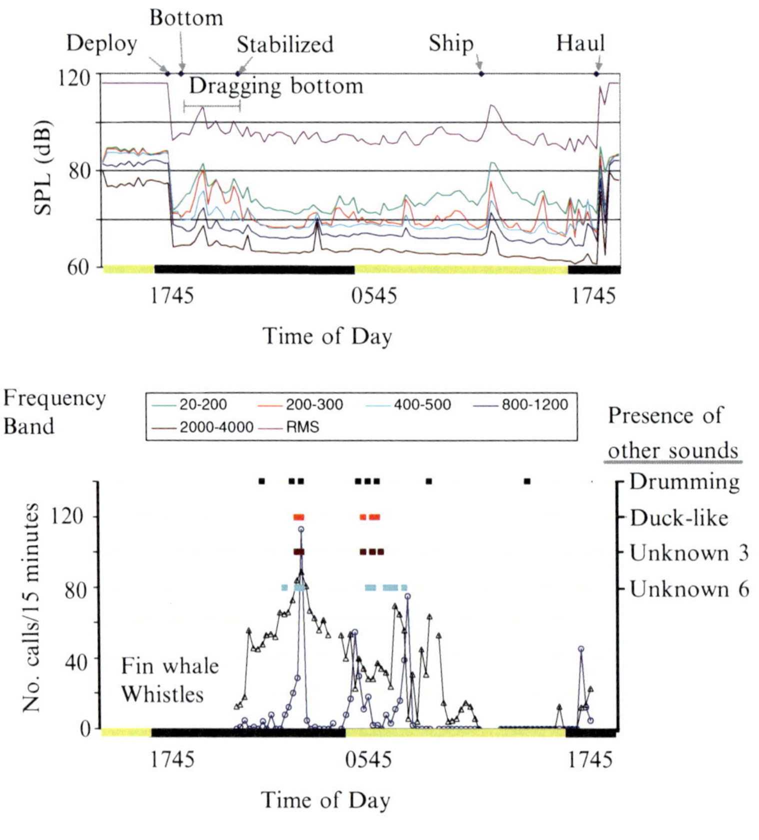

Fig. 1 Temporal pattern of spectral density of selected frequency bands (top) and number presence of selected sounds (bottom). SPL, sound pressure level; RMS, root mean square

midfrequencies resulted from periodic movement of the traps (dragging noise), whereas peaks in higher frequency bands resulted from passing ships. Ships that passed nearby produced strong peaks at all frequencies and completely masked biological sounds. Temporal patterns in received sound levels (Fig. 1, top) provided unexpected data on fishing operations. The DAULS took ~1 h to reach the seafloor and then was dragged along the bottom for another 4-5 h before coming to a stop, followed by periodic dragging throughout the recording, thus contradicting fishermen's assumptions of little trap movement after placement. Although chorus activity was not detected, biological sounds exhibited strong temporal patterns (Fig. 1, bottom). Low-frequency (<30 Hz) fin whale and high-frequency delphinid (3-5 kHz) sounds dominated the recording and peaked at night. Unidentified calls could not be enumerated but were present throughout the recordings, except when masked by ship or dragging noises (e.g., gap just before 0500h). Most individual biological sound types were infrequent, but several distinct sounds labeled "drumming," "ducklike," "unknown 3," and "unknown 6" occurred throughout the night and early morning hours (Fig. 1, bottom). Most unidentified sounds that we attributed to biological sources had fundamental frequencies below 1,200 Hz, well within the range expected for fishes. However, at this time, we have insufficient data to distinguish between fish sounds and not described low-frequency cetacean sounds. In addition, we noted that most of these sounds exhibited very low received source levels and were often difficult to detect above the background noise. Samples of the most common unidentified sounds together with their spectrograms can be obtained at http://www.fishecology.org.

4 Conclusions

More than 50 years after Marshall hypothesized the importance of sound production in deep-sea fishes, we still have little understanding of the importance of fish sounds and the undersea soundscape in the ecology of the deep sea. However, our observations suggest that sound production by deep sea fishes is likely to be important. Furthermore, if deep-sea sounds tend to be low amplitude, as suggested here, then continued increases in anthropogenic noise may be particularly problematic.

References

Cato DH (1978) Marine biological choruses observed in tropical waters near Australia. J Acoust Soc Am 64:736–743.

Griffin DR (1955) Hearing and acoustic orientation in marine animals. Deep-Sea Res 3 (Suppl):406–417.

Kelly L, Kewley DJ, Burgess AS (1985) A biological chorus in deep water northwest of Australia. J Acoust Soc Am 77:508–511.

Locascio JV, Mann DA (2008) Diel periodicity of fish sound production in Charlotte Harbor, Florida. Trans Am Fish Soc 137:606–615.

Mann DA, Jarvis SM (2004) Potential sound production by a deep-sea fish. J Acoust Soc Am 115:2231–2233.

Marshall NB (1954) Aspects of deep sea biology. Philosophical Library, New York.

Marshall NB (1967) Sound-producing mechanisms and the biology of deep-sea fishes. In: Tavolga WN (ed) Marine bio-acoustics, vol 2. Pergamon Press, Oxford, UK, pp 123–133.

McDonald MA, Hildebrand JA, Wiggins SM (2006) Increases in deep ocean ambient noise in the Northeast Pacific west of San Nicolas Island, California. J Acoust Soc Am 120:711–718.

Rountree RA (2008) Do you hear what I hear? Future technological development and needs in passive acoustics underwater observation. Mar Tech Rep 51:40–46.

Rountree RA, Gilmore RG, Goudey CA, Hawkins AD, Luczkovich J, Mann DA (2006) Listening to fish: Applications of passive acoustics to fisheries science. Fisheries 31:433–446.

Unusual and Unexpected Biological Noisemakers in the Irish Sea and St. George's Channel

Rodney Coates, Gillian A. Coates, Rachel Porter, Andrew Davies, and Dylan Evans

1 Introduction

This paper might perhaps be described as the prologue to a detective story that will reach its denouement some years in the future. The authors are aware of the presence of click-making animals in our local marine environment. They sound like snapping shrimp but probably are not. We have alternative suspects and we have research in progress to help unravel the mystery.

2 Background

During the summer of 2003, Seiche Ltd. had occasion to make underwater recordings of the noise signatures of commercial ferries entering and leaving Holyhead Harbour, Anglesey, UK. While recording the signature of the ferry Stena Forwarder, it was noted that each record was contaminated by what sounded like the characteristic noise produced by snapping shrimp. As the closest approach was attained, noise from the vessel completely swamped the "snapping shrimp noise." The recordings were made with a pair of hydrophones deployed at 3-m water depth from the extreme outer end of the Holyhead breakwater.

Although snapping shrimps are ubiquitous in warm-temperate and tropical seas, the northern limit of their range is usually considered to be the Channel Islands. It is not beyond possibility that snapping shrimp may have been introduced to Holyhead Harbour by a foreign fishing vessel flushing ballast water. The presence of clicking sounds having been observed, however, we attempted to investigate further. A careful underwater visual search of the seafloor at the end of the breakwater revealed no snapping shrimp. This was not particularly surprising because these crustacea, living as they do at the end of mud burrows that may be several meters in length, are hard to uncover even in

R. Coates (✉) • G.A. Coates
Seiche Ltd., Swallow Court, Trefor, Holyhead, Anglesey LL65 4TA, UK
e-mail: rcoates@seiche.com

R. Porter • A. Davies
School of Marine Science, University of Wales, Bangor, Menai Bridge, Anglesey LL59 5AB, UK

D. Evans
Anglesey Sea Zoo, Brynsiencyn, Ynys Mon LL61 6TQ, UK

A.N. Popper and A. Hawkins (eds.), *The Effects of Noise on Aquatic Life*,
Advances in Experimental Medicine and Biology 730, DOI 10.1007/978-1-4419-7311-5_42,

tropical regions where they are relatively common. In the years that followed the 2003 observations, Seiche Ltd. has maintained at Holyhead Harbour a yacht that we have used to tow hydrophone arrays within and beyond the harbor. We find click makers along the breakwater but only toward its extreme end. We find their habitat to extend perhaps 50 m outward into the harbor, which appears to be the limit at which their noise can be detected. They can also be heard on the inshore rock reefs in Holyhead Bay and beyond. Most regions of the harbor and the local bays are, however, devoid of them. Consequently, we believe that there is a need for longer term acoustic observations and visual surveys.

3 The Nature of Marine Animal Clicks

From an engineering viewpoint, a click is an impulsive time-domain waveform. It is a central tenet of Fourier methods that the narrower a temporal function, the broader is its spectrum in the frequency domain. The snapping shrimp "click" is not, as was long believed, produced by a "hammer-and-anvil" mechanism. Instead, it is a consequence of the generation and collapse of a cavitation bubble, itself caused by the production of a tiny but rapid jet of water as the snapping claw is triggered to shut. At least one other marine animal also produces a cavitation "click," the mantis prawn (order Stomatopoda). These crustaceans employ their "praying mantis" claws to spear prey. So fast is the stab that the claw spear point trails a cavitation bubble as it moves toward its target. Once again, a sharp "click" is heard as the bubble collapses and a very broadband signature results. Often, the pulse width is described as being as short as, perhaps, 5 ms. This figure is probably more a consequence of the bandwidth of the measurement hydrophone than of an intrinsic property of the bubble collapse.

There are other click-making candidates that do not employ a cavitation bubble-generating mechanism. One large group of click makers are the syngnathid ("fused-jaw") fish. This family includes the sea horses. When a sea horse feeds, it sucks in a microscopic planktonic organism through a very tiny mouth. The sucking action is produced by a rapid flexing of the opercular plates. The flexing generates what, to the human ear, is a distinct click. Because the click is of mechanical origin, its bandwidth must, inevitably, be much smaller than that of the "singularity" of theoretical physics, which results when a cavitation bubble is snapped out of existence by external water pressure. If the bandwidth of the mechanical click is much smaller than that of the cavitation click, its duration must be much longer. So we have, in principle, a test to establish whether the Holyhead click maker is a cavitation bubble maker, and thus possibly a snapping shrimp, or whether it is a mechanical click maker. The test is to measure, using standard procedures, the bandwidth of the click. This is simple to state in principle, yet far from easy to achieve in practice because of the corrupting influences of local reflecting and reverberant surfaces on the click waveform.

Why introduce sea horses into the debate? If snapping shrimp, with their normally tropical habitat, are alien invaders, should not the same be true of sea horses? As it happens, there are native species of sea horsed in UK waters (long snouted, *Hippocampus guttulatus;* short snouted, *Hippocampus hippocampus*). It is true that they are rare in the waters of North Wales (the short snouted is probably not present), but *H. guttulatus* could potentially be a suspect. However, there is another reason for considering the Syngnathidae because the family also includes the pipefishes. Pipefish feed in the same manner as sea horses and they also click when they feed. Although sea horses are rare in Welsh waters, pipefish are very common (great, *Syngnathus acus*; worm, *Nerophis lumbriciformis*; straight nosed, *Nerophis ophidion)*. Their preferred habitat consists of beds of thong weed. Such beds occur around most of the inshore rock reefs of Holyhead Bay.

4 Exotic Species in Welsh Waters

In May 2004, Dr. Ivor Rees (School of Ocean Sciences, University of Wales, Bangor) caught a specimen of *Alpheus glaber*, a red snapping shrimp, during scientific trawling somewhere to the east of the Isle of Man from RV Prince Madog, the research vessel of the school. If there is one snapping shrimp near the Isle of Man, there are likely to be more, and it is by no means impossible that they are to be found around the coast of Anglesey, some 70 km to the south. Nor would it be remarkable if they had been introduced into the harbor as a consequence of trawling or, perhaps more likely, commercial dredging for common whelks (*Buccinum undatum*), which represents a not-inconsiderable local fishery on the Island of Anglesey.

If snapping shrimp are a viable suspect for our click sounds, and it is right that they should be, then so also must the mantis prawn. One report available through the MarLIN Web site states that colonies of the stomatopod *Rissoides demaresti* have been found near the southern tip of the Lleyn Peninsula, some 30 km from Holyhead Harbour (http://www.marlin.ac.uk/speciesinformation.php?speciesID=4255).

5 Experimental Procedures and Conclusions

We intend, first, to obtain as near-perfect acoustic signatures of as many of the "vocalizing" suspects as we can. This will enable us to formulate a procedure for acoustic identification. The easier species to acquire and stimulate are our various species of pipefish and other possible local click-maker fish, including several species of wrasse known to be common in Holyhead Harbour (goldsinny, *Ctenolabrus rupestris*; corkwing, *Crenilabrus melops*; ballan, *Labrus bergylta*). Anglesey Sea Zoo is also, as it happens, the sole successful UK breeder of native sea horses. Snapping shrimp and mantis prawns are available through the aquarium trade and, with the help of Anglesey Sea Zoo, we shall resource suitable specimens.

High-quality signature acquisition will be achieved by the field use of reverberation-free specimen containment. The prototype container is currently under construction at the School of Ocean Science, Menai Bridge. It is our intention to provide, within the container, a brine-shrimp drip-feeding capability to stimulate feeding and thus click making. Video cameras and Seiche wideband acoustic sensors are also being installed on the container. The acoustic sensors employ broadband, uniform sensitivity 1-3 composite hydrophones specifically designed for this task. The sensor output is digitally transmitted over plastic or glass fiberoptic cable to a cabin in the support boat. Fiberoptic transmission eliminates wharfside and shipboard electrical interference. It greatly reduces the expense, weight, handling and corrosion issues consequent on the use of copper cable. Finally, it eliminates the problem of work hardening and fracture of a copper conductor when in long-term use in buoyed and anchored equipment. The only penalty is the need for battery power in the seafloor equipment. By using 3.7-V Li-Ion batteries and modern circuit techniques, excellent deployment lifetimes can be achieved.

We are undertaking the construction of shallow landers employing video camera recording. It is our intention to add to existing lander technology developed at the School of Ocean Science wideband acoustic detectors developed by Seiche Ltd. together with their associated recorders. This will permit long-term monitoring of regions of the sea floor around the Holyhead breakwater and near selected inshore reefs. Our objective will be to assess, on a continuous basis, the variability of average click rate with time. We hope to identify diurnal, lunar, or longer cycles in the click-rate analysis.

It may well transpire that it is not snapping shrimps or mantis prawns that are the Holyhead click makers. Nonetheless, species within the genera *Alpheus* and *Synalpheus* and the family of stomatopods are, as has been mentioned, available to us locally. It is therefore also important that we attempt to monitor click rate, commencing with the most readily accessible stomatopod colony near the Lleyn Peninsula. As time and resources permit, we also propose to deploy landers at the *Alpaheus glaber* site, to the east of the Isle of Man. This will be possible during normal cruises of the RV Prince Madog. Hopefully, click-rate measurement and click analysis will allow us to home in on what we, for the moment, presume must be a significant colony of these animals.

Although the investigation of unexpected click makers may be thought to have little core scientific value, the key aspect of what we are engaged in is the devising of passive acoustic monitoring equipment and methods that will help us acquire long-term environmentally sensitive data. This, in turn, may be employed to establish the viability of certain classes of marine ecosystems.

A major area of concern where such methodology would prove massively beneficial is in the passive acoustic monitoring of mangrove. Throughout the developing Far East, mangrove is, quite literally, a dying habitat. Snapping shrimp are widespread and plentiful within the mangrove. Monitoring average click rate may yet prove to be a key indicator of habitat quality.

Part IV
Physiological Effects of Sounds

Noise-Induced Hearing Loss: From Animal Models to Human Trials

Colleen G. Le Prell

1 Introduction

Exposure to noise results in neural swelling, hair cell loss, and strial damage. Toxic-free radical accumulation also contributes to cell death in the inner ear. The type and extent of pathology and amount and permanence of hearing loss all vary with the level and duration of the insult as well as the dynamic level change. New data indicate that temporary changes in hearing are potentially much more harmful than previously believed. The potential for the use of free radical scavengers as therapeutic agents varies with the noise insult; new studies are essential for determining the potential to prevent the previously unknown long-term sequellae of noise insults that induce temporary changes in hearing. Evaluating the effect of noise on the human inner ear and the potential for protection using novel therapeutic agents presents multiple challenges.

2 Relationship of Temporary Threshold Shift and Permanent Threshold Shift

Some data suggest that temporary threshold shift (TTS) and permanent threshold shift (PTS) deficits fall along a functional continuum. PTS increases with TTS as TTS increases above 50 dB (Henderson et al. 1991), and TTS measured 24 h postnoise is correlated with PTS (Hamernik et al. 2002). Other data reveal a morphological continuum of damage. Elegant survival-fixation techniques show that PTS is best predicted by hair cell loss and neural degeneration, not TTS deficits (Nordmann et al. 2000). Anatomical changes after noise have been well reviewed (Ohlemiller 2008; Wang et al. 2002). Unlike functional and morphological changes after noise, molecular responses to TTS- and PTS-inducing sounds are different. The Bcl-2 gene family includes both proapoptotic genes (Bax, Bak) and antiapoptotic genes (Bcl-2, Bcl-xl) (for a review, see Danial 2007). PTS exposure induces Bak gene expression and cell death, whereas TTS exposure induces Bcl-xl expression (Yamashita et al. 2008).

C.G. Le Prell (✉)
University of Florida, Gainesville, FL 32611, USA
e-mail: colleeng@phhp.ufl.edu

A.N. Popper and A. Hawkins (eds.), *The Effects of Noise on Aquatic Life*,
Advances in Experimental Medicine and Biology 730, DOI 10.1007/978-1-4419-7311-5_43,

2.1 *Clinical Relevance of TTS*

The issue of whether repeat TTSs result in a PTS is clinically relevant; indirect evidence from real-world populations supports such a relationship. For example, TTS and PTS are detected at the same frequencies in motorcycle riders tested for a TTS postride or an existing PTS (McCombe et al. 1995). Other new data indicate a direct link between TTS and PTS. Noise exposure that resulted in a TTS accelerated hearing loss as a result of progressive neural degeneration with age (Kujawa and Liberman 2006). Kujawa and Liberman (2009) recently revisited the issue of neural degeneration after a TTS. They found a rapid, extensive loss of synaptic contacts between hair cells and nerve fibers within 24 h of noise as well as progressive long-term neural degeneration. Neural loss occurred even though the hair cell population was intact and normal threshold function had returned. These data suggest that a TTS may be more harmful than previously believed. Given this, it may be clinically beneficial to reduce the TTS.

3 Relationship of Chronic Noise and Impulse Noise: Kurtosis

Loud sounds can be present for hours, such as in some occupational settings or at some recreational events; other loud sounds are brief and impulsive (such as gunfire or fireworks). The equal energy hypothesis states that two sounds with equal energy should be equally harmful (with the exception of extreme impulse noise, which can induce rupture of the tympanic membranes and fracture of the ossicular chain). However, hearing loss after noise is not necessarily well correlated with the total energy of the exposure. We now know that another major dimension of sound is rapid amplitude change or kurtosis. Hamernik et al. (2002) have systematically explored a continuum of noise exposures from Gaussian to Gaussian+impact to pure impact noise. The kurtosis statistic of the amplitude distribution, describing "peakedness," contributes importantly to the extent and distribution of sensory cell death and hearing loss induced by noise exposure, with greater trauma associated with higher kurtosis values (for recent review, see Davis et al. 2009).

3.1 *Gaussian Noise*

Gaussian noise is spectrally flat, with all frequency components at relatively equal levels. Broadband noise has been used to induce a TTS; increasing the sound level or the exposure duration results in a PTS (for one example, see Gao et al. 1992). The extent of anatomical damage and its relationship to exposure time and level has been carefully explored (Spoendlin 1971, 1976).

3.2 *Octave Band Noise*

Octave band noise centered at 4 kHz is widely used to induce sensory cell death and hearing loss in guinea pig and chinchilla subjects; noise trauma tends to be a higher frequency octave band (8–16 kHz) when rats or mice serve as subjects. There is a trade-off between exposure level and duration, and exposures show a wide range across studies: from 86-dB sound pressure level (SPL) × 5 days (Salvi et al. 1982) to 120-dB SPL × 5 hours (e.g., see Le Prell et al. 2007a).

3.3 *Kurtotic Noise*

Kurtotic noise is a Gaussian-like broadband noise that is distinguished from Gaussian noise by the presence of brief narrowband impulses. This exposure is a potential model for industrial noise exposure, in which high background noise levels are punctuated by brief noise bursts.

3.4 *Impulse Noise*

Impulse noise typically consists of brief noise pulses modeled after the sound of gunfire. Impulse noise exposures can produce a TTS or PTS based on the number of pulses (Duan et al. 2002). The effects of varying pulse rate and peak SPL have been systematically explored in chinchillas (Henderson et al. 1991).

3.5 *Conditioning Noise*

Sound conditioning was initially described by Canlon et al. (1988) as a phenomenon in which long-term exposure to continuous low-level sound reduces deficits associated with exposure to a subsequent sound that otherwise induces significant deficits in auditory sensitivity. Sound exposures that occur for hours rather than days more closely model occupational noise conditions, and a now common conditioning paradigm is a 4–6 h/day exposure to a noise band for multiple days (see, e.g., McFadden et al. 1997). The protection conferred by conditioning exposures does not depend on intact middle ear muscles (Ryan et al. 1994). Instead, the efferent system was implicated when disruption of the olivocochlear efferents eliminated protection against a PTS with prior lower-level exposure (Zheng et al. 1997). We now know that the dopaminergic lateral olivocochlear system is involved in conditioning (Niu and Canlon 2002), whereas the medial olivocochlear pathway does not appear to contribute (Yamasoba and Dolan 1998). Increased levels of endogenously produced antioxidant enzymes (in the stria vascularis as well as the organ of Corti) may also play a role (Jacono et al. 1998).

4 Potential for Protection

Until the mid-1990s, we believed that most, if not all, noise-induced hearing loss (NIHL) occurred largely as a consequence of mechanical trauma that induced morphological damage. Because NIHL was assumed to result from direct mechanical destruction, mechanical devices (ear plugs, ear muffs) that reduce sound coming into the ear were assumed to be the only strategies for reducing NIHL. Although some impulse noise exposures can cause mechanical damage, we now know that cell death after a noise insult is largely driven by oxidative stress (for reviews, see Henderson et al. 2006; Le Prell et al. 2007b). Free radical formation in the inner ear is well characterized, with immediate noise-induced free radical production (i.e., within 1–2 h of noise exposure; see Ohlemiller et al. 1999; Yamashita et al. 2004) and continued free radical production for 7–10 days postnoise (Yamashita et al. 2004). These key findings have opened the door for the potential use of free radical scavengers or antioxidants to reduce NIHL after noise.

Early work with antioxidant agents revealed that glutathione (GSH) and GSH-related molecules reduce PTS, and multiple groups quickly extended these findings to other free radical scavengers (for reviews, see Henderson et al. 2006; Le Prell et al. 2007b). Delivery of individual dietary antioxidants that directly scavenge free radicals has been effective (for a review, see Le Prell et al. 2007b) and nutrient combinations have the potential for additive and/or synergistic effects, with combinations proving effective even with a short prenoise treatment onset (Le Prell et al. 2007a). Recent reviews are available for several therapeutics that increase the availability of the GSH precursor cysteine (D-methionine, see Campbell et al. 2007; N-acetylcysteine, see Kopke et al. 2007) or catalyze the formation of GSH (ebselen, see Lynch and Kil 2009). Of particular relevance, there is a window of up to 3 days during which postnoise rescue treatment is feasible (Yamashita et al. 2005).

Antioxidant therapy has been perhaps surprisingly effective in reducing hearing loss induced by impulse noise exposure (for examples, see Henderson et al. 1999; Kopke et al. 2005). Data such as these suggest metabolic stress after impulse noise importantly contributes to a PTS postnoise. The extent to which antioxidant therapy can effectively attenuate long-term neural degeneration after TTS-inducing noise exposure remains to be determined; however, long-term auditory nerve protection achieved via antioxidant vitamins delivered during an ototoxic drug insult (Maruyama et al. 2007) suggests the potential for enhanced neural survival with antioxidant treatment.

Given data suggesting the potential to protect the ear against NIHL, human clinical trials are a critical next step. Evaluating the potential for protection using novel therapeutics presents multiple challenges, such as the identification of appropriate subject populations, assurance that human subject hearing is not put at an increased risk of harm, and selection of appropriate test metrics that are both sensitive to small changes in auditory function and clinically relevant. Several groups are now evaluating a variety of agents in human trials; these data have the potential to drive changes in evidence-based clinical practice.

References

Campbell KCM, Meech RP, Klemens JJ, Gerberi MT, Dyrstad SSW, Larsen DL, Mitchell DL, El-Azizi M, Verhulst SJ, Hughes LF (2007) Prevention of noise- and drug-induced hearing loss with D-methionine. Hear Res 226:92–103.

Canlon B, Borg E, Flock A (1988) Protection against noise trauma by pre-exposure to a low level acoustic stimulus. Hear Res 34:197–200.

Danial NN (2007) BCL-2 family proteins: Critical checkpoints of apoptotic cell death. Clin Cancer Res 13:7254–7263.

Davis RI, Qiu W, Hamernik RP (2009) Role of the kurtosis statistic in evaluating complex noise exposures for the protection of hearing. Ear Hear 30:628–634.

Duan M, Qiu J, Laurell G, Borg EG, Olofsson A (2002) Susceptibility to impulse noise trauma is different among species. Assoc Res Otolaryngol Abstr 25:104.

Gao WY, Ding DL, Zheng XY, Ruan FM, Liu YJ (1992) A comparison of changes in the stereocilia between temporary and permanent hearing losses in acoustic trauma. Hear Res 62:27–41.

Hamernik RP, Ahroon WA, Patterson JH Jr, Qiu W (2002) Relations among early postexposure noise-induced threshold shifts and permanent threshold shifts in the chinchilla. J Acoust Soc Am 111:320–326.

Henderson D, Bielefeld EC, Harris KC, Hu BH (2006) The role of oxidative stress in noise-induced hearing loss. Ear Hear 27:1–19.

Henderson D, McFadden SL, Liu CC, Hight N, Zheng XY (1999) The role of antioxidants in protection from impulse noise. Ann NY Acad Sci 884:368–380.

Henderson D, Subramaniam M, Gratton MA, Saunders SS (1991) Impact noise: The importance of level, duration, and repetition rate. J Acoust Soc Am 89:1350–1357.

Jacono AA, Hu B, Kopke RD, Henderson D, Van De Water TR, Steinman HM (1998) Changes in cochlear antioxidant enzyme activity after sound conditioning and noise exposure in the chinchilla. Hear Res 117:31–38.

Kopke R, Bielefeld E, Liu J, Zheng J, Jackson R, Henderson D, Coleman JK (2005) Prevention of impulse noise-induced hearing loss with antioxidants. Acta Otolaryngol (Stockh) 125:235–243.
Kopke RD, Jackson RL, Coleman JKM, Liu J, Bielefeld EC, Balough BJ (2007) NAC for noise: From the bench top to the clinic. Hear Res 226:114–125.
Kujawa SG, Liberman MC (2006) Acceleration of age-related hearing loss by early noise exposure: evidence of a misspent youth. J Neurosci 26:2115–2123.
Kujawa SG, Liberman MC (2009) Adding insult to injury: Cochlear nerve degeneration after "temporary" noise-induced hearing loss. J Neurosci 29:14077–14085.
Le Prell CG, Hughes LF, Miller JM (2007a) Free radical scavengers vitamins A, C, and E plus magnesium reduce noise trauma. Free Radical Biol Med 42:1454–1463.
Le Prell CG, Yamashita D, Minami S, Yamasoba T, Miller JM (2007b) Mechanisms of noise-induced hearing loss indicate multiple methods of prevention. Hear Res 226:22–43.
Lynch ED, Kil J (2009) Development of ebselen, a glutathione peroxidase mimic, for the prevention and treatment of noise-induced hearing loss. Semin Hear 30:47–55.
Maruyama J, Miller JM, Ulfendahl M (2007) Effects of antioxidants on auditory nerve function and survival in deafened guinea pigs. Neurobiol Dis 25:309–318.
McCombe AW, Binnington J, Davis A, Spencer H (1995) Hearing loss and motorcyclists. J Laryngol Otol 109:599–604.
McFadden SL, Henderson D, Shen YH (1997) Low-frequency 'conditioning' provides long-term protection from noise-induced threshold shifts in chinchillas. Hear Res 103:142–150.
Niu X, Canlon B (2002) Activation of tyrosine hydroxylase in the lateral efferent terminals by sound conditioning. Hearing Research 174:124–132.
Nordmann AS, Bohne BA, Harding GW (2000) Histopathological differences between temporary and permanent threshold shift. Hear Res 139:13–30.
Ohlemiller KK (2008) Recent findings and emerging questions in cochlear noise injury. Hear Res 245:5–17.
Ohlemiller KK, Wright JS, Dugan LL (1999) Early elevation of cochlear reactive oxygen species following noise exposure. Audiol Neurootol 4:229–236.
Ryan AF, Bennett TM, Woolf NK, Axelsson A (1994) Protection from noise-induced hearing loss by prior exposure to a nontraumatic stimulus: Role of the middle ear muscles. Hear Res 72:23–28.
Salvi RJ, Perry JW, Hamernik RP, Henderson D (1982) Relationships between cochlear pathologies and auditory nerve and behavioral responses following acoustic trauma. In: Hamernik RP, Henderson D, Salvi RJ (eds) New perspectives in noise-induced hearing loss. Raven Press, New York, pp 165–188.
Spoendlin H (1971) Primary structural changes in the organ of Corti after acoustic overstimulation. Acta Otolaryngol (Stockh) 71:166–176.
Spoendlin H (1976) Anatomical changes following various noise exposures. In: Henderson D, Hamernik R, Dosanjh S, Mills JH (eds) Effects of noise on hearing. Raven Press, New York p 69–89.
Wang Y, Hirose K, Liberman MC (2002) Dynamics of noise-induced cellular injury and repair in the mouse cochlea. J Assoc Res Otolaryngol 3:248–268.
Yamashita D, Jiang H-Y, Le Prell CG, Schacht J, Miller JM (2005) Post-exposure treatment attenuates noise-induced hearing loss. Neuroscience 134:633–642.
Yamashita D, Jiang H, Schacht J, Miller JM (2004) Delayed production of free radicals following noise exposure. Brain Res 1019:201–209.
Yamashita D, Minami SB, Kanzaki S, Ogawa K, Miller JM (2008) Bcl-2 genes regulate noise-induced hearing loss. J Neurosci Res 86:920–928.
Yamasoba T, Dolan DF (1998) The medial cochlear efferent system does not appear to contribute to the development of acquired resistance to acoustic trauma. Hear Res 120:143–151.
Zheng XY, Henderson D, McFadden SL, Hu BH (1997) The role of the cochlear efferent system in acquired resistance to noise-induced hearing loss. Hear Res 104:191–203.

Auditory Effects of Underwater Noise in Odontocetes

James J. Finneran

1 Introduction

A common result of noise exposure is a noise-induced threshold shift, defined as an increase in auditory threshold that persists after cessation of a noise exposure. If the hearing threshold returns to normal after some period of time, the threshold shift is called a temporary threshold shift (TTS). If the threshold does not return to normal, the remaining amount of threshold shift is called a permanent threshold shift (PTS). Observed PTS/TTS may be the result of a variety of mechanical and biochemical processes, including physical damage or distortion of the tympanic membrane and cochlear hair cell stereocilia, oxidative stress-related hair cell death, changes in cochlear blood flow, and swelling of cochlear nerve terminals resulting from glutamate excitotoxicity (Henderson et al. 2006; Kujawa and Liberman 2009). Although the outer hair cells are the most prominent target for noise effects, severe noise exposures may also result in inner hair cell death and loss of auditory nerve fibers (Henderson et al. 2006).

Threshold shifts are determined by first measuring preexposure hearing thresholds at one or more discrete frequencies, exposing the subject to a "fatiguing" noise exposure, and then measuring postexposure thresholds at one or more frequencies. The amount of threshold shift (in dB) is determined by subtracting the preexposure threshold (in dB) from the postexposure threshold (in dB). If the threshold eventually returns to normal, the measured threshold shift is denoted as a TTS. TTS measured at a specific postexposure time is denoted with a subscript, e.g., TTS_4 indicates the amount of TTS measured 4 min after cessation of the fatiguing noise exposure.

A large number of TTS/PTS studies have been performed in humans and terrestrial mammals to identify the relationships between noise exposure and induced threshold shifts and to develop safe exposure guidelines for people working in noisy environments. Although much information has been learned, the applicability of these data to marine mammals is limited, in part because of the differences between the peripheral auditory systems of marine and terrestrial animals but also because the types of noise exposures most relevant for people (e.g., 8-h exposure to broadband noise) may not be relevant to marine mammals exposed to shorter duration, intermittent sources such as sonar systems, pile driving, and seismic air guns. For these reasons, a number of TTS measurements have been conducted with marine mammals to determine noise-exposure conditions

J.J. Finneran (✉)
US Navy Marine Mammal Program, Space and Naval Warfare Systems Center Pacific, San Diego, CA 92152, USA
e-mail: james.finneran@navy.mil

A.N. Popper and A. Hawkins (eds.), *The Effects of Noise on Aquatic Life*,
Advances in Experimental Medicine and Biology 730, DOI 10.1007/978-1-4419-7311-5_44,

necessary for TTS in these animals. This paper reviews some of the major findings arising from TTS experiments with *Tursiops truncatus* (bottlenose dolphins) and *Delphinapterus leucas* (beluga whales).

2 Major Findings

The major findings to arise from TTS experiments with *Tursiops truncatus* and *Delphinapterus leucas* parallel findings from terrestrial mammal experiments. As in terrestrial mammals, the most significant factors that affect TTS in *Tursiops truncatus* and *Delphinapterus leucas* are the hearing test frequency, exposure sound pressure level (SPL), exposure duration, exposure frequency, temporal pattern, and recovery time.

2.1 Hearing Test Frequency

The amount of TTS measured will vary with the hearing test frequency, with the maximum TTS after tonal exposures not occurring at the exposure frequency but instead occurring one-half to one octave above the exposure frequency (Finneran et al. 2007; Schlundt et al. 2000). The overall spread of TTS from tonal exposures can thus extend over an extended frequency range, i.e., narrowband exposures can produce broadband (greater than one octave) TTS. Figure 1 illustrates the frequency spread of TTS after a tonal exposure. For impulsive sounds, there has not been a systematic effort to explore the relationship between exposure frequency content and the resulting frequency spread of TTS. TTS in a *Delphinapterus leucas* exposed to a water gun impulse occurred at frequencies above the predominant energy in the exposure, suggesting an upward shift in TTS as one would expect based on terrestrial mammal data.

2.2 Exposure SPL and Duration

As in terrestrial mammals, the amount of TTS increases with exposure SPL and the exposure duration, but the relationships are not monotonic. At low-exposure levels, there will exist some SPL

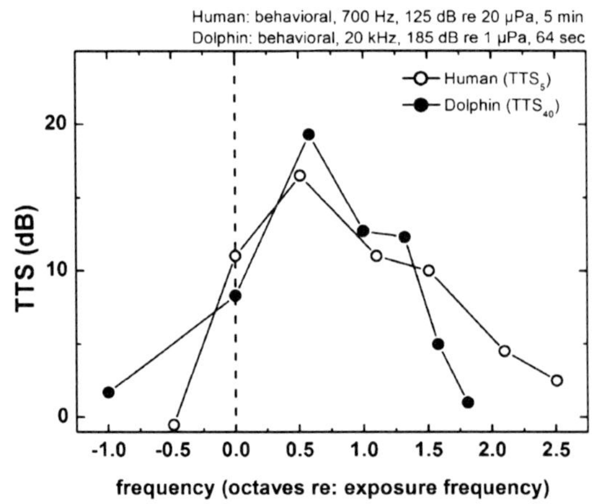

Fig. 1 Upward frequency spread of temporary threshold shift (TTS) after tonal exposures in a human and *Tursiops truncatus*. Human data from Ward (1962). *Tursiops truncatus* data from Finneran et al. (2007)

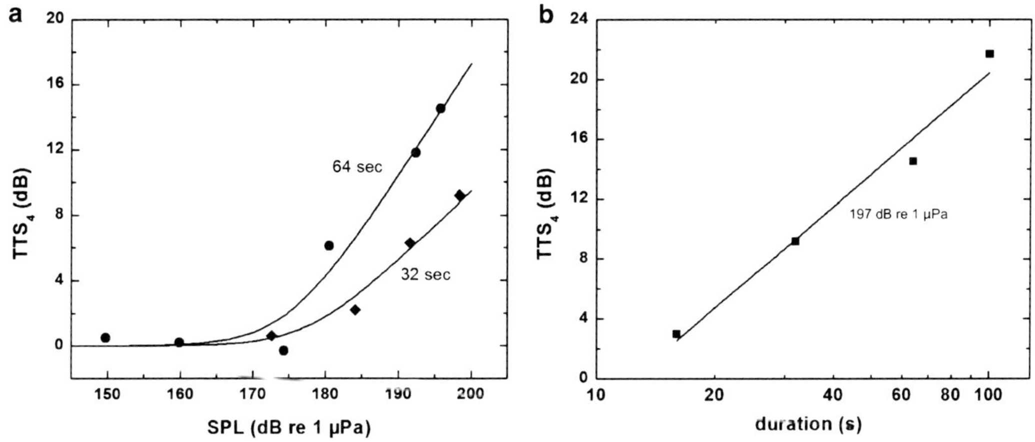

Fig. 2 Growth of *Tursiops truncatus* TTS_4 with exposure sound pressure level (SPL; **a**) and exposure duration (**b**) for 3-kHz tones. Modified from Finneran et al. (2010a)

below which no TTS will occur regardless of exposure duration; this level is called "effective quiet." To date, there have been no studies performed to measure effective quiet in a marine mammal. As the exposure SPL increases, the amount of TTS at a particular frequency will grow exponentially, then approach a linear relationship with SPL as the amount of TTS increases (Fig. 2a). Above effective quiet, TTS will also increase with duration, approaching a linear relationship with the logarithm of time (Fig. 2b). Although not yet demonstrated in marine mammals, terrestrial mammal data have shown that if the exposure SPL and duration are increased sufficiently, the amount of TTS will reach a plateau where further increases in exposure do not result in additional threshold shift. This region is called asymptotic threshold shift (ATS).

Because TTS is a function of both the exposure SPL and exposure duration, the use of sound exposure level (SEL) as a single metric has become increasingly common. TTS is correlated with SEL, especially if the range of exposure durations is relatively small. As the exposure duration increases, however, the relationship between TTS and SEL begins to break down. Specifically, duration has a more significant effect on TTS than would be predicted on the basis of SEL alone (Finneran et al. 2010a; Mooney et al. 2009a). This means that if two exposures have the same SEL but different durations, the exposure with the longer duration will tend to produce more TTS. Despite this, SEL continues to be used in many situations because it is simple and more accurate than SPL alone.

2.3 *Exposure Frequency*

Early TTS data did not reveal significant differences in TTS onset at 3, 10, and 20 kHz despite significant changes in *Tursiops truncatus* hearing thresholds over this range of frequencies (Schlundt et al. 2000). For this reason, most acoustic impact criteria have used similar effects threshold for the onset of TTS regardless of exposure frequency (e.g., Southall et al. 2007). More recent data, however, have revealed large differences (~15 dB) between TTS onset at 3 kHz compared with 20 kHz (Finneran and Schlundt 2010; Finneran et al. 2007; Fig. 3). TTS growth rates in *Tursiops truncatus* have also been shown to increase with exposure frequency above 3 kHz (Finneran and Schlundt 2010). These data demonstrate the need for frequency-specific criteria for TTS onset and/or more accurate auditory weighting functions.

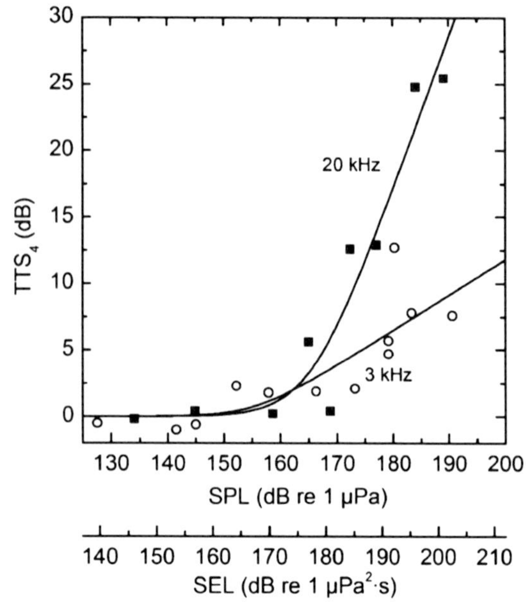

Fig. 3 Growth of TTS_4 in a *Tursiops truncatus* after 16-s exposures at 3 and 20 kHz. SEL, sound exposure level. Modified from Finneran and Schlundt (2010)

2.4 Exposure Temporal Pattern

Most marine mammal TTS experiments have featured single continuous or impulsive exposures; however, some preliminary studies have been conducted to examine the effects of intermittency and temporal pattern on TTS (Finneran et al. 2010b; Mooney et al. 2009b). These studies have shown that TTS can accumulate across multiple exposures but that the resulting TTS will be less than the TTS from a single, continuous exposure with the same SEL (or the same SPL and exposure "on time"). This means that predictions based on total, cumulative SEL will overestimate the amount of TTS from intermittent exposures. For the specific exposure conditions used by Finneran et al. (2010b), the modified power law model (Humes and Jesteadt 1989) fit the growth of TTS across multiple exposures; however, it is unknown to what extent this approach would apply to other exposure conditions.

2.5 Recovery Time

The amount of observed TTS tends to decrease with increasing time after the exposure; however, the relationship is not monotonic. The amount time required for complete recovery of hearing depends on the magnitude of the initial shift; for relatively small shifts, recovery may be complete in a few minutes, whereas large shifts (e.g., 40 dB) require several days for recovery. TTS recovery functions in *Tursiops truncatus* exposed to 3-kHz tones have exhibited complex patterns (Fig. 4) that often contain regions where TTS is linear with the logarithm of time but also typically contain regions with varying slopes. Double exponential functions used to fit human TTS recovery data (Keeler 1968; Patuzzi 1998) also fit the *Tursiops truncatus* recovery data and, for the 3-kHz exposures with durations from 1 to 128 s, the recovery functions can be described using TTS_4 and

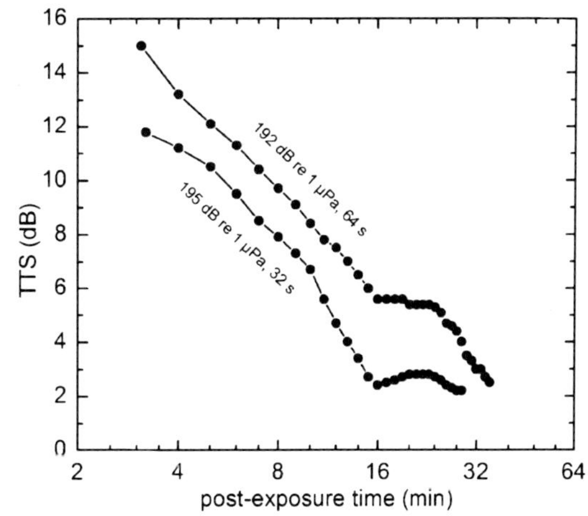

Fig. 4 Recovery of TTS in a *Tursiops truncatus* after exposure to 3-kHz tones. Modified from Finneran et al. (2010a)

recovery time only; i.e., recovery functions did not depend on the specific SPL and duration but only on the resulting TTS_4. The extent to which this result may be extrapolated to other exposure conditions is unknown.

3 Conclusions

TTS data remain one of the few direct measures of the effects of noise exposure and are therefore an important component for predicting and mitigating the effects of anthropogenic noise on marine mammals. Although much progress has been made in understanding the auditory effects of intense sounds on odontocetes, data gaps still exist regarding the effects of exposure frequency, TTS growth and recovery after exposure to intermittent noise, and the extent and manner in which TTS data from *Tursiops truncatus* and *Delphinapterus leucas* may be extrapolated to other species.

References

Finneran JJ, Carder DA, Schlundt CE, Dear RL (2010a) Growth and recovery of temporary threshold shift (TTS) at 3 kHz in bottlenose dolphins (*Tursiops truncatus*). J Acoust Soc Am 127:3256–3266.

Finneran JJ, Carder DA, Schlundt CE, Dear RL (2010b) Temporary threshold shift in a bottlenose dolphin (*Tursiops truncatus*) exposed to intermittent tones. J Acoust Soc Am 127:3267–3272.

Finneran JJ, Schlundt CE (2010) Frequency-dependent and longitudinal changes in noise-induced hearing loss in a bottlenose dolphin (*Tursiops truncatus*). J Acoust Soc Am 128:567–570.

Finneran JJ, Schlundt CE, Branstetter B, Dear RL (2007) Assessing temporary threshold shift in a bottlenose dolphin (*Tursiops truncatus*) using multiple simultaneous auditory evoked potentials. J Acoust Soc Am 122:1249–1264.

Henderson D, Bielefeld EC, Harris KC, Hu BH (2006) The role of oxidative stress in noise-induced hearing loss. Ear Hear 27:1–19.

Humes LE, Jesteadt W (1989) Models of the additivity of masking. J Acoust Soc Am 85:1285–1294.

Keeler JS (1968) Compatible exposure and recovery functions for temporary threshold shift-mechanical and electrical models. J Sound Vib 2:220–235.

Kujawa SG, Liberman MC (2009) Adding insult to injury: Cochlear nerve degeneration after "temporary" noise-induced hearing loss. J Neurosci 29:14077–14085.

Mooney TA, Nachtigall PE, Breese M, Vlachos S, Au WWL (2009a) Predicting temporary threshold shifts in a bottlenose dolphin (*Tursiops truncatus*): The effects of noise level and duration. J Acoust Soc Am 125:1816–1826.

Mooney TA, Nachtigall PE, Vlachos S (2009b) Sonar-induced temporary hearing loss in dolphins. Biol Lett 5:565–567.

Patuzzi R (1998) Exponential onset and recovery of temporary threshold shift after loud sound: Evidence for long-term inactivation of mechano-electrical transduction channels. Hear Res 125:17–38.

Schlundt CE, Finneran JJ, Carder DA, Ridgway SH (2000) Temporary shift in masked hearing thresholds of bottlenose dolphins, *Tursiops truncatus*, and white whales, *Delphinapterus leucas*, after exposure to intense tones. J Acoust Soc Am 107:3496–3508.

Southall BL, Bowles AE, Ellison WT, Finneran JJ, Gentry RL, Greene CR Jr, Kastak D, Ketten DR, Miller JH, Nachtigall PE, Richardson WJ, Thomas JA, Tyack PL (2007) Marine mammal noise exposure criteria: initial scientific recommendations. Aquat Mamm 33:411–522.

Ward WD (1962) Damage-risk criteria for line spectra. J Acoust Soc Am 34:1610–1619.

Implementation of Acoustic Dosimeters With Recoverable Month-Long GPS/TDR Tags to Interpret Controlled-Exposure Experiments for Large Whales

Bruce R. Mate

1 Introduction

Obtaining extended and detailed information from instrumented whales is important for understanding whale behavior and interpreting responses to anthropogenic noise but has been difficult to achieve with tags because of short attachment times, poor spatial/temporal resolution, or lack of adequate acoustic data. In 2007–2008, sperm whales were tagged in the Gulf of California with Wildlife Computers TDR-PAT-MK-10 tags. While attached, the tags sent Fastloc-GPS locations and summary dive data (shape, duration, and depth) via ARGOS after dives >10 min and >10 m. Tags released from the whales floated to the surface and were recovered. Downloaded data revealed high-resolution time-depth recorder (TDR; 1-s and 2-m dive-depth increments) and GPS data within 60 m (Mate 2008). The tags provided insights into resting, foraging, and traveling behaviors, diving to depths over 1,200 m for periods of up to 28 days. Fast zigzag patterns during dives likely represented foraging attempts on Humboldt squids. Despite coordinated travels, whales in the same social unit did not usually dive synchronously or to the same depth and displayed considerable variability at the same position and time.

A principal components analysis (PCA) was used to detect changes in dive behavior during a sperm whale's GPS/TDR dive record over a 28-day period as it moved from 14 days in a deep-water habitat to 14 days in a shallow-water habitat (Fig. 1). The technique was sufficiently sensitive to identify changes in the first 5 dives of the shallow-water habitat as substantially different from those of the previous 5 or 14 days of deep-water dives. The whale did not consistently dive to the bottom, so these differences were not just a reflection of bottom depth. This environmental change probably affected the behavior of squid, sperm whale prey, and therefore sperm whale dive behavior. The detection of this dive change was possible from the ARGOS-relayed summary dive data (and thus feasible with near real-time monitoring) rather than having to wait for the detailed dive data obtainable only by recovering the tag at the end of the experiment. This could be important for a controlled-exposure experiment (CEE) or behavioral-response study (BRS) strategy designed to expose animals to sequentially higher levels of sound so there is a means of recognizing changes in dive behavior in real time to avoid extreme noise exposures, which may be injurious or problematic for the subject animals. The capability of maintaining tags on animals for longer periods of time and

B.R. Mate (✉)
Marine Mammal Institute, Department of Fisheries and Wildlife,
Oregon State University, Newport, OR 97365, USA
e-mail: bruce.mate@oregonstate.edu

A.N. Popper and A. Hawkins (eds.), *The Effects of Noise on Aquatic Life*,
Advances in Experimental Medicine and Biology 730, DOI 10.1007/978-1-4419-7311-5_45,

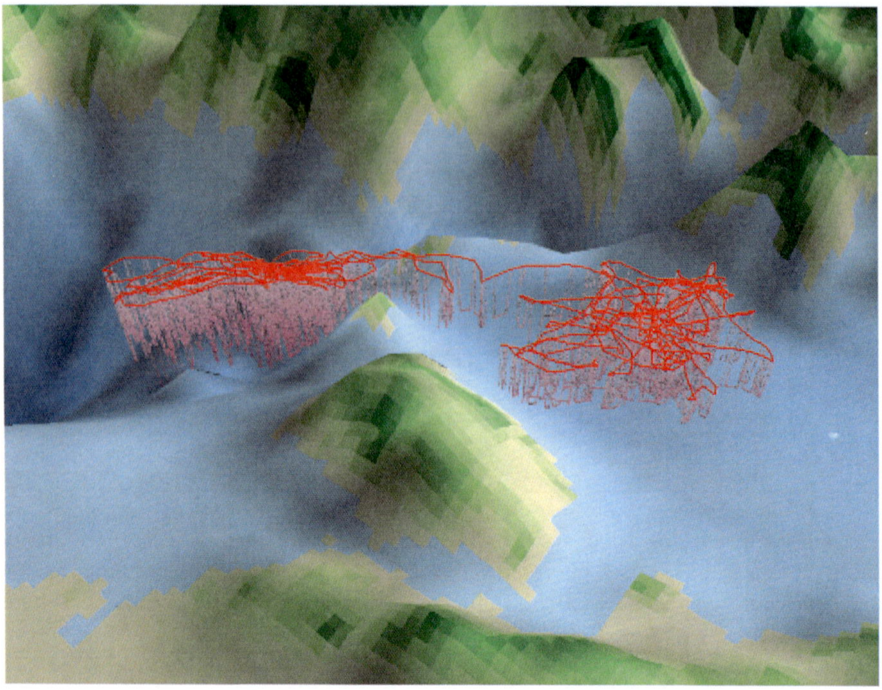

Fig. 1 The movement of a sperm whale equipped with a Wildlife Computers GPS/TDRPAT-MK-10 tag in the central Gulf of California during 14 days of dives in deep water to the south (left) and 14 days of dives in shallower water to the north (right). The view is from the east looking west over Tiburon Island in the foreground

monitoring their responses to sound exposures allows for adaptive modifications in the experimental exposure protocol, which can potentially document recovery from temporary effects, changes in response to repeat exposures, and possible habituation.

Although we used an environmental change rather than a noise as a "change proxy," these results provide evidence that the tag technologies (attachments, GPS, and dive-sensing records), data gathered, and the PCA tools can inform a dose-response experiment involving noise and dive behavior changes.

The extent to which whales change their behavior in the presence of anthropogenic sound is still in doubt for most whale species. Although it seems intuitive that loud sounds like military sonar systems and seismic surveys might affect the habits of whales, their underwater behaviors are difficult to study. It is also hard to understand why whales do not always respond to the noises of ships by getting out of the way (Berman-Kowalewski et al. 2010) The development of long-term satellite tracking of whales (Mate et al. 2007) and sophisticated geographic information system analyses (Bailey et al. 2010) now make it possible to do both broad-scale habitat characterization (Cotte et al. 2009) and the collection of detailed diving behavior (Mate 2008).

The development of an acoustic dosimeter is underway at Cornell University, Ithaca, NY (C. Clark, personal communication) and will be incorporated into the next generation of the GPS/TDR tag. It will provide detailed records of up to five predetermined acoustic events designed to provide insight into how anthropogenic noise affects whale behavior during future CEEs or BRSs lasting over a month. The dosimeter is being incorporated into the GPS/TDR tag structure but is totally self-sufficient. It measures the sound exposure level (SEL) for echolocation clicks, codas, creaks (a proxy for a foraging attempt), vessel noise, and seismic/sonar sources during 5-min summary periods. The GPS/TDR-acoustic dosimeter tags will record GPS locations and the depth of whale dives every 2 s as well as the SEL and time stamp of the preselected acoustic events. Thus the tags will provide a huge database including pre- and postexposure dive behavior during "control" periods and noise-exposure events. Potential responses to experimental noises can then be viewed in the context of nonexposure (control) variability. Analyses will be undertaken to look for possible effects of just vessel noise in the absence of intentional exposures, which may help inform if and how whales respond to the approach of vessels.

The highest accumulated sound level will be measured for each acoustic event category occurring every 5 min, integrating over full 1-s periods, not just the highest sound pressure level value. Before field deployment, the dosimeter will be tested in conjunction with a calibrated acoustic buoy to affirm the quality of its measurements.

The tags come off 1) at a preset date and time, 2) if they don't change water depth for an extended period of time (suggesting they have come off the animal and sunk to the bottom in their attachment sleeve), or 3) when the tag batteries are down to 25% of their capacity. The final criterion ensures sufficient battery power for the tag to send its GPS location so it can be located for recovery. The tag-recovery vessel uses the ARGOS-relayed or tag-uplink GPS message in a Fastloc proprietary software system and the vessel's own GPS system to display the relative position of the vessel to the tag.

The first application of these amalgamated technologies is likely to be on blue and fin whales in the southern Gulf of California during the 2010 Southern California BRS. For this study, the measured wave forms will be 1) low-frequency social vocalizations, 2) ship noise, 3) naval sonar, and 4) pseudonoise of the same frequency range as the sonar source but with different time characteristics. Although we do not expect social vocalizations to be very routine in August when we start this experiment with blue/fin whales, we will determine if their vocalization rate changes in response to dive depths, noise exposure, or time of day. During that experiment, other team participants will be applying B-probes, and it is hoped that the double tagging of individual whales will provide a cross-correlation between the two techniques.

2 Conclusions

The long-term attachment of GPS/TDR tags can provide detailed data over extended periods before, during, and after the CEE/BRS. The incorporation of an acoustic dosimeter into this tag provides the information on SELs for biologic sources (vocalizations, clicks, codas, and creaks) and anthropogenic noises, both controlled (seismic or sonar) as well as uncontrolled (ship noise). PCA techniques can recognize changes in dive behavior when whales move between different habitat types from both ARGOS-relayed summary dive data and detailed TDR data on tag recovery, suggesting that significant noise-induced charges will also be detectable.

References

Bailey H, Mate BR, Palacios DM, Irvine L, Bograd SJ, Costa DP (2010) Behavioural estimation of blue whale movements in the Northeast Pacific from state-space model analysis of satellite tracks. Endang Species Res 10:93–106.

Berman-Kowalewski M, Gulland F, Wilkin S, Calambokidis J, Mate BR, Cordaro J, Rotstein D, St. Leger J, Collins P, Fahy K, Dover S (2010) Association between blue whale mortality and ship strikes along the California coast. Aquat Mamm 36:59–66.

Cotte C, Guinet C, Taupeir-Letage I, Mate B, Petiau E (2009) Scale-dependent habitat use by a large free-ranging predator, the Mediterranean fin whale. Deep-Sea Res Part I Oceanogr Res Pap 56:801–811.

Mate BR (2008) Technical instrumentation issues related to the design and execution of a controlled-exposure experiment for larger cetaceans to assess possible behavioural responses and potential impacts. Bioacoustics 17:334–336.

Mate B, Mesecar R, Lagerquist B (2007) The evolution of satellite-monitored radio tags for large whales: One laboratory's experience. Deep-Sea Res Part II Top Stud Oceanogr 54:224–247.

Marine Mammal Auditory System Noise Impacts: Evidence and Incidence

Darlene R. Ketten

1 Introduction

Sound is an inevitable element of every human activity in the oceans. Some, like exploration and military sonar exercises, produce impulse sounds that are intense but infrequent; others, like shipping, generate nonimpulsive, less intense, but continuous noise. A recent National Research Council review (2003) found that the ocean's acoustic budget has increased by 3 dB, i.e., doubled, per decade in the last half century. In effect, in some ocean areas, and particularly along our fragile coasts, we are creating an environment akin to that of human workplaces.

It is reasonable that we are concerned that any sound added to the marine environment may adversely impact a species within its "acoustic reach." Our concern for marine mammals is particularly acute because many species are endangered and hearing is arguably their primary sense. In fact, the important issues are multifactorial: what species will be exposed, for how long, to what frequencies, and at what levels, and then how do these parameters compare with an animal's hearing abilities. Only with all these factors in hand can we reliably determine the probability of adverse impacts affecting fitness or endangering populations.

Before we can have a useful perspective, much less responsibly impose regulations and sanctions to prevent impacts from anthropogenic oceanic noise, it is first necessary to understand the susceptibility to noise damage in marine mammal stocks and the current status of their hearing. In the last decade, we have gained substantial ground in testing and documenting both normal and impaired hearing in some marine mammals, but we must still infer hearing characteristics for the majority of species. This paper summarizes our knowledge of sound impact mechanisms in land mammals, the current evidence for marine mammal hearing loss, and, finally, in the context of these data, the implications for when and how marine mammal ears may suffer noise damage.

D.R. Ketten (✉)
Woods Hole Oceanographic Institution, Woods Hole, MA 02543, USA

Harvard Medical School, Boston, MA 02115, USA
e-mail: dketten@whoi.edu

A.N. Popper and A. Hawkins (eds.), *The Effects of Noise on Aquatic Life*,
Advances in Experimental Medicine and Biology 730, DOI 10.1007/978-1-4419-7311-5_46,

2 Noise-Induced Hearing Loss

Noise is not synonymous with sound. Sound is a physical phenomenon perceived through hearing, whereas noise is defined essentially as an aperiodic signal that interferes with the perception of sound and has a negative physiological impact. Experimental and human noise effect data reviewed by Davis et al. (2003), Kryter (1996), and Slepecky (1986) are summarized in the following sections.

2.1 Human Incidence

Humans are, in one sense, an on-going, natural experiment for noise impacts. Noise-induced hearing loss (NIHL) is second only to aging effects as a cause of loss among humans. We have long been aware that repeated exposure to loud noises may result in hearing loss. Early-stage NIHL in humans manifests itself as a "notch," or preferential loss, near 4 kHz but may extend as high as 6 kHz for extensive impulse exposures (Humes 2010). As NIHL progresses, distinctive threshold increases occur near the peak frequency and at partial-octave intervals of the offending signal as well as frequencies above 4 kHz. Current OSHA regulations permit chronic exposures at an average sound pressure level (SPL) of 85 dB (A-weighted) over an 8-hour period, with a mandatory halving of exposure time for each 3-dB increment in SPL. European standards are comparable. Recent statistics indicate that ~15% of people over 20 years of age in the United States have high-frequency hearing loss attributed solely to noise exposure. By age 45, 20% have substantial NIHL; by 75, ~50% of the population has profoundly impaired hearing from presbycusic, i.e., age-related, loss that is the result of long-term noise exposure compounded by diminished cellular recovery (http://www.nidcd.nih.gov).

Although other mammalian species are often used to investigate NIHL mechanisms, natural hearing loss is not commonly studied in any other species. Consequently, we know little about the incidence and nature of long-term noise effects in most mammals, making it difficult to estimate the state of health of "natural" marine ears and the probable risks from anthropogenic sources for wild marine mammal populations.

2.2 Mechanisms of NIHL

There is no simple single factor or formula for estimating loss from noise. Any noise exposure will not necessarily result in a measureable hearing loss, but every ear has tolerance limits. Sounds within an individual's hearing range may be noisome or damaging depending on the synergistic effect of several factors, e.g., intensity, frequency, duration, whether the signal profile is impulsive or continuous, and the subject's sensitivity at that frequency.

The fundamental cause of NIHL is overstimulation of the inner ear sensory cells, which results in metabolic exhaustion of the hair cells, organ of Corti support cell damage, and, in severe cases, retrograde ganglion cell and axonal degeneration. First-order damage, the protracted bending and shearing of auditory hair cell stereocilia, prevents production of neurochemical releasers that initiate auditory fiber impulses. In effect, the impacted area of the inner ear becomes chemically "silent" as a result of the loss of the stereociliary triggers. If the stereocilia recover to any extent, the function returns but may require greater energy than previously to respond and thresholds increase. Longitudinal and radial variations in cell structure along the cochlea also produce microregions with different vulnerabilities. Finally, adjunct conditions, such as exposure to ototoxins, heavy metals, hypertension, or stress hormones, may accelerate or exacerbate losses.

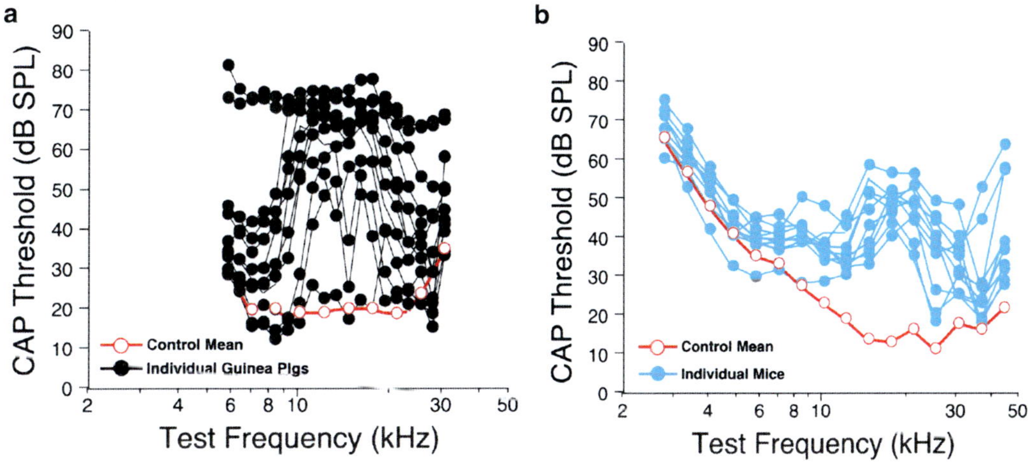

Fig. 1 (**a**) Guinea pigs exposed to 12-kHz narrowband noise for 4 h, 109 dB sound pressure level (SPL). (**b**) CBA/CaJ mice exposed to 1- to 16-kHz octave-band noise for 2 h, 103 dB SPL. CAP, compound action potential. Updated from Yoshida and Liberman (2000), courtesy of C. Liberman

Whatever the incipient cause of damage, some structural correlates for loss types are now fairly well understood. Damage to inner hair cells results in a total lack of response, whereas the loss of outer hair cells produces elevated thresholds. If hair cells recover from noise insults, the attendant loss is a temporary threshold shift (TTS). In experiments, threshold elevations as large as 50 dB returned to baseline sensitivity, although recovery sometimes required as much as 30 days. Shifts over 50 dB are often permanent (PTS). In humans, they most often result from an extreme, acute exposure or from accumulated TTS insults to the ear, some of which may occur without allowing recovery from a prior TTS. It is generally assumed that TTS represents a full recovery with no attendant physical injury, but it has been difficult to explain how even a reversible loss occurs without at least submicroscopic physical damage. Recent data (Kujawa and Liberman 2009) show that in TTS, sensory cells appear to remain intact but that there can be acute afferent nerve terminal damage and eventually cochlear nerve degeneration. Thus TTS to PTS may not have fundamentally different mechanisms but rather are a graded, although nonlinear, continuum.

There are two intriguing features in threshold shifts. One is that continuous high SPL exposures result in losses at the center frequency (CF) and at higher loci at half-octave intervals that are explained by cochlear nonlinearities. However, this does not account for anomalous damage near the hook nor for the notch phenomena that may be related to spiral topography. The second is that individual responses to identical exposures can be as great as interspecific differences except in genetically identical individuals as demonstrated in mice (Fig. 1), suggesting that there is a large genetic element in NIHL susceptibility.

2.3 NIHL Summary

Common findings across species for noise effects are as follows: inner ear damage locations and severity correlate with the power spectrum of the signal but higher frequencies may also be impacted; intensity and duration can act synergistically to broaden the loss; there is a critical limit beyond which shifts grow rapidly; continuous exposures over time are asymptotic; impulse noise produces more profound effects than continuous noise at equivalent levels; onset limits for TTS are the same for normal and hearing-impaired individuals, thus there is a smaller "shift window" for impaired individuals; effects spread more to higher frequencies from any stimulus, possibly because the tonotopic structure of the basilar membrane means all incoming signals first traverse higher frequency encoding regions at the base of the cochlea before reaching lower frequency regions. Temporal integration is reduced, but frequency discrimination is often preserved in both TTS and PTS. Signal rise time and duration of peak pressures are significant factors in PTS but not in TTS.

3 Marine Mammal Hearing Loss: Evidence

Potential impacts from noise in marine mammals, just as in land mammals, may be physiological, pathological, acute or chronic, and even subclinical or largely behavioral. Many papers in this volume provide details on all these aspects; therefore, this review is confined to the key points of the physiological and anatomical elements of hearing loss.

It is not news that some marine mammals may be hearing impaired. This is evident in hearing curves published over the last 50 years (Fig. 2). In the last 10 years, awareness of preexisting loss and testing to determine the onset of TTS via behavioral and noninvasive auditory evoked potential (AEP) techniques have both increased.

Published results are now available for 12 species of odontocetes and pinnipeds for tonal, impulse, and octave-band noise (OBN) stimuli. With the exception of tests on a few stranded juvenile whales and dolphins, all data were obtained from captive animals with prior test experience, some of which are older subjects with high-frequency losses. Southall et al. (2007) provides a detailed discussion and original citations for the studies summarized below. All SPLs are in decibels re 1 μPa and SELs are in decibels re 1 μPa^2-s unless otherwise noted.

For two odontocete species, *Tursiops truncatus* (bottlenose dolphin) and *Delphinapterus leucus* (beluga whale), tested in a natural harbor, a 6-dB or greater shift for single, short-duration pulses required exposures of 160 kPa peak (SPL 226 dB peak-peak; SEL 186 dB), whereas for pure-tone stimuli (3–80 kHz) with short exposures (up to 8 s), the mean shift onset was 195 dB SPL (SEL 192–201 dB). The lowest onset was at 182 dB SPL for 1 subject at 75 kHz. With longer exposures (up to 130 s), greater shifts occurred (23 dB) at equal or lower exposures. As seen in land mammals, shifts occurred also at octave intervals and at frequencies above the probe stimulus. Equivalent shifts were obtained at similar received SELs under more controlled pool conditions, implying that masking from harbor noise was not a factor. Studies with longer stimuli (30 min) using OBN reported 11-dB shifts with stimuli of 179 dB SPL and 212–214 dB SEL, but exact exposures required monitoring the subject for surface breaths. Parallel AEP studies for *Tursiops* using longer exposures at surface (50-min OBN) found 4- to 8-dB shift onsets at 160 dB and 193–195 dB SEL. AEP and auditory steady-state response (ASSR) studies show the same trends as behavioral studies but often report shifts 10–20 dB greater. SEL growth is the most consistent finding among these studies, with a typical rate of 1 dB TTS/dB SEL. Recovery rates vary somewhat from 1–2 dB per doubling of time for short exposures to low- or midfrequency signals versuss 5–6 dB per doubling of exposure time for frequencies closer to peak sensitivities.

TTS studies in pinnipeds have been conducted in air and in water for 6 species with parameters similar to those for cetaceans. TTS, like basic hearing, among pinnipeds covaries with aquatic versus aerial adaptation and body mass; i.e., smaller phocids are impacted at lower exposures underwater than larger species and more air-adapted otariids. At 152 dB SPL (SEL 183 dB), *Phoca vitulina* (harbor seals) sustain a TTS of 6–8 dB after a 30-min exposure to 2.5-kHz OBN compared with TTS onsets near 170 dB (SEL 205 dB) for *Zalophus californianus* (California sea lion) and an adult *Mirounga angustirostris* (northern elephant seal). In air, measures of the same subjects and regimens required 99 dB re 20 μPa (SEL 131 dB) for 6 dB of TTS in the harbor seal versus

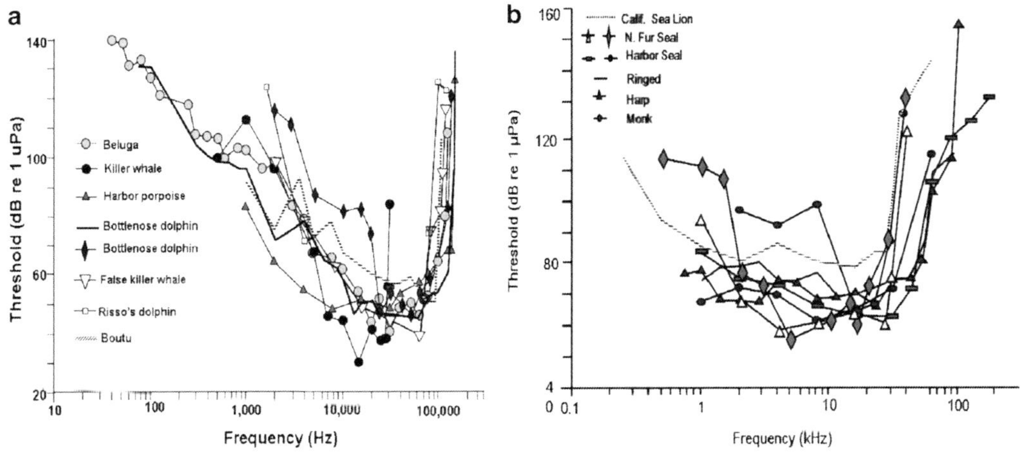

Fig. 2 Odontocete (**a**) and pinniped (**b**) audiograms. Elevated thresholds for one of the bottlenose dolphins, fur seals, and harbor seals indicate hearing deficits. Modified from Wartzok and Ketten (1999)

121–122 dB re 20 μPa (SEL 154–163 dB) in the sea lion and elephant seal. As in cetaceans, recovery times were relatively short, but longer exposures of 50 min required 3 days recovery, in common with land mammal data. SEL values were lower than in cetaceans, with a growth rate of ~2.5 dB TTS/dB noise and ~2.5 dB/doubling.

4 Conclusions

Undeniably, there have been serious consequences from noise exposures for marine mammals, including mass strandings. However, despite the importance of such events in bringing underwater noise to our attention, ironically, to date, there has been no demonstrable evidence of acute, traumatic, disruptive, or profound auditory damage in any marine mammal as the result anthropogenic noise exposures, including sonar (D'Amico et al. 2009; Ketten et al. 2003). This does not negate our concern but rather underscores our need for a better understanding of the many facets and consequences of sound use.

Although we are still uncertain about how robust or fragile these ears are, we have clear evidence that despite adaptations for diving and high-pressure environments, they are not impervious to permanent noise damage. Longitudinal studies report broad hearing losses and steep notches in both odontocetes and cetaceans (Ridgway and Carder 1997; Schusterman et al. 2002). Recent postmortem studies of ears from some of these subjects (Ketten et al. 2008) found evidence of sensorineural hearing loss in the form of extensive ganglion cell and auditory nerve fiber degeneration consistent with profound hearing deficits. Ears from strandings also show NIHL and age-related changes as well as other ear pathologies, including labyrinthitis ossificans, parasitic infestations, trauma, and chronic otitis media.

Despite these indicators that marine mammal hearing losses likely occur by the same mechanisms and etiologies as in land mammals, current data are insufficient to formulate population-level hazard criteria. Southall et al. (2007) attempted a risk-matrix approach for noise exposures using available audiometric and anatomical data and weighted SEL and peak SPL criteria but concluded that data gaps prevent assigning definitive exposure criteria. Furthermore, the fact that marine mammals sustain natural hearing loss from aging, trauma, and disease complicates definitive attribution of hearing loss from anthropogenic sources based on sampling a few individuals in coastal populations. Multiple possible loss causes should be considered and eliminated in any animal for which there is little or no history; therefore, the simple finding of a single animal with a hearing decrement cannot be taken as a clear indicator of a population-level hazard from any source without corroborative evidence. Protracted exposures and TTS to PTS conversions rather than acute individual impacts may be our major concern for populations. A high incidence of NIHL is unlikely to occur naturally across genders and ages in any wild population. Such a finding coincident with long-term or frequent intense exposures would be an appropriate cause for significant concern and action.

Acknowledgments Work for this paper was supported by the Office of Naval Research. I am grateful for many helpful discussions, particularly with Paul Nachtigall, Robert Gisiner, Colleen Reichmuth, Bill Ellison, and Sam Ridgway. Above all, I would like to acknowledge the invaluable contributions of Dr. Ron Schusterman to this field. He will be sorely missed, intellectually and personally, by all who were honored to call him colleague, mentor, and friend.

References

D'Amico A, Gisiner RC, Ketten DR, Hammock JA, Johnson C, Tyack PL, Mead J (2009) Beaked whale strandings and naval excercises. Aquat Mamm 35:452–472.

Davis RR, Kozel P, Erway LC (2003) Genetic influences in individual susceptibility to noise: A review. Noise Health 5:19–28.

Humes L (2010) Noise-induced loss in the U.S. military: A review. J Acoust Soc Am 127:1747.

Ketten DR, Arruda J, Cramer S, Hammock J (2008) Non-anthropogenic deafness in marine mammals: Hearing that is going, going, gone. J Acoust Soc Am 124:2505.

Ketten DR, Rowles T, Cramer S, O'Malley J, Arruda J, Evans PGH (2003) Cranial trauma in beaked whales. Proceedings of the Workshop on Active Sonar and Cetaceans, 17th Annual Conference of the European Cetacean Society, 8 March 2003, Las Palmas, Gran Canaria, pp 21–27.

Kryter KD (1996) Handbook of hearing and the effects of noise. Academic Press, New York.

Kujawa SG, Liberman MC. (2009) Adding insult to injury: Cochlear nerve degeneration after "temporary" noise-induced hearing loss. J Neurosci 29:14077–14085.

National Research Council (2003) Ocean noise and marine mammals. National Academies Press, Washington, DC.

Ridgway SH, Carder DA (1997) Hearing deficits measured in some *Tursiops truncatus* and the discovery of a deaf/mute dolphin. J Acoust Soc Am 101:590–594.

Schusterman RJ, Southall BL, Kastak D, Reichmuth C (2002) Age-related hearing loss in sea lions and their scientists. J Acoust Soc Am 111:2342–2343.

Slepecky N. (1986) Overview of mechanical damage to the inner ear: Noise as a tool to probe cochlear function. Hear Res 22:307–321.

Southall BL, Bowles AE, Ellison WT, Finneran JJ, Gentry RL, Greene CR Jr, Kastak D, Ketten DR, Miller JH, Nachtigall PE, Richardson WJ, Thomas JA, Tyack PL (2007) Marine mammal noise exposure criteria: Initial scientific recommendations. Aquat Mamm 33:411–521.

Wartzok D, Ketten DR (1999) Marine mammal sensory systems. In: Reynolds JE, Rommel SA (eds) Biology of marine mammals. Smithsonian Institution Press, Washington, DC, pp 117–175.

Yoshida N, Liberman MC (2000) Sound conditioning reduces noise-induced permanent threshold shift in mice. Hear Res 148:213–219.

Effect of Broadband Sounds on the Auditory Evoked Potential Thresholds in the Beluga Whale

Vladimir V. Popov and Alexander Y. Supin

1 Introduction

Intensive human activity in the ocean brings up the important issue of the negative impact of man-made noises on the behavior and physiology of marine animals. All living organisms are exposed to hazards, but cetaceans call for special attention. These highly developed marine mammals show complex behavior, their hearing is very sensitive, and the spectrum of perceived frequencies is very wide. Presently, researchers are focusing on disturbances of behavior and the parameters of auditory perception of some fairly intensive sounds simulating man-made noises (Southall et al. 2007). Also of importance is the effect of noises on perception thresholds at sound levels that do not cause lasting threshold shifts. The noises may mask the perception of communication signals, and in toothed whales, the operation of the sonar may be disturbed, which is the main tool for distance orientation in those animals. It appears of interest to investigate the relationship between the auditory thresholds and the level of broadband auditory signals.

2 Methods

The methods are based on the registration of the compound response (the set of evoked potentials) on the combination of several series of tone pips, with various frequencies and various rates of the pips in the series. The use of amplitude-modulated auditory stimuli in combination with recordings of overall evoked potentials proved to be an effective method for testing the auditory system of toothed whales (Finneran and Houser 2007; Popov et al. 1997). The magnitude of response at each frequency can be estimated on Fourier transformation of the compound response. By varying the parameters of tone components in a complex stimulus, we can estimate the magnitude of the response to signals with different carrier frequencies.

The experiments were performed on a mature *Delphinapterus leucas* (beluga whale) at the Utrish Marine Station, Institute of Ecology and Evolution, Russian Academy of Sciences. All the experiments were performed in conformity with the regulations for the use of animals in biomedical

V.V. Popov (✉) • A.Y. Supin
Severtsov Institute of Ecology and Evolution, Russian Academy of Sciences,
Moscow 119071, Russian Federation
e-mail: popov.vl.vl@gmail.com

A.N. Popper and A. Hawkins (eds.), *The Effects of Noise on Aquatic Life*,
Advances in Experimental Medicine and Biology 730, DOI 10.1007/978-1-4419-7311-5_47,

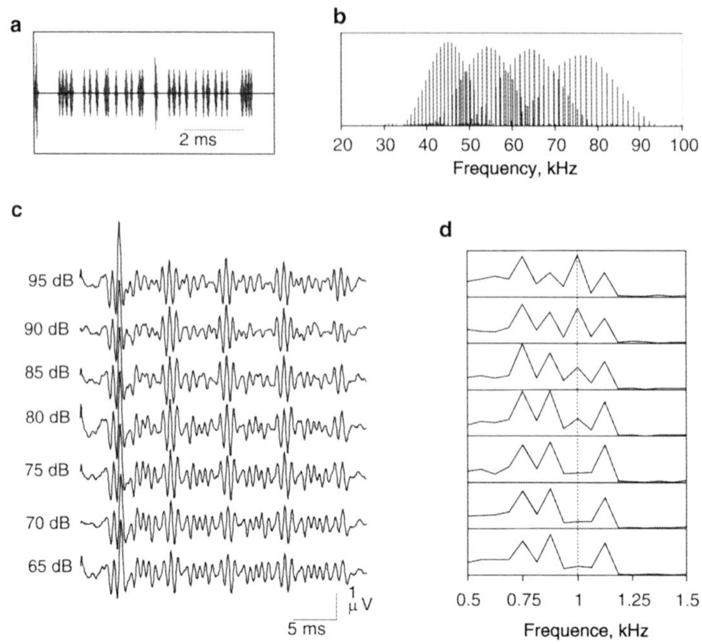

Fig. 1 (**a**) Oscillograms of an integral stimulus. (**b**) Amplitude spectrum of the stimulus. (**c**) Evoked potentials recorded in response to the integrated stimulus. The intensity of the component with a carrier frequency of 64 kHz and a pip rate of 1 kHz is indicated at the oscillograms, the intensity of the integrated stimulus being 85 dB. (**d**) Amplitude spectra of the responses. Dashed line shows the component with a carrier frequency of 64 kHz and a pip rate of 1 kHz

research of the Ministry of Science and Education, Russian Federation. For noninvasive evoked-potential recordings, suction-cup electrodes consisting of a 15-mm stainless-steel disk mounted within a 60-mm silicon suction cup were used. The electrical potentials were amplified within a range of 200 to 5,000 Hz; the amplified signal was digitized to be sent to the computer. To distinguish the signal from the noise, the digitized signals were coherently averaged.

The digital synthesis of auditory stimuli with an update rate of 512 kHz was used. The signals were converted into an analog form to be presented to the animals through a piezoceramic transducer (B&K 8104). The transducer was located 1 m in front of the animal's head. The signal level was controlled by a calibrated hydrophone (B&K 8103). The whale was exposed to a mixture of four series of tone pips with definite carrier frequencies. The envelope of each signal was a single cosinusoid; the ratio of the number of cycles of the carrier frequency to the envelope was invariably 8:1. Series were formed with pip rates of 0.875, 1, 1.125, and 1.25 kHz for the carrier frequencies of 45, 54, 64, and 90 kHz, respectively. The number of pips in each series was 40. All four series were summed up. A fragment of the resultant stimulus is represented in Figure 1a. The respective amplitude spectrum of that stimulus is shown in Figure 1b. The spectrum of that mixture was fairly broadband, with a considerable overlap of the spectra of the individual series. The intensities of all series of signals were leveled according to the frequency characteristics of the transducer, different durations of the tone pips, and different pip rates in the series. In the course of the experiment, both the intensity of the entire mixture and the intensities of individual components (series) could be controlled.

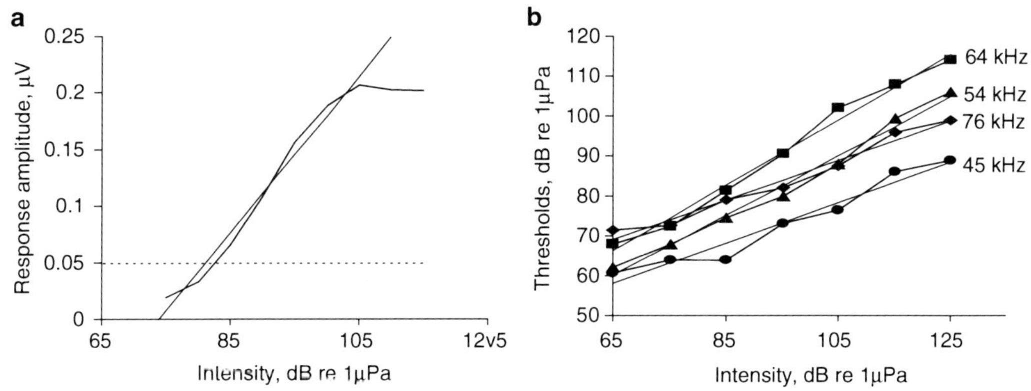

Fig. 2 (**a**) Relationship between the amplitude of the spectral peak at a frequency of 1 kHz (carrier frequency of 64 kHz) and the intensity. (**b**) Dependence of thresholds on the intensity of the three other components

3 Results

A compound response was recorded from the surface of the animal's head in response to a combination of four series of tone pips (Fig. 1c).

In that experimental series, the intensity of the integrated stimulus was a constant 85 dB, and the intensity of the component with a carrier frequency of 64 kHz at a 1-kHz signal sequence varied. The component intensities are indicated near the oscillograms. Some portion of the overall response, from 6 to 36 ms, was Fourier transformed. The spectra obtained contain four spectral peaks at the frequencies of the pips rate in each series: 0.875, 1, 1.125, and 1.25 kHz (Fig. 1d). The decrease in the intensity of the component with a carrier frequency of 64 kHz caused a decline of the spectral peak amplitude at a frequency of 1 kHz. The diagram of the relationship between the amplitude of the spectral peak at a frequency of 1 kHz (carrier frequency of 64 kHz) and the intensity also demonstrate a decrement of the amplitude with a reduction in intensity (Fig. 2a).

The intersection of the regression line with a near-threshold amplitude value of 0.05 μV was assumed to be the threshold of response. Figure 2b shows the dependence of the thresholds of responses to carrier frequencies of 45, 54, 64, or 76 kHz on the integrated intensity of the three other components. The increase in the intensity of the integrated stimulus results in an increase in the threshold of response. The threshold values obtained are approximated by regression lines. For the central frequencies of the integrated stimulus (54 and 64 kHz), an increase in the complex intensity by 1 dB resulted in an increase in the threshold of 0.7 and 0.8 dB, respectively. For the lateral frequencies of 45 and 76 kHz, the thresholds increased 0.5 dB/dB. These data are in good conformity with those on the effect of broadband noise on thresholds of the cat auditory nerve response (Costalupes et al. 1984).

Acknowledgments The study was supported by Grant No. 09-04-00688 from the Russian Foundation for Basic Research, Moscow, Russian Federation.

References

Costalupes JA, Young ED, Gibson DJ (1984) Effects of continuous noise backgrounds on rate of auditory nerve fibers in cat. J Neurophysiol 51:1326–1344.

Finneran JJ, Houser DS (2007) Bottlenose dolphin (*Tursiops truncatus*) steady-state responses to multiple simultaneous sinusoidal amplitude modulated tones. J Acoust Soc Am 121:981–992.

Popov VV, Supin AY, Klishin VO (1997) Paradoxical lateral suppression in the dolphin's auditory system: Weak sounds suppress response to strong sounds. Neurosci Lett 234:51–54.

Southall BL, Bowles AE, Ellison WT, Finneran JJ, Gentry RL, Greene CR Jr, Kastak D, Ketten DR, Miller JH, Nachtigall PE, Richardson WJ, Thomas JA, Tyack PL (2007) Marine mammal noise exposure criteria: Initial scientific recommendation. Aquat Mamm 33:411–522.

The ESME Workbench: Simulating the Impact of Anthropogenic Sound on Marine Mammals

David C. Mountain, David Anderson, Andrew Brughera, Matthew Cross, Dorian S. Houser, Nael Musleh, Michael Porter, and Martin Siderius

1 Introduction

The Effects of Sound in the Marine Environment (ESME) Workbench (http://esme.bu.edu) is a software tool designed to predict the impact of anthropogenic sounds on marine mammals. The ESME Workbench allows the user to select site-specific environmental data such as bathymetry, sound-speed profiles, sediment type, and average wind speed to predict sound propagation in a wide range of scenarios and to record the sound exposures received by virtual animals. The Workbench provides access to raw exposure information as well as summarized exposure information at the end of the simulation run. These data are made available in formats suitable for postprocessing utilizing a variety of data analysis tools.

2 Modules

The current version of the ESME Workbench is based on a series of interacting modules: Environment Builder, Acoustic Simulator, Scenario Simulator, and Data Viewer.

D.C. Mountain (✉) • D. Anderson • A. Brughera • N. Musleh
Boston University Hearing Research Center, Boston, MA 02215, USA
e-mail: dcm@bu.edu; da@bu.edu; brughera@bu.edu; naelmusleh@gmail.com

M. Cross • D.S. Houser
Biomimetica Inc., Santee, CA 92071, USA
e-mail: matthewjancross@gmail.com; dorian.houser.ctr@navy.mil

M. Porter
Heat, Light and Sound Research Inc., La Jolla, CA 92037, USA
e-mail: mikeporter@hlsresearch.com

M. Siderius
Department of Electrical and Computer Engineering, Portland State University, Portland, OR 97201, USA
e-mail: siderius@pdx.edu

A.N. Popper and A. Hawkins (eds.), *The Effects of Noise on Aquatic Life*,
Advances in Experimental Medicine and Biology 730, DOI 10.1007/978-1-4419-7311-5_48,

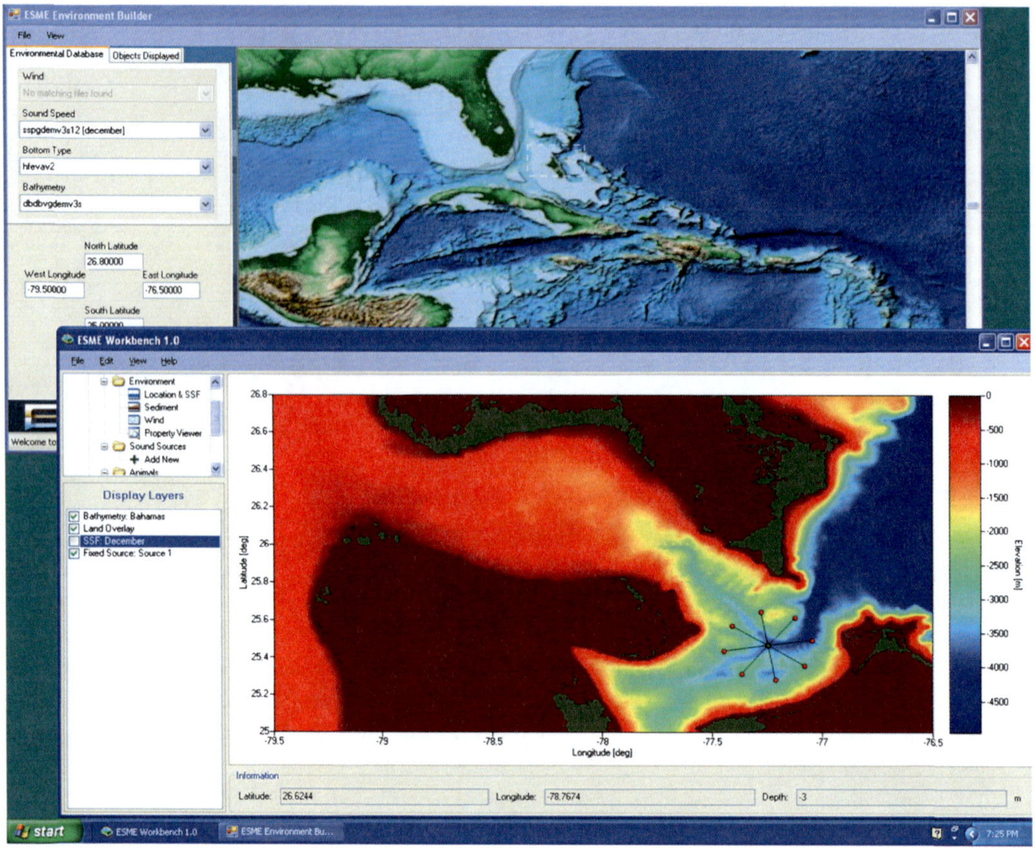

Fig. 1 The Effects of Sound in the Marine Environment (ESME) Environment Builder and Workbench modules

2.1 Environment Builder

The Environment Builder module allows the user to choose a region for the simulation by either specifying the coordinates or clicking and dragging a box on a map. The user then can browse the ESME Environmental Database and choose the data sources for environmental parameters (e.g., wind, sound-speed profile, bottom type, and bathymetry).

2.2 Acoustic Simulator

The acoustic propagation models use range-dependent depth profiles and depth-dependent sound-speed profiles to compute the received sound level for the simulated animal from each simulated source. The propagation models use bottom and sea surface characteristics to account for losses that occur during reflection at these boundaries. Sound sources are specified through parameters such as source location and depth, frequency, intensity, and beam pattern.

The current version of the simulator uses the Bellhop algorithm (Porter and Liu 1994). Bellhop uses a ray-tracing algorithm that computes relatively quickly and produces good results for water depths greater than 20 wavelengths. Source code, executables, and documentation can be downloaded from the Ocean Acoustics Library (http://oalib.hlsresearch.com/).

2.3 Scenario Simulator

The ESME Scenario Simulator consists of three closely interacting subsystems: Animal Behavior, Source Behavior, and Acoustic Exposure.

The Animal Behavior subsystem uses a Markov model for each individual. Animals are assumed to be in one of a finite number of behavioral states (descending, foraging, ascending, surface travel) and to transition from one state to another in a probabilistic manner (Houser 2006). Animal location as well as instantaneous and cumulative sound exposure is logged for subsequent analysis.

The Source Behavior subsystem is a simple movement model intended for surface and submerged sources moving in a deterministic manner. Initial location, depth, course, and speed are provided (or are defined as being randomized within specified limits), and (optionally) a closed, simple polygon is provided for the source to bounce around inside of. Acoustic duty cycle information (ping duration and repetition rate) is also specified. Source speed is presumed to be constant throughout the duration of the simulation.

The Acoustic Exposure subsystem combines the output of the Animal Behavior, Source Behavior, and Acoustic Simulator modules to predict animal exposure to each active source, once per second of simulated time. The exposure data are logged in both raw and summarized forms, which are output to a text file suitable for postprocessing using any of a number of commercial or custom analysis packages.

2.4 Data Viewer

The Data Viewer module allows users of the ESME Workbench to visualize different data that are used by or produced by the various modules described above. For example, environmental parameters such as sound-speed profiles and surface loss can be displayed as line graphs. When a parameter data layer (e.g., sound-speed profile) is the active layer, the sound-speed graph is displayed when the user clicks on one of the sound-speed profile locations in the user console window.

Sound fields can be plotted as transmission loss (TL) in decibel attenuation or as sound level (dB re 1 μPa). The range and offset of the color scale can be adjusted by clicking and dragging the mouse over the color bar. Images can be exported in a number of standard formats (BMP, GIF, JPEG, or PNG), and data can be exported to comma-separated text (CSV) files. A cursor tool tip that shows numerical data values can be used to explore the color plots.

References

Houser DS (2006) A method for modeling marine mammal movement and behavior for environmental impact assessment. IEEE J Ocean Eng 31:76–81.

Porter MB, Liu YC (1994) Finite-element ray tracing. In: Lee D, Shultz MH (eds) Theoretical and computational acoustics, vol. 2. World Scientific Publishing, Singapore.

Using EarLab to Study Masking Due to Anthropogenic Sound

David C. Mountain, David Anderson, and Andrew Brughera

1 Introduction

The mammalian auditory system is a highly evolved acoustic signal-processing system that performs well even in highly reverberant and cluttered acoustic environments. In cetaceans, the auditory system is even more highly evolved and is generally more important than vision for navigation, foraging, and social communication. As humans inject more and more acoustic energy into the marine environment, these important acoustic functions may become compromised. Unfortunately, little is known about the hearing capabilities of many marine species in quiet and much less is known about the impact of anthropogenic sounds that could mask biologically significant signals.

Biophysical computer models based on physiological and behavioral experiments performed on easily studied species can be extrapolated to those species that are not easily studied. We have used the EarLab (http://earlab.bu.edu) desktop simulation environment to study masking effects in a variety of scenarios. Model parameters for species of interest are estimated from behavioral audiograms and other available data. The models can be used to predict how different types of biologically significant sounds are represented in neural firing patterns and how the neural representation degrades in the presence of anthropogenic noise.

2 Methods

To create a corpus of waveforms with known signal-to-noise ratios (SNRs), we used cetacean vocalizations recorded under relatively quiet conditions and added scaled recordings of shipping noise. Because animal vocalizations are commonly transient in nature while shipping noise represents a more continuous signal, we have defined the SNR in terms of the peak-to-peak amplitude of the signal (V_S) and the noise (V_N)

$$SNR = 20\log\left(\frac{V_S}{V_N}\right) \tag{1}$$

D.C. Mountain (✉) • D. Anderson • A. Brughera
Hearing Research Center and Department of Biomedical Engineering,
Boston University, Boston, MA 02215, USA
e-mail: dcm@bu.edu

A.N. Popper and A. Hawkins (eds.), *The Effects of Noise on Aquatic Life*,
Advances in Experimental Medicine and Biology 730, DOI 10.1007/978-1-4419-7311-5_49,

For each vocalization-noise pair, several files were prepared with different SNRs. The SNR values were chosen to bracket the SNR for which human listeners found the vocalization to be barely detectable.

To simulate the cetacean auditory system, EarLab modules representing the middle ear, cochlear mechanics, inner hair cells, and the auditory nerve were used. If audiograms for the species of interest were available, they were used to estimate the middle ear cutoff frequency and the range of the cochlear frequency-place map. If no audiogram was available, a human model was scaled to match the putative frequency range for the species. The output of the model was the average instantaneous firing rate for 256 populations of auditory nerve fibers, with each population representing fibers tuned to a specific frequency range. The spacing between the best frequencies of these populations were chosen to represent equal spacing along the length of the cochlea.

3 Results

For data-visualization purposes, cochleagrams were created. These plots are similar to spectrograms in that they are a form of time-frequency representation and color is used to represent the response amplitude. One difference is that the frequency axis is scaled to represent equal increments in cochlear position rather than equal increments in frequency. We should also point out that the bandwidth of the cochlear filters changes with best frequency, whereas in the classical spectrogram, the filter bandwidths are constant.

Figure 1 illustrates the cochleagrams for a *Balaenoptera physalus* (fin whale) vocalization in quiet and in the presence of shipping noise (0-dB SNR). A scaled human model was used for this simulation, and only frequency channels corresponding to the vocalization frequencies are shown.

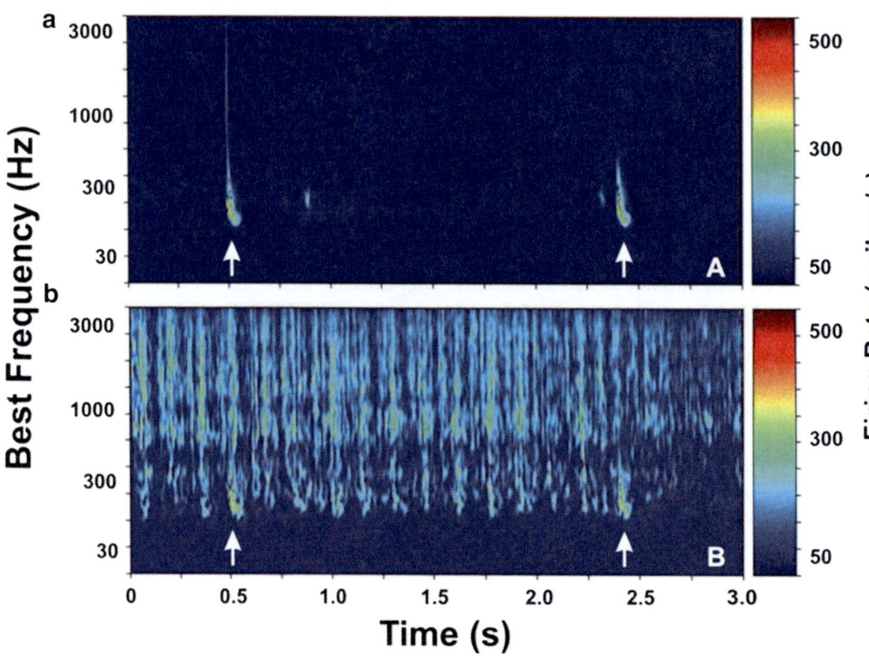

Fig. 1 Simulated auditory nerve activity for a fin whale call in quiet (**a**) and the same call with background shipping noise (**b**)

The brief, frequency-modulated vocalizations (Fig. 1, arrows) are clearly visible in the quiet condition but barely detectable visually in the noise background. At this SNR, the vocalization was barely detectable acoustically.

Biologically plausible detectors (e.g., energy detection, cochleagram correlation) are currently being tested with the simulated auditory nerve output for their ability to replicate human detection performance for these signals. Our preliminary results indicate that model performance is similar to that for human listeners.

4 Conclusions

Physiologically based models provide a method for more realistic estimates of masking effects in complex acoustic environments than conventional approaches because they include phenomena such as the upward spread of masking and forward masking that are not included in conventional techniques.

Portable Auditory Evoked Potential System to Assess Odontocete Hearing

Aude F. Pacini, Paul E. Nachtigall, and Laura N. Kloepper

1 Introduction

The hearing of marine mammals has been extensively studied in the last decades and has focused primarily on species available in captivity such as the bottlenose dolphin *Tursiops truncatus*. Recent work has shown that mass stranding events could be related to anthropogenic sound exposure such as naval sonar activities, seismic surveys, or oil drilling exploration. Although little is known about the hearing abilities of most odontocete species, it is primordial to be able to obtain a rapid hearing assessment of stranded animals. A portable system was designed to be easily transported and used during stranding events, at rehabilitation facilities, and in laboratory settings. This overview provides a description of this system as well as an overview of the data collected so far.

2 Materials and Methods

The system was first presented by Taylor et al. (2007), and additional modifications have been implanted to provide more flexibility and portability to the system.

2.1 Background Noise Measurements

For any new facility or stranding, background noise measurements were collected with a RESON TC-4032 (−170 dB re 1 V/mPa; Slangerup, Denmark) connected to a Microtrak II two-channel digital recorder (M-Audio, Irwindale, CA). Noise files were recorded at a 96-kHz sampling rate with one channel with no gain and the second channel with variable gain. Files of 1-s duration were then extracted, averaged, and analyzed with Adobe Audition 3.0 and a custom MATLAB program.

A.F. Pacini (✉) • P.E. Nachtigall • L.N. Kloepper
Marine Mammal Research Program, Hawai'i Institute of Marine Biology,
Kane'ohe, HI 96744, USA
e-mail: aude@hawaii.edu

A.N. Popper and A. Hawkins (eds.), *The Effects of Noise on Aquatic Life*,
Advances in Experimental Medicine and Biology 730, DOI 10.1007/978-1-4419-7311-5_50,

2.2 Acoustic Stimulus Presentation

At the start of each data session, the animal was stationed at the surface and three electrodes encased in soft latex rubber suction cups were positioned on the animal's back. The acoustic stimuli were presented through a hydrophone positioned 1 or 2 m in front of the animal's ears at a depth of 50 cm. Different hydrophones were used depending on the range of frequencies tested: an ITC 1032 (Santa Barbara, CA) for frequencies between 4 and 50 kHz and a RESON 4013 for frequencies above 50 kHz. Because of the limited time usually available during stranding events, the hydrophones were normally calibrated after the data collection.

Sound stimuli were sinusoidally amplitude-modulated (SAM) tone bursts. The tones were usually modulated at 1,000 Hz based on modulation rate transfer functions obtained in the past or before the hearing measurements. For frequencies below 50 kHz, the update rate was 256 kHz and was raised to 512 or 800 kHz for frequencies above 50 kHz. The tones were digitally synthesized with a custom LabVIEW program and a National Instrument PCMIA-6062 E DAQ card (Austin, TX) implemented in a laptop. The tone bursts were normally 19 ms in duration followed by 30 ms of silence so that the acoustic stimuli were presented at a rate of 20 ms^{-1}. Output voltages were measured peak to peak with a Tektronix TPS 2014 oscilloscope (Beaverton, OR) and were then converted to equivalent root mean square voltages (peRMS) to calculate the sound pressure level (SPL) for each individual frequency. SPL was varied in 1- to 10-dB steps with a Hewlett-Packard P-350D (Palo Alto, CA) attenuator.

2.3 Auditory Brain Stem Response Recording

Three gold human EEG electrodes embedded in latex suction cups were used to collect the animal's neurological responses to the acoustic stimulus. The first electrode was positioned 5 cm posterior to the blowhole, the second electrode was on the back, and the third ground electrode was on the dorsal fin of the subject. The three suction cups were connected to a Grass CP-511 bioamplifier (West Warwick, RI), and the signal was amplified 10,000 times and filtered from 300 to 3,000 Hz. Additional filtering was performed by a Krohn-Hite 3384 bypass filter (Brockton, MA) with similar settings. The response signal was then digitized at a 16-kHz rate with the same PCMIA computer card that generated the acoustic stimulus. A complete record consisted of collecting and averaging 1,000 responses, which were 26 ms long and triggered with the acoustic stimulus.

2.4 Data Analysis

During a stranding event, a complete audiogram can be collected in less than an hour and can include up to 10 frequencies ranging from 4 to 128 kHz. For each frequency, an average of seven stimulus-level trials was necessary to obtain a threshold. SPL was decreased until no response was visible for at least two trials. Previous work has shown that SAM tone bursts generate an envelope following response (EFR) (Nachtigall et al. 2007; Popov and Supin 2007). For each frequency and SPL, a 256-point fast Fourier transform (FFT) was performed on a 16-ms window of the EFR. Each FFT provided a frequency spectrum, and the peak response at 1,000 Hz was used to estimate the response of the subject to the acoustic stimulus. For a given frequency, the peak responses were the repetition rate and were then plotted as a function of the stimulus SPL. A linear regression was used to calculate the hypothetical zero value that was used as the approximation of the threshold for that frequency. An audiogram consisted of compiling all the thresholds for each frequency and the corresponding SPL threshold (Fig. 1).

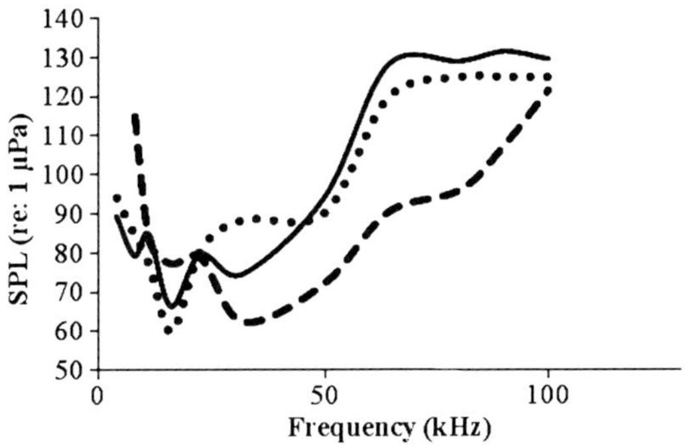

Fig. 1 Audiograms of three captive adult *Tursiops truncatus* collected in French Polynesia. SPL, sound pressure level

3 Conclusions

Absolute thresholds cannot be obtained using the auditory evoked potential (AEP) because of the inherent biological electrical noise. However, previous studies have indicated that behavioral audiograms are very similar to hearing measurements obtained with the AEP technique (Yuen et al. 2005) Therefore, this method seems to be ideal to measure hearing in stranded animals either during their rehabilitation or soon after the stranding event if portable pools are available. In addition to the basic system, additional measurements can be collected. The use of a jaw phone to produce sound in localized areas directly on the head of the subject has been used to examine sound paths with the described system (Mooney et al. 2008). The modulation rate transfer function can easily be collected by varying the modulation rate of the acoustic stimulus (Mooney et al. 2009). This system has been used in many situations including in the laboratory, in collaboration with marine mammal parks, in long-term rehabilitation facilities, in portable pools after stranding events, and during capture and release of wild animals. Continuous improvements will provide a reliable and adaptable platform to use on new species and perhaps on mysticetes.

References

Mooney TA, Nachtigall PE, Castellote M, Taylor KA, Pacini AF, Esteban JA (2008) Hearing pathways and directional sensitivity of the beluga whale *Delphinapterus leucas*. J Exp Mar Biol Ecol 362:108–116.

Mooney TA, Nachtigall PE, Taylor KA, Rasmussen MH, Miller LA (2009) Auditory temporal resolution of a wild white-beaked dolphin (*Lagenorhynchus albirostris*). J Comp Physiol A 195:375–384.

Nachtigall PE, Mooney TA, Taylor KA, Yuen ML (2007) Hearing and auditory evoked potential methods applied to odontocete cetaceans. Aquat Mamm 33:6–13.

Popov VV, Supin AY (2007) Analysis of auditory information in the brains of cetaceans. Neurosci Behav Physiol 37:285–291.

Taylor KA, Nachtigall PE, Mooney TA, Supin AY, Yuen MML (2007). A portable system for the evaluation of the auditory capabilities of marine mammals. Aquat Mamm 33:93–99.

Yuen ME, Nachtigall PE, Supin AY, Breese M (2005) Behavioral and auditory evoked potential audiograms of a false killer whale (*Pseudorca crassidens*). J Acoust Soc Am 118:2688–2695.

Barotrauma in Fish and Barotrauma Metrics

Thomas J. Carlson

1 Introduction

In general terms, barotrauma is defined as an injury or disorder resulting from the establishment of a pressure difference across the wall of an anatomical structure or an injury of a body part or organ as a result of changes in pressure. In fish, barotrauma is physiological damage to nonauditory tissue. Barotrauma may be expressed as physical injury or changes in behavior and may result in immediate or delayed direct or indirect mortality.

Barotrauma injury in fish can be quite variable in both cause and effect depending on details of the exposure to pressure change and the physiology and physiological state of exposed fish. Barotrauma end points include mortal injury (McKinstry et al. 2007) and a number of injuries with varying severity and impact not linked to immediate or delayed mortality. Injuries that are not mortal injuries, such as temporary stunning (Sutherland 1972), loss of buoyancy, or any of many other temporary or permanent injuries which reduce fitness, can, nonetheless, result in mortality by predation.

Changes in pressure that can cause barotrauma in fish can be divided into two general categories: decompression and mechanical. Decompression can be subdivided into rapid and impulsive decompression. These categories of barotrauma are not mutually exclusive but may co-occur depending on exposure.

2 Decompression

Essentially all decompressive barotrauma is caused by changes in the state (in solution or free) and volume expansion of air-containing structures (e.g., swim bladder). The most important driving physical principals are 1) the relationship between the volume of a bubble or structure (swim bladder) and the change in pressure acting on the air-filled bubble or structure (Boyle's law) and 2) the relationship between the solubility of gas in blood and other body fluids and the pressure acting on the fish (Henry's law). Both the magnitude and direction of change in pressure are important.

T.J. Carlson (✉)
Pacific Northwest National Laboratory, Battelle, Richland, WA 99352, USA
e-mail: Thomas.Carlson@pnl.gov

A.N. Popper and A. Hawkins (eds.), *The Effects of Noise on Aquatic Life*,
Advances in Experimental Medicine and Biology 730, DOI 10.1007/978-1-4419-7311-5_51,

In general, for the pressure changes fish experience during rapid decompression, the compression of air-filled structures and bubbles does not cause barotrauma; it is expansion that causes injury.

There is an increase in the volume of any air-filled enclosure within the body of the fish when external pressure decreases. The change in volume (V) is proportional to the magnitude of the change in pressure (P; Boyle's law: $P_1V_1 = P_2V_2$). The rate of change in volume is essentially the same as the rate of change in pressure. Simultaneously, an increase in blood volume is caused by the formation of bubbles from gas released from solution in the blood (Henry's law: $P = k_gC$, where k_g is Henry's law constant for a gas and C is concentration, or, for a particular gas, $P_1C_2 = P_2C_1$). The increased blood volume puts pressure on arteries and veins and can disrupt the function of internal organs and biological processes essential for the survival of the fish. These physical phenomena are the causes of commonly observed decompressive barotrauma injuries that include, but are not limited to, rupture of blood vessels, bruising, severe physical damage to organs, swim bladder rupture, and occlusion of the circulatory system (Beyer et al. 1976; Cramer and Oligher 1964; Rummer and Bennett 2005; Tsvetkov et al. 1972).

For physoclistous fish, it has long been known that the depth of acclimation before pressure change exposure is a critical factor influencing the magnitude of injury and mortality rates (Abernethy et al. 2001, 2002, 2003; Beyer et al. 1976; Cada 1990; Cada et al. 1997; Rummer and Bennett 2005).

The susceptibility of physoclistous fish to barotrauma during hydroturbine passage, given acclimation to depth, has been found to be proportional to the ratio of absolute pressure at their acclimation depth to the lowest absolute pressure experienced during turbine passage. Recent evidence suggests that not including physiological state as an element of treatment most likely results in a bias in assessing the risk of barotrauma from exposure to a higher rate of change of rapid decompression (Carlson and Abernethy 2005). It is likely that the same caution is true for barotrauma assessment for exposure to near-field impulsive sound.

3 Rapid Decompression

Rapid decompression, which is sometime referred to in the published literature as "catastrophic decompression" (Rummer and Bennett 2005), typically refers to a decrease in the static pressure acting on a fish that may be more or less rapid. Rapid decompression is almost always caused by the forced movement of a fish from a greater to a lesser depth, thereby from higher to lower static pressure. It occurs when a fisherman brings a fish at depth to the surface more rapidly than the fish can remove gas from its swim bladder (D'Aoust 1973; D'Aoust and Smith 1974; Feathers and Knable 1983). Fish without swim bladders are at a low risk of injury from rapid decompression, whereas the risk to fish with swim bladders depends on their physiology and physiological state.

Another source of rapid decompression, the passage of fish through hydroturbines, has been extensively studied. In this case, water flowing through a hydroturbine experiences changes in pressure associated with a transfer of energy to turn a rotor, thereby generating electricity. Figure 1 shows how pressure changes as water flows through a hydroturbine. Depending on the nature of the hydroturbine, the rate of change in pressure experienced by a fish carried with flow through the machine can occur at rates very similar to those occurring during impulsive decompression.

The response of juvenile Chinook salmon to rapid decompression during passage through a hydroturbine is shown in Figure 2 (Skalski and Seaburg 2010). The *x*-axis is in terms of a derived variable, which is the natural log of the ratio of acclimation to nadir (minimum) pressures. The *y*-axis is the probability of mortal injury.

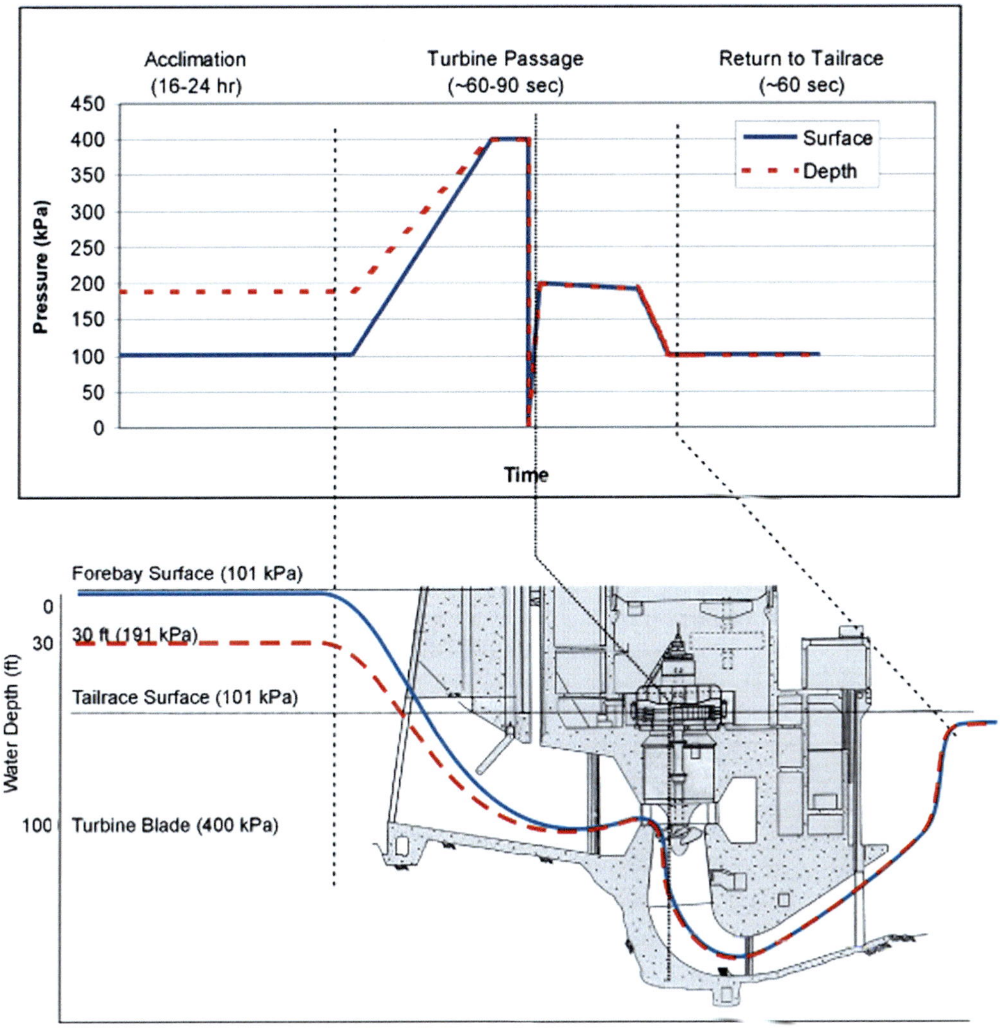

Fig. 1 Two examples of the pressure change a fish might experience while passing through a Kaplan hydropower turbine. One is a pressure-time series for a fish acclimated to near-surface static pressure (blue line) and the other is for a fish acclimated to the static pressure at a depth of 30 ft (red dashed line). From Abernethy et al. (2002)

Fig. 2 Response of juvenile Chinook salmon to rapid decompression. The independent variable is the natural log of the ratio of acclimation to nadir (lowest exposure) pressures. The dependent variable is the probability of mortal injury. From Skalski and Seaburg (2010)

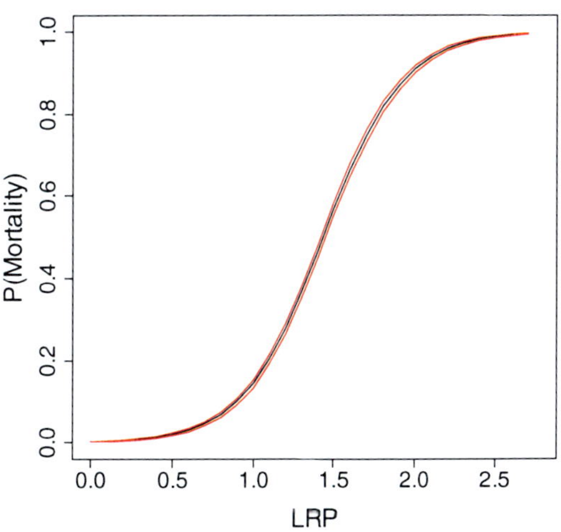

4 Mechanical (Impact) Injury From Impulsive Sound

A major mechanism of injury to fish from impulsive signals is believed to be impact experienced by the fish as a significant difference in pressure across tissue. The difference in pressure, if large enough, could cause tearing and other mechanical injury of almost any organ or other tissue in the body of the fish. Almost all observations of impact injury come from studies of exposure of fish to high-energy impulsive signals caused by explosions (Keevin and Hempen 1997; Weston 1960; Yelverton et al. 1975). The severity of damage, in addition to the characteristics of the impulsive signal, is a function of the physiology of the exposed fish, which can vary widely with species, age, and condition factors.

5 Barotrauma Metrics

The most common barotrauma metric is direct, immediate mortality. However, research indicates that there are a number of barotrauma injuries that do not result in immediate mortality and may not even be severe enough for direct delayed mortality but may increase the risk to the fish of indirect mortality by predation. For rapid decompression studies, we derived a new response metric, mortal injury.

In a study of rapid decompression (Brown et al. 2007), juvenile Chinook salmon were exposed to rapid decompression. Those fish that died during exposure were necropsied after removal from a test chamber. The surviving fish were held for 48 hours after exposure. Those that died during holding were immediately necropsied. At the end of the holding period, all surviving fish were necropsied.

A series of statistical tests were conducted to identify observed injuries that had high odds of occurring in fish that died during the period from exposure through the end of the holding period. These tests found that a set of 22 barotrauma injuries had high odds of occurring in fish that died. A logistic model-building exercise found that 8 of the 22 higher odd barotrauma injuries were sufficient to satisfy goodness-of-fit measures for a model predicting mortality given the occurrence of immediate and delayed mortal injuries in test fish (McKinstry et al. 2007). For subsequent studies, the mortal injury metric permitted necropsy of test fish immediately after exposure to rapid decompression without sacrificing consideration of delayed mortality. Mortal injury also permits the test fish to be classified into one of two classes depending on observation of a mortal injury during necropsy. This binomial classification of test fish permit use of logistic regression and similar analysis tools to derive stimulus-response functions that facilitate quantitative assessment of exposure risk.

Studies currently underway of barotrauma in juvenile salmonids after exposure to pile driving and blasting impulsive sounds are exploring additional barotrauma metrics.

References

Abernethy CS, Amidan BG, Cada GF (2001) Laboratory studies of the effects of pressure and dissolved gas supersaturation on turbine-passed fish. PNNL-13470, Pacific Northwest National Laboratory, Richland, WA.

Abernethy CS, Amidan BG, Cada GF (2002) Simulated passage through a modified kaplan turbine pressure regime: A supplement to Laboratory studies of the effects of pressure and dissolved gas supersaturation on turbine-passed fish. PNNL-13470-A, Pacific Northwest National Laboratory, Richland, WA.

Abernethy CS, Amidan BG, Cada GF (2003) Fish passage through a simulated horizontal bulb turbine pressure regime: A supplement to "Laboratory studies of the effects of pressure and dissolved gas supersaturation on turbine-passed fish." PNNL-13470-B, Pacific Northwest National Laboratory, Richland, WA.

Beyer DL, D'Aoust BG, Smith LS (1976) Decompression-induced bubble formation in salmonids: Comparison to gas bubble disease. Undersea Biomed Res 3:321–338.
Brown RS, Carlson TJ, Welch AE, Stephenson JR, Abernethy CS, Mckinstry CA, Theriault MH (2007) Assessment of barotrauma resulting from rapid decompression of depth-acclimated juvenile Chinook salmon bearing radio telemetry transmitters. PNNL-17690, Pacific Northwest National Laboratory, Richland, WA.
Cada FG (1990) A review of studies relating to the effects of propeller-type turbine passage on fish early life stage. N Am J Fish Manage 10:418–426.
Cada GF, Coutant CC, Whitney RR (1997) Development of biological criteria for the design of advanced hydropower turbines. Report DOE/ID-10578, Idaho Operations Office, US Department of Energy, Idaho Falls, ID.
Carlson TJ, Abernethy CS (2005) Pilot study of the effects of simulated turbine passage pressure on juvenile Chinook salmon acclimated with access to air at absolute pressures greater than atmospheric. PNNL-15011, Pacific Northwest National Laboratory, Richland, WA.
Cramer FK, Oligher RC (1964) Passing fish through hydraulic turbines. Trans Am Fish Soc 92:243–259.
D'Aoust BG (1973) Experiments in the physiology of the Swim bladder. Exp Physiol Biochem 6:33–46.
D'Aoust BG, Smith LS (1974) Bends in fish. Comp Biochem Physiol 49A:311–321.
Feathers MG, Knable AE (1983) Effects of depressurization upon largemouth bass. N Am J Fish Manage 3:86–90.
Keevin TM, Hempen GL (1997). The environmental effects of underwater explosions with methods to mitigate impacts. A Manual Published by the US Army Corps of Engineers, St. Louis District, St. Louis, MO.
McKinstry C, Carlson TJ, Brown, RS (2007) Derivation of mortal injury metric for studies of rapid decompression of depth acclimated physostomous fish. PNWD-1708, Battelle-Pacific Northwest Division, Richland, WA.
Rummer JL, Bennett WA (2005) Physiological effects of swim bladder overexpansion and catastrophic decompression on red snapper. Trans Am Fish Soc 134:1457–1470.
Skalski JR, Seaburg AG (2010) The comprehensive analysis of mortal injury for hatchery yearling and subyearling Chinook salmon smolts subject to simulated pressure changes associated with turbine passage. Columbia Basin Research, School of Aquatic and Fishery Sciences, University of Washington, Seattle, WA.
Sutherland DF (1972) Immobilization of fingerling salmon and trout by decomposition. Technical Report NMFS SSRF-655, National Oceanic and Atmospheric Administration, Seattle, WA.
Tsvetkov VI, Pavlov DS, Nezdoliy VK (1972) Changes in hydrostatic pressure lethal to the young of some freshwater fish. J Ichthyol 12:307–318.
Weston DE (1960) Underwater explosions as acoustic sources. Proc Phys Soc 76:233–249.
Yelverton JT, Richmond DR, Hicks W, Saunders K, Fletcher ER (1975) The relationship between fish size and their response to underwater blast. Report DNA 3677T, Defense Nuclear Agency, Washington, DC.

Assessment of Barotrauma Injury and Cumulative Sound Exposure Level in Salmon After Exposure to Impulsive Sound

Michele B. Halvorsen, Brandon M. Casper, Thomas J. Carlson, Christa M. Woodley, and Arthur N. Popper

1 Introduction

Concern continues about the potential effects the sounds produced during pile-driving activities have on fish (e.g., Popper and Hastings 2009). We investigated the effects of high-intensity sound on juvenile Chinook salmon (*Oncorhynchus tshawytscha*). State-of-the-art equipment was developed to expose fish to pile-driving signals in the laboratory.

2 High-Intensity Controlled Impedance–Fluid-Filled Wave Tube

The equipment is called a high-intensity controlled impedance–fluid-filled wave tube (HICI-FT) that was designed and built by Peter Rogers and James Martin at the Georgia Institute of Technology, Atlanta, GA. The HICI-FT is made of a stainless-steel tube with an electromagnetic shaker at each end. Adaptations to the HICI FT were vibration isolation feet to decouple the system from the building and a water-conditioning system, including a chiller, for the health of the fish and compliance of water inside the HICI-FT. Also the HICI-FT needed to be cooled to function properly; therefore, cool air was brought to the electromagnetic shakers and adaptations were made to remove heat from the room.

3 Pile-Driving Impact Signals

Eight pile-driving signals, recorded during driving of steel shell pile and provided by the Washington Department of Transportation, were used in these studies. These eight signals differed slightly in their spectral content and their time waveform; however, they were normalized to have the same

M.B. Halvorsen (✉) • B.M. Casper • A.N. Popper
Department of Biology, University of Maryland College Park, College Park, MD 20742, USA
e-mail: mb_halvorsen@yahoo.com

T.J. Carlson
Pacific Northwest National Laboratory, Battelle, Portland, OR 97204, USA

C.M. Woodley
Pacific Northwest National Laboratory, Battelle, Richland, WA 99352, USA

A.N. Popper and A. Hawkins (eds.), *The Effects of Noise on Aquatic Life*, Advances in Experimental Medicine and Biology 730, DOI 10.1007/978-1-4419-7311-5_52,

single-strike sound exposure level (SEL). These 8 signals were randomized into a block of 96 strikes in which each signal was represented exactly 12 times randomly within the block. The block of 96 was then repeated to achieve the desired number of pile strikes, i.e., the block of randomized signals was repeated 10 times to produce 960 strikes.

4 SEL Target

Each data point (i.e., exposure-treatment group) had a specific cumulative SEL target. Some data points were paired in that the elements of the pair had the same cumulative SEL target but were obtained with a different number of pile strikes, i.e., a different single-strike SEL for the signals within a block. This pairing was done to test whether the same cumulative SEL caused similar barotrauma injuries regardless of the number of pile strikes and different single-strike SEL.

5 Physiological Condition

At the start of each experiment, after handling, juvenile salmon were allowed to acclimate in the HICI-FT acrylic entry chamber to fill their swim bladder and achieve neutral buoyancy. Experimental findings from rapid decompression studies indicated that negatively buoyant physostomous fish (i.e., salmonids) at the time of exposure were at lower risk of barotrauma than those that were neutrally buoyant (Brown et al. 2007). These findings support other experiments that found that fish without a swim bladder were at a much lower risk of barotrauma than fish with a swim bladder (Yelverton et al. 1975).

6 Data Management

It is important to perform and demonstrate quality control of data. Quality was achieved by utilizing a one-over system where one person digitally entered the data and another person double-checked each cell for accuracy. Often the data were further ensured with a third check of each cell along with inspection for compliance with experimental protocols. Quality assurance was applied to every part of the experiment, which included observations of the sound presented to the test fish in the HICI-FT, observation of fish during acclimation, fish physical metrics, and the score for each barotrauma injury. After quality assurance was completed, the data for each fish went through a rigorous ranking assignment with justification for the ranking. The rankings used were 1) full acceptance of the data for a fish, 2) conditional acceptance of the data for a fish, and 3) deletion of data for a fish from study. The criteria for these rankings included fish physiological condition, buoyancy at exposure, food in gut, weight/length ratio, and compliance with experimental protocols during experimental procedures. Finally, after removal of unacceptable fish, the data were analyzed, including the number of injuries per test fish and the frequency of occurrence of injuries for the sample of fish within each treatment.

7 Barotrauma Scoring

After exposure to a block of pile-driving impact signals, fish were individually examined for barotrauma injuries. Exposed fish were examined using a panel of 63 potential barotrauma injuries. The occurrence of an injury was scored as 1 and absence by 0. In the first step of analysis, physiologically significant injuries and those that occurred with the highest frequency throughout the entire study regardless of SEL or strike number exposure were identified. The result was a group of ~20 injuries that had the highest frequency of occurrence and were also physiologically significant for fish health. Further statistical analysis was conducted using the smaller grouping of injuries.

Data analysis continues. Initial results indicate a strong trend in increase in the number of injuries per fish and the frequency of occurrence of injuries within treatment with increasing severity of exposure.

References

Brown RS, Carlson TS, Welch AE, Stephenson JR, Abernethy CS, Mckinstry CA, Theriault MH (2007) Assessment of barotrauma resulting from rapid decompression of depth-acclimated juvenile Chinook salmon bearing radio telemetry transmitters: A white paper. PNNL-17690, Pacific Northwest National Laboratory, Battelle, Richland, WA.

Popper AN, Hastings MC (2009) The effects on fish of human-generated (anthropogenic) sound. Integr Zool 4:43–52. Available via http://www.wsdot.wa.gov/NR/rdonlyres/0B027B4A-F9FF-4C88-8DE0-39B165E4CD94/61426/BA_HumanSoundonFish.pdf.

Yelverton JT, Richmond DR, Hicks W, Saunders K, Fletcher ER (1975) The relationship between fish size and their response to underwater blast. A white paper. Topical Report DNA 3677T, Defense Nuclear Agency, Department of Defense, Washington, DC.

Shipboard Assessment of Hearing Sensitivity of Tropical Fishes Immediately After Exposure to Seismic Air Gun Emissions at Scott Reef

Mardi C. Hastings and Jennifer Miksis-Olds

1 Introduction

A shipboard system for measurement of auditory evoked potentials (AEPs) in fish was developed to investigate the effects on hearing in tropical reef fish after exposure to emissions from an air gun array used in a three-dimensional marine seismic survey at Scott Reef, Western Australia (Woodside Energy Ltd. 2007). Scott Reef is a coral atoll located in the Indian Ocean ~430 km north of Broome. The species studied included *Myripristis murdjan* (pinecone soldierfish), *Chromis viridis* (blue green damselfish), *Sargocentron spiniferum* (sabre squirrelfish), and *Lutjanus kasmira* (bluestripe seaperch).

2 Methods

Fish were collected in the field by professional divers using standard methods developed by aquarium collectors and marine scientists. All fish were caught in water depths less than 20 m, with the majority of fish caught in depths between 5 and 15 m. The fish were placed in covered buckets with small holes drilled in the lids and stored at holding locations in a sand habitat at the northwest edge of South Scott Reef.

2.1 Sound Exposure Sites

Three exposure sites were identified in the center of South Scott Reef lagoon. Sites were selected to be representative of bathymetry and habitat types within the area. The three sites were located on a line at distances from 45 to 2,743 m from a planned seismic survey vessel track. Water depth at these sites was between 25 and 45 m.

At each site, two groups of three 1.0-m^3 soft net cages were suspended 5 m below the water surface and tethered to the seabed. The fish were transferred by boat in their holding buckets to the

M.C. Hastings (✉)
Georgia Institute of Technology, Atlanta, GA 30332, USA
e-mail: mardi.hastings@gatech.edu

J. Miksis-Olds
The Pennsylvania State University, State College, PA 16804, USA

A.N. Popper and A. Hawkins (eds.), *The Effects of Noise on Aquatic Life*,
Advances in Experimental Medicine and Biology 730, DOI 10.1007/978-1-4419-7311-5_53,

sites where specimens were placed in the three cages based on species (*Myripristis murdjan* and *Sargocentron spiniferum* replicates were placed in the same cage). Two three-cage groups at each location allowed for single-pass and double-pass exposure testing without disturbing the separate replicates.

The experimental fishes were then exposed to discharges from a seismic air gun array with a combined capacity of 2,055 in^3 and a shot point interval of 18.75 m as it was towed along a single seismic survey line. Exposures to either one or two air gun array passes occurred over five days. After exposure, the experimental animals were immediately removed from the cages and transported in buckets to the holding location before testing.

Sound pressure was recorded using underwater noise loggers from the Centre for Marine Science and Technology (CMST) at Curtin University, Perth, Western Australia. Loggers were attached to the fish cages to record the sound pressure received by the fish.

Control fishes were treated exactly as experimental animals, other than for the actual sound exposure from the air gun array. They were placed in cages 45 m from the seismic survey track line, held for the same amount of time as fishes later exposed to one and two passes of the air gun array, and removed and transported to the holding location before testing.

2.2 Assessment of Hearing Thresholds

A test chamber suitable for measurement of AEPs in fish was designed for shipboard use and fabricated at CMST. A tank was constructed using a 1-m length of 400-mm-diameter Schedule 60 steel pipe. The pipe was flanged to a steel plate at the bottom and oriented upright. An underwater sound transducer (Lubell Labs LL9162T) was placed at the bottom and embedded in sand to minimize rattle and structural flanking. The tank and baseplate were mounted inside a plywood enclosure lined with foil to reduce electromagnetic interference and with damping material to minimize airborne noise interference. A rectangular PVC pipe frame with four legs filled with sand for stabilization was placed inside the enclosure and isolated from contact with the tank to support the fish. In addition, the enclosure sat on top of four automobile tires to minimize interference from structural vibration through the deck of the boat. The enclosure had a hinged lid for access to the test tank from the top. The tank was filled with local seawater to a height of 800 mm and then drained and refilled daily. Seawater temperature was 21–22°C throughout the study. Hearing sensitivity of each specimen was determined by measuring AEPs. Baseline animals were held on deck in tanks with continuous freshwater flow but were not transported and placed in cages.

Each fish specimen was held in a plastic mesh sling secured with a clamp across the dorsal fin and suspended in the test tank 200 mm below the water surface. Slings were cut to size for each individual so as to minimize movement while allowing for normal respiration. Stainless-steel needle electrodes were used to record the AEP signal. A recording electrode was inserted subdermally into the dorsal surface of the fish directly over the brain stem and a reference electrode was inserted subdermally either between the nares or into the dorsal musculature near the tail. A ground electrode was placed directly in the water near the fish.

Sound stimuli were produced and AEP waveforms recorded using EVoked REsponse Study Tool (EVREST) software (Finneran 2009). This system was implemented on a ruggedized computer with an expansion chassis for a 16-bit data-acquisition board and an integrated custom programmable electronic output attenuator to control the level of acoustic stimuli. Stimuli consisted of 50-ms pulsed tones with 4-ms rise-and-fall times for frequencies below 1,200 Hz, and 22-ms pulsed tones with 1-ms rise-and-fall times for frequencies above 1,200 Hz. The phase of the tone was flipped between presentations to eliminate any electrical artifacts in the AEP signal. Stimulus tones were amplified using an automotive power amplifier and 12-V battery.

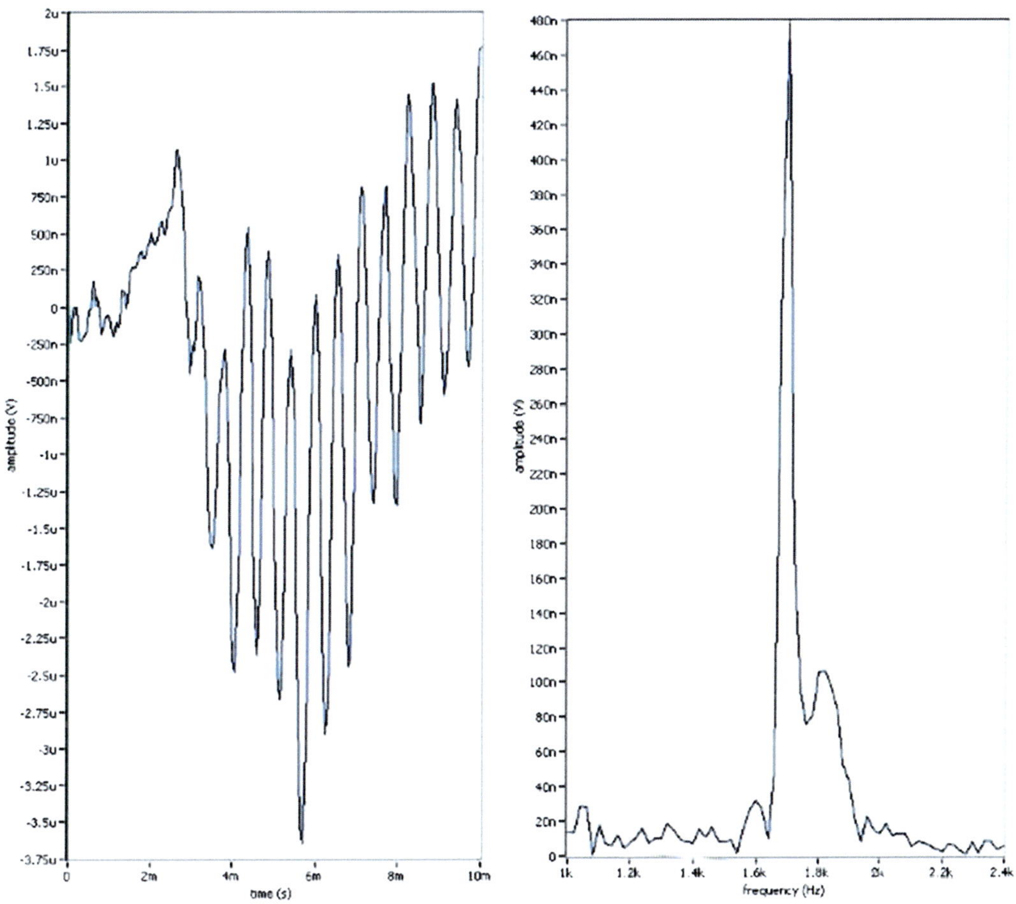

Fig. 1 Auditory evoked potential (AEP) recording from a *Chromis viridis* specimen for a 140-dB (re 1 μPa) stimulus at 845 Hz

Acoustic stimuli were calibrated and background acoustic noise was recorded each day with a Reson TC 4033 hydrophone (sensitivity −202.2 dB re 1 V/μPa) connected to a Reson VP1000 preamplifier set to provide a gain of 6 dB. Discrete test frequencies between 100 and 2,000 Hz were chosen a priori to minimize distortion. During each trial, the stimulus was presented at a level high enough to elicit a clear evoked potential and then decreased in 5-dB steps until a threshold level was no longer visible.

2.3 Data Reduction

Hearing thresholds were determined by visual inspection of AEP waveforms and power spectra. Signal levels at each frequency were changed in minimum 5-dB steps until a typical AEP waveform and corresponding spectral signal amplitude could not be seen above the noise. AEP waveforms were examined for latency and periodic response, and then AEP power spectra were analyzed for signal amplitude at twice the stimulus frequency, typical of the AEP recordings in fish (Suga et al. 2006). Figure 1 displays a typical AEP waveform and spectrum.

Data at each test frequency from the baseline, control, and exposure groups for each species were compared using a one-way analysis of variance (ANOVA) with $\alpha = 0.05$ to determine any differences in threshold. In each case, the Bonferroni adjustment was applied to the *p* values of statistical significance. Any differences were verified using Tukey's honestly significant difference (HSD) post hoc comparison ($\alpha = 0.05$).

3 Results

AEP measurements were made on 51 specimens of *Chromis viridis* with an average standard body length of 39 ± 7 mm and mass of 2.6 ± 1.0 g; 47 *Lutjanus kasmira* with a standard length of 164 ± 14 mm and mass of 123 ± 29 g; 10 *Sargocentron spiniferum* and 10 other holocentrid fishes, primarily from the genus *Sargocentron*, with standard lengths of 237 ± 25 mm and 152 ± 18 mm and mass of 400 ± 107 g and 103 ± 31 g, respectively; and 20 *Myripristis murdjan* with a standard length of 164 ± 22 mm and mass of 192 ± 75 g.

No significant differences in auditory thresholds were found among exposure groups or between exposure groups and baseline or control thresholds at any test frequency for any species, except for the group of *Chromis viridis* exposed to a single pass of air gun emissions. This exposure group suffered from a bacterial infection that was not apparent until they were retrieved from the cages. There was no difference between thresholds measured 0-6 h postexposure and baseline thresholds; however, because of deterioration of these fish over time, thresholds measured 6–12 h postexposure were significantly higher than thresholds measured during the 0- to 6-h period: 18 dB higher at 225 Hz and 16 dB higher at 455 Hz. These specimens were dying at an exponential rate after being brought onboard for testing, so the threshold shifts were not related to sound exposure.

Myripristis murdjan had the lowest AEP thresholds above 400 Hz of any species tested at Scott Reef. Even though sample sizes were small, no differences in AEP thresholds were found among exposure groups or between exposure groups and baseline or control thresholds at any test frequency for *Myripristis murdjan* specimens. The exposure groups included 3 specimens tested within 0–6 h after a cumulative exposure of at least 190 dB re 1 μPa^2-s. Figure 2 shows the baseline and control AEP thresholds along with the overall average thresholds for exposed specimens.

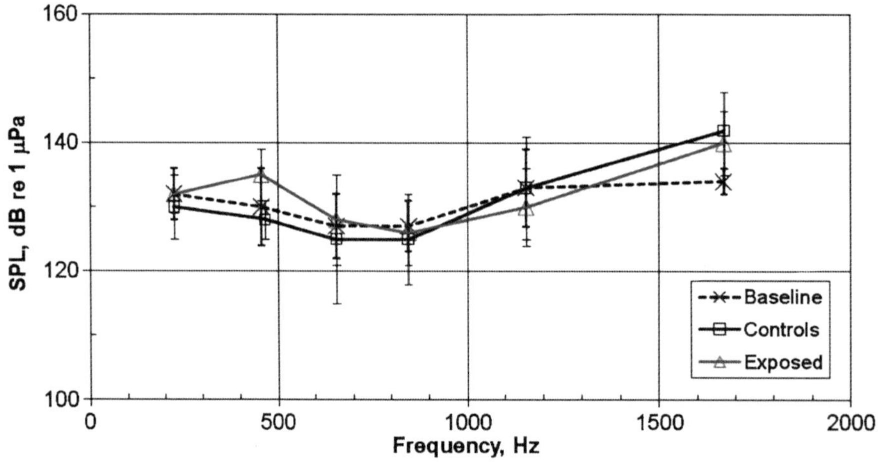

Fig. 2 AEP thresholds measured for *Myripristis murdjan*. SPL, sound pressure level. No significant difference was found among baseline, control, and exposed groups

4 Summary and Conclusions

This is the first study to examine the potential effects of emissions from an air gun array on hearing of tropical reef fish using AEP thresholds measured in the field. Over 7 days, 51 *Chromis viridis*, 47 *Lutjanus kasmira*, 20 *Myripristis murdjan*, and 10 *Sargocentron spiniferum* were tested. AEP thresholds were determined before and after exposure to emissions from one or two passes of an air gun array. Temporary threshold shift resulting from sound exposure was not found in any species, even when cumulative sound exposure levels reached 190 dB re 1 μPa^2-s.

Acknowledgments This project was sponsored by Woodside Energy Ltd., Perth, WA, through Environmental Resources Management Australia Pty. Ltd., Perth, WA, Australia.

References

Finneran JJ (2009) Evoked response study tool: A portable, rugged system for single and multiple auditory evoked potential measurements. J Acoust Soc Am 126:491–500.

Suga T, Akamatsu T, Hiraishi T, Yamamoto K (2006) Contamination of auditory evoked potential of goldfish *Carassius auratus* with microphonic potential. J Mar Acoust Soc Jpn 33:85–88.

Woodside Energy Ltd. (2007) Environmental protection statement: Maxima 3D seismic survey, Scott Reef. Woodside Energy Ltd., Perth, WA, Australia.

A Lack of Correlation Between Air Gun Signal Pressure Waveforms and Fish Hearing Damage

Robert D. McCauley and Chandra Salgado Kent

1 Introduction

Offshore marine petroleum seismic surveys involve the repetitive use of intense, short bursts of low-frequency noise, the reflections of which are used to image subsea geology. The seismic signal is produced by a spatial array of sources, usually air guns that violently release high-pressure compressed air into the water column. Although the signal produced by a single air gun is largely omnidirectional at low frequencies (typically, most energy over 10–150 Hz), the signal received by a spatial array of individual air guns is highly directional and dependent on the array configuration and receiver orientation.

There have been several studies that have looked at the behavioral response of fish to air gun signals and that have attempted to quantify damage to fish hearing systems and changes in hearing sensitivity from air gun signal exposure. Popper and Hastings (2009) present the most recent review of these works. For hearing damage, McCauley et al. (2003b) found significant damage to the hearing epithelia of *Pagrus auratus* (pink snapper) 58 days after they were exposed in an approach-departure fashion, multiple times to a single 20-cui air gun at ranges from 5 to 300 m in Jervoise Bay, Western Australia (termed JB). A vigorous behavioral response was observed from these fish on the first set of seismic passes that led to the hearing damage, but 58 days later, the same fish failed to respond to a second set of seismic passes (McCauley and Fewtrell 2008; McCauley et al. 2003a), supporting the histological observations of hearing damage occurring after the first set of passes. In contrast, Popper et al. (2005) exposed several fish species to a small seismic array in a shallow-river system and found limited loss of hearing in two species that recovered hearing within 18–24 h and that showed no damage to sensory hair cells (Song et al. 2008).

Using experiments carried out in Australia, we briefly present further observations of hearing damage experiments in fish that support the case of Popper et al. (2005) and Song et al. (2008) for no or limited hearing damage to fish ears from seismic array passes. The pressure waveforms of received seismic signals from the Australian experiments are examined to explore what differences occur between the small single air gun used in the JB experiments and larger arrays of air guns used to expose shallow-placed fish that do not appear to show significant damage to fish ears.

R.D. McCauley (✉) • C.S. Kent
Centre for Marine Science and Technology, Curtin University, Perth, WA 6845, Australia
e-mail: R.McCauley@cmst.curtin.edu.au

A.N. Popper and A. Hawkins (eds.), *The Effects of Noise on Aquatic Life*,
Advances in Experimental Medicine and Biology 730, DOI 10.1007/978-1-4419-7311-5_54,

2 Ear Damage Based on Caged-Fish Observations in Northern Australia

Two sets of observations of damage to the hearing epithelia of caged fish exposed to air gun signals have been made in northern Australia recently, one in the Timor Sea in 2006 (10°15′ S) and one on a coral atoll complex in the Kimberley area (14° S) of northern Western Australia in 2007. These works are unpublished, but a brief summary is presented here. In the Timor Sea, *Pristipomoides multidens* (gold band snapper, around 55-cm standard length) were removed from 2.5-m-long × 1.5-m-wide × 1.5-m-high fish traps set on the seabed and that had captured fish on a prior seismic pass. The fish hearing structures were fixed, preserved for histology at sea, and processed ashore. A 3,090-cui air gun array towed at 5-m depth passed the traps set at 112-m depth on the seabed at ranges of 370 m, 2.1 km, and 58 km from the closest air gun pass. Damage to the hair cells lying on the sensory epithelia surrounding the sagittal otolith was quantified as per McCauley et al. (2003b). There was an exponentially increasing amount of hair cell damage with decreasing range from seismic pass or increasing cumulative sound exposure, although the sample size was low, making the results indicative only. This experiment is termed Timor Sea (TS). In 2007, an experiment was conducted in a large coral reef lagoon (averaging 45-m depth at the study site) where several fish species were held in 1-m^3 cages at 7-m depth at ranges from 45 m to 2.7 km abeam a seismic vessel track, with various exposure regimens and control passes. Fish were sampled in the field by fixing and preserving the cranial structure, with the sensory epithelia surrounding the sagittal otolith removed in the laboratory and quantified for damage to hair cells as per McCauley et al. (2003b). Exposed fish were sampled in a time series to 60 days postexposure. The primary species sampled was *Lutjanis kasmira* (sea perch or snapper) of ~150-mm standard length. In these experiments, a 2,055-cui air gun array towed at 5-m depth ran past the cages. When data of hair cell damage for seismic-exposed fish was lumped and compared with those for control fish, a statistical increase in damage to hair cells was found, but this damage covered much less than 1% of the hair cell-populated portion of the sensory epithelia and was considered to be small. No relationships could be found between hair cell damage and the number of seismic passes (one or two), intensity of signals received, or time after exposure, indicating that the damage observed was marginal and not a clear-cut response to the seismic passes. This experiment is termed Kimberley (KI).

In summary, four sets of observations of damage to the sensory epithelia of fishes exposed to seismic air gun array passes exist, with one (McCauley et al. 2003b) showing significant damage to passes of a single air gun, one (Popper et al. 2005; Song et al. 2008) showing a small amount of hearing loss and no hair cell damage to passes of a small air gun array in a shallow river, one (TS) showing an increasing amount of hair cell damage with increasing cumulative seismic energy received but with low sample size, and one (KI) showing low levels of hair cell damage to passes of a 2,055-cui array but no clear patterns based on exposure type.

3 Nature of Seismic Sources

Seismic air gun arrays comprise a planar array of elements, the air guns in several strings aligned in the tow direction. Based on the configuration of seventeen 1,500–4,000 cui, mostly 3-dimensional seismic survey arrays available to the author, each gun string is from 14 to 20 m long, with gun elements on average 2.7 m apart; most arrays have two strings of guns, the separation across the array is from 10 to 20 m, and the guns are towed at 5- to 7-m depth. Typically, a wide size range of air guns is used in an array to spread the frequency content of the primary pulse with gun elements ranging from 20 to 40 cui up to 300 to 500 cui. The larger gun elements are usually clusters of two or three air guns, with these located at the tow end of the string and the smaller guns at the rear. Some arrays are asymmetrical in design. The resulting array will be directional in noise energy

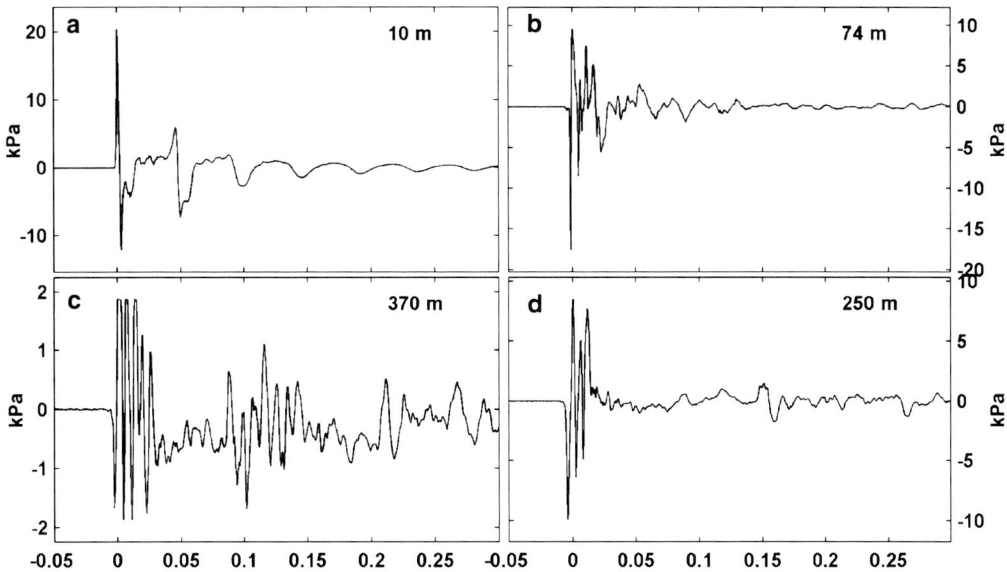

Fig. 1 Four air gun waveforms as a maximum signal received in experiments by McCauley et al. (2003a,b; **a**); a maximum signal from a 2,055-cui air gun array at a cage 7 m deep at a 74-m range (1.5° below array plane; **b**); a signal of 3,090-cui array received at a 370-m horizontal range for a receiver on the seabed at 112-m water depth (16° below array plane, signal has clipped on positive pressures; **c**); and a signal of a 3,040-cui array measured by a receiver 250 m abeam the array on the seabed at 152-m water depth (30° below array plane; **d**)

output, designed to focus energy downward for subsea imaging. Signals received in any orientation other than directly below will contain multiple time-offset air gun signals that will interfere with each other.

With respect to the results available for caged fish exposed to commercial seismic arrays, the directionality patterns inherent in the spatial grid of air gun sources imply that 1) the caged-fish experiments carried out to date have not been in the focal zone of any air gun array (directly below the array) and 2) the signal received at the fish cages has been the sum of a complex mix of signals from multiple air gun elements. Figure 1 shows the air gun waveforms for the highest level signals received at the caged fish experiments using a single 20-cui air gun at approximately the same depth as the fish (fish 2–3 m, air gun 5-m depth, range 5–10 m at closest approach; McCauley et al. 2003a; Fig. 1a); for cages at 7-m depth, 74 m abeam a 2,055-cui passing air gun array in an approximately 45-m-deep coral lagoon (cage 1.5° below array horizontal; Fig. 1b); and for a fish trap on the seabed at 112-m depth, 370 m abeam a 3,090-cui source towed at 5-m depth (cage 16° below array horizontal; Fig. 1c). For comparison, Figure 1d shows the signal received by a 3,040-cui array towed at 6-m depth passing 250 m abeam a receiver on the seabed at 152-m depth (receiver 30° below array horizontal). These signals were collected under various programs, not detailed here, using calibrated sea noise logging systems set up for intense air gun signals, with receivers at the fish cages or on the seabed (Fig. 1d). The waveforms for the Popper et al. (2005) and Song et al. (2008) fish exposures were not available, but the cages were believed to be close to the same plane as the air gun array, implying interference from the different air guns at the received location and waveforms possibly similar to those shown in Figure 1b-d.

There are some differences in the signal shapes received when comparing the maximum signal received in the McCauley et al. (2003b) experiment and the maximum signals received from commercial air gun arrays passes at elevations of ≤30° from the array horizontal plane. Details of

Table 1 Maximum air gun signal exposures received in caged-fish experiments

Site	Max p-p	Max SEL	Time between max peaks, ms	>190 p-p	>195 p-p	>200 p-p	>205 p-p	>210 p-p	CSEL
JB	210	181	1.972 ± 0.3742	83	33	8	2	1	188
TS	198	175	8.864 ± 5.0922	36	9	0	0	0	188
KI	209	179	8.023 ± 3.6897	43	22	11	2	0	190
3040	206	178	8.955 ± 3.1462	37	17	5	1	0	189

Legend: JB, Jervoise Bay, Western Australia (McCauley et al. 2003a,b); TS, Timor Sea; KI, Kimberley; 3040, measurements from a 3,040-cui array passing a receiver; Max p-p, maximum peak-peak (dB re 1 μPa); Max SEL, maximum sound exposure level (dB re 1 μPa^2-s); time between max peaks, time between maximum +ve/−ve peak and the following maximum −ve/+ve peak, with 95% confidence limits based on the highest 11 signals received in each dataset; 190, 195, 200, 205, 210 p-p, number of signals greater than the thresholds for peak-peak (dB re 1 μPa); CSEL, cumulative sound exposure level (sum of all linear SEL values of received air gun signals; dB re 1 μPa^2-s).

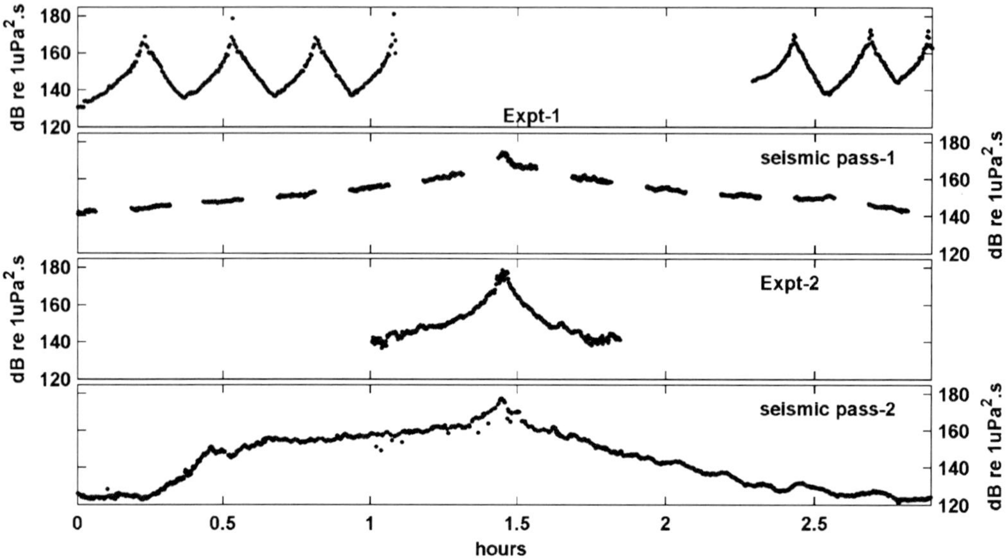

Fig. 2 Time sequence of air gun signals (sound exposure level units). Expt-1, McCauley et al. (2003a,b); seismic pass-1, equivalent to measurements from the Timor Sea fish cages but made during a seismic line (this noise logger was on a duty cycle and did not sample continuously); Expt-2, Kimberley experiments for the 2,055-cui array passing; seismic pass-2, the 3,040-cui array passing. The datasets are arbitrarily aligned by time to match the Expt-1 duration

the maximum signal levels reached during the JB, TS, and KI sets of fish-exposure experiments and the 3,040-cui array passing are listed in Table 1. The time sequence of exposures aligned over the same time frame as the McCauley et al. (2003b) experiment duration are shown in Figure 2. The magnitudes of the maximum signals experienced in peak-peak (p-p) and sound exposure level (SEL), as given by the pressure waveforms at the JB and KI experimental passes and the 3,040-cui array passing given in Table 1, are similar in decibel values (the JB p-p is twice that of the KI exposures in linear terms), whereas the TS measurements are ~10 dB down from p-p and 3 dB down in SEL. The number of signals >200 dB re 1 μPa p-p experienced was similar for the JB, KI, and 3,040-cui measurements, whereas for the >190 dB re 1 μPa category, the JB exposure had approximately double the number of signals (the same values in SEL units were similar in trend to p-p shown). This was due to the nature of exposures delivered. Figure 2 shows the JB exposures varying from a seismic vessel passing, where the signal ramps up, reaching a peak often slightly before the array passes abeam, and then dropping rapidly away on leaving. The JB exposures used multiple short approach-departures rather than a single pass-by because the experiments were designed to capture fish behavior (McCauley et al. 2003a). The cumulative sound exposures received in all measurements were within 2 dB of each other. Although the TI exposures had lower maximum signal pressures, they had a similar cumulative SEL exposure because the air gun was not being operated in a data-collection mode but in a free run at a 3.3-s shot spacing compared with 8.2 s for the KI experiments, 15 s for JB, and 10.4 s for the 3,040-cui array.

A feature that showed significant differences between signal types at the fish cage experiments was the time between the first maximum signal peak (+ve or –ve) and the following maximum signal peak of opposite sign. All signals at short range measured from the single 20-cui air gun had a leading +ve peak with the time between this and the following –ve peak of ~2 ms or approximately one quarter of that experienced from the air gun arrays (8–9 ms). The signals received from the commercial arrays at <30° elevation from the array plane had a mix of leading positive and negative peaks followed by a peak of the opposite sign (at the TS traps, it was 8/3 leading +ve/–ve peaks; at the KI cages, it was 6/5 leading +ve/–ve peaks; for the 3040 measurements, it was 0/11 or all leading –ve peaks).

4 Conclusions

The pressure waveforms of maximum air gun signals received during experiments with caged fish, which have and have not caused apparent significant hearing damage, differ in their shape and time between minimum and maximum pressure peaks but do not differ greatly systematically in exposure magnitudes as individual signal or cumulative energy delivered when measured as pressure parameters. The measurements were of signal pressure, not particle motion, particularly particle acceleration, which is what drives the fish hearing system. Although involving only a single air gun, the experiment that has shown considerable damage to fish ears involved short-range exposures (<10 m), well inside the near field where particle motion is not proportional to the pressure magnitude and was not measured. The experiments with air gun arrays have all taken place at elevations <30° from the array horizontal plane, out of the array focal zone, at elevations with considerable interference from different array elements, and at ranges where the particle motion may have been proportional to the measured pressure, although because it has never been measured, this is not known. This implies that as yet the issue of hearing damage in fish resulting from exposure to intense impulse signals is a poorly resolved issue.

What has been resolved is that in at least some instances, reasonably short-range exposures (down to 45 m; KI experiments here; Song et al. 2008) to air gun arrays up to 2,500 cui at shallow elevations below the array horizontal plane in shallow water (<50 m depth) have not shown significant damage to the hearing epithelia of several fish species.

Acknowledgments Santos Ltd., Adelaide, SA, Australia, supported the fish hearing experiments in the Timor Sea and Woodside Energy Ltd., Perth, WA, Australia, supported the experiments in the Kimberley. Andrew Levings, Horst Fischer, Julie Lloyd, and Alex Beatty supported the Timor Sea field work. Malcolm Perry, Frank Thomas, and Michael Archer supported the Kimberley experiments.

References

McCauley RD, Duncan AJ, Fewtrell J, Jenner C, Jenner M, Penrose JD, Prince RIT, Adhitya A, Murdoch J, McCabe K (2003a) Marine seismic surveys: Analysis and propagation of air-gun signals; and effects of exposure on humpback whales, sea turtles, fishes and squid. In: Environmental implications of offshore oil and gas development in

Australia: Further research. Australian Petroleum Production Exploration Association, Canberra, ACT, Australia, pp 364–521.
McCauley RD, Fewtrell J (2008) Experiments and observations of fish exposed to seismic survey pulses. Bioacoustics 17:205–207.
McCauley RD, Fewtrell J, Popper AN (2003b) High intensity anthropogenic sound damages fish ears. J Acoust Soc Am 113:638-642.
Popper AN, Hastings MC (2009) The effects of human-generated sound on fish. Integr Zool 4:43–52.
Popper AN, Smith ME, Cott PA, Hanna BW, MacGillivray AO, Austin ME, Mann DA (2005) Effects of exposure to seismic air gun use on hearing of three fish species. J Acoust Soc Am 117:3958–3971.
Song J, Mann DA, Cott PA, Hanna BW, Popper AN (2008) The inner ears of northern Canadian freshwater fishes following exposure to seismic air gun sounds. J Acoust Soc Am 124:1360–1366.

Effects of Short-and Long-Term Exposure to Boat Noise on Cortisol Levels in Juvenile Fish

Ilaria Spiga, Joe Fox, and Robert Benson

1 Introduction

Anthropogenic noise can elicit a significant elevation in plasma cortisol levels in fish (Smith et al. 2004; Wysocki et al. 2006). In response to a stressor, fish attempt to compensate by using a series of biochemical and physiological changes that start with a neuroendocrine response. This response includes the release of cortisol into the circulatory system (Wendelaar Bonga 1997).

Sciaenops ocellatus (red drum) and *Cynoscion nebulosus* (spotted sea trout) are highly vocal fish that commonly inhabit Texas bays and estuaries. These areas are widely traveled by motor-driven recreational and fishing boats, and the number of fishing boats is increasing (Tseng et al. 2006). This study evaluates the short- and long-term effect of boat noise on cortisol levels in *Sciaenops ocellatus* and *Cynoscion nebulosus*.

2 Materials and Methods

2.1 Noise Acquisition

Noise recordings were made of a sport fishing vessel approaching and passing a stationary hydrophone in Corpus Christi Bay, TX. A 15-s segment of the recording was digitalized and projected via an amplifier into fish holding tanks using underwater speakers. In the long-term study, both continuous and intermittent sounds were carried by a dual-channel transmission.

3 Experimental Setup and Protocol

3.1 Short-Term Study

Juvenile fish of both species were exposed to continuous boat noise (180 dB re 1 μPa) for periods of 0, 15, 30, and 60 min. Each tank was divided into eight cylindrical units, and curtainlike structures were built to prevent fish in different groups from interacting with each other. At the

I. Spiga (✉) • J. Fox • R. Benson
Texas A&M University-Corpus Christi, Corpus Christi, TX 78412, USA
e-mail: ilaria.spiga@gmail.com; Joe.Fox@tamucc.edu; Robert.Benson@tamucc.edu

A.N. Popper and A. Hawkins (eds.), *The Effects of Noise on Aquatic Life*,
Advances in Experimental Medicine and Biology 730, DOI 10.1007/978-1-4419-7311-5_55,

end of the designated treatment period, fish were removed, flash frozen in liquid nitrogen, and analyzed using an ELISA assay.

3.2 Long-Term Study

Juvenile fish of both species were exposed to ambient background noise (120 dB re 1 μPa), continuous noise (175 dB re 1 μPa), and intermittent noise (175 dB re 1 μPa, 30-min periodicity). A baseline subsample of 50 fish was collected with a net for initial cortisol analysis followed by immediate flash-freezing in liquid nitrogen. The duration of noise exposure was 8 wk. At the end of the trial, an additional 50 fish were collected for cortisol analysis using an ELISA assay.

4 Results

4.1 *Sciaenops ocellatus*

Fish exposed to either of the two noise treatments in the long-term study did not show an elevation in the whole body cortisol level at the end of 8 wk. In the short-term exposure study, 30 min of noise exposure significantly increased plasma cortisol levels ($P < 0.01$). Plasma cortisol declined thereafter to a level lower than that of the no-noise control group after 60-min exposure time.

4.2 *Cynoscion nebulosus*

Fish exposed to either of the two noise treatments in the long-term study did not show an elevation in the whole body cortisol level at the end of 8 wk. A significant elevation in whole body cortisol concentration was shown after 15 min ($P < 0.001$) and after 30 min ($P < 0.001$) in the short-term study. The concentration of whole body cortisol decreased after 60 min of noise exposure.

5 Discussion

When fish are exposed to stressors, there is an immediate release of catecholamines followed by the activation of the hypothalamic–pituitary–interrenal axis, which stimulates the synthesis and secretion of glucocorticoid hormones (cortisol) in teleosts (Wendelaar Bonga 1997). We found that short-time boat noise had the potential of being a stressor for fish. However, long-term exposure to noise did not elicit elevation to either continuous or intermittent noise. Moreover, we noted that fish in the long-term study showed behavioral changes at the onset of noise, including startling and swimming in tight circles with increased acceleration. This response diminished within 1 wk. This suggests that fish of both species became acclimated to both noise treatments over time. We therefore infer that acclimation to noise under the present experimental conditions is relatively rapid. This is further substantiated by the observation that both species of fish reacted to short-term noise exposition by increasing whole body cortisol concentrations. *Sciaenops ocellatus* showed a significant elevation in cortisol after 30 min; however, the cortisol concentration declined after 60 min of noise exposure.

Cortisol levels in *Cynoscion nebulosus* peaked after 15 min of exposure and remained significantly higher after 30 min but, as with *Sciaenops ocellatus* and *Cynoscion nebulosus*, returned to lower values.

6 Conclusions

Boat noise similar to that projected by recreational fishing boats elicits in juvenile *Sciaenops ocellatus* and *Cynoscion nebulosus* a rapid increase in cortisol levels; however, these higher levels do not persist for more than 60 min. Despite persistence of stressful sound, acclimation in the species of fish studied occurs rapidly.

References

Smith ME, Kane AS, Popper AN (2004) Noise-induced stress response and hearing loss in goldfish (*Carassius auratus*). J Exp Biol 207:427–435.

Tseng YP, Wolber NR, Ditton RB (2006) Demographics, participation, attitudes, and management preferences of Texas anglers. Human Dimensions of Fisheries Research Laboratory Technical Document HD-631, Texas A&M University, College Station, TX.

Wendelaar Bonga SE (1997) The stress response in fish. Physiol Rev 77:591–625.

Wysocki LE, Dittami JP, Ladich F (2006) Ship noise and cortisol secretion in European freshwater fishes. Biol Conserv 128:501–508.

Potential Effects of Long-Term Exposure to Boat Noise on the Growth, Survival, and Nutrient Retention in Juvenile Fish

Ilaria Spiga, Joe Fox, and Robert Benson

1 Introduction

Fishing and recreational boats are common vessels in coastal waters and contribute to underwater noise, with levels and frequencies related to their size. Texas bays and estuaries are widely traveled by motor-driven recreational and fishing boats and the number of them has increased in the past few years (National Marine Manufacturers Association 2008). Studies have shown that fish react to underwater noise from boats by displaying abnormal behavior (Fuiman et al. 1999; Popper et al. 2004) and reduced growth (Davidson et al. 2009; Wysocki et al. 2007). Fish react to stressful conditions by modulating metabolic rate and repartitioning energy, with the subsequent conversion of stored to available energy at the expense of growth (Wendelaar Bonga 1997).

Sciaenops ocellatus (red drum) and *Cynoscion nebulosus* (spotted sea trout) are sport fish that commonly inhabit Texas bays and estuaries. There is substantial concern that increased anthropogenic noise could have a negative impact on the growth and development of these fish. In these preliminary experiments, we examine the effects of long-term exposure of *Cynoscion nebulosus* and *Sciaenops ocellatus* to boat noise on growth, survival, and metabolic balance.

2 Materials and Methods

2.1 Noise Acquisition

Noise recordings were obtained of a sport fishing vessel approaching and passing a stationary hydrophone in Corpus Christi Bay, TX. A 15-s segment of the recording was digitalized and projected via an amplifier into fish holding tanks using underwater speakers. The dual-channel digital sound file contained both continuous and intermittent noise.

I. Spiga (✉) • J. Fox • R. Benson
Texas A&M University-Corpus Christi, Corpus Christi, TX 78412, USA
e-mail: ilaria.spiga@gmail.com; Joe.Fox@tamucc.edu; Robert.Benson@tamucc.edu

A.N. Popper and A. Hawkins (eds.), *The Effects of Noise on Aquatic Life*,
Advances in Experimental Medicine and Biology 730, DOI 10.1007/978-1-4419-7311-5_56,

2.2 Experimental Setup and Protocol

Juvenile *Sciaenops ocellatus* and *Cynoscion nebulosus* were provided by a local hatchery and stocked in six fiberglass tanks. The tanks were organized by species according to the following treatments: one control tank with ambient background noise (120 dB re 1 μPa), one tank with continuous noise (175 dB re 1 μPa), and one with intermittent noise (175 dB re 1 μPa, 30-min periodicity). The duration of noise exposure was 8 wk, with a mid-trial sample obtained after 4 wk for determination of total length. At the end of the trial, the fish were enumerated for survival and measured for total length, mass, and derivation of other growth factors (e.g., condition factor, specific growth rate). Whole fish were then individually lyophilized, ground to a dry powder, and analyzed for protein and energy determinations with an automated nitrogen analyzer and bomb calorimetry, respectively.

3 Results

Both species of fish reacted to the initial playback of noise, either intermittent or continuous, by swimming in tight circles and by accelerating. In the intermittent-noise tank, fish startled when the noise was present. These behaviors were not presented after 1 wk of exposure to noise. There were no significant differences among treatments ($P \geq 0.05$) for the dependent variables determined in this experiment for either species of fish. Regardless, some of the initial assumptions were confirmed by the results.

3.1 *Sciaenops ocellatus*

The percent survival of fish was similar among treatments at the end of the 8 wk, although it was numerically lower in fish subjected to noise treatments. The rate of growth, expressed as mean percent weight gain, was lower during the first month than at termination, whereas the specific growth rate and mean total length were higher. An analysis of the frequency distribution of condition factor indicated a shift toward higher values for treatment fish compared with those in the ambient noise control tank. Mean percent whole body protein of the fish subjected to boat noise treatments was lower than that of the control fish at the end of the 8-wk study period. The protein-to-energy ratio declined in the fish exposed to boat noise. Higher feed conversion ratio values were shown by the fish subjected to continuous noise.

3.2 *Cynoscion nebulosus*

The percent survival was highest in fish under intermittent-noise conditions. Mean weight gain was lower during the first month than at the end of the trial. Specific growth rate and mean total length were higher after 4 wk than at the end of the experiment. Fish showed lower mean condition factor values when exposed to continuous noise than those subjected to ambient or intermittent noise. Fish subjected to intermittent noise had the highest mean whole body protein levels, whereas fish exposed to continuous noise were the lowest. The protein-to-energy ratio increased in a manner similar to that of protein concentration. Fish showed the lowest feed conversion values when exposed to continuous noise.

4 Discussion

Most studies evaluating the effect of noise on fish have shown a disruption of normal behavior and physiology, all having the potential to impair the ability of fish to forage, escape predators, and reproduce (Richardson et al. 1995). In this experiment, *Sciaenops ocellatus* and *Cynoscion nebulosus* showed an initial response to noise, which was not presented after 1 wk of treatment. It can be inferred that both of these species habituated to increased noise levels. Habituation might be a consequence of their living in noisy habitats such as estuaries and bays and, therefore, is a manifestation of their capacity to maintain normal activities in variable conditions. Neither species of fish showed long-term impairment of growth due to noise exposure, although both grew slowly during the first month of noise exposure (e.g., lower mean percent weight gain). Future research should attempt to further characterize this decrease in growth rate and the physiological responses associated with it. As a source of stress, underwater noise can induce a relocation of energy. The stress response to noise in this experiment was determined by measuring whole body protein and energy retention after 8 wk. Protein content and the protein-to-energy ratio declined for *Sciaenops ocellatus* exposed to both continuous and intermittent noise. On the other hand, *Cynoscion nebulosus* exposed to continuous noise had the lowest whole body protein content, and those exposed to intermittent noise had the highest whole body protein values and protein-to-energy ratio values. This suggests that both species are affected by long-term noise exposure and display different abilities to use protein as energy. The overall feed conversion rate was high in both species of fish when subjected to either noise treatment, suggesting that noise and the stress it imparts could induce a reduction in feed intake.

5 Conclusions

Long-term exposure to boat noise similar to that projected by recreational fishing boats does not appear to influence the growth, survival, and nutrition of *Sciaenops ocellatus* or *Cynoscion nebulosus*, although behavioral observations indicated an initial response to noise (either intermittent or continuous). Further studies should include replications and growth trials to assess when the acclimation happens.

References

Davidson J, Bebak J, Mazik P (2009) The effects of aquaculture production noise on the growth, condition factor, feed conversion, and survival of rainbow trout, *Oncorhynchus mykiss*. Aquaculture 288:337–343.

Fuiman LA, Smith ME, Malley V (1999) Ontogeny of routine swimming speed and startle responses in red drum, with a comparison of responses to acoustic and visual stimuli. J Fish Biol 55A:215–226.

National Marine Manufacturers Association (2008) 2007 U.S. recreational boat registration statistics. Market Statistics Department, National Marine Manufacturers Association, Chicago, IL.

Popper AN, Fewtrel J, Smith ME, McCauley RD (2004) Anthropogenic sound: Effects on the behavior and physiology of fishes. J Mar Technol Soc 37:35–40.

Richardson WJ, Greene CR Jr, Malme CI, Thompson DH (1995) Marine mammals and noise. Academic Press, San Diego, CA.

Wendelaar Bonga SE (1997) The stress response in fish. Physiol Rev 77:591–625.

Wysocki LE, Davidson JW III, Smith ME, Frankel AS, Ellison TE, Mazik PM, Popper AN, Bebak J, (2007) The effects of aquaculture production noise on hearing, growth, and disease resistance of rainbow trout, *Oncorhynchus mykiss*. Aquaculture 272:687–697.

Predicting Hearing Loss in Fishes

Michael E. Smith

1 Introduction

Exposure to intense sounds can damage the auditory hair cells of terrestrial vertebrates, and this damage and/or loss results in reduced hearing sensitivity. This is also true for aquatic vertebrates such as fishes exposed to underwater sounds. Exposure to sounds at high intensities has been shown to induce hearing loss in a number of fish species (e.g., goldfish *Carassius auratus* [Smith et al. 2004a]; fathead minnows *Pimephales promelas* [Scholik and Yan 2001]; northern pike *Esox lucius* and lake chub *Couesius plumbeus* [Popper et al. 2005]). Because data on the physiological effects of sound on most fish species is generally lacking, models predicting sound-induced hearing loss in fishes would be useful. The application of such models would be of tremendous value for policy decisions concerning anthropogenic noise exposure, especially when threatened or endangered species are involved.

2 Linear Threshold Shift Hypothesis

Mammalian models have long been used to predict the effects of noise on humans. Temporary threshold shifts (TTSs) increase with the duration of sound exposure until an asymptotic threshold shift (ATS) is reached (Clark 1991). Once the ATS is reached, further noise exposure no longer increases TTS, but ATS increases linearly with the sound pressure level (SPL) of the stimulus (Carder and Miller 1972).

Smith et al. (2004b) tested the hypothesis that noise-induced threshold shifts in fishes would increase linearly with increasing sound pressure differences (SPDs) between the exposure noise and baseline hearing thresholds across multiple frequencies in the fish audiograms. This relationship was coined the linear threshold shift (LINTS) hypothesis. The advantage of using SPDs instead of raw SPLs is that it accounts for differences in hearing loss at different frequencies. In other words, if a sound stimulus has much of its energy near 400 Hz, one might expect that the greatest hearing

M.E. Smith (✉)
Department of Biology, Western Kentucky University, Bowling Green, KY 42101, USA
e-mail: michael.smith1@wku.edu

A.N. Popper and A. Hawkins (eds.), *The Effects of Noise on Aquatic Life*,
Advances in Experimental Medicine and Biology 730, DOI 10.1007/978-1-4419-7311-5_57,

loss would also be near 400 Hz. Using SPDs also accounts for differential hearing sensitivity across frequencies. For example, goldfish exposed to white noise exhibited the greatest threshold shifts between 800 and 1,000 Hz, which is near the frequency where their hearing is most sensitive (Smith et al. 2004b).

Threshold shift data were compiled for five species of fishes (bluegill sunfish *Lepomis macrochirus* [Scholik and Yan 2002]; fathead minnow *Pimephales promelas* [Scholik and Yan 2001]; catfish *Pimelodus pictus* [Amoser and Ladich 2003]; goldfish *Carassius auratus* and tilapia *Oreochromis niloticus* [Smith et al. 2004b]). When threshold shifts were plotted as a function of SPD above baseline hearing thresholds, a significant linear relationship was evident (TTS = 0.23(SPD) − 2.44; $r^2 = 0.62$). The LINTS relationship predicts no or minimal hearing loss when the exposure stimulus is <20 dB above baseline hearing levels and ~20 dB of hearing loss when the exposure is 100 dB above baseline levels.

Although the LINTS hypothesis is robust and predictive on many levels, it has limited utility. This model was developed using long-term continuous-noise exposures in which fish were exposed for at least 24 h, and, presumably, an ATS was reached. Shorter exposures would result in lower threshold shifts than the LINTS model would predict. For example, goldfish exposed to 160–170 dB re 1 µPa white noise for 10 min, 1 h, and 24 h exhibited mean threshold shifts of ~5, 12, and 27 dB, respectively (Smith et al. 2004a). In addition, most anthropogenic noises that have the potential to damage fish auditory systems are not long-term continuous sounds but impulsive or repetitive sounds that are temporary or sporadic. Thus a better measure of cumulative acoustic exposure is sound exposure level (SEL), the squared weighted sound pressure integrated over time instead of peak or root mean square (RMS) SPL per se.

3 Equal Energy Hypothesis

The concept of using the total amount of energy received by an organism's auditory system as a means of predicting hearing loss has been conceptualized and defined by Elredge and Covell (1958) as the equal energy hypothesis (EEH). It states that equal amounts of sound-induced hearing loss is expected from varying sound sources as long as their sum amounts of energy are equal. In other words, the EEH states that the frequency, level, and duration of a given sound are not important as long as their total energy is equivalent. The EEH has been shown to have predictive values in mammals (Elredge and Covell 1958) but has only been recently tested in fish.

Smith and Gilley (2008) exposed three species of fishes to five equal-energy combinations of exposure times (from 6 to 1,440 min.) and SPL (from 155 to 179 dB re 1 µPa) following a 3-dB trade-off rule for a doubling of exposure duration. They exposed *Carassius auratus* and channel catfish *Ictularus punctatus* to a 1,000-Hz tone and rainbow trout *Onchorynchus mykiss* to a 400-Hz tone. Although all exposure combinations resulted in threshold shifts in all three species, there were significant differences between the five duration-intensity combinations. Despite the variability found across the exposure combinations, there was some degree of consistency found between the threshold shifts of *C. auratus* and *I. punctatus* that were exposed for intermediate intensities and durations (i.e., 23 and 90 min at 173 and 169 dB re 1 µPa, respectively). This suggests that the EEH may have some predictive value in a limited range. At higher SPLs, the shorter duration times may not be long enough for the fish to exhibit the physiological effects of acoustic trauma because there is likely a lag between noise exposure and damage to the auditory sensory cells that would lead to greater hearing loss. Lower SPLs may not have enough energy to cause auditory damage no matter how long the exposure. Thus this preliminary data suggest that the EEH is not a very predictive model of hearing loss in fishes.

4 Particle Motion

In the models discussed above, the sound stimuli were quantified in terms of SPL (dB re 1 μPa), but fish can also detect particle displacement (Fay 1988). In the past, fish species were often characterized as either hearing "specialists" or "generalists," with specialists having anatomical structures that connect or bring an air bubble (e.g., swim bladder) in close proximity to the inner ear. Fish with such specializations are more sensitive to sound pressure and can detect higher frequencies than those without (i.e., generalists). In reality, there is a continuum of both anatomical structures and hearing abilities in fishes. Popper and Fay (2011) proposed placing fish species on a continuum of pressure-detection methods, with fish that have extensive use of pressure (like Otophysan fishes) on one extreme and fish that only use motion detection (such as fish with no swim bladders like sharks and flatfish) on the opposite extreme.

In both of these extremes, sensory hair cells in the inner ear are deflected due to shear forces produced by the relative motion of the sensory epithelia and the overlying otolith (or otoconia), which is much denser than the surrounding tissue. Larger relative motions will presumably produce greater shear forces with a greater probability of damaging hair cells. Hastings (2010) recently proposed using a lumped-parameter mechanical and fluid mathematical model to predict hearing loss in fishes. Inputs into this model include swim bladder geometry, otolith size, anatomical connections between the swim bladder and inner ear, number of hair cells, and frequency and amplitude of the sound stimuli. Preliminary data from five fish species suggest that the model predicts excessive relative motion in the ear that is correlated with hair cell damage (Hastings 2010). Although this type of modeling shows great promise in our understanding of the relationships between peripheral auditory structures of various fish species and their susceptibility for noise-induced damage to the auditory system, the large number of parameters needed as model inputs may not be practical for most managers attempting to mitigate the effects of anthropogenic sound.

5 Conclusions

Early models of hearing loss in fishes were dependent on measurements of pressure level of sound sources. Although the LINTS model is predictive for pressure-detecting fishes, it is not appropriate for all species. Thus models that incorporate particle motion, of both the sound stimulus and the resulting vibration in the inner ears of fish, will be important to our understanding of the effects of noise on hearing loss in fishes.

References

Amoser S, Ladich F (2003) Diversity in noise-induced temporary hearing loss in otophysine fishes. J Acoust Soc Am 113:2170–2179.
Carder HM, Miller JD (1972) Temporary threshold shifts from prolonged exposure to noise. J Speech Hear Res 15:603–623.
Clark WW (1991) Recent studies of temporary threshold shift (TTS) and permanent threshold shift (PTS) in animals. J Acoust Soc Am 90:155–163.
Eldredge DH, Covell WP (1958) A laboratory method for the study of acoustic trauma. Laryngoscope 68:465–477.
Fay RR (1988) Hearing in vertebrates: A psychophysics databook. Hill-Fay Associates, Winnetka, IL.
Hastings MC (2010) A model for predicting effects of sound on auditory sensory cells in fishes. J Acoust Soc Am 127:1755.
Popper AN, Fay RR (2011) Rethinking sound detection by fishes. Hear Res 273:25–36. Doi:10.1016/j.heares.2009.12.023.

Popper AN, Smith ME, Cott PA, Hanna BW, MacGillivray AO, Austin ME, Mann DA (2005) Effects of exposure to seismic airgun use on hearing of three fish species. J Acoust Soc Am 117:3958–3971.

Scholik AR, Yan HY (2001) Effects of underwater noise on auditory sensitivity of a cyprinid fish. Hear Res 152:17–24.

Scholik AR, Yan HY (2002) The effects of noise on the auditory sensitivity of the bluegill sunfish, *Lepomis macrochirus*. Comp Biochem Physiol A 133:43–52.

Smith ME, Gilley RR (2008) Testing the equal energy hypothesis in noise-exposed fishes. Bioacoustics 17:343–345.

Smith ME, Kane AS, Popper AN (2004a) Noise-induced stress response and hearing loss in goldfish (*Carassius auratus*). J Exp Biol 207:427–435.

Smith ME, Kane AS, Popper AN (2004b) Acoustical stress and hearing sensitivity in fishes: Does the linear threshold shift hypothesis hold water? J Exp Biol 207:3591–3602.

Effects of Pile-Driving Noise on *Oncorhynchus mykiss* (Steelhead Trout)

Richard Oestman and Christopher J. Earle

1 Introduction

Underwater sounds from anthropogenic sources may affect the behavior and physiology of aquatic organisms, including a variety of salmonid fishes protected under the US Endangered Species Act. One mechanism of effect is underwater sound generated by pile driving in and near aquatic habitats (Popper and Hastings 2009). In 2008, several state and US federal agencies convened the Fisheries Hydroacoustic Working Group (FHWG) to evaluate the available scientific evidence and determine interim criteria for the onset of injury to fish from underwater sound generated by pile driving. Interim criteria were subsequently agreed to by the National Marine Fisheries Service, United States Fish and Wildlife Service, Federal Highway Administration, California Department of Fish and Game, and the California (Caltrans), Oregon, and Washington State Departments of Transportation (FHWG 2008). The interim criteria for the onset of injury were set at a peak sound pressure level of 206 dB re 1 μPa and accumulated sound exposure levels (SELs) of 187 dB re 1 μPa^2-s for fish >2 gm body weight and 183 dB for fish <2 gm body weight. The accumulated SEL is a measure of the cumulative energy to which a fish is exposed to over the course of a pile-driving event (<1 day).

Laboratory data for a variety of sound sources have been used to estimate the thresholds of physical effects of pile driving on fish (Popper and Hastings 2009). However, there have been only a few experiments that evaluate pile-driving sound propagation and the attendant physical effects of pile-driving sound on fish in natural environments, particularly in riverine systems. This study was undertaken to evaluate whether exposure to underwater sound from pile driving in excess of the interim criteria can result in damage to juvenile salmonid tissues.

R. Oestman (✉)
ICF International, Seattle, WA 98104, USA
e-mail: roestman@icfi.com

C.J. Earle
ICF International, Olympia, WA 98501, USA
e-mail: cearle@icfi.com

A.N. Popper and A. Hawkins (eds.), *The Effects of Noise on Aquatic Life*,
Advances in Experimental Medicine and Biology 730, DOI 10.1007/978-1-4419-7311-5_58,

2 Methods

Caltrans is in the process of replacing the two Highway 101 bridges that cross the Mad River in McKinleyville, CA. The bridge replacements involve driving several 2.2-m-diameter cast-in-steel shell piles immediately adjacent to the river with a Pileco D225 diesel impact hammer. This study used caged-fish deployments within the Mad River to expose juvenile *Onchorhynchus mykiss* to a variety of peak sound pressures levels and cumulative SELs during pile driving. Four experimental trials were conducted (on 1, 6, 8, and 10 July 2009). Each trial consisted of driving 1 pile section (20-24 m long). During each trial, cages containing juvenile *Oncorhynchus mykiss* sourced from a nearby fish hatchery were placed at 4 exposure locations at various distances from the pile-driving activity (from 35 to 150 m away) and at an in-river control location (350 m away). Equipment deployed at each exposure location included identical, paired cages; one contained a hydrophone and the other contained 10 fish. The cages were submerged to a depth of 0.5 m in 1- to 3-m-deep water. Underwater sound (peak and SEL) was recorded at each location during the experiments.

Hatchery, transport, and overnight control groups of fish were necropsied before each pile-driving event. After cessation of each pile-driving event, a necropsy was performed on each exposed and in-river control fish. Gross observations and necropsy included condition of the skin, eyes, fins, mesenteric vasculature, and swim bladder as well as general observations for parasites on the skin and gills.

Tissue samples were collected from all fish. For each trial, fish from the cage closest to the pile driving, fish from any other cage that received an accumulated SEL of >187 dB, and the respective in-river control groups underwent histopathology. Histopathology included examination of tissues from the head, gill, liver, and a body wedge comprising the swim bladder, trunk kidney, spinal cord, vertebrae, skeletal muscle, and skin. All on-site necropsies and subsequent histopathology of fish were conducted by Gary D. Marty, DVM, Fish Pathology Services, Abbotsford, BC, Canada.

3 Results

During pile driving, fish were exposed to underwater maximum peak sound pressure levels ranging from 163 to 188 dB, well below the interim peak criterion of 206 dB (Caltrans 2010a). Cumulative SELs ranged from 178 to 194 dB. The cumulative SEL exceeded the cumulative SEL criterion of 187 dB during 2 trials. In both trials, the criterion was exceeded in the 2 cages closest to the pile being driven, which were at distances of 35 and 50 m from the driven pile. Thus 4 exposure groups experienced cumulative SELs in excess of 187 dB. Control fish in a cage 350 m from the pile driving experienced total accumulated SELs of 159–166 dB.

A total of 159 exposed fish and 156 control fish were necropsied, and histopathology was performed on tissue from 60 exposed fish and 37 in-river control fish (Caltrans 2010b,c). Necropsy and histopathology identified no significant differences between the incidence of pathological conditions in exposed and control fish. Relatively few of the 35 pathological conditions evaluated were detected in either the necropsied fish or the fish that underwent histopathology. The principal detected conditions included fin fraying (which is fairly common in hatchery fish), enlarged liver cells (hepatocellular megalocytosis), foci of inflammation in the liver (focal/multifocal parenchymal leukocytes), inflammation around the ducts that carry bile from the liver to the gallbladder (cholangitis/pericholangial leukocytes), and granulomatous inflammation in several organs. The observed frequency of these conditions was consistent with the low incidence of such lesions common in any juvenile salmonid population.

References

California Department of Transportation (Caltrans) (2010a) Mad River bridges replacement project: Hydroacoustic monitoring report for bridge piles driven in 2009. Prepared by Illingworth & Rodkin, Petaluma, CA.

California Department of Transportation (Caltrans) (2010b) Effects of pile driving sound on juvenile steelhead. Prepared by ICF Jones & Stokes, Seattle, WA.

California Department of Transportation (Caltrans) (2010c) Necropsy and histopathology of steelhead trout exposed to steel pile driving at the Mad River bridges, US Highway 101, July 2009. Prepared by Gary D. Marty, DVM, PhD, Fish Pathology Services, Abbotsford, BC, Canada.

Fisheries Hydroacoustic Working Group (FHWG) (2008) Memorandum – Agreement in principal for interim criteria for injury to fish from pile driving activities. Available via Caltrans' Fish Bioacoustic Web page at http://www.dot.ca.gov/hq/env/bio/files/fhwgcriteria_agree.pdf. Accessed April 25, 2010.

Popper AN, Hastings MC (2009) The effects of anthropogenic sources of sound on fishes. J Fish Biol 75:455–489.

Part V
Anthropogenic Sounds and Behavior

Controlled Exposure Study of Dolphins and Sea Lions to Midfrequency Sonarlike Signals

Dorian Houser, Laura Yeates, Daniel Crocker, Stephen W. Martin, and James J. Finneran

1 Estimates of Acoustic Impacts to Marine Mammals

Many nations that conduct acoustic activities within the oceans are required to address the potential impacts to marine mammals resulting from those actions. In the United States, federal government agencies must comply with the regulations established under the National Environmental Policy Act (NEPA), the Marine Mammal Protection Act (MMPA), and the Endangered Species Act (ESA). Under the MMPA, the determination of potential impacts to marine mammals requires estimates of both behavioral and physiological responses that result from sound exposure. Both determinations are fraught with difficulty and uncertainty.

Temporary threshold shift (TTS), a temporary reduction in hearing sensitivity, has been used as a criterion for the onset of physiological responses resulting from sound exposure. The sound exposure level (SEL; dB re 1 μPa^2-s) required to induce TTS in *Tursiops truncatus* (bottlenose dolphins) exposed to midfrequency tonal signals has served as the threshold for the physiological impacts to both odontocete and mysticete whales in US Navy environmental compliance efforts involving midfrequency active sonar (MFAS). The TTS data have been acquired over more than a decade (Finneran et al. 2005, 2010; Schlundt et al. 2000), yet there is still uncertainty as to appropriate thresholds for exposures to higher and lower frequencies and from multiple sound exposures. Nevertheless, data regarding the sound exposures that result in TTS are far more grounded than are data for predicting the behavioral responses to sound exposure.

Behavioral responses to sound are potentially impacted by the sound source (type of sound), duration and amplitude of the exposure, the animal's prior experience with the sound source, and the exposed animal's state of motivation. The significance of the response is dependent on the behavior at the time of the exposure (e.g., foraging or breeding), the magnitude and duration of

D. Houser (✉) • L. Yeates
National Marine Mammal Foundation, San Diego, CA 92106, USA
e-mail: dorian.houser@nmmfoundation.org; laura.yeates@nmmfoundation.org

D. Crocker
Sonoma State University, Rohnert Park, CA 94928, USA
e-mail: crocker@sonoma.edu

S.W. Martin • J.J. Finneran
US Navy Marine Mammal Program, Space and Naval Warfare Systems Center Pacific, San Diego, CA 92152, USA
e-mail: stephen.w.martin@navy.mil; james.finneran@navy.mil

A.N. Popper and A. Hawkins (eds.), *The Effects of Noise on Aquatic Life*, Advances in Experimental Medicine and Biology 730, DOI 10.1007/978-1-4419-7311-5_59,

the response, and the life history stage of the animal. Although many studies have been conducted exploring the relationship between sound exposure and the behavioral reactions of wild marine mammals (for a review, see Richardson et al. 1995), the complexities of animal behavior, unknowns about animal motivation and experience, and inherent uncertainties regarding the sound field to which animals are exposed have made interpretation of these studies difficult and subjective. Still, federal agencies within the United States are required to estimate behavioral impacts to marine mammals that result from the noise generated by their activities.

The evolving approach to behavioral harassment estimates in US Navy environmental compliance documents is based on the development of a risk function, which predicts the probability of a behavioral response as a function of the sound pressure level (SPL; dB re 1 μPa) received by the animal (i.e., the received level). The current US Navy risk function for odontocetes exposed to midfrequency sound (1-10 kHz) is derived from 3 datasets: an analysis of behavioral observations made during TTS experiments with trained *Tursiops truncatus* (Finneran and Schlundt 2004); reported alterations in *Orcinus orca* (killer whales) behavior from animals exposed to MFAS in the coastal waters of Washington State; and behavioral responses of *Eubalaena glacialis* (North Atlantic right whales) exposed to alarm signals within the frequency range of 500-4,500 Hz (Nowacek et al. 2004). Each of these datasets has, to some extent, been criticized for its relevancy. TTS studies utilized positive reinforcement to increase the tolerance of subjects to high-exposure levels, and these studies were not designed as a behavioral response studies. Observations of *Orcinus orca* exposed to MFAS are complicated by the proximity of whale-watching vessels and no direct measures of sound exposure. *Eubalaena glacialis* are mysticetes, and their response to alarm signals (i.e., signals intentionally designed to elicit a response) may not be representative of odontocete species and/or other sound types. Nevertheless, at the time the risk function was created, these studies provided the only data suitably related to the midfrequency range of US Navy tactical sonar.

2 Behavior Response Studies in Wild Marine Mammals

Risk functions, which are dose-response functions, provide an intuitive means by which to explore relationships between sound exposure and the behavioral reactions of marine mammals. Ideally, data for these functions would be collected from wild animals across a broad range of exposure conditions. Exposures of different species within and across phylogenetic groups could potentially provide insight into the degree of tolerance of phylogenetically related species to sound type and novelty. A number of behavioral response studies are currently underway with wild marine mammals. Most notable of these are those involving sound exposure to marine mammals previously tagged with dive and acoustic recorders. Eventually, these studies will provide the bulk of the information for use in determining the relationships between sound exposure and the behavioral responses of marine mammals. However, the logistical difficulties of deploying tags on wild marine mammals, particularly cetaceans, and exposing them to desired sound levels will take time. The acquisition of a robust dataset for the development of a risk function for any species or sound source will likely take many years to build.

3 Stress Responses of Marine Mammals Exposed to Anthropogenic Sound

The relevance of a risk function to an individual or population of marine mammals is limited in its broader management application. Whether certain types of behavioral responses are of biological significance is currently under debate. In this regard, physiological measures of the generalized stress response coupled to observations of sound-induced behavioral reactions are warranted. Both acute stress markers (e.g., catecholamines) and chronic stress markers (e.g., cortisol and aldosterone) can provide a more comprehensive view of the impact to the exposed animal, potentially shedding light on the animal's allostatic load. Several investigators have attempted to link the generalized stress response to sound exposure in marine mammals held in collections (Romano et al. 2004; Thomas et al. 1990), but the few results obtained have been variable, inconclusive, and/or hampered by small sample sizes.

4 Controlled Exposure Studies With Animals of the US Navy Marine Mammal Program

The US Navy Marine Mammal Program (MMP) maintains a collection of *Tursiops truncatus* and *Zalophus californianus* (California sea lions) for defense purposes. The availability of these animals for research provides an opportunity for a large sample size study of marine mammal behavioral responses to midfrequency sonarlike signals. In addition, because many of the *Tursiops truncatus* are trained to wear harnesses and voluntarily present themselves for blood draws, there is potential to relate behavioral reactions to measures of the generalized stress response.

A group of *Tursiops truncatus* (n = 30) and *Zalophus californianus* (n = 15) have been trained to leave a station (A), travel to another station (B) and touch a paddle, and then return to the original station (A) for a fixed fish reward. Subjects are trained on the task until there is 100% completion across a 10-trial block. On meeting the nominal performance criterion, the subjects perform a 10-trial control session followed by a 10-trial sound-exposure session. Control and exposure sessions are performed in a 9- × 18-m floating pen with an underwater sound projector placed 1 m behind station B. Trials are limited to 30 s. The fixed reward for completing a trial is a number of fish equal to ~1% of the weight of the daily allotment of fish for each animal. For control sessions, trials are performed without an acoustic exposure. During an exposure session, the animal receives an acoustic exposure in each trial at the midpoint along its trajectory to station B. Acoustic exposures consist of a 0.5-s upward frequency-modulated sweep (center frequency ~3,250 Hz) followed by a 0.5-s tone (~3,450 Hz). Subjects are assigned to groups, and the SPL received by subject groups ranges from ~115 to 185 dB re 1 μPa. The study is designed so that a subject receives the same exposure across all trials and is part of a group that receives the same exposure level.

Sessions are video recorded, and intercom and underwater recordings are mixed with the video for use in the analysis of behavioral responses. Behavioral responses anticipated to occur have been a priori scored for severity by an independent group of scientists to avoid potential biases in the behavioral analysis. At the completion of data collection, the behavioral responses will be scored, and the results will be used to investigate the dose-response relationship between sound exposure and the severity and occurrence of behavioral reactions. The series of 10 trials will be used to investigate whether acute sensitization/desensitization occurs during the trial sequence.

Blood samples are collected from *Tursiops truncatus* 1 wk before and immediately after the exposure session and 1 wk after testing. Samples are analyzed via radioimmunoassay to assess circulating levels of cortisol, aldosterone, and epinephrine. A subset of *Tursiops truncatus* wear an ECG system to record heart rate during the sessions. Hormone levels and heart rates will be compared with exposure levels to determine whether acute or chronic stress responses resulted from the sound exposures.

4.1 Utility of the Study

Behavioral response studies with wild marine mammals are needed to enable the prediction of behavioral disturbances resulting from exposure to anthropogenic noise. However, collection animals can also be studied and can facilitate a more rapid acquisition of both behavioral and physiological data. Although information obtained from behavioral-response studies with collection animals will not supplant data obtained with wild populations, it can serve as a useful interim source for seeding risk functions used in the quantitative prediction of impacts to marine mammals exposed to anthropogenic sound.

References

Finneran JJ, Carder DA, Schlundt CE, Dear RL (2010) Growth and recovery of temporary threshold shift (TTS) at 3 kHz in bottlenose dolphins (*Tursiops truncatus*): Experimental data and mathematical models. J Acoust Soc Am 127:3256–3266.

Finneran JJ, Carder DA, Schlundt CE, Ridgway SH (2005) Temporary threshold shift (TTS) in bottlenose dolphins (*Tursiops truncatus*) exposed to mid-frequency tones. J Acoust Soc Am 118:2696–2705.

Finneran JJ, Schlundt CE (2004) Effects of intense pure tones on the behavior of trained odontocetes. Technical Report 1913, SPAWAR Systems Center San Diego, San Diego, CA.

Nowacek DP, Johnson MP, Tyack PL (2004) North Atlantic right whales (*Eubalaena glacialis*) ignore ships but respond to alerting stimuli. Proc Biol Sci 271:227–231.

Richardson WJ, Greene CR Jr, Malme CI, Thomson DH (1995) Marine mammals and noise. Academic Press, San Diego, CA.

Romano T, Keogh M, Kelly C, Feng P, Berk L, Schlundt CE, Carder DA, Finneran JJ (2004) Anthropogenic sound and marine mammal health: Measures of the nervous and immune systems before and after intense sound exposures. Can J Fish Aquat Sci 61:1124–1134.

Schlundt CE, Finneran JJ, Carder DA, Ridgway SH (2000) Temporary shift in masked hearing thresholds of bottlenose dolphins, *Tursiops truncatus*, and white whales, *Delphinapterus leucas*, after exposure to intense tones. J Acoust Soc Am 107:3496–3508.

Thomas JA, Kastelein RA, Awbrey FT (1990) Behavior and blood catecholamines of captive belugas during playbacks of noise from an oil drilling platform. Zoo Biol 9:393–402.

A Direct Comparison of Bottlenose Dolphin and Common Dolphin Behavior During Seismic Surveys When Air Guns Are and Are Not Being Utilized

Sarah B. Barry, Anna C. Cucknell, and Nicola Clark

1 Introduction

The majority of noise produced by air guns from seismic surveys are recorded at sound pressure levels focused in the low frequencies (0 to 300 Hz) of ~240 dB re 1 μPa at 1 m. Bottlenose (*Tursiops truncatus*) and common (*Delphinus delphis*) dolphins are sensitive to a broad range of frequencies, from just 50 (Turl 1993) to 300,000 Hz and from 100 to 150,000 Hz (Richardson et al. 1998), respectively. This paper uses combined data collected during 21 offshore seismic surveys over 8 years (2001–2008) from Europe, West Africa, and Australia to investigate the observed behavioral variations caused by seismic air gun activity on bottlenose and common dolphins.

2 Methodology

Behavioral observations were split into six categories: "foraging," "milling," "traveling," "close to boat," "surface-active displays," and "traveling with active displays."

The relative proportions of behaviors when the air guns were firing were statistically analyzed with the χ^2 test, assuming the expected levels to be the same across all behaviors. This analysis was also carried out for the sightings collected during no air gun activity. After this, each behavioral category was examined using a z-test, directly comparing the proportions of each behavior when the air guns were and were not firing. The z-test takes into account both the proportion and sample size of each behavioral category for this comparison. A confidence level of 95% (P = 0.05) was employed for all statistical analysis.

S.B. Barry (✉) • N. Clark
Gardline Environmental Limited, Great Yarmouth, Norfolk NR30 3NG, UK
e-mail: sarah.barry@gardline.co.uk

A.C. Cucknell
The Cork Initiative, Penryn, Cornwall TR10 8BL, UK

A.N. Popper and A. Hawkins (eds.), *The Effects of Noise on Aquatic Life*,
Advances in Experimental Medicine and Biology 730, DOI 10.1007/978-1-4419-7311-5_60,

Table 1 Behavioral definitions

Foraging	Foraging and lunge foraging, diving, fluking
Milling	Logging, milling, slow swimming
Traveling	Normal and fast swimming, shallow diving, traveling, transiting, porpoising
Close to boat	Bow riding, close to boat, spy hopping
Surface-active displays	Breaching, leaping, head/tail slaps, fin slapping, lobtailing, sexual
Traveling/active displays	When traveling and active displays were listed together

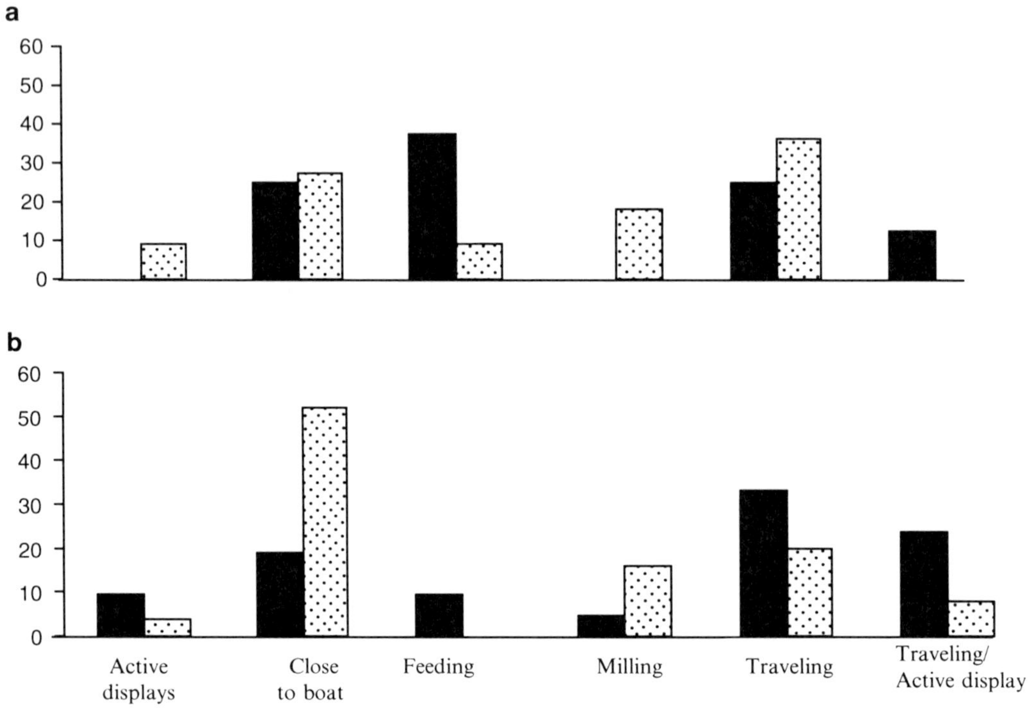

Fig. 1 Relative percentage of behaviors during seismic air gun activity (black bars) and inactivity (dotted bars) for bottlenose (**a**) and common (**b**) dolphin sighting data

3 Results

Analysis of both species with the χ^2 test showed significant variations in behavior from the expected equal chance of sighting all behavioral categories. Bottlenose dolphins sighted during air gun activity demonstrated predominant proportions of "foraging" and "traveling" behaviors ($\chi^2 = 67.28$; $df = 5$; $P < 0.001$). Conversely, the predominant behaviors while the air guns were inactive were "close to boat" and "traveling" ($\chi^2 = 55.04$; $df = 5$; $P < 0.001$). However, a direct statistical comparison between each behavioral category with a z-test demonstrated no significant differences between the bottlenose dolphin behavior in regard to air gun activity.

Common dolphins demonstrated predominant behaviors of "close to boat" ($\chi^2 = 106.4$; $df = 5$; $P < 0.001$) when the air guns were not firing and "traveling" and "traveling with active displays" ($\chi^2 = 31.16$; $df = 5$; $P < 0.001$) when the air guns were firing. In addition, in contrast to the bottlenose dolphin results, common dolphins were found to show a significant difference in behavior with air gun activity, with more "close to boat" behavior when the air guns were not firing (z-test statistic 2.31; $P = 0.021$).

4 Discussion

Our results show some interesting contradictions to the current findings. During air gun activity, common dolphins demonstrated predominant behaviors of "traveling" and "traveling with active displays." Interpreting the purpose of surface behaviors in wild cetaceans is continually subject to a variety of speculative hypotheses. It is now generally accepted that elements of active displays are a form of nonvocal communication between individuals advertising a motivational and/or an intentional state (Lusseau 2006), such as to transfer information about the direction of travel (Herzing 2000).

Although the air guns were inactive, both bottlenose and common dolphin observations demonstrated predominantly "close to boat" behavior. Goodwin and Cotton (2004) grouped behaviors into classes, listing approaching the vessel, bow riding, or rubbing alongside vessels as "positive" behaviors, all of which would be included in the "close to boat" behavioral category. Despite the use of such behavioral categories being potentially biased (Slooten 1994), it is generally accepted that if a cetacean approaches a vessel, it is doing so in a relaxed manner and does not feel threatened by the vessel.

Stone's (2003) report on cetacean observations during seismic surveys highlights a decreased tendency for cetaceans to engage in "foraging" activity during periods of air gun firing. Conversely, our results show that bottlenose dolphins were observed predominantly displaying "foraging" behavior when the guns were firing. Bottlenose dolphins are well known to be opportunistic feeders (dos Santos et al. 2007) with a variety of foraging strategies worldwide (Perrin et al. 2009). With this in mind, it would not be unreasonable to suggest that dolphins may be feeding opportunistically on fish that may be suffering from temporary threshold shift (TTS), a startle response (McCauley et al. 2000), or hearing damage due to the air gun activity (Popper et al. 2005).

Bottlenose dolphins are the most cosmopolitan in distribution of all dolphin species; their opportunistic nature has resulted in them becoming highly adaptable to a great variety of niches worldwide (Perrin et al. 2009). We speculate that such adaptability may explain why the bottlenose dolphin does not appear to show any reaction, agitation, or distress to firing seismic air guns. Common dolphins demonstrated significantly more "close to boat" behavior when the air guns were not firing compared with sightings during air gun activity. Monitoring studies have shown that various small and medium-sized odontocetes are sufficiently tolerant of air gun sounds (Richardson et al. 2004); however, these results suggest that different species of small cetaceans differ in their sensitivity to air gun sounds, with common dolphins affected more than bottlenose dolphins.

5 Conclusions

The data suggest that not all species of cetaceans are affected by seismic air guns as previously speculated. Some species may even use seismic air guns to their own advantage. The results also show consistencies with studies and reports already published on the impact of seismic surveys on cetaceans. It appears that different species adapt to seismic air gun emissions in different ways. However, it is important to keep in mind that there is still a long way to go in terms of understanding and interpreting what is "normal" and unaffected cetacean behavior regardless of anthropogenic activities.

References

dos Santos ME, Coniglione C, Louro S (2007) Feeding behaviour of the bottlenose dolphin, *Tursiops truncatus*, in the Sado estuary, Portugal. Rev Bras Zoociênc 9:31–39.

Goodwin L, Cotton PA (2004) Effects of boat traffic on the behaviour of bottlenose dolphins (*Tursiops truncatus*). Aquat Mamm 30:279–283.

Herzing DL (2000) Acoustics and social behavior of wild dolphins: Implications for a sound society. In: Au WWL, Popper AN, Fay RR (eds) Hearing by whales and dolphins. Springer, New York.

Lusseau D (2006) Why do dolphins jump? Interpreting the behavioural repertoire of bottlenose dolphins (*Tursiops* sp.) in Doubtful Sound, New Zealand. Behav Processes 73:257–265.

McCauley RD, Fewtrell J, Duncan AJ, Jenner C, Jenner MN, Penrose JD, Prince RIT, Adhitya A, Murdoch J, McCabe K (2000) Marine seismic surveys–A study of environmental implications. APPEA J 40:692–708.

Perrin WF, Wursig B, Thewissen JGM (2009) Encyclopedia of marine mammals. Academic Press, San Diego, CA.

Popper AN, Smith ME, Cott PA, Hanna BW, MacGillivray AO, Austin ME, Mann DA (2005) Effects of exposure to seismic airgun use on hearing of three fish species. J Acoust Soc Am 117:3958–3971.

Richardson JW, Greene CR Jr, Malme C, Thomson D (2004) Marine mammals and noise. Academic Press, San Diego, CA.

Slooten E (1994) Behavior of Hector's dolphin: Classifying behavior by sequence analysis. J Mammal 75:956–964.

Stone CJ (2003) The effects of seismic activity on marine mammals in UK waters, 1998-2000. Joint Nature Conservation Committee (JNCC) Report 323, Peterborough, UK.

Turl C (1993) Low-frequency sound detection by a bottlenose dolphin. J Acoust Soc Am 94:3006–3008.

Behavioral Reactions of Harbor Porpoise to Pile-Driving Noise

Jakob Tougaard, Line A. Kyhn, Mats Amundin, Daniel Wennerberg, and Carolina Bordin

1 Introduction

Pile driving of large steel monopiles in offshore waters has increased rapidly in recent years due to the expanding development of offshore wind energy. In particular, *Phocoena phocoena* (harbor porpoise) has been the focus of attention with respect to a possible negative impact. Impact pile driving, where a large steel monopile is driven 20-30 m into the seabed, is capable of generating very loud sound pressures, exceeding 230 dB re 1 μPa peak-peak in source levels and detectable at distances of tens of kilometers (Bailey et al. 2010). Such high sound pressures, coupled with the repetitive emission of sounds (1–2 strokes/s) at a high duty cycle (10%) gives the potential for exposing nearby animals to very high and potentially damaging sound exposure levels (Gordon et al. 2009). Besides the potential to inflict acute injury, the pile-driving noise has the potential to affect behavior of marine mammals over an even larger area.

2 Regulation of Pile Driving and Other Loud Sound Sources

Focus on the regulation of pile driving has so far been mainly on acute injury, i.e., measures to reduce the risk that marine mammals are exposed to sound levels that could damage their auditory system. Mitigation is primarily in the form of deterrent devices (pingers and seal scarers) deployed before pile driving and soft start (ramp up) of the piling procedure. Deterrent devises and soft start protocols (ideally) allow animals to vacate the zone of injury before the full-power pile driving commences, but they do not reduce the size of the zone of injury nor the size of the zone of behavioral reaction.

Thus, even with a carefully designed protocol to protect against injury, there may still be an impact due to behavioral reactions to the sound. Because the zone of behavioral reactions for pile driving could be very large, the number of individuals affected by behavioral reactions is likely to be considerably larger than the number of individuals potentially exposed to injuring sound levels.

J. Tougaard (✉) • L.A. Kyhn
National Environmental Research Institute, Aarhus University, DK-4000 Roskilde, Denmark
e-mail: jat@dmu.dk

M. Amundin • D. Wennerberg • C. Bordin
Kolmården Djurpark, SE-618 92 Kolmården, Sweden

A.N. Popper and A. Hawkins (eds.), *The Effects of Noise on Aquatic Life*, Advances in Experimental Medicine and Biology 730, DOI 10.1007/978-1-4419-7311-5_61,

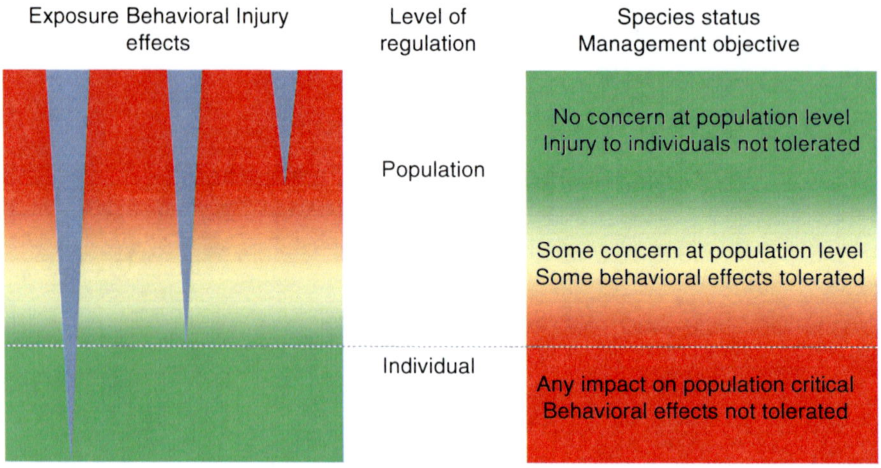

Fig. 1 The accepted exposure levels depend not only on the physiology and behavior of the species in question but also on management objectives. Left: Increasing exposure leads to larger effects. Right: The less concern there is for significant impact at population level, the larger the behavioral impact that can be tolerated

The large difference between the size of zone of injury and the zone of behavioral reaction highlights a central question in terms of mitigation: Is it sufficient to protect only against acute injury or must measures be taken to reduce the impact on behavior? The simple answer is that it depends. It depends on the species involved and, more importantly, on the management objectives for that species in the relevant area. This is illustrated in Figure 1.

In general, the protection of marine mammals is regulated at two levels: individuals and populations. In most cases, individual marine mammals are specifically protected from deliberate injury. At the same time, the species are generally protected, with the aim of maintaining favorable living conditions and viable populations (or appropriate management units). Population-based management clearly allows for some impact on the individuals as long as this impact does not affect the overall management objective. This means that although injury to individuals should be minimized whenever possible, the level of behavioral impact tolerated from a particular activity should depend on the status of the relevant species. The greater concern there is for a population, the less the behavioral impact should be tolerated. On the other hand, for species where there is little concern over the long term, the need for mitigating behavioral effects is small and the focus should be on acute injury.

3 Assessing the Impact on Population Parameters

The difficulty of assessing behavioral effects is that they can only rarely be observed directly. Even when a behavioral reaction can be quantified, the real impacts will most often only manifest themselves later in the life of the affected individuals through changes in their survival and reproductive success and ultimately in the size of the population (Fig. 2). In most cases, the impact must be inferred indirectly from behavioral observations, which requires a firm understanding of the links between behavior and population parameters (National Research Council 2005).

Evident from Figure 2 is that the commonly seen practice of quantifying impact in terms of habitat exclusion is problematic. Because habitat exclusion is not a population effect per se, a reduction in

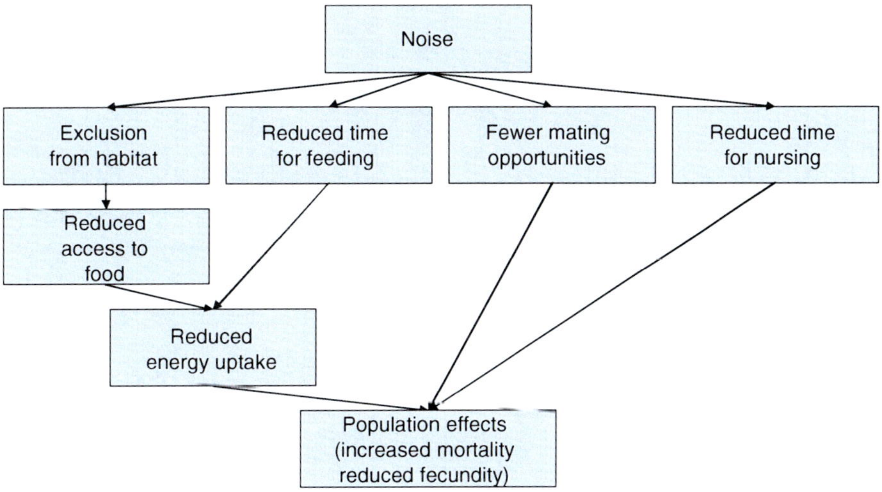

Fig. 2 Behavioral reactions to noise can have effects on population parameters directly and indirectly

abundance in the impact area does not in itself imply a significant effect. At the same time, because there are other effects than exclusion, a lack of response (no reduction in abundance) cannot be taken as evidence for no effect.

4 Behavioral Effects of Pile-Driving Noise

Several studies have demonstrated the behavioral reactions of harbor porpoises to pile driving at distances up to 25 km from the pile-driving site by use of passive acoustic monitoring (Brandt et al. 2009; Diederichs et al. 2009; Tougaard et al. 2009). The observed reaction was a reduction in acoustic activity of porpoises within the impact area during pile driving and up to 48 h after. A reduction in acoustic activity can be interpreted as animals vacating the area or, alternatively, a reduction could arise because animals remained in the area but with an altered behavior leading to fewer echolocation clicks being emitted or fewer clicks being picked up by the detectors.

To address the question, Do the animals leave the area or remain but with changed behavior?, a controlled-exposure study was conducted. Pile-driving sounds were played back at reduced levels from underwater loudspeakers (Lubell 9162) located close to shore at Fyns Hoved, Great Belt, Denmark, a location where porpoises can be tracked visually by theodolite from a nearby cliff top. Results in Figure 3 clearly show that porpoises avoided a zone with a distance of~200 m around the loudspeakers when sound was transmitted. Received levels of sound were, on average, 140 dB re 1 μPa (peak-peak) at a distance of 200 m from the loudspeakers. This threshold level for reactions is consistent with the results from the real pile driving. Bailey et al. (2010) measured sound pressures of ~150 dB at distances 10-30 km from a pile-driving site. Thus, even though the source levels in this controlled-exposure study was 50-60 dB lower than a real pile driving and hence that the size of the impact area was greatly reduced, the thresholds for reaction are consistent. This gives confidence to concluding that porpoises likely reacted in a similar way to real pile driving, i.e., by vacating the impact area.

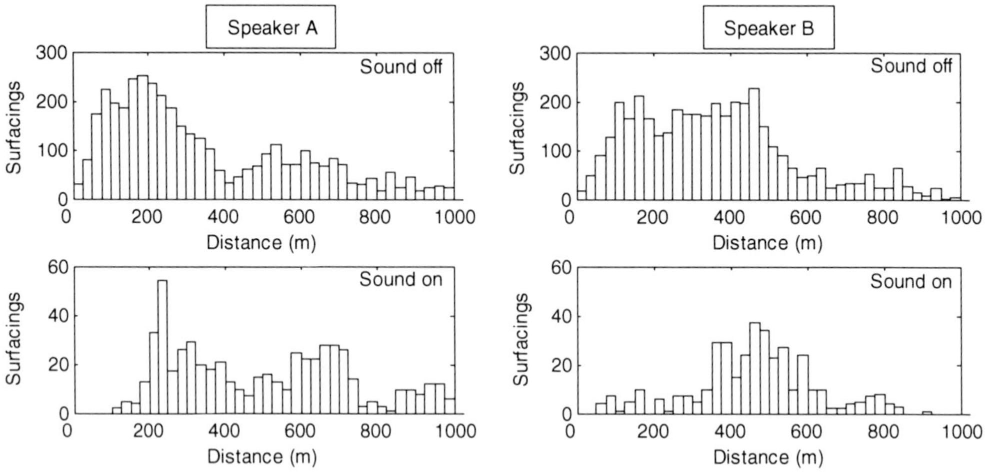

Fig. 3 Results of controlled-exposure experiments. Harbor porpoises were exposed to simulated pile-driving noise at reduced levels and tracked visually by theodolite. Two different underwater sound sources were used.

References

Bailey H, Senior B, Simmons D, Rusin J, Picken G, Thompson PM (2010) Assessing underwater noise levels during pile-driving at an offshore windfarm and its potential effects on marine mammals. Mar Pollut Bull 60:888–897.

Brandt MJ, Diederichs A, Nehls G (2009) Harbour porpoise responses to pile driving at the Horns Rev II offshore wind farm in the Danish North Sea. Final Report to DONG Energy, BioConsult SH, Husum, Germany.

Diederichs A, Brandt MJ, Nehls G (2009)Effects of construction of the transformer platform on harbor porpoises at the offshore test field "alpha ventus." Report to Stiftung Offshore-Windenergie, BioConsult SH, Husum, Germany.

Gordon J, Thomson DH, Lonergan M, Gillespie D, Hastie GD, Calderan SV, Todd VLG, Jaffey B (2009) Acute risks to marine mammals from pile driving: A role of aversive signals in mitigation. International Workshop on Underwater Noise and Offshore Wind Farms, 2–3 June 2009, Hamburg, Germany..

National Research Council (2005) Marine mammal populations and ocean noise: Determining when noise causes biologically significant effects. National Academies Press, Washington, DC.

Tougaard J, Carstensen J, Teilmann J, Skov H, Rasmussen P (2009) Pile driving zone of responsiveness extends beyond 20 km for harbor porpoises (*Phocoena phocoena*, (L.)). J Acoust Soc Am 126:11–14.

Effects of Offshore Pile Driving on Harbor Porpoises (*Phocoena phocoena*)

Miriam J. Brandt, Ansgar Diederichs, Klaus Betke, and Georg Nehls

1 Introduction

The world's growing demand for sustainable and environmentally friendly energy has led a growing number of countries to explore the options for the installation of offshore wind farms. In particular, noise emissions during the construction phase, when, in many cases, steel foundations are driven into the seafloor, are expected to cause temporal avoidance of the area by marine mammals and even have the potential to inflict physical damage to their sensory system (Madsen et al. 2006).

The harbor porpoise (*Phocoena phocoena*) is the only regularly occurring cetacean species in the German North Sea. Due to its wide distribution, all wind farm constructions in the North Sea inevitably affect this species to a certain extent. To assess these impacts, a profound knowledge of the behavior of the species in relation to noise levels created by offshore pile driving is essential. The main task is to describe the temporal and spatial extent of disturbance and thereby assess the scale of habitat exclusion.

During two different wind farm construction projects in the North Sea, we examined the impacts of offshore pile driving on harbor porpoises using passive acoustic monitoring (T-PODs).

2 Methods

The responses of harbor porpoises to wind farm construction were monitored by continuous registration of echolocation clicks using hydrophones with data loggers (T-PODs, version 4, www.chelonia.demon.co.uk). The T-POD is accompanied by the software package T-POD.exe (version 7.41) that uses a train detection algorithm (version 3.0) to discriminate cetacean trains from other sources. Clicks are then appointed to different probability classes depending on the likelihood of being of porpoise origin. We only used the two highest probability classes for analyses.

M.J. Brandt (✉) • A. Diederichs • G. Nehls
BioConsult SH, 25813 Husum, Germany
e-mail: m.brandt@bioconsult-sh.de

K. Betke
itap GmbH, 26129 Oldenburg, Germany
e-mail: klaus.betke@web.de

A.N. Popper and A. Hawkins (eds.), *The Effects of Noise on Aquatic Life*,
Advances in Experimental Medicine and Biology 730, DOI 10.1007/978-1-4419-7311-5_62,

T-PODs were placed in the water column ~1 m above the sea bottom. At Horns Rev II, six T-PODs were deployed along a transect line reaching from inside the wind farm area to a maximum distance of ~22 km to the southeast. The distance of the T-POD positions to single wind turbines ranged from 0.5 to 25 km. Water depth was 9-18 m. At Horns Rev II, data were recorded during the construction of 95 monopile foundations in 2009. Pile-driving events lasted on average 46 min. At Alpha Ventus offshore wind farm, seven T-PODs were deployed at a mean distance of 1.7–3.1 km, two at 8.3–9.1 km, and three at 15.6–19.6 km to single turbines. Water depth was ~30 m. Here four piles were rammed into the seabed during construction of the transformer platform in 2008. In 2009, 42 piles were driven into the seabed during the constructions of 6 jacket and 6 tripod foundations. Ten pile-driving events (separated by at least 60 min) lasting on average 5.5 h could be identified during construction of the tripod foundations, and 64 pile-driving events lasting on average 60 min could be identified during construction of the jacket foundations.

T-POD data were analyzed using GAM procedures where the parameter "porpoise-positive minutes per hour (PPM/H)" was used as the response variable; hour after pile driving, distance to pile, and time of day were entered as continuous nonlinear predictor variables; and in the case of Alpha Ventus data, T-POD position and year and in the case of Horns Rev II, month were entered as factors. One model was calculated for each of the three distance categories at Alpha Ventus and for each T-POD position at Horns Rev II. The duration of the effect was then visually defined as the time between the points when porpoise activity reached the first local maximum.

3 Results

At Horns Rev II, hour after pile driving had a significant effect on PPM/H at all positions. The curve that the GAM fitted to the data was of different shapes at the different T-POD positions (Fig. 1). At position 1, PPM/H steadily increased after the pile-driving event. PPM/H was substantially below the overall mean up to 24 h after pile driving. However, PPM/H continued to increase, with a narrow confidence interval, until leveling off at ~72 h after pile driving. At positions 2 and 3, the patterns are similar. At position 2, the effect lasted between 18 and 40 h; at position 3, it was between 17 and 42 h. At positions 4 and 5, the effects were substantially shorter: 9–21 h and 10–31 h, respectively. At position 6, the shape of the curve differed: PPM/H was higher than the overall average up to ~35 h after pile driving while decreasing and fluctuating around the overall mean afterward.

A similar pattern was found at Alpha Ventus. Here the effect of hour after pile driving was significant at 1.7–3.1 and 8.3–9.1 km but not at 15.6–19.6 km from the pile-driving site. The effect lasted between 20 and 35 h at 1.7–3.1 km and 9–12 h at 15.6–19.6 km (Fig. 1).

4 Discussion

We found a clear negative impact of pile driving during wind farm construction on porpoise acoustic activity. Porpoise activity measured as PPM/H was temporarily reduced during and after pile driving at a minimum distance of up to 17.8 km at Horns Rev II, whereas no such effect was found at a mean distance of 21.7 km. At the closest distance studied (2.5 km), porpoise activity was reduced between 24 and 70 h after pile driving. Results at Alpha Ventus were similar, with an effect still being detectable up to 9 km and no effect between 16 and 20 km. In the near vicinity, porpoise activity was reduced for 20–35 h after pile driving. Sound pressure levels during pile driving were

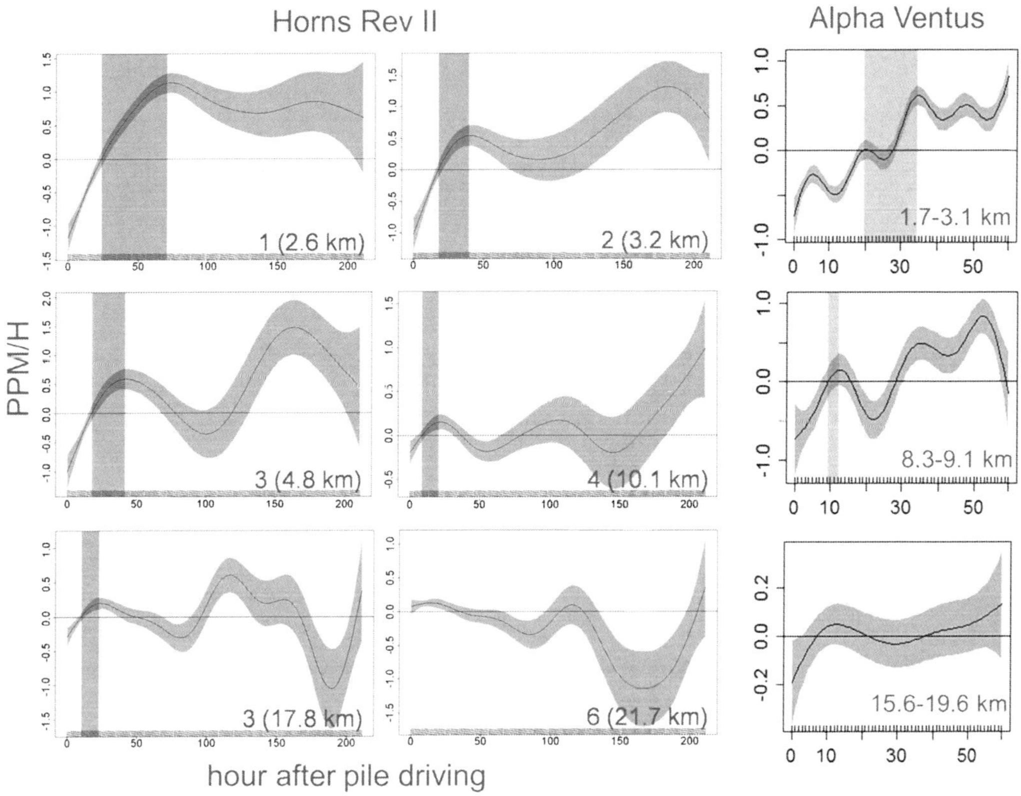

Fig. 1 Deviance of the overall mean of porpoise-positive minutes per hour (PPM/H) at different hours after pile driving at the offshore wind farms Horns Rev II and Alpha Ventus as calculated by GAM. Gray shaded areas, confidence intervals; gray vertical bars, duration of the negative effect as defined in Methods. The Figures from Horns Rev II are taken from Brandt et al. (2011)

higher at Horns Rev II than at Alpha Ventus. At Horns Rev II, 176 dB re 1 μPa (sound exposure level [SEL]) were measured 720 m from the pile driving. At Alpha Ventus, a sound pressure level of between 167 and 170 dB re 1 μPa (SEL) was calculated at 750 m based on measurements at greater distances. During both studies, the duration of the negative effect on porpoise activity decreased with distance. The mean time between pile-driving events was 38 h during both projects. This is within the time it took for porpoise activity to recover in the near vicinity to the construction site. Thus porpoise activity was lower for the whole construction period lasting 5 mo at Horns Rev II and 4 mo at Alpha Ventus.

Our results partly confirm findings by Tougaard et al. (2009) who found an effect up to a similar distance of ~20 km. However, the effect we found at both construction sites lasted considerably longer then the increase from 5.9 to 7.5 h between porpoise encounters after pile driving that they found. Unlike them, we also found a spatial gradient in the duration of the effect during both projects.

To keep negative effects on harbor porpoises at a minimum, these results should be taken into account for future spatial and temporal planning of pile-driving activities in the North Sea.

References

Brandt MJ, Diederichs A, Betke K, Nehls G (2011) Responses of harbour porpoises to pile driving at the Horns Rev II offshore wind farm in the Danish North Sea. Mar Ecol Prog Ser 421:205–216.

Madsen PT, Wahlberg M, Tougaard J, Lucke K, Tyack PL (2006) Wind turbine underwater noise and marine mammals: Implications of current knowledge and data needs. Mar Ecol Prog Ser 309:279–295.

Tougaard J, Carstensen J, Teilmann J (2009) Pile driving zone of responsiveness extends beyond 20 km for harbor porpoises (*Phocoena phocoena* (L.)). J Acoust Soc Am 126:11–14.

Evaluating the Effects of Offshore Pile Driving on *Phocoena phocoena* (Harbor Porpoises) by Using Passive Acoustic Monitoring

Klaus Lucke, Michael Dähne, Sven Adler, Anja Brandecker, Kathrin Krügel, Janne K. Sundermeyer, and Ursula Siebert

1 Introduction

The German offshore wind energy production will be largely expanded in the next years. As a test. the first offshore wind turbines in German waters were installed in 2009 at the "alpha ventus" wind farm site off the island of Borkum in the southern German Bight. The whole process is accompanied by environmental impact assessments (EIAs) covering all marine taxa and other relevant aspects such as noise emissions. A regulatory framework for conducting these investigations, the so-called StUK 3 (Bundesamt für Seeschifffahrt und Hydrographie [BSH] 2007), is set by the permitting agency (BSH [Federal Maritime and Hydrographic Agency]). To evaluate whether these requirements are appropriate and lead to scientifically robust results, a comprehensive research study was conducted simultaneously at a larger scale and by using methods additional to the ones of the EIA studies. The aim of these investigations was to validate the existing requirements or improve methodologies and the scope of their use wherever appropriate.

A key species in this context is *Phocoena phocoena* (harbor porpoise), the only resident odontocete species in German waters. The presence, habitat use, and behavior of this species is studied by passive acoustic monitoring using C-PODs (Chelonia Ltd., Long Rock, Cornwall, UK). A focus is put on the construction period and the effects of pile driving on the presence and distribution of harbor porpoises in the study area.

2 Materials and Methods

The C-POD is a fixed autonomous logging device designed to passively detect the presence of *Phocoena phocoena*. It uses digital waveform characterization to select clicks and registers the time, frequency, intensity, envelope, and bandwidth of cetacean clicks. This information is used as input to

K. Lucke (✉) • M. Dähne • J.K. Sundermeyer • U. Siebert
Research and Technology Centre (FTZ) Westcoast, University of Kiel, D-25761 Büsum, Germany
e-mail: lucke@ftz-west.uni-kiel.de

M. Dähne • A. Brandecker • K. Krügel
German Oceanographic Musueum, D-18435 Stralsund, Germany

S. Adler
Institute for Biological Sciences, University of Rostock, 18051 Rostock, Germany

A.N. Popper and A. Hawkins (eds.), *The Effects of Noise on Aquatic Life*,
Advances in Experimental Medicine and Biology 730, DOI 10.1007/978-1-4419-7311-5_63,

an automated train detection and classification algorithm. Its detection range varies depending on the existing background noise level, with a maximum of several hundred meters. The output format was chosen as "detection positive 10 minutes" (dp10m; i.e., 10-min periods with at least 1 detection).

A total of 22 devices were installed, and data were regularly retrieved from them starting in August 2008 over an 80- × 30-km area stretching from the Dutch waters in the west toward shipping lanes leading toward Wilhelmshaven, Germany, in the east. The northern and southern boundaries were set by two shipping lanes for large commercial vessels. This design allowed determination of potential gradients in the presence and habitat use of *Phocoena phocoena* along an east-west transect (i.e., parallel to the coastline and depth contour) as well as from north to south (i.e., with increasing water depth). All positions were placed at comparable distances to the shipping lanes to eliminate bias due to shipping activity. The C-PODs (V0 and 1) were calibrated in a tank and deployed 10 m above the ground in water depths of 25.5–34.5 m.

3 Results

The C-POD data revealed that during the pile-driving activities for the construction of the transformer platform as well as for the first offshore wind turbines, the detection rate for *Phocoena phocoena* decreased significantly. This effect occurred over a distance of at least 8 km but possibly ranging up to 25 km, a range comparable to results from other wind farm studies (Tougaard et al. 2009). Data allowed for discriminating seasonal variation in the presence of *Phocoena phocoena* from effects driven by external factors. A comparison of mean dp10m values per day with the standard variation of these values showed that seasonality can be ruled out as a cause for the observed decrease.

A decrease in harbor porpoise abundance before the start of pile-driving activity, as observed in some instances, indicates that other events, most likely of an acoustic nature, have affected the animals' presence negatively in the area. Underwater sound recordings were conducted only over a limited period of time during the pile-driving period that lasted from April to September 2009. Consequently, these data were insufficient to conduct a comprehensive assessment of the acoustic underwater scenario during the whole period and identify potential external sound sources (i.e., unrelated to the pile driving). Therefore, all available information on such additional disturbance factors, such as the use of deterrent devices (pinger, seal scarer), military sonar, research sonar, and other piling activities were collected and statistically analyzed for correlation with the C-POD data.

Deterrence devices were shown to have a short range effect, but due to incoherent use of these devices before, during, and after the pile driving, the resulting correlation is statistically not unequivocal. There was no correlation with the use of a fisheries sonar used for scientific purposes as well as military sonar (as far as relevant information was made available).

The C-POD data showed an increase in the presence of *Phocoena phocoena* near the pile-driving site (<3 km) toward the end of the piling-driving period in August-September 2009, i.e., during the installation of the last of the 12 wind turbines.

4 Discussion

The most likely explanation for the premature decrease in dp10m values in the vicinity of the piling-driving site is the arrival of the construction ships, including large barges equipped with heavy gear to lift the components of the turbines and vibratory pile driving (used to initially install some pile foundations). These activities would explain the decrease in some but not all of the incidents when the dp10m values dropped to zero before the impact pile driving was commenced. Therefore, the analysis of Automatic Identification System data (i.e., information on shipping activity in the area) and the use of noise loggers will probably provide some declarative strength in this respect. The other activities analyzed in this context can be ruled out as the main cause for the absence of *Phocoena phocoena* before the pile driving.

The increasing presence of *Phocoena phocoena* in late summer 2009 cannot be unequivocally explained; it could be due to either seasonal variation, disturbance, or deteriorating conditions in adjacent areas; habituation; or other factors. However, the detection of *Phocoena phocoena* close to the active pile driving gives reason for severe concern about the acoustic exposure of these animals from the piling impulses. It took >15,000 piling strikes on average to install each turbine (a tripod or jacket construction installed on three to four pile foundations), with source levels calculated to be in excess of 230 dB re 1 μPa at 1 m (Matuschek and Betke 2009). Each animal that returned to the construction site and remained during the construction period was consequently exposed to a high number of piling impacts, thereby accumulating the acoustic energy in their hearing system. Although a temporary threshold shift limit of 200 dB peak to peak re 1 μPa at 1 m was determined for *Phocoena phocoena* (Lucke et al. 2009) for an exposure to single impulses, the received sound energy from a high number of piling strikes at high received intensities is clearly above any safe exposure limit.

To avoid any negative effect for the animals, mitigation measures such as a soft start (even though its efficacy is not proven) and the use of sound reduction methods such as an air bubble curtain are required. The best solution in this respect would, of course, be the use of an alternative, less noisy installation technique.

References

Bundesamt für Seeschifffahrt und Hydrographie (Federal Maritime and Hydrographic Agency; BSH) (2007) Standard. Investigation of the impacts of offshore wind turbines on the marine environment. BSH No. 7003, Bundesamt für Seeschifffahrt und Hydrographie, Hamburg, Germany.

Lucke K, Siebert U, Lepper PA, Blanchet MA (2009) Temporary shift in masked hearing thresholds in a harbor porpoise (*Phocoena phocoena*) after exposure to seismic airgun stimuli. J Acoust Soc Am 125:4060–4070.

Matuschek R, Betke K (2009) Measurement of construction noise during pile driving of offshore research platforms and wind farms. Proceedings of the NAG/DAGA 2009 International Conference on Acoustics, 23–26 March 2009, Rotterdam, The Netherlands, pp 262–265.

Tougaard J, Carstensen J, Teilmann J, Skov H, Rasmussen P (2009) Pile driving zones of responsiveness extends beyond 20 km for harbor porpoises (*Phocoena phocoena* (L)). J Acoust Soc Am 126:11–14.

Effects of Underwater Explosions on Presence and Habitat Use of Harbor Porpoises in the German Baltic Sea

Janne K. Sundermeyer, Klaus Lucke, Michael Dähne, Anja Gallus, Kathrin Krügel, and Ursula Siebert

1 Introduction

The impacts of high-energy impulsive sounds generated by underwater explosions on individual marine mammals are diverse, ranging from behavioral reactions to injury or even death.

Unexploded ordnance (ammunition) in the marine environment and its safe disposal is a recurring problem in German waters. After World War II, large amounts of ammunition, mainly mines and chemical weapons, were dumped into the Baltic and North Seas. One of these ammunition dump sites is located in shallow water in the Kiel Bight (German Baltic Sea), which is a protected area (EU-Code: DE 1528–39), and is inhabited by the endangered Baltic Sea population of *Phocoena phocoena* (harbor porpoise). Porpoises, like other odontocetes, use sound for echolocation to navigate and locate prey (Verfuss et al. 2005). They have an acute sense of hearing; thus they are highly susceptible to the effects of high-energy impulsive sound emissions.

In 2006, the "explosive ordnance disposal service" of Schleswig-Holstein began removing the more than 100 mines and warheads dumped in the Kiel Bight by detonating them. After concern was raised about the lack of an impact assessment for this work, the work was stopped for one year to establish a monitoring program and to allow for appropriate mitigation measures to be investigated. In April and June 2008, air bubble curtains (ABCs) were tested as an alternative mitigation measure by detonating a series of small charges in controlled experiments. To systematically investigate the efficacy of ABCs in reducing the acoustic power of the explosive shock waves, a series of 25 explosive charges were exploded underwater 2.5 km off the coast of Northern Germany (Kiel Bight, Baltic Sea). This paper describes the results of the acoustic monitoring program.

2 Static Acoustic Monitoring

Static acoustic monitoring devices (T-PODs) were used to monitor the porpoise echolocation activity before, during, and after blowing of test charges. The T-POD (Chelonia Ltd., Penzance, UK) is a self-contained data logger that logs echolocation clicks from *Phocoena phocoena* and other cetaceans.

J.K. Sundermeyer (✉) • K. Lucke • U. Siebert • M. Dähne
Research and Technology Centre (FTZ) Westcoast, University of Kiel, 25761 Büsum, Germany
e-mail: sundermeyer@ftz-west.uni-kiel.de

A. Gallus • K. Krügel • M. Dähne
German Oceanographic Museum, 18439 Stralsund, Germany

A.N. Popper and A. Hawkins (eds.), *The Effects of Noise on Aquatic Life*,
Advances in Experimental Medicine and Biology 730, DOI 10.1007/978-1-4419-7311-5_64,

The calibrated T-PODs were deployed at two positions close to the dumping site and at five reference positions from June 2007 onward. A 5-yr T-POD monitoring had previously been performed at the reference stations. Data were analyzed using T-POD.exe software, which detects harbor porpoise click trains. The output format was chosen as "detection-positive 10 minutes" (dp10m; i.e., 10-min periods with at least 1 detection). The parameters "waiting time" (number of minutes in a silent period >10 min) and "porpoise-positive days per month" (dpd/mo) have been used to assess a change in the presence of *Phocoena phocoena.*

The analysis of T-POD data from 2005 to 2008 revealed a seasonal variation in porpoise presence in the Kiel Bight. The long-term observations from 2005 to 2008 showed the lowest percentage of dpd/mo from February to April. From July to November, porpoises were registered nearly every day.

The dp10min showed that *Phocoena phocoena* utilizes the area continuously if undisturbed by explosions but that detonations of the test charges in April and June 2008 led to a prolonged absence of porpoises. The results indicate that porpoises avoided the test site over a range of at least 10 km for 12 h after the detonations. Waiting times showed similar results.

The calculated sound pressure level of the explosions, derived from actual measurements, falls below the limit for a temporary threshold shift (TTS) for *Phocoena phocoena* of 200 dB peak to peak re 1 μPa (Lucke et al. 2009) at a distance of 14 km (Nützel 2008).

3 Mitigation

During the test detonations, three ABC configurations were deployed on the seabed in 10-m water depth. These were 1) a single perforated pipe and 2) a double and (3) a triple pipe forming half-circles of increasing diameters (7.5–11.5 m). The half-circles were facing away from the coastline to reduce the sound emitted into the open water. A total air volume of up to 20 m^3/min was pumped into the pipes. During these tests, deterrent devices were used and the area was also surveyed for *Phocoena phocoena* by visual observers.

Sound measurements conducted at different water depths showed a reduction in peak sound pressure level between 11.1 and 17.3 dB re 1 μPa (Nützel 2008). However, comparing the attenuation achieved with different numbers of ABCs led to unequivocal results because the highest attenuation was reached when two ABCs were used.

4 Conclusions

This is the first study using T-PODs to investigate effects of underwater explosions on *Phocoena phocoena*. The results show that *Phocoena phocoena* was avoiding the impact area. The consequences for individuals or the porpoise population cannot be assessed yet.

In comparison to the rather small test charges, the detonation of the sea mines and warheads (up to 350 kg) represent an even greater threat for porpoises due to their higher source levels. At very short ranges, death, physical damage, or impairment of the auditory system may occur (TTS, permanent threshold shift [PTS], or deafness), whereas at greater distances from the site of an explosion, a range of psychophysical or behavioral effects including stress, masking, and habitat exclusion may be observed (Ketten 2000; Richardson et al. 1995; Southall et al. 2007). Severe physical effects can only be avoided if either the animals can be effectively excluded from the risk zone or the emitted levels can be reduced substantially. Blasting of the ammunition was resumed recently with only the ABCs, visual monitoring, and the use of deterrent devices as mitigation measures.

However, the results of test detonations indicate that a more effective set of mitigation measures would be required. To reduce negative effects on porpoises to the lowest possible extent, further removal of ammunition by blasting should be avoided under any circumstances.

References

Ketten DR (2000) Cetacean ears. In: Au WWL, Popper AN, Fay RR (eds) Hearing by whales and dolphins. Springer, New York, pp 43–108.

Lucke K, Lepper PA, Blanchet MA, Siebert U (2009) Temporary shift in masked hearing thresholds in a harbor porpoise (*Phocoena phocoena*) after exposure to seismic airgun stimuli. J Acoust Soc Am 125:4060–4070.

Nützel B (2008) Untersuchungen zum Schutz von Schweinswalen vor Schockwellen. Technischer Bericht TB 2008–7, Forschungsanstalt der Bundeswehr für Wasserschall und Geophysik (FWG), Kiel, Germany.

Richardson WJ, Greene CR Jr, Malme CI, Thomson DH (1995) Marine mammals and noise. Academic Press, San Diego, CA.

Southall BL, Bowles AE, Ellison WT, Finneran JJ, Gentry RL, Greene CR Jr, Kastak D, Ketten DR, Miller JH, Nachtigall PE, Richardson WJ, Thomas JA, Tyack PL (2007) Marine mammal noise exposure criteria: Initial scientific recommendations. Aquat Mamm 33:411–521.

Verfuss UK, Miller LA, Schnitzler HU (2005) Spatial orientation in echolocating harbour porpoises (*Phocoena phocoena*). J Exp Biol 208:3385–3394.

Behavioral-Response Studies: Problems With Statistical Power

Rebecca A. Dunlop, Michael J. Noad, and Douglas H. Cato

1 Problems With Behavioral-Response Study Experimental Design

The behavioral-response study (BRS) is an experimental design used by field biologists to determine the function and/or behavioral effects of conspecific, heterospecific, or anthropogenic stimuli. Although this has been used for many years in studies of acoustic stimuli and whales, many of these studies have been limited by lack of true replication, otherwise known as "pseudoreplication" (Hurlbert 1984; Kroodsma 1989; Searcy 1989).

The term "pseudoreplication" covers a specific range of experimental errors in which the sample size (n) used in a particular statistical test is not sufficient to test the hypothesis (McGregor 2000); either the stimulus is not representative (i.e., a true replicate) of the class of stimuli or the subject animals are not representative of the class or animals to which they belong. In a review of playback studies with marine mammals, Deeke (2006) found that only 17 of the 46 studies reviewed avoided pseudoreplication. A frequently used experimental design is a "before, during, and after" (BDA) design in which the subject is monitored before, during, and after presentation of a stimulus. One of the most common errors in the analysis of BDA data is to treat subjects within each BDA period as independent samples rather than as a repeated-measures design. Also, subjects may interact biologically within an experimental BDA (which relates to internal validity errors) or each subject may be repeatedly presented with the stimulus; therefore, samples are not independent, although they are treated as independent samples in the analysis. A number of variations of a stimulus may be used within the experiment, but the analysis wrongly pools the samples and treats these multiple recordings as one treatment. Finally, conclusions may be wrongly drawn about the function of, or response elicited by, a specific stimulus without true replication of the experiment. Invalid conclusions are then made about how applicable the result is to the population or species.

Part of the problem is balancing the need to obtain adequate sample size to ensure statistically significant results against the high cost and logistical difficulty of such experiments. Many previous studies have been inconclusive through inadequate sample size, but this was not evident until after

R.A. Dunlop (✉) • M.J. Noad
Cetacean Ecology and Acoustics Laboratory, School of Veterinary Science, Gatton, QLD 4343, Australia
e-mail: r.dunlop@uq.edu.au; mnoad@uq.edu.au

D.H. Cato
University of Sydney Institute of Marine Science, University of Sydney, NSW 2006, Australia

Defence Science and Technology Organisation, Pyrmont, NSW 2009, Australia
e-mail: doug.cato@sydney.edu.au; Doug.Cato@dsto.defence.gov.au

A.N. Popper and A. Hawkins (eds.), *The Effects of Noise on Aquatic Life*,
Advances in Experimental Medicine and Biology 730, DOI 10.1007/978-1-4419-7311-5_65,

the data were analyzed. Hence it is important to make a reliable estimate of the sample size required as part of the experimental design, such as by using a statistical power analysis. A power analysis calculates the number of samples needed to detect a change based on the variance in the response, but this requires some initial information about the statistical variability expected. Ideally, the power analysis should be carried out as part of a pilot study and, based on the results of this analysis, the sample size needed for the full experiment should be determined. In reality, field experiments are limited by time, money, or logistics, and such an ideal situation rarely exists.

2 BRS With Australian Humpback Whales

Experiments were carried out during the humpback whale acoustic research collaboration (HARC) project in September-October 2004 and 2008 during the humpback whale southward migration. The study site was located at Peregian Beach, which is 150 km north of Brisbane, on the east coast of Australia (26°29′ S, 153°06′ E). Acoustic observations were made from a fixed array of hydrophones moored offshore. Visual data collection (from an elevated survey point) involved two platforms of observation: "scan sampling" (all groups in the area were sampled as often as possible) and "focal follow" (one group was focally followed throughout the experimental period and all surfacing behaviors were captured from this group).

2.1 *Experimental Design*

This experiment followed a typical BDA design. The "B" period consisted of a preexposure (stimulus-off) control; the "D" period was when the stimulus was turned on and was followed by a postexposure (stimulus-off) "A" period. Exposure treatments were one of three: a silent control, conspecific social vocalizations, and artificial tones. The silent control was a recording of "silence" to provide a control scenario in which the recording equipment was deployed from the boat but with no audible stimulus. To increase the sample size of this treatment, groups that migrated within 2 km of a stationary vessel (the research boat or a similar-size vessel) were also included in this control category. The social-sound recording was a random collection of social vocalizations spliced together to make a 204-s recording. Social sounds were taken from a DTAG recording of a mother-calf-escort group. This 204-s recording was repeatedly played for 20 min. The artificial-tone recording consisted of a 1.5-s tone ascending from 2 to 2.1 kHz, repeated every 8 s, for 20 min. Nonexposed groups were focally followed during times when no playback experiments were underway. These groups were randomly selected from groups that were migrating through the study area and were focally followed for at least 1 h.

2.2 *Response-Model Development*

Each response variable (e.g., direction and speed of travel, dive profile, surface behavior) was analyzed separately using general linear mixed models (GLMMs) that included the random effect of group (and associated variance) and takes into account covariance due to repeated sampling of a group (eliminating a common source of pseudoreplication error). The response model initially tested two independent factors: experimental period (before, during, and after) and stimulus

Table 1 Significant factors dictating changes in course-traveled and mean deep-dive times in humpback whale groups during the Australian behavioral-response study

Dependent Measure	Platform	Significant Factors Included in the Final Model
Mean course traveled per experimental period	Focal	Stimulus
	Scan	Stimulus, experimental period, proximity to the source vessel (within or beyond 3.5 km from the vessel during exposure)
Mean deep-dive time per experimental period	Focal	Stimulus, experimental period, background noise levels (including noise from singers)
	Scan	Biased sampling, no result

Table 2 Number of groups used in the behavioral-response study analysis for each platform of observation

Platform of Observation	Sample Size: Nonexposed	Silent control	Social-sound exposure	Tone exposure
Focal follow	6	0	8	6
Scan sampling	15	5	24	22

(nonexposed, silent control, social sounds, and tones). Various response models were developed but due to the small sample size of focal-follow data, only a few of the covariates such as "proximity of group from source vessel," "received level of stimulus," and "background noise levels" could be added in to this model. For scan-sampled data (where there were more data points), additional independent parameters such as group composition and the "nearest neighbor" could be added to, and eliminated from, the response model based on their significance to the response variability. The final models were developed separately for both scan-sampled and focally followed groups. Table 1 illustrates some of the results obtained for the movement-response parameter "course traveled" and behavioral-response parameter "deep-dive time."

2.3 Problems With Power

A power analysis assesses the probability of accepting the null when it is false (beta error) and rejecting the null when it is correct (α error), otherwise known as a type I and a type II error, respectively. Typically, α is set at 0.05 and β at 0.2, which means that the power of the experiment (which is calculated as $1 - \beta$) should be at least 0.8. The analysis calculates the sample size necessary (from means and variances generated from the results) to achieve this power. Table 2 illustrates the sample size for the Australian behavioral-response study.

The power calculation presented here used "course traveled" as a response variable and used only focally followed groups as the sample population. It is based on a balanced design, where the number of controls equals the number of exposures. The power-analysis graph (Fig. 1) generated for the response to tones indicates that a sample size of 10 per treatment (that is 10 controls and 10 exposures) was needed to detect a significant change in course traveled. With 6 replicates (the number used in this experiment), the power of the experiment is only 0.55. For the response to the social-sound stimulus (Fig. 1), 8 replicates produce a power of only 0.45, and the analysis indicates that 15 samples (15 control and 15 exposed groups) are required for significant response detection with minimal chance of a type I or type II error.

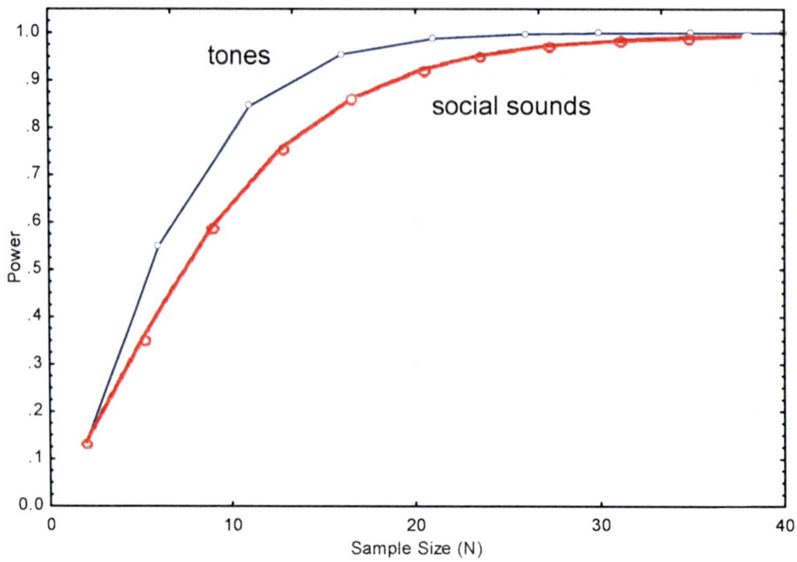

Fig. 1 Power analysis generated for focal groups exposed to "tones" ($n = 6$) and groups exposed to "social sounds" ($n = 8$) using the results of the behavioral-response study. Graph illustrates the sample size and the associated power based on the mean and variance of course traveled within the "during" period for exposed and nonexposed groups

3 Conclusions

An improvement in modeling techniques over the years and the development of more sophisticated statistical-analysis packages have allowed a more complex analysis of data to be carried out, and, therefore, some of the analysis issues can be addressed. Modeling techniques can also incorporate other environmental effects that may influence the response of whales to the experimental stimulus (as shown in the Australian BRS results) and can incorporate a random effect in the statistical model (which accounts for the "within-group" variance and therefore repeated measurements within the dataset).

This paper illustrates that, despite the improved statistical ability to model data, the power of the focal-follow result is low due to small sample size. It is difficult to achieve reliable results in these experiments due to the high cost and logistical difficulties balanced with the need to obtain an adequate sample size to ensure statistically significant results. However, by having two platforms of observation, we were able to retest the result using the larger sample size from the scan-sampled data. There is a trade-off between data quality and sample size; the focal-follow data are more concentrated in terms of the quality of data collected for each group (higher resolution), but the increased effort to collect these data results in a smaller sample size in term of numbers of groups of whales for which data are collected. The scan-sampled data could be viewed as lower quality (because there is a reduced effort of data collection for each group), the trade-off being that sample size is greater. However, it is not possible to use scan sampling to assess some behavioral responses (e.g., deep-dive time). To accurately calculate this, a focal follow of the group is required. Future experiments will increase focal-follow effort and use the scan-sampling platform as a way of putting the focal-follow groups into social and environmental context.

Acknowledgments This work was supported by the US Office of Naval Research, Arlington, VA; the Australian Defence Science and Technology Organisation, Canberra, ACT, Australia; the Australian Marine Mammal Centre, Hobart, TAS, Australia; and the Exploration and Production Sound and Marine Life Joint Industry Programme, International Association of Oil and Gas Producers, London, UK.

References

Deeke VB (2006) Studying marine mammal cognition in the wild: A review of four decades of playback experiments. Aquat Mamm 32:461–482.
Hurlbert SH (1984) Pseudoreplication and the design of ecological field experiments. Ecol Monogr 54:187–211.
Kroodsma DE (1989) Suggested experimental designs for song playbacks. Anim Behav 37:600–609.
McGregor PK (2000) Playback experiments: Design and analysis. Acta Ethol 3:3–8.
Searcy WA (1989) Pseudoreplication, external validity and the design of playback experiments. Anim Behav 38:715–717.

Detection and Tracking of Whales Using a Shipborne, 360° Thermal-Imaging System

Elke Burkhardt, Lars Kindermann, Daniel Zitterbart, and Olaf Boebel

1 Introduction

Shipborne, quasi-continuous visual marine mammal observations for mitigation or research require, while being restricted to daylight hours, the utmost concentration by observers as well as large teams when conducted during month-long cruises. To overcome such limitations, the use of thermal imaging has first been examined in the context of offshore oil exploration by Greene and Chase (1987). The approach exploits the thermal signature of a whale's blow, which, at least at high latitudes, is warmer than the environment. Their study and additional research by Cuyler et al. (1992) recorded thermal images of both odontocete and mysticete blows at ranges up to 100 m. Significantly larger detection ranges were achieved by Perryman et al. (1999) who detected gray whale blows at several kilometers range in thermal images taken from ashore. Most recently, Baldacci et al. (2005) tested a handheld naval infrared (IR) camera in the Mediterranean Sea, reporting detections of various species at typically 1–2 nautical miles.

However, the limited field of view of IR cameras used in these studies poses a significant constraint when continuous monitoring of a ship's entire perimeter is required. This problem is overcome in this study by using a shipborne, 360° cooled thermal imager, FIRST-Navy, mounted on an actively stabilized gimbal. The sensor, developed by Rheinmetall Defense Electronics (Bremen, Germany) provides a continuous 5-Hz video stream of the ship's perimeter with a horizontal resolution of 0.05°/pixel and a vertical resolution of 0.03°/pixel. The system was installed on the crow's nest of the *R/V Polarstern* in autumn 2009 and was tested during two expeditions to the Greenland Sea and the Southern Ocean, generating thermal-imaging streams that allow detection, location, and track of whales in the ship's vicinity.

2 Results

The FIRST-Navy sensor was operated for a total of 837 h. Because the significant data rates (3.5 TB/day) prohibit continuous saving of data, independent information on the presence of marine mammals was used to trigger archiving of relevant video sequences (spanning the speed-dependent

E. Burkhardt (✉) • L. Kindermann • D. Zitterbart • O. Boebel
Alfred Wegener Institute, 27570 Bremerhaven, Germany
e-mail: Elke.Burkhardt@awi.de

A.N. Popper and A. Hawkins (eds.), *The Effects of Noise on Aquatic Life*,
Advances in Experimental Medicine and Biology 730, DOI 10.1007/978-1-4419-7311-5_66,

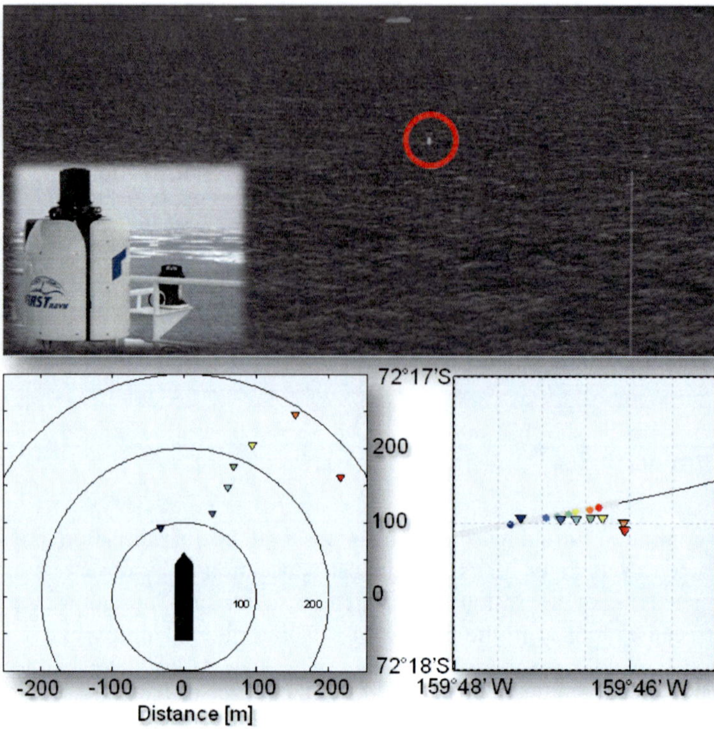

Fig. 1 Top: Infrared signature of a minke whale blow at ~250-m distance. Inset: FIRST-Navy sensor. Bottom left: Relative locations of whale blows. Bottom right: Geo-referenced blow (triangles) and ship (circles) positions. Colors express time, progressing from blue to red

relevant period of ±3 to 60 min around each sighting). This ensured the acquisition of video data sequences with whales known to be present. Retrospectively, the ensuing dataset of 35-h duration was visually screened for IR signatures of whale spouts, so far revealing over 300 spouts (i.e., Fig. 1, top) at ranges up to 3 km.

From a blow's position within a given thermal image, the relative distance of the blow to the ship was calculated using simple triangulation, similar to the methods used with distance sampling techniques (Lerczack and Hobbs 1998). The bearing under which the blow was detected can be extracted from the image with an extremely high resolution of 0.05 angular degrees. The vertical resolution of 0.03 angular degrees surpasses those of "Big Eye" binocular (small) reticules (0.06–0.08 angular degrees) and is comparable to angular resolutions that may be achieved by visual interpolation when using "Big Eye" under ideal conditions. However, the fact that the FIRST-Navy sensor is stabilized against the ship's roll and pitch and the possibility to scroll forth and back in the recorded video material allows avoiding overlooking blows while providing highly accurate and objective relative blow positions (Fig. 1, bottom left). Concurrent 1-s-resolution records of the ship's navigational data then allow calculation of absolute, geo-referenced whale tracks (Fig. 1, bottom right). Figure 1 exemplifies such results for an encounter with a minke whale. Similar tracks have so far been processed for 19 events for humpback, sperm, minke, fin, and bowhead whales. This provides a unique opportunity of documenting in detail the locomotive behavior and blow rates of cetaceans in the near vicinity of ships equipped with such systems.

Acknowledgments This project was funded by German Federal Ministry of Education and Research (Bonn, Germany) Grant BMBF 03F0479I and Federal Ministry for the Environment, Nature Conservation, and Nuclear Safety (Bonn, Germany) Grant BMU 370891101-01. The responsibility for the contents of this publication rests with the authors.

References

Baldacci A, Carron M, Portunato N (2005) Infrared detection of marine mammals. Technical Report SR-443, North Atlantic Treaty Organization Undersea Research Center, La Spezia, Italy.
Cuyler LC, Wiulsrød R, ØRitsland NA (1992) Thermal infrared radiation from free living whales. Mar Mamm Sci 8:120–134.
Greene CR, Chase SC (1987) Infrared detection of whale spouts: A report on a feasibility study. Prepared for Shell Western E&P, Inc., by Greeneridge Sciences, Inc., Santa Barbara, CA.
Lerczack JA, Hobbs RC (1998) Calculating sighting distances from angular reading during shipboard, aerial, and shore-based marine mammal surveys. Mar Mamm Sci 14:590–599.
Perryman WL, Donahue LA, Laake JL, Martin TE (1999) Diel variation in migration rates of Eastern Pacific gray whales measured with thermal imaging sensors. Mar Mamm Sci 15:426–445.

Distribution of Bowhead Whale Calls Near an Oil Production Island With Fluctuating Underwater Sound

W. John Richardson, Trent L. McDonald, Charles R. Greene, Jr., Susanna B. Blackwell, and Bill Streever

1 Introduction

Bowhead whales (*Balaena mysticetus*) migrate west parallel to the north coast of Alaska during the late summer and autumn. In 2000, BP Exploration (Alaska) Inc. (BPXA), Anchorage, AK, began constructing its Northstar oil production facility on an artificial gravel island 5 km seaward of the natural barrier islands in the Beaufort Sea, AK, and a few kilometers inshore of the southern edge of the migration corridor of bowheads. Construction of Northstar continued during the 2001 bowhead migration season. Oil production and gas injection began in late 2001 and continue. Gas turbines operate continuously on the island to generate electricity. During the summer and autumn, vessels, helicopters, and since 2003 a hovercraft travel to and from Northstar.

There has been concern about the effects of underwater noise emanating from Northstar and its support vessels both on migrating whales themselves and on their accessibility to Inupiat subsistence hunters based on Cross Island, 27 km east of Northstar (Streever et al. 2008). Hunters were concerned that migrating whales might be deflected offshore.

We expected that the level of underwater sound emanating from Northstar would fluctuate as industrial operations on and around the island varied, and we hypothesized that this fluctuating sound could affect the distribution of bowheads in the southern part of their migration corridor. Aerial surveys have been a standard method of monitoring bowhead whale distribution, but an a priori statistical power analysis showed that aerial surveys would not provide sufficient sightings to detect or characterize a deflection of the small size that was expected. Instead, for four migration seasons (2001–2004), we used an acoustical approach to determine whether the closest detectable

W.J. Richardson (✉)
LGL Limited environmental research associates, King City, ON L7B 1G3, Canada
e-mail: wjr@LGL.com

T.L. McDonald
WEST Inc., Cheyenne, WY 82001, USA
e-mail: tmcdonald@west-inc.com

C.R. Greene, Jr. • S.B. Blackwell
Greeneridge Sciences Inc., Santa Barbara, CA 93117, USA
e-mail: cgreene@greeneridge.com

B. Streever
BP Exploration (Alaska) Inc., Anchorage, AK 99519, USA
e-mail: bill.streever@bp.com

A.N. Popper and A. Hawkins (eds.), *The Effects of Noise on Aquatic Life*,
Advances in Experimental Medicine and Biology 730, DOI 10.1007/978-1-4419-7311-5_67,

whale calls tended to be farther offshore when levels of underwater sound produced by Northstar and its support vessels were above average. If so, that would be evidence of a Northstar effect on whale distribution, calling behavior, or both.

2 Methods

Throughout 4 field seasons, each ~1 mo in duration, low-frequency underwater sounds were recorded via a bottom-mounted cabled hydrophone or autonomous recorder located ~450 m seaward of Northstar (Blackwell and Greene 2006). From the recordings, we extracted the overall broadband level (10–450 Hz), the 28- to 90-Hz level (mainly from industrial sources), and the occurrence and levels of transient sounds (mainly from vessel traffic) and tonal sounds (from machinery and vessels).

Throughout the same four field seasons, bowhead calls were recorded by an array of 10 directional autonomous seafloor acoustic recorders (DASARs) located 7–22 km offshore of Northstar (Greene et al. 2004). There were two overlapping hexagonal DASAR arrays, with each DASAR spaced 5 km from its nearest neighbors. Each DASAR incorporated the directional sensor from a DIFAR sonobuoy to provide bearings to recorded calls. When two or more DASARs detected the same call, the calling whale's location was plotted by triangulation. The offshore distance of each call was determined relative to a baseline through Northstar and parallel to the coast.

Weighted quantile regression (Koenker 2005) was used to relate the 5th quantile of the offshore distances of calls to the various measures of anthropogenic sound near Northstar after allowing for the apparent effects of natural environmental covariates (McDonald et al., in press). Call location data were weighted, with weights inversely proportional to both the probability of detection and uncertainty in locations of the calls. Because individual whales could not be distinguished and some whales called repeatedly, offshore distances of calls were not all independent. To overcome this difficulty, cluster analysis was used to identify groups of potentially interdependent calls. Following cluster analysis, block permutation of uncorrelated call clusters was used to assign significance levels to coefficients in the quantile regression model. The magnitude, duration, and type of anthropogenic sound that would induce deflection or a change in calling behavior were unknown before data collection. Hence several measures of industrial sounds were computed over 7 sound-averaging times (15, 30, 45, 60, 70, 90, and 120 min preceding each call), and statistical model selection was used to determine the best combination of predictors. Normal within-season variation in the migration corridor's apparent southern edge was accommodated by treating day–night changes, distance of the call east or west of Northstar, and date as covariates.

3 Results

Simultaneous data on Northstar sounds and locations of calling whales were obtained for 31, 24, 29, and 28 days during the autumns of 2001–2004, respectively (Blackwell et al. 2007). In those periods, 2,325, 5,124, 25,176, and 43,932 call locations, respectively, were determined in circumstances with detection probability >10%.

Although the overall distribution of calling bowhead whales each year appeared similar at times with low versus higher underwater noise near Northstar, detailed analysis indicated that after allowance for natural covariates, the apparent southern (proximal) edge of the call distribution was significantly ($P < 0.01$) associated with industrial sound output each year. In 2001, the 5th quantile of offshore distance was an estimated 1.58 km farther offshore when sound at industrial frequencies (28–90 Hz), measured 440 m from Northstar and averaged over 45 min preceding the call, increased from 94.3 to 103.7 dB re 1 μPa. (In 2001, 94.3 dB was the median level in the absence of appreciable Northstar sound; 103.7 dB was the median level near Northstar when the level there exceeded 94.3 dB.) In 2002, the 5th quantile of the call distribution was an estimated 1.66 km farther offshore during times when transient sounds associated with boat traffic were present during the preceding 2 h. In 2003 and 2004, the 5th quantile was estimated to be farther offshore when tones were recorded in the 10- to 450-Hz band just before the call. In 2003, the apparent offshore shift was an estimated 0.67 km when tones were present within the preceding 15 min. In 2004, the apparent shift was 2.24 km when tones were present within the preceding 2 h.

Thus, during each of the four seasons studied, there was an apparent offshore displacement of the southern edge of the distribution of whale calls when underwater sound levels from Northstar and associated vessels were above average.

4 Discussion

Visual observations of bowheads and other baleen whale species have shown that some individual whales are displaced on exposure to continuous industrial sound, with received sound pressure level exceeding roughly 120 dB re 1 μPa (Richardson et al. 1995). In this study, measured levels of Northstar sound ~450 m from Northstar rarely exceeded 120–125 dB re 1 μPa. The closest bowheads were several kilometers farther offshore, where received levels of sound (Northstar plus background) rarely exceeded 105–110 dB re 1 μPa. Based on previously available data, one would not expect to detect much, if any, displacement of bowheads by the weak Northstar sounds typically occurring in the southern part of the bowhead migration corridor. At the whale locations, those sounds are either below ambient levels or barely above ambient levels. However, in each of the four seasons of study, the southernmost calls tended to be slightly but significantly farther offshore when Northstar sounds increased. This finding in part demonstrates the high statistical power of methods that provide a large number of observations (here, thousands or tens of thousands of calls per season) and of statistical methods that account for natural covariates before assessing the effect of main interest (here, the effect of industrial operations).

The acoustic data alone cannot distinguish whether the apparent displacement of calls was attributable to actual displacement of whales, to noise-induced changes in bowhead calling behavior, or both. It would be desirable to quantify the relative contributions of actual displacement versus the change in calling behavior in causing the observed change in whale call distribution, but that cannot be done from passive acoustic monitoring alone. The advantages of a passive acoustic approach over visual methods in detecting and monitoring cetaceans are widely known and were the reason for adopting an acoustic approach in this study. However, even though acoustic monitoring may (as in this study) be able to provide a very large sample of data collected continuously over a large area, acoustic monitoring alone may not show whether distribution of cetaceans is affected by a human activity. For that, additional data are needed, such as whale distribution data collected via a method not dependent on calling behavior or data on the effect of human activity on calling rate and on the source levels of the calls.

Acknowledgments We thank BP Exploration (Alaska), Inc., Anchorage, AK, for active participation in design and implementation of the project as well as financial support. Alaska Clean Seas, Anchorage, AK, was very helpful in providing vessels. For advice on design and interpretation as well as encouragement, we thank the peer/stakeholder group including representatives of US National Marine Fisheries Service, Silver Spring, MD; Department of Wildlife Management and Science Advisory Committee, North Slope Borough, Barrow, AK; Alaska Eskimo Whaling Commission, Barrow, AK; US Bureau of Ocean Energy Management, Regulation and Enforcement, Washington, DC; and various external reviewers. Also, we thank the many additional participants from Greeneridge Sciences Inc., Santa Barbara, CA; LGL Alaska, Anchorage, AK; LGL Limited environmental research associates, King City, ON, Canada; and WEST Inc., Cheyenne, WY.

References

Blackwell SB, Greene CR Jr (2006) Sounds from an oil production island in the Beaufort Sea in summer: Characteristics and contribution of vessels. J Acoust Soc Am 119:182–196. Doi: 10.1121/1.2140907.

Blackwell SB, Richardson WJ, Greene CR Jr, Streever B (2007) Bowhead whale (*Balaena mysticetus*) migration and calling behaviour in the Alaskan Beaufort Sea, autumn 2001–04: An acoustic localization study. Arctic 60:255–270. Available via http://pubs.aina.ucalgary.ca/arctic/Arctic60-3-255.pdf. Accessed 2 May 2010.

Greene CR Jr, McLennan MW, Norman RG, McDonald TL, Jakubczak RS, Richardson WJ (2004) Directional frequency and recording (DIFAR) sensors in seafloor recorders to locate calling bowhead whales during their fall migration. J Acoust Soc Am 116:799–813. Doi: 10.1121/1.1765191.

Koenker R (2005) Quantile regression. Cambridge University Press, Cambridge, UK.

McDonald TL, Richardson WJ, Greene CR Jr., Blackwell SB, Nations C, Nielson RW, Streever B (in press) Detecting changes in the distribution of calling bowhead whales exposed to fluctuating underwater sounds. J. Cetac. Res. Manage.

Richardson WJ, Greene CR Jr, Malme CI, Thomson DH (1995) Marine mammals and noise. Academic Press, San Diego, CA.

Streever B, Angliss RP, Suydam R, Ahmaogak M, Bailey C, Blackwell SB, George JC, Greene CR Jr, Jakubczak RS, Lefevre J, McDonald TL, Napageak T, Richardson WJ (2008) Progress through collaboration: A case study examining effects of industrial sounds on bowhead whales. Bioacoustics 17:345–347.

Assessment of the Effect of Natural and Anthropogenic Aquatic Noise on Vaquita (*Phocoena sinus*) Through a Numerical Simulation

Giovanni Castellazzi, Petr Krysl, Lorenzo Rojas, and Ted W. Cranford

1 Introduction

Phocoena sinus (vaquita) is a small marine harbor porpoise endemic to the Gulf of California that is believed to be the most endangered cetacean in the world as reported by the International Union for Conservation of Nature (IUCN) red list (see Rojas-Bracho et al. 2008). Simultaneous use of the same habitat by fishermen and *Phocoena sinus* has led to the precipitous decline of the porpoise. *Phocoena sinus* is easily entangled in fishing nets, resulting in drowning.

Our challenge was twofold: first, to understand the pathway by which *Phocoena sinus* receives sound and second, whether *Phocoena sinus* should be able to detect nets using their echolocation system. Recently, research cruises have focused attention on recording sounds present in the *Phocoena sinus* habitat. We used numerical analysis to understand how sound propagates through the anatomy of *Phocoena sinus*. Finite element modeling (FEM) offers the capacity to simulate what happens when anatomic geometry interacts with sound pressure waves. An example of a similar application of FEM can be found in Cranford et al. (2008) where they constructed a numerical simulation of the acoustic pathways in the head of a Cuvier's beaked whale. This kind of simulation is valuable because marine mammal hearing is not fully understood and contradictory suggestions have been put forth to explain it. It is clear that toothed whales have a well-developed sense of hearing, probably as a result of selective pressures that compensate for the diminishing penetration of light in water. It is not yet clear how toothed whales detect, receive, filter, or amplify sounds.

G. Castellazzi (✉)
DISTART, University of Bologna, 40126 Bologna, Italy
e-mail: giovanni.castellazzi@unibo.it

P. Krysl
Department of Structural Engineering, University of California, San Diego, La Jolla, CA 92093, USA
e-mail: pkrysl@ucsd.edu

L. Rojas
San Instituto Nacional de Ecología (INE), Fraccionamiento Zona Playa, Ensenada, BC 22860, Mexico
e-mail: lrojas@cicese.mx

T.W. Cranford
Department of Biology, San Diego State University, San Diego, CA 92182, USA
e-mail: tcranfor@mail.sdsu.edu

A.N. Popper and A. Hawkins (eds.), *The Effects of Noise on Aquatic Life*,
Advances in Experimental Medicine and Biology 730, DOI 10.1007/978-1-4419-7311-5_68,

We focused our attention on simulating interactions with pinger sounds because they have been used in entanglement mitigation efforts around the globe with other porpoise species. It is worth emphasizing that the interaction between porpoise anatomy and high-frequency sounds has not been investigated previously.

2 Materials and Methods

We had access to two postmortem *Phocoena sinus* specimens thanks to the generosity of the Government of Mexico. The specimens were scanned using X-ray computed tomography (CT). The 3-dimensional image data provide a detailed map of anatomic structure, which is the starting point for building an acoustic simulation.

The material parameters needed for the isotropic constitutive equation are density, bulk modulus, shear modulus, and dynamic viscosity. The sample density can be mapped from the CT image using a conversion from the Hounsfield units. Because the dynamic viscosity is not available as a map of the Hounsfield units and needs to be estimated from the literature, we used data from Krysl et al. (2006) to assign representative average mechanical properties to tissues in the following groups: hard bone, soft bone, connective tissue, muscle, and acoustic fats/blubber.

2.1 Numerical Simulation

The CT data provide the anatomic geometry for the model: a voxel in the CT image corresponds to a hexahedral height-node finite element. No approximations or interpolation is made on the CT image resolution. For simplicity, the computational box is assumed initially at rest and unstressed. Such an assumption makes it easy to reconcile the boundary and the initial conditions. It also implies that the specimen and its bounding box are near the sea surface and that the system is not exposed to significant hydrostatic pressure. The three velocity components are prescribed along all the bounding surfaces of the computational box. In particular, these initial and boundary conditions correspond to plane sound waves propagating in the left-to-right direction (see Fig. 1), with an exponential ramp-up from a rest/unstressed state to full power within a fraction of a millisecond. In this paper, we consider a preliminary set of traveling sound waves of frequencies from 80 to 120 kHz directed along the line connecting the tip of the nose to the first vertebra of the animal. The subsequent step, which we will investigate in the future, is to consider different angles for the traveling sound waves to realize whether or not the anatomy of the animal can filter out the entering sound waves.

3 Results

These preliminary results are given in terms of pressure map distribution. For the sake of brevity, we focus our attention on pressures over the bony ear complex (tympanoperiotic complex [TPC]). Figure 1 shows the distribution of the pressure due to a traveling 80-kHz frequency-pressure wave. Figure 1a shows a close-up of the left TPC and Figure 1b and c shows the whole body model. The lateral view of the left TPC is shown in Figure 2, where the pressure map is overlaid on the TPC in detail. Figure 2 shows the pressure resulting from a set of frequencies from 80 to 120 kHz, where warm colors indicate high pressure and cold colors indicate low pressure.

These maps show that the various acoustic frequencies illuminate the TPC in different ways. For example, Figure 2a shows a low frequency that is strongly amplified at the ear. Going toward

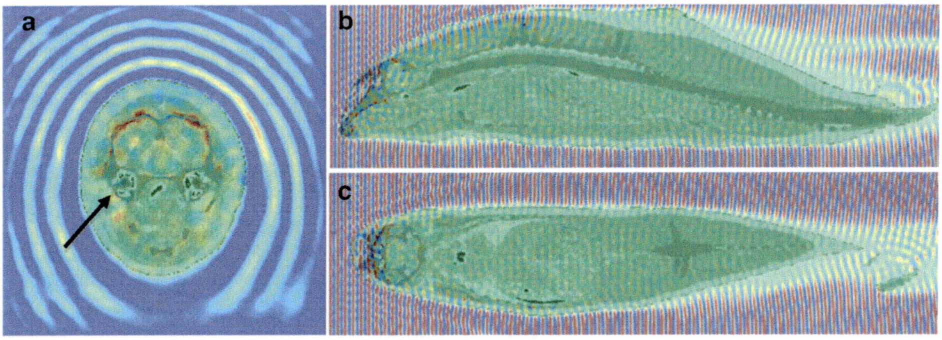

Fig. 1 Sagittal view of the left bony ear complex (**a**) and coronal (**b**), and axial (**c**) views of the whole body of the porpoise. Pressures are shown for an 80-kHz frequency. Arrow indicates position of the left ear

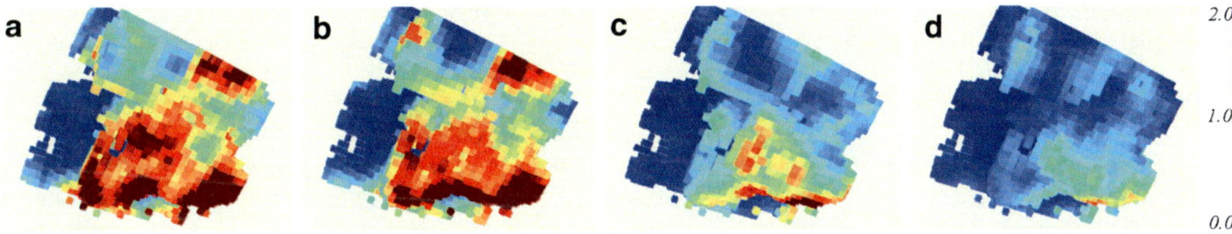

Fig. 2 Pressure results for frequencies of 80 (**a**), 95 (**b**), 110 (**c**), and 120 (**d**) kHz at the left bony ear complex. The pressures are normalized using the external pressure measured at the tip of the porpoise nose. Red indicates twice the pressure at the tip of the rostrum; green shows a pressure that is equal to that at the tip of the rostrum; blue indicates half and lower pressures relative to the tip of the rostrum

120 kHz (upper limit of *Phocoena sinus* hearing; see Gregory 1991), the pressure is focusing, as expected, on a specific spot, presumably responsible for the mechanical functioning of the TPC.

4 Conclusions

This paper addresses *Phocoena sinus* conservation through better understanding of their interaction with the acoustic components of their habitat. An advanced numerical simulation of the interaction of sound with an entire toothed whale is presented here for the first time. The unique FEM model provides a window of understanding into the acoustic functioning. This information may allow us to propose conservation actions to help protect this species. This approach produced the capability to model the acoustic pathways for sound entering the head and the body of these animals.

References

Cranford TW, Krysl P, Hildebrand JA (2008) Acoustic pathways revealed: Simulated sound transmission and reception in Cuvier's beaked whale (*Ziphius cavirostris*). Bioinspir Biomim 3:1–10.

Gregory KS (1991) Acoustic signals of the vaquita (*Phocoena sinus*). Aquat Mamm 17.3:130–131.

Krysl P, Cranford TW, Wiggins SM, Hildebrand JA (2006) Simulating the effect of high-intensity sound on cetaceans: Modeling approach and a case study for Cuvier's beaked whale (*Ziphius cavirostris*). J Acoust Soc Am 120: 2328–2339.

Rojas-Bracho L, Reeves RR, Jaramillo-Legorreta A, Taylor BL (2008) *Phocoena sinus*. In: International Union for Conservation of Nature (IUCN) red list of threatened species, version 2010.4, IUCN, Gland, Switzerland. Available via http://www.iucnredlist.org.

Impacts of Anthropogenic Sounds on *Phocoena phocoena* (Harbor Porpoise)

Ronald Kastelein and Nancy Jennings

1 Introduction

Anthropogenic underwater sound in the seas and oceans is increasing. *Phocoena phocoena* (harbor porpoise) is sensitive to underwater sound because of its very acute hearing, wide hearing frequency range, and high responsiveness to sounds. The detection of sounds by animals and the degree to which sounds have an effect on animals involve the characteristics of sounds at the source, the propagation of sound between the source and the receiver (the animal), and the hearing characteristics of the receiver. Here, we focus on the hearing properties of *Phocoena phocoena*, factors affecting sound detection, and effects of detected sounds on the physiology, behavior, and echolocation ability of the species.

2 Hearing and the Basic Audiogram

For a sound to have an impact on marine organisms, it has to be heard. The audibility of a sound depends on the frequency range of the hearing of an animal, which can be established experimentally in the form of hearing thresholds for several frequencies (audiograms). For *Phocoena phocoena*, audiograms have been established by using the psychophysical (behavioral) technique (Andersen 1970; Kastelein et al. 2002, 2010) and the electrophysiological technique (by measuring the auditory brainstem response; Bibikov 1992; Lucke et al. 2007; Popov et al. 1986). The psychophysical technique provides more realistic hearing thresholds than the electrophysiological technique but is much more time consuming. The electrophysiological technique is best suited to the quick establishment of relative changes in hearing thresholds, e.g., during temporary threshold shift (TTS) studies.

It is incorrect simply to use the basic audiogram to calculate detection ranges of sounds by *Phocoena phocoena* at sea. Basic audiograms, because they are determined under very quiet background conditions, can realistically be used to determine whether a species is physically able to

R. Kastelein (✉)
Sea Mammal Research Company (SEAMARCO), Harderwijk 3843 CC, The Netherlands
e-mail: Researchteam@zonnet.nl

N. Jennings
Dotmoth, Dundry, Bristol BS41 8LN, UK
e-mail: nancy@dotmoth.co.uk

A.N. Popper and A. Hawkins (eds.), *The Effects of Noise on Aquatic Life*,
Advances in Experimental Medicine and Biology 730, DOI 10.1007/978-1-4419-7311-5_69,

detect a sound of a particular frequency or not. Actual detection of a sound at sea involves many other factors, e.g., the directionality of hearing, masking of the sound, and signal duration. Detected sounds can have many, and varied, adverse effects.

2.1 Factors Affecting Sound Detection

Hearing is directional: its sensitivity varies depending on the angle from which a sound reaches the listener; therefore, the location of a sound source relative to the listener affects the audibility of the sound. This phenomenon is frequency dependent because the higher the frequency of the sound, the more pronounced is its directionality. The hearing of almost all marine mammal species has only been tested for sounds coming from directly in front of them. This method usually provides the highest hearing sensitivity. In *Phocoena phocoena*, the directionality of hearing in the horizontal plane has been established for sounds of three frequencies (Kastelein et al. 2005a), and based on these data, the directivity index has been calculated for these frequencies. Individuals of *Phocoena phocoena* can use their directional hearing, combined with head movements, to determine the location of a sound source or to reduce the perceived sound level of an annoying sound.

The sound source localization ability of the species depends on the characteristics of the sounds as well as on the hearing properties of *Phocoena phocoena*. The longer a signal and the higher the received signal-to-noise ratio, the greater the ability of *Phocoena phocoena* to localize it (Kastelein et al. 2007). Being able to localize a sound is important if an individual wants to reduce the received level by swimming away from a sound.

The background noise level and spectrum also affect the audibility of a sound. Masking occurs when one sound (the background noise) interferes with the detection of another sound (the signal). The degree of interference depends on the amplitudes of the two sounds and on the difference in frequency of signal and noise; masking is greatest when the two sounds have a similar spectrum. A pure tone is masked mainly by noise at adjacent frequencies in a critical bandwidth around the frequency of the tone. Noise at frequencies outside this masking band has little influence on detection of the signal unless the noise level is very high. Some information exists on the critical bandwidths of *Phocoena phocoena* (Popov et al. 2006). The lowest signal-to-noise ratio at which a subject can detect a tonal signal in broadband masking noise is defined as the critical ratio (CR; Fletcher 1940). The CR of *Phocoena phocoena* is slightly lower than that of other odontocetes in which the CR has been measured (Kastelein and Wensveen 2008; Kastelein et al. 2009). This suggests that porpoises can detect signals in noise relatively well.

The duration of a signal affects its audibility because the mammalian ear integrates energy over time until an integration time threshold is reached. Therefore, the amplitude of a signal required for detection increases as its duration decreases below the integration time threshold. The integration time for tonal signals over the entire hearing range of *Phocoena phocoena* has been measured; the lower the frequency, the higher the integration time (Kastelein et al. 2010). This information is important for the calculation of audibility ranges of sounds such as those made by pile driving, seismic surveys, and detonations, which have a high source level but are of short duration.

2.2 Hearing Damage Caused by Detected Sounds

Loud sounds may cause direct physical injury in *Phocoena phocoena* or permanent hearing damage in the form of permanent threshold shifts (PTSs). Loud sounds of sufficient duration may cause permanent hearing loss to animals at close range and TTSs to animals further away. Permanent hearing loss reduces the ability of an animal to perceive sounds that are important for its survival, fitness, and ability to reproduce, and exposure to sounds at levels causing permanent damage needs to be avoided. However, the levels at which permanent hearing damage occurs in marine mammals are not known. For a variety of reasons (e.g., governmental regulations, practical issues, and ethical issues), studies designed to cause permanent hearing loss in marine mammals have not been conducted. TTS, which is followed by complete recovery, has been studied. The course and time of recovery generally depend on the amount of exposure to noise and the amount of shift incurred. Determining the amplitudes and durations of noises that begin to initiate a TTS will allow the estimation of sound levels at which interference with the animal's life history events and behavior may occur. TTS reduces the acoustic communication range of an animal and lessens its ability to detect certain important sounds for a certain period of time. TTS in *Phocoena phocoena* has been measured after exposure to seismic air gun stimuli, although the results of the hearing tests may have been influenced by very high levels of background noise (Lucke et al. 2009).

3 Behavioral Effects of Detected Sounds

Once a sound has been heard by an individual of *Phocoena phocoena*, the effect of the sound on its behavior depends on the characteristics of the sound (e.g., level, spectrum, duty cycle, signal duration), the characteristics of the animal (e.g., responsiveness, age, sex, history, reproductive state, hunger level), and the context (e.g., animal alone or in a group, water depth, location, weather). Behavioral effects occur in steps, which show gradients in severity.

3.1 *Startle Response*

After simply hearing a sound without responding behaviorally, the first step on the gradient of behavioral effects is the occurrence of a startle response to a single sound stimulus. In recent startle-response studies, individuals of *Phocoena phocoena* were subjected to approximately six received sound levels that caused startle responses in 0–100% of study sessions. From the psychometric functions, the received sound levels causing a startle response in 50% of sessions could be derived (Kastelein, Hoek, de Jong, in preparation).

3.2 *Increased Anxiety*

Increased anxiety, measured as an increase in respiration and surfacing rate, occurs when individuals of *Phocoena phocoena* are subjected to several sound stimuli in a series (Kastelein et al. 1995, 1997a,b, 2000, 2001, 2006). Increased respiration and surfacing rates lead to higher energy expenditure, which may have negative impacts on the health, survival, and reproductive success of *Phocoena phocoena*.

3.3 *Deterring Effect*

If received sound levels are increased further, *Phocoena phocoena* may avoid the vicinity of, or be deterred away from, the sound source. Avoidance threshold levels have been determined for specific stimuli (Kastelein et al. 2005b, 2008a,b). In recent experiments, deterring stimuli were projected in various levels of background noise mimicking sea states 0–8. The higher the sea state, the smaller the deterring effect of the stimuli (Kastelein et al. 2011). It is important to quantify sound levels that deter *Phocoena phocoena* so that such sound levels can be avoided in ecologically important areas used by the species for feeding, breeding, and suckling.

4 Effects of Sounds on Echolocation Ability

It is possible to compromise the ability of *Phocoena phocoena* to echolocate by increasing the background noise level in the frequency range of its echolocation clicks (Kastelein, Verboom, de Haan, Au, in preparation). However, because *Phocoena phocoena* produces narrowband echolocation signals of high frequency (120–130 kHz; Verboom and Kastelein 1995) and because anthropogenic underwater noise has most of its energy in the low-frequency range, it is unlikely that current anthropogenic noise will reduce the echolocation ability of *Phocoena phocoena* (Kastelein et al. 1999).

5 Conclusions

Knowledge of the hearing of *Phocoena phocoena* is gradually increasing as research continues. Information on the audibility of anthropogenic sounds and their effect on the detection of biologically relevant sounds by *Phocoena phocoena* has increased. Knowledge on the effect of anthropogenic sound on the behavior of *Phocoena phocoena* is also increasing, although the effects of many broadband sounds, resembling sounds from various human activities at sea, still need to be quantified. The next step is to useg behavioral data to predict the effects of anthropogenic sounds on the size, dynamics, and health of populations of *Phocoena phocoena*.

References

Andersen S (1970) Auditory sensitivity of the harbour porpoise, *Phocoena phocoena*. In: Pilleri G (ed) Investigations on cetacea, vol 2. Institute for Brain Research, Bern, Switzerland, pp 255–259.

Bibikov NG (1992) Auditory brainstem responses in the harbor porpoise (*Phocoena phocoena*). In: Thomas JA, Kastelein RA, Supin AY (eds), Marine mammal sensory systems. Plenum Press, New York, pp 197–211.

Fletcher H (1940) Auditory patterns. Rev Mod Phys 12:47–65.

Kastelein RA, Au WWL, Rippe HT, Schooneman NM (1999) Target detection by an echolocating harbor porpoise (*Phocoena phocoena*). J Acoust Soc Am 105:2493–2498.

Kastelein RA, Bunskoek P, Hagedoorn M, Au WWL, de Haan D (2002) Audiogram of a harbor porpoise (*Phocoena phocoena*) measured with narrow-band frequency-modulated signals. J Acoust Soc Am 112:334–344.

Kastelein RA, de Haan D, Goodson AD, Staal C, Vaughan N (1997a) The effects of various sounds on a harbour porpoise (*Phocoena phocoena*). In: Nachtigall PE, Lien J, Au WWL, Read AJ (eds) Harbour porpoises: Laboratory studies to reduce bycatch. De Spil Publishers, Woerden, The Netherlands, pp 367–383.

Kastelein RA, de Haan D, Staal C, Goodson AD (1997b) The response of a harbour porpoise (*Phocoena phocoena*) to nets of various mesh sizes, with and without deterring sound. In: Read AJ, Wiepkema PR, Nachtigall PE (eds) The biology of the harbour porpoise. De Spil Publishers, Woerden, The Netherlands, pp 385–409.

Kastelein RA, de Haan D, Vaughan N, Staal C, Schooneman NM (2001) The influence of three acoustic alarms on the behaviour of harbour porpoises (*Phocoena phocoena*) in a floating pen. Mar Environ Res 52:351–371.

Kastelein RA, de Haan D, Verboom WC (2007) The influence of signal parameters on the sound source localization ability of a harbor porpoise (*Phocoena phocoena*). J Acoust Soc Am 122:1238–1248.

Kastelein RA, Goodson AD, Lien J, de Haan D (1995) The effects of acoustic alarms on harbour porpoise (*Phocoena phocoena*) behaviour. In: Nachtigall PE, Lien J, Au WWL, Read AJ (eds) Harbour porpoises: Laboratory studies to reduce bycatch. De Spil Publishers, Woerden, The Netherlands, pp 157–167.

Kastelein RA, Hoek L, Wensveen PJ, Terhune JM, de Jong CAF (2010) The effect of signal duration on the underwater hearing thresholds of two harbor seals (*Phoca vitulina*) for single tonal signals between 0.2 and 40 kHz. J Acoust Soc Am 127:1135–1145.
Kastelein RA, Janssen M, Verboom WC, de Haan D (2005a) Receiving beam patterns in the horizontal plane of a harbor porpoise (*Phocoena phocoena*). J Acoust Soc Am 118:1172–1179.
Kastelein RA, Jennings N, Verboom WC, de Haan D, Schooneman NM (2006) Differences in the response of a striped dolphin (*Stenella coeruleoalba*) and a harbour porpoise (*Phocoena phocoena*) to an acoustic alarm. Mar Environ Res 61:363–378.
Kastelein RA, Rippe T, Vaughan N, Schooneman NM, Verboom WC, de Haan D (2000) The effects of acoustic alarms on the behavior of the harbor porpoises (*phocoena phocoena)* in a floating pen. Mar Mamm Sci 16:46–64.
Kastelein RA, Steen N, de Jong C, Wensveen PJ, Verboom WC (2011) Effect of broadband-noise masking on the behavioral response of a harbor porpoise (*Phocoena phocoena*) to 1-s duration 6–7 kHz sonar up-sweeps. J Acoust Soc Am 129:2307–2315.
Kastelein RA, Verboom WC, Jennings N, de Haan D (2008a) Behavioral avoidance threshold level of a harbor porpoise (*Phocoena phocoena*) for a continuous 50 kHz pure tone. J Acoust Soc Am 123:1858–1861.
Kastelein RA, Verboom WC, Jennings N, de Haan D, van der Heul S (2008b) The influence of 70 and 120 kHz tonal signals on the behavior of harbor porpoises (*Phocoena phocoena*) in a floating pen. Mar Environ Res 66:319–326.
Kastelein RA, Verboom WC, Muijsers M, Jennings NV, van der Heul S (2005b) The influence of acoustic emissions for underwater data transmission on the behaviour of harbour porpoises (*Phocoena phocoena*) in a floating pen. Mar Environ Res 59:287–307.
Kastelein RA, Wensveen PJ (2008) Effect of two levels of masking noise on the hearing threshold of a harbor porpoise (*Phocoena phocoena*) for a 4.0 kHz signal. Aquat Mamm 34:420–425.
Kastelein RA, Wensveen PJ, Hoek L, Au WWL, Terhune JM, de Jong CAF (2009) Critical ratios in harbor porpoises (*Phocoena phocoena*) for tonal signals between 0.315 and 150 kHz in random Gaussian white noise. J Acoust Soc Am 126:1588–1597.
Lucke K, Lepper PA, Hoeve B, Everaarts E, van Elk N, Siebert U (2007) Perception of low-frequency acoustic signals by a harbour porpoise (*Phocoena phocoena*) in the presence of simulated offshore wind turbine noise. Aquat Mamm 33:55–68.
Lucke K, Siebert U, Lepper PA, Blanchet M (2009) Temporary shift in masked hearing thresholds in a harbor porpoise (*Phocoena phocoena*) after exposure to seismic airgun stimuli. J Acoust Soc Am 125: 4060–4070.
Popov VV, Ladygina TF, Supin AY (1986) Evoked potentials of the auditory cortex of the porpoise, *Phocoena phocoena*. J Comp Physiol A 158:705–711.
Popov VV, Supin AY, Wang D, Wang K (2006) Nonconstant quality of auditory filters in the porpoises *Phocoena phocoena* and *Neophocaena phocaenoides* (Cetacea, Phocoenidae). J Acoust Soc Am 119:3173–3180.
Verboom WC, Kastelein RA (1995) Acoustic signals by harbour porpoises (*Phocoena phocoena*). In: Nachtigall PE, Lien J, Au WWL, Read AJ (eds) Harbour porpoises: Laboratory studies to reduce bycatch. De Spil Publishers, Woerden, The Netherlands, pp 1–39.

Changes in Vocal Behavior of North Atlantic Right Whales in Increased Noise

Susan E. Parks, Mack P. Johnson, Douglas P. Nowacek, and Peter L. Tyack

1 Introduction

Social communication requires the detection and recognition of a signal from a sender by a receiver. Sources of noise in the environment can limit the range for successful detection of signals (Brumm and Slabbekoorn 2005; Wiley and Richards 1978). A number of compensation mechanisms are used by individuals producing acoustic communication signals in increased noise to increase the probability of detection. These types of changes are called "vocal adjustments," indicating that changes in sound production occur over short time scales, with individuals actively changing their signals in response to exposure to changing noise levels (Patricelli and Blickley 2006). This plasticity enables animals to compensate actively for changing noise spectra and levels in their environment. A diverse range of vertebrate taxa, ranging from small bird species to marine mammals, exhibit similar "vocal adjustment" capabilities (Brumm and Slabbekoorn 2005; Tyack 2008).

Four general types of vocal adjustment have currently been described that increase the signal-to-noise ratio of the signal. These include changes to the 1) amplitude of the signal, 2) frequency content of the signal, 3) temporal structure within the sound, and/or 4) the timing of sound production (Patricelli and Blickley 2006). For these changes to occur, there has to be plasticity in the vocal production mechanism of the species (Tyack 2008).

Most marine mammals rely on sound for communicating, navigating, and foraging, and the potential impact of anthropogenic sources of sound on marine mammals has received increasing attention in the past two decades (Nowacek et al. 2007; Richardson et al. 1995). Several studies have demonstrated a pronounced increase in sound levels below 100 Hz in deep water for the Northeast Pacific (Andrew et al. 2002; McDonald et al. 2006) and in shallow coastal waters in the North Atlantic on the Canadian continental shelf (Zakarauskas et al. 1990) since the 1960s. Distant shipping has been suggested as the most likely source for these observed increases in low-frequency noise. If ambient sound levels increase due to human activities, the range over which individuals

S.E. Parks (✉)
Applied Research Laboratory, The Pennsylvania State University, State College, PA 16804, USA
e-mail: sep20@psu.edu

M.P. Johnson • P.L. Tyack
Woods Hole Oceanographic Institution, Woods Hole, MA 02543, USA

D.P. Nowacek
Duke University Marine Laboratory, Beaufort, NC 28516, USA

A.N. Popper and A. Hawkins (eds.), *The Effects of Noise on Aquatic Life*,
Advances in Experimental Medicine and Biology 730, DOI 10.1007/978-1-4419-7311-5_70,

can locate one another will be reduced unless marine mammals can modify their calling behavior to improve the probability of detection of these signals.

Many studies have investigated short-term changes in the vocal behavior of cetaceans related to increased background noise levels from vessels. These studies indicate that marine mammals employ all four of the currently described vocal adjustments to improve the signal-to-noise ratio of their calls, including increasing the intensity of their calls, shifting the frequencies (Hz) of their calls to a frequency band with lower noise levels, increasing the duration of their calls, increasing the repetition rate of their signals, or waiting until the noise decreases before calling (Tyack 2008). Here we investigate how one species of baleen whale, *Eubalaena glacialis* (North Atlantic right whale), modifies its communication system to compensate for increases in low-frequency noise in the environment.

2 Right Whale Communication in Noise

Eubalaena glacialis is a highly endangered species of baleen whale (Kraus et al. 2005) that uses sound for social communication (Clark 1983; Parks and Tyack 2005). The primary habitat for this species is off the east coast of the United States, an area with high levels of shipping traffic. Concern has been raised over the impact of increased environmental noise on this species, with potential impacts ranging from limiting the communication range for critical behavioral functions such as locating and selecting mates and reuniting with offspring after separation to chronic increase in stress levels (Parks and Clark 2007). The primary contact call in *Eubalaena glacialis*, the upcall, is produced in a frequency range that overlaps with the peak noise from distant shipping (50–350 Hz; Parks et al. 2009). In this low-frequency, long-duration noise regimen, the most effective compensation mechanism would be to increase the intensity or shift the frequency of the contact calls (Fig. 1).

A recent study indicates that there may be long-term population-level changes occurring in response to shifts in ambient noise in some species (Parks et al. 2007). Comparisons were made between the call parameters of *Eubalaena* upcalls in both low- and high-noise environments by collecting both modern (high-noise) and historical (lower noise) recordings of upcalls from *Eubalaena* populations in the North Atlantic (high noise) and South Atlantic (lower noise). The results of these comparisons indicated that right whale contact calls were produced at higher frequencies in higher noise habitats, but no measures were available for individual call-intensity responses to increasing noise (Parks et al. 2007).

Individual call variation of *Eubalaena glacialis* in response to noise was investigated using the DTAG, an archival acoustic recording tag, that was attached to individual whales using suction cups in the Bay of Fundy, Canada, during the summers of 2001, 2002, and 2005 (Johnson and Tyack 2003). Calls produced by the tagged whales were identified by high-intensity received levels, the presence of multiple high-frequency harmonics, and visual observations confirming that the whale was not with other right whales. Eleven individual whales produced at least two upcalls during tag attachment to compare differences in call parameters in different noise conditions. Measurements were made of the background noise level and the received level, duration, and minimum frequency of the calls. Analyses indicate that the received level of calls was significantly affected by the noise level, whereas frequency changes were not consistent, with some individual whales increasing and others decreasing their call frequency in increasing noise.

3 Discussion

Selection should favor modifications in calling behavior that maximize the detectability of a signal in noise while minimizing the costs associated with signal production. New tools provide insight into the studies of individual acoustic behavior of free-ranging marine mammals. We are now able

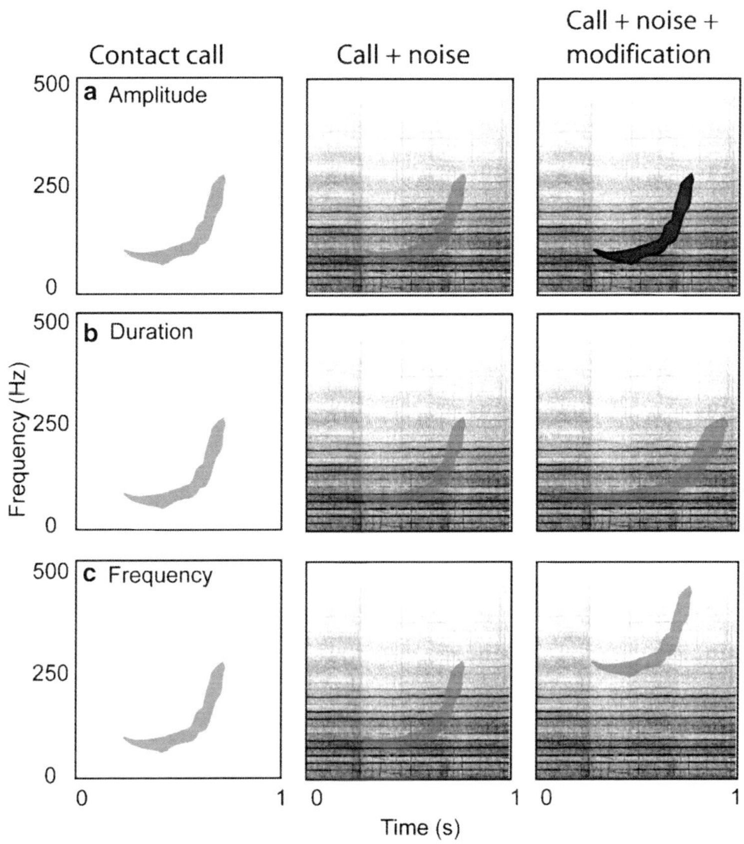

Fig. 1 Illustration of potential contact call modifications by right whales to improve signal-to-noise ratio. (**a**) Increased intensity; (**b**) increased duration; (**c**) increased minimum frequency

to ask questions concerning how individual whales respond to acoustic cues in their environment while submerged beneath the ocean's surface. These studies are significant as they demonstrate that *Eubalaena glacialis*, a highly endangered species, is capable of modification of its long-distance contact calls in moderately increased levels of noise. Several questions remain to be answered, including the cost of compensation, whether these compensation mechanisms are sufficient to allow *Eubalaena glacialis* to communicate in their industrialized environment, how much noise is too much for them to compensate for, and at what noise levels will their acoustic communication start to break down? Answers to these questions will help us to assess what impact noise may be having on limiting the recovery of this species.

References

Andrew RK, Howe BM, Mercer JA (2002) Ocean ambient sound: Comparing the 1960s with the 1990s for a receiver off the California coast. Acoust Res Lett Online 3:65–70.

Brumm H, Slabbekoorn H (2005) Acoustic communication in noise. In: Slater PJB, Snowdon CT, Roper TJ, Brockmann HJ, Naguib M (eds) Advances in the study of behavior, vol. 35. Academic Press, San Diego, CA, pp 151–209.

Clark CW (1983) Acoustic communication and behavior of the southern right whale. In: Payne RS (ed) Communication and behavior of whales. American Association for the Advancement of Science Selected Symposium 76, Westview Press, Boulder, CO, pp 163–198.

Johnson MP, Tyack PL (2003) A digital acoustic recording tag for measuring the response of wild marine mammals to sound. IEEE J Ocean Eng 28:3–12.

Kraus SD, Brown MW, Caswell H, Clark CW, Fujiwara M, Hamilton PK, Kenney RD, Knowlton AR, Landry S, Mayo CA, McLellan WA, Moore MJ, Nowacek DP, Pabst DA, Read AJ, Rolland RM (2005) North Atlantic right whales in crisis. Science 309:561–562.

McDonald MA, Hildebrand JA, Wiggins SM (2006) Increase in deep ocean ambient noise in the Northeast Pacific west of San Nicolas Island, California. J Acoust Soc Am 120:711–718.

Nowacek DP, Thorne LH, Johnston DW, Tyack PL (2007) Responses of cetaceans to anthropogenic noise. Mammal Rev 37:81–115.

Parks SE, Clark CW (2007) Acoustic communication: Social sounds and the potential impacts of noise. In: Kraus S, Rolland R (eds) The urban whale: North Atlantic right whales at the crossroads. Harvard University Press, Cambridge, MA, pp 310–332.

Parks SE, Clark CW, Tyack PL (2007) Short and long-term changes in right whale calling behavior: The potential effects of noise on communication. J Acoust Soc Am 122:3725–3731.

Parks SE, Tyack PL (2005) Sound production by North Atlantic right whales (*Eubalaena glacialis*) in surface active groups. J Acoust Soc Am 117:3297–3306.

Parks SE, Urazghildiiev I, Clark CW (2009) Variability in ambient noise levels and call parameters of North Atlantic right whales in three habitat areas. J Acoust Soc Am 125:1230–1239.

Patricelli GL, Blickley JL (2006) Avian communication in urban noise: Causes and consequences of vocal adjustment. The Auk 123:639–649.

Richardson WJ, Greene CR Jr, Malme CI, Thomson DH (1995) Marine mammals and noise. Academic Press, San Diego, CA.

Tyack PL (2008) Implications for marine mammals of large-scale changes in the marine acoustic environment. J Mammal 89:549–558.

Wiley RH, Richards DG (1978) Physical constraints on acoustic communication in atmosphere: Implications for evolution of animal vocalizations. Behav Ecol Sociobiol 3:69–94.

Zakarauskas P, Chapman DMF, Stall PR (1990) Underwater acoustic ambient noise levels on the eastern Canadian continental shelf. J Acoust Soc Am 87:2064–2071.

Acoustic Compensation to Shipping and Air Gun Noise by Mediterranean Fin Whales (*Balaenoptera physalus*)

Manuel Castellote, Christopher W. Clark, and Marc O. Lammers

Seafloor recorders were deployed in the western Mediterranean Sea and adjacent Atlantic waters during 2006–2009 to monitor noise levels and fin whale presence. Acoustic parameters of 20-Hz pulses (pulse duration, pulse bandwidth, interpulse interval, and center and peak frequencies) were compared for areas with different shipping noise levels and different shipping intensities in the Strait of Gibraltar and during seismic air gun events. Significant differences were detected between noise contexts. In general, acoustic parameter values decreased with increased noise levels. In high-noise conditions, 20-Hz pulse duration shortened, bandwidth decreased, and center and peak frequencies decreased. Similar results were obtained in the presence of air gun events, and bearings to singing whales indicated that whales moved away from the air gun source and out of our detection area for a time period that extended well beyond the duration of the air gun activity. This study provides evidence that fin whales modify their acoustic behavior to compensate for increased ambient noise and shows that under some conditions they will leave an area for an extended period. Sensitization and habituation processes may play a role in these behaviors and are discussed in the context of these results. The biological consequences of these behavioral changes are not known, but the potential negative effect in their reproductive success and survival is discussed.

M. Castellote (✉)
National Marine Mammal Laboratory, Alaska Fisheries Science Center, National Oceanic and Atmospheric Administration, Seattle, WA 98115, USA
e-mail: manuel.castellote@noaa.gov

C.W. Clark
Bioacoustics Research Program, Cornell Laboratory of Ornithology, Cornell University, Ithaca, NY 14850, USA
e-mail: cwc2@cornell.edu

M.O. Lammers
Hawai'i Institute of Marine Biology, University of Hawai'i, Kane'ohe, HI 96744, USA
e-mail: lammers@hawaii.edu

A.N. Popper and A. Hawkins (eds.), *The Effects of Noise on Aquatic Life*, Advances in Experimental Medicine and Biology 730, DOI 10.1007/978-1-4419-7311-5_71,

Long-Term Monitoring of Anthropogenic Noise and Its Relationship to Bottlenose Dolphin (*Tursiops truncatus*) Distribution in the Cres–Lošinj Archipelago, Northern Adriatic, Croatia

Nikolina Rako, Marta Picciulin, Peter Mackelworth, Draško Holcer, and Caterina Maria Fortuna

1 Introduction

The Cres-Lošinj archipelago represents an important nursing and feeding ground for the resident *Tursiops truncatus* (bottlenose dolphin) population. Scientific research on *Tursiops truncatus* in this area has been conducted since 1987, and the population abundance is currently estimated to be 113 individuals (95% CI = 107–121; SE = 6.967), showing a significant decline of 39% between 1995 and 2003 (Fortuna 2006). Being top predators, dolphins represent the symbol of a healthy environment; hence a reduction in their abundance is considered an indication of the endangered status of this population. The low *Tursiops truncatus* density in the region may be related to the habitat degradation of the northern Adriatic waters, particularly to the scarce availability of food resources in an environment subject to high anthropogenic pressure (Bearzi et al. 2004).

The Cres-Lošinj archipelago is characterized by a strong nautical tourism that constitutes the primary source of underwater anthropogenic noise that causes an increase in sea ambient noise (SAN) over low frequencies (below 1 kHz). This is mainly because of the increased number and high mobility of fast-moving recreational vessels that colonize the area during the summer season.

Exposure to loud noise can have various effects on marine animals. The most common one is behavioral disturbance, with reactions of the disturbed species strongly dependent on the connections between costs and benefits associated with the environmental conditions.

N. Rako (✉) • P. Mackelworth
Blue World Institute of Marine Research and Conservation, 51551 Veli Lošinj, Croatia
e-mail: nikolina.rako@blue-world.org

M. Picciulin
CSEE-Department of Life Sciences, University of Trieste, 34127 Trieste, Italy
e-mail: marta.picciulin@gmail.com

D. Holcer
Blue World Institute of Marine Research and Conservation, 51551 Veli Lošinj, Croatia

Croatian Natural History Museum, HR-10000 Zagreb, Croatia
e-mail: drasko.holcer@hpm.hr

C.M. Fortuna
Blue World Institute of Marine Research and Conservation, 51551 Veli Lošinj, Croatia

Institute for Environmental Protection and Research (ISPRA), 00144 Rome, Italy
e-mail: caterina.fortuna@isprambiente.it

A.N. Popper and A. Hawkins (eds.), *The Effects of Noise on Aquatic Life*,
Advances in Experimental Medicine and Biology 730, DOI 10.1007/978-1-4419-7311-5_72,

Studies over the past few decades have demonstrated that dolphins tend to avoid highly disturbed places (Richardson et al. 1995; Simmonds and Mayer 1997), and changes in habitat use have already been reported as potentially related to noise disturbance (Evans et al. 1992; Foote et al. 2004). Recent work on the critical habitats and factors affecting *Tursiops truncatus* distribution within the Cres-Lošinj archipelago has suggested the possibility of anthropogenic noise being a long-term factor impacting dolphin habitat use (Fortuna 2006).

The aim of this study was to identify critical areas in terms of noise and vessel presence and to correlate these types of anthropogenic pressure to *Tursiops truncatus* distribution within the Cres–Lošinj archipelago.

2 Materials and Methods

The Cres-Lošinj archipelago is situated in the Kvarneri (northern Adriatic Sea, Croatia). The study area extends over 2,000 km^2 along the eastern side of the islands of Cres and Lošinj. In 2005–2008, 380 boat surveys were undertaken, covering 6,515 nautical miles (nmi) in adequate survey conditions. During these surveys, a long-term monitoring of SAN was carried out twice a month at 10 predefined locations, giving a total of 710 5-min SAN samples. The 10 acoustic stations were put into 3 groups of different anthropogenic impact, high, medium, and low, characterized by different proximities to tourist and municipal locations on land.

Acoustic recordings were made from a 5.70-m inflatable research vessel using a RESON TC4032 hydrophone and a calibrated Pioneer DC-88 DAT recorder. Collected data were analyzed using SPECTRA RTA software calibrated with a signal of 100 mV root mean square (RMS) @ 1 kHz and hydrophone sensitivity, obtaining a one-third octave band analysis for each acoustic sample.

During the acoustic sampling, data on vessel presence, type, and distance from the monitoring location were collected. Different vessel types observed in the study area were scored for later comparison of their produced noise. At the same time, data on *Tursiops truncatus* distribution were collected by recording the presence of each dolphin encounter coordinates (using GPS) and photo identification techniques. Data were analyzed by calculating the overall and seasonal encounter rates weighted by the research effort inside the area. All the graphic outputs were obtained using GIS software.

3 Results

The results of the acoustic analysis gave a description of the local background noise in the study area, indicating a progressive increase over the years. Furthermore, the results showed that critically noisy areas may bear some relationship to the spatial and temporal distribution of dolphins. In particular, the area around three marine petrol stations and tourist beach camps situated along the coast of the islands of Lošinj and Cres was highlighted as important. In this area of high anthropogenic impact, we found a significant difference in the SAN intensities measured over frequencies below 1 kHz compared with the other two groups of medium and low impact during the summer seasons.

Data collected on bottlenose dolphin distribution showed a concomitant season-dependent avoidance of this area, with the lowest dolphin encounter rates over the summer months. Analysis of the underwater emissions produced by different vessel types indicated that fast-moving recreational vessels are particularly noisy, producing the highest noise intensities over the low range of frequencies. Additionally, data on vessel presence within the study area indicated that a significantly higher number of fast-moving recreational vessels are found in the high-anthropogenic impact area during the summer months.

4 Discussion

To identify the critically noisy areas, long-term monitoring of SAN was undertaken within the study area. Results of this study indicated the high-anthropogenic impact group as the noisiest area, showing significantly higher levels of underwater noise compared with the other two impact groups. This increment of the noise was shown over the range of low frequencies, which is consistent with the data obtained from monitoring the vessel. In fact, different vessel types recorded in the study area emitted most of the noise energy in the range below 1 kHz. Of all the recorded vessels, the noisiest were fast-moving recreational vessels, producing the highest levels of noise over the low range of frequencies. They were found in the highest number in the group of high-anthropogenic impact during the tourist season. It is therefore more than likely that the observed increase in the SAN was correlated to the intense human activities in this particular area.

Studies have shown that a large percentage of important marine habitats are exposed to high levels of anthropogenic noise (Allen and Read 2000; Gisiner 1998; Perry 1998). Coastal dolphins can tolerate some degree of chronic exposure to man-made noise, but localized displacement has been reported when they are exposed to a particularly strong disturbance as in the case of a simultaneous presence of multiple noise sources (Richardson and Wursig 1995). This seems consistent with data gathered on *Tursiops truncatus* distribution within the study area. In fact, the particularly low encounter rate found in the area of high-anthropogenic impact may suggest avoidance of this critically noisy area. Alternately, the low number of dolphin sightings, despite a high search effort, may indicate that dolphins cross this area using longer dives to avoid the proximity of sources of loud noise. This would then reflect on the possibility of detection of the animals by researchers.

Based on our overall results, we conclude that there is a negative impact on the distribution of bottlenose dolphins within the study area associated with the presence and noise of fast-moving vessels. In this case, SAN proved to be a good indicator of this anthropogenic pressure.

References

Allen MC, Read AJ (2000) Habitat selection of foraging bottlenose dolphins in relation to boat density near Clearwater, Florida. Mar Mamm Sci 16:815–824.

Bearzi G, Holcer D, Notarbartolo di Sciara G (2004) The role of historical dolphin takes and habitat degradation in shaping the present status of northern Adriatic cetaceans. Aquat Conserv Mar Freshw Ecosyst 14:363–379.

Evans PGH, Canwell PJ, Lewis E (1992) An experimental study of the effects of pleasure craft noise upon bottlenose dolphins in Cardigan Bay, West Wales. In: Evans PGH (ed) European Research on Cetaceans 6, Proceedings of the Sixth Annual Conference of the European Cetacean Society, 20–22 February 1992, San Remo, Italy, pp 43–46.

Foote AD, Osborne RW, Hoelzel AR (2004) Whale-call response to masking boat noise. Nature 428:910.

Fortuna CM (2006) Ecology and conservation of bottlenose dolphins (*Tursiops truncatus*) in the north eastern Adriatic Sea. PhD thesis, University of St. Andrews, Fife, UK.

Gisiner RC (ed) (1998) Proceedings of the workshop on the effects of anthropogenic noise in the marine environment. Office of Naval Research, 10–12 February 1998, Arlington, VA.

Perry C (1998) A review of the impact of anthropogenic noise on cetaceans. Report SC/50/E9 to the Scientific Committee of the International Whaling Commission, 27 April-8 May 1998, Oman.

Richardson WJ, Greene CR Jr, Malme CI, Thompson DH (1995) Marine mammals and noise. Academic Press, San Diego, CA.

Richardson WJ, Würsig B (1995) Influences of man-made noise and other human actions on cetacean behaviour. Mar Freshw Behav Physiol 29:183–209.

Simmonds MP, Mayer SJ (1997) An evaluation of environmental and other factors in some recent marine mammal mortalities in Europe: Implications for conservation and management. Environ Rev 5:89–98.

Does Vessel Noise Affect the Use of Sound by Foraging *Orcinus orca* (Killer Whales)?

Marla M. Holt, Dawn P. Noren, and Candice K. Emmons

1 Introduction

Investigations on the behavioral responses of cetaceans to a variety of anthropogenic sounds are numerous, with a large proportion of these studies focused on responses to vessels (Nowacek et al. 2007; Richardson et al. 1995). Behavioral responses that affect activities associated with survival and reproduction are of particular concern because they have clearer repercussions on life functions. Furthermore, populations that are repeatedly exposed to vessels such as those that frequent urban ports and/or those targeted for vessel-based whale-watching activities also get particular attention given the consistency of the exposure (Clark et al. 2009; Jensen et al. 2009). Behavioral responses of killer whales to vessels include a decrease in foraging activity, an increase in dive duration and swim speed, and an increase in some surface-active behaviors (Lusseau et al. 2009; Noren et al. 2009; Williams et al. 2009). However, it remains unclear whether the physical presence and/or the sound emitted from motorized vessels is associated with the observed behavioral effects because information about received sound levels and other sound exposure variables were not reported in previous investigations.

The waters surrounding the San Juan Islands, WA, are areas that endangered southern resident killer whales (SRKWs) use extensively for feeding (Hanson et al. 2010). This region is also an area where recreational and commercial vessel traffic is common, and in particular, there is a well-developed whale-watching industry, with an average number of ~20 boats surrounding SRKWs during summer daylight hours (Koski et al. 2006). There are ~85 individual SRKWs divided among three (J, K, and L) pods, and as part of their Endangered Species Act (ESA) recovery plan, several risk factors that could hinder SRKW population recovery were identified. One is prey availability and/or quality because, at least in their summer range, these killer whales target *Oncorhynchus tshawytscha* (Chinook salmon), many of whose stocks are also depleted (Hanson et al. 2010). Another risk factor is disturbance by vessel presence and/or noise from vessels that can result in behavioral and acoustic responses because killer whales rely on sound for many important life functions.

Fish-eating killer whales produce echolocation clicks for sensing prey in the water and population-specific pulsed calls and whistles that serve communicative functions (Ford 1989). Calls likely

M.M. Holt (✉) • D.P. Noren • C.K. Emmons
Marine Mammal Ecology Team, Northwest Fisheries Science Center, National Marine Fisheries Service, National Oceanic and Atmospheric Administration, Seattle, WA 98112, USA
e-mail: Marla.Holt@noaa.gov

A.N. Popper and A. Hawkins (eds.), *The Effects of Noise on Aquatic Life*, Advances in Experimental Medicine and Biology 730, DOI 10.1007/978-1-4419-7311-5_73,

function to maintain cohesion and coordinate movement among killer whales within matrilineally related groups (Ford 1989), and they commonly share prey among members of their matriline (Ford and Ellis 2006). Ford (1989) classified the pulsed calls of killer whales as discrete, aberrant, or variable. Discrete calls are structurally distinct, stereotyped, and repeated. Ford (1989) reported that the proportion of discrete call types observed among J pod members of the SRKW community differed depending on whether the whales were foraging or traveling. Hoezel and Osborne (1986) also reported that the density (calls per minute) and diversity of SRKW calls increased during foraging and milling (both involved in feeding) compared with traveling, but these data were not adjusted for group size.

Given the uses of sound associated with foraging in killer whales, there is concern that noise levels from nearby vessels are high enough in the whales' summer range to result in masking of these sounds, which has implications for foraging success. Previous research efforts have shown that as the number of vessels surrounding this killer whale population increased, so too did background noise levels (Holt et al. 2009). Furthermore, SRKWs vocally compensated for changes in vessel noise by increasing their call source levels (Holt et al. 2009) and call duration (Foote et al. 2004). These results indicate that masking of communicative sounds by vessel noise is a potential auditory challenge to the whales.

The objectives of the present study were to determine how sound-type usage and vessel noise levels vary with group activity states in the current population of SRKWs. We placed a particular emphasis on foraging behavior given the risk factors of this population and determining from an acoustic perspective why they might decrease foraging activities in the presence of vessels.

2 Methods

We collected data in waters surrounding the San Juan Islands, WA, off an 8-m research vessel, *RV Noctiluca*, during the summer months of three years (2007–2009). When SRKWs were sighted, the research vessel was positioned ahead and in the general path of the whales (~1,000 m), the motor was shut down, and recording equipment was set up. We collected call and background sound data continuously while also conducting 1-min observations every 10 min of the following: total vessel number within 1,000 m; pod identity; composition and group size; group activity state (forage, travel, social, and rest; based on Ford 1989); group spatial distribution (tight, loose, spread, flank, or nonlinear); group swim speed (fast or slow); and group direction (directional or nondirectional).

Details about our calibrated acoustic recording system and measurements of call and background sound levels are described in Holt et al. (2009). We determined the occurrence of different sound types (fast clicks, slow clicks, buzzes, calls, whistles, and other sounds) produced by SRKWs for each 1-min observation, and we collected behavioral data when the group was within 1 km of our research vessel. Within each of these sound-type categories, we statistically compared their occurrence (presence or absence) and also call rate (call per minute per whale) and discrete call-type diversity (number of call types per minute per whale) among activity states.

3 Results and Discussion

Results based on data collected in 2007 and 2008 are reported here and those from 2009 are still being analyzed. The majority of data we collected in 2007 and 2008 were in the presence of J pod members (either in single pod, subpod, or multiple pod groups). Forage and travel were the most often observed activity states. Because we rarely observed social and resting whales, we did not include these data in the subsequent analyses. The occurrence of both fast and slow clicks and (marginally) calls were significantly higher when the whales were foraging compared with traveling. Call rate and call-type diversity were also significantly higher when the whales were foraging compared with traveling. These results illustrate that the production of calls as well as clicks are integral for foraging and that a broader expression of their vocal repertoire occurred while foraging.

In contrast, both the total number of vessels and background sound levels were significantly higher when the whales were traveling compared with foraging. These differences to some extent reflect differences in spatial distribution of the whale-watching fleet that mirror the spatial distribution of the whales within a group (i.e., whales are more dispersed when foraging). However, the rate at which the J pod individuals increased the amplitude of their calls (S1 call type) as background levels increased was lower when they were foraging compared with traveling. These preliminary results suggest that when the whales were foraging, they were less effective at compensating for higher noise levels, and, consequently, vessel noise may limit the range at which calls can be exchanged. Lusseau et al. (2009) reported that SRKWs spent less time foraging and more time traveling in the presence of vessels. One very plausible reason why is because the active space of their communicative sounds when foraging (and dispersed) is reduced from the masking effects of vessel noise.

Acknowledgments We thank Anna-Maria Seibert, Christine Froschl, and Teresa Mongillo for field support. Val Veirs, Scott Veirs, and the students of Beamreach.org provided valuable support and assistance with hydrophone calibration, sound propagation, and localization error assessments.

Financial support was provided by the Northwest Fisheries Science Center, National Marine Fisheries Service (NMFS), National Oceanic and Atmospheric Administration (NOAA), Seattle, WA; the NOAA NMFS Ocean Acoustics Program, Silver Spring, MD; and the National Research Council Postdoctoral Associateship Program, Washington, DC.

We thank the conference organizers and sponsors for travel support to attend the Conference on the Effects of Noise on Aquatic Life in Cork, Ireland.

Data were collected under Scientific Research Permit No. 781-1824-00 from the US NMFS Office of Protected Resources, Silver Spring, MD, and Species at Risk Act/Marine Mammal Licence MML 2007-08/SARA-68 and MML 2008-05/SARA-85, Department of Fisheries and Oceans, Ottawa, ON, Canada.

References

Clark CW, Ellison WT, Southall BL, Hatch L, Van Parijs S, Frankel A, Ponirakis DW (2009) Acoustic masking in marine ecosystems: Intuition, analysis, and implications. Mar Ecol Prog Ser 395:201–222.

Foote AD, Osborne RW, Hoelzel AR (2004) Whale-call response to masking boat noise. Nature 428:910.

Ford JKB (1989) Acoustic behaviour of resident killer whales (*Orcinus orca*) off Vancouver Island, British Columbia. Can J Zool 67:727–745.

Ford JKB, Ellis GM (2006) Selective foraging by fish-eating killer whales. Mar Ecol Prog Ser 316:185–199.

Hansen MB, Baird R, Ford JKB, Hempelmann-Halos J, Van Doornik DM, Candy JR, Emmons CK, Schorr GS, Gisborne B, Ayres KL, Wasser SK, Balcomb KC, Balcomb-Bartok K, Sneva JG, Ford MJ (2010) Species and stock identification of prey consumed by endangered southern resident killer whales in their summer range. Endang Species Res 11:69–82.

Hoezel AR, Osborne RW (1986) Killer whale call characteristics: Implications for cooperative foraging strategies. In: Kirkevold BC, JS Lockard JS (eds) Behavioral biology of killer whales. Alan R. Liss Inc., New York, pp 373–403.

Holt MM, Noren DP, Veirs V, Emmons CK, Veirs S (2009) Speaking up: Killer whales (*Orcinus orca*) increase their call amplitude in response to vessel noise. J Acoust Soc Am 125:EL27–EL31.

Jensen FH, Bejder L, Wahlberg M, Aguilar Soto N, Madsen PT (2009) Vessel noise effects on delphinid communication. Mar Ecol Prog Ser 395:161–175.

Koski K, Osborne R, Tallmon R (2006) Soundwatch public outreach/boater education project 2004-2005 final program report. National Marine Fisheries Service Contract No. AB133F-04-SE-0835 (Req. No. NFFP 5000-4-00026), The Whale Museum, Friday Harbor, WA.

Lusseau D, Bain DE, Williams R, Smith JC (2009) Vessel traffic disrupts the foraging behavior of southern resident killer whales. Endang Species Res 6:211–221.
Noren DP, Johnson AH, Rehder D, Larson A (2009) Close approaches by vessels elicit surface active behaviors by southern resident killer whales. Endang Species Res 8:179–182.
Nowacek DP, Thorne LH, Johnston DW, Tyack PL (2007) Response of cetaceans to anthropogenic noise. Mamm Rev 37:81–115.
Richardson WJ, Greene CR Jr, Malme CI, Thomson DH (1995) Marine mammals and noise. Academic Press, San Diego, CA.
Williams R, Bain DE, Smith JC, Lusseau D (2009) Effects of vessels on behaviour patterns of individual southern resident killer whales, *Orcinus orca*. Endang Species Res 6:199–209.

Do Silent Ships See More Fish? Comparison of a Noise-Reduced and a Conventional Research Vessel in Alaska

Alex De Robertis, Christopher D. Wilson, and Neal J. Williamson

1 Introduction

Under some circumstances, fish detect and avoid approaching vessels, often well before the vessel passes over the fish (Mitson 1995). Such vessel-induced avoidance behavior is potentially a major source of error in surveys of fish populations. Sound propagates a long distance in water compared with other stimuli, and fish are generally most sensitive to sound in the frequency range at which the underwater-radiated noise from ships is most intense (Mann et al. 2009). Thus the primary stimulus for this avoidance behavior is thought to be auditory (Mitson 1995). The concern that vessel noise causes fish avoidance has led to the formulation of recommendations for maximum low-frequency (<1-kHz) underwater-radiated noise levels for fisheries research vessels (Mitson 1995). These recommendations, made under the auspices of the International Council for the Exploration of the Seas (ICES), are based on the hearing capabilities of *Clupea harengus* (Atlantic herring) and *Gadus morhua* (Atlantic cod), two species with sensitive hearing; the recommendation is therefore expected to minimize noise-induced vessel avoidance for other species as well.

Several nations have constructed vessels that comply with the ICES radiated-noise limits. Specialized vessel designs, including diesel-electric propulsion, fixed-pitch propellers, and quiet hull designs have resulted in substantial reductions in noise levels over a wide frequency range. Despite the existence of and investment in noise-reduced vessels, little is known about the impacts of noise reduction on vessel avoidance (reviewed in De Robertis et al. 2010).

In the United States, the National Oceanic and Atmospheric Administration (NOAA) has built four noise-reduced fisheries research ships. The first of these, the NOAA ship *Oscar Dyson* (OD), is now being used to conduct a long time series of acoustic-trawl surveys of *Theragra chalcogramma* (walleye pollock) in Alaska that have previously been conducted with the conventional (i.e., not noise-reduced) NOAA ship *Miller Freeman* (MF). Because the OD emits much less radiated noise than the MF (see De Robertis et al. 2008), there is concern that the survey abundance estimates used to manage fisheries, derived from the two vessels, will differ due to differences in fish reactions to the vessels. Thus a series of field experiments were conducted to establish if *Theragra chalcogramma* differentially avoid the two ships.

A. De Robertis (✉) • C.D. Wilson • N.J. Williamson
Alaska Fisheries Science Center, National Marine Fisheries Service, National Oceanic and Atmospheric Administration, Seattle, WA 98115, USA
e-mail: Alex.DeRobertis@noaa.gov; Chris.Wilson@noaa.gov; Neal.Williamson@noaa.gov

A.N. Popper and A. Hawkins (eds.), *The Effects of Noise on Aquatic Life*, Advances in Experimental Medicine and Biology 730, DOI 10.1007/978-1-4419-7311-5_74,

2 Methods

Experiments comparing the acoustic estimates of *Theragra chalcogramma* abundance derived from the OD and MF were conducted concurrently with established abundance surveys. Five experiments were conducted in four survey areas. The vessels traveled at 11–12 knots in a side-by-side arrangement (separated by 900 m) to generate acoustic abundance measurements with a vertical 38-kHz echo sounder. These paired measurements were used to estimate the ratio of fish abundance observed by the two vessels as detailed by De Robertis et al. (2008).

In three locations, an echo sounder mounted in a free-floating buoy (cf. De Robertis and Wilson 2010) was used to characterize the behavioral responses of *Theragra chalcogramma* when approached by the ships. The vessels took turns passing the buoy at 15-min intervals. Each vessel approached the buoy at ~11.5 knots and passed within <10 m of the buoy.

3 Results

The side-by-side vessel ratio exhibited strong contrasts among study areas (Fig. 1a). In the eastern Bering Sea, the area where the fish were shallowest (<140 m), there was no significant difference in acoustic abundance estimates between vessels during the day (i.e., the 95% confidence interval includes 1.0) in experiments in 2006 and 2008. However, there was a pronounced and significant vessel difference at night. Significant differences were also observed in the Shumagin Islands and Shelikof Strait, although diel differences were not detected. No significant differences were detected in the Bogoslof area where the fish were deepest (400–700 m). At each location, the differences in acoustically measured fish abundance were greater for shallower fish. The acoustic records from the

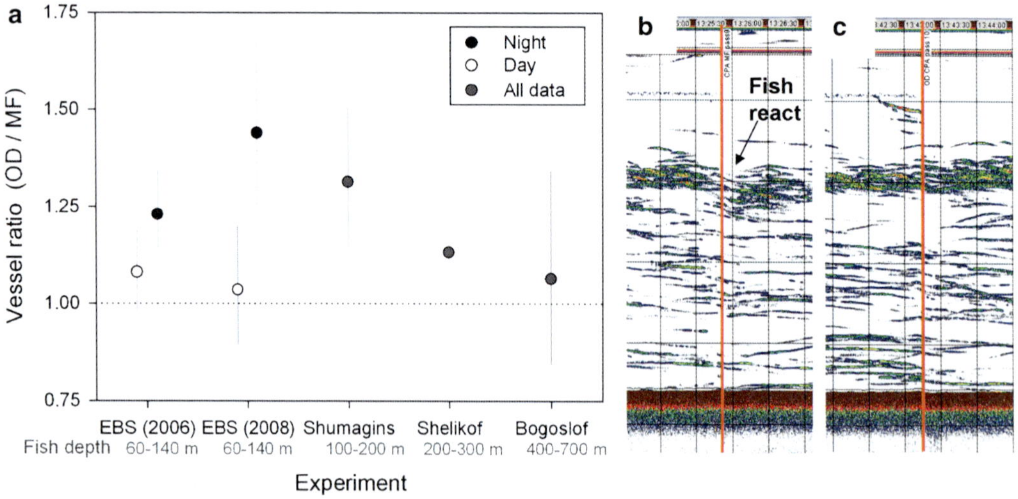

Fig. 1 (**a**) Summary of comparisons of acoustic fish abundance measurements from the *Oscar Dyson* (OD) and *Miller Freeman* (MF). Values are means ± 95% CI. Day and night results are shown separately for the eastern Bering Sea (EBS) because there is a significant diel difference. (**b**) and (**c**) Echograms from the acoustic buoy during vessel approach where the layer of *Theragra chalcogramma* at ~75 m depth is disturbed by passage of the MF (**b**) but not by the OD, which passes 15 min later (**c**). Orange line, vessel's closest approach to the buoy

buoy confirmed that *Theragra chalcogramma* exhibited a stronger avoidance response to the MF than to the OD, as evidenced by a deepening of the fish layer and a reduction in acoustic backscatter when the MF passed overhead (Fig. 1, b and c).

4 Conclusions

The noise-reduced vessel detected more fish than the conventional (i.e., non-noise-reduced) vessel under some survey conditions. More backscatter from *Theragra chalcogramma*, in some situations as much as 44%, was observed with the OD compared with the MF due to a differential behavioral response to the vessels. These vessel-specific reactions were independently confirmed with the acoustic buoy. The buoy observations also indicated that in an absolute sense, walleye pollock responses to the OD were small. The vessel discrepancy in each area was stronger for shallower fish, consistent with the expectation of a stronger response for fish closer to the vessels where acoustic transmission loss is lower. However, there was substantial variability among areas and a strong diel effect in the eastern Bering Sea. Thus fish depth (i.e., range to the vessel) is not the only factor influencing fish reactions to vessels and other aspects of fish behavior must be considered. The ICES vessel noise recommendations were designed such that one-third octave band underwater-radiated noise from the vessel should not exceed 30 dB above the hearing threshold of fish (Mitson 1995). Therefore, fish including *Theragra chalcogramma* can perceive acoustic stimuli from compliant vessels at distances of hundreds of meters (Mann et al. 2009). Thus the factors influencing the decision to react rather than perceptual limits likely play a key role in determining avoidance reactions.

This and other studies indicate that vessel avoidance behavior is variable and that the current understanding of the processes influencing the behavior is insufficient to accurately predict when vessel avoidance will occur and what the strength of the response will be. For example, one could not have predicted the diel difference in the eastern Bering Sea a priori. Uncertainty also remains regarding the impact of vessel noise reduction; for example, in the one other direct comparison of a noise-reduced vessel to a conventional vessel, Ona et al (2007) observed a stronger response by *Clupea harengus* to a noise-reduced rather than a conventional vessel. Although the present study was not designed to identify the stimuli triggering vessel avoidance responses, radiated noise is an obvious candidate as the OD is substantially quieter than the MF. Nonetheless, other potential stimuli such as near-field particle acceleration or infrasound should not be discounted.

The vessel-specific differences in acoustic survey results have fishery management implications. For example, the acoustic survey-based estimates of abundance from the OD are expected to be higher than those from the MF in several of the study areas. The resulting biases could be introduced into the fish abundance time series used to manage the fishery by switching vessels. Thus consideration of fish reactions to vessel noise is of great importance for fish abundance surveys.

References

De Robertis A, Hjellvik V, Williamson NJ, Wilson CD (2008) Silent ships do not always encounter more fish: Comparison of acoustic backscatter recorded by a noise-reduced and a conventional research vessel. ICES J Mar Sci 65:623–635. Doi:10.1093/icesjms/fsn025.

De Robertis A, Wilson CD (2010) Silent ships sometimes do encounter more fish. 2. Concurrent echosounder observations from a free-drifting buoy and vessels. ICES J Mar Sci 67:996–1003. Doi:10.1093/icesjms/fsp301.

De Robertis A, Wilson CD, Williamson NJ, Guttormsen MA, Stienessen S (2010) Silent ships sometimes do encounter more fish. 1. Vessel comparisons during winter pollock surveys. ICES J Mar Sci 96:985–995. Doi:10.1093/icesjms/fsp299.

Mann DA, Wilson CD, Song J, Popper AN (2009) Hearing sensitivity of the walleye pollock. Trans Am Fish Soc 138:100–1008. Doi:10.1577/T08-052.1.

Mitson RB (ed) (1995) Underwater noise of research vessels: Review and recommendations. ICES Cooperative Research Report 209.

Ona E, Godø OR, Handegard NO, Hjellvik V, Patel R, Pedersen G (2007). Silent research vessels are not quiet. J Acoust Soc Am 121:145–150.

Behavioral Observations of Baleen Whales in Proximity of a Modern Fishing Vessel

Matteo Bernasconi, Ruben Patel, and Leif Nøttestad

1 Introduction

Many marine species have evolved mechanisms to use sound to communicate, interact, and hunt in aquatic environments (Fudge and Rose 2009; Tyack and Clark 2000). For these reasons, human activities that introduce noise to the world's oceans may be of concern for marine life. It is the opinion among many scientists that whales and dolphins in particular may be sensitive to elevated sound levels and that noise may potentially result in negative physical and/or behavioral impacts (Hildebrand 2005). For example, mass strandings and serious injuries found in groups of marine mammals have been linked to the operation of low-frequency military sonar systems (Bernasconi et al. 2009). Concerns that the impact of sound may be widespread and detrimental has led to "bad press" about the application of active acoustic techniques to marine life. Nonetheless, ad hoc groups of experts have concluded that sonar used in fishery research is not a source of high potential risk for marine mammals (O'Brien 2004) and thus blanket "bad press" is unjustified. We have collected data on the acoustic target strength (TS) of cetaceans since 2007 and analyze here the observational data collected in conjunction with the acoustic data to see if any impact and disturbance of the vessel or its sonar systems on the animals under investigation can be detected.

2 Materials and Methods

Acoustic observations of whales were included as an integrated subgoal of annual acoustic surveys of pelagic fish stocks that covered substantial areas (7,395 nautical square miles [nmi^2]) of the Norwegian Sea between 62°30′-75.00° N and 18° W-22° E. This region is a well-known marine mammal feeding and migration area. The data were collected in summer 2008 onboard the Norwegian fishing vessel *Eros*, a 72-m combined purse seiner and pelagic trawler adapted and

M. Bernasconi (✉)
Pelagic Ecology Research Group, Scottish Oceans Institute, University of St. Andrews, St. Andrews, Fife KY16 8LB, Scotland, UK

Institute of Marine Research, 5817 Bergen, Norway
e-mail: matteo@imr.no

R. Patel • L. Nøttestad
Institute of Marine Research, 5817 Bergen, Norway

A.N. Popper and A. Hawkins (eds.), *The Effects of Noise on Aquatic Life*, Advances in Experimental Medicine and Biology 730, DOI 10.1007/978-1-4419-7311-5_75,

equipped as a modern scientific platform. Its acoustic instrument suite included 2 omnidirectional sonar systems (20 and 110 kHz; sound pressure level [SPL] 210 dB re 1 μPa) and a scientific echo sounder with 5 frequencies (18, 38, 70, 120, and 200 kHz; SPL 185 dB re 1 μPa). From a platform 13 m above the sea surface, 2 marine mammal observers collected whale data during the breathing cycles and recorded the distance, position of the animals relative to the ship, and swimming direction. Observations were logged using a GPS radio-linked software synchronized by time to the acoustic data logging. Sonar data processing, whale position, speed, and directional changes were computed with dedicated MATLAB scripts. The main goal of the observations was to measure whale TS in situ from different body aspects (head, side, and tail). The vessel interactions with the whales were in two modes: 1) the animal interacting freely with the ship (TS experiment [TSE]) at distances between 50 and 400 m and 2) the ship attempting encircling maneuvers (EMs) around the animal at a distance of ~200 m. We hypothesized that during the TSE, the whales would not be perturbed by the operating sonar systems. However, EMs may cause a disturbance to the whales due to the combined vessel and acoustic noise influencing the whales' swimming pattern. Variability in whale breathing interval, speed, and heading between these two observational modes were examined using an ANOVA type II Wald test to analyze possible behavioral changes.

3 Results and Discussion

A total of six individuals of the Balenopteridae family were observed, but EMs were attempted with only four individuals. The whales were exposed to a maximum source level of 183.9 dB re 1 μPa at 20 kHz (see Table 1 for details). Collection of data on cetaceans' body reflectivity was the main goal of the experiments, with no specific protocol for behavioral data analysis being in place a priori. Our observations can nevertheless provide some insight into potential behavioral changes by animals in the vicinity of a vessel operating standard fishery acoustic instruments commonly used by the Norwegian fishing fleet. The breathing intervals, with an average of 62 s during EMs, did not increase at all compared with an average of 60 s observed during the TSE, and examination with ANOVA gave a resulting $P = 0.12$ ($N = 284$). We believe that this parameter has limitations for behavioral inference. However, persistent light in summer at high latitudes influences the depth distribution of main prey such as herring, mackerel, and krill, and whales do not need to dive deeper than ~50 m to reach aggregations of suitable prey (Skjoldal 2004). Short dive durations and concomitantly short recovery periods are thus needed by whales. No significant changes in swimming speed were observed, with an average of 5.3 knots during the TSE and 6.6 knots while attempting EMs, with a resulting $P = 0.15$ ($N = 280$). More consistent changes in headings in response to the ship maneuvers were observed (Fig. 1), with an average of 19.7° during the TSE and an average of 35.8° during EMs ($P = 0.014$; $N = 229$).

Table 1 Single whale observations

Whale	fw01	fw02	fw03	hw01	hw02	bw01
Time exposure, min	57:49	25:51	41:22	48:35	33:51	26:11
Max SPL exposure at 20 kHz (dB re 1 μPa)	165.2	176.0	175.3	183.9	176.0	160.9
EM attempts	No	Yes	Yes	Yes	Yes	No
Blow intervals (TSE/EM), s	24/NA	51/56	30.5/29.0	51/32	28/26	82.7/NA
Speed (TSE/EM), knots	3.9/NA	7.0/5.3	3.5/6.3	3.6/5.4	0.8/0.8	5.2/NA
Directional changes (TSE/EM), degrees	13.0/NA	20/21	22/40	5/20	0/20	28.5/NA
Distance from ship after dive,* m	350.0	205.0	182.0	140.0	95.0	438.0

Values are medians. Max SPL, maximum sound pressure level; EM, encircling maneuver; TSE, target strength experiment; fw, fin whale; hw, humpback whale; bw, blue whale. *Apnea longer than 90 s (NA 37)

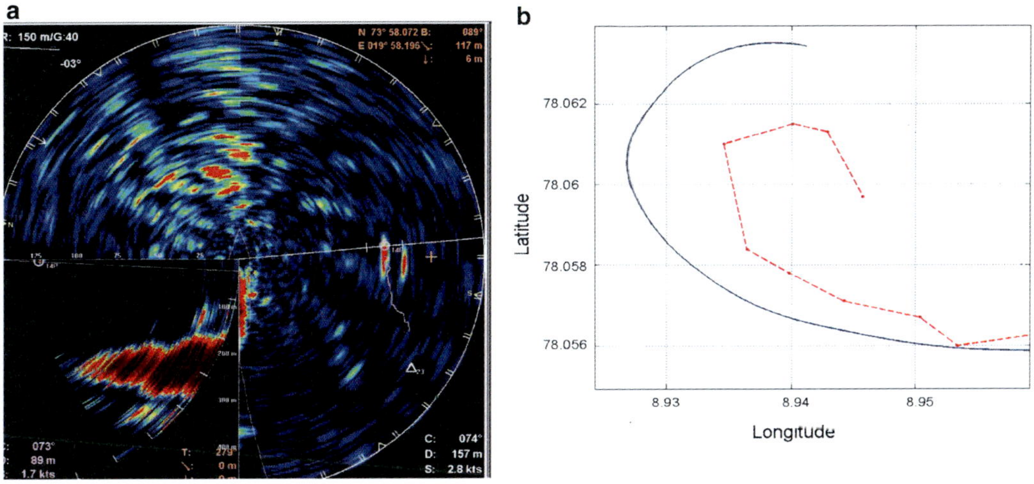

Fig. 1 (**a**) Sonar screen showing humpback whale (hw) 01 swimming during the target strength experiment (TSE) and acoustically tracked automatically (purple line). (**b**) Plot of direction changes by a whale (red line) during an encircling maneuver (EM).Blue line, vessel track

Individual whales swam most of the time in line with the main axis of our sonar transducers. However, whale heading changes were observed when an EM was attempted. Based on surface observations, it seems likely that the vessel and/or the active acoustic instrumentation onboard did not have any apparent impact on baleen whales. The whales did not leave the area, increase dive duration, or swimming speed. The whales only kept some distance to the vessel, which can be looked on as a minimum safe "predator" distance. We believe that whales gradually adapt over time to operating fishing vessels and sonar systems and identify noises that do not represent any immediate threat. This is not to say that noise from fishing vessels and their sonar do not represent any disturbance, but they can probably live properly with such noise; adaptation to a noisy environment has been observed for many bird species. Future studies should focus on testing hypotheses containing more reliable parameters that can be translated into clear quantities. In theory, animals that communicate in the infrasound bandwidth should not be affected by the fishery sonar frequencies we used. Using fishery sonar with applicable frequencies and low source levels as tools for whale mitigation measures onboard seismic ships seems promising. This may lead the way toward more operational measures on when, where, and why seismic shooting should be allowed or not.

References

Bernasconi M, Patel R, Nøttestad L, Knudsen F, Brierley A (2009) Use of active sonar for cetacean conservation and behavioral-ecology studies: A paradox? Proc Inst Acoust 31:112–118.

Fudge SB, Rose GA (2009) Passive- and active-acoustic properties of a spawning Atlantic cod (*Gadus morhua*) aggregation. ICES J Mar Sci 66:1259–1263.

Hildebrand JA (2005) Impacts of anthropogenic sound. In: Reynolds JE III, Perrin WF, Reeves RR, Montgomery S, Ragen TJ (eds) Marine mammal research: Conservation beyond crisis. The Johns Hopkins University Press, Baltimore, MD, pp 101–124.

O'Brien PE (2004) Report on the SCAR ad hoc group on marine acoustic technology and the environment workshop, September 2001. Polarforschung 72:69.
Skjoldal HR (2004) The Norwegian sea ecosystem. Tapir Academic Press, Trondheim, Norway.
Tyack PL, Clark CW (2000) Communication and acoustical behavior in dolphins and whales. In: Au WWL, Popper AN, Fay RR (eds) Hearing by whales and dolphins. Springer, New York, pp 156–224.

Assessing Effects of an Acoustic Marine Geophysical Survey on the Locomotive Behavior of Bottlenose Dolphins, *Tursiops truncatus*

Sylvia E. van der Woude

1 Introduction

This presentation provides observations on the effects of noise from a seismic and bathymetric survey on the behavior of bottlenose dolphins (*Tursiops truncatus*). It is a counterpart of the study presented in 2007 at the 1st International Conference on the Effects of Noise on Aquatic Life, Nyborg, Denmark (van der Woude 2008). While the previous study focused on the dolphins' acoustic responses to the survey, the present study focuses on the visually observed responses. Both studies include detailed data on the noise source and the receiver. Because the animals under observation were captives, the main objective of the study was to contribute methodologically to future field studies.

2 Methods

The geophysical survey covered the northern tip of the Gulf of Aqaba (Red Sea), an area ~6 × 7 km that includes an extensive open-sea enclosure in Eilat, Israel, and can accommodate 10 dolphins. Five different devices producing pulses with peak frequencies between 1 and 375 kHz were simultaneously applied in different combinations (see Table 1 in van der Woude 2008). GPS log files of the devices supplied information on their distance to the dolphin enclosure (varying between 5 km and 50 m), relative direction, and relative speed.

The dolphins' behavior was monitored visually and acoustically. Commented video recordings were made from an observation tower and linked to acoustic recordings obtained from a spacious three-hydrophone array. To relate visual data to acoustic data, only simultaneous recordings were analyzed and examined in 10-s intervals. Visual data were examined for the locomotive behavior of individual dolphins, i.e., quantitatively scored whereabouts, swimming associations, and speeds.

S.E. van der Woude (✉)
Institute for Biology, Free University of Berlin, 14195 Berlin, Germany

International Laboratory for Dolphin Behavior Research (ILDBR), 88100 Eilat, Israel
e-mail: s_vdw@web.de

A.N. Popper and A. Hawkins (eds.), *The Effects of Noise on Aquatic Life*,
Advances in Experimental Medicine and Biology 730, DOI 10.1007/978-1-4419-7311-5_76,

3 Results and Discussion

In contrast to continuous acoustic recordings, visual recordings were discontinued because animals submerged out of sight. However, similar to vocal activity, locomotive behavior clearly differed between survey days (15 h on 9 days) and control days (13 h on 8 days). This effect was more pronounced the closer the noise-producing devices were and the faster they were moving. Detailed results will be presented in a poster and published in a peer-reviewed paper (van der Woude, in preparation). Methodological limitations will be discussed as well as the interpretation and biological significance of behavioral modifications and possible long-term effects. Although there were no indications for injuries like deafness (temporary and/or permanent threshold shift), most locomotive changes are considered costly to the animals and some may have profound consequences in the wild.

References

van der Woude SE (2008) Assessing effects of an acoustic marine geophysical survey on the behavior of bottlenose dolphins, *Tursiops truncatus*. Bioacoustics 17:188–190.

Acoustic Monitoring of Beluga Whales (*Delphinapterus leucas*) in Cook Inlet, Alaska

Marc O. Lammers, Robert Small, Shannon Atkinson, Manuel Castellote, Justin Jenniges, Anne Rosinski, Sue Moore, Chris Garner, and Whitlow W.L. Au

1 Introduction

Beluga whales (*Delphinapterus leucas*) in Cook Inlet, Alaska, are listed as endangered and share habitats with a variety of anthropogenic activities including coastal development, oil and gas exploration, shipping, and military activities. Their population has declined from an estimated 653 animals in 1994 to 321 in 2009 (Hobbs et al. 2009). As a result, there is an urgent need for data that will help regulatory agencies such as the National Oceanic and Atmospheric Administration (NOAA) and Alaska's Department of Fish and Game (ADF&G) implement effective management and recovery plans. Among the principal types of information needed are quantifiable measures of seasonal presence in the inlet, temporal and spatial patterns of habitat preference, and the occurrence of animals in areas impacted or considered for industrial development.

To address these information needs, a research partnership was formed in 2007 between the University of Alaska Fairbanks (UAF), ADF&G, and the Hawai'i Institute of Marine Biology (HIMB) to apply an acoustic-monitoring strategy to study the occurrence of beluga whales in Cook Inlet.

M.O. Lammers (✉) • A. Rosinski • W.W.L. Au
Hawai'i Institute of Marine Biology, Kane'ohe, HI 96744, USA
e-mail: lammers@hawaii.edu

R. Small
Alaska Department of Fish and Game, Juneau, AK 99811, USA
e-mail: bob.small@alaska.gov

S. Atkinson
Juneau Fisheries, School of Fisheries and Ocean Sciences, University of Alaska Fairbanks, Juneau, AK 99801, USA
e-mail: atkinson@sfos.uaf.edu

M. Castellote • S. Moore
National Marine Mammal Laboratory, National Oceanic and Atmospheric Administration, Seattle, WA 98115, USA
e-mail: Manuel.Castellote@noaa.gov

J. Jenniges
Alaska SeaLife Center, Seward, AK 99664, USA
e-mail: justinj@alaskasealife.org

C. Garner
United States Army, Fort Richardson, AK 99505, USA
e-mail: christopher.d.garner@us.army.mil

A.N. Popper and A. Hawkins (eds.), *The Effects of Noise on Aquatic Life*,
Advances in Experimental Medicine and Biology 730, DOI 10.1007/978-1-4419-7311-5_77,

This approach is based on the use of ecological acoustic recorders (EARs), digital, low-power systems that record ambient sounds at frequencies up to 40 kHz on a recording schedule (Lammers et al. 2008).

2 Methods

Between June and December 2009, a network of 10 EARs was deployed in both upper and lower Cook Inlet (Fig. 1). These units were coupled with echolocation click detectors called C-PODs (Chelonia Ltd., Cornwall, UK) into an acoustic-mooring package. The objectives of colocating the two instruments were to compare the effectiveness of the two beluga detection methods and to improve detection range and efficiency during behavioral states when the animals might only produce certain types of signals (e.g., echolocation clicks, best recorded by the C-POD) or transit in the distance and produce faint calls (best recorded with the EAR). Here, only results obtain by EARs are presented because the C-POD data are being analyzed separately by M. Castellote.

3 Results

Beluga whale acoustic signals (Fig. 2) were detected at the following locations: Eagle River, Beluga River, Fire Island, and Trading Bay. No belugas were heard at Eagle Bay, Cairn Point, Tuxedni Bay, Kenai River, and Homer Spit. The unit deployed at North Eagle Bay was lost and not recovered. All

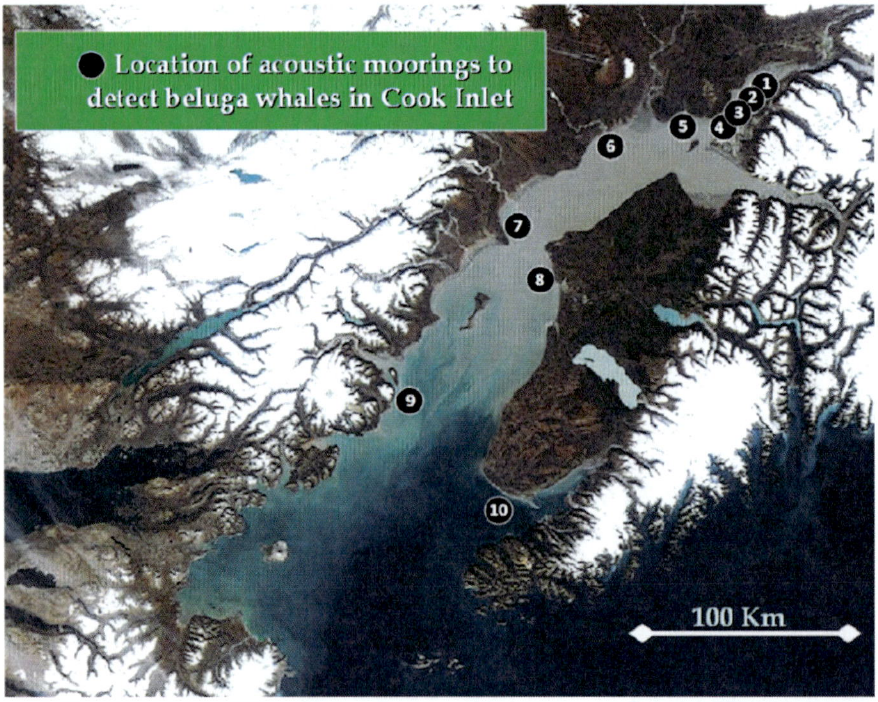

Fig. 1 Location of ecological acoustic recorder (EAR) deployment sites in Cook Inlet, Alaska. 1: North Eagle Bay; 2: Eagle River; 3: South Eagle Bay; 4: Cairn Point; 5: Fire Island; 6: Beluga River; 7: Trading Bay; 8: Kenai River; 9: Tuxedni Bay; 10: Homer Spit

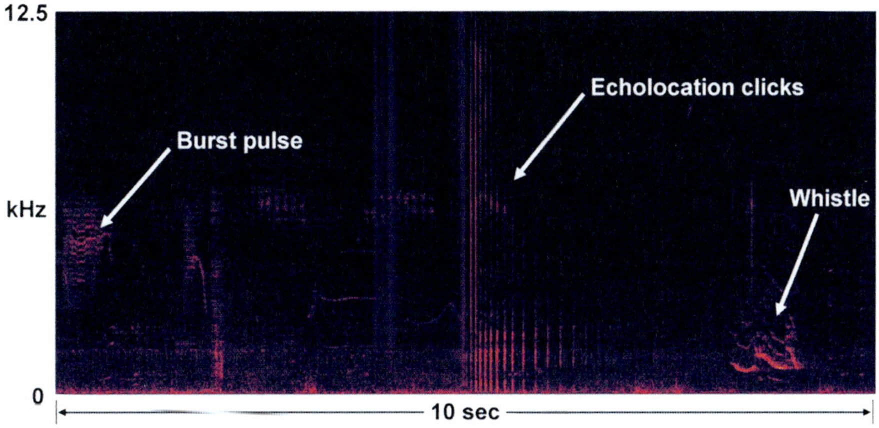

Fig. 2 Sonogram of beluga signals recorded on the EAR at the Beluga River mooring site

confirmed that beluga detections occurred in the upper Cook Inlet. Belugas were not heard at any lower inlet locations south of Trading Bay, but signals produced by killer whales (*Orcinus orca*) were recorded on 6, 9, and 15 October at Homer Spit and on 11 November at Kenai River. These killer whale calls were tentatively identified as belonging to resident killer whales of the AB clan (C. Matkin, personal communication).

Beluga detections were highest during two multiday episodes at Beluga River and Eagle River. At Beluga River, signals were recorded nearly continuously between 7 July, when the EAR was deployed, and 13 July. These sounds were highly variable and included many forms of whistles, calls, buzzes, and echolocation pulse trains. At Eagle River, a similar episode of nearly continuous detections occurred between 16 August, when the EAR was first deployed, and 22 August, when the EAR stopped recording due to a hardware malfunction. Of the EARs that recorded throughout their deployment, belugas were detected most consistently at Fire Island, where signals were recorded on 31 of the 118 (26.2%) deployment days. Of note, however, is that the Eagle River EAR detected belugas on all six days that it recorded.

4 Discussion

The results of this initial phase of the acoustic monitoring project reveal that during the summer and fall months belugas appear to restrict their occurrence to the upper part of Cook Inlet. Although only six days of recordings were obtained at Eagle River, the high number of detections made during this time suggests that this is likely an important part of their range. The fact that similarly high numbers of detections occurred episodically at Beluga River suggests that this too, at times, is an important part of their habitat.

The more sparse but consistent level of detections at Fire Island suggests a different pattern of occurrence in the area. No episodes of sustained high-acoustic activity were noted at Fire Island, but belugas were heard on more than one-fourth of the days that the EAR was deployed. This pattern suggests that animals were frequently nearby and transited by Fire Island. Lack of any directional information in the data precludes speculation about in which direction the belugas moved.

The pattern of occurrence of belugas at Trading Bay resembled that observed at Fire Island. Detections were sparse but relatively consistent after the beginning of September. This may indicate that belugas gradually transitioned further down the inlet in the fall. If this is the case, they did not go very far because no detections were made at the four lower inlet EARs well into November.

Interestingly, killer whales were heard several times at two of the lower inlet EARs. These were likely fish-eating resident killer whales and therefore not predators of belugas. The lack of detection of both belugas and killer whales at any of the EAR locations suggests some degree of resource partitioning may be happening, with belugas occupying mostly upper inlet waters to avoid competing for resources with killer whales. Alternatively, belugas may choose to remain in heavily sedimented, high-current upper inlet waters to avoid predation by their main predators, marine mammal-eating transient killer whales.

The detection of beluga calls in the recordings was a challenge because of considerable amounts of vessel and water flow noise within the frequency range of beluga calls. Consequently, the automated analysis algorithms designed to detect calls required significant manual ground truthing, which was time and labor intensive. However, the results to date are encouraging and validate the acoustic-monitoring approach as a method of learning about beluga whale occurrence and distribution in Cook Inlet. Continuing work will focus on the analysis of data from the EARs that have been deployed over the winter, on further characterizing habitat use by belugas, and on quantifying the natural and anthropogenic noise characteristics of Cook Inlet.

References

Hobbs RC, Sims CL, Shelden KEW, Rugh DJ (2009) Estimated abundance of beluga whales in Cook Inlet, Alaska, from aerial surveys conducted in June 2009. Unpublished Report, National Marine Mammal Laboratory, National Marine Fisheries Service, Seattle, WA.

Lammers MO, Brainard RE, Au WWL, Mooney TA, Wong K (2008) An ecological acoustic recorder (EAR) for long-term monitoring of biological and anthropogenic sounds on coral reefs and other marine habitats. J Acoust Soc Am 123:1720–1728.

Soundscape of a Nearshore Coral Reef Near an Urban Center

Whitlow W.L. Au, Michael Richlen, and Marc O. Lammers

1 Introduction

The School of Ocean and Earth Science Technology (SOEST) of the University of Hawai'i, Honolulu, HI, installed a nearshore coral reef observatory in 2005 to study the geochemistry and physics of bentic processes, wave boundary layer processes, water quality, and sediment pore water processes. Kilo Nalu ("observe the wave" in Hawai'ian) provides underwater nodes for data connectivity to the shore via a fiberoptic link and the availability of 24-V DC power so that a suite of observational instruments can be used to resolve waves, tides, currents, and nearshore water quality. The observatory is in the offshore waters of the Kaka'ako Waterfront Park in Honolulu, HI. In 2008, an acoustic-monitoring system was added to monitor the soundscape for biological entities. The Kilo Nalu acoustic-monitoring sensor (Fig. 1, red dot) is mounted on the bottom on a coral reef at a depth of 20 m.

2 Acoustic Sensor Package

The acoustic sensor consists of a Sensor Technology Limited SQ-26 cylindrical hydrophone that has a sensitivity of approximately −193 dB re 1 V/μPa and a receiving response that is flat from 1 Hz to 28 kHz and is usable to 40 kHz. The hydrophone is attached to the front face of a 10-cm-diameter aluminum cylinder that houses the electronics as shown in Figure 2. The electronics consist of an amplifier-filter board of an ecological acoustic recorder (EAR), a remote autonomous acoustic recorder developed jointly by the Hawai'i Institute of Marine Biology and the Coral Reef Ecology Division, National Marine Fisheries Service, Honolulu, HI (Lammers et al. 2008). For our application, the CF-2 microcontroller was replaced by a "Rabbit" microcontroller that controls the frequency cutoff of an eight-order Bessel anti-aliasing filter and the sampling rate of a serial analog-to-digital (A/D) converter and interfaces the output of the A/D converter to an Ethernet converter module driving a fiberoptic cable back to shore. A custom-built low-noise DC-to-DC converter converts the 24 V provided by the node to the 5 V required by the sensor electronics. With the electronic gain set at 47 dB, the total sensitivity of the acoustic sensor package is approximately

W.W.L. Au (✉) • M. Richlen • M.O. Lammers
Hawai'i Institute of Marine Biology, Kane'ohe, HI 96744, USA
e-mail: wau@hawaii.edu

A.N. Popper and A. Hawkins (eds.), *The Effects of Noise on Aquatic Life*,
Advances in Experimental Medicine and Biology 730, DOI 10.1007/978-1-4419-7311-5_78,

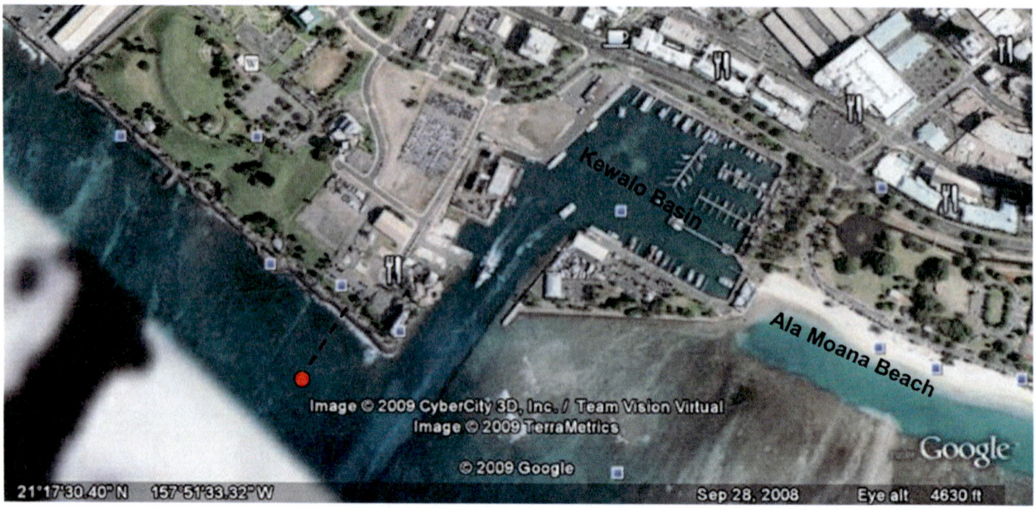

Fig. 1 Location of the Kilo Nalu acoustic sensor in a coral reef environment near a rural center

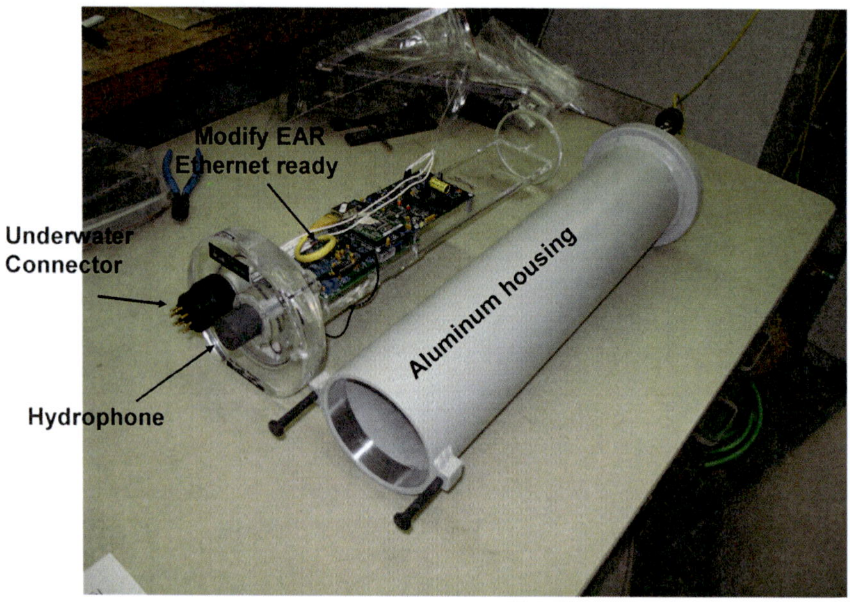

Fig. 2 Photograph of the components of the acoustic sensor package. EAR, ecological acoustic recorder

−146 dB re 1 V/μPa. A sample rate of 24 kHz and a anti-aliasing cutoff frequency of 9.6 kHz are used. The sensor package is mounted on the bottom at 20 m depth ~300 m from shore. A laptop computer on shore collects and stores the acoustic data on a portable disk drive.

3 The Soundscape: Biological Sounds

The predominant biological sound at this and all other coral reef environments in the tropics and subtropics is the sound produced by snapping shrimp. The frequency spectrum of a single snap of any individual shrimp is extremely broad (Fig. 3a for *Synalpherus paraneomeris*) and is the broadest

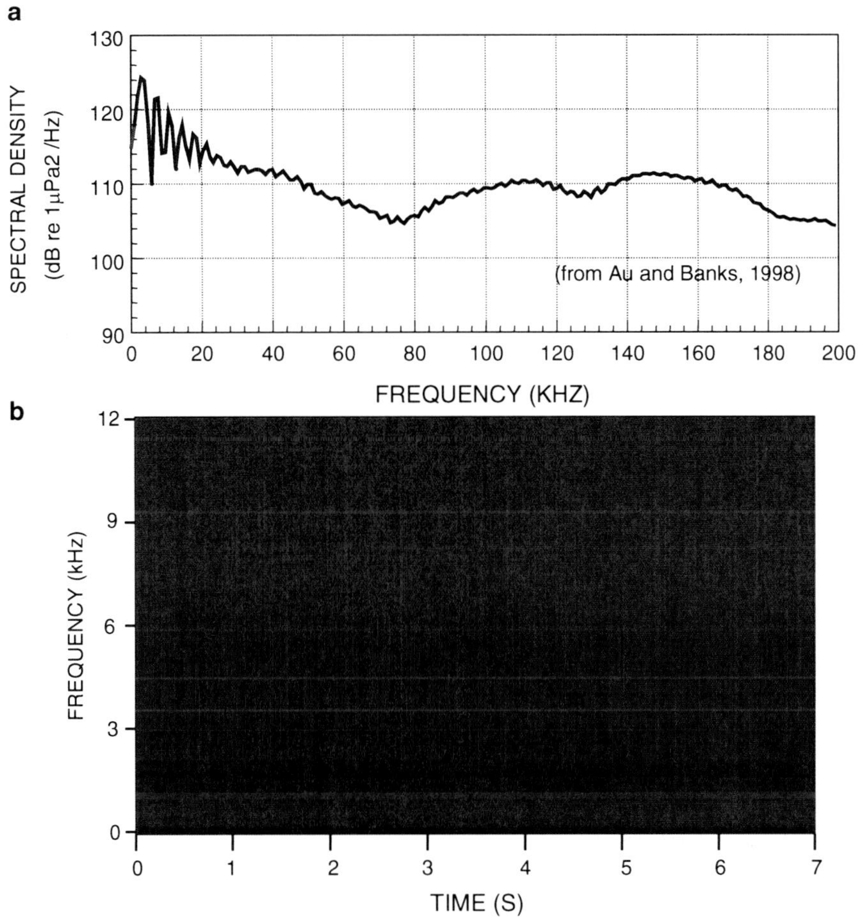

Fig. 3 (**a**) Frequency spectrum of a single snap from a *Synalpherus paraneomeris*. (**b**) Spectrogram of an ensemble of snapping shrimp

of any animal (Au and Banks 1998). The spectrogram of an ensemble of snapping shrimp sounds is shown in Figure 3b.

Fish sounds are also a part of the biological soundscape at the Kilo Nalu site. These sounds tend to be detected during the sunrise and sunset periods. An example of the sounds produced by a squirrelfish (*Sargocentron xantherythrum*) is shown in Figure 4. The sounds consist of a series of broadband clicks typically produced by stridulation (Au and Hastings 2008). There is a main click followed by three much smaller clicks, and the interaction of the four clicks caused the spectrum in Figure 4c to have local minima and maxima or a rippled pattern. Most of the energy of the fish sound is between 0.5 and 1.4 kHz, slightly below the peak frequency of the snapping shrimp, which is ~2.5 kHz.

During the winter months, humpback whales (*Megaptera novaeanglaie*) migrate to Hawai'i from Alaska and other high latitude areas. Songs can be detected from about early December until the end of April at the Kilo Nalu site. Humpback whale songs consist of different units arranged in phrases that are, in turn, arranged in a pattern to form themes and the themes are themselves formed in a pattern to produce a song (Payne and McVay 1971). An example of a portion of a humpback whale song is shown in Figure 5. Most of the energy of humpback whale songs is between 200 Hz

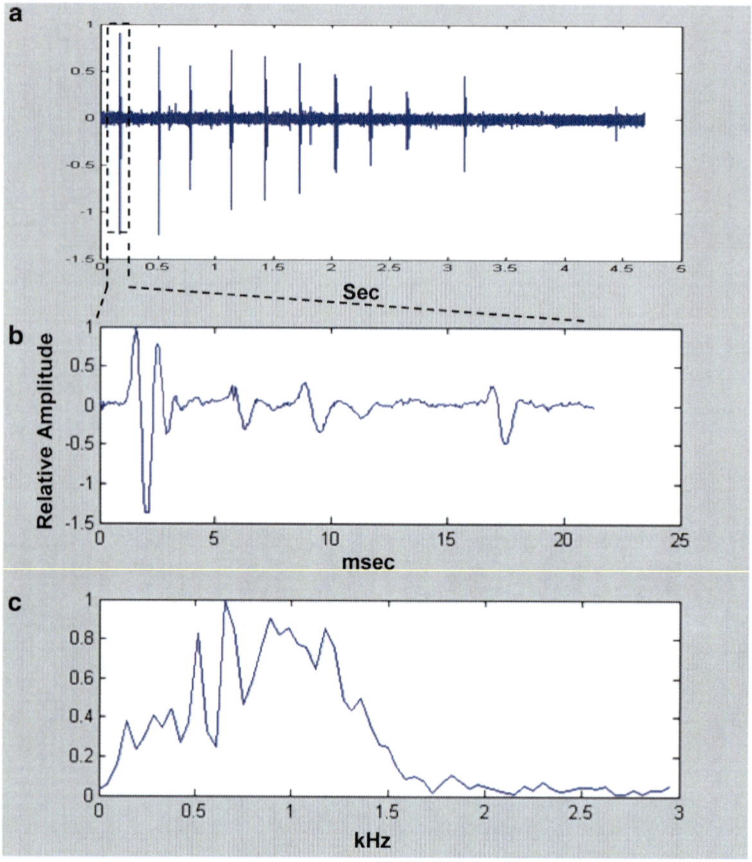

Fig. 4 (**a**) Broadband clicks produced by a squirrelfish. (**b**) Waveform of a single click. (**c**) Spectrum of a single click

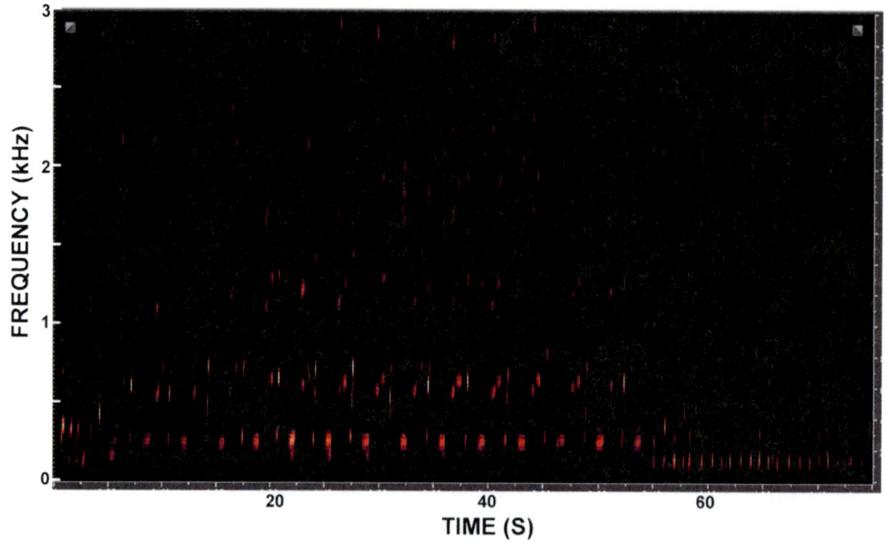

Fig. 5 Portion of a humpback whale song detected at Kilo Nalu

and 2 kHz, although harmonics can extend to over 24 kHz (Au et al. 2006). There are also many instances in which the songs of several humpback whales sound like a chorus, and these chorusing sounds can be detected by the acoustic sensor.

4 The Soundscape: Anthropogenic Sounds

The dominant sounds during the day are boat sounds. The Kilo Nalu site shown in Figure 1 is close to the opening of a small boat harbor (Kewalo Basin), and all sorts of boats go in and out of the harbor during the day. Tour boats, boats associated with sunset dinner cruises, dive boats, fishing boats, pleasure craft, and sport fishing boats move in and out of the harbor and produce noise that is part of the daylight soundscape. An example of noise from a boat is shown in Figure 6. Different boats will produce different sounds so that during any given day, a wide variety of boat sounds will be present during the daylight hours. The frequency spectrum of most of the boats covers a broad range, and all produce sounds in the frequency range of the biological sounds. Although sounds from an individual boat do not last long as it transits through the area, there are a sufficient number of boats to produce disturbing daylight-hour noise to interfere with the acoustic functioning of biological sounds.

Examples of the soundscape for a typical day and for a typical week are shown in Figure 7 based on the root mean square (RMS) sound pressure level (SPL). As is obvious, boat sounds dominate the soundscape during the day and are considerably louder than the snapping shrimp sounds by ~10–15 dB. The snapping shrimp sounds are lower in amplitude during the day than at night by ~4 dB in contrast to the 2 dB that Lammers et al. (2008) observed at the Waikiki Marine Life Conservation District that is only several miles away.

The only biological sound that boat sounds will not drown out are songs of humpback whales. An example of humpback whale sounds mixed with boat sounds during the day is shown in Figure 8. The levels of humpback whale sounds can be as high as the levels of boat sounds as can be seen in the nighttime display of humpback whale and snapping shrimp sounds in Figure 8. When listening to files containing boat and humpback whale sounds, there are frequency bands that

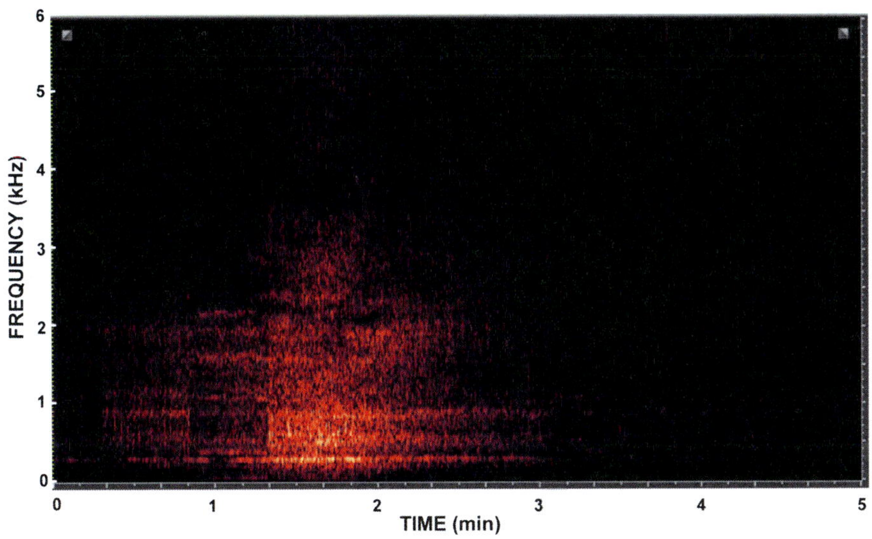

Fig. 6 Spectrogram of a boat sound passing in the vicinity of the Kilo Nalu site

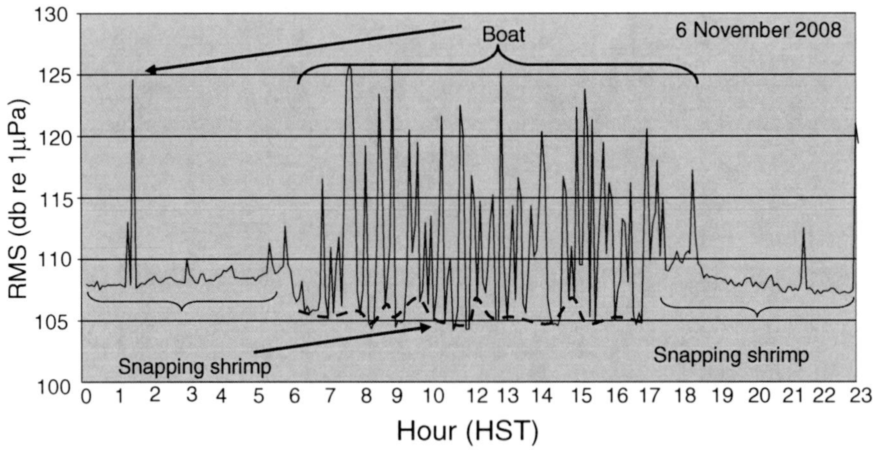

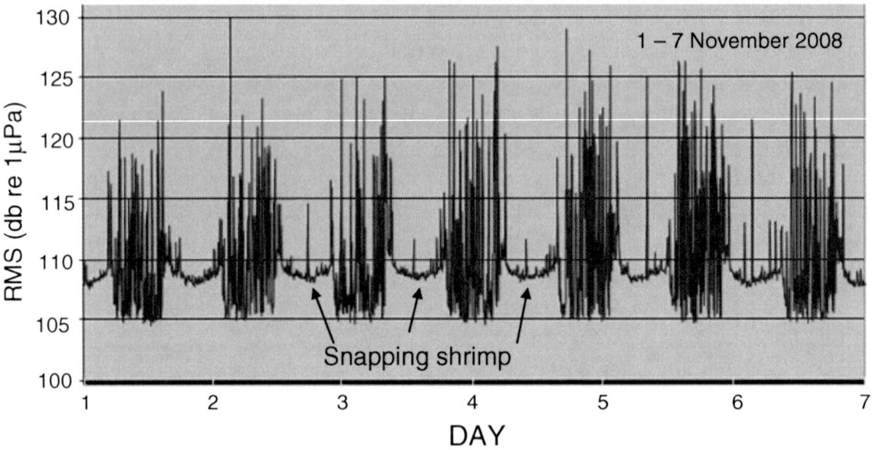

Fig. 7 The root mean square (RMS) sound pressure level for a typical day (top) and for a typical week (bottom). HST, Hawai'ian standard time

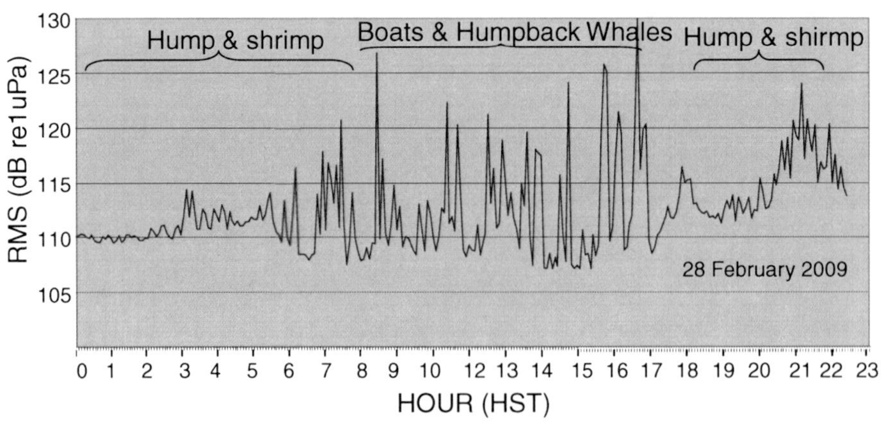

Fig. 8 RMS sound pressure levels including boats, humpback whales, and snapping shrimp over 24 h

contain mainly humpback whale sounds without any components of boat sounds and vice versa. Also, the comparable SPLs of humpback whale songs with boat noises make it possible to hear both simultaneously.

5 Conclusions

There can be no doubt that anthropogenic noise in the form of boat noise continues to impact the marine life in the vicinity of the Kilo Nalu observatory. However, precise and definitive statements on the severity of this impact cannot be made mainly because of our poor understanding on how various marine life uses sounds and a complete lack of data of the soundscape of this area when it was more "pristine." Our understanding of the role of sound emission by different species of fish, snapping shrimp, and marine mammals has to improve substantially. Whether these sounds are used for territorial purposes, for mating purposes, or merely for contact purposes we do not know. What we do know is that the boat noise will limit the range at which conspecifics will be able to detect emitted signals. Whatever the functioning of these signals are, their effectiveness will no doubt be affected during the daylight hours.

References

Au WWL, Banks K (1998) The acoustics of snapping shrimp in Kaneohe Bay. J Acoust Soc Am 103:41–47.

Au WWL, Hastings M (2008) Principles of marine bioacoustics. Springer Science+Business Media, New York.

Au WWL, Pack AA, Lammers MO, Herman L, Deakos M, Andrews K (2006) Acoustic properties of humpback whale songs. J Acoust Soc Am 120:1103–1110.

Lammers MO, Brainard RE, Au WWL, Mooney TA, Wong K (2008) An ecological acoustic recorder (EAR) for long-term monitoring of biological and anthropogenic sounds on coral reefs and in nearby waters. J Acoust Soc Am 123:1720–1728.

Payne RS, McVay S (1971) Songs of humpback whales. Science 173:585–597.

Gap Durations in Ambient Noise Impact the Communication Range of Harp Seals: A Pilot Study

John M. Terhune

1 Introduction

Pagophilus groenlandicus (harp seal) underwater vocalizations have sound durations that average 385 ± 550 (SD) ms but 51% are ≤200 ms (data reanalyzed from Perry and Terhune 1999). The detection thresholds for short duration sounds increase when sounds are <780 cycles (Kastelein et al. 2010), thus reducing the effective source level of such a call and its communication range.

Within *Pagophilus groenlandicus* breeding herds, call rates often exceed 90 calls/min, and this determines the ambient noise level (Terhune and Ronald 1986). A common description of ambient noise presents the sound pressure levels (SPLs) that are exceeded 95, 50, and 5% of the time. This description does not take into account the "fine structure" of the noise, however. For communication to occur, individual calls must be unmasked for some minimum period of time. Thus the temporal distribution of the noise levels will be important because communication will be possible during quieter times that exceed the duration of individual vocalizations.

The goal of this pilot study was to measure the variability of the underwater ambient (biotic and abiotic combined) noise levels within the *Pagophilus groenlandicus* breeding herd to determine if the durations of quieter gaps interacted with the higher detection thresholds of short duration calls when estimating the harp seal communication range. Limited sets of data were examined to determine if there was sufficient evidence to warrant a more detailed study.

2 Methods

A calibrated recording of *Pagophilus groenlandicus* underwater calls (Sony TDC-100 digital audiotape recorder and Vemco VHLF hydrophone) obtained in the Gulf of St. Lawrence on 1 March 2003 was examined. Five 240-s segments, widely spaced throughout the recording, were sampled. Each sample was played back through a Krohn-Hite 3364 filter set to bandpass an approximately one-third octave bandwidth centered at 1 kHz. The 3-dB down points were at 860 and 1,140 Hz and the slopes were −35 dB/octave. The amplitudes of consecutive 100-ms durations were measured using

J.M. Terhune (✉)
University of New Brunswick, Saint John, NB E2L 4L5, Canada
e-mail: terhune@unbsj.ca

A.N. Popper and A. Hawkins (eds.), *The Effects of Noise on Aquatic Life*,
Advances in Experimental Medicine and Biology 730, DOI 10.1007/978-1-4419-7311-5_79,

the energy contour function of the Multi-Speech Signal Analysis Workstation model 3700, version 2.5.2, at a digitization rate of 11,025 Hz. The maximum amplitudes per 100 and consecutive 200, 300, etc., to 2,000 ms durations were determined. This measures the duration over which a new call could be made at a particular masking level. Thus if the highest noise amplitude of 3 consecutive 100-ms periods was 90 dB re 1 μPa, then a 300-ms duration call that occurred in this time period would be masked at a level of 90 dB, even if the noise level was lower for the other two 100-ms periods. This procedure enables measuring the potential for seals to call for differing durations at different masking levels. An assumption of this analysis is that a call would be masked by the highest background noise level that occurs momentarily during that signal. Thus long duration signals would be more subject to masking than shorter signals.

The amplitudes were entered into a column of a spreadsheet and copied to 19 columns, each with a lag of one 100-ms step. Then the maximum amplitudes that occurred over different durations of time ("window width" in 100-ms units) were determined. This was accomplished by determining the maximum amplitudes per case for 1, 2, 3, 4, etc., up to 20 lagged columns in the data sheet. To avoid replication, the data were subsampled to include only measurements obtained every x time units. Thus when examining the maximum noise levels over a 600-ms duration, every sixth case was selected. The lagged variables would not permit a complete analysis at the beginning and end of the 240-s sample so, for analysis purposes, 2,000 consecutive cases selected from the middle of each dataset from each of the 5 samples were used in the analyses.

A communication range model was constructed using the assumptions that the sound spreading was spherical, that high-frequency absorption was negligible, and that the "almost certain" detection threshold of the seal would be 20 dB above the highest one-third octave noise level that occurred during the time window. The detection threshold for short duration sounds was increased by the amounts determined using *Phoca vitulina* (harbor seal) data (Kastelein et al. 2010). Communication range calculations for a call with a source level of 160 dB re 1 μPa at 1 m, using ambient noise levels that are exceeded 95, 50, and 5% of the time, were performed for sound window durations of 100–2,000 ms and for the simple average of the noise levels measured in 100-ms segments. This latter calculation did not assume a threshold increase due to the shortness of the call.

3 Results

Overall, the mean ambient SPL was 92.1 ± 3.6 dB re 1 μPa, with the 95 and 5% levels at 87.4 and 98.7 dB re 1 μPa, respectively. The mean SPL differences between 2 and 3 adjacent 100-ms samples were 2.6 ± 2.2 and 3.3 ± 2.9 dB, respectively. The maximum differences were 26.0 and 26.8 dB, respectively, but these large shifts were infrequent and the 5% values were 6.9 and 8.8 dB, respectively. Only 44 of 9,995 cases exhibited more than a 12-dB shift between 2 adjacent 100-ms segments and 140 of 9,990 cases exhibited more than a 12-dB shift between 2 measurements separated by 100 ms.

The interaction of the distribution of noise windows of various durations and the impact of threshold increases associated with short duration sounds resulted in the longest communication ranges occurring for 300-ms windows for the ambient levels that are exceeded 95, 50, or 5% of the time (Fig. 1). Using only the average noise levels and not taking the higher detection thresholds for short duration sounds into account resulted in communication range estimations that were 40% longer than the 300-ms calculations.

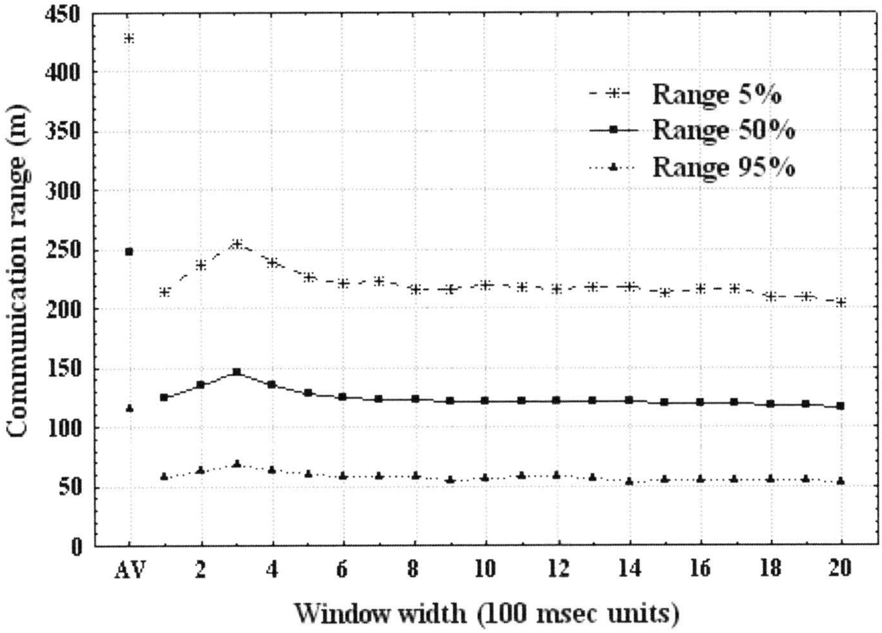

Fig. 1 Preliminary model of the underwater communication ranges of a *Pagophilus groenlandicus* (harp seal) call (1 kHz, 160 dB re 1 μPa at 1 m) in the presence of ambient noise levels that are exceeded 95, 50, and 5% of the time. The noise levels were measured as the highest level over durations of 100–2,000 ms or as a statistical average (AV) of 100-ms measurements. No short duration detection threshold correction was made for the AV range calculations (see text)

4 Discussion

The communication range calculations for ambient noise levels that are exceeded 95, 50, and 5% of the time indicate that the use of the distribution of the ambient noise levels alone will result in ranges that exceed all of the calculations based on the "noise windows" concept. This is because the model for the ambient level alone assumes that the hearing threshold of the listener is constant and does not take into account the 6-dB increase when the sound duration is shortened to 100 ms (Kastelein et al. 2010). This effect would be further increased if even shorter durations were used when measuring the average ambient noise levels.

This preliminary analysis indicates that the "fine structure" of the masking noise (for irregular noises at least, as opposed to the rhythmic noise of a ship's propeller or a pile driver) is important. If we assume that the communication range of a *Pagophilus groenlandicus* call would be determined by the highest background noise when the call is made and we measure the noise of a *Pagophilus groenlandicus* herd over consecutive 100-ms intervals, the actual masking level is higher than would be predicted by just the statistical distribution of the noise levels. Noise measures should be analyzed over periods similar to the call durations of the species concerned, and the occurrence of short duration calls must be considered. The distribution of quieter periods when communication can occur cannot be assessed by examining just the statistical distribution of the noise levels, especially if very short duration samples were measured. Finally, for short duration sounds, the higher detection thresholds must be considered. This is especially true for *Pagophilus groenlandicus* calls, the majority of which are very short sounds.

References

Kastelein RA, Hoek L, Wensveen PJ, Terhune JM, de Jong CAF (2010) The effect of signal duration on the underwater hearing thresholds of two harbor seals (*Phoca vitulina*) for single tonal signals between 0.2 and 40 kHz. J Acoust Soc Am 127:1135–1145.

Perry EA, Terhune JM (1999) Variation in harp seal underwater vocalisations among three breeding locations. J Zool (Lond) 249:181–186.

Terhune JM, Ronald K (1986). Distant and near range functions of harp seal underwater calls. Can J Zool 64: 1065–1070.

Importance of Underwater Sounds for Migration of Fish and Aquatic Mammals

Ilse van Opzeeland and Hans Slabbekoorn

1 Introduction: The Possible Role of Sound in Underwater Migration

Aquatic mammals and fishes often migrate over large distances through oceans, seas, and river systems. Many aquatic mammals migrate seasonally between mating and feeding areas (Stevick et al. 2002). Many species of fishes show a similar seasonality with spring and autumn migration between spawning and feeding areas. Some species also have a distinct migration cycle related to phases in life, which can include marine and freshwater stages, covering distances of more than 6,000 km such as the eel (e.g., *Anguilla anguilla*; van den Thillart et al. 2009).

Sound can play an important role in underwater migration because light conditions are often a constraint and water has excellent sound transmission properties. There are three possible ways in which sound can play a critical role in migratory behavior. First, sounds of biotic and abiotic origin can be exploited as acoustic cues for orientation. Second, echolocation sounds can be actively applied for probing the environment during migration. And third, social calls of conspecifics can synchronize and coordinate movements of members of more or less coherent groups.

The first way in which sound can play a role, as an acoustic beacon, depends on distinct acoustic variation along the migratory routes. Local water conditions and local animal communities that make both abiotic and biotic sound sources vary spatially. Consequently, different underwater areas can have distinct acoustic signatures (Radford et al. 2010) with habitat-specific spectral profiles, much in the same way as in above-water habitats (e.g., Slabbekoorn 2004). Although aquatic mammals can show avoidance responses or slow down in response to playbacks of sounds of predator species and certain industrial sounds (Moore and Clarke 2002), there are no data on the use of sounds for soundscape orientation (Slabbekoorn and Bouton 2008). Similarly, there are little or no data on adult fish using environmental sounds for orientation or navigation (Simpson et al. 2008; Sonny et al. 2006). However, data on pelagic larvae of reef fishes attracted to reef sounds are accumulating (Montgomery et al. 2006).

In the context of the second way in which sound can be important for migration, it is well known that toothed whale and dolphin species use echolocation clicks for orientation and localization of

I. van Opzeeland (✉)
Ocean Acoustics Lab, Alfred Wegener Institute, D-27568 Bremerhaven, Germany
e-mail: Ilse.van.Opzeeland@awi.de

H. Slabbekoorn
Behavioural Biology, Institute of Biology, Leiden University, 2333 BE Leiden, The Netherlands
e-mail: H.W.Slabbekoorn@biology.leidenuniv.nl

A.N. Popper and A. Hawkins (eds.), *The Effects of Noise on Aquatic Life*,
Advances in Experimental Medicine and Biology 730, DOI 10.1007/978-1-4419-7311-5_80,

prey, whereas baleen whales have also been suggested to use reflections of their low-frequency calls from natural boundaries as a form of echo ranging (Clark and Ellison 2003). Humpback whales produce clicks during nighttime foraging that are also thought to generate echoes that provide information about the seafloor and other large objects (Stimpert et al. 2007). Except for some species sensitive to environmental influences on their own electric fields, we are unaware of fishes exploiting some sort of echolocation with sounds.

For the third way in which sound can be important for traveling underwater, there is again more evidence for mammals than for fishes. Aquatic mammals migrating in groups often rely on acoustic cues for group cohesion. Harp seals (*Pagophilus groenlandicus*), for example, are thought to use vocalizations to coordinate herd movement during migration (Serrano and Miller 2000). Many baleen whale species produce sounds that are also likely to function in keeping track of each other during migration. Also, nonvocal sounds such as those generated by specific aerial displays in dolphins have been suggested to convey information about diving or travel intentions (Lusseau 2006). Many fish species also produce sounds that could serve in activating and aggregating, but we currently have very little empirical evidence. Trains of low-frequency "thumping" sounds have, for example, been attributed to eel passing through the river Dee in Scotland (Hawkins 2006), and, theoretically, such sounds could influence synchronization and coordination of migratory activity.

2 Conclusions: The Possible Impact of Anthropogenic Noise on Underwater Migration

Underwater migration may be affected by noisy human activities in several ways. Deterrence from high-level noise sources could lead to changes in optimal migration routes, obstruction of passage ways through river systems, or disturbance of group coherence. Deterioration of signal-to-noise ratios can cause masking of critical orientation cues, reflections from echolocation clicks, and social calls (Slabbekoorn et al. 2010). Furthermore, artificial noise conditions may further raise physiological stress levels in periods of already increased pressure due to migratory swimming expenses. The presence of anthropogenic noise may therefore result in considerable detrimental effects on migration efficiency in both aquatic mammals and fishes. However, a thorough understanding of the impact of noise requires more insight into the role of sounds in migration. There is especially a lack of insight into the importance of environmental sounds for orientation in both aquatic mammals and fishes and into whether communicative sounds from conspecifics also play a role in migratory decisions in fish.

References

Clark CW, Ellison WT (2004) Potential use of low-frequency sound by baleen whales for probing the environment: Evidence from models and empirical measurements. In: Thomas JA, Moss CF, Vater M (eds) Echolocation in bats and dolphins. University of Chicago Press, Chicago, IL, pp 564–581.

Hawkins A (2006) Assessing the impact of pile driving upon fish. In: Irwin CL, Garrett P, McDermott KP (eds) Proceedings of the 2005 International Conference on Ecology and Transportation, 29 August-2 September 2005, Center for Transportation and the Environment, North Carolina State University, Raleigh, NC, pp 21–22.

Lusseau D (2006) Why do dolphins jump? Interpreting the behavioural repertoire of bottlenose dolphins (*Tursiops* sp.) in Doubtful Sound, New Zealand. Behav Processes 73:257–265.

Montgomery JC, Jeffs A, Simpson SD, Meekan M, Tindle C (2006) Sound as an orientation cue for the pelagic larvae of reef fishes and decapod crustaceans. Adv Mar Biol 51:143–196.

Moore SE, Clarke JT (2002) Potential impact of offshore human activities on gray whales (*Eschrichtius robustus*) J Cetacean Res Manage 4:19–25.
Radford CA, Stanley JA, Tinde CT, Montgomery JC, Jeffs AG (2010) Localised coastal habitats have distinct underwater sound signatures. Mar Ecol Prog Ser 401:21–29.
Serrano AS, Miller EH (2000) How vocal are harp seals (*Pagophilus groenlandicus*)? A captive study of seasonal and diel patterns. Aquat Mamm 26:253–259.
Simpson SD, Jeffs A, Montgomery J (2008) Nocturnal relocation of adult and juvenile coral reef fishes in response to reef noise. Coral Reefs 27:97–104.
Slabbekoorn H (2004) Habitat-dependent ambient noise: Consistent spectral profiles in two African forest types. J Acoust Soc Am 116:3727–3733.
Slabbekoorn H, Bouton N (2008) Soundscape orientation: A new field in need of sound investigation. Anim Behav 76:e5–e8.
Slabbekoorn H, Bouton N, van Opzeeland I, Coers A, ten Cate C, Popper AN (2010) A noisy spring: The impact of globally rising underwater sound levels on fish. Trends Ecol Evol 25:419–427.
Sonny D, Knudsen FR, Enger PS, Kvernstuen T, Sand O (2006) Reactions of cyprinids to infrasound in a lake and at the cooling water inlet of a power plant. J Fish Biol 69:735–748.
Stevick PT, McConnell BJ, Hammond PS (2002) Patterns of movement. In: Hoelzel AR (ed) Marine mammal biology: An evolutionary approach. Blackwell Science, Oxford, UK, pp 185–216.
Stimpert AK, Wiley DN, Au WWL, Johnson MP, Arsenault R (2007) 'Megapclicks': Acoustic click trains and buzzes produced during night-time foraging of humpback whales (*Megaptera novaeangliae*). Biol Lett 3:467–470.
van den Thillart G, Dufour S, Rankin JC (eds) (2009) Spawning migration of the European eel. Springer, New York.

Behavioral Responses of Diamondback Terrapins (*Malaclemys terrapin terrapin*) to Recreational Boat Sounds

Lori A. Lester, Edward A. Standora, Walter F. Bien, and Harold W. Avery

1 Introduction

Anthropogenic sound caused by recreational boat traffic is a major concern for many marine animals because it may alter their behavior, mask sounds necessary for survival, and cause hearing loss. These alterations could potentially lower the chance of survival for individuals and lead to population declines. In this study, the diamondback terrapin (*Malaclemys terrapin terrapin*) in Barnegat Bay, NJ, is used as a model organism to understand how boat engine sound influences behavior. Previously, we used the auditory brain stem response (ABR) technique to determine that terrapins can hear a limited range of low-frequency tones less than 1,000 Hz. Most anthropogenic activities such as recreational boating also produce sound with low-frequency components (Richardson and Würsig 1997).

During the summer months, Barnegat Bay has one of the highest densities of recreational boating in the world (BBNEEP 2002). Terrapins are likely exposed to high levels of anthropogenic sound in Barnegat Bay. Diamondback terrapins are listed as a species of special concern in New Jersey (Hart and Lee 2006); population declines have been caused by anthropogenic impacts such as habitat destruction, roadkill, bycatch in crab traps, commercial harvesting, and increased recreational boating. Of over 3,000 terrapins that have been captured as part of a population study in Barnegat Bay, 14% have scars caused by boat propellers. Thus our research goal was to determine whether terrapins behaviorally respond to playback recordings of approaching recreational boats.

2 Methods

An underwater recording was made with a C54XRS hydrophone and a Sound DSA ST191 digital-recording computer of an approaching Lowe Boat (14 ft, 9.9 hp, 22.9 km/h). The experimental site consisted of a 60-m section of a mosquito ditch. Mosquito ditches are dug to drain marsh surface waters (especially intermittent pools that are used by mosquitoes as breeding habitat) to decrease the mosquito population. Each terrapin was exposed to a 1-min playback recording of boat engine sound using

L.A. Lester (✉) • W.F. Bien • H.W. Avery
Department of Biology, Drexel University, Philadelphia, PA 19104, USA
e-mail: lal56@drexel.edu; wfb22@drexel.edu; haltort@aol.com

E.A. Standora
Department of Biology, Buffalo State College, Buffalo, NY 14222, USA
e-mail: standoea@buffalostate.edu

A.N. Popper and A. Hawkins (eds.), *The Effects of Noise on Aquatic Life*,
Advances in Experimental Medicine and Biology 730, DOI 10.1007/978-1-4419-7311-5_81,

an underwater speaker (Lubell LL9816) that was suspended in the water at the center of the experimental site during high tide.

Terrapins were exposed to playback recordings of boat sounds in situ. An individual terrapin was placed in the mosquito ditch at one end and allowed to swim 60 m. The playback recording was started when the terrapin was 10 m from the speaker. Six experimental trials were run for each terrapin: three sound trials and three control trials in which no sound was played. Flags were placed at 10-m increments along the side of the mosquito ditch to divide the ditch into 6 sections. The amount of time that the terrapin spent swimming in each section was monitored as an estimate of its speed before, during, and after sound exposure. A DST milli-L temperature and depth data logger was secured with epoxy on each terrapin's carapace to measure its depth every second during the trials. The depth data logger allowed us to determine if the terrapin was attempting to escape the sound by sticking its head out of the water, burying in the substrate, or climbing out of the ditch.

3 Results

There was high interindividual variation of behavioral responses to the playback recordings of boat sound. Some terrapins spent more time near the surface during sound exposure, whereas others spent more time buried in the substrate at the bottom of the mosquito ditch. The swimming rate of the terrapins before, during, and after sound exposure also varied considerably among the individuals. Some terrapins increased swimming speed in response to the sound; however, other terrapins' swimming speed slowed or remained constant.

4 Discussion

Boat injuries are prevalent in terrapins captured in Barnegat Bay, NJ, and may be detrimental to long-term population viability. Kemp's ridley sea turtles (*Lepidochelys kempii*) alter behavior in response to anthropogenic sound by increasing the mean submergence time (Samuel 2004). By increasing the mean submergence time, *L. kempii* minimize the risk of boat strike. Although terrapins are sensitive to boat sound frequencies, many of the individuals in this study did not alter behavior during and after exposure to the playback recordings. The failure of many terrapins to behaviorally respond to anthropogenic sounds may be detrimental to survival in areas with intense boating. Partial or complete closure of some estuarine areas to boating may be necessary to decrease injury and mortality rates of terrapins and possibly other wildlife.

Acknowledgments This research was funded by the Earthwatch Institute, Boston, MA, and the Betz Chair of Environmental Science at Drexel University, Philadelphia, PA.

References

BBNEEP (2002) Barnegat Bay national effective estuary program: Comprehensive conservation and management plan. US Environmental Protection Agency Region 2 and New Jersey Department of Environmental Protection.

Hart KM, Lee DS (2007) The diamondback terrapin: The biology, ecology, cultural history, and conservation status of an obligate estuarine turtle. Stud Avian Biol 32:206–213.

Richardson WJ, Würsig B (1997) Influences of man-made noise and other human actions on cetacean behavior. Mar Freshw Behav Physiol 29:183–209.

Samuel Y (2004) Underwater low-frequency noise and anthropogenic disturbance in a critical sea turtle habitat. Masters Thesis, Cornell University, Ithaca, NY.

Assessing Disturbance From Under-Ice Noise on Fishes in Boreal Lakes

Peter A. Cott, David A. Mann, Dennis M. Higgs, Tom A. Johnston, and John M. Gunn

1 Introduction

Resource development is increasing in northern Canada and with it comes the potential for a variety of impacts to aquatic environments (Birtwell et al. 2005; Cott et al. 2003), e.g., impacts from anthropogenic noise (Mann et al. 2009). Compared with what is known about aquatic environments in southern areas, there is a dearth of information on the ecology of fishes living in northern boreal environments (Birtwell et al. 2005). In the absence of northern-specific information, southern models are often used to assess environmental impacts of northern developments. Yet, northern and southern systems likely behave differently due to large differences in climate and productivity and therefore likely differ in their sensitivity to potential impacts. As such, unsuitable or incorrect resource management decisions may be made if northern-based information is not used.

Sound is important to fishes in several ways, such as in the detection of predators and prey, defense, courtship, and reproduction (Hawkins 1993). Since little is known about how fishes in boreal ecosystems use sound or the potential impacts anthropogenic sound may have on them, the assessments of noise-related impacts from northern developments may be difficult. The level of impact from disturbance is dependent not only on the proximity to and the type of sound source (Mann et al. 2009; Popper et al. 2005) but also on the physical environment, the life stage of the fish (Popper et al. 2005), and the life process the fish is undergoing at a specific point in time (e.g., spawning) (Finstad and Nordeide 2004).

P.A. Cott (✉)
Department of Fisheries and Oceans, Yellowknife, NWT X1A 1E2, Canada

Cooperative Freshwater Ecology Unit, Laurentian University, Sudbury, ON P3E 2C6, Canada
e-mail: Pete.Cott@dfo-mpo.gc.ca

D.A. Mann
College of Marine Science, University of South Florida, St. Petersburg, FL 33701, USA

D.M. Higgs
Department of Biology, University of Windsor, Windsor, ON N9B 3P4, Canada

T.A. Johnston
Ontario Ministry of Natural Resources, Sudbury, ON P3G 1E7, Canada

Cooperative Freshwater Ecology Unit, Laurentian University, Sudbury, ON P3E 2C6, Canada

J.M. Gunn
Cooperative Freshwater Ecology Unit, Laurentian University, Sudbury, ON P3E 2C6, Canada

A.N. Popper and A. Hawkins (eds.), *The Effects of Noise on Aquatic Life*,
Advances in Experimental Medicine and Biology 730, DOI 10.1007/978-1-4419-7311-5_82,

Here we identify some sources of under-ice noise and their potential to disturb these fishes and describe what is known about the hearing of northern boreal fishes. Also, we present the potential for a widespread boreal piscivore, *Lota lota* (Burbot), to vocalize.

2 Under-Ice Noise

In northern Canada, the majority of industrial activity occurs during the winter when frozen waters and terrain allow access into remote areas. Activities are often conducted on ice (e.g., ice roads) over frozen lakes and rivers (Cott et al. 2008). Such activities generate sounds that can enter under-ice environments, changing the soundscapes of aquatic ecosystems and potentially disturb fishes (Mann et al. 2009). An assessment of noise generated from a diamond exploration camp was conducted in response to public and regulatory concern over the impacts of anthropogenic noise on fishes in ice-covered lakes. Sound from a variety of sources, typical of northern developments, was measured, including large- and small-diameter drills, ice road traffic and maintenance (trucks, graders), snowmobiles, aircraft (airplanes and helicopters), ice augers, ice chisels, walking, ice cracks, and ambient conditions. Although the natural under-ice soundscape is quiet (with the exception of ice cracks; peak sound pressure level [SPL] > 145 dB re 1 μPa), anthropogenic noise can be loud under ice (peak SPL 120–155 dB re 1 μPa depending on the source). Unlike ice cracks, which are intermittent and transient, anthropogenic noises are often persistent. Most sounds measured were within the hearing range of northern fishes (50–1,600 Hz), particularly by fishes with sensitive hearing such as *Couesius plumbeus* (lake chub) (Mann et al. 2007). However, away from the source, sounds that attenuate rapidly are not likely to cause physical harm to resident fishes (Mann et al. 2009).

3 Hearing of Boreal Fishes

Anthropogenic noise has been shown to impact the physiology (e.g., McCauley et al. 2003; Popper et al. 2005) and behavior (e.g., Finstad and Nordeide 2004; Hawkins 1993) of fishes. The effects of noise on fish can be correlated to the hearing sensitivity of the fish (e.g., Popper et al. 2005), making it important to understand the hearing capability of the fishes impacted at the life stages at which they may be. Recently, the hearing of eight species of northern boreal fish was assessed through auditory evoked potential of the lowest detectible sound using the auditory brain stem response technique. The research showed that the hearing sensitivity of fishes varied depending on the frequency investigated (Mann et al. 2007). In general, all species assessed had the most sensitive hearing at <400 Hz. Fish possessing Weberian ossicles, in this case, *Couesius plumbeus* and *Catostomus catostomus* (longnose sucker), had the most sensitive hearing and over the broadest frequency range (100–1,600 Hz). *Cottus ricei* (spoonhead sculpin) and *Percopsis omiscomaycus* (trout-perch) had sensitive hearing but in a narrow frequency range (>200 Hz) (Mann et al. 2007). Adult *Esox lucius* (northern pike) had more sensitive hearing than young-of-the-year *Esox lucius* (Popper et al. 2005), suggesting ontogenetic shifts in hearing ability.

4 Fish Vocalizations

Sound is central to the behavior of many animals including fishes, particularly in relation to reproduction (Hawkins 1993). Gadids are well represented among fishes that produce sounds. *Gadus morhua* (Atlantic cod) have been observed making "hum" or "grunt" sounds during spawning courtship

(Finstad and Nordeide 2004; Hawkins 1993; Rowe and Hutchings 2006), and it has been suggested that their spawning vocalization is linked to spawning success (Rowe and Hutchings 2008). The mechanism by which *Gadus morhua* produce sounds is by movement of sonic muscles that are attached to their swim bladder (Hawkins, 1993; Rowe and Hutchings 2006; 2008). *Lota lota*, a close relative of *Gadus morhua*, also possess sonic musculature on their swim bladders (Hawkins 1993), which may enable them to vocalize.

Anthropogenic sounds may interfere with fish vocalizations, interfering with and masking their communication (Hawkins 1993). For instance, Finstad and Nordeide (2004) suggest that noise from commercial fishing activities may disturb spawning vocalizations and behavior of *Gadus morhua*. Such disruptions may interfere with mate selection, spawning success, or egg viability (Rowe and Hutchings 2006). *Lota lota* may be more sensitive to under-ice noise than other boreal fishes because they spawn under ice in the winter (Scott and Crossman 1973) when the majority of northern development occurs. The under-ice noise produced by industrial activities may disturb spawning *Lota lota* directly or may mask their prespawning vocalizations, reducing their ability to reproduce successfully.

To assess the potential for *Lota lota* vocalization, a pen net (10 m × 10 m × 10 m) dubbed "the *Lota*-tron" was positioned in Great Slave Lake, NWT, Canada, with the top of the *Lota*-tron frozen into the ice and anchored to the substrate. Adult prespawn *Lota lota* ($n = 31$) were collected in the vicinity and placed in the *Lota*-tron. An autonomous multichannel acoustic recorder (AMAR), capable of continuous recording of low-frequency sound (at levels produced by *Gadus morhua*), was deployed within the *Lota*-tron. The AMAR was deployed in mid-December 2009 and retrieved in mid-April 2010. This spanned the spawning period (mid-to-late February) for *Lota lota* in this location. These sound data are being analyzed for evidence of *Lota lota* vocalizations (P.A. Cott, unpublished observations).

5 Summary

There are many potential sources of anthropogenic noise that can manifest under ice in boreal lakes that are within the hearing ranges of northern boreal fishes. Impacts of noise on fish can be correlated to the fish's hearing sensitivity. In general, boreal fishes have most sensitive hearing < 400 Hz, but this varies by species and life stage. By assessing the hearing capabilities and vocalizations of fish species and sound signatures from development activities, it may be possible to forecast potential impacts resulting from expected under-ice noise exposure.

There is still the need for basic information to facilitate the assessment and identification of potential impacts (e.g., the hearing capabilities of many species of boreal fishes at different life stages and the sound signatures of various anthropogenic noise sources). Additionally, the impacts to fish from a particular noise source (e.g., pile driving) or the potential for a sound-related disturbance at a particular time in a species life history, such as courtship or spawning, warrants further study.

References

Birtwell IK, Samis SC, Khan NY (2005) Commentary on the management of fish habitat in northern Canada: Information requirements and policy considerations regarding diamond, oil sands and placer mining – Summary Report. Canadian Technical Report of Fisheries and Aquatic Sciences 2607, Fisheries and Oceans Canada, West Vancouver, BC, Canada.

Cott PA, Hanna BW, Dahl JA (2003) Discussion on seismic exploration in the Northwest Territories 2000–2003. Canadian Manuscript Report for Fisheries and Aquatic Sciences 2648, Department of Fisheries and Oceans, Yellowknife, NWT, Canada.

Cott PA, Sibley P, Gordon A, Bodaly RA, Mills KH, Somers WM, Fillatre GA (2008) The effects of water withdrawal from ice-covered lakes on oxygen, temperature and fish. J Am Water Resour Assoc 44:328–342.

Finstad JL, Nordeide JT (2004) Acoustic repertoire of spawning cod, *Gadus morhua*. Environ Biol Fishes 70: 427–433.

Hawkins AD (1993) Underwater sound and fish behaviour. In: Pitcher TJ (ed) Behaviour of teleost fishes, 2nd edn. Chapman & Hall, London, pp 129–169.

Mann D, Cott P, Horne B (2009) Under-ice noise generated from diamond exploration in a Canadian sub-arctic lake and potential impacts on fishes. J Acoust Soc Am 126:2215–2222.

Mann DA, Cott PA, Hanna BW, Popper AN (2007) Hearing in eight species of northern Canadian freshwater fishes. J Fish Biol 69:1–12.

McCauley RD, Fewtrell J, Popper AN (2003) High intensity anthropogenic sound damages fish ears. J Acoust Soc Am 113:638–642.

Popper AN, Smith ME, Cott PA, Hanna BW, MacGillivray AO, Austin ME, Mann DA (2005) Effects of exposure to seismic airgun use on hearing of three fish species. J Acoust Soc Am 117:3958–3971.

Rowe S, Hutchings JA (2006) Sound production by Atlantic cod during spawning. Trans Am Fish Soc 135: 529–538.

Rowe S, Hutchings JA (2008) A link between sound producing musculature and mating success in Atlantic cod. J Fish Biol 72:500–511.

Scott WB, Crossman EJ (1973) Freshwater fishes of Canada. Fisheries Research Board of Canada Bulletin 184.

Impacts of River-Based Air Gun Seismic Activity on Northern Fishes

Peter A. Cott, Arthur N. Popper, David A. Mann, John K. Jorgenson, and Bruce W. Hanna

1 Introduction

Air guns are an energy source commonly used by the oil and gas industry for seismic exploration, particularly in marine environments. Exposure to air gun noise has been shown to negatively impact marine fishes (e.g., McCauley et al. 2003); however, little is known about the potential impacts on fishes in freshwater systems. A proposal to use air guns for a seismic survey along the Mackenzie and Liard Rivers (~1,200 km) in the Northwest Territories (NWT) led to considerable public and regulatory concern. As a result, the Canadian Department of Fisheries and Oceans (DFO) required the proponent to validate their prediction that the project will have no impacts on fish. The proponent conducted a study that showed no acute mortality associated with air gun use, although some stunning was observed in caged fish closest (2 m) to the air gun blasts (Cott et al. 2003). Outstanding data gaps associated with sublethal impacts of air gun noise (e.g., physical damage, stress, herding, and hearing loss) prompted the DFO to initiate a study, in collaboration with academic partners, to assess these impacts on northern fish in a riverine setting. The project was split into two components: 1) physiological impacts on fish hearing and ear damage and 2) behavioral impacts on free-swimming fishes.

2 Methods

The study was conducted in the Mackenzie Delta near Inuvik, NWT, during late July to early August 2004. A 730-in.^{3} air gun array with an acoustic output analogous to that proposed for the river seismic exploration program was used as the sound source (Popper et al. 2005).

P.A. Cott (✉) • B.W. Hanna
Department of Fisheries and Oceans, Yellowknife, NWT X1A 1E2, Canada
e-mail: Pete.Cott@dfo-mpo.gc.ca

A.N. Popper
Department of Biology, University of Maryland, College Park, MD 20742, USA

D.A. Mann
College of Marine Science, University of South Florida, St. Petersburg, FL 33701, USA

J.K. Jorgenson
Department of Fisheries and Oceans, Winnipeg, MB R3T 2N6, Canada

A.N. Popper and A. Hawkins (eds.), *The Effects of Noise on Aquatic Life*, Advances in Experimental Medicine and Biology 730, DOI 10.1007/978-1-4419-7311-5_83,

Three species of fish native to the Mackenzie River were selected for assessment of impacts on hearing and inner ear damage: *Couesius plumbeus* (lake chub), a hearing specialist; *Esox lucius* (northern pike), a hearing generalist (both juvenile and adults); and *Coregonus nasus* (broad whitefish), a species important to the aboriginal subsistence fishery (Mann et al. 2007; Popper et al. 2005). Fish were captured in the Mackenzie River at Inuvik. Treatment fish were placed in a mesh holding cage in the river ~15 m from the air gun array and exposed to 5 or 20 shots. The air gun array was positioned broadside to the fish cage to attain maximum noise exposure (205–209 dB re 1 μPa) (Popper et al. 2005). Fish hearing was assessed using the auditory brain stem technique for evidence of threshold shifts in hearing ability. Exposed fishes were tested after exposure and over time (18 and 24 h) to assess for delayed hearing loss or recovery (Popper et al. 2005). The inner ears of these fish were subsequently removed and examined for physical trauma, focusing on the sensory epithelia, using scanning electron microscopy (Song et al. 2008). The hearing and inner ear structure of several other species of Mackenzie River fish were also investigated to address data gaps associated with northern fish biology (Mann et al. 2007; Song et al. 2008).

To determine behavioral impacts, free-swimming fish in the Mackenzie Delta were observed using hydroacoustics to detect evidence of herding or startle response when subjected to air gun noise (Jorgensen and Gyselman 2009). The herding response of fish to oncoming air gun noise was assessed by tracking fishes using an acoustic launch anchored over fish targets as the seismic barge approached with air guns firing. The startle response of individual fish to a single air gun shot was assessed by drifting the acoustic launch over fish targets while the seismic barge was anchored, with no engine noise from either vessel (Jorgensen and Gyselman 2009).

3 Results and Discussion

The temporary threshold shift (TTS) in hearing in *Couesius plumbeus* exposed to both 5 and 20 shots was significantly different from that in unexposed fish, with the threshold shifts in fish exposed to 20 shots being greater than that of those exposed to 5 shots. Trends suggest recovery in *Couesius plumbeus* hearing after 18 h. There was a significant difference in threshold shifts in hearing in adult *Esox lucius* exposed to 5 air gun shots, but hearing recovered after 24 h. Both of these species showed the greatest threshold shifts at 400 Hz, with *Couesius plumbeus* also showing large threshold shifts at 200 Hz. No evidence of hearing loss was detected in young-of-the-year *Esox lucius* exposed to either 5 or 20 shots, possibly due to incomplete auditory development compared with the adults. The fish with the poorest hearing ability, *Coregonus nasus*, showed no effect from air gun noise (Popper et al. 2005). There was no damage observed in the sensory epithelia in any of the fish exposed to air gun noise, including those held 18-24 h postexposure (Song et al. 2008). Hearing loss or hearing damage from exposure to air gun noise is unlikely in free-swimming riverine fishes because they would not be exposed to as much noise as in this study (Popper et al. 2005; Song et al. 2008).

No evidence of herding behavior from air gun noise was observed, i.e., fish were not being "pushed" by the air gun noise. There was no significant difference between experimental and reference fish with respect to horizontal direction, horizontal speed, vertical direction, vertical speed, or tortuosity (a measure of the deviation from random movement). Similarly, there was no significant difference in the startle response between experimental and reference fish with the above-mentioned parameters. Fish did not respond to air gun sound by changing their swimming behavior. The fish observed were likely coregonids such as *Coregonus nasus* (Jorgensen and Gyselman 2009), a species shown by Popper et al. (2005) and Mann et al. (2007) to have poor hearing relative to other species.

4 Conclusions

The results of our study indicate that air gun noise can cause temporary hearing loss for some species of riverine fish. However, there was no damage to the inner ears of any of the noise-exposed fish examined. Furthermore, no evidence of startle or herding behavior associated with air gun noise was detected. Due to the limited air gun array size and the single-pass nature of a riverine seismic program, sound exposure to riverine fishes is transient relative to a marine program and adverse impacts are unlikely. This information allowed regulators to advise project proponents on the means to mitigate the effects on fishes (such as avoiding spawning areas where impacts are unknown) and provided new insights on the effects of sound on fishes in lotic environments. Note that this study looked at the impacts of air gun-generated noise on northern riverine fish species and is not directly representative of marine seismic programs due to differences in sound attenuation in rivers.

References

Cott PA, Hanna BW, Dahl JA (2003) Discussion on seismic exploration in the Northwest Territories 2000–2003. Canadian Manuscript Report for Fisheries and Aquatic Science 2648.

Jorgensen JK, Gyselman EC (2009) Hydroacoustic measurements of the behavioral response of arctic riverine fishes to seismic airguns. J Acoust Soc Am 126:1598–1609.

Mann DA, Cott PA, Hanna BW, Popper AN (2007) Hearing in eight species of northern Canadian freshwater fishes. J Fish Biol 69:1–12.

McCauly RD, Fewtrell J, Popper AN (2003) High intensity anthropogenic sound damages fish ears. J Acoust Soc Am 113:638–642.

Popper AN, Smith ME, Cott PA, Hanna BW, MacGillivray AO, Austin ME, Mann DA (2005) Effects of exposure to seismic airgun use on hearing of three fish species. J Acoust Soc Am 117:3958–3971.

Song J, Mann DA, Cott PA, Hanna BW, Popper AN (2008) Ear structure in seven species of northern Canadian freshwater fishes, including effects of seismic exposure on ear tissues in select species. J Acoust Soc Am 124: 1360–1366.

Effects of Underwater Noise on Larval settlement

Jenni A. Stanley, Craig A. Radford, and Andrew G. Jeffs

1 Introduction

Many benthic marine organisms possess a pelagic larval phase that typically results in dispersal away from the parental habitat and ends in the larva selecting a suitable benthic habitat in which to settle (O'Connor and Gregg 1998). Settlement and metamorphosis often involve a specific cue or combination of chemical and/or physical cues (Gebauer et al. 2003). The larvae of many marine organisms are known to be capable of extending their larval phase, often for considerable periods, until suitable settlement cues or habitats are detected. Some larvae will spontaneously metamorphose or even die without metamorphosing in the absence of specific settlement cues (Gebauer et al. 2003; Pechenik 1990). Brachyuran crabs seem to lack the ability to delay metamorphosis indefinitely because they appear to have a temporal threshold beyond which settlement and metamorphosis occur even in the absence of settlement cues (Weber and Epifanio 1996).

The focus of research into the effects of anthropogenic noise on the marine environment has mostly been on its direct impact on marine mammals and adult fish. However, anthropogenic noise may have significant impacts if it interferes with key biological processes involved in maintaining populations, such as settlement and recruitment. Numerous experimental studies now indicate that ambient underwater sound plays an important role in the orientation and settlement of the pelagic larvae of many ecologically and economically important coastal organisms such as fish and crabs (Jeffs et al. 2003; Leis and Lockett 2005; Montgomery et al. 2006; Radford et al. 2007; Simpson et al. 2005; Stobutzki and Bellwood 1998; Tolimieri et al. 2004). Therefore, the aim of our research has been to investigate the potential for underwater sound to trigger settlement behavior and/or shorten TTM in settlement stage larvae (megalopae) of common species of brachyuran crabs in both temperate and tropical waters.

2 Induction of Metamorphosis by Ambient Reef Sound

A number of species of megalopae have been shown to orientate and swim toward ambient underwater reef sound (Radford et al. 2007); however, the role of underwater sound as a cue for settlement and metamorphosis has not been examined. Experiments in both laboratory tanks and the field

J.A. Stanley (✉) • C.A. Radford • A.G. Jeffs
Leigh Marine Laboratory, University of Auckland, Warkworth 0941, New Zealand
e-mail: ja.stanley@auckland.ac.nz

A.N. Popper and A. Hawkins (eds.), *The Effects of Noise on Aquatic Life*,
Advances in Experimental Medicine and Biology 730, DOI 10.1007/978-1-4419-7311-5_84,

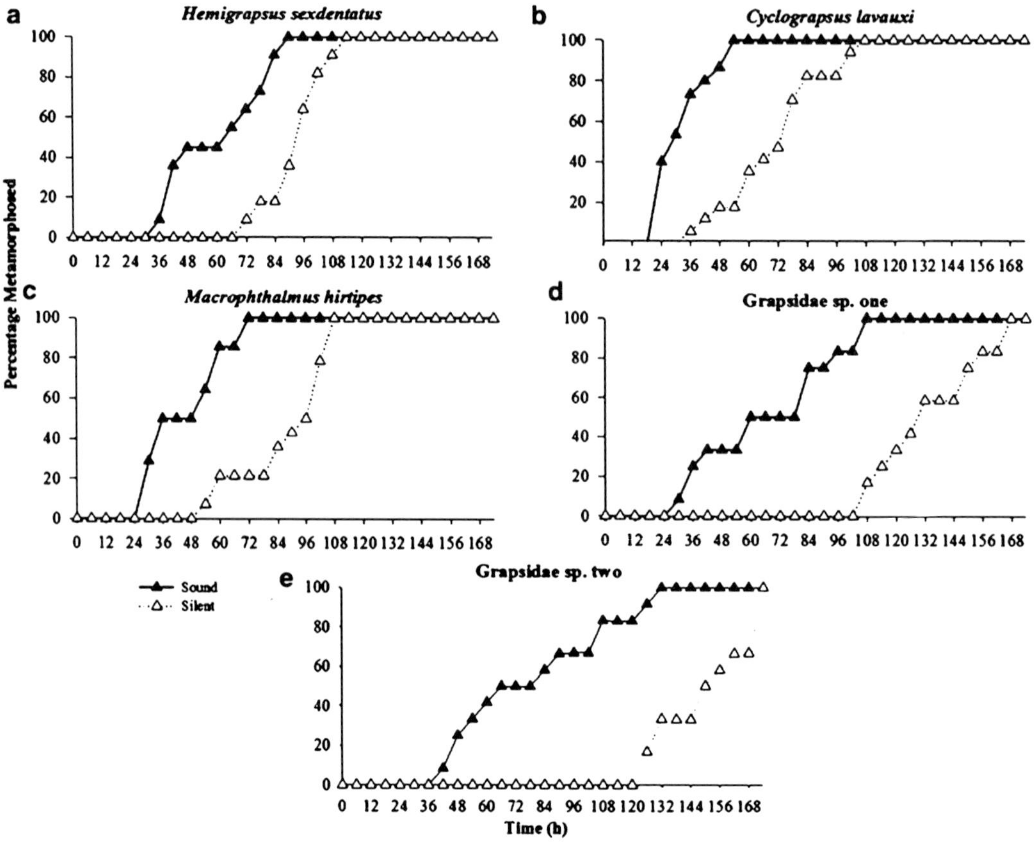

Fig. 1 Percentage of total number of megalopae metamorphosed against time. From Stanley et al. (2010)

have shown that the time to metamorphosis in the larvae of several species of temperate and tropical crabs is markedly reduced (34–60%) when exposed to ambient underwater reef sound compared with a silent (control) treatment (Fig. 1). These results appear to greatly extend the role that ambient underwater sound plays in triggering and advancing the behavioral and physiological changes in settling larval marine organisms (Stanley et al. 2010).

3 Anthropogenic Impacts

In addition to natural ambient biotic and abiotic sounds in the ocean, there is also human-generated noise from shipping traffic, oil and gas exploration, and coastal construction. Due to the demand of renewable energy sources, there has also been an increase in the construction of offshore wind farms in the marine environment. Wind farms are also being planned for the shallow waters of inshore areas into the medium-term future (Madsen et al. 2006; Nedwell and Howell 2004). There are many underwater noise sources related to wind farms; these include vessel movements, pile driving, drilling, dredging, rock laying, trenching, and the ongoing operational noise of the wind turbine (Nedwell and Howell 2004). Underwater noise in a wide range of frequencies and intensities are generated by these activities. Research on existing wind farms and associated activities have shown that the operation of wind farms have added to the ambient background noise at frequencies up to

2,000 Hz and peak pressures up to 200 dB re 1 μPa at 1 m (Nedwell and Howell 2004; Thomsen et al. 2006). Activities such as pile driving, seismic air guns, and boat engine noise have also been shown to significantly increase the auditory threshold (Scholik and Yan 2001) and damage the inner ear sensory hair cells of fish (McCauley et al. 2003; Popper et al. 2004). These activities also have the potential to damage the analogous structures in crustaceans, the statocysts, which to date have been the only organ with significant evidence for sound reception (Montgomery et al. 2006).

Given our findings on the significant influence of natural sound in the settlement and metamorphosis of larval crabs, it would seem likely that anthropogenic noise, especially of a continuous nature, has the potential to interfere with the critical settlement and recruitment processes of many coastal organisms using ambient underwater to locate and settle into suitable habitats. This anthropogenic noise is occurring at biologically relevant frequencies that have the potential to act as orientation and settlement cues. Also, some of these sources of anthropogenic sound have large peak pressure levels that may mask natural levels of ambient underwater reef sound used for natural orientation and settlement cues (Cato 1992; Radford et al. 2008).

4 Conclusions

Research has shown that ambient underwater emanating from coastal habitats in many parts of the world is important to the early recruitment processes of a wide range of coastal organisms. Furthermore, the settlement physiology and behavior of some larvae are mediated by ambient underwater sound. Consistent results from a wide range of species and locations suggest that underwater sound plays a significant role in the recruitment to crab populations. It also raises the possibility that anthropogenic noise could interfere with recruitment processes by disrupting these important settlement cues and leading to premature or reduced settlement.

References

Cato DH (1992) The biological contribution to the ambient noise in waters near Australia. Acoust Aust 20:76-80.

Gebauer P, Paschke K, Anger K (2003) Delayed metamorphosis in decapod crustaceans: Evidence and consequences. Rev Chil Hist Nat 76:169–175.

Jeffs A, Tolimieri N, Montgomery JC (2003) Crabs on cue for the coast: The use of underwater sound for orientation by pelagic crab stages. Mar Freshw Res 54:841–845.

Leis JM, Lockett MM (2005) Localization of reef sounds by settlement-stage larvae of coral-reef fishes (Pomacentridae). Bull Mar Sci 76:715–724.

Madsen PT, Wahlberg M, Tougaard J, Lucke K, Tyack P (2006) Wind turbine underwater noise and marine mammals: Implications of current knowledge and data needs. Mar Ecol Prog Ser 309:279–295.

McCauley RD, Fewtrell J, Popper AN (2003) High intensity anthropogenic sound damages fish ears. J Acoust Soc Am 113:638–642.

Montgomery JC, Jeffs AG, Simpson SD, Meekan M, Tindle C (2006) Sound as an orientation cue for the pelagic larvae of reef fishes and decapod crustaceans. Adv Mar Biol 51:143–196.

Nedwell J, Howell D (2004) A review of offshore windfarm related underwater noise sources. Report No. 544 R 0308 prepared by Subacoustech Ltd., Southampton, UK, for the Collaborative Offshore Wind Energy Research into the Environment (COWRIE), London, UK.

O'Connor NJ, Gregg AS (1998) Influence of potential habitat cues on duration of the megalopal stage of the fiddler crab *Uca pugnax*. J Crustac Biol 18:700–709.

Pechenik JA (1990) Delayed metamorphosis by larvae of benthic marine invertebrates: Does it occur: Is there a price to pay? Ophelia 32:63–94.

Popper AN, Fewtrell J, Smith ME, McCauley RD (2004) Anthropogenic sound: Effects on the behavior and physiology of fishes. Mar Technol Soc J 37:35–40.

Radford CA, Jeffs AG, Montgomery JC (2007) Directional swimming behavior by five species of crab postlarvae in response to reef sound. Bull Mar Sci 80:369–378.
Radford CA, Jeffs AG, Tindle CT, Montgomery JC (2008). Temporal patterns in ambient noise of biological origin from a shallow water temperate reef. Oecologia 156:921–929.
Scholik AR, Yan HY (2001) The effects of underwater noise on auditory sensitivity of fish. Proc IOA 32:27–36.
Simpson SD, Meekan M, Montgomery J, McCauley R, Jeffs A (2005). Homeward sound. Science 308:221.
Stanley JA, Radford CA, Jeffs AG (2010) Induction of settlement in crab megalopae by ambient underwater reef sound. Behav Ecol 21:113–120.
Stobutzki IC, Bellwood DR (1998) Nocturnal orientation to reefs by late pelagic stage coral reef fishes. Coral Reefs 17:103–110.
Thomsen F, Lüdemann K, Kafemann R, Piper W (2006) Effects of offshore wind farm noise on marine mammals and fish. Prepared by Biola, Hamburg, Germany, for the Collaborative Offshore Wind Energy Research into the Environment (COWRIE), London, UK.
Tolimieri N, Haine O, Jeffs A, McCauley R, Montgomery J (2004) Directional orientation of pomacentrid larvae to ambient reef sound. Coral Reefs 23:184–191.
Weber JC, Epifanio CE (1996) Response of mud crab (*Panopeus herbstii*) megalopae to cues from adult habitat. Mar Biol 126:655–661.

Does Vessel Noise Change the Calling Rate and Intensity of Soniferous Fishes?

Joseph J. Luczkovich, Cecilia S. Krahforst, and Mark W. Sprague

1 Introduction

Fishes of the family Sciaenidae (drums and croakers) are well-known for their abilities to produce sounds using both sonic muscles and the swim bladder (Luczkovich et al. 2008a,b; Rountree et al. 2006; Sprague and Luczkovich 2004). Calls of sciaenid fishes like *Micropogonias undulatus* (Atlantic croaker) can be heard with hydrophones throughout the day, producing sounds when disturbed, during aggression, and during spawning (male advertisement calls). In this study, we examined if the noise associated with coastal vessels (ferry boats and tugboats) that operated daily during the early morning through early evening had any effect on the seasonal and daily calling rate of *Micropogonias undulatus*.

2 Methods

Fishes were recorded in situ using passive recorders (long-term acoustic recording system [LARS], Loggerhead Instruments, Inc., Sarasota, FL). The recordings were time-stamped 10-s wave files (<10 kHz) recorded to a compact flash disk at 15-min intervals from March through December 2008 at a site (Potash Corporation of Saskatchewan site; 35°23.207' N latitude and 76°44.673' W) in the Pamlico River near Aurora, NC. The fishes were exposed on a regular basis to vessel noises from a North Carolina State Department of Transportation ferryboat making sixteen 0.5-h trips each day, beginning at 0530 and ending at 2015 EDT. In addition, large tugboats pushing barges from the phosphate mine pass by the site intermittently. Passive recordings were analyzed using MATLAB

J.J. Luczkovich (✉)
Department of Biology, East Carolina University, Greenville, NC 27858, USA

Institute for Coastal Science and Policy, East Carolina University, Greenville, NC 27858, USA
e-mail: luczkovichj@ecu.edu

C.S. Krahforst
Department of Biology, East Carolina University, Greenville, NC 27858, USA

Coastal Resources Management Doctoral Program, East Carolina University, Greenville, NC 27858, USA

M.W. Sprague
Department of Physics, East Carolina University, Greenville, NC 27858, USA

A.N. Popper and A. Hawkins (eds.), *The Effects of Noise on Aquatic Life*,
Advances in Experimental Medicine and Biology 730, DOI 10.1007/978-1-4419-7311-5_85,

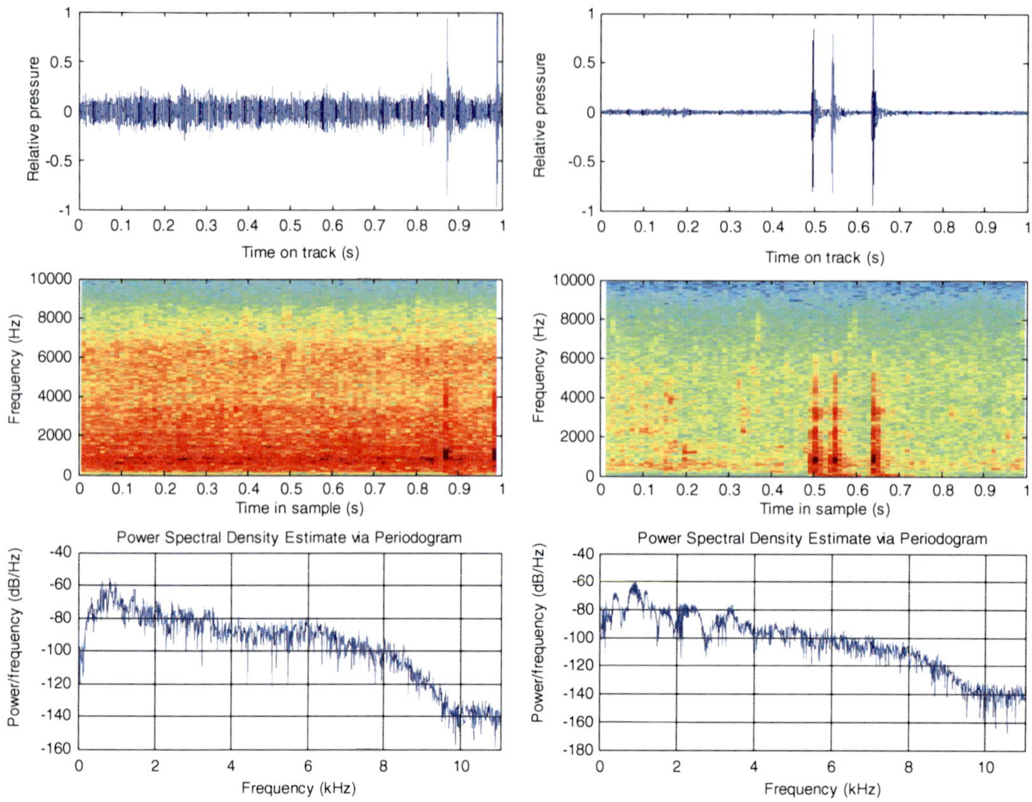

Fig. 1 Oscillograms (top panels), spectrographs (middle panels), and average power spectra (bottom panels) of a ferry with a single *Micropogonias undulatus* (Atlantic croaker) call (left panels) and three *M. undulatus* calls with no ferry sounds (right panels)

spectral analysis to estimate ferry and vessel noise (1,000-8,000 Hz) and fish sound (200-1,000 Hz) frequencies. We created automated sound event detectors in XBAT (Extensible Bioacoustics Tools, see http://www.xbat.org; Fig. 1), which was used to count fish calls and vessel noises and estimate minimum, maximum, and mean frequencies of vessel noises. Variation in salinity (10–23 practical salinity units [psu]), temperature (5-30°C), dissolved oxygen (1-13 mg/l), and turbidity (1-3,000 nephlometric turbidity units [NTU]) were measured daily at both sites with automated water quality meters (see Krahforst et al., Chpater 38). We compared the number of fish calls detected in 60 s (six 10-s recordings) each hour that ferries or other vessels (phosphate mine tugboats and barges) passed by the recording station to the number of calls detected in each hour without ferries present. We analyzed these data using ANOVA (SYSTAT, $\log_{10}$-transformed number of *M. undulatus* detections) with factors month (March, April, July, August, September, October, November, and December) and vessel noise (ferry or no ferry operating during recording). Finally, a plot of the vessel noise midfrequencies (halfway between the minimum and maximum frequencies reported by the XBAT detector), which shifted temporally due to varying vessel traffic, was compared with a plot of a calling index for *M. undulatus* during the month of October. The calling index was computed by dividing the calling rate for each observation by the maximum calling rate recorded times 100.

3 Results

The vessel sounds were broadband, ranging from 200 to 8,000 Hz (Fig. 1, left panels), and variable, especially at the lower frequencies that often overlapped the *M. undulatus* Atlantic croaker calls (300-1,000 Hz; Fig. 1, right panels). We recorded 5,926 *M. undulatus* calls at this station, made during 906 hourly measurements from March through December 2008. Sounds were made by *M. undulatus* both day and night regardless of the presence of the ferry (Fig. 2). However, fish sounds were less common (the calling index declined) when large vessels with low midfrequencies passed by the recorder on 13-15 October 2008 (Fig. 2, left). These vessel noise frequencies were close to the *M. undulatus* calling frequency range, perhaps causing the fish calling index to decline (Fig. 2, right). The rate of fish sound production varied significantly with month (ANOVA, $F_{7,892}$ = 22.4359, $P < 0.0001$), with the greatest number of *M. undulatus* calls occurring in a 3-mo period in the fall (August, September, and October; Fig. 3). There was no overall significant difference in the number of calls detected when a ferry was either present or absent in the recording area (ANOVA, $F_{1,892} = 0.0478$, $P = 0.8269$).

4 Discussion

The number of detections with and without ferry noise suggested that vessel noise has a limited effect on *M. undulatus* sound production. Instead, the variation in seasonal changes of acoustic sound production may be related to seasonal behavior and migration patterns, water quality parameters, and photoperiod. Spawning time for *M. undulatus* occurs from August through December, when the sound production, which is typically associated with spawning, increases. Sound production by *M. undulatus* persists despite the significant amount ferry noise present at the site. When large vessels pass by, the calling rate appears to fall. It is likely that these large vessels may prevent sound

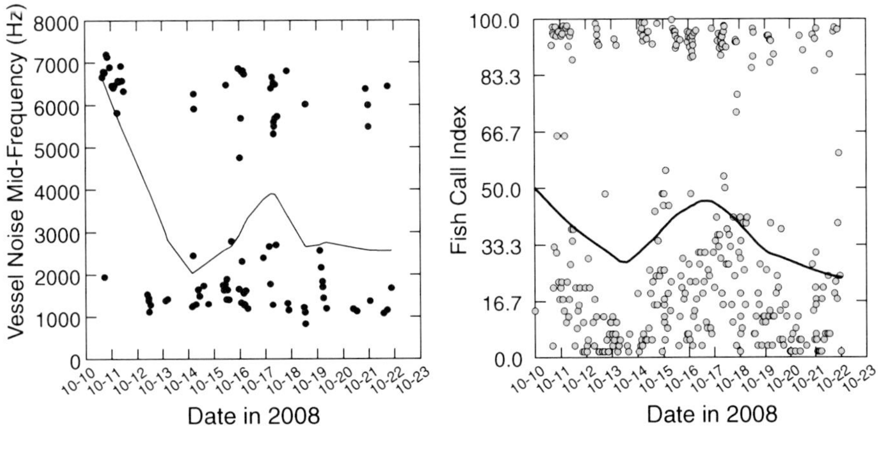

Fig. 2 Vessel and fish sounds detected by the passive acoustic recorder during two weeks in October 2008 at the Potash Corporation of Saskatchewan (PCS) site in the Pamlico River Estuary in North Carolina. Left: Temporal change in midfrequencies (minimum + maximum frequencies/2) associated with each vessel noise detected (points). Solid line is locally weighted (LOWESS) fit to the data. Right: Calling index (number of calls/60 s, scaled as a percentage of the maximal number of calls) of *Micropogonias undulatus*

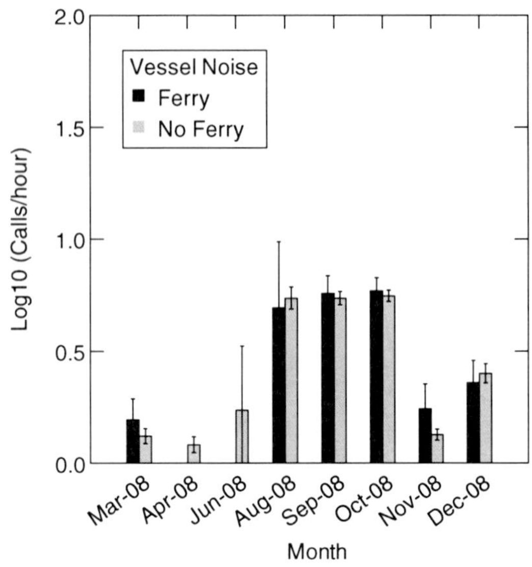

Fig. 3 Calling rate (number of calls/hour, $\log_{10}$ transformed) variation of *Micropogonias undulatus* at the PCS site near a ferry route in the Pamlico River Estuary in North Carolina by month with and without the presence of ferry noise

communication in this species at this location due to masking of the calls as well as a change in calling rate. Some vessel sounds (the phosphate tugboat and barge) overlapped the frequency of fish calls; when this occurred, calling rate declined. These low-frequency sounds would also prevent another fish nearby from hearing the sounds. The impact of noise on the ability of these fishes to attract a mate and successfully spawn needs to be investigated.

References

Luczkovich JJ, Mann DA, Rountree RA (2008a) Passive acoustics as a tool in fisheries science. Trans Am Fish Soc 137:533–541.

Luczkovich JJ, Pullinger RC, Johnson SE, Sprague MW (2008b) Identifying sciaenid critical spawning habitats by the use of passive acoustics. Trans Am Fish Soc 137:576–605.

Rountree RA, Gilmore RG, Goudrey CA, Hawkins AD, Luczkovich JJ, Mann DA (2006) Listening to fish: Applications of passive acoustics to fisheries science. Fisheries 31:433–446.

Sprague MW, Luczkovich JJ (2004) Measurement of an individual silver perch *Bairdiella chrysoura* sound pressure level in a field recording. J Acous Soc Am 116:3186–3191.

Behavioral Responses of Captive Herring to Sonar Signals (1-1.6 kHz) of a Naval Frigate Throughout a Yearly Cycle

Lise Doksæter, Petter H. Kvadsheim, Nils O. Handegard, Nina Nordlund, and Olav R. Godø

1 Introduction

Modern long-range antisubmarine warfare active sonar systems are typically operating in the 1- to 10-kHz frequency band (OSPAR 2009) and within the hearing range of most clupeid fish such as the ecologically and commercially important Atlantic herring (*Clupea harengus*) (Enger 1967). Herring behavior in response to such sonar systems have been studied in the wild during the winter (Doksæter et al. 2009) without revealing any significant behavioral responses. Detecting small-scale changes in the wild may be difficult, and herring behavior is also changing over the seasons due to differences in physiological and motivational states (Fernø et al. 1998). Therefore, we studied fish behavior during sonar exposure in captivity, allowing close-up monitoring of detailed changes in behavioral dynamics over extended time periods.

2 Methods

Herring were held in net pens at an aquaculture facility for a year. Four experiments were conducted in different seasons when fish were transferred to a movable, deep net pen located in a sheltered fjord. Under controlled conditions, herring were exposed to naval sonar signals from a passing frigate. Two types of signals were used: a 1- to 1.6-kHz hyperbolic frequency-modulated (FM) upsweep and a 1-kHz weighted continuous-wave (CW) signal. The frigate followed a predetermined transect, passing the pen at a closest point of approach (CPA) of 500 m. The FM signals were introduced to the herring in two ways: "gradual" and "sudden." A gradual introduction involved transmission of the signals starting 1 nautical mile (nmi) away from the net pen, thus gradually increasing the received level for the fish. During a sudden introduction, the first signal was transmitted at the CPA, thus giving maximal levels. Experiments were conducted as a block design, with each block consisting of four runs across the transect, including the two types of FM transmission (gradual and sudden), CW transmission, and one control run without any transmission in randomized order.

L. Doksæter (✉) • N.O. Handegard • O.R. Godø
Institute of Marine Research (IMR), NO-5817 Bergen, Norway
e-mail: lise.doksaeter@imr.no

P.H. Kvadsheim • N. Nordlund
Norwegian Defence Research Establishment (FFI), NO-3191 Horten, Norway

A.N. Popper and A. Hawkins (eds.), *The Effects of Noise on Aquatic Life*, Advances in Experimental Medicine and Biology 730, DOI 10.1007/978-1-4419-7311-5_86,

Additionally, positive control experiments were conducted to verify that herring were able to perform an avoidance response within the net pen as well as for testing the ability of the experimental setup to detect and document such a reaction. During these experiments, herring were exposed to engine noise from a two-stroke outboard engine and an impulse sound, both known to induce avoidance responses in herring. Herring behavior was monitored by an upward-looking echo sounder mounted on the bottom of the pen that measured the vertical structure of the fish as well as by video monitoring to document fine-scale behavioral changes. A hydrophone was placed in the middle of the pen to measure the received sound pressure levels within the pen. The echo sounder recordings were used to determine the median depth structure of the herring as a function of time. The difference in depth before and during exposure was compared, revealing potential vertical escape reactions. The behavior of the herring in the video recordings for all runs was classified according to standard criteria with respect to group behavior, vertical behavior, number of fish reacting, and overall response.

3 Results

Some minor startle responses by a few herring were seen on the video recordings in 3 of 14 FM "sudden" exposures. None of the other sonar transmission types caused any form of reaction. Both the engine noise and the impulse sound generated a typical avoidance response involving strong schooling, an increase in school density, and a rapid downward movement by all fish.

No significant vertical movement was recorded by the echo sounder for any of the sonar transmission types (CW, FM gradual/sudden) compared with the controls. The positive control experiments, however, showed a significant distribution change toward deeper water.

The highest measured received sound pressure level was 168 dB_{RMS} (root mean square) re 1 µPa during a FM sudden run.

4 Discussion

This study has documented that herring in a net pen do not react with any significant avoidance responses when exposed to naval sonar signals in the frequency range of 1–1.6 kHz and received sound pressure levels up to 168 dB_{rms} re 1 µPa. The herring did, however, produce a strong vertical avoidance response to the sound of a two-stroke outboard engine and an impulse sound at sound pressure levels much lower than the sonar. The effects of received levels higher than those tested cannot be excluded. However, the total volume covered by such levels will be relatively small due to spherical spreading that rapidly reduces the signal level (e.g., the 500-m distance from frigate to net pen gave a >50-dB transmission loss). For herring to be affected at the population level, the sonar transmission must be conducted at very high source levels in an area of extremely dense concentrations of fish in combination with very low stock levels (Kvadsheim and Sevaldsen 2005). The present results are in agreement with other military sonar effect studies on adult fish conducted on the rainbow trout (Popper et al. 2007) and free-living herring (Doksæter et al. 2009). Jørgensen et al. (2005) documented mortality in juvenile herring at exposure levels >180 dB_{rms} re 1 µPa. This is probably related to the combination of the juvenile being exposed to resonance-frequency, high-source levels as well as juveniles generally being more sensitive than adult fish.

The lack of reactions to the military sonar was consistent throughout all seasons, but the reaction to the engine noise tended to be stronger in winter than in summer/autumn. This reflects the situation in the wild because the herring are generally more responsive during the winter between

November and February and least sensitive to stimuli such as noise or predators in spring and summer after spawning has ended (Fernø et al. 1998; Kvamme et al. 2003; Nøttestad et al. 1996).

References

Doksæter L, Kvadsheim PH, Godø OR, Handegard NO, Donovan C, Lam FP, Miller PJO (2009) Behavioral response of herring (*Clupea harengus*) to 1-2 and 6-7 kHz sonar signals and killer whale calls. J Acoust Soc Am 125:554–564.

Enger PS (1967) Hearing in herring. Comp Biochem Physiol 22:527–538.

Fernø A, Pitcher TJ, Melle W, Nøttestad L, Mackingson S, Hollingworth C, Misund OA (1998) The challenge of the herring: Making optimal collective spatial decisions. Sarsia 83:149–167.

Jørgensen R, Olsen KK, Falk-Pettersen IB, Kanapthippilai P (2005) Investigation of potential effects of low frequency sonar signals on survival, development and behavior of fish larvae and juveniles. Report from the Norwegian College of Fishery Science, University of Tromsø, Tromsø, Norway. Available via http://triton.nfh.uit.no/dok/lydrapport_m_vedlegg_28_04-05.pdf. Accessed 30 April 2010.

Kvadsheim PH, Sevaldsen E (2005) The potential impact of 1-8 kHz active sonar on stocks of juvenile fish during sonar exercises. Norwegian Defence Research Establishment (FFI) Report 2005/01027. Available via http://rapporter.ffi.no/rapporter/2005/01027.pdf. Accessed 30 April 2010.

Kvamme C, Nøttestad L, Fernø A, Misund OA, Dommasnes A, Axelsen BE, Dalpadado P, Melle W (2003) Migration patterns in Norwegian spring spawning herring: Why young herring swim away from the wintering area in late summer. Mar Ecol Prog Ser 247:197–210.

Nøttestad L, Aksland M, Beltestad A, Fernö A, Johannessen A, Misund OA (1995) Schooling dynamics of Norwegian spring spawning herring (*Clupea harengus L.*) in a coastal spawning area. Sarsia 80:277–284.

OSPAR (2009) Overview of the impact of anthropogenic underwater sound in the marine environment. Report 441 to the OSPAR Commission, London, UK.

Popper AN, Halvorsen MB, Kane A, Miller DL, Smith ME, Song J, Stein P, Wyscocki LE (2007) The effects of high-intensity, low-frequency active sonar on rainbow trout. J Acoust Soc Am 122:623–635.

Acoustic Tagging: A Suitable Method for the Study of Natural Herring Behavior Around Spawning?

Lise Langård, Arne Johannessen, Anders Fernö, Leif Nøttestad, Georg Skaret, Aril Slotte, Jostein Røttingen, and Jan T. Øvredal

1 Introduction

Herring schools can be studied by standard acoustic methods, but to understand the relationship between the individual and school levels as well as the interactions between subpopulations of herring, it is essential to learn more about individual herring behavior. Acoustic tagging has successfully been used to track individual fish of various species, but herring are highly sensitive to being handled and are therefore not easy to tag. In this study, we tagged herring with acoustic transmitters to track prespawning and spawning individuals. The conditions for a tagging experiment were ideal because the herring remained in a single school (>100,000 fish) for more than 1 mo within an area of ~200 m^2 in the sheltered semienclosed Lindåspollene basins (Johannessen et al. 2009). This is a small (7-km^2) well-defined ecosystem in western Norway, which comprises 3 distinct 60- to 90-m-deep basins containing numerous islets along with several sheltered sites. Here we evaluate whether acoustic tagging is a suitable method for the study of the natural behavior of herring around spawning.

2 Tagging Process

Herring were caught individually by jigging from the *R/V Hans Brattstrøm* in February in two seasons. This demanded a high effort lasting for ~4 days because herring seldom feed before spawning (Nøttestad et al. 1996). The fish were kept in lidded holding tanks (500 l) onboard the vessel for ~3 h before being tagged. Fish with signs of injuries or abnormal swimming were sorted out. To minimize handling time, three to four persons took part in the tagging process: 1) each fish was

L. Langård (✉)
Department of Biology, University of Bergen, NO-5020 Bergen, Norway

Institute of Marine Research, 5817 Bergen, Norway
e-mail: lise.langaard@imr.no

A. Johannessen • A. Fernö
Department of Biology, University of Bergen, NO-5020 Bergen, Norway

L. Nøttestad • G. Skaret • A. Slotte • J.Røttingen • J.T. Øvredal
Institute of Marine Research, 5817 Bergen, Norway

A.N. Popper and A. Hawkins (eds.), *The Effects of Noise on Aquatic Life*, Advances in Experimental Medicine and Biology 730, DOI 10.1007/978-1-4419-7311-5_87,

captured by a portable net fitted with soft plastic material to avoid damage; 2) it was then held carefully to avoid scale loss (eyes were covered); and 3) the tag was inserted surgically in the abdomen just behind the pelvic fins. The wounds were closed with histoacryl (0.1 ml). The whole procedure took a maximum of 1 min. The fish were left to recover for 3-4 h in a lidded tank (500 l) with circulating seawater and were monitored continuously. Total length of all tagged fish was measured and lost scales were sampled for age determination; for fish that were sorted, the stage of maturity was also determined (see Mjanger et al. 2007). If neither signs of injury nor abnormal swimming was observed, the fish were released at the site where they had been caught. A total of 45 herring (length 28-35 cm, age 4-12 yr) were tagged in the course of the 2 experiments.

2.1 Acoustic Tags and Receivers

Two types of acoustic transmitters were used during the two seasons: 1) 17 V9 coded transmitters (V9-1H: length 24 mm, weight in water 2.2 g; V9-2H: length 29 mm, weight in water 2.9 g; Vemco Ltd., Halifax, NS, Canada; http://www.vemco.com/products/receivers/vrap.php), and 2) 28 Thelma Biotel (Trondheim, Norway) acoustic transmitters, of which 19 were LP-7.3 mm (length 18 mm, weight in water 1.2 g) and 9 were depth transmitters (ADT-9-SHORT, length 34 mm, weight in water 3.3 g).

On the basis of the results from pilot studies, VR2 receiver units (Vemco Ltd.; see Skilbrei et al. 2009) were moored at 6 sites in Lindåspollene, covering a total range of 1,500 m from the innermost to the outermost location. The receivers were moored to a weight on the seabed and buoyed to a depth of ~10 m. The depth at the receiver sites ranged from 10 to 70 m. The circular detection area of the VR2 had a maximum practical range of 200-400 m.

For more detailed tracking of behavior, a Vemco radio acoustic position (VRAP) real-time system was used (Løkkeborg et al. 2002), with the 3 buoys separated by ~150-250 m. The system is estimated to detect signals from acoustic transmitters at a range of 200-800 m and enables individual tracks to be updated every 5-180 s, with an accuracy of up to 1-2 m (http://www.vemco.com) depending on the number of fish tracked simultaneously, type of transmitter, hydrographical conditions, and background noise. The buoys were deployed in a herring spawning area, which was identified by observations of diving eider ducks, herring roe in the stomachs of sampled cod and haddock, and web cameras.

3 Evaluation of the Method

During first season, 17 of ~50 herring caught were tagged and released, including 2 pressure tags. The proportion of tagged fish rose in 2010, with 28 fish tagged and released out of 36 fish caught, including 9 pressure tags. The highest discard rate took place after capture. Of the total number of discarded herring (33 in experiment 1 and 8 in experiment 2), 30 and 6 fish, respectively, were discarded before tagging and only a few fish were discarded after tagging (3 and 2, respectively).

The acoustic tags enabled prespawning herring to be reliably tracked to a maximum distance of 400 m from the buoys for up to 90 days. In the first season, we evaluated whether the recordings were in accordance with the expected natural movement pattern of herring. Several types of activity patterns were identified. Five tags that repeatedly transmitted signals from a fixed position and depth were interpreted as representing dead fish. Three tags transmitted signals for ~1 mo, after which contact was lost, and these were interpreted as representing either dead fish or signals lost in other ways. The rest of the tagged fish appeared to behave normally. Nine fish remained within the relatively stationary main prespawning school for 1-3 mo after being tagged, with shorter or longer excursions to nearby localities, including the spawning site at depths of 0-20 m. This suggests that the fish had spawned, with the high level of activity in the upper layers displayed by some fish late in the period indicating postspawning feeding (Nøttestad et al. 1996). The tagged herring also performed vertical migrations (VRAP and VR2 data) covering the whole recorded water depth.

4 Conclusions

To the best of our knowledge, this is the first time that herring have been successfully tagged with internal acoustic transmitters of these sizes. The condition of the fish should always be evaluated before and after tagging, and scale loss must be minimized. Most discards were after capture and a few fish were discarded after tagging. The prespawning period in February, with large and developing gonads, is believed to be a vulnerable period for the tagging of herring, but the low temperatures (1-3°C) may have been favorable for tagging (see Krieger 1982). Although some individuals died or their signals were lost for other reasons, behavior that was interpreted as representing natural herring behavior was monitored in more than half of the fish for several weeks after tagging. These data will have to be analyzed in greater detail to evaluate whether the behavior was affected in any way by the tagging process. Acoustic receivers ought to be located at sites where there is a low risk of interfering sources of sound, and the distances between receivers need to be finely tuned and optimized to the distance of detection to avoid double detections. This new miniaturized acoustic tagging methodology may help us better understand the relationships between the behavior of individual herring and of schools.

Acknowledgments We thank Ragnar Bergfjord and the crew of *R/V Hans Brattstrøm* for assistance with the field work. The work was funded by the University of Bergen, the Institute of Marine Research, and the Meltzer Foundation, Bergen, Norway.

References

Johannessen A, Nøttestad L, Fernö A, Langård L, Skaret G (2009) Two components of north-east Atlantic herring within the same school during spawning: Support for the existence of a metapopulation? ICES J Mar Sci 66:1740–1748.

Krieger KJ (1982) Tagging herring with coded-wire microtags. Mar Fish Rev 44:18–21.

Løkkeborg S, Fernö A, Jørgensen T (2002) Effect of position-fixing interval on estimated swimming speed and movement pattern of fish tracked with a stationary positioning system. Hydrobiology 483:259–264.

Mjanger H, Hestenes K, Olsen E, Svendsen BV, Wenneck TDL, Aanes S (2007) Håndbook for prøvetaking av fisk og krepsdyr, versjon 3.16. Institute of Marine Research, Bergen, Norway.

Nøttestad L, Aksland M, Beltestad A, Fernö A, Johannessen A, Misund OA (1996) Schooling dynamics of Norwegian spring spawning herring (*Clupea harengus* L.) in a coastal spawning area. Sarsia 80:277–284.

Skilbrei OT, Holst JC, Asplin L, Holm M (2009) Vertical movements of "escaped" farmed Atlantic salmon (*Salmo salar* L.)—a simulation study in a western Norwegian fjord. ICES J Mar Sci 66:278–288.

Effects of Pile Driving on the Behavior of Cod and Sole

Frank Thomsen, Christina Mueller-Blenkle, Andrew Gill, Julian Metcalfe, Peter K. McGregor, Victoria Bendall, Mathias H. Andersson, Peter Sigray, and Daniel Wood

1 Introduction

Studies on the effects of offshore wind farm construction on marine life have focused on behavioral reactions in porpoises and seals (Thomsen et al. 2006). The effects on fish have only very recently come into the focus of scientists, regulators, and stakeholders (Popper and Hastings 2009). Pile-driving noise during construction is of particular concern because the very high sound pressure levels (see Thomsen et al. 2006) could potentially prevent fish from reaching breeding or spawning sites, finding food, and acoustically locating mates that could result in long-term effects on reproduction and population parameters. There is also the possibility that avoidance reactions might displace fish away from potential fishing grounds that could lead to reduced catches (see, e.g., Engås et al. 1996). However, the nature and extent of behavioral reactions of marine fish due to pile driving have not been studied in controlled experiments. Therefore, the impacts of pile driving on marine fish remain unknown.

2 Methods

We performed playbacks of pile-driving noise to cod and sole held in two large (40-m) net pens located in a quiet bay in West Scotland. Movements of the fish were recorded and analyzed using a novel acoustic tracking system. Received sound pressure level and particle motion were measured during the experiments (see Mueller-Blenkle et al., this volume).

F. Thomsen (✉) • C. Mueller-Blenkle • J. Metcalfe • V. Bendall • D. Wood
Cefas, Lowestoft, Suffolk NR33 0HT, UK
e-mail: frank.thomsen@cefas.co.uk

A. Gill
Natural Resources Department, Cranfield University, Cranfield, Bedfordshire MK43 0AL, UK

P.K. McGregor
Cornwall College Newquay, Newquay, Cornwall TR7 2LZ, UK

M.H. Andersson
Department of Zoology, Stockholm University, S-106 91 Stockholm, Sweden

P. Sigray
Department of Meteorology, Stockholm University, S-106 91 Stockholm, Sweden

A.N. Popper and A. Hawkins (eds.), *The Effects of Noise on Aquatic Life*,
Advances in Experimental Medicine and Biology 730, DOI 10.1007/978-1-4419-7311-5_88,

3 Results

Both species showed a significant movement response to the pile-driving stimulus at relatively low received sound levels (sole: 144–156 dB re 1 μPa peak; cod: 140–161 dB re 1 μPa peak; particle motion between 6.51×10^{-3} and 8.62×10^{-4} m/s^2 peak). The observed sole significantly increased their swimming speed during the playback period compared with before and after playback. Cod exhibited a similar reaction, yet the results were not significant. Cod showed a freezing response that was significant at the onset and cessation of playback. There were indications that both species moved away from the speaker at the start of the playbacks. There was a high variability in the behavioral reactions across individuals and a decrease in response with multiple exposures.

4 Discussion and Conclusions

This study is the first to document the behavioral response of marine fish due to playbacks of pile-driving sounds. The results indicate that a range of received sound pressure and particle motion levels will trigger behavioral responses in sole and cod. Our study further implies a relatively large zone of behavioral response to pile-driving sounds in marine fish. Yet the exact nature and extent of the behavioral response needs to be investigated further. Some of our results point toward habituation to the sound.

The implications for regulatory advice and the implementation of mitigation measures in the construction of offshore wind farms are threefold. First, the concerns raised about the effects of pile-driving noise on fish were well founded (Thomsen et al. 2006). Second, the costs imposed by some mitigation measures that have so far been applied following the precautionary principle go part of the way in addressing a real problem. Third, we suggest that our behavioral thresholds should be considered in assessments of the impacts of offshore wind farms. Mitigation measures should be further developed and, if meaningful, applied, especially if they could lead to a reduction in acoustic energy that is emitted into the water column.

References

Engås A, Løkkeborg S, Ona E, Soldal AV (1996) Effects of seismic shooting on local abundance and catch rates of cod (*Gadus morhua*) and haddock (*Melanogrammus aeglefinus*). Can J Fish Aquat Sci 53:2238–2249.

Popper AN, Hastings MC (2009) The effects of anthropogenic sources of sound on fish. J Fish Biol 75:455–489.

Thomsen F, Lüdemann K, Kafemann R, Piper W (2006) Effects of offshore wind farm noise on marine mammals and fish. Prepared by Biola, Hamburg, Germany, for the Collaborative Offshore Wind Energy Research into the Environment (COWRIE), London, UK.

A Novel Field Study Setup to Investigate the Behavior of Fish Related to Sound

Christina Mueller-Blenkle, Andrew B. Gill, Peter K. McGregor, Mathais H. Andersson, Peter Sigray, Victoria Bendall, Julian Metcalfe, and Frank Thomsen

1 Introduction

There is an urgent need to obtain information on the effects of underwater sound on marine fish due to imminent policy drivers, e.g., the European Union Marine Strategy Framework Directive, on one hand and the increasingly noisy activities in the marine environment on the other. Yet studying the influence of sound, particularly on the behavior of fish, is a challenging task. Studies in tanks can suffer problems with the reflection of sound, especially at the low frequencies that are most important for fish. Studies in the field are often limited because the observation of fish is very complicated.

This paper presents a novel setup for studying fish behavior related to sound under almost natural but yet controlled conditions. It also includes measurements of particle motion that are of particular importance for hearing in fish but are rarely measured in other studies.

2 Experimental Setup

The setup was composed of two large net pens (mesocosms), a sound production and monitoring system including a particle motion sensor, and an acoustic tracking system for fish installed on the seabed in a bay. Pile-driving sound was played back from two locations on either side of the two mesocosms, producing different sound fields and thus different environmental conditions for the experimental fish.

C. Mueller-Blenkle (✉) • V. Bendall • J. Metcalfe • F. Thomsen
Cefas, Lowestoft, Suffolk, NR33 0HT, UK
e-mail: christina.mueller@cefas.co.uk

A.B. Gill
Natural Resources Department, Cranfield University, Cranfield, Bedfordshire MK43 0AL, UK

P.K. McGregor
Cornwall College Newquay, Newquay, Cornwall TR7 2LZ, UK

M.H. Andersson
Department of Zoology, Stockholm University, S-106 91 Stockholm, Sweden

P. Sigray
Department of Underwater Research, Swedish Defence Research Institute, S-164 90 Stockholm, Sweden

A.N. Popper and A. Hawkins (eds.), *The Effects of Noise on Aquatic Life*,
Advances in Experimental Medicine and Biology 730, DOI 10.1007/978-1-4419-7311-5_89,

2.1 Mesocosms

Two mesocosms 40 m in diameter and 5 m in height were installed in Loch Ceann Traigh, a shallow, quiet bay in West Scotland. The mesocosms were located 15 m apart from each other in a water depth between 10 and 15 m with a 4.2-m tidal range. The mesocosms were sitting on and were open toward the seabed. Zippers allowed divers easy access to the mesocosms to get the experimental fish in and out of the cages.

2.2 Sound Playback and Recording System

In the experiments, fish were exposed to recordings of pile-driving sound (provided by the Institute for Applied and Technical Physics [ITAP], Oldenburg, Germany). A J11 loudspeaker, manufactured in the United States, was used because this device can adequately produce frequencies below 1 kHz (range 32–10,000 Hz) and thus was most suitable for the playback of pile-driving sound. The loudspeaker was connected to two amplifiers and a transformer to produce high sound levels and to a laptop that held the sound files. Source sound pressure levels of up to 170 dB re 1 μPa_{peak} could be produced using this system. The sound playback system was powered by a leisure battery and stored in a waterproof container.

The recording system for the playback trials was composed of four Reson TC4013 hydrophones and TC4013–12/VP1000 amplifiers connected to a Dell Inspiron Mini 10 laptop running “Raven” sound-recording software. A large waterproof case located on a floating platform contained the preamplifiers and amplifiers for the hydrophones, a leisure battery as the power supply, and the laptop to record the data.

The hydrophones were attached to the net located on either side of each mesocosm and measured the sound pressure levels at distances of 5, 45, 60, and 100 m from the sound source, providing data on transmission loss during playback.

2.3 Particle Motion Sensor System

A novel instrument measuring particle acceleration that was developed by the Department of Meteorology at Stockholm University and the Swedish Defence Research Agency (Sigray et al. 2009) was used. The system was designed to measure particle acceleration in the frequency range of 0.1–360 Hz, with a sampling frequency of 800 Hz employed in the experiments. An underwater unit containing amplifiers, filters, and line drivers was placed on the seafloor, with the accelerometers suspended 0.9 m above the substrate at a distance of 5–10 m from the sound source. The dry unit consisted of power amplifiers, receiver, analog-to-digital converter, and a recording device (laptop). The whole system was powered by a 12-V marine battery. For detailed information, see Sigray and Andersson, Chapter 111.

2.4 Acoustic Tracking System

During the experiments, the movements of fish were recorded using an acoustic tracking system (Vemco radio acoustic positioning [VRAP]). The VRAP system uses three acoustic tracking buoys that detect acoustic pulses from tagged fish, triangulates the fish position, and then relays the data to a base station via a radio link. Fish were equipped with VEMCO V9 acoustic pingers/tags transmitting on 1 of 8 frequencies at a range between 63 and 84 kHz. Tags were programmed to transmit on 1 day in an 8-day cycle so, e.g., tags 1-8 transmitted on days 1, 9, 17, 25, 33, and so on while tags 9–16 transmitted on days 3, 11, 19, 27, and so on. Therefore, different fish were observed on different experiment days. Additionally, the system could be programmed to observe certain fishes located in both mesocosms in one experiment and switch to observe other fish in the next trial. Using the VRAP system, the position of one fish could be monitored about every 22 s. With 4 fish being monitored during the trial, the position of a single fish was taken about every 90 s.

3 Conclusions

The setup has proven to be feasible for behavioral studies on fish related to sound. The acoustic field in both mesocosms differed significantly in both sound pressure and particle motion levels, exposing the fish in the mesocosms to varying sound conditions depending on the position of the loudspeaker. But the mesocosms were also large enough for the fish to move away from the sound source to avoid higher sound levels if they chose to. Programming tags to observe fish at different times allowed larger numbers of fish to be monitored without introducing and removing fish for every experiment. Introducing fish on different days allowed the reaction of the fish to the first sound stimulus to be observed.

An experimental plan with different observed fish, a change in the loudspeaker position, and variable sound playback stimuli (different parts of the same sound recording) was chosen to avoid the problem of pseudoreplication (McGregor 2008).

References

McGregor PK (2008) Designing experiments to test for behavioral effects of sound. International Conference on the Effects of Noise on Aquatic Life, 13–17 August 2007, Nyborg, Denmark.

Sigray P, Andersson MH, Fristed T (2009) Partikelrörelser i vattnet vid ett vindkraftverk - Akustisk störning på fisk. Rapport 5963–7, Naturvårdsverket, Stockholm, Sweden.

Do Ocean-Based Wind Farms Alter the Migration Pattern in the Endangered European Silver Eel (*Anguilla anguilla*) Due to Noise Disturbance?

Mathias H. Andersson, Ingvar Lagenfelt, and Peter Sigray

1 Introduction

Numerous offshore wind farms have been built and will increase in the near future in the coastal areas of northern Europe. These locations are often in the direct path of migrating *Anguilla anguilla* (European silver eel) on their 5,000- to 6,000-km journey from Europe to the Sargasso Sea. At a certain time in their life that occurs between the ages of 6 and 20, *A. anguilla* start their long spawning migration. When this occurs, an irreversible physiological transformation starts in which the eyes and pectoral fins are enlarged, the skin color changes, and the digestive organs are regressed. Additionally, they stop feeding during the migration phase, which gives *A. anguilla* a limited amount of stored energy. If the fish are disturbed or hindered, the energy reserves might not be enough for them to reach their destination or it could, at least, result in limited spawning success. This effect could be devastating to the already highly threatened *A. anguilla* population that is listed as "critically endangered" in the International Union for Conservation of Nature (IUCN) red list.

One of several possible disturbances is the operational noise from wind turbines that has been measured at several wind farms in Europe. The results show that turbines generate noise below 1,000 Hz and at intensities, in terms of sound pressure, well above ambient levels (Wahlberg and Westerberg 2005). These frequencies and intensities will be detectable by *A. anguilla* when they pass the wind farm area because their hearing range is between 10 and 300 Hz (Jerkö et al. 1998). Only recently has particle motion been measured from an operating wind turbine (Sigray and Andersson, Chapter 111). The authors showed that the turbines generated particle acceleration with an amplitude of 0.0019 m/s^2 in the frequency spectrum of 2–200 Hz at a distance of 10 m. This level is not high enough to startle or disturb *A. anguilla* during migration (Sand et al. 2000). Therefore, the fish are most likely only perceiving sound pressure, whereas particle motion is "inaudible" when the fish pass the wind farm unless they pass the wind turbine foundations within a distance of 1–10 m.

The sound between Sweden and Denmark in the southern Baltic Sea is a shallow connection to Kattegat, important for the water exchange in the Baltic Sea, and is an important migrating route for

M.H. Andersson (✉)
Department of Zoology, Stockholm University, S-106 91 Stockholm, Sweden
e-mail: mathias.andersson@zoologi.su.se

I. Lagenfelt
Swedish National Board of Fisheries, S- 401 26 Gothenburg, Sweden

P. Sigray
Department of Underwater Research, Swedish Defence Research Institute, S-164 90 Stockholm, Sweden

A.N. Popper and A. Hawkins (eds.), *The Effects of Noise on Aquatic Life*,
Advances in Experimental Medicine and Biology 730, DOI 10.1007/978-1-4419-7311-5_90,

A. anguilla living in the Baltic Sea area (Westin 1990). In 2007, 48 wind turbines were erected in the sound on concrete gravitation foundations, constituting the Lillgrund wind farm. Additionally, the sound is heavily trafficked by commercial ships, and 36,900 ships were recorded passing the strait in 2008 (Swedish Maritime Administration 2008). These anthropogenic activities increased the ambient level in the area, potentially creating a noise barrier for the migrating *A. anguilla*. The aim of this study was to investigate if the migration pattern of *A. anguilla* was affected by the wind farm and if an effect could be linked to the noise generated by the wind farm during operation.

2 Methods

Two sets of methods were used to monitor the migration pattern of wild-caught *A. anguilla* through the sound: active and passive telemetry. For the active monitoring, 23 fish were tagged during 2001–2004 with ultrasonic tags. The fish were released 1-2 km south of the planned wind farm area (Fig. 1) and tracked actively using a hydrophone (Vemco model VH40) mounted on a boat. Passive telemetry was used in the baseline year 2005 and during the operational phases in 2008 and 2009, when 241 fish were tagged (Vemco tag V13) and released during October and November each year. The fish were released 11 km south of the wind farm. A receiver array containing 19 receivers (Vemco model VR2) was positioned on the seafloor in a 10-km line (maximum 500 m apart) passing from the Danish border in the west, through the wind farm, and continuing toward the Swedish mainland to the east (Fig. 1). All *A. anguilla* in the study were metamorphosed (silver eel) females with a total length (TL) of 60–100 cm and weighed between 0.5 and 1.6 kg.

Noise was measured using the battery-powered hydrophone system DSG-Ocean (Loggerhead Instruments) with a hydrophone from High Tech, Inc. The DSG was deployed in October to November 2009 1 km south of the wind farm (Fig. 1). In addition, measurements at 160 and 400 m from the southern tip of the wind farm were performed in May to June 2010. All locations were in the path of the fish migration route. The DSG recorded sound for 5 min every 30 min for 2 wk with a sample rate of 50 kHz. Data were stored on a 16-GB memory card. Data on wind speed, wind direction, and production from the wind farm were obtained from the wind farm operator, Vattenfall AB.

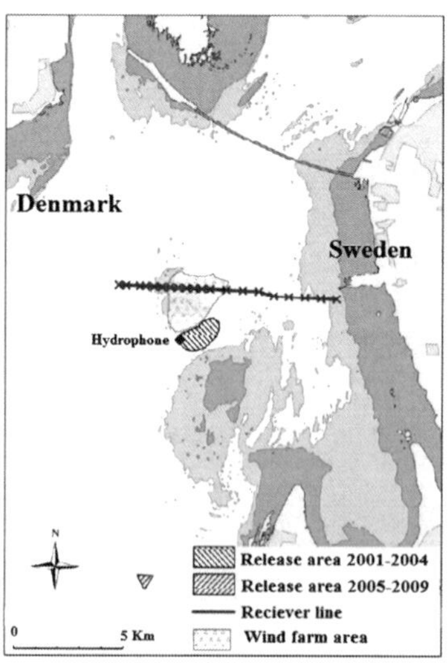

Fig. 1 Map of the sound with the location of the release areas, receivers, and hydrophone in 2009

This study is an ongoing project, part of the monitoring program for the Lillgrund wind farm (cf. Bergström et al. 2010), and a research project, supported by Vindval, of the Swedish Environmental Protection Agency, Stockholm.

3 Results and Discussion

During the baseline study (2001-2005), 30% of released fish were detected by the receivers, and results showed that most of the fish had swum to the west of the shallow area where the wind farm was built the following year (Fig. 2). During operation (2008-2009), again ~30% of the tagged fish were detected, and the results showed that the fish passage had shifted more toward the east side (Fig. 2). Several individuals was recaptured on the Danish side, south of Copenhagen, suggesting that a large number of eels swam more to the west, thus avoiding the receiver array. Few individuals passed directly through the wind farm area both before and after construction. This is most likely due to the shallow depth in the area. Noticeable is the time of passage from release to detection where the fish to the west were faster on average (143 h) than fish passing to the east (270 h) or through the wind farm (257 h). However, the difference was not statistically different due to the large variation (Kruskal-Wallis test, $P > 0.01$). In addition, no correlation between swimming time and total production of the wind farm was noticed (Spearman correlation coefficient, $P = 0.35$).

Sound measurements showed that the wind farm generated a few dominating tones (24, 130, and 520 Hz). However, the 520-Hz tone was sliding over several tones in frequency and should not be regarded as a stationary frequency and was therefore excluded in the analysis. This had little effect on the possible impact on the eel because it has limited hearing ability above 300 Hz (Jerkö et al. 1998). A transmission loss of 15 log(R) was estimated based on the measurements. During maximum production of 2.3 MW/turbine, the source levels of the 24- and 130-Hz tones were 126 and 142 dB re 1 μPa at 1 m, 93 and 109 dB re 1 μPa at 160 m, 87 and 103 dB re 1 μPa at 400 m, and

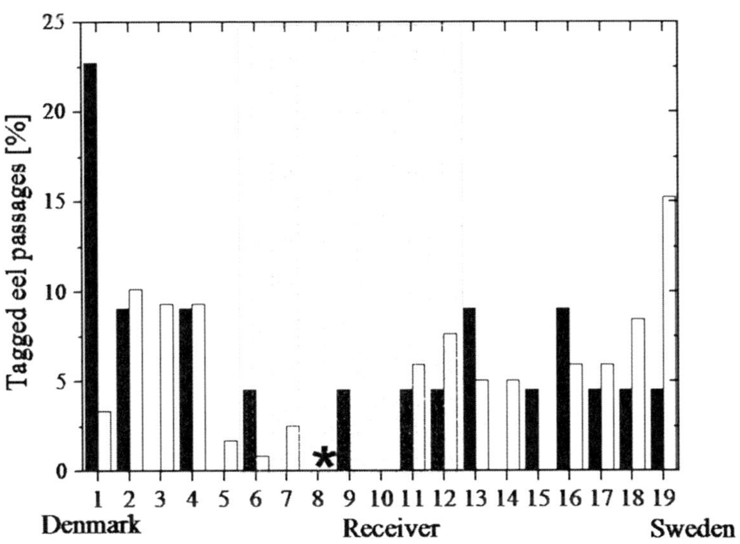

Fig. 2 Distribution of tagged *Anguilla anguilla* (European silver eel) passages through the receiver line from the west (Denmark) to the east (Sweden) at baseline (2001–2005; open bars) and during production (2008–2009; solid bars). Shaded area, those receivers located within the Lillgrund wind farm; solid star, receiver not yet recovered

81 and 96 dB re 1 μPa at 1,000 m, respectively. The measurements also showed intense ship traffic in the area, corresponding well with the previously recorded passages (Swedish Maritime Administration 2008) of four ships every hour. This might result in a masking effect of the wind farm operational noise, and the eel will most likely not perceive the wind farm until it is a few hundred meters from the wind farm because the ambient noise in the sound is higher than measured (ship traffic included). In conclusion, our results suggest that migrating *A. anguilla* seems not to be affected by the noise from the wind farm in the sound. The eels did not shift their migration path before or after construction, and the swimming speed did not differ significantly between the monitored areas. *A. anguilla* appear not to be disturbed by the wind farm-generated noise because no correlation could be seen between park production and swimming speed. These results are site specific because the sound has intense ship traffic and studies in other areas with less noise might show different results.

References

Bergström L, Lagenfelt I, Sundqvist F, Andersson MH, Sigray P (2010) Fishery study at Lillgrund: Control program for Lillgrund wind farm in 2009 (in Swedish). Board of Fisheries Annual Report 2009, Göteborg, Sweden.

Sand O, Enger PS, Karlsen HE, Knudsen FR, Kvernstuen T (2000) Avoidance response to infrasound in downstreaming migrating European silver eels. Environ Biol Fishes 57:327–336.

Swedish Maritime Administration (2008) Annual report 2008. Swedish Maritime Administration, Norrköping, Sweden. Available via http://www.sjofartsverket.se/en/About-us/Finances/Annual-Report/. Accessed 14 April 2010.

Wahlberg M, Westerberg H (2005) Hearing in fish and their reactions to sounds from offshore wind farms. Mar Ecol Prog Ser 288:295–309.

Westin L (1990) Orientation mechanisms in migrating European silver eel (*Anguilla anguilla*): Temperature and olfaction. Mar Biol 106:175–179

Effects of Vessel Engine Noise on the Acoustic Signaling Behavior of *Dascyllus albisella* (Hawaiian Damselfish)

Pollyanna I. Fisher-Pool, Marc O. Lammers, and Lisa M. Munger

1 Introduction

Visual communication in aquatic environments is limited by light, depth, and turbidity. As a result, sounds are a vital method of communication for a multitude of marine organisms such as shrimp, marine mammals, and fish. Fish possess well-developed auditory systems and can discriminate sounds produced in specific frequency bands (Fay et al. 2008; Kasumyan 2008). Hundreds of fish species are known to produce specialized sounds (Rountree et al. 2003), with many well-studied species found in coral reef ecosystems. Signals of reef fishes are used for communication during agonistic responses (Lobel 1992), territorial defense (Mann and Lobel 1998), feeding (Kasumyan 2008), spawning (Lobel and Mann 1995; Luczkovich et al. 1999, 2008), and courtship (Kaatz and Lobel 1999; Lobel 1992; Lobel and Kerr 1999; Maruska et al. 2007). However, marine environments are becoming subjected to increasing amounts of anthropogenic noise, particularly from shipping and vessel traffic. Little is known about how vessel-generated noise affects the communication and behavior of fish and many other marine species within coral reef ecosystems.

Studies have shown that the introduction of anthropogenic noise into fishes' environments can have physiological and behavioral impacts. It has been demonstrated that elevated cardiac output and other physiological stress indicators in *Micropterus salmoides* (largemouth bass) occur in response to boating disturbances, with recovery times for these events varying relative to the magnitude of disturbance (Graham and Cooke 2008). Vasconcelos et al. (2007) reported impairment in Lusitanian toadfishes' auditory sensitivity caused by ship noise and noted that communication is negatively impacted in noisy coastal environments. Temporary hearing loss in goldfish and catfish in response to exposure to 12- and 24-h periods of white noise (158 dB re 1 μPa) was seen to severely affect communication between fishes (Amoser and Ladich 2003). In some cases, the effects may be long lasting; McCauley et al. (2002) demonstrated permanent damage to the epithelia of pink snappers after exposure to seismic air gun noise.

P.I. Fisher-Pool (✉)
Joint Institute of Marine and Atmospheric Research, University of Hawai'i at Manoa, Honolulu, HI 96822, USA
e-mail: Pollyanna.fisher-pool@noaa.gov

M.O. Lammers • L.M. Munger
Hawai'i Institute of Marine Biology, Kane'ohe, HI 96744, USA
e-mail: lammers@hawaii.edu; lmunger@hawaii.edu

A.N. Popper and A. Hawkins (eds.), *The Effects of Noise on Aquatic Life*, Advances in Experimental Medicine and Biology 730, DOI 10.1007/978-1-4419-7311-5_91,

The domino damselfish (*Dascyllus albisella)* is a relatively well-studied species and is known to produce sounds in a variety of behavioral contexts, including aggressive "pop" and "chirp" signals as described by Mann and Lobel (1998). Mann and Lobel (1997) studied *Dascyllus albisella* sound propagation at the Johnston atoll and found that signals attenuate quickly and that sounds therefore are used for communication over distances less than 12 m. As a result, this species may be vulnerable to behavioral disruption by anthropogenic noise sources. We designed an in vivo experiment to examine the effects of vessel noise on the acoustic behavior of *Dascyllus albisella*. We hypothesized that masking of *Dascyllus albisella* signals by vessel engine noise would affect their behavior by resulting in one or more of the following: changes in calling rates, frequency, and timing.

2 Methods

We exposed a colony of wild *Dascyllus albisella* to controlled vessel engine noise stimuli. The colony was located in a marine reserve in Kane'ohe Bay, Oahu, HI. The sound source was the University of Hawai'i's 43-ft *R/V Miriam*, which is regularly moored at a dock ~10 m from a *Dascyllus albisella* colony. This vessel is equipped with twin 305-HP 8-cylinder diesel engines and is typical of many vessels that operate along the Hawaiian coast in reef environments. Data were collected using a seafloor-placed ecological acoustic recorder (EAR), which records ambient sounds at frequencies up to 40 kHz on a programmable schedule (Lammers et al. 2008). The EAR was deployed at ~4 m depth next to a *Porites compressa* coral head with a resident *Dascyllus albisella* colony and programmed to sample at 25 kHz, providing a recording bandwidth of 12.5 kHz. During exposure experiments, only 1 engine was used and the duty cycle was set to record for 60 s every 120 s for a period of 30 min before noise stimulus, 30 min during the noise stimulus, and 30 min after cessation of the stimulus, 3 times each day for 1 wk.

Data were analyzed aurally and visually to determine the number of *Dascyllus* signals per recording. Periods with no exposure to engine noise were compared with stimulus periods. Calling rates (calls per minute) were obtained for 30 min immediately before boat noise stimulus ("before") and 30 min immediately after the stimulus ("after") and were compared using a paired Wilcoxon signed-rank test (Hollander and Wolfe 1999).

3 Results

The fish in the study produced a series of pulse sounds with 90% of signal energy between 100 and 1,100 Hz (Fig. 1). Vessel noise from the *R/V Miriam* covered the same frequency band (Fig. 2).

A total of 14 noise stimulus trials were conducted (4 morning, 5 midday, and 5 afternoon). Results from trials are given in Table 1. Calling rates did not differ significantly between the "before" and "after" periods ($P = 0.328$) (Fig. 3).

4 Discussion

The sounds produced by *Dascyllus albisella* in this study were similar to the aggressive "pops" and "chirps" described by Mann and Lobel (1998), with frequencies within the reported range of a 146-Hz bandwidth around a 442-Hz mean peak frequency. This is expected because we did not record

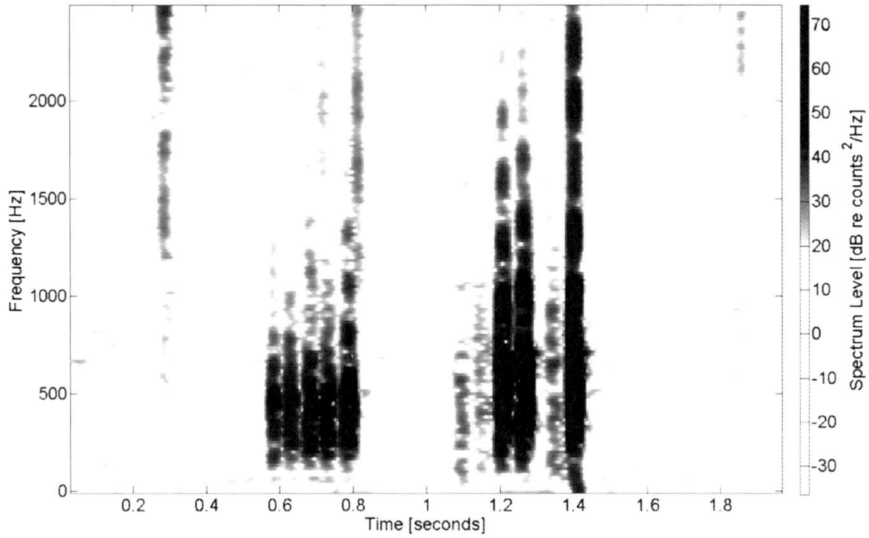

Fig. 1 *Dascyllus albisella* spectrogram recorded by an ecological acoustic recorder (EAR), 2,000-point fast Fourier transform (FFT), and 90% overlap

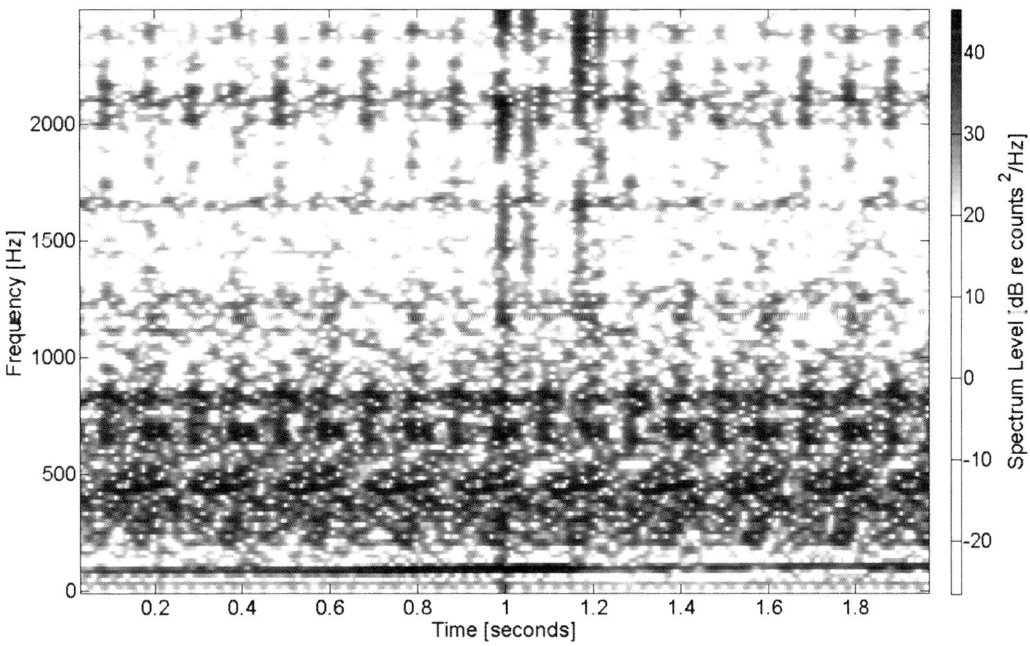

Fig. 2 Vessel noise spectrogram, 2,000-point FFT, and 90% overlap

Table 1 Summary of *Dascyllus albisella* noise-exposure trial results

	Number of Trials	Minutes Recorded	Median Call Rate (Calls/Minute)	Minimum Call Rate	Maximum Call Rate
"Before"	14	222	0.35	0	3.2
"After"		201	0.29	0	2.6

Before, before boat noise stimulus; after, after boat noise stimulus

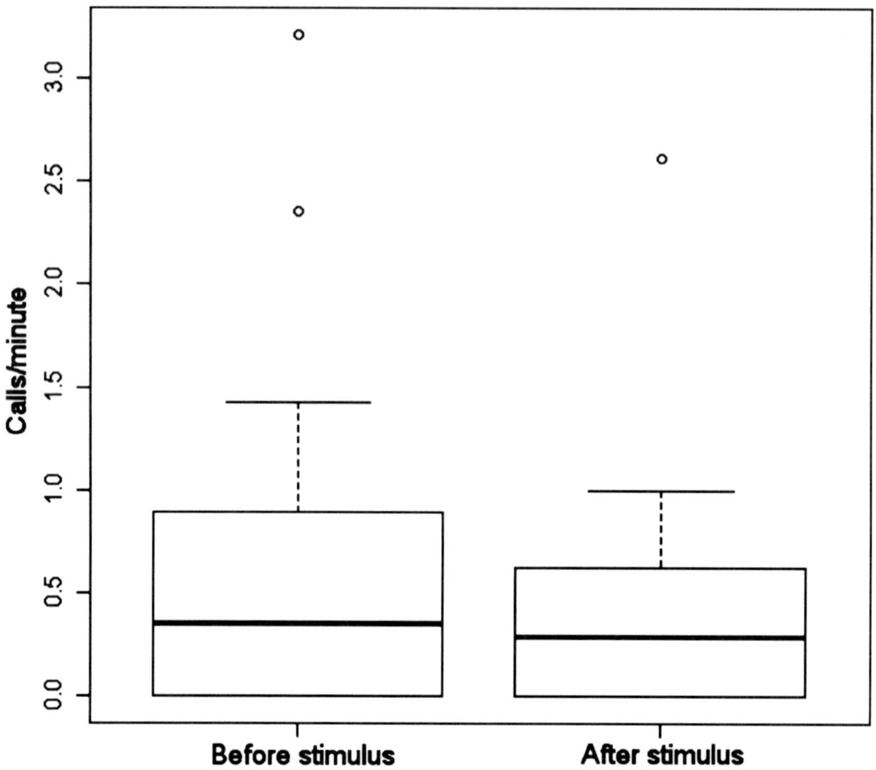

Fig. 3 *Dascyllus albisella* calling rates before and after noise stimulus (N = 14 trials). Box boundaries represent lower and upper quartiles; bold lines are median calling rates; small circles are outliers. Whiskers are drawn to data values within 1.5 times the box range

during the known spawning season and sounds recorded during our experiment were therefore likely aggressive signals rather than the courtship-associated "signal jumps" reported by Mann and Lobel (1998).

Initial analysis of the results led us to believe that stimuli of vessel engine noise did not cause a significant change in fish acoustic behavior. However, the environment within the marine reserve was exposed to uncontrolled anthropogenic noise that may have confounded the results; sound from other small boats traveling from a nearby pier provided intermittent background noise. A spatially diverse experiment of fish colonies in remote areas compared with colonies near boat traffic would provide further insight on *Dascyllus albisella* acoustic response to vessel noise in reef ecosystems. In addition, an analysis of recordings from periods without ship noise throughout the day and night would be useful in determining baseline acoustic behavior and temporal patterns in sound production.

Acknowledgments We thank Michael Richlen and Anne Rosinsky, Marine Mammal Research Program, Hawai'i Institute of Marine Biology, Kane'ohe, HI, for their assistance in the deployment of field instrumentation and noise stimuli production.

References

Amoser S, Ladich F (2003) Diversity in noise-induced temporary hearing loss in otophysine fishes. J Acoust Soc Am 113:2170–2179.

Fay RR, Popper AN, Webb JF (2008) Introduction to fish bioacoustics. In: Webb JF, Fay RR, Popper AN (eds) Fish bioacoustics. Springer, New York, pp 1–15.

Graham AL, Cooke SJ (2008) The effects of noise disturbance from various recreational boating activities common to inland waters on the cardiac physiology of a freshwater fish, the largemouth bass (*Micropterus salmoides*). Aquat Conserv Mar Freshw Ecosyst 18:1315–1324.

Hollander M, Wolfe DA (1999) Nonparametric statistical methods. John Wiley & Sons, New York.

Kaatz IM, Lobel PS (1999) Acoustic behavior and reproduction in five species of *Corydoras* catfishes (Callichthydae). Biol Bull 197:241–242.

Kasumyan AO (2008) Sounds and sound production in fishes. J Ichthyol 48:981–1030.

Lammers MO, Brainard RE, Au WWL, Mooney TA, Wong K (2008) An ecological acoustic recorder (EAR) for long-term monitoring of biological and anthropogenic sounds on coral reefs and other marine habitats. J Acoust Soc Am 123:1720–1728.

Lobel PS (1992). Sounds produced by spawning fishes. Environ Biol Fishes 33:351–358.

Lobel PS, Kerr LM (1999) Courtship sounds of the Pacific damselfish, *Abudefduf sordidus* (Pomacentridae). Biol Bull 197:242–244.

Lobel PS, Mann DA (1995) Spawning sounds of the damselfish, *Dascyllus albisella* (Pomacentridae), and relationship to male size. Bioacoustics 6:187–198.

Luczkovich JJ, Pullinger RC, Johnson SE, Sprague MW (2008) Identifying sciaenid critical spawning habitats by the use of passive acoustics. Trans Am Fish Soc 137:576–605.

Luczkovich JJ, Sprague MW, Johnson SE, Pullinger RC (1999) Delimiting spawning areas of weakfish *Cynoscion regalis* (family Sciaenidae) in Pamlico Sound, North Carolina, using passive hydroacoustic surveys. Bioacoustics 10:143–160.

Mann DA, Lobel PS (1997) Propagation of damselfish (*Pomacentridae*) courtship sounds. J Acoust Soc Am 101:3783–3791.

Mann DA, Lobel PS (1998) Acoustic behavior of the damselfish *Dascyllus albisella*: Behavioral and geographic variation. Environ Biol Fish 51:421–428.

Maruska KP, Boyle KS, Dewan LR, Tricas TC (2007) Sound production and spectral hearing sensitivity in the Hawaiian sergeant damselfish, *Abudefduf abdominalis*. J Exp Biol 210:3990–4004.

McCauley RD, Fewtrell J, Popper AN (2003) High intensity anthropogenic sound damages fish ears. J Acoust Soc Am 113:638–642.

Rountree R, Goudey C, Hawkins T (2003) Listening to fish: Passive acoustic applications in marine fisheries. Massachusetts Institute of Technology Sea Grant Technical Report MITSG 03–02, Cambridge, MA.

Vasconcelos RO, Amorim MCP, Ladich F (2007) Effects of ship noise on the detectability of communication signals in the Lusitanian toadfish. J Exp Biol 210:2104–2112.

Distribution and Potential Impact of Boat Noise on Fish Sound Production From an Autonomous Acoustic Array in the Eastern Gulf of Mexico

Carrie C. Wall, Peter Simard, and David A. Mann

1 Introduction

Numerous species of soniferous fish, such as *Epinephelus morio* (red grouper) and *Opsanus beta* (gulf toadfish), have been recorded in the Gulf of Mexico off west-central Florida. These waters are also used by vessels ranging from small recreational craft to large commercial ships. The low-frequency (50- to 6,000-Hz) sounds produced by vessels can mask the low-frequency (50- to 500-Hz) sounds produced by fishes (Bradbury and Vehrencamp 1998; Webb et al. 2007). Communication ranges for fish may thus be impacted. Therefore, the spatial and temporal patterns of both boat noise and fish sounds are an important consideration to understand the potential impact of boat noise on fish communication.

2 Acoustic Array

From June to September 2008, 23 autonomous acoustic recorders were deployed in the eastern Gulf of Mexico covering ~3,600 km^2 (Fig. 1). Each Digital SpectroGram (DSG) acoustic recorder was bottom mounted in a trawl-resistant housing in waters up to 30 m in depth. The DSGs recorded 10 s every hour at a 50-kHz sample rate for up to 4 mo. Sound was recorded from 10 June to 15 September 2008, producing over 40,000 acoustic files. Both fish sounds and boat noise were commonly found in these recordings.

C.C. Wall (✉) • P. Simard • D.A. Mann
College of Marine Science, University of South Florida, St. Petersburg, FL, USA
e-mail: cwall@mail.usf.edu; psimard@mail.usf.edu; dmann@marine.usf.edu

A.N. Popper and A. Hawkins (eds.), *The Effects of Noise on Aquatic Life*,
Advances in Experimental Medicine and Biology 730, DOI 10.1007/978-1-4419-7311-5_92,

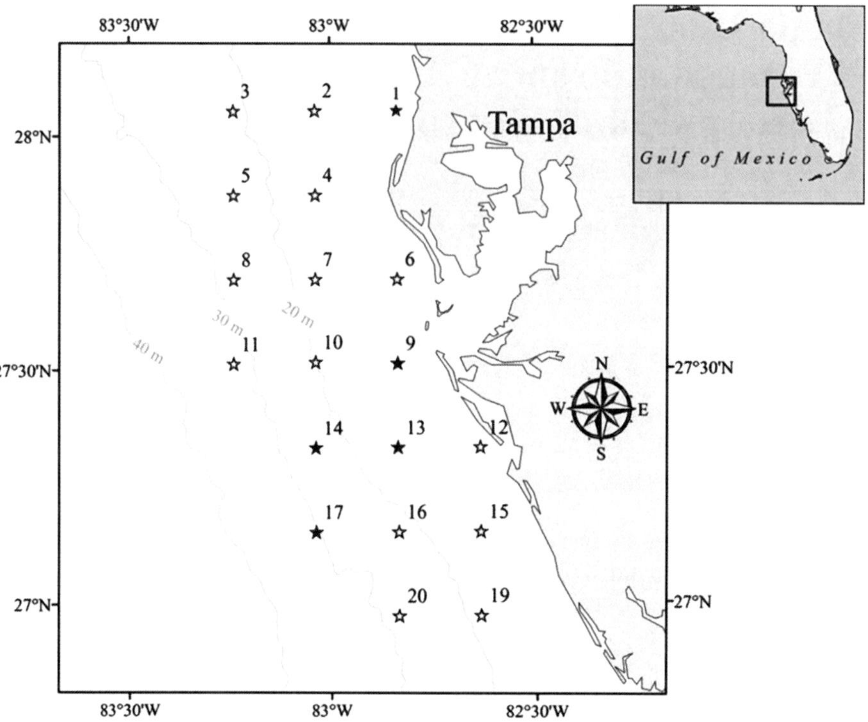

Fig. 1 Acoustic array deployed from June to September 2008. Open stars are locations where data were collected from July to September 2008 only. Numbers indicate locations of the array in the eastern Gulf of Mexico

3 Sound Production

The immense number of files produced during these deployments precludes manually verifying every file. However, analysis of a subset of files has identified sounds produced by *Opsanus tau*, *Epinephelus morio*, *Bairdiella chryosoura* (silver perch), more than six unknown species, and verified boat noise.

Composite spectrograms were used to examine the duration and frequency range of boat noise (Fig. 2). Two frequency bands were used to identify potential boat and fish noise (100–300 Hz) and potential boat and ambient noise (1,800–2,000 Hz), which serves as a baseline, over a 24-h period (Fig. 3). The values were calculated by subtracting the 1,800- to 2,000-Hz band median root mean square (RMS) per hour for all days from the 100- to 300-Hz band median RMS per hour for all days. A peak in the sound level at 2000h was observed for all sites in Figure 3, suggesting a high likelihood of fish communication during this time. To identify the frequency spectrum and potential overlap of boat noise at this time, band sound pressure levels were calculated for 30 July 2008 at 2000h (Fig. 4).

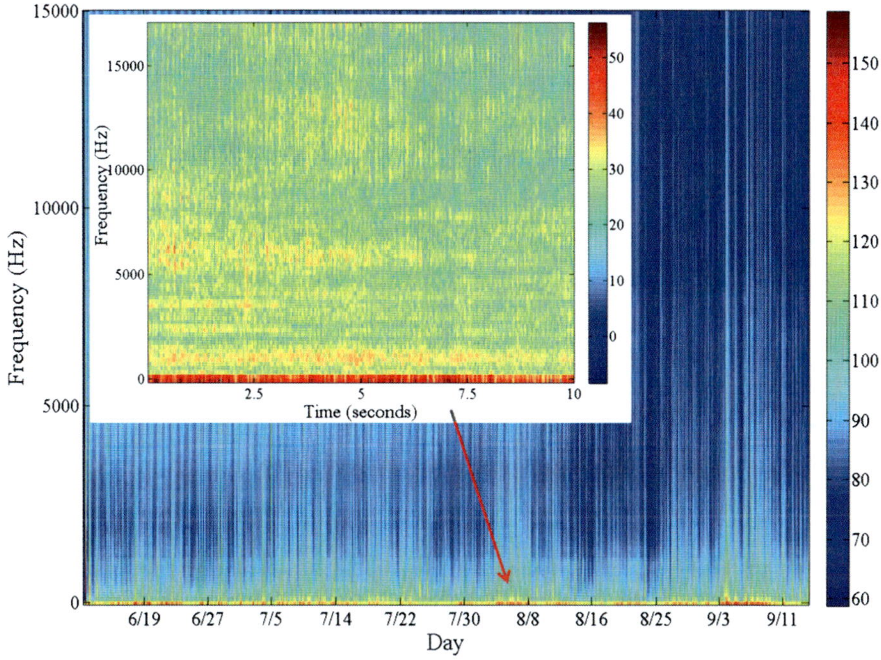

Fig. 2 Composite spectrogram showing band sound pressure level (100-Hz resolution) from 11 June to 15 September 2008, at Station 1. Inset is a spectrogram (200-Hz resolution) on 7 August 2008, illustrating higher sound intensities associated with boat presence below 500 Hz and near 2,000 Hz

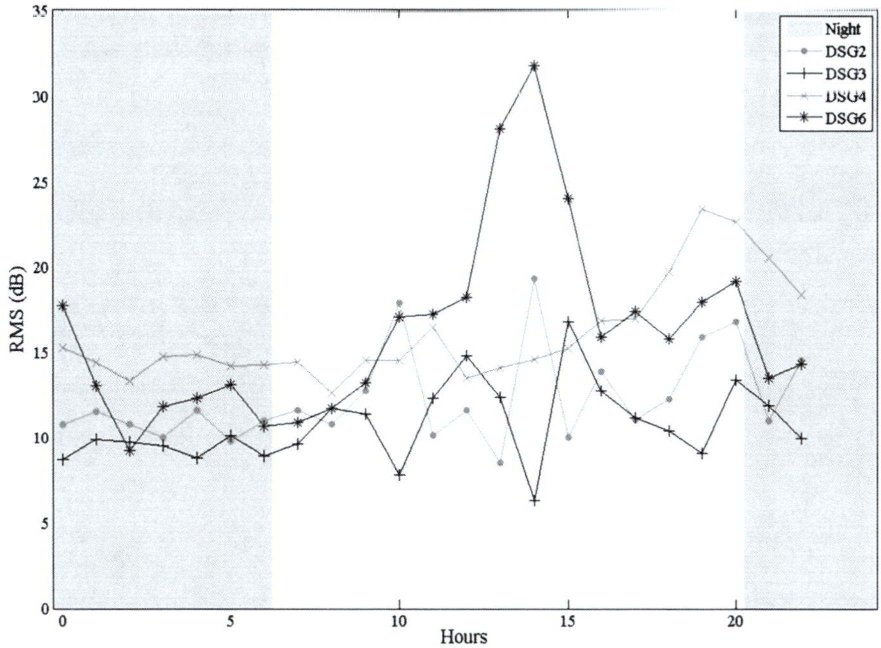

Fig. 3 Median hourly difference between the 100- to 300-Hz band and the 1,800- to 2,000-Hz band sound pressure level at 4 sites from 10-26 June 2008. RMS, root mean square, DSG, Digital SpectroGram

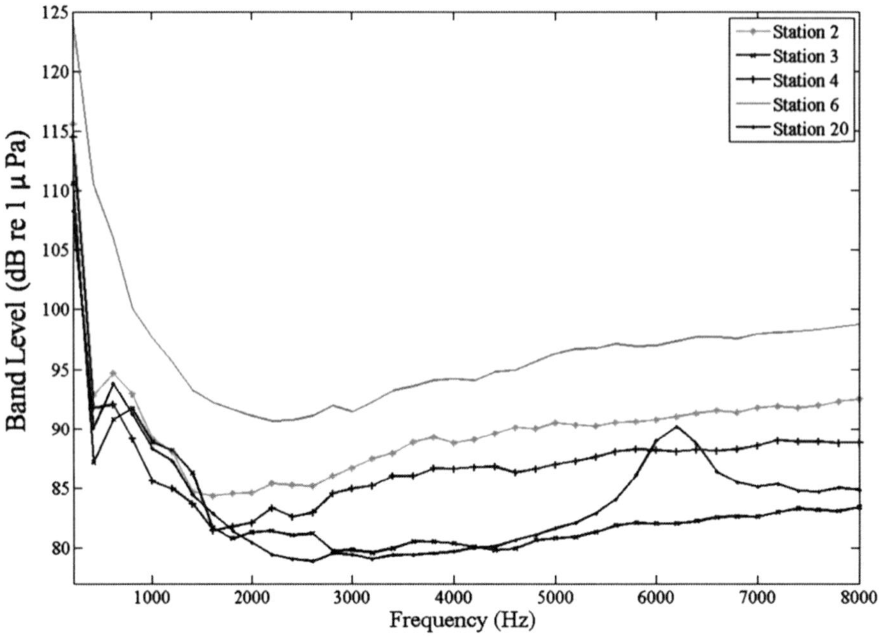

Fig. 4 Band sound pressure levels (200-Hz resolution) for 30 July 2008 at 2000h

4 Conclusions

Peaks in band levels between 500 and 1,500 Hz indicate broadband low-frequency sounds during a period of suspected high fish sound production. Future research will determine the extent of temporal overlap and possible impact on communication range for fishes due to the presence of boats across the entire study area.

Acknowledgments This research was funded by National Oceanographic Partnership Program Award OCE-0741705 and a University of South Florida (USF)/United States Geological Survey graduate assistantship to C. Wall. We thank Mote Marine Laboratory, Sarasota, FL, and the USF Center for Ocean Technology, St. Petersburg, FL, among numerous colleagues, for assistance in the field.

References

Bradbury J, Vehrencamp S (1998) Principles of animal communication. Sinauer Associates, Sunderland, MA.
Webb JF, Fay RR, Popper AN (eds) (2007) Fish bioacoustics. Springer, New York.

Cichlid Courtship Acoustics: Signals and Noise Influence Reproductive Behavior

Hans Slabbekoorn, Machteld Verzijden, and Christina May

1 Introduction: Cichlid Sounds and Sexual Selection

Cichlid fish are well known for their diversity and the role of sexual selection in speciation. The species-specific mate preferences that have received the most attention in this context are based on visual (e.g., Couldridge and Alexander 2002; Seehausen et al. 2008) and chemical signals (Plenderleith et al. 2005). However, it has been known for a long time that cichlids also generate species-specific sounds during courtship (Amorim et al. 2008; Lobel 1998; Myrberg et al. 1965; Verzijden et al. 2010). Therefore, sounds may also have been an important factor in the rapid evolutionary diversification of cichlids in African lakes (Kocher 2004; Kornfield and Smith 2000).

All fish can hear sound, and various fish species have been reported to be able to discriminate between their own and other species' sounds or synthetic stimuli (Lugli et al. 2004; McKibben and Bass 1998; Myrberg and Spires 1972; Rollo and Higgs 2008) and, in some cases, even seem to recognize individuals acoustically (Myrberg and Riggio 1985). We performed a series of recordings and experiments (Fig. 1) that elucidate the importance of sound to courtship behavior in the African cichlid *Pundamilia nyererei* (Verzijden et al. 2010; Estramil et al., unpublished data; May et al., unpublished data).

We used laboratory-reared fish that were descendents of wild-caught individuals from Makobe Island in Lake Victoria, Tanzania (Seehausen 1997). We used recordings of male cichlid sounds made during courtship displays to test the impact on female cichlids. The experimental tank was divided into three compartments with the use of two grids, which allowed females to pass through but confined the larger males to their compartments. We placed two males, matched for size and nuptial coloration, in the outer compartments and a female cichlid in the middle. Opaque dividers initially obscured both males from the female. We played back a sequence of sounds from one side three times, after which the blinds were removed and the female could see both males and interact with the male of her choice.

We scored the courtship interactions between the female and the males that are predictive of mate choice in East African cichlids (Seehausen 1997; Verzijden and ten Cate 2007). The female

H. Slabbekoorn (✉) • C. May
Behavioural Biology, Institute of Biology, Leiden University, 2333 BE Leiden, The Netherlands
e-mail: h.w.slabbekoorn@biology.leidenuniv.nl

M. Verzijden
Behavioural Biology, Institute of Biology, Leiden University, 2333 BE Leiden, The Netherlands

Department of Biology, Lund University, SE-22364 Lund, Sweden

A.N. Popper and A. Hawkins (eds.), *The Effects of Noise on Aquatic Life*,
Advances in Experimental Medicine and Biology 730, DOI 10.1007/978-1-4419-7311-5_93,

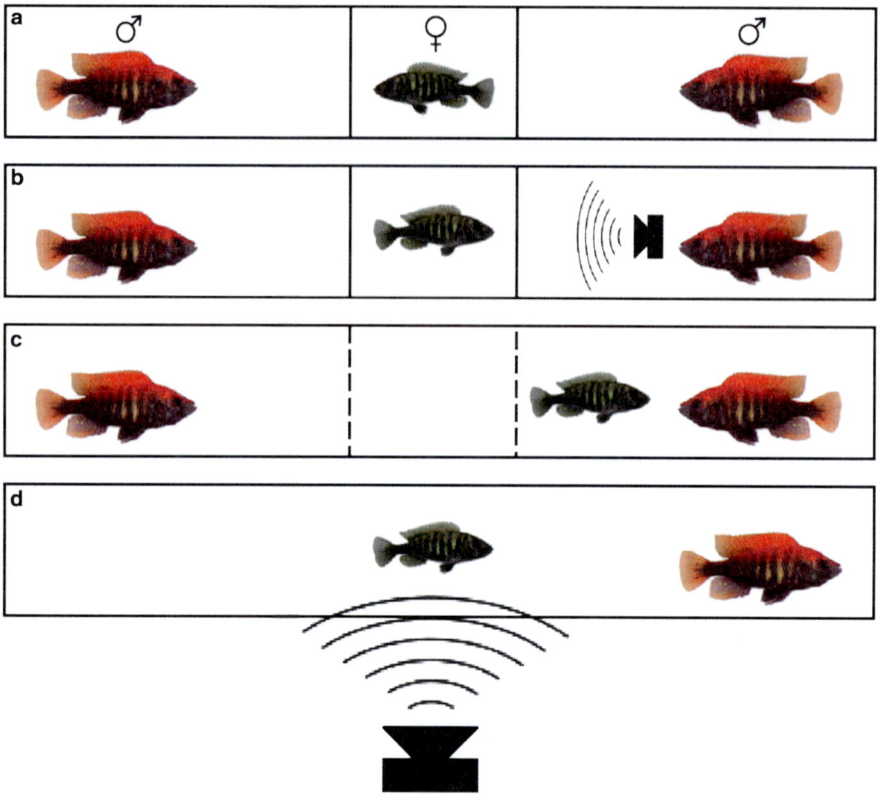

Fig. 1. Schematic representation of the experimental setup for the mate choice trials (**a–c**) and for the experimental exposure to artificially increased noise levels (**d**). Pictures courtesy of Martine Maan

approach response to male quiver was the most stereotypic interaction and was used to quantify response strength and side preferences. We found that the presence of conspecific sounds influenced the female choice when selecting one male over another to court. Males that were associated with the playback of sounds were preferred over males that were not associated with sound. These playback results provided the first experimental evidence of sounds affecting mate preferences in any fish species (Verzijden et al. 2010).

In follow-up experiments, we were able to confirm that the cichlids modify their behavior especially when hearing conspecific sounds and that they do not respond to just any sound (Estramil et al., unpublished data). We also tested the impact of more noisy water conditions on the natural swimming behavior of male cichlids in a solitary enclosure and the intensity of courtship behavior of male cichlids that encounter an unfamiliar female. We used an external loudspeaker placed beneath the fish tank and observed the fish behavior under low and high noise levels under dim and bright light conditions. Both the solitary and the social context revealed significant reductions in activity levels and aberrant swimming behavior related to increased noise levels (May et al., unpublished data).

2 Conclusions: Acoustic Signals and Artificial Noise Matter

Our results provide evidence for conspecific sound to affect mate choice decisions of female cichlids interacting with two live male cichlids. We confirmed that directional hearing and side preferences can be tested in relatively small fish tanks (cf. Rollo and Higgs 2008). We believe that our setup is also suitable for testing the communicative significance of acoustic variation in fish sounds related to species, size, context, and motivation. The impact of noise on the swimming and reproductive behavior in a fish tank could be indicative of a similar impact of artificial sounds on free-living fish. However, field studies are needed because hardly any data exist on the impact of anthropogenic noise on the behavior of free-living fish, although human activities are making the underwater world increasingly noisy on a global scale.

References

Amorim MCP, Simoes JM, Fonseca PJ, Turner GF (2008) Species differences in courtship acoustic signals among five Lake Malawi cichlid species. J Fish Biol 72:1355–1368.

Couldridge VCK, Alexander GJ (2002) Color patterns and species recognition in four closely related species of Lake Malawi cichlid. Behav Ecol 13:59–64.

Kocher TD (2004) Adaptive evolution and explosive speciation: The cichlid fish model. Nat Rev Gen 5:288–298.

Kornfield I, Smith PF (2000) African cichlid fishes: Model systems for evolutionary biology. Ann Rev Ecol Syst 31:163–196.

Lobel PS (1998) Possible species specific courtship sounds by two sympatric cichlid fishes in Lake Malawi, Africa. Environ Biol Fish 52:443–452.

Lugli A, Pavan G, Torricelli P (2004) The response of the male freshwater goby to natural and synthetic male courtship sound playback following exposure to different female sexual stimuli. Ethol Ecol Evol 16:55–70.

McKibben JR, Bass AH (1998) Behavioral assessment of acoustic parameters relevant to signal recognition and preference in a vocal fish. J Acoust Soc Am 104:3520–3533.

Myrberg AA, Kramer E, Heinecke P (1965) Sound production by cichlid fishes. Science 149:555–558.

Myrberg AA, Riggio RJ (1985) Acoustically mediated individual recognition by a coral-reef fish (*Pomacentrus partitus*). Anim Behav 33:411–416.

Myrberg AA, Spires JY (1972) Sound discrimination by bicolor damselfish, *Eupomacentrus partitus*. J Exp Biol 57:727–735.

Plenderleith M, van Oosterhout C, Robinson RL, Turner GF (2005) Female preference for conspecific males based on olfactory cues in a Lake Malawi cichlid fish. Biol Lett 1:411–414.

Rollo A, Higgs D (2008) Differential acoustic response specificity and directionality in the round goby, *Neogobius melanostomus*. Anim Behav 75:1903–1912.

Seehausen O (1997) Distribution of and reproductive isolation among color morphs of a rock-dwelling Lake Victoria cichlid (*Haplochromis nyererei*). Ecol Freshw Fish 6:59–66.

Seehausen O, Terai Y, Magalhaes IS, Carleton KL, Mrosso HD, Miyagi R, van der Sluijs I, Schneider MV, Maan ME, Tachida H, Imai H, Okada N (2008) Speciation through sensory drive in cichlid fish. Nature 455:620–623.

Verzijden MN, ten Cate C (2007) Early learning influences species assortative mating preferences in Lake Victoria cichlid fish. Biol Lett 3:134–136.

Verzijden MN, van Heusden J, Bouton N, Witte F, ten Cate C, Slabbekoorn H (2010) Sounds of male Lake Victoria cichlids vary within and between species and affect female mate preferences. Behav Ecol 21:548–555. Doi:10.1093/beheco/arq018.

Using Catch Statistics to Investigate Effects of Seismic Activity on Fish Catch Rates

Aud Vold, Sven Løkkeborg, and Maria M. Tenningen

Norway has been a petroleum nation for several decades, with most of its offshore activities originally concentrated in the North Sea. As these reserves diminish, focus is shifting to areas further north, some of which, particularly the Lofoten/Vesterålen area, are crucial for the reproduction of major fish stocks such as *Gadus morhua* (northeastern arctic cod), *Melanogrammus aeglefinus* (haddock), and *Clupea harengus* (herring) in the Barents Sea. Cod stocks are the basis for enormous seasonal fisheries during the spawning season in late winter, but other species such as *Pollachius virens* (saithe), *Melanogrammus aeglefinus*, *Brosme brosme* (tusk), *Molva molva* (ling), *Sebastes marinus* (golden redfish), and *Lophius piscatorius* (angler) provide a living for local fishermen the whole year-round. Although the area has not yet been opened for exploitation, seismic surveys for petroleum reserves have been carried out by the Norwegian authorities during the summers of 2008 and 2009 (Fig. 1). Seismic air gun investigations are major sources of low-frequency noise in areas of petroleum industry activity. The sound from the guns has its peak energy in the most sensitive hearing range of fish. It has already been documented that air gun activity can have a negative impact on fish catch rates (Engås et al. 1996; Skalski et al. 1992). Fishermen in the area are, quite naturally, anxious about the effect of seismic activity on their livelihoods. To determine whether the seismic exploration carried out in the Lofoten/Vesterålen area has had an effect on the local fisheries, data from official databases of landed catches have been analyzed.

The fisheries in the Lofoten/Vesterålen area are dominated by coastal vessels less than 28 m in overall length fishing with gill nets, longlines, jigs, and Danish seines. The catch data were taken from the official databases of landed catches managed by the Norwegian Directorate of Fisheries, Bergen, Norway. Records of vessel identification, date of catch delivery, catch location, type of fishing gear used, and the weight of delivered catch by species were selected from the databases. These suffered from several shortcomings for our analyses. 1) The recording of catch locations is coarse (given the International Council for the Exploration of the Sea [ICES] location with an accuracy of 0.5 × 1°; Fig. 1), and in any case, some fish buyers are known to register catch location rather arbitrarily. 2) Fishing dates are not supplied, only delivery dates. Fishermen occasionally accumulate their catches from several sea trips for one delivery. 3) There is a lack of fishing effort data because only gear type is given, with no information on gear quantity or effort changes. In our analyses, we have used catch per delivery as a measure of catch per unit effort (CPUE) and assumed that there

A. Vold (✉) • S. Løkkeborg • M.M. Tenningen
Institute of Marine Research, 5817 Bergen, Norway
e-mail: aud.vold@imr.no

A.N. Popper and A. Hawkins (eds.), *The Effects of Noise on Aquatic Life*,
Advances in Experimental Medicine and Biology 730, DOI 10.1007/978-1-4419-7311-5_94,

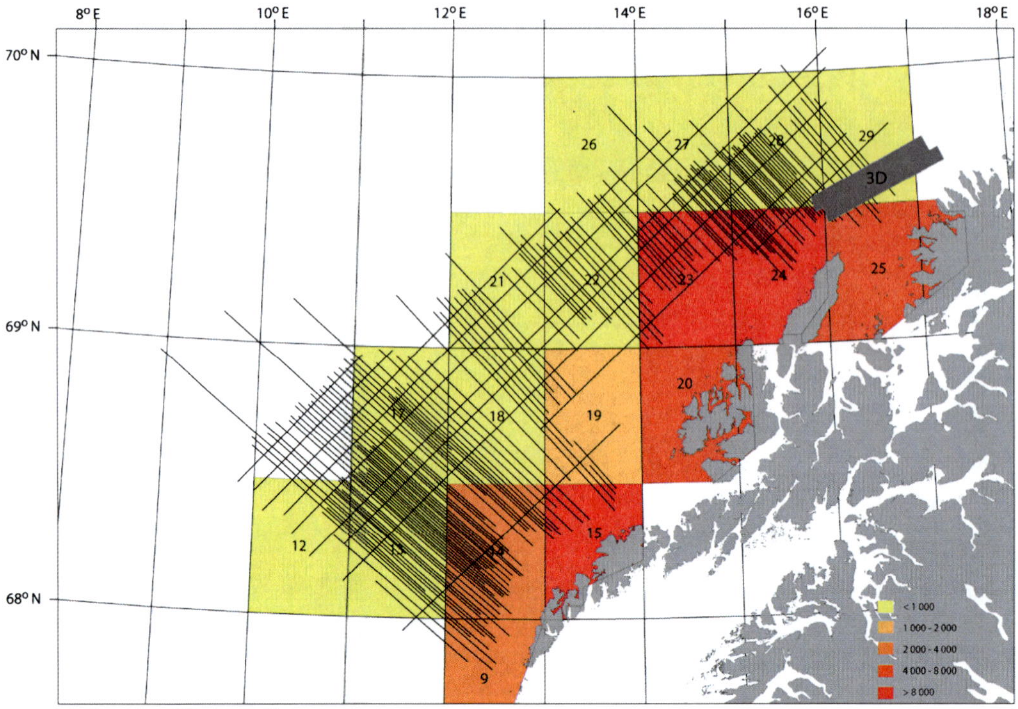

Fig. 1 The study area was located in International Council for the Exploration of the Sea (ICES) statistical area 05, northern Norway. Subarea Lofoten refers to locations 9, 13, 14, 15, 17, and 18 and Vesterålen to locations 19, 21, 22, 23, 24, and 28. The intensity of the color reflects the number of catches taken in each location. Black lines, survey transects of the seismic vessel

were no systematic changes in fishing effort during the period analyzed. The spatial and temporal identification of the seismic vessel transects were much more accurate. We considered the catch rates within one ICES location as possibly affected by air gun sound if the seismic vessel was operating in that location simultaneously with fishing. However, the distance from the seismic vessel to the fishing ground could have been as much as 30 nautical miles if the vessels were operating in the opposite corners of the same ICES location. To be revealed by our analyses, the effects of seismic sound would therefore had to have been severe.

The seismic surveys in summer 2008 were a series of two-dimensional surveys lasting from May to September, covering a large area (Fig. 1). Some locations were only affected by occasional visits by the seismic vessel scattered over the whole survey period. The effects of the air gun sound were analyzed in three ways. 1) Catch rates from individual fishing vessels using stationary gears (gill nets or longlines) close to seismic survey transects before, during, and after seismic activity were compared. 2) Average catch rates from all fishing vessels operating in each location before, during, and after seismic activity in summer 2008 were compared. 3) Catch rates during the seismic survey in 2008 were compared with the previous 5 years. The data were analyzed under the null hypothesis that catch rates did not differ between years or periods (before, during, or after shooting).

The analyses showed that seismic air gun activity can influence fish catch rates, but in this study, the direction of the changes was not consistent. The effects varied between subareas, species, and fishing gears used, and the daily variation in catch rates of single vessels was too wide to discover any effect on vessel level. When analyzing the average catches of all vessels operating within each location, a decline was found in delivered catches of *Pollachius virens* and *Melanogrammus aeglefinus* caught by gill nets in the Vesterålen subarea (Fig.1), but a similar effect was not seen for other fishing methods (longline, hook and line, or Danish seine) or in the Lofoten subarea. A 45% decrease in the quantity of *Pollachius virens* delivered from the whole study area in summer 2008 compared with 2006 and 2007 was found. However, the catches were not less than in the period from 2003 to 2005. The total stock of *Pollachius virens*, and thus the fishing quotas, however, were higher in 2008 than in the previous years, giving an expectation of increasing catch rates, which is the opposite of what was found in these analyses. On the other hand, the catch rates of *Sebastes marinus* and *Lophius piscatorius* caught by gill nets increased significantly during the air gun shooting period, whereas the catch rates of *Gadus morhua*, *Molva molva* and *Brosme brosme* did not appear to be significantly altered. In the cases where no effects of seismic sound were observed, this may be either because there was no effect or because the data used in the analyses were not adequate for the purpose. Earlier experiments (Engås et al. 1996; Løkkeborg and Soldal 1993) have clearly demonstrated that trawl and longline catches of *Gadus morhua* were negatively affected by air gun sound. It is therefore important to bear in mind that during the season analyzed in this study, *Gadus morhua* was not a target species but was only caught incidentally as bycatch. The small and varying catches may have obscured evidence of effects of seismic activity.

The alterations in catch rates indicate that air gun sound affects fish behavior in several ways. Different fishing gears use different behavioral traits to catch fish. Whereas gill nets only catch moving fish that accidentally encounter the net, longlines exploit the feeding motivation of fish by attracting them to baited hooks. There is reason to believe that stationary bottom-dwelling species such as *Lophius piscatorius* increase their swimming activity when scared by air gun sound, thus making them more available to gill nets. This may also be true for *Sebastes* spp., but it has also been documented (Pearson et al. 1992) that *Sebastes* swim above the bottom dive when scared by air gun sound, making them more available for bottom-set fishing gears. *Pollachius virens* and *Melanogrammus aeglefinus* are more mobile species. The reduced catch rates of these species in gill nets may indicate that they migrate out of the investigation area.

References

Engås A, Løkkeborg S, Ona E, Soldal AV (1996) Effects of seismic shooting on local abundance and catch rates of cod (*Gadus morhua*) or haddock (*Melanogrammus aeglefinus*). Can J Fish Aquat Sci 53:2238–2249.

Løkkeborg S, Soldal AV (1993) The influence of seismic exploration with air guns on cod (*Gadus morhua*) behaviour and catch rates. ICES J Mar Sci 196:62–67.

Pearson WH, Skalski JR, Malme CI (1992) Effects of sounds from a geophysical survey device on behavior of captive rockfish (*Sebastes* spp.). Can J Fish Aquat Sci 49:1343–1356.

Skalski JR, Pearson WH, Malme CI (1992) Effects of sound from a geophysical survey device on catch-per-unit-effort in a hook-and-line fishery for rockfish (*Sebastes* spp.). Can J Fish Aquat Sci 49:1357–1365.

Effects of Sounds From Seismic Air Guns on Fish Behavior and Catch Rates

Svein Løkkeborg, Egil Ona, Aud Vold, and Are Salthaug

1 Introduction

Seismic surveys are performed to explore for oil and gas resources. During these surveys, air guns are discharged about every 10-15 s and sound pulses with a short rise time and very high peak sound pressure level (SPL) are emitted (source levels at ~250 dB re 1 μPa). Concerns have been raised about the impacts of seismic air gun emissions on marine life, and fishermen in particular claim that seismic surveys have a serious negative influence on fish distribution and commercial catch rates due to avoidance responses.

Research on the effects of seismic air gun sounds on marine fish includes studies on behavioral effects (e.g., Hassel et al. 2004; Pearson et al. 1992) and how catch rates are affected (Engås et al. 1996; Løkkeborg and Soldal 1993; Skalski et al. 1992). Here we provide an overview of the main findings of these studies and focus on their implications for commercial fisheries.

A comprehensive field study of this issue was recently carried out in the Norwegian gillnet and longline fishery, and the objectives and experimental design of this study are described and some preliminary results are presented. A thorough analysis of these data and findings will be presented elsewhere.

2 Behavioral Effects

Behavioral changes in fish exposed to human-generated sounds range from startle and avoidance responses to more subtle reactions such as changes in swimming activity, vertical distribution, and schooling behavior. A relationship between sound source level and the strength of the response is often observed, leading to behavioral responses of stronger intensity when a sound stimulus is progressively increased (Blaxter et al. 1981; Pearson et al. 1992).

A startle response (also called the C-start response) is a stereotyped response in which the fish's body forms a C-shape that usually points away from the sound source. The startle response is the most intense behavioral reaction to sounds observed in fish and often occurs after brief loud noises with a rapid rise time (see Blaxter et al. 1981; Eaton et al. 2001). Startle responses have been observed in several species exposed to seismic air gun discharges (*Sebastes* spp. [rockfish],

S. Løkkeborg (✉) • E. Ona • A. Vold • A. Salthaug
Institute of Marine Research, 5005 Bergen, Norway
e-mail: svein.lokkeborg@imr.no

A.N. Popper and A. Hawkins (eds.), *The Effects of Noise on Aquatic Life*,
Advances in Experimental Medicine and Biology 730, DOI 10.1007/978-1-4419-7311-5_95,

Pearson et al. 1992; *Dicentrarchus labrax* [European sea bass], Santulli et al. 1999; *Gadus morhua* [cod], *Pollachius pollachius* [Pollack], *Pollachius virens* [saithe], and *Merlangius merlangus* [whiting], Wardle et al. 2001; Ammodytes marinus [sandeel], Hassel et al. 2004). This response pattern may initiate an avoidance reaction to a noxious source, and observations of startle responses in studies of enclosed fish may thus be an indication of an avoidance reaction that would have taken place had the fish not been confined but observed in their natural environment.

Detailed observations of behavioral changes have been made in *Sebastes* spp. exposed to air gun sounds (Pearson et al. 1992). Fish held in a field enclosure showed changes in both swimming pattern and depth distribution during 10-min exposures to sounds from a single air gun. These observations suggested that subtle changes in behavior (changes in depth distribution and shifts to active behaviors such as eddying and milling) to sounds became evident at 161 dB re 1 μPa, and that changes in these behaviors became more extensive as sound level rose. The threshold for alarm responses (increases in activity and changes in schooling and position in the water) was observed at ~180 dB re 1 μPa, and the threshold for startle responses appeared to be between 200 and 205 dB re 1 μPa. Differences in response patterns were observed between the five *Sebastes* species studied.

This field experiment was performed on fish confined in enclosures and avoidance reactions would not have been observed. However, the changes in schooling behavior observed in *Sebastes mystinus* (blue rockfish) and *Sebastes melanops* (black rockfish) after air gun discharges may be analogous to the alarm responses seen in schools under attack by predators (Pearson et al. 1992).

Free-swimming gadoids (*Gadus morhua*, *Pollachius pollachius*, *Pollachius virens*, and *Merlangius merlangus*) inhabiting an inshore reef were observed during air gun firing (Wardle et al. 2001). All fish observed showed a C-start response at all ranges tested, with the maximum range being 109 m, giving a sound level of 195 dB re 1 μPa at the reef. The movement pattern of the fish was slightly changed, but no movements away from the air gun or the reef were observed. This lack of directional avoidance reaction raises some interesting points of difference between this study and several others, which demonstrate that gadoids leave the survey area during seismic shooting (see below). In the study by Wardle et al. (2001), the gun was stationary so that the sequences of shots showed no increase in intensity as with an approaching seismic vessel towing an array. The authors suggest that the air gun shot on its own is either too complex or variable in composition or too short to provide directional information that will allow the fish to respond with directed movements away from the sound source. In addition, this study was performed at a shallow inshore site that may cause reflections and give rise to a complex combination of pressure waves. Furthermore, a resident fish population inhabiting a reef may show marked site fidelity, whereas fish in the open sea may respond more freely to stimuli presented by seismic activities (Wardle et al. 2001).

Studies on behavioral changes in free-swimming fish exposed to air gun sounds have also been carried out in offshore waters. The vertical distribution of *Merlangius merlangus* was found to change in deeper waters during a seismic survey (Chapman and Hawkins 1969). *Sebastes* spp. and *Micromesistius poutassou* (blue whiting) were found in deeper waters in periods with seismic air gun shooting than during periods without shooting (Skalski et al. 1992; Slotte et al. 2004). Horizontal movements away from seismic survey areas have been observed in both demersal (Engås et al. 1996) and pelagic (Slotte et al. 2004) species.

3 Effects on Commercial Catches

Reduced catches on fishing grounds exposed to seismic survey activities have been demonstrated in three scientific studies (Engås et al. 1996; Løkkeborg and Soldal 1993; Skalski et al. 1992). These studies demonstrated pronounced catch reductions during periods of air gun sound emissions compared with preshooting levels, and the findings also indicate size- and species-specific differences in response pattern to air gun sounds.

Engås et al. (1996) adopted an experimental approach to investigate the spatial and temporal extent of the effects of seismic survey activities on local fish abundance and commercial catch rates in the Barents Sea. Continuous 3-dimensional (3-D) seismic shooting using an 18-gun array was performed over a 5-day period within an area of 3 × 10 nautical miles (nmi). Trawl catches of *Gadus morhua* and *Melanogrammus aeglefinus* (haddock) and longline catches of *Melanogrammus aeglefinus* in an area of 40 × 40 nmi centered around the shooting area declined by ~50% during the shooting period compared to a 7-day preshooting period, whereas longline catches of *Gadus morhua* were reduced by 21%. The catch reductions were most pronounced within the shooting area where trawl catches of both species and longline catches of *Melanogrammus aeglefinus* were reduced by ~70% and longline catches of *Gadus morhua* by 45%. The local abundances of *Gadus morhua* and *Melanogrammus aeglefinus* in the experimental area were estimated from acoustic mapping and found to decline in line with the catch reductions. A relatively greater reduction in catches and acoustic estimates was found in large (>60 cm) compared with small *Gadus morhua*. This comprehensive investigation demonstrated that seismic survey activity caused pronounced reductions in local abundance and catch rates of *Gadus morhua* and *Melanogrammus aeglefinus* within an area of at least 18 nmi from the shooting area, and these effects lasted for at least 5 days.

Similar reductions in catch rates caused by seismic activity were founded in an analysis of catch data obtained from commercial fishing vessels that happened to be operating on fishing grounds where seismic surveys were being carried out (Løkkeborg and Soldal 1993). This analysis found a 55-80% reduction in longline catches of *Gadus morhua* and a reduction of 80-85% in the bycatch of *Gadus morhua* in shrimp trawling.

Skalski et al. (1992) examined how single air gun emissions affected catchability in the *Sebastes* spp. hook-and-line fishery on the coast of California. A survey vessel traversed over fish aggregations on rock pinnacles at depths of 82-183 m and produced sound levels of 186-191 dB re 1 μPa at the base of *Sebastes* aggregations. There was an average decline in total catch rates of 52% during periods of sound emissions. This overall decline was reflected in the individual catches of three of the five most abundant *Sebastes* species caught. There was no sign of fish dispersing from the pinnacles, and the reduced catchability was explained by decreased responsiveness to baits and behavioral changes because fish schools were observed descending in the water column.

4 Effects on Norwegian Gillnet and Longline Fisheries: New Knowledge

A research project aimed at improving our understanding of how seismic surveys affect fish distribution and commercial fisheries was carried out in connection with the Norwegian Petroleum Directorate's seismic survey off the coast of Vesterålen (northern Norway) in the summer of 2009. The main objective of the project was to study the degree to which the species of fish occupying this area were affected by seismic shooting activity, whether by leaving the area or changing their behavior in other ways that might affect the fisheries.

Seismic 3-D data were collected during a period of 38 days within an area of 8 × 46 nmi, which overlapped with traditional fishing grounds for *Reinhardtius hippoglossoides* (Greenland halibut), *Sebastes marinus* (golden redfish), *Pollachius virens*, and *Melanogrammus aeglefinus*. Two local gillnet boats that fished for *Reinhardtius hippoglossoides* and *Pollachius virens*/*Sebastes marinus* were chartered as were two local longline boats that fished for *Reinhardtius hippoglossoides* and *Melanogrammus aeglefinus*. The fishing trials commenced 12 days before the start of seismic shooting and continued until 25 days after termination of the seismic survey. A research vessel carried out an acoustic survey of the distributions of fish to determine whether these changed between before, during, and after the seismic survey. Sound measurements were also made at a range of depths and distances from the seismic air gun array (see Øvredal and Totland, Chapter 109).

The project revealed that the sound of the air guns affected the fisheries in the study area in different ways, including both increased and reduced catch rates for the different species and types of gear. Gillnet catches of *Reinhardtius hippoglossoides* and *Sebastes marinus* increased during seismic shooting and remained higher after the end of the seismic shooting compared with preshooting catches. On the other hand, longline catches of *Reinhardtius hippoglossoides* fell during the seismic survey. The results for *Pollachius virens* showed a decline in gillnet catches both during and after seismic shooting. This decline in gillnet catches of *Pollachius virens* was in agreement with the acoustic survey estimates, indicating that *Pollachius virens* left the area in response to the seismic survey.

There were large day-to-day variations in longline catches of *Melanogrammus aeglefinus*, and differences in catch rates from before and during the seismic survey could not be demonstrated. The area in which the *Melanogrammus aeglefinus* fishing took place was less affected by the sound of the air guns than the fishing grounds for the other species because there was no direct overlap between this area and the seismic vessel transects. Nevertheless, there was a declining trend in *Melanogrammus aeglefinus* catches towards the end of the period of seismic activity when the distance to the survey vessel diminished.

The acoustic survey of the distribution of demersal fishes in this area largely confirms the results of the fishing experiments. During seismic shooting, lower densities of *Pollachius virens* were measured in the area, whereas no changes were demonstrated in the abundance or distribution of the other demersal species that could be ascribed to the seismic survey.

The results of this study provide clear indications that the fish reacted to the sound of the air guns in that catch rates changed (rose or fell) during the period of seismic shooting. These results can be explained by the fish raising their level of swimming activity, thus making the *Reinhardtius hippoglossoides* and *Sebastes marinus* more vulnerable to be taken by gillnets, whereas *Pollachius virens* may have migrated out of the area.

5 Conclusions

The studies cited above demonstrate that fish show behavioral responses to seismic air gun sounds by increasing their swimming activity and changing depth distribution and avoidance reactions. The hearing abilities of fish differ, and sensitivity and responses to seismic sounds are thus likely to vary among different species.

Observations indicate that fish that inhabit different habitat types behave differently to seismic sound. Avoidance responses have been observed in pelagic fish and fish staying close to relative smooth and featureless seabeds (e.g., Chapman and Hawkins 1969; Engås et al. 1996; Slotte et al. 2004), whereas fish associated with underwater structures (e.g., reefs, rock pinnacles) tend to be more stationary and are less likely to disperse during seismic air- gun emissions (Skalski et al. 1992; Wardle et al. 2001). Resident fish populations inhabiting a reef may thus show strong site fidelity, whereas fish in the open sea or on featureless banks may respond more freely to sound stimuli. An optimal strategy for fish associated with underwater structures should be to stay in shelter in response to loud sounds that may comprise a threat.

The recent Norwegian study on the effects of seismic surveys on gillnet and longline catches showed that the species studied all reacted to the air gun sound but that this effect was not reflected in movements out of the area and reduced catches of all species. Earlier studies, however, have revealed pronounced reductions in catch rates of several species for trawl, longline, and hook-and-line (Engås et al. 1996; Løkkeborg and Soldal 1993; Skalski et al. 1992). In these studies, the seismic shooting was concentrated within much smaller areas, which meant that the fish were exposed to stronger and more continuous sounds (number of air gun shots per unit area and period of time) than was the case in the recent study where the seismic vessel operated within a large survey area (i.e. along 46 nmi-long transects). Seismic air gun emissions distributed over a large area may thus produce lower sound exposure levels and thus have less impact on commercial fisheries.

References

Blaxter JHS, Gray JAB, Denton EJ (1981) Sound and startle responses in herring shoals. J Mar Biol Assoc UK 61:851–870.

Chapman CJ, Hawkins AD (1969) The importance of sound in fish behaviour in relation to capture by trawls. FAO Fish Rep 62:717–729.

Eaton RC, Lee RKK, Foreman MB (2001) The Mauthner cell and other identified neurons of the brainstem escape network of fish. Prog Neurobiol 63:467–485.

Engås A, Løkkeborg S, Ona E, Soldal AV (1996) Effects of seismic shooting on local abundance and catch rates of cod (*Gadus morhua*) and haddock (*Melanogrammus aeglefinus*). Can J Fish Aquat Sci 53:2238–2249.

Hassel A, Knutsen T, Dalen J et al. (2004) Influence of seismic shooting on the lesser sandeel (*Ammodytes marinus*). ICES J Mar Sci 61:1165–1173.

Løkkeborg S, Soldal AV (1993) The influence of seismic exploration with airguns on cod (*Gadus morhua*) behaviour and catch rates. ICES Mar Sci Symp 196:62–67.

Pearson WH, Skalski JR, Malme CI (1992) Effects of sounds from a geophysical survey device on behavior of captive rockfish (*Sebastes* spp.). Can J Fish Aquat Sci 49:1343–1356.

Santulli A, Modica A, Messina C, Ceffa L, Curatolo A, Rivas G, Fabi G, D'Amelio V (1999) Biochemical responses of European sea bass (*Dicentrarchus labrax* L.) to the stress induced by off shore experimental seismic prospecting. Mar Pollut Bull 38:1105–1114.

Skalski JR, Pearson WH, Malme CI (1992) Effects of sound from a geophysical survey device on catch-per-unit-effort in a hook-and-line fishery for rockfish (*Sebastes* spp.). Can J Fish Aquat Sci 49:1357–1365.

Slotte A, Kansen K, Dalen J, Ona E (2004) Acoustic mapping of pelagic fish distribution and abundance in relation to a seismic shooting area off the Norwegian west coast. Fish Res 67:143–150.

Wardle CS, Carter TJ, Urquhart GG, Johnstone ADF, Ziolkowski AM, Hampson G, Mackie D (2001) Effects of seismic air guns on marine fish. Cont Shelf Res 21:1005–1027.

Part VI
Population Effects

A Bioenergetics Approach to Developing a Population Consequences of Acoustic Disturbance Model

Daniel P. Costa

1 Introduction

In an effort to know how measurable short-term responses result in biologically meaningful changes in populations, a National Research Council Committee developed the population consequences of acoustic disturbance (PCAD) framework (National Research Council 2005). This framework detailed how behavioral responses to sound may affect life functions, how life functions are linked to vital rates, and how changes in vital rates cause population change through a series of transfer functions. However, many of these transfer functions are poorly understood. Here a bioenergetics model is described that can be used to parameterize these transfer functions and can identify species and/or particular life history characteristics that are likely to be sensitive or resilient to acoustic disturbance.

2 Energetic Model

Energetic models provide a means to quantitatively assess the effort animals spend acquiring resources as well as the relative way in which they allocate those resources. Measurements of energy acquisition and allocation provide a quantitative assessment of how animals organize their daily or seasonal activities and how they prioritize their behaviors. Thus, energy flow can be described as what goes into the animal as food and what comes out in the form of growth, reproduction, repair, waste, and metabolic work (Fig. 1). Survival and reproduction require a positive balance between the costs of maintenance and the acquisition of food energy. If a marine mammal cannot compensate for decreases in energy acquisition, it must either reduce its overall rate of energy expenditure or utilize its stored energy reserves. Conversely, to grow and reproduce, animals must obtain more energy than is needed to survive. The rate of prey energy acquired is directly related to the availability and quality of prey. As prey becomes less available, the cost of finding it increases and a greater proportion of time and therefore energy is expended searching for it (Fig. 2a).

D.P. Costa (✉)
Department of Ecology and Evolutionary Biology, University of California, Santa Cruz, Santa Cruz, CA 95060, USA
e-mail: costa@ucsc.edu

A.N. Popper and A. Hawkins (eds.), *The Effects of Noise on Aquatic Life*, Advances in Experimental Medicine and Biology 730, DOI 10.1007/978-1-4419-7311-5_96,

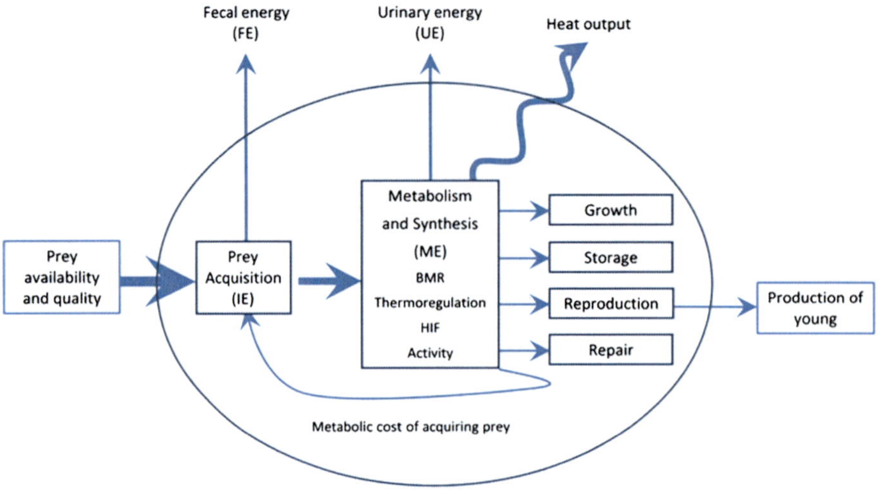

Fig. 1 Energy-flow diagram showing the relationship between energy acquisition and allocation. IE, ingested energy

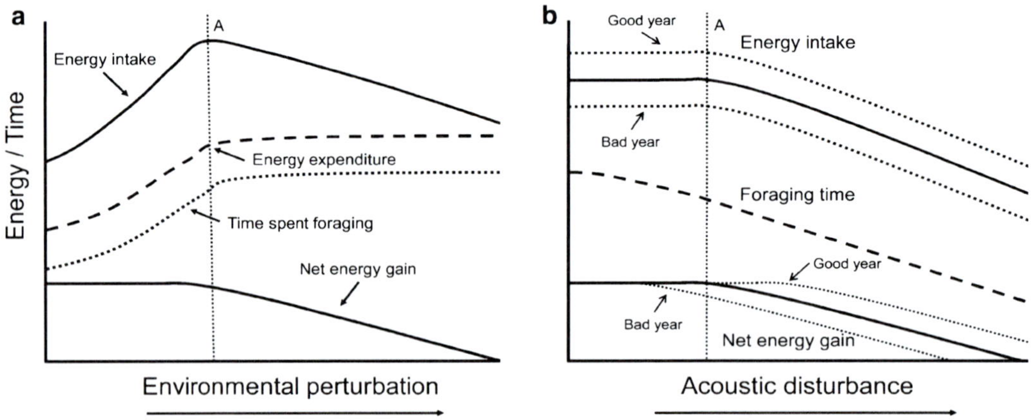

Fig. 2 (**a**) Response of an animal to reductions in prey availability due to environmental perturbation such as an El Niño-Southern Oscillation (ENSO) event. (**b**) Response to acoustic disturbance. Vertical dotted line (A) is the point beyond which the animal can no longer adjust its behavior to accommodate the perturbation and goes into an energy deficit. In b, in a bad year, animals have less ability to accommodate to acoustic disturbance, whereas in a better than average year (good year), animals have a greater capacity

Eventually, there is a threshold when more energy is spent searching for prey than is obtained and the animal goes into negative energy balance (Fig. 2a). Many marine mammals (especially capital breeders) undergo profound variations in this feast-or-famine dynamic equilibrium as they gain significant amounts of energy while feeding in highly productive environments followed by prolonged negative energy balance while fasting during migration or reproduction (Brodie 1975; Costa 1993; Lockyer 2007). This ability to store energy for later use provides some marine mammals with an ability to withstand periodic reductions in foraging time and thus may provide tolerance to acoustic disturbance. This is particularly relevant if the disturbance occurs during a period when foraging is less important, e.g., during migration. However, these species may also be more sensitive to acoustic disturbance if it occurs during these short intense foraging periods, especially if the available foraging habitat is spatially limited. Finally, this model does not account for all forms of cost because some activities may expose an animal to different risks of predation or disease. For example, an increase in foraging time may also result in a greater risk of predation.

It follows that we should be able to quantify the impact of acoustic disturbance by observing the change in the time-activity budget associated with a specific acoustic disturbance or, in the worst case scenario, we can assume that the animal does not forage over the entire period of disturbance. A similar model can then be developed where the acoustic disturbance results in a reduction in foraging time. Similar to the previous model, there is a range of disturbance that can be compensated for except here the animal compensates by working harder, spending less time in other activities, or increasing the intensity of its foraging effort. As long as the animal reduces its activity during the exposure period, its energy expenditure won't increase and the net energy intake will remain relatively constant. However, the ability of the animal to compensate will be lower during a "bad" year when resources are low and will be higher during a "good" year when resources are more abundant than normal.

The parameters needed to inform such an energetic model can be obtained from studies of animals responding to natural environmental fluctuations (e.g., El Niño-Southern Oscillation [ENSO] events; Costa 2008; Crocker et al. 2006; Le Boeuf and Crocker 2005; Testa et al. 1991). The response to these natural fluctuations can be used to quantify the linkage between changes in the time-energy budget (and associated energy deficit) and reductions in pup growth, weaning mass, and subsequent juvenile survival.

Because marine animals have evolved to live in a dynamic environment, it is likely that they have some ability to respond to environment perturbation. There are data to support the idea that some species are quite capable of accommodating to perturbations in their environment whether they are natural or human induced. For example, Goebel (2002) found that the weaning mass of northern fur seals was not tightly linked to success over an individual foraging tripbecause females could compensate for the difficulties encountered during a single trip on subsequent foraging trips. This suggests that success averaged over a series of foraging events may be more important than what occurs over a single foraging trip or period. Other studies indicate that some species are already working at their maximum capability and have little, if any, capacity to increase their foraging effort and thus any environmental perturbation is likely to result in reductions in offspring growth, weaning mass, and subsequent survival (Costa 2008). Finally, although animals may be resilient when resources are abundant, the opposite is true: animals will be more sensitive to acoustic disturbance during a bad year when resources are less available (Fig. 2b).

The ability to respond to environmental change appears to differ both between individuals and between species and has also been linked to differences in foraging behavior (Costa et al. 2004; Forcada et al. 2008). There are also striking differences in the ability of some animals to adjust their foraging tactics because they are already operating at or near their physiological limits (Costa 2008; Costa et al. 2004; Fig. 2, line A). This suggests that there may differences in the intrinsic ability of animals to respond to environmental fluctuations and human disturbance. That is, some species may be more tolerant to acoustic disturbance and others more sensitive, but we should be able to gain insight into these differences by examination of the plasticity in the animals' foraging behavior and ability from their response to natural environmental perturbations.

3 Life History and Behavioral Correlates

Although marine mammals exhibit an array of life history traits, they can be described as either a capital or an income breeder. Mysticete cetaceans and most phocid seals, like elephant seals, are capital breeders (Costa 1993; Lockyer 2007). Capital breeders accumulate and store the energy and resources needed for lactation before giving birth. On parturition, the female fasts during lactation to provision their young with milk that is derived from stored body reserves. The primary advantage of this life-history pattern is that the foraging grounds can be spatially and temporally separate from breeding grounds. All other marine mammals are income breeders. This is a more conventional life-history pattern where the female forages more or less continuously while she is lactating. These strategies confer different benefits and costs. Capital breeding disassociates reproductive success from local food availability. The nutritional provisioning of young by capital breeders (phocid and mysticete cetaceans) is thus largely unconstrained by traveling time to and from the foraging grounds, thereby allowing them to utilize prey that are more dispersed, patchy, unpredictable, or distant from the breeding grounds. The necessity of feeding during lactation constrains income breeders to forage closer to the rookery, thus linking reproductive success and local prey abundance (Costa 1993) and thereby potentially connecting population status to localized environmental changes (e.g., ENSO; Boyd and Murray 2001; Costa 2008; Testa et al. 1991; Trillmich et al. 1991).

4 Conclusions

A bioenergetic model can be used to assess the potential impact of an acoustic disturbance. Quantitative data on the range of disturbance that animals can tolerate and the linkage to population processes can be obtained from studies of animals responding to natural environmental variation. The PCAD working group, which is sponsored by the Office of Naval Research, is working on developing such models.

Acknowledgments This work was supported by a contract from the Exploration and Production Sound and Marine Life Joint Industry Programme, administered by the International Association of Oil and Gas Producers, London, UK.

References

Boyd IL, Murray AWA (2001) Monitoring a marine ecosystem using responses of upper trophic level predators. J Anim Ecol 70:747–760.

Brodie PF (1975) Cetacean energetics, an overview of intraspecific size variation. Ecology 56:152–161.

Costa DP (1993) The relationship between reproductive and foraging energetics and the evolution of the Pinnipedia. Symp Zool Soc Lond 66:293–314.

Costa DP (2008) A conceptual model of the variation in parental attendance in response to environmental fluctuation: foraging energetics of lactating sea lions and fur seals. Aquat Conserv Mar Freshw Ecosyst 17:S44–S52.

Costa DP, Kuhn CE, Weise MJ, Shaffer SA, Arnould JPY (2004) When does physiology limit the foraging behaviour of freely diving mammals? Int Congr Ser 1275:359–366.

Crocker DE, Costa DP, Le Boeuf BJ, Webb PM, Houser DS (2006) Impact of El Niño on the foraging behavior of female northern elephant seals. Mar Ecol Prog Ser 309:1–10.

Forcada J, Trathan PN, Murphy EJ (2008) Life history buffering in Antarctic mammals and birds against changing patterns of climate and environmental variation. Glob Change Biol 14:2473–2488.

Goebel ME (2002) Northern fur seal lactation, attendance and reproductive success in two years of contrasting oceanography. Dissertation, University of California, Santa Cruz, Santa Cruz, CA.

Le Boeuf BJ, Crocker DE (2005) Ocean climate and seal condition. BMC Biol 3:9.

Lockyer C (2007) All creatures great and smaller: A study in cetacean life history energetics. J Mar Biol 87: 1035–1045.

National Research Council (2005) Marine mammal populations and ocean noise: Determining when noise causes biologically significant effects committee on characterizing biologically significant marine mammal behavior. National Academy Press, Washington, DC.

Testa JW, Oehlert G, Ainley DG, Bengtson JL, Siniff DB, Laws RM, Rounsevell D (1991) Temporal variability in Antarctic marine ecosystems: Periodic fluctuations in the phocid seals. Can J Fish Aquat Sci 48:631–639.

Trillmich F Ono KA (1991) Effects of El Niño on pinnipeds. Springer-Verlag, Berlin.

Energetic Cost of Behaviors Performed in Response to Vessel Disturbance: One Link in the Population Consequences of Acoustic Disturbance Model

Dawn P. Noren, Robin C. Dunkin, Terri M. Williams, and Marla M. Holt

1 Introduction

Several studies have shown that cetaceans respond to the physical presence and/or acoustic emissions from marine vessels. For example, cetaceans perform surface-active behaviors (SABs) in response to an increase in the number of and/or close approaches by vessels (Lusseau 2006; Noren et al. 2009; Williams et al. 2002, 2009). SABs are often performed in bouts of one or more behaviors performed sequentially, and the majority of SABs provide both visual and acoustic signals that are important to social marine mammals. Indeed, the use of sound is essential to the survival and reproduction of cetaceans (National Research Council 2003), and because of this, anthropogenic sound exposure in marine mammals is a concern. Individuals may compensate for increased vessel noise by changing the amplitude (Holt et al. 2009; Scheifele et al. 2005), duration (Foote et al. 2004), repetition rate, and/or frequency of the sounds they produce.

Although many studies have described changes in the performance of SABs and acoustic signals in cetaceans relative to changes in vessel presence and background noise, it is difficult to quantify whether these changes in behavior impact marine mammal populations. The population consequences of acoustic disturbance (PCAD) model provides a framework to assess the biological significance of behavioral responses to disturbance (National Research Council 2005). This model describes several stages required to relate acoustic disturbance to the effects on a marine mammal population. Data on the physiological effects of sound exposure, including the energetic costs of performing behaviors in response to increased exposure to vessel noise, are critical to the PCAD model.

The aim of this study is to fill a data gap in the PCAD model by determining the metabolic cost of behaviors performed in response to vessel presence and associated noise. Specifically, in terms relative to the PCAD model (National Research Council 2005, Fig. 3–1), this study provides data on transfer function 2 (metabolic cost) of behavioral changes (performing SABs and vocalizations)

D.P. Noren (✉) • M.M. Holt
Marine Mammal Ecology Team, Northwest Fisheries Science Center, National Marine Fisheries Service, National Oceanic and Atmospheric Administration, Seattle, WA 98112, USA
e-mail: dawn.noren@noaa.gov

R.C. Dunkin • T.M. Williams
Department of Ecology and Evolutionary Biology, University of California, Santa Cruz, Santa Cruz, CA 95064, USA

A.N. Popper and A. Hawkins (eds.), *The Effects of Noise on Aquatic Life*, Advances in Experimental Medicine and Biology 730, DOI 10.1007/978-1-4419-7311-5_97,

that can immediately affect life functions (e.g., feeding rates). For example, if metabolic costs associated with these behavioral changes are significant, it is possible that cetaceans would need to increase their daily food consumption to meet their increased metabolic demands.

It is interesting to note that vessels can also disrupt foraging behavior in cetaceans. For example, *Orcinus orca* (killer whales) switch from foraging behavior to travel in the presence of vessels (Lusseau et al. 2009). Thus foraging opportunities and, consequently, energy acquisition can be reduced in the presence of vessels. If energetic demand is also increased due to the performance of SABs and vocalizations, then vessel impacts to cetaceans' daily energy budgets could be exacerbated.

2 Methods

Two trained adult male *Tursiops truncatus* (Atlantic bottlenose dolphins) were used in experiments to determine the metabolic cost of behavioral responses (SABs and vocalizations) to vessels by free-ranging animals. To determine the metabolic costs of performing SABs, oxygen consumption from the two dolphins was measured via flow-through respirometry immediately after they swam the length of the research pool, had bouts of tail slaps, and had bouts of bows (a proxy for breaches; Fig. 1). Oxygen consumption values were recorded continuously until levels reached resting values. To assess the energetic cost of sound production, oxygen consumption from the same dolphins was measured via flow-through respirometry before performing distinct stereotypical vocalizations, during the performance of the vocalizations, and during recovery after the performance of vocalizations. For both studies, respiration rates were also measured before, during, and after the performance of behaviors.

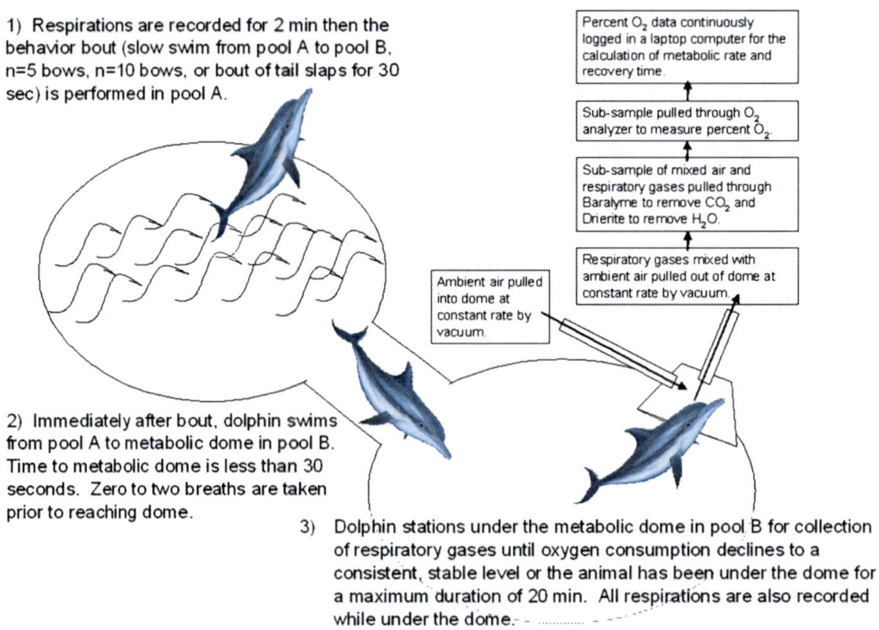

Fig. 1 Schematic of experimental trial to measure the energetic cost of performing surface-active behaviors

3 Results

Respiration and oxygen consumption rates after the performance of behaviors differed across trial types for both dolphins. Not surprisingly, respiration rates were highest after bouts of 10 bows and lowest after swimming. Metabolic rates measured after 10 bow bouts were higher than those after 5 bow bouts, whereas metabolic rates after tail slaps and slow swimming were the lowest. Energetic costs of slow swimming and bouts of tail slaps were similar to metabolic rates measured during rest. Recovery periods required for metabolic rates to return to resting values were greatest for the most energetically costly SAB bouts (10 bows and 5 bows). Furthermore, recovery periods after bouts of 10 bows were surprisingly long (≥14 min), particularly compared with the time required for the performance of these bouts (≤1 min). Data on the metabolic cost of sound production are still being analyzed so the results are not yet available. However, results from studies on sound production in other organisms suggest that these costs could be significant (e.g., Oberweger and Goller 2001).

4 Discussion

Experimental studies on trained marine mammals can be used to determine the energetic costs of behaviors performed in response to increased vessel presence and anthropogenic noise. These studies can provide useful data to populate information gaps in the PCAD model. Quantifying the energetic costs of behaviors performed in response to vessel disturbance will help us determine if short-term behavioral responses to disturbance have long-term individual- and/or population-level impacts.

The results of this study show that behaviors performed in response to increased vessel presence and anthropogenic noise can increase metabolic rates in cetaceans. As a consequence, the performance of energetically expensive SABs (e.g., breaches) in response to vessel presence and/or close approaches by vessels may impact daily energy requirements. For example, if the frequency of disturbance and the resulting behavioral responses are great enough to increase daily metabolic rates, daily food consumption will need to increase to meet these higher energetic demands. Furthermore, because cetaceans utilize sound to find food and/or coordinate feeding activities with conspecifics, increased noise due to vessel activity can also have the confounding effect of masking cetacean acoustic signals during foraging, consequently reducing the efficiency of foraging efforts. Moreover, some cetaceans actually cease foraging behavior in the presence of vessels (Lusseau et al. 2009). Thus vessel presence and the associated increase in ambient noise levels have the potential to increase energetic demand, reduce the efficiency of energy intake, and/or reduce opportunities for energy intake.

Acknowledgments This research was funded by Northwest Fisheries Science Center, National Marine Fisheries Service, National Oceanic and Atmospheric Administration, Seattle, WA, and Office of Naval Research (ONR), Arlington, VA, grants to T. M. Williams, D. P. Noren, and M. M. Holt. These experiments were made possible through the efforts of dolphin trainers Traci Kendall, Jen Walsh, Brett Long, and Beau Richter from the Mammalian Physiology Laboratory, University of California, Santa Cruz, Santa Cruz, CA. All protocols were approved by the University of California, Santa Cruz, Institutional Animal Care and Use Committee.

References

Foote AD, Osborne RW, Hoelzel AR (2004) Whale-call response to masking boat noise. Nature 428:910.

Holt MM, Noren DP, Veirs V, Emmons C, Veirs S (2009) Speaking up: Killer whales (*Orcinus orca*) increase their call amplitude in response to vessel noise. J Acoust Soc Am 125:EL27–EL32.

Lusseau D (2006) The short-term behavioral reactions of bottlenose dolphins to interactions with boats in Doubtful Sound, New Zealand. Mar Mamm Sci 22:802–818.

Lusseau D, Bain DE, Williams R, Smith JC (2009) Vessel traffic disrupts the foraging behavior of southern resident killer whales *Orcinus orca*. Endang Species Res 6:211–221.

National Research Council (2003) Ocean noise and marine mammals. National Academies Press, Washington, DC.

National Research Council (2005) Marine mammal populations and ocean noise. National Academies Press, Washington, DC.

Noren DP, Johnson AH, Rehder D, Larson A (2009) Close approaches by vessels elicit surface active behaviors by southern resident killer whales. Endang Species Res 8:179–192.

Oberweger K, Goller F (2001) The metabolic cost of birdsong production. J Exp Biol 204:3379–3388.

Scheifele PM, Andrew S, Cooper RA, Darre S, Musiek FE, Max L (2005) Indication of a Lombard vocal response in the St. Lawrence River beluga. J Acoust Soc Am 117:1486–1492.

Williams R, Bain DE, Smith JC, Lusseau D (2009) Effects of vessels on behaviour patterns of individual southern resident killer whales *Orcinus orca*. Endang Species Res 6:199–209.

Williams R, Trites AW, Bain DE (2002) Behavioral responses of killer whales (*Orcinus orca*) to whale-watching boats: Opportunistic observations and experimental approaches. J Zool 256:255–270.

Part VII
Anthropogenic Sound Sources and Their Measurement

A Common Sense Approach to Source Metrics

William T. Ellison and Adam S. Frankel

1 Introduction

The analysis and assessment of the impact of anthropogenic sound on the ocean environment require a clear understanding of the spatial, spectral, and temporal properties of the sources that generate the sounds and the animals that are exposed. The mantra "spatial, spectral, and temporal" is one that applies to all acoustic assessment problems and should serve as the underlying basis for the analysis toolbox anyone brings to bear on these issues. Table 1 correlates the salient aspects of the three features of sound production, transmission, and reception with this mantra.

Recently proposed criteria for the onset of physical injury from sound (Southall et al. 2007) has suggested levels potentially as high as 230 dB re 1 μPa (peak) or 215 dB re 1 μPa^2-s (frequency weighted) for selected combinations of sound type and marine mammal species. These levels are near to the maximum sound level that some sources can generate even at quite close distances and pose the issue of needing a better understanding of the properties of these sources and how the resultant sound field changes with distance, frequency, and time. Most sound sources used for military, resource exploration, or scientific purposes use an array concept to focus the resultant sound field in a given direction as well as to magnify the level of sound created, and the underlying principles of this source type is addressed below. The objective here is to provide the nonengineering professional the ability to make a first estimate on array source properties as well as simple formulas for some of the underlying terms.

2 Spatial Properties of Sources

The spatial character of typical underwater sound sources would appear to be the most intuitive of the three, yet is probably the most difficult. The typical symbolic representation of a sound source is a point in space from which sound emanates and progresses out into the surrounding medium, yet most productive sound sources, such as seismic air gun arrays, military sonar systems, and multibeam depth sounders, are not pointlike at all but rather a precisely distributed and operated array of transducers, each operating at a specified time and amplitude sequence. The two most important features of an array are its near-field character and its final beam-formed far field.

W.T. Ellison (✉) • A.S. Frankel
Marine Acoustics, Inc., Middletown, RI 02842, USA
e-mail: bill.ellison@marineacoustics.com; adam.frankel@marineacoustics.com

A.N. Popper and A. Hawkins (eds.), *The Effects of Noise on Aquatic Life*,
Advances in Experimental Medicine and Biology 730, DOI 10.1007/978-1-4419-7311-5_98,

Table 1 Sound production, transmission, and reception features partitioned into spatial, spectral, and temporal categories

	Source Characteristics	Sound Propagation	Sound Reception
Spatial	• Beam forming	• Ducted	• Receiver directivity (DI) and threshold (DT)
	• Near field/far field	• Refractive	• Noise field
	• Source level (SL)	• Bottom interacting	• Signal excess [SE = RL – (NL– DT)]
	• Particle velocity	• Transmission loss (TL)	
		• Receive level (RL = SL – TL)	
Spectral	• Narrowband	• Interface losses	• Hearing spectrum
	• Broadband	• Absorption	• Sound spectrum
			• Noise spectrum
			• Weighting
Temporal	• Pulse	• Multipath	Metrics (see Southall et al. 2007, Appendix)
	• Nonpulse		
	○ Tone		• SPL (root mean square)
	○ Frequency modulated (FM)		• Peak
	○ Continuous		• SEL
			• CSEL

Understanding the underlying physical principles governing the sound field in these two regions is needed to assess the acoustic impact within each region. Fortunately, the most important features of these two regions may be illustrated using a very simple linear array of equally spaced point sources and are further simplified by the fact that they are well described by a nondimensional variable, the ratio of all dimensions to the acoustic wavelength (λ = sound velocity/frequency).

2.1 Near Field of an Acoustic Array

An array example is shown below in Figure 1, where the array is made up of four individual transducers (or elements), each with a known source level, SL_E in dB re 1 μPa at 1 m, typically measured in a calibrated facility and known quite accurately as a function of input signal strength and time-frequency characteristics.

Line arrays are designed to create a focused beam in a direction normal to the axis of the array. Thus a vertical array, such as surveillance towed array sensor system low-frequency active (SURTASS LFA), produces an axis (also called a beam) of sound horizontally, and a horizontal array, such as a seismic air gun array, produces a vertical beam. Used in this manner, the amplification capable of such an array magnifies the output of a single element (SL_E) by a factor of 20 log(N_E), e.g., 12 dB for a 4-element array. If the elements of the array are spaced a fraction of a wavelength apart, say λ/b, then the calculation of the far-field "on-axis" transition, R_{FF}, for any array of (N_E + 1) elements is quickly determined. First, the length, $L = N_E\lambda/b$, and the far-field transition point must be greater than $(N_E/b)^2\lambda/2$. Thus, the far-field transition point grows as the square of the number of elements in a simple array and grows as well with increasing space between elements. A well-known misinterpretation of the sound levels associated with array sources comes from the fact that the published source levels are those needed to calculate the sound field in the far field. In reality, as one approaches the near field and closer to the source, the sound field defocuses, and very near the source, the field is actually dominated only by the sound field from the nearest source elements.

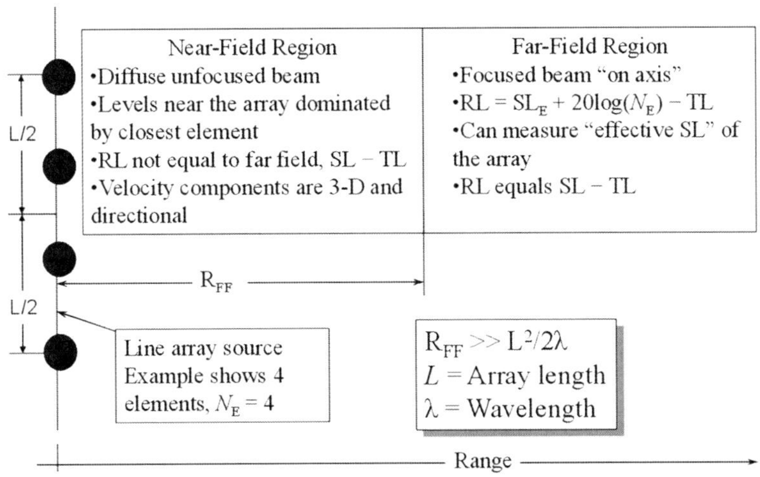

Fig. 1 Characteristics of the near and far field of a simple four-element line array

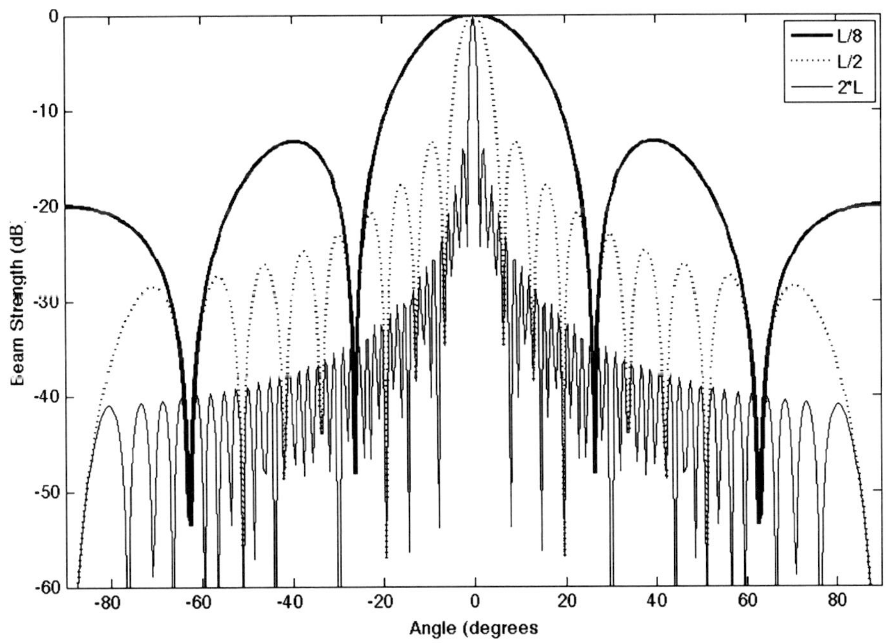

Fig. 2 Beam patterns for a 10-element line array spaced at λ/8, λ/2, and 2λ

2.2 *Beam Pattern of a Simple Line Array*

As with the near-field calculations, the beam-forming attributes of an array can also be formulated in terms of the nondimensional array-spacing term, λ/b. Figure 2 shows the resultant beam pattern for a 10-element line array at three different nondimensional element spacings, λ/8, λ/2, and 2λ.

There are several key features that are observed from this plot. First, the amplitude of the main lobe and each of the succeeding "numbered" side lobes are the same independent of spacing.

Recall that the main axis amplification is primarily a function of the number of elements, $20\log(N_E)$, but so are the side-lobe levels. However, they vary at what angle (relative to the axis) at which they occur. So the first side lobes (at ~40° for λ/8, 12° for λ/2, and 2° for 2λ) for each spacing example are each ~14 dB less than the main beam. The underlying pattern here is a classical $[\sin(x)/x]^2$ pattern where the side lobes occur at odd multiples of $x = [n\pi/2]$, $n = 3,5,7,9,\ldots.$ and $20\log[1/(np/2)] = -13.5, -17.9, -20.8, -23.0\ldots$.dB as roughly depicted in Figure 2. (See Chapter 3 in Urick [1983] for a more detailed exposition of different array configurations and formulas). Although some arrays are two and even three dimensional, knowing the spacing and number of elements in any given direction can provide a quick and simple and relatively accurate estimate of both the near-field extent, source amplification, and resultant beam pattern and side lobe levels at any given transmitting wavelength.

2.3 *Particle Velocity*

Particle velocity (U) is an integral component of any acoustic field. In the derivation of the wave equation for propagation of acoustic waves, one must consider the necessary elasticity (bulk modulus; *B*) and density (ρ) of the medium and the mechanical nature of the rarefaction and compression resulting from the presence of an acoustic field. Particle velocity is directional in nature. An easy way to visualize particle velocity is to place oneself in the far field of a sound source. Looking back toward the source, the particle velocity can be felt coming from that direction, and its local amplitude at distance is equal to the local acoustic pressure divided by the acoustic impedance of the medium, which is equal to the density times the sound velocity (*c*), or $U = P/(\rho c)$. In this idealized case and in the absence of other sources, the radial toward the source will be the only direction in which the particle velocity will be felt. However, as one approaches the source (and enters into the near-field range), the particle velocity increases faster than $P/(\rho c)$, and depending on the spatial extent of the source and its nearness to other boundaries (sea surface or bottom), particle velocity from other directions may also arise. Figure 3 provides a visual depiction of a simple one-element source in the presence of the ocean surface modeled as a dipole (Junger and Feit 1972, Eq. 3.10 et seq). The pressure field, radial particle velocity, and the tangential particle velocity are graphically illustrated at 50 Hz for a single-element array.

2.4 *Absorption*

The absorption of sound in the ocean is a well-documented effect, and although the primary governing variable is frequency, its effect is primarily noticed with propagation range because the loss is directly proportional to range, i.e., decibels/kilometer, and is primarily a dominant effect at frequencies greater than 10 kHz. Figure 4 shows the frequency dependence of absorption on transmission loss, calculated as $20\log(R) + \alpha R$, where R is range in kiloyards and α represents the absorption loss in decibels/kiloyard.

3 Spectral and Temporal Properties of Sources

Just as frequency (wavelength) and distance combine in a useful nondimensional way (λ/distance), so too do frequency and time combine in a very useful nondimensional construct termed time (*T*) and bandwidth (*W*) product, or *TW*. A limiting form of this term, $TW = 1$, actually provides a wealth of information about a number of common source signals.

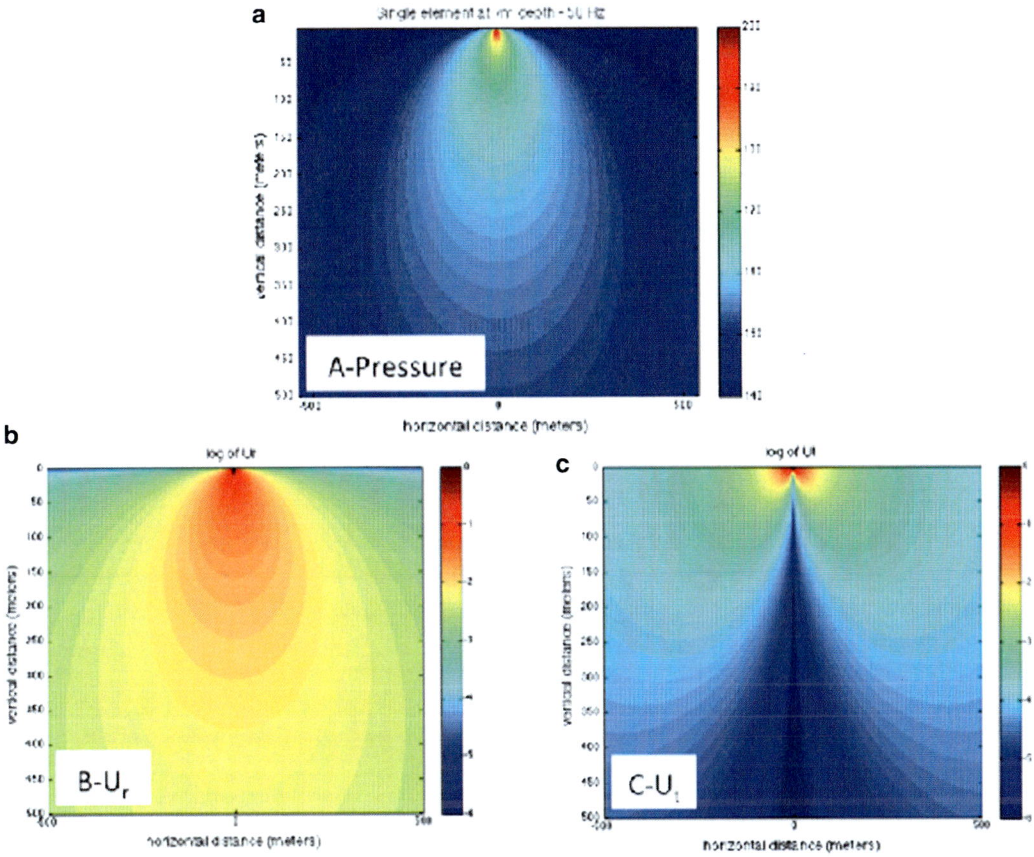

Fig. 3 Pressure field (**a**), radial particle velocity field (**b**), and velocity field in a tangential direction normal to the radial (**c**) from a single element at 7-m depth, broadcasting at 50 Hz

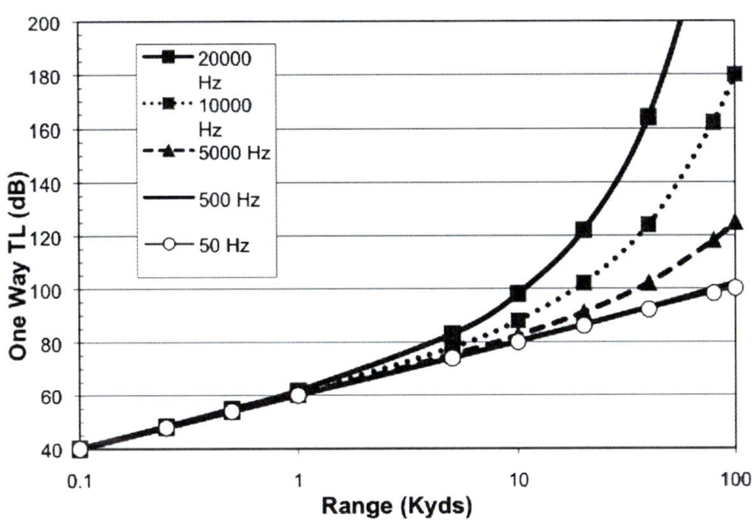

Fig. 4 Effect of absorption on transmission loss (TL) of different frequencies. Absorption increases with frequency, increasing TL. Note that the 50-Hz and 500-Hz curves barely differ

3.1 Pulsed Sounds

In Southall et al. (2007), the properties used to define pulsed sounds included a transient nature, high peak pressure values (both positive and negative), fast rise time, short duration (possibly non-linear), and broadband. Common source types fitting these characteristics included explosions, pile driving, and air guns. Almost any text on signal processing will contain a mathematical definition of an impulse as it is an elegant function in itself as well as a building-block function in transform theory. The following definition from Burdic's (1984) text is typical. He defines an impulse of strength A as

$$A\delta(t) = \lim(\mathrm{a} \to 0)\left[(A\ /\ Aa\ - a)\mathrm{rect}(t\ /\ ta\ - a)\right]$$

where

$$\begin{aligned}\mathrm{rect}(t\,/\,a) &= 1, \ -a\,/\,2 \le t \le a\,/\,2 \\ &= 0, \text{ elsewhere}\end{aligned}$$

Or the impulse exists only for a vanishingly short time as t approaches zero. In the frequency domain, achieved by a Fourier transform, this same result reveals the impulse as having infinite and uniform bandwidth. So from a very simple view, this is the ultimate $TW = 1$ function: short, peaky, and very broadband. And a useful rule of thumb on short pulse sounds of duration T is that the resultant bandwidth is well approximated by $W = 1/T$.

3.2 Tonal and Frequency-Modulated Sounds

The second $TW = 1$ set of sounds are pure tones, with vanishingly narrow bandwidth and infinitely long duration. Tonals have long played a major role in evaluating hearing thresholds. They are also used in sonar systems as the primary signal for discerning the Doppler shift from a moving target but are poor at resolving the distance accurately. Very short pulse sounds are good at range resolution but poor at Doppler. The most interesting signal is the frequency-modulated (FM) signal; it has the best range resolution and can carry extra energy so that it can be discerned in noise better than either tones or pulses.

References

Burdic WS (1984) Underwater acoustic system analysis (see Sec. 6.4.2). Prentice-Hall, Englewood Cliffs, NJ.

Junger M, Feit D (1972) Sound, structures and their interaction. MIT Press, Cambridge, MA.

Southall BL, Bowles AE, Ellison WT, Finneran JJ, Gentry RL, Greene CR Jr, Kastak D, Ketten DR, Miller JH, Nachtigall PE, Richardson WJ, Thomas JA, Tyack PL (2007) Marine mammal noise exposure criteria: Initial scientific recommendations. Aquat Mamm 33:411–522.

Urick RJ (1983) Principles of underwater sound for engineers. McGraw Hill, New York.

Underwater Sounds From Unattenuated and Attenuated Marine Pile Driving

James Reyff

1 Introduction

In 2009, the California Department of Transportation (known as Caltrans) published a guidance manual known as the *Fish Guidance Manual* on how to evaluate noise impacts from pile driving on fish (ICF Jones and Stokes and Illingworth and Rodkin, Inc. 2009). This manual provides detailed information and guidance in the process of assessing and mitigating potential impacts to fish. The manual includes as an appendix a detailed compendium of the underwater pile-driving sound data collected.

Underwater sound data of pile installation noise for the *Fish Guidance Manual* have been compiled for measurements collected over the period 2000 to 2006 during marine pile driving in coastal and river environments of northern California. Since that time, numerous other measurements have been conducted. The Washington State Department of Transportation (WSDOT) has conducted numerous field measurement studies during pile driving and has made the reports available on their Web site at http://www.wsdot.wa.gov/Environment/Air/PileDrivingReports.htm. Many projects included attenuation systems, such as bubble curtains, dewatered or bubbled casings, and cofferdams used to reduce underwater sound. Information on the effectiveness of the attenuation systems is presented in the compendium of the *Fish Guidance Manual* and reports provided by WSDOT. These reports describe the measurement and data-analysis methods as well as methods to attenuate sounds during the acoustical measurements.

The *Fish Guidance Manual* offers the most extensive set of underwater sound measurements made for pile driving. Much of the data collected were made near the source (e.g., at 10 m) to represent source levels. Many projects described included measurements at further distances and differing water depths. These measurements demonstrate the complexity of predicting underwater sound levels from these relatively shallow water environments. Predicting sound from pile-driving activity is difficult, but the compendium provides valuable data to assess these impacts.

J. Reyff (✉)
Illingworth and Rodkin, Inc, Petaluma, CA 94952, USA
e-mail: jreyff@illingworthrodkin.com

A.N. Popper and A. Hawkins (eds.), *The Effects of Noise on Aquatic Life*,
Advances in Experimental Medicine and Biology 730, DOI 10.1007/978-1-4419-7311-5_99,

2 Discussion

For all of the projects described in the *Fish Guidance Manual*, peak sound pressures were always measured along with root mean square (RMS) sound pressure levels (measured over the duration of a sound pulse). Since marine pile driving became an issue in the United States, peak sound pressures were suspected of being associated with fish injuries, so initial protective criteria were peak-pressure based. The RMS sound pressure level for a pulse has been used by the National Oceanic and Atmospheric Administration (NOAA) as criteria for assessing the exposure of marine mammals to sound. After an extensive review of studies involving fish exposure to underwater sounds, Hastings and Popper (2005) and Popper and Hastings (2009) identified the unweighted sound exposure level (SEL) as an indicator of injuries to fish. Carlson et al. (2007) documented the concept of accumulated SEL that was used to identify interim criteria to assess the dose of acoustical energy that a fish would receive from repetitive pile-driving sounds during a day. Currently, the descriptors peak pressure and accumulated SEL are used to assess possible adverse impacts to fish from pile-driving sounds (Fisheries Hydroacoustic Working Group 2008). The National Marine Fisheries Services use of peak sound pressures and SEL as criteria for assessing impacts on fisheries is described by Stadler and Woodbury (2009). The RMS sound pressure level continues to be used by NOAA to establish safety zones to protect marine mammals from these sound effects.

Underwater sound pressures from pile driving depend primarily on the size of the pile and the size of the hammer. Other factors, however, can cause large variations in measured sound pressures at a particular project site or from project site to project site. These factors primarily include water depth, tidal conditions or currents (if sound attenuation systems are used), geotechnical conditions that determine how difficult it is to drive the pile, and the contribution of groundborne sound.

Data from many of the projects that are described in the *Fish Guidance Manual* are summarized in Table 1 for impact hammers and vibratory installation. Not included in this table are sound levels associated with use of attenuation systems. Results from these projects were highly variable and cannot be summarized into one level for a certain type of pile or pile size. The *Fish Guidance Manual* includes an extensive set of pile-driving sounds from various projects. Information includes

Table 1 Summary of near-source (10-m) unattenuated sound pressures for in-water pile driving

Pile Type and Approximate Size	Relative Water Depth, m	Average Sound Pressure, dB		
		Peak	RMS*	SEL**
Impact pile driving				
0.30-m Steel H type – Thin	<5	190	175	160
0.6-m AZ steel sheet	~15	205	190	180
0.61-m Concrete pile	~15	188	176	166
0.36-m Steel pipe pile	~15	200	184	174
0.61-m Steel pipe pile	~15	207	194	178
0.8-m Steel pipe pile	~10	210	193	183
1.5-m Steel CISS	<5 m	210	195	185
2.4-m Steel CISS	~10	220	205	195
Vibratory pile installation				
0.30-m Steel H type	<5	165	150	150
0.30-m Steel pipe pile	<5	171	155	155
0.8-m Steel pipe pile	~5	180	170	170
0.6-m AZ steel sheet	~15	175	160	160
1-m Steel pipe pile - loudest	~5	185	175	175
1.8-m Steel pipe pile	~5	183	170	170

*RMS, root mean square; impulse level (35 ms average)

**SEL, sound exposure level for 1 s of continuous driving. CISS, cast-in steel shell the pile type; pile size; location of the project; water depth; distance from the pile where the data were collected; measured peak, RMS, and SEL levels where available; and an approximation of the attenuation rate. These data can be used as a ready reference and for comparative purposes when predicting sound levels for an upcoming project.

3 Signal Analysis

The *Fish Guidance Manual* includes selected acoustical analysis of recorded pile-driving signals. This includes pressure plotted over time (waveforms), and narrowband frequency spectra are usually provided. In addition, the accumulation of unweighted SEL is provided over a majority of the duration of the sound pulse or signal. From these data, measures of peak pressure, RMS, and SEL of the pulse can be obtained.

An example of these data are shown in Figure 1 for measurements at 0 m from a 0.76-m-diameter steel pipe pile driven with an air bubble curtain operating. This type of pile has an unattenuated sound level of ~210-dB peak, 193-dB RMS, and 183-dB SEL. The data presented in Figure 1 indicate the air bubble curtain system provided ~10 dB of sound reduction.

4 Accumulated SEL

Not included in the *Fish Guidance Manual* are measurements of accumulated SEL. This descriptor was not used before 2008 when most data were acquired. Since an interim agreement was reached with US west coast agencies, the accumulated SEL for pile-driving events are now measured or

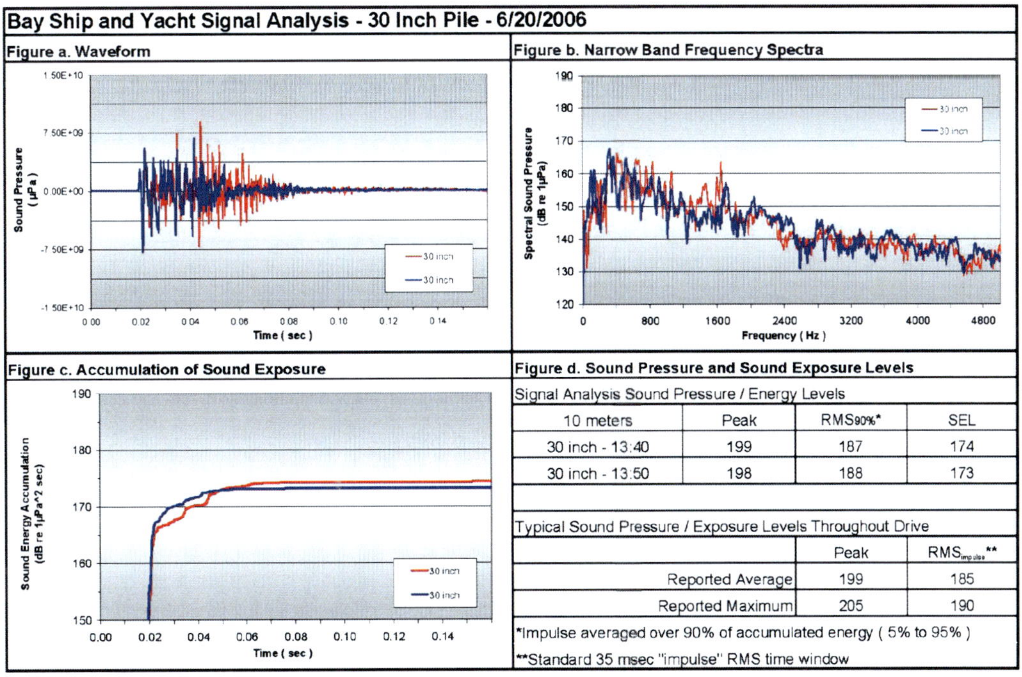

Signal Analysis Sound Pressure / Energy Levels			
10 meters	Peak	RMS90%*	SEL
30 inch - 13:40	199	187	174
30 inch - 13:50	198	188	173

Typical Sound Pressure / Exposure Levels Throughout Drive		
	Peak	$RMS_{impulse}$**
Reported Average	199	185
Reported Maximum	205	190

*Impulse averaged over 90% of accumulated energy (5% to 95%)

**Standard 35 msec "impulse" RMS time window

Fig. 1 Analysis of pile strike pulses for a 0.76-m-diameter pile with an air bubble curtain system. RMS, root mean square

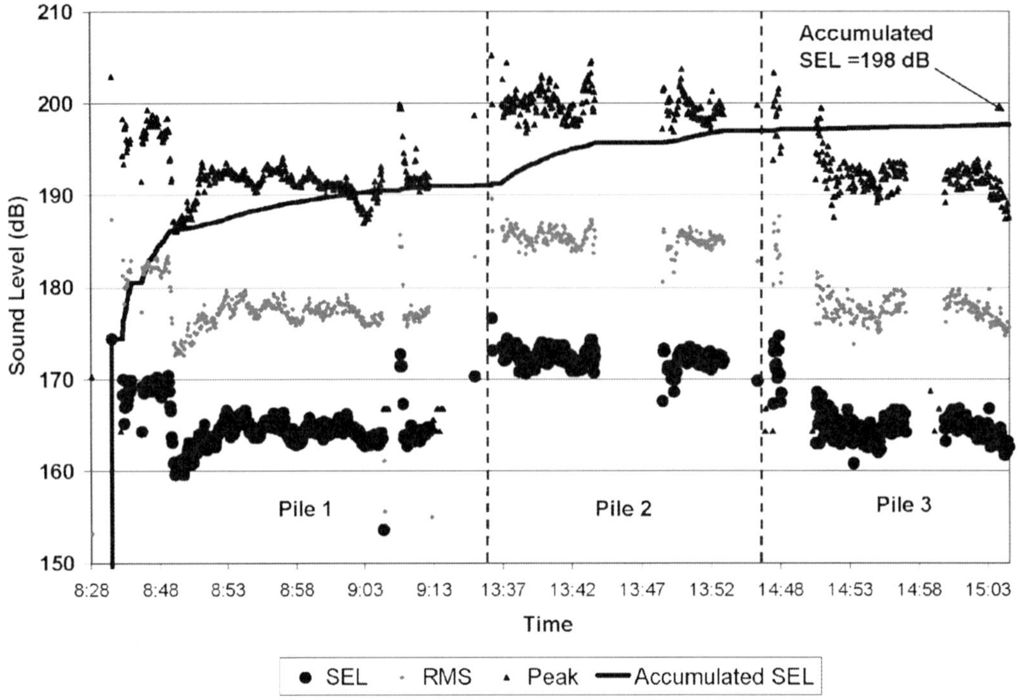

Fig. 2 Time history of sound pressure levels during the driving of 3 piles in 1 workday. SEL, sound exposure level

calculated for pile driving. Figure 2 shows the daily history of pile-driving sound during the installation of three piles for one project. These data are based on measurements at 10 m from each pile. In this case, the accumulated SEL for all 3 piles was 198 dB, whereas the single-strike SEL ranged from 160 to 175 dB. The actual duration of pile driving that day was less than 60 min or 2,400 pile strikes.

5 Vibratory Pile Installation

Vibratory hammers are routinely used to install piles before impact driving. Although peak sound levels can be substantially less than those produced by impact hammers, the total sound energy imparted can be comparable to impact driving because the vibratory hammer operates continuously and requires more time to install the pile. To meet or demonstrate pile-resistance requirements for some projects, piles need to be struck multiple times with an impact hammer; this can preclude the use of vibratory hammers in many cases.

6 Attenuation Systems

Various measures have been developed to reduce underwater sound generated by in-water pile driving. These measures fall into two general categories: 1) treatments that reduce the transmission of sound through the water, and 2) treatments to reduce the sound generated by the pile. The first category includes simple unconfined air bubble curtains, multiple-stage unconfined air bubble curtains, confined air bubble curtains, and cofferdams. The second category includes alternative hammer types such as vibratory hammers and oscillating, rotating, or press-in systems. The use of wood, nylon, and micarta pile caps also would fall in the second category. Information is currently available on the general effectiveness of various air bubble curtain systems and cofferdams. The WSDOT (2006) has tested the effectiveness of pile caps.

7 Conclusions

The *Fish Guidance Manual* offers the single most extensive set of underwater sound measurements made for pile driving. Much of the data collected were in accordance with agency requirements, which is near the source (e.g., at 10 m), representing source levels. Many projects described included measurements at further distances and differing water depths. The measurement data demonstrate the complexity of predicting underwater sound levels from these relatively shallow water environments. Predicting sound from pile-driving activity is difficult, but the dataset in the *Fish Guidance Manual* provides valuable data to assess these impacts.

The predictions of possible sound impacts are conducted during the design phase of projects in order to protect fish resources. This information is used by designers, engineers, and resource agencies to develop strategies to reduce the potentially harmful effects of pile-driving sounds to fish. Methods to predict the resulting underwater sound levels from pile driving are described in the *Fish Guidance Manual*. These methods are similar to those developed by NOAA's National Marine Fisheries Services, which is based on a spreadsheet calculator. Inputs to both of these methods require knowledge of the source levels and possible transmission losses over distance. Both of these data inputs may be estimated from the large volume of measurement data contained in the *Fish Guidance Manual*.

It is hoped that the compendium of sound data contained in the *Fish Guidance Manual* will be a continuously expanding document because additional data are almost constantly being acquired. For instance, underwater sound data for pile-driving activities in or near shallow rivers have been collected extensively since the compendium of sound data was developed. There are several projects that have provided a wealth of information on sound levels associated with these activities in these environments.

References

Carlson T, Hastings M, Popper AN (2007) Update on recommendations for revised interim criteria for fish during pile driving activities. Memorandum to California and Washington Departments of Transportation. Available via http://www.dot.ca.gov/hq/env/bio/fisheries_bioacoustics.htm. Accessed 28 April 2010.

Fisheries Hydroacoustic Working Group (2008) Agreement in principle for interim criteria for injury to fish from pile driving activities. Memorandum to National Oceanic and Atmospheric Administration's Fisheries Northwest and Southwest Regions; US Fish and Wildlife Service Regions 1 and 8; California, Washington, and Oregon Departments of Transportation; California Department of Fish and Game; and Federal Highway Administration. Available via http://www.dot.ca.gov/hq/env/bio/fisheries_bioacoustics.htm. Accessed 28 April 2010.

Hastings MC, Popper AN (2005) Effects of sound on fish. Report to California Department of Transportation. Available via http://www.dot.ca.gov/hq/env/bio/fisheries_bioacoustics.htm. Accessed 28 April 2010.

ICF Jones and Stokes and Illingworth and Rodkin, Inc. (2009) Technical guidance for assessment and mitigation of the hydroacoustic effects of pile driving on fish. Prepared for the California Department of Transportation. Available via http://www.dot.ca.gov/hq/env/bio/fisheries_bioacoustics.htm Accessed 28 April 2010.

Popper AN, Hastings MC (2009) The effects of anthropogenic sources of sound on fishes. J Fish Biol 75:455–489.

Stadler JH, Woodbury DP (2009) Assessing the effects to fishes from pile driving: Application of new hydroacoustic criteria. Proceedings of Inter-Noise 2009, 38th International Congress and Exposition on Noise Control Engineering, 23-26 August 2009, Ottawa, ON, Canada. Available via http://www.itre.ncsu.edu/ADC30/10WinterWrkshop_presentation_links.htm. Accessed 28 April 2010.

Washington State Department of Transportation (WSDOT) (2006) Underwater sound levels associated with pile driving at the Cape Disappointment boat launch facility, wave barrier project. Available via http://www.wsdot.wa.gov/Environment/Air/PileDrivingReports.htm. Accessed 28 April 2010.

What is the Source Level of Pile-Driving Noise in Water?

Michael A. Ainslie, Christ A.F. de Jong, Stephen P. Robinson, and Paul A. Lepper

1 Introduction

To meet the growing demand for carbon-free energy sources, the European Union (EU) has ambitious plans to increase its capacity for generation of offshore wind power. The United Kingdom and The Netherlands, for example, plan to increase their offshore power-generating capacity to 33 and 6 GW, respectively, by the year 2020. Assuming that this power is generated entirely by wind and that a single wind turbine can generate up to 10 MW, at least 3,900 offshore turbines would be required by these two states alone to achieve this goal. A popular turbine construction method known as "pile driving" involves the use of hammering a steel cylinder (a "monopile") into the seabed. A concern has arisen for the possible effect on mammals (Southall et al. 2007) and fish (Popper and Hastings 2009) of the sound produced by the succession of hammer impacts required to sink the pile to its required depth (tens of meters).

The EU plans to meet this concern by monitoring the sound of impulsive sound sources, including pile drivers, although a consensus has not yet developed over the most appropriate acoustic metric to be used. It is impractical to measure sound at every point where an animal might be so it makes sense instead to characterize the source in such a way that its impact can be estimated by modeling. It is conventional to parameterize a source of underwater sound by means of its "source level (SL)," which is a measure of its radiated power or energy. We describe the difficulties associated with both the meaning and measurement of source level in this context. The environmental impact is addressed in a companion paper (Lepper, Robinson, Theobald, Ainslie, and de Jong, Chapter 102).

M.A. Ainslie (✉)
TNO Sonar Department, The Hague, NL-2597 AK, The Netherlands
e-mail: michael.ainslie@tno.nl

C.A.F. de Jong
TNO Monitoring Systems, Delft, NL-2628 CK, The Netherlands
e-mail: christ.dejong@tno.nl

S.P. Robinson
National Physical Laboratory, Teddington, Middlesex, TW11 3TU, UK
e-mail: stephen.robinson@npl.co.uk

P.A. Lepper
Loughborough University, Loughborough, Leicestershire, LE11 3TU, UK
e-mail: p.a.lepper@lboro.ac.uk

A.N. Popper and A. Hawkins (eds.), *The Effects of Noise on Aquatic Life*,
Advances in Experimental Medicine and Biology 730, DOI 10.1007/978-1-4419-7311-5_100,

2 Definitions

We adopted the standard definitions (Morfey 2001) in terms of the acoustic pressure $p(t)$ during a time period (T) of sound exposure level: $\mathrm{SEL}(T) \equiv 10\log_{10}\left[E(T)E(T)\left(\mu\mathrm{Pa}^2\mathrm{s}\right)-\left(\mu\mathrm{Pa}^2\mathrm{s}\right)\right]$, where E is the sound exposure $E(T)=\int_0^T p^2 \mathrm{d}t$; of sound pressure level: $\mathrm{SPL}(T) \equiv 10\log_{10}\left[\left(E(T)/E(T)T-T\right)/\left(E(T)/E(T)T-T\right)\left(\mu\mathrm{Pa}^2\right)-\left(\mu\mathrm{Pa}^2\right)\right]$; and of peak pressure: $p_{\mathrm{peak}} \equiv \max\left|p(t)\right|$. No frequency weighting is applied. SL is sometimes defined as the SPL at a distance of 1 m from the source. Alternatively, it can be defined in terms of the product of the distance (s) from the source and the root mean square (RMS) pressure at that distance measured in far-field and free-field conditions [$p_{\mathrm{FF}}(s)$; Morfey 2001]: $\mathrm{SL}_{\mathrm{RMS}} \equiv 10\log_{10}\left[p_{\mathrm{FF}}(s)^2 s^2 / p_{\mathrm{FF}}(s)^2 s^2\left(\mu\mathrm{Pa}^2\mathrm{m}^2\right)-\left(\mu\mathrm{Pa}^2\mathrm{m}^2\right)\right]$. Because of the s^2 scaling and its far-field nature, this quantity is more closely related to the free-field radiated power than to intensity or mean square pressure. It is equal to SPL at 1 m only in very special conditions (de Jong et al. 2010). For transient sources, such as a pile driver, the averaging time for $\mathrm{SL}_{\mathrm{RMS}}$ is not well defined, so it is useful instead to define an energy SL (SL_{E}) in terms of the far-field and free-field sound exposure (E_{FF}) and scaled by s^2 in the same way, i.e., $\mathrm{SL}_{\mathrm{E}} \equiv 10\log_{10}\left[E_{\mathrm{FF}}(s)s^2 / E_{\mathrm{FF}}(s)s^2\left(\mu\mathrm{Pa}^2\mathrm{sm}^2\right)-\left(\mu\mathrm{Pa}^2\mathrm{sm}^2\right)\right]$. The definition of propagation loss (PL) to an arbitrary position **x** relevant to a transient signal is

$$\mathrm{PL}(\mathbf{x}) \equiv \mathrm{SL}_{\mathrm{E}} - \mathrm{SEL}(\mathbf{x}). \tag{1}$$

3 Methods and Measurements

Measurements are made of SEL in one-third octave bands as a function of distance from the pile driver. Use of Equation 1 makes it possible to estimate SL_{E} from a measurement of SEL and a model calculation of PL. If the duration (dt) of the transmitted pulse is known (at the sound source), $\mathrm{SL}_{\mathrm{RMS}}$ can be estimated using $\mathrm{SL}_{\mathrm{RMS}} \approx \mathrm{SL}_{\mathrm{E}} - 10\log_{10}\left[\delta t/\delta t 1\mathrm{s}-1\mathrm{s}\right]$. However, it is not clear how this duration can be estimated. It cannot be measured at the source. Similarly, no simple and unambiguous conversion to an SL defined in terms of peak pressure ($\mathrm{SL}_{\mathrm{peak}}$) is known to the authors. We therefore limited our attention to the SL_{E}, which we calculated by rearranging Equation 1.

Various models are available for the calculation of underwater acoustic propagation loss. The sound source is commonly modeled as a point monopole that, for simplicity, is assumed not to be in contact with either the sea surface or the seabed. While hiding the real problem under the carpet (the real source passes through the sea surface and is in firm contact with the seabed), this pragmatic approach is adopted here to make some progress. Uncertainty in the estimated SL associated with imperfectly known conditions is estimated by means of parameter variations. The risk of modeling error is mitigated by checking selected calculation results against the results of high-fidelity propagation models.

Inputs for the SL calculations are data from measurements of offshore piling activities in The Netherlands (NL; de Jong and Ainslie 2008) and the United Kingdom (Robinson et al. 2007). In these studies, a similar hydraulic piling hammer was used at the same nominal energy of 800 kJ/stroke. The pile diameters (f) were 4 m (NL) and 2 m (UK). At the UK site, the water depth (H)

varied between 8 and 15 m depending on local variations and the tide and the sediment mostly consisted of chalk. The water depth at the NL site (Q7) was ~21 m with a relatively flat, sandy bottom.

4 Results

Comparing the underwater noise produced at various piling sites does not require an estimation of SL. Interpolation or extrapolation of measurement data to the received SEL at a standard distance, e.g., 750 m (the reference distance for evaluating piling noise that is currently applied by the German government), introduces less uncertainty than the SL estimation. However, the advantage of SL estimations is that these can be used as input for prediction models, e.g., to produce noise maps.

Figure 1 gives an estimate of the SL_E spectrum for NL site Q7 for various choices of environmental conditions using an implementation of Weston's (1976) flux method, with all measurement points in the "mode-stripping" region. The curves represent the power averaged levels found from applying Equation 1 to the various measurement results. The large variations at higher frequencies are relevant for the prediction of the impact of piling noise on marine species that have a high-frequency hearing sensitivity (especially "high-frequency cetaceans"; Southall et al. 2007). The total broadband SL_E of these spectra varies between 215 and 220 dB re 1 μPa²m²s, with most of the energy in the frequency range of 100-500 Hz. Measurement distances from the pile were between 0.9 and 5.6 km.

At the UK site, broadband SELs of 178 and 164 dB re 1 μPa²-s were observed at distances of 57 and 1,850 m, respectively. Interpolation between measurement results at Q7 led to an estimated SEL of 168 dB re 1 μPa²-s at 1,850 m, i.e., 4 dB above the SEL observed at the UK site. Applying Equation 1 to the UK measurements yields SL_E in the range of 204.5 to 213.5 dB for f of 2 m and hammer energy of 800 kJ.

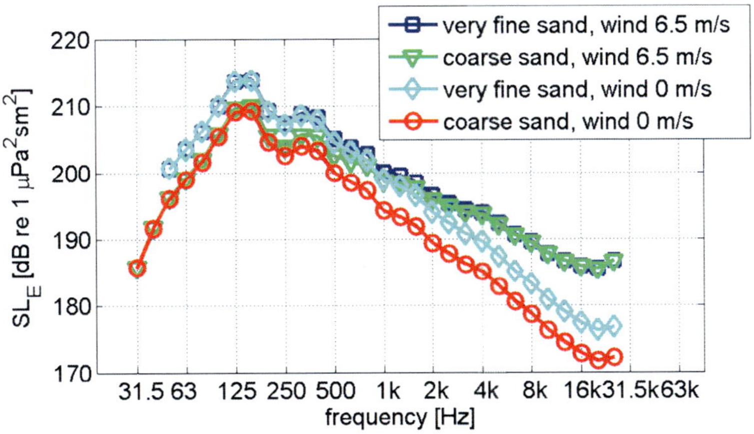

Fig. 1 One-third octave energy source level (SL_E) spectra estimated for The Netherlands site Q7. Pile diameter (f) = 4 m; hammer energy = 800 kJ

5 Conclusions

We sidestep fundamental questions concerning the definition of "SL" by idealizing the monopile as a single-point monopole that is not in contact with either boundary. Applying the energy conservation principle, we obtain values for the SL_E between 204.5 and 213.5 dB re μPa²m²s for the UK site (f = 2 m, H = 21 m, chalk) and between 215 and 220 dB for the NL site (f = 4 m, H = 8–15 m, sand), both for a hammer energy of 800 kJ. This range of SL_E estimations converts (de Jong and Ainslie 2008) to an acoustic source energy of 2.3 to 18 kJ/piling stroke for the UK site and 26 to 82 kJ/piling stroke (NL site). Thus, for the situations considered, the energy radiated as sound is between 0.3 and 10% of the total stroke energy.

An estimate of SL_{RMS} is possible but only if the duration of the transmitted pulse is known or estimated. An estimate of SL_{peak} is beyond the present scope because this requires a detailed understanding of the sound-generation mechanism and time-domain modeling of the radiated waveform.

Further research is needed to study radiation mechanisms. An internationally accepted standard definition of SL is urgently needed in order to facilitate comparison between measurements made using different methods, especially if SLs are expressed in the form of RMS or peak pressure.

References

de Jong CAF, Ainslie MA (2008) Underwater radiated noise due to the piling of the Q7 Offshore Wind Park. In: Zakharia M (ed) Proceedings of the European Conference on Underwater Acoustics, Acoustics'08, 29 June to 4 July 2008, Paris, France, pp 117–122.

de Jong CAF, Ainslie MA, Blacquière G (2010) Measuring underwater sound: Towards measurements standards and noise descriptors. Technical Report TNO-DV 2009 C613, TNO, The Hague, Netherlands.

Morfey CL (2001) Dictionary of acoustics. Academic Press, San Diego, CA.

Popper AN, Hastings MC (2009) The effects of anthropogenic sources of sound on fishes. J Fish Biol 75:455–489.

Robinson SP, Lepper PA, Ablitt J (2007) The measurement of the underwater radiated noise from marine piling including characteristics of a "soft start" period. Proceedings of the IEEE Oceans 2007 Conference, 18-21 June 2007, Aberdeen, UK, pp 1–6.

Southall BL, Bowles AE, Ellison WT, Finneran JJ, Gentry RL, Greene CR Jr, Kastak D, Ketten DR, Miller JF, Nachtigall PE, Richardson WJ, Thomas JA, Tyack PL (2007) Marine mammal noise exposure criteria: Initial scientific recommendations. Aquat Mamm 33:411–522.

Weston DE (1976) Propagation in water with uniform sound velocity but variable-depth lossy bottom. J Sound Vib 47:473–483.

Ground Roll Waves as a Potential Influence on Fish: Measurement and Analysis Techniques

Richard A. Hazelwood

1 Introduction

The influence of acoustic noise in the environment is well known, but ground vibration has been less studied, especially underwater. A program measuring anthropogenic noise and vibration started in 2010, led by the National Physical Laboratory (NPL), Teddington, UK. Unlike acoustic pressure waves, waves in solids take various forms, one of which is usefully described as a "ground roll," including the large waves that damage buildings even when well away from earthquake epicenters.

For this work, a theoretical simulation has been made of offshore wind-farm piling, a likely source of major vibration in the seabed. This understanding helps design instrumentation for the measurement task. Water pressure waves are usually measured in pascals (or dB re 1 μPa), but Hawkins and Johnstone (1978) described how particle motion provides a better measure of the impact on fish. Human sensitivity is given as velocity, a vector measured in meters per second. For acoustic waves in bulk water, pressures and velocities are linked by a substantially fixed acoustic impedance (pascals per [meter per second]), but this link changes dramatically near a boundary.

Some sensor designs are described, with details of the simple system currently being evaluated.

2 Simulation of Ground Roll Waves by Finite Element Analysis

In 1887, Rayleigh showed how waves propagated along the surface of a solid. He considered an idealized infinite half-space where a surface divides the solid below from a vacuum above. His analytic solution shows how elastic energy is exchanged between the vertical and horizontal motions and deformations of the material, described by a wave in which the particles describe an ellipsoidal "rolling" motion. There is an analogy with the rolling motion of sea waves where the particles, along with a floating cork, follow circular paths, with a radius that reduces with depth. These waves suffer little absorption and thus dominate the mechanical energy of the restless seas.

An exact solution for the speed of Rayleigh waves was published recently by Mechkour (2003). Their speed depends on the material properties, in particular the "Poisson ratio" (ν). This indicates how a rod increases in diameter when reduced in length. The figure shows a shear wave front for $\nu = 1/3$, with the Rayleigh wave moving 7% slower. Finite element analysis (FEA) allows ν to vary

R.A. Hazelwood (✉)
R & V Hazelwood Associates, LLP, Guildford, Surrey GU2 8UT, UK
e-mail: dick@r-vhazelwood.co.uk

A.N. Popper and A. Hawkins (eds.), *The Effects of Noise on Aquatic Life*,
Advances in Experimental Medicine and Biology 730, DOI 10.1007/978-1-4419-7311-5_101,

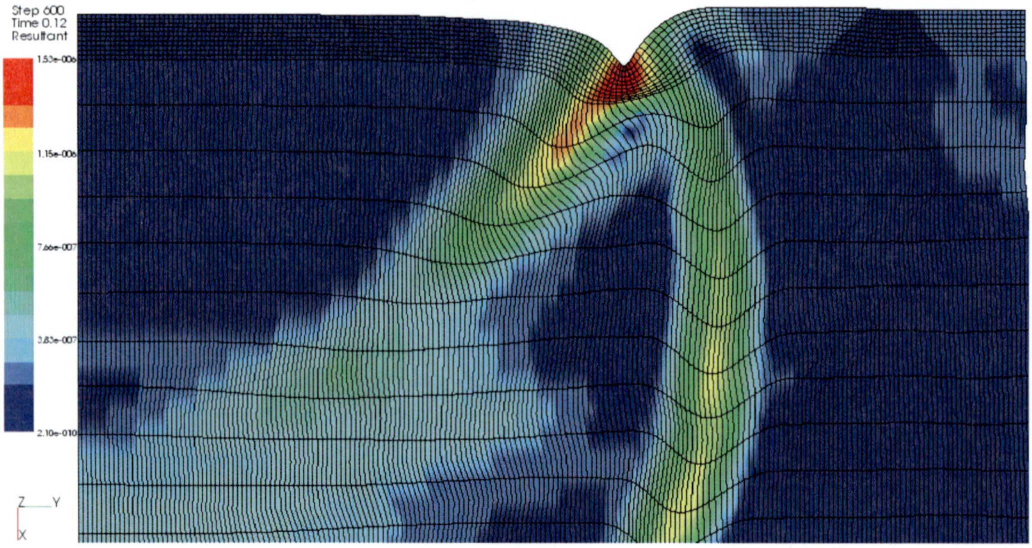

Fig. 1 Exaggerated deformation of a Rayleigh half-space 0.12 s after excitation by a 1-MN force

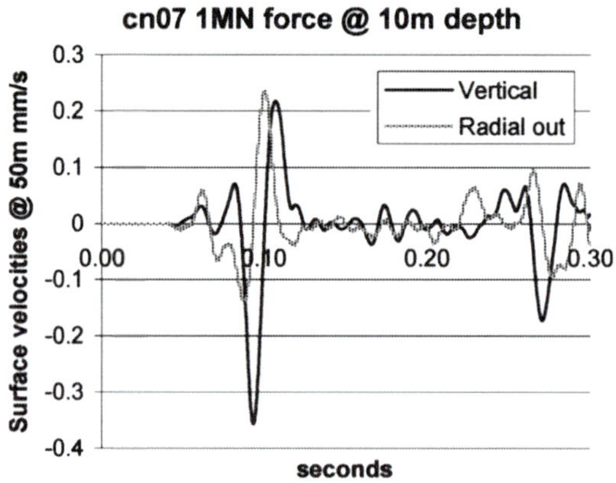

Fig. 2 Single-layer response to a 20-ms force pulse peaking at 1 MN

through the loosely compacted layers of seabed sediment. For a fluid, $\nu = ½$, and saturated clays have been shown to have similar values.

The FEA package from PACSYS Ltd., Nottingham, UK, gives this "pafVu," one frame (time step 600) of the animation produced (Fig. 1). Later models will simulate a water layer and a pile to study the interaction with acoustic propagation to longer ranges. A disadvantage of FEA is that models are limited in range if the computation is to be kept within acceptable timescales. Here the model is made axisymmetric about the x-axis. This simplification assumes that the source radiates equally in all azimuthal directions so that the model can be represented by this 2-dimensional (2-D) diagram. Data from actual piles show that they are maintained at less than 1° tilt as driven.

The meganewton downward force creates a ground roll wave moving in ever increasing circles.

At a 50-m radius on the surface, a small upward motion precedes the major dip. Little subsequent motion occurs until energy arrives from a bottom reflection at 0.17 s (Fig. 2), followed by a reflection from the outer 100-m radius at 0.27 s. These are artifacts of the finite model, to be discounted.

The analysis of response at the forcing point (2.5-m radius, 10 m deep) shows that the work done over the 20-ms pulse was 16 J, small compared with a typical piling-hammer blow where energy is measured in hundreds of kilojoules. But a force of 100 MN would inject 160 kJ of energy, more comparable with anticipated pile blow energies, and give a very large peak downward velocity of 35 mm/s. The efficiency of energy transfer by the pile will be investigated in the larger model.

This predicted scaling is comparable with data for construction piling given in BS 5228–2:2009 (British Standards Institution 2009).

3 Measuring Real Seabed Motion

Whereas acoustic pressure waves are recorded with hydrophones, the seismic industry uses large numbers of geophones. These are directional, resolving the vector components of the velocity, often used in sets of three, to measure all three components, much as is done with triaxial accelerometers. Unlike the typical piezoelectric accelerometer, they are magnetic and similar in structure to moving coil loudspeakers. Both vertical and horizontal types with the industry standard sensitivity of 20 V/(m/s) were purchased from China, and check calibrations were made using a laser vibrometer at NPL. The three geophones were accompanied by an inclinometer (all in one housing) and a separate hydrophone, all mounted on a stainless steel "sledge" (Figs. 3 and 4). The inclinometer module was used to report the sledge attitude because the high geophone sensitivity means they must be correctly orientated. A magnetic compass would be affected by the geophone stray fields so the sledge is designed to be oriented by a short tow across the seabed, also helping to embed its runners in the sediment to improve the coupling.

Figs. 3 and 4 Tests conducted at the National Physical Laboratory's (NPL) Wraysbury calibration facility used a triple geophone set. The additional plastic "hull" minimizes the risk of it sinking too far into the soft Wraysbury sediment

Fig. 5 With a graded layer, the seabed oscillates at 40 Hz

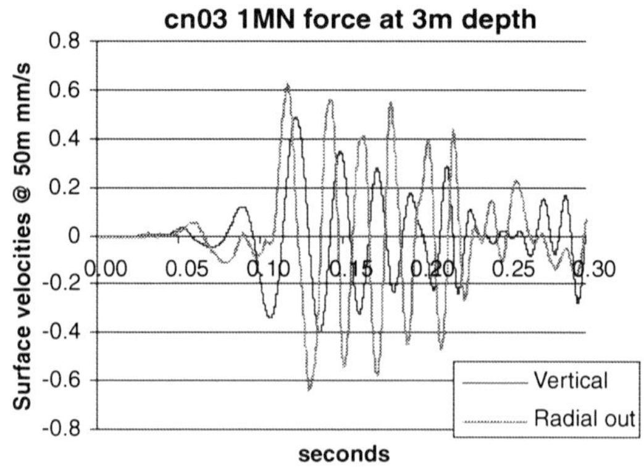

4 Modeling the Sediment

Reality will involve a sediment layer with lower stiffness than the deeper material. This more difficult problem can be considered by FEA, and graded material layers were tested using rough estimate properties designed to demonstrate principles rather than the absolute levels at this stage.

The ν varied from 0.5 at the surface to 0.33 in the bulk. The shear wave speed then increases from zero while the compression wave speed also rises from that in water. Typical results show an oscillatory response as the seabed wobbles after the impact. Similar "wavelet" responses are seen in saturated clay soils on land. The peak radial velocity seen in Figure 5 of 0.6 mm/s is now greater than the upward velocity, although the energy injected also doubled to 32 J.

As well as exceeding the human perception thresholds given by BS 5228–2:2009 (Brititsh Standards Institution 2009), it is orders of magnitude greater than the sensitivity of the salmon as measured by Hawkins and Johnstone (1978). If the same horizontal motion occurred in bulk seawater, the acoustic pressure would peak at 900 Pa, a sound pressure level (root mean square) of 170 dB re 1 μPa. However, it should be noted that results here are not as yet loaded by water.

For a wavelet, energy may be found a more appropriate metric than the peak velocity, in analogy to the use of energy integral criteria elsewhere (equivalent continuous sound exposure level [L_{eq}]). Robinson et al. (2007) have shown that the acoustic radiation is proportional to piling energy and hammer energy is widely quoted in kilojoules for larger machines.

Acknowledgments Dr. Patrick Macey (PACSYS Ltd., Nottingham, UK) provided considerable help in model building. This National Physical Laboratory (Teddington, UK) contract work also benefited from discussions with Steve Robinson and Pete Theobald.

References

British Standards Institution (2009) Code of practice for noise and vibration control on construction sites. British Standard 5228–2:2009.

Hawkins AD, Johnstone ADF (1978) The hearing of the Atlantic salmon, J Fish Biol 13:655–673.

Mechkour H (2003) The exact roots of the Rayleigh wave equation. Balkan Society of Geometers (BSG) Proceedings 8, Second International Colloquium of Mathematics in Engineering and Numerical Physics, 22–27 April 2002, Bucharest, Romania, pp 96–104.

Robinson SP, Lepper PA, Ablitt J (2007) The measurement of the underwater radiated noise from marine piling including characterisation of a "soft start" period. Proceedings of IEEE Oceans 2007 Conference, 18–21 June 2007, Aberdeen, UK, pp 1–6.

Assessment of Cumulative Sound Exposure Levels for Marine Piling Events

Paul A. Lepper, Stephen P. Robinson, Michael A. Ainslie, Pete D. Theobald, and Christ A.F. de Jong

1 Introduction

The installation of offshore wind farms in European waters and the scale of the planned activity have led to concern over the generation of noise and its potential impact on marine life. Much of this concern is centered around the noise generated by pile driving, which is used for the installation of the turbine foundations, and its potential impact on marine life (Thomsen et al. 2006). The noise generated by pile driving has the potential to cause injury, induce temporary or permanent hearing loss, and evoke avoidance reactions. One injury criterion for marine mammals is defined as the onset of auditory permanent threshold shift (PTS) (Southall et al. 2007), which is governed by either an instantaneous peak pressure or an integrated sound exposure level. The latter is the total noise energy to which the mammal is exposed during a given duration that, for a pile-driving source, would be either the duration of the piling or the time over which the mammal is in auditory range and is known as sound exposure level (SEL). In this case, cumulative exposure can be a useful parameter. This paper considers a summation of the SELs to which the animal is exposed during the entire piling sequence.

P.A. Lepper (✉)
Loughborough University, Loughborough, Leicestershire LE11 3TU, UK
e-mail: p.a.lepper@lboro.ac.uk

S.P. Robinson • P.D. Theobald
National Physical Laboratory, Teddington, Middlesex TW11 0LW, UK
e-mail: stephen.robinson@npl.co.uk

M.A. Ainslie
TNO Sonar Department, The Hague NL-2597, AK, The Netherlands
e-mail: michael.ainslie@tno.nl

C.A.F. de Jong
TNO Monitoring Systems, Delft NL-2628, CK, The Netherlands
e-mail: christ.dejong@tno.nl

A.N. Popper and A. Hawkins (eds.), *The Effects of Noise on Aquatic Life*,
Advances in Experimental Medicine and Biology 730, DOI 10.1007/978-1-4419-7311-5_102,

2 Fleeing Animal Model

The levels at the receptor (unweighted sound exposure level [SEL_0] received levels for a single hammer strike) used in this paper are based on the predictions calculated from a typical piling sequence measured in UK coastal waters. This allows the calculated cumulative exposures to be compared with the thresholds obtained from the literature, e.g., from the criteria published by Southall et al. (2007). To do this, a trajectory is chosen for each animal whereby the animal swims away (fleeing) from the source in a straight line at constant speed, heading, and depth. To calculate the cumulative SEL (SEL_{cum}), the energy received level is calculated for each individual hammer strike (Madsen 2005) and the animal's potential position at that time is then summed over the entire piling sequence.

Figure 1 shows a recorded SEL_0 received level at a single location for a complete piling sequence of 4,362 hammer strikes for a monopile in ~15 m of water. In this case, the total piling sequence took around 2 h 20 min with an ~8-dB increase in received level from the start to the maximum observed SEL_0 received level of ~155 dB re 1 μPa^2-s about two-thirds of the way through the sequence. Using range-dependent modeling and taking into account bottom bathymetry, the transmission loss on a bearing from the source at various ranges can then be estimated. Figure 2 shows an estimate of received level at a specific depth for a given source level as a two-dimensional

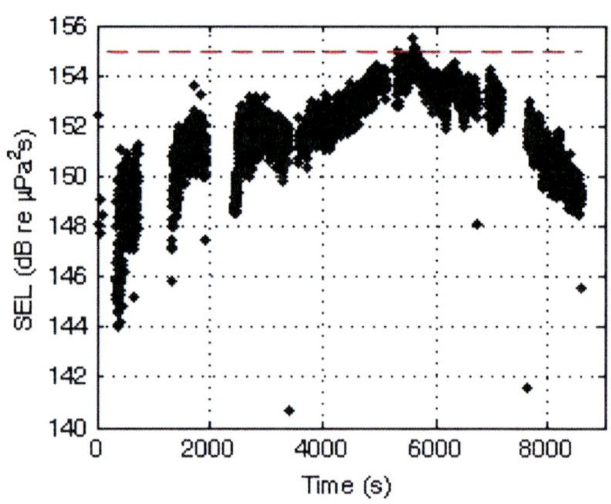

Fig. 1 Single-strike unweighted sound exposure level (SEL_0) received level for a piling sequence at a fixed location for a marine monopile in shallow water. Dashed line: maximum received level of 155 dB re 1 μPa^2-s

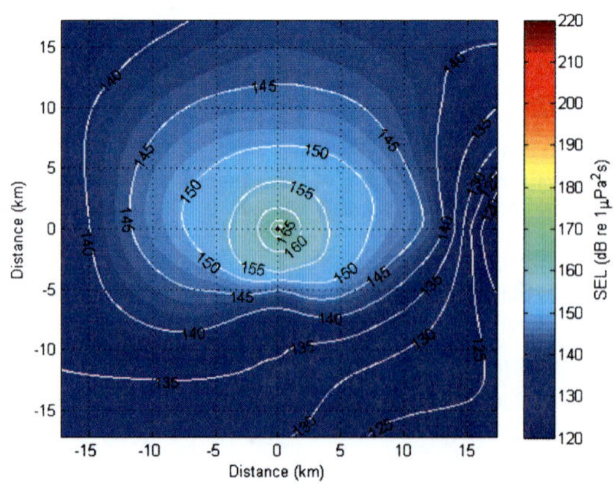

Fig. 2 Two-dimensional model of SEL_0 received level at a given depth surrounding a monopile source in a range-dependent bathymetry

profile around a source. Using this approach and the source variation data taken from Figure 1, the likely received level at the animal can be estimated for each hammer strike at any range and bearing from the pile location.

3 Cumulative Exposure Calculated for Marine Piling

Using the methodology described in Section 2, the fleeing mammal model has been used to calculate the cumulative exposure assuming a number of conditions. Figure 3 shows an example estimate of the unweighted SEL_{cum} for a maximum energy source level of 210 dB re 1 $\mu Pa^2 \cdot s \cdot m^2$ (Ainslie et al., Chapter 100) for the sequence given in Figure 1, a specific start distance from the source in this example of 100 m and an animal swim speed of 1.5 m s^{-1}.

Using the sequence time and amplitude data, the variation in source level for each hammer strike was calculated as representing the changes in source levels seen over time (soft start) or gaps (slow start) in specific piling sequences (Fig. 3, black trace and dots). The individual SEL_0 received level (Fig. 3, blue trace and dots) at the animal is then estimated for an animal swimming away from the source. The total exposure for each successive strike (Fig. 3, red trace and dots) was then added to give the total cumulative exposure for the entire piling sequence. This figure can then be compared with the known impact criteria threshold for a cumulative exposure.

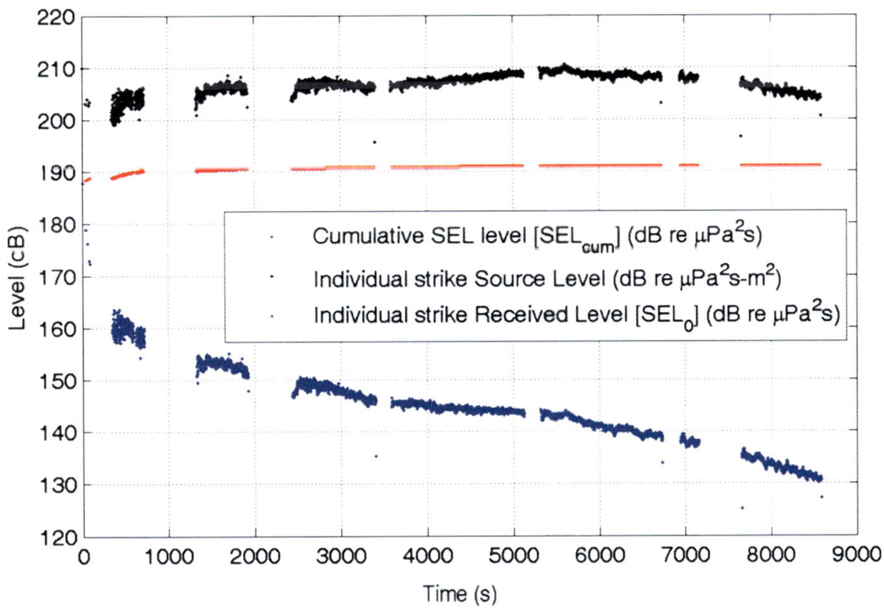

Fig. 3 Individual strike source level, SEL_0 received level, and cumulative exposure at receptor SEL_{cum} for a given piling sequence. The receptor was assumed to start 100 m from the source and swim away at a constant speed of 1.5 m s^{-1}

4 Impact Zone Prediction

The range from a source at which an animal starts, remains, or transects through and the area allowing an exposure in excess of predefined impact criteria often form the basis of impact assessments. In the case of a fleeing animal, the total cumulative exposure can be estimated for a given piling sequence on a known transect and start position. These models are then used to find a start range outside of which the total exposure is kept below a predefined threshold. Figure 4 shows the effect of start range on total SEL_{cum} (weighted and unweighted) for the piling sequence example shown in Figure 1, with a maximum example source level of 210 dB re 1 $\mu Pa^2 \cdot s \cdot m^2$ and a swim speed of 1.5 m s^{-1} applied to frequency-weighted functional hearing groups for marine mammals in both static and fleeing animal models as outlined by Southall et al. (2007). In this case, the difference in the static and fleeing animal models shows a marked increase in minimum start range to avoid exposure.

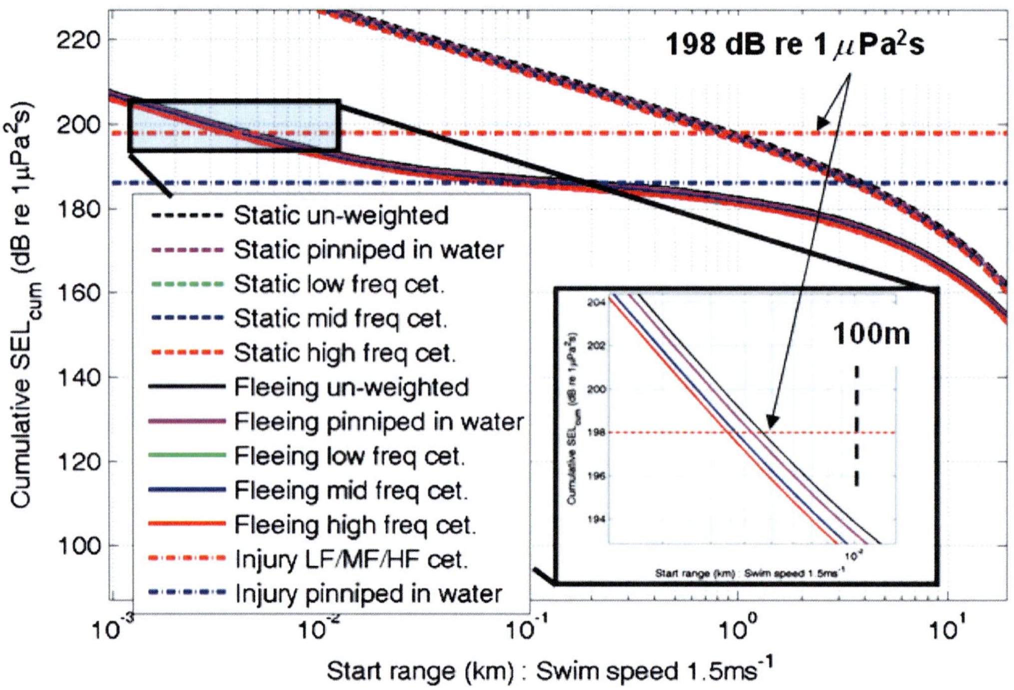

Fig. 4 Total SEL_{cum} versus start range for a typical piling sequence applied to different marine mammal functional hearing groups for both static and fleeing animal scenarios. LF, low frequency; MF, midfrequency; HF, high frequency

5 Conclusions

Both the fleeing and static model methods have been used to calculate the cumulative exposure/ SEL_{cum} for a typical piling event during the installation of a wind turbine monopile in shallow water. The actual sequence timing, number of hammer strikes, and variation in source level and shallow-water propagation loss properties are considered. Total exposure examples for functional hearing groups proposed by Southall et al. (2007) are given for each functional hearing group. This approach has also been applied to model variation in total source level (use of barrier methods) and effectiveness of soft start as an aid to development of mitigation strategies of various marine operations.

References

Madsen PT (2005) Marine mammals and noise: Problems with root mean square sound pressure for transients. J Acoust Soc Am 117:3952–3957.

Southall BL, Bowles AE, Ellison WT, Finneran JJ, Gentry RL, Greene CR Jr, Kastak D, Ketten DR, Miller JH, Nachtigall PE, Richardson WJ, Thomas JA, Tyack PL (2007) Marine mammal noise exposure criteria: Initial scientific recommendations. Aquat Mamm 33: 411–522.

Thomsen F, Ludemann K, Kafemann R, Piper W (2006) Effects of offshore wind farm noise on marine mammals and fish. Prepared by Biola, Hamburg, Germany, for the Collaborative Offshore Wind Energy Research into the Environment (COWRIE), London, UK.

Modeling the Propagation of Transient Sounds in Very Shallow Water Using Finite Difference Time Domain (FDTD) Calculations

Mark W. Sprague and Joseph J. Luczkovich

I ntroduction

Many soniferous fishes live in very shallow water (less than 20 m) close to shore and in estuaries, rivers, and lakes. Often these shallow-water fishes produce transient, or pulsed, sounds containing many frequency components (Fish and Mowbray 1970; Mann et al. 1997; Sprague and Luczkovich 2001).

Sound propagation in very shallow water is dominated by interactions of the sound waves with the water surface and bottom. These boundary interactions cause sound energy to spread cylindrically instead of the spherical spreading characteristic in deep water (Urick 1983). Multiple reflected propagation paths between the source and receiver result in a superposition of waves forming a combined sound that is a jumble of reflections arriving at different times and phases.

In this paper, we use the finite difference time domain (FDTD) method to model transient sound propagation in shallow water. This technique is useful for predicting the propagation of transient pulses in confined and irregular environments.

2 FDTD Method

The FDTD was first used to model electromagnetic wave propagation (Yee 1966), but it has been adapted for acoustics to model sound propagation in confined spaces such in ducts (Botteldooren 1994) and indoor spaces (Sakamoto et al. 2002). In the FDTD, the acoustic propagation equations for pressure and particle velocity are converted from differential to finite difference equations on a space and time grid. Spatial variations in pressure are used to calculate the changes in particle velocity, and spatial variations in particle velocity are used to calculate the changes in pressure in an alternating sequence known as leapfrogging.

M.W. Sprague (✉)
Department of Physics, East Carolina University, Greenville, NC 27858, USA
e-mail: spraguem@ecu.edu

J.J. Luczkovich
Department of Biology and Institute for Coastal Science and Policy,
East Carolina University, Greenville, NC 27858, USA
e-mail: luczkovichj@ecu.edu

A.N. Popper and A. Hawkins (eds.), *The Effects of Noise on Aquatic Life*,
Advances in Experimental Medicine and Biology 730, DOI 10.1007/978-1-4419-7311-5_103,

3 Shallow-Water FDTD Calculations

We adapted the FDTD to underwater acoustics to demonstrate its usefulness for modeling shallow-water propagation. To reduce computation time, we used a cylindrically symmetric geometry with a perfectly matched layer (PML) to eliminate reflections from the end of the spatial grid (Teixeira and Chew 1997). We used a simplified source-receiver geometry with constant depth and horizontally stratified acoustic parameters. Our shallow water had a sound speed of 1,536 m/s and a density of 1,024 kg/m^3, and the seafloor had a sound speed of 1,700 m/s and a density of 2,035 kg/m^3. The seafloor was at a depth of 3 m, and the sound source was at a depth of 2.38 m. We used the FDTD to calculate the propagation of a pressure impulse (Sakamoto et al. 2002) at the source to the receiver locations. Then, we performed a convolution of our impulse results at the receiver locations with a recorded *Cynoscion regalis* (weakfish) pulse to obtain the propagated pulse waveform. Figure 1 shows the plots of the waveform and power spectrum of the pulse as it propagates in the 3-m deep water. The pulse shape and the frequency components change significantly as the pulse propagates away from the source.

4 Discussion and Conclusions

This calculation shows how a transient pulse evolves as it propagates in very shallow water. The dominant frequency of the pulse does not change, but the pulse has very different time and frequency characteristics at each distance. This calculation demonstrates the usefulness of the FDTD

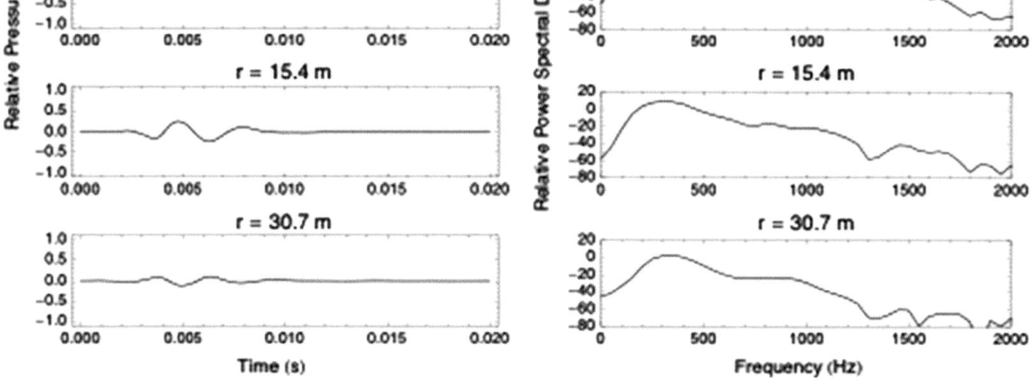

Fig. 1 Finite difference time domain (FDTD) calculations of the propagation of a single *Cynoscion regalis* pulse in 3-m deep water. The source and receivers were at a depth of 2.38 m. Left: Waveforms; right: corresponding power spectra. r, Radius

for modeling pulse propagation in very shallow water. The FDTD can be used in more complicated noncylindrical geometries such as those with asymmetrically varying depths (e.g., channels and sandbars) to predict pulse propagation in realistic geometries.

References

Botteldooren D (1994) Acoustical finite-difference time-domain simulation in a quasi-Cartesian grid. J Acoust Soc Am 95:2313–2319.

Fish MP, Mowbray WH (1970) Sounds of the western North Atlantic fishes. The Johns Hopkins University Press, Baltimore, MD.

Mann DA, Bowers-Altman J, Rountree RA (1997) Sounds produced by the striped cusk-eel *Ophidion marginatum* (Ophidiidae) during courtship and spawning. Copeia 1997:610–612.

Sakamoto S, Seimiya T, Tachibana H (2002) Visualization of sound reflection and diffraction using finite difference time domain method. Acoust Sci Technol 23:34–39.

Sprague MW, Luczkovich JJ (2001) Do striped cusk-eels *Ophidion marginatum* (Ophidiidae) produce the "chatter" sound attributed to weakfish *Cynoscion regalis* (Sciaenidae)? Copeia 2001:854–859.

Teixeira F, Chew W (1997) PML-FDTD in cylindrical and spherical grids. IEEE Microwave Guided Wave Lett 7:285–287.

Urick RJ (1983) Propagation of sound in the sea: Transmission loss, 1. In: Principles of underwater sound, 3rd edn. McGraw-Hill, New York, pp 99–146.

Yee KS (1966) Numerical solution of initial boundary value problems involving Maxwell's equations in isotropic media. IEEE Trans Antennas Propagation 14:302–307.Doi: 10.1109/TAP.1966.1138693.

Proximate Measurements of Acoustic Emissions Associated with the Installation and Operation of an Exploration Jackup Drilling Rig in the North Sea

Victoria L.G. Todd and Paul R. White

Underwater sound recordings were obtained from the Noble Kolskaya jackup gas-exploration drilling rig in the North Sea on the Dogger Bank. The aim was to document received levels, characteristics, and range dependence of sounds produced by the rig's site installation and drilling during the winter. Sound pressure levels (SPLs) generated by the Kolskaya were similar to previous measurements from metal-legged bottom-founded rigs both in level (120 dB re 1 μPa) and in frequency range of dominant tonalities (2-1,400 Hz). Received levels were highly variable over short periods and generally varied by 15-20 dB between quietest (holding) and loudest (drilling) operations. The rig was significantly quieter than its associated support vessels at low frequency, although radiated noise levels were higher above 2 kHz. Rig high-frequency SPLs dropped rapidly above 8 kHz. Noise is discussed with reference to the hearing capabilities of the harbor porpoise (*Phocoena phocoena*).

V.L.G. Todd
Ocean Science Consulting Ltd., Ocean House, Belhaven, Dunbar, East Lothian EH42 1PD, UK

Institute of Sound and Vibration Research, University of Southampton, Southampton, Hampshire SO17 1BJ, UK

P.R. White
Institute of Sound and Vibration Research, University of Southampton, Southampton, Hampshire SO17 1BJ, UK

A.N. Popper and A. Hawkins (eds.), *The Effects of Noise on Aquatic Life*,
Advances in Experimental Medicine and Biology 730, DOI 10.1007/978-1-4419-7311-5_104,

Measurement of Underwater Noise Arising From Marine Aggregate Operations

Stephen P. Robinson, Pete D. Theobald, Paul A. Lepper, Gary Hayman, Victor F. Humphrey, Lian-Sheng Wang, and Samantha Mumford

1 Introduction

As of 2009, there were 75 licensed areas within UK waters for marine aggregate extraction, with dredging activity taking place at any given point in time. In 2008, the dredged area totaled 137.9 km^2, extracting 21.24 million tons of sand and gravel for the building and construction industry. There is concern that the extraction of marine aggregate has the potential to generate noise levels that could have a negative impact on marine species in or around the dredging area. Although this paper deals only with the underwater noise generated and the potential impact it might have, there are many ways in which dredging can have an impact on marine life. For marine aggregate extraction, the type of dredger used is a trailing suction hopper dredger (TSHD). This type of dredger lowers a drag head and suction pipe to the sea floor, in water depths of up to 50 m, to extract the sand or gravel, depositing it in a hopper on the vessel. The vessel will often screen the dredged material for granular size and return the unwanted material and water over the side of the vessel. Such an operation can take from as little as 3 h to up to 12 h, concentrated in a relatively small area. The vessels usually dredge in tight lanes, usually less than 2 km in length and up to 100 m in width.

There has been very little work in the United Kingdom on assessing the noise generated by marine aggregate extraction operations and its potential impact on aquatic life. The most extensive measurements of dredging activity was undertaken in the Beaufort Sea during oil exploration activities in the 1980s, which included a number of examples of suction dredgers that are summarized in publications by Greene (1987) and Richardson et al. (1995). Although mostly lower frequency measurements and not strictly comparable with dredgers used in the United Kingdom for marine

S.P. Robinson (✉) • P.D. Theobald • G. Hayman
National Physical Laboratory, Teddington, Middlesex TW11 0LW, UK
e-mail: stephen.robinson@npl.co.uk

P.A. Lepper
Loughborough University, Loughborough, Leicestershire LE11 3TU, UK
e-mail: p.a.lepper@lboro.ac.uk

V.F. Humphrey • L.-S. Wang
Institute of Sound and Vibration, University of Southampton, Southampton, Hampshire SO17 1BJ, UK
e-mail: vh@isvr.soton.ac.uk

S. Mumford
Gardline Environmental Ltd., Great Yarmouth, Norfolk NR30 3NG, UK
e-mail: samantha.mumford@gardline.com

A.N. Popper and A. Hawkins (eds.), *The Effects of Noise on Aquatic Life*,
Advances in Experimental Medicine and Biology 730, DOI 10.1007/978-1-4419-7311-5_105,

aggregate extraction, the measurements did indicate that suction-dredging activities can generate source levels similar to those of larger tanker vessels when underway. Other measurements around Sakhalin Island, which were compared by Ainslie et al. (2009) with other vessels, including the Overseas Harriette measured by Arveson and Vendittis (2000), do support this finding that dredgers while extracting aggregate generate noise levels in the lower frequency region similar to those of larger tanker or cargo carriers while underway, as do other measurements in the United Kingdom that are not reported in the open literature.

This paper reports some initial results from measurements of one of the largest TSHD vessels in the UK fleet, the Sand Falcon, and considers the noise it generates under different operating conditions.

2 Generation of Noise From TSHDs

The potential sound sources during a dredging activity are numerous and their relative contributions to the overall noise generally are mostly unknown. Due to the suction pipe, drag head, and return of high volumes of excess water from the vessel's hopper over the sides from both spillways and screening towers, the noise generated by this type of vessel is very different from that of conventional ship noise. Even though their transit speed is very slow while dredging, typically less than 1.5 knots, they have been shown to generate one-third octave source levels at lower frequencies that are comparable to larger tankers and cargo vessels while underway at speed (Ainslie et al. 2009; Richardson et al. 1995).

The possible source mechanisms for a TSHD vessel while dredging will be propeller/thruster noise (very low speed), general hull-radiated hull noise (this could include internal pumps), drag head noise, overboard pump noise, suction pipe noise, and water and sediment discharge noise. The first two of these are common to other surface vessels, possibly with the exclusion of the large internal pumps on some vessels, but the others are unique to this type of dredging vessel and could radiate sound into water at higher frequencies than those normally associated with surface vessels. The water and sediment discharge has the potential to generate bubbles in the water and therefore broadband noise. The suction pipe and overboard pump also have the potential to generate broadband noise through friction and cavitation, respectively. The effect of the drag head is more difficult to postulate, but it is possible that it could generate some vibration in the seabed.

3 Measurement Methodology

To measure the noise generated from the dredging vessel, a series of static measurement locations were used while the dredger passed by dredging with screening, pumping only water (drag head lifted), and dragging its drag head while not pumping (pumps off).

The static measurement locations were provided by noise-monitoring buoys, designed and manufactured by Loughborough University, Leicestershire, UK, positioned at ~20-40 and 500 m and by a survey vessel positioned ~100 m off the dredging lane. The buoys were aligned to form a normal intersect with the dredging lane while the measurement vessel was set off this transect slightly. Both the buoys and the measurement vessel were positioned to measure the dredger along the center portion of its dredging lane length, which was ~1 km in length. The positions of the buoys and measurement vessel were marked with GPS while the complete track for the dredging vessel, the Sand Falcon, was provided by the vessel operators, Cemex, Surrey, UK, for the entire duration of dredging.

The survey vessel was used to deploy a broadband acoustic measurement system and a conductivity, temperature, and depth (CTD) sensor to determine the sound-speed profile. The acoustic measurement system deployed from the survey vessel consisted of two Reson TC4032 low-noise hydrophones on a weighted line with an integral bungee cord length, subsurface resistor, and surface float arrangement to reduce the low-frequency influence of wave motion. The water depth was ~30 m and the hydrophones were each at a depth of ~6.5 and 10 m. The hydrophones were attached to a B&K Pulse system and sampled with a 24-bit resolution at a sample rate of 200 kHz on each channel, providing a measurement bandwidth of 100 kHz. The hydrophones were deployed throughout the measurement period for around 6 h while the survey vessel was anchored and silent. The tidal flow peaked at ~3 knots during the measurements, and although some of the measurements were performed over a slack tide, local conditions resulted in relatively high-flow conditions even around the slack tide. The sea state was relatively flat during the measurements at around Force 2. The water temperature was measured to be ~4.6°C and showed negligible changed over a depth of 20 m.

4 Results and Discussion

The data gathered were extensive, and so only a small sample of the data is presented here. Given that full analysis has not been completed at this stage in the project, only the one-third octave band received levels at the measurement locations are presented. To assess the characteristics of the noise generated by the dredging activity, the received levels are shown (see Fig. 1) for the different operational conditions of the Sand Falcon. These are full dredging (sucking sand and gravel from the seabed), pumping only water (with drag head lifted but all pumps still running), and no pumping but still dragging (pumps off with the drag head on the seabed). Background noise measurements were also performed on the following day once the Sand Falcon had left the area and these are also included in Figure 1.

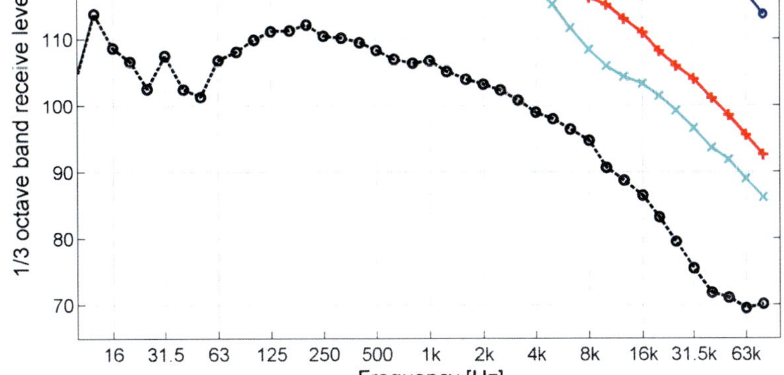

Fig. 1 One-third octave band spectra for the Sand Falcon full dredging, pumping only water (drag head lifted), drag head on seabed with pump off, and background noise measured at ~100 m

The results shown in Figure 1 are from when the Sand Falcon was at its position of closest approach to the survey vessel and are taken using a 4-s spectral average over ~90 s of data in each case, except for the background noise that was from 7 min of measurement data. This was measured from the survey vessel using the shallower of the two Reson TC4032 hydrophones. The data clearly show differences in the higher frequency noise levels (above 1 kHz), with full dredging approaching levels that are 20 dB higher above 16 kHz compared with pumping water only or dragging the drag head with no pumping. Although this has not been fully analyzed, this does indicate that it is the effect of the sand and gravel passing through the suction pipe and pump that generates much of the higher frequency noise. It should be noted that dragging the drag head with no pumping is not an actual operational state for dredging activities and was done as part of this exercise to help identify the different sources that contribute to the spectra.

To assess the impact of the noise generated from marine aggregate extraction operations, it is necessary to analyze the existing data to calculate the one-third octave source level data and then use these, with detailed knowledge of the transmission loss in the area, to estimate the zones of potential impact on different marine species. The work reported here is just the initial results from the first part of such a study.

5 Conclusions

A comprehensive set of acoustic measurements has been completed of the underwater noise radiated from a large trailing suction hopper dredger in shallow UK coastal waters while extracting sand and gravel from the seabed. Initial analysis of the received level data in one-third octave bands shows significant variation between operational dredger conditions at higher frequencies, which also indicates that the dominant noise source at frequencies above 2 kHz might be caused by the sand and gravel passing through the suction pipe and pump. The peak levels, however, do occur below 500 Hz and are consistent with ship noise and hull noise.

Acknowledgments We acknowledge the financial support of the Marine Aggregate Levy Sustainability Fund, administered by the Marine Environmental Protection Fund, and the support and cooperation of the British Marine Aggregate Producers Association and Cemex UK.

References

Ainslie MA, de Jong CAF, Dol HS, Blacquière G, Marasini C (2009) Assessment of natural and anthropogenic sound sources and acoustic propagation in the North Sea. Report TNO-DV 2009 C085, TNO, The Hague, The Netherlands.

Arveson PT, Vendittis DJ (2000) Radiated noise characteristics of a modern cargo ship. J Acoust Soc Am 107:118–129.

Greene CR (1987) Characteristics of oil industry dredge and drilling sounds in Beaufort Sea. J Acoust Soc Am 82:1315–1324.

Richardson WJ, Greene CR Jr, Malme CI, Thomson DH (1995) Marine mammals and noise. Academic Press, San Diego, CA.

Svein Vaage Broadband Air Gun Study

Anders Mattsson, Gregg Parkes, and David Hedgeland

1 Introduction

Marine seismic exploration is a method for collecting geophysical data that offer an opportunity for a detailed look at the geological structure beneath the seabed. The product of a seismic survey can be either a two-dimensional or a three-dimensional image, which can then be used to identify potential areas for oil and gas exploration and production. Seismic imaging is analogous to the ultrasound technology that is commonly used in the medical profession for imaging the human body.

A typical marine seismic operation uses a source that consists of several acoustic source elements and a receiver, usually a cable or streamer that houses many hydrophone sensors. Both source and receiver arrays are deployed from and towed behind the seismic vessel. Today, the high-pressure air source or air gun is used almost exclusively during marine seismic operations. The source releases a pulse of energy that travels downward through the water column and into the seabed and the subsurface. When it reaches the various sediment formations, some of the energy is reflected and travels back to the receivers, where the data are then digitally transferred and recorded onboard the survey vessel. For typical seismic operations, the useful frequency band is in the 0- to 100-Hz range.

The time taken to complete a seismic survey is affected by factors such as obstructions, tides, and weather as well as the turns between sail lines. Therefore, although a seismic vessel can operate on a 24-h basis, the seismic source is not continuously active all day every day. Fair weather conditions and wave heights below 4 m are usually needed to ensure data quality and image of the subsurface.

In many areas of the world, governments and regulatory authorities strive to embrace the latest scientific data or best available science/technology as the basis for implementing up-to-date and practical mitigation and monitoring methods for protection of the marine environment. In 2006, the Exploration and Production (E&P) Sound and Marine Life Joint Industry Programme (JIP) was established. The JIP is a multimillion-dollar multiyear commitment to support research activities that will improve our understanding of the potential interactions between marine life and E&P operations offshore. The JIP is administered by the International Association of Oil and Gas Producers (OGP; London, UK) and is supported by multinational exploration and production

A. Mattsson (✉)
Petroleum Geo-Services, 1366 Lysaker, Norway
e-mail: anders.mattsson@pgs.com

G. Parkes • D. Hedgeland
Petroleum Geo-Services, Weybridge, Surrey KT13 0NY UK
e-mail: gregg.parkes@pgs.com; david.hedgeland@pgs.com

A.N. Popper and A. Hawkins (eds.), *The Effects of Noise on Aquatic Life*,
Advances in Experimental Medicine and Biology 730, DOI 10.1007/978-1-4419-7311-5_106,

companies and the International Association of Geophysical Contractors (IAGC; Houston, TX) representing the geophysical industry. There are five broad subject categories in which the JIP supports research activities: sound source characterization and propagation; physical, physiological, and hearing effects of sound; behavioral reactions and biological significance; monitoring and mitigation; and research tools. Under the sound source characterization and propagation research category, Petroleum Geo-Services (Lysaker, Norway) conducted the Svein Vaage broadband air gun study.

2 Measurements

The emission characteristics of a seismic air gun below 200 Hz are well known and have been well documented by the industry. There has also been significant research at frequencies up to 500 Hz; however, so far there has never been a thorough investigation at higher frequencies. Extensive new measurements, up to 50 kHz, have now been made of the most commonly used air gun types in the industry. These new data will help improve the theoretical models of air guns, particularly at very high frequencies, and will also greatly improve our understanding of how the seismic air gun might affect the marine environment (Mattson 2008).

3 Test Site and Duration

A test site was specifically designed and constructed in a fjord on the west coast of Norway. Both near-field and far-field measurements of single air guns and clusters (two air guns located side by side) were conducted from June to October 2007 and June 2009 to June 2010. To minimize any risk to nearby fish farms, the nearest fish farm was monitored by a local research group. There was also a fishery representative onboard the test platform at all times during the measurements (Mattson 2008).

4 Test Setup

Measurements had to be conducted in a test area with a water depth of more than 400 m to ensure that the water bottom reflection did not degrade the recorded pulse signal (Mattson 2008). The test platform was a converted barge containing a compressor, high-pressure system, generators, crane, workshop, work boat, and an instrument room with a high-resolution recording system. Highly sensitive hydrophones were positioned on a floating collar connected to the barge. Two hydrophone arrays were built up. One array was positioned in the corner of the floating collar and one array was centered below the air gun. A set of near-field hydrophones was deployed at fixed positions around the seismic source. The source was positioned with a crane to allow different firing depths (Mattson 2008; Fig. 1).

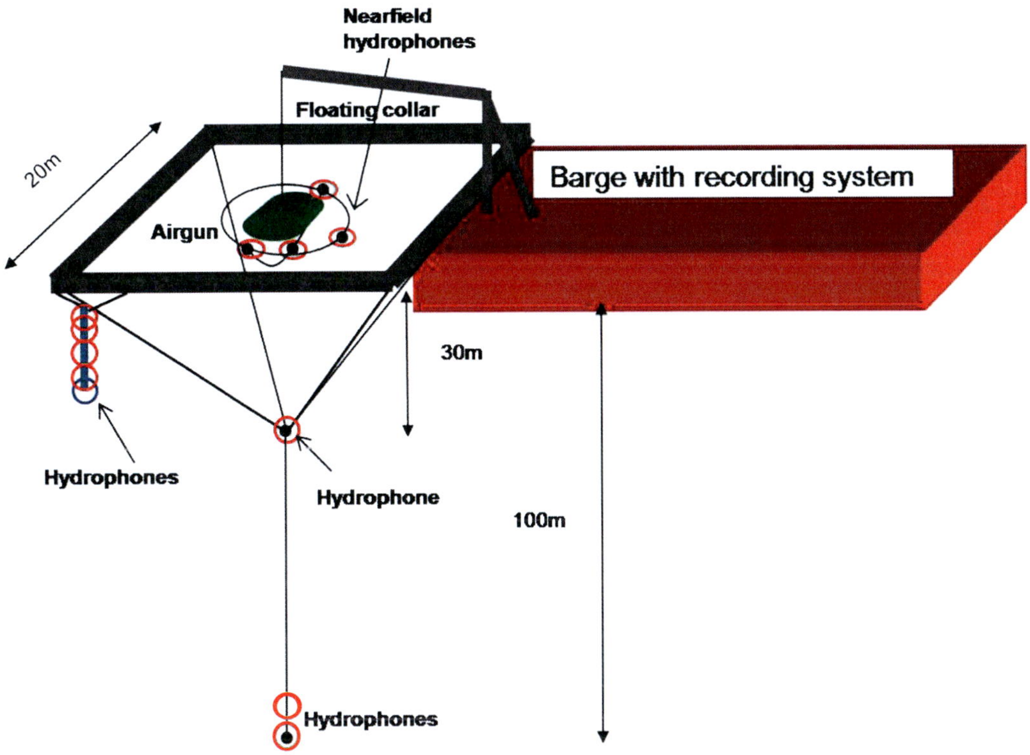

Fig. 1 Test setup with hydrophones and air gun

5 Test Program

The most commonly used air guns and two-element cluster configurations within the geophysical industry were measured. Air gun type, volume size, operating pressure, and depth of the source were varied in the experiment. The effect of varying the separation between individual air guns within clusters was also measured (Mattson 2008). At the end of the test program, particle velocity measurements at different positions relative to the seismic source were conducted.

6 Conclusions

The test setup was able to produce very consistent measurements. The ambient noise conditions at the barge were such that energy at 50 kHz could be seen above the noise floor (Mattson 2008). An initial quality control of the data has been done, but all data still have to be thoroughly analyzed. Development of software codes and implementation of new source modeling capability in existing software tools will follow.

References

Mattsson A (2008) Single airgun and cluster measurement project. Exploration and Production Sound and Marine Life Joint Industry Programme Review Meeting, 28–30 October 2008, Houston, TX, p 22.

Cetacean Hearing-Damage Zones Around a Seismic Source

Robert Laws

1 Introduction

In deep marine seismic surveying, a vessel tows a large array of hydrophones that measure the echoes from the subsurface resulting from the periodic firing of the seismic source. These echoes are processed to generate images many kilometers deep in the earth so that geologists can try to locate oil and gas. Such surveys typically collect over five terabytes of data per day. Typically, the vessel moves at 2.5 m/s. The seismic source array uses ~5,000 in.[3] (~80 l) of compressed air at 2,000 psi (~140 bar) distributed among 18 air guns at a depth of 6 m and emits almost all of its energy below 200 Hz. In three-dimensional (3-D) surveying, there are usually two source arrays fired alternately 25 m to the port and starboard of the sail line. Each one typically has an in-line interval of 37.5 m. In two-dimensional (2-D) surveying, there is a single source array that is normally fired at an in-line interval of 25 m. The source wave field can either be calculated from hydrophone measurements made close to the array following the method of Ziolkowski et al. (1982) be modeled using the method given by Laws et al. (1990). In this paper, I use the former with data that are sampled at 4,000 Hz. I assume reflection coefficients of −1 and +0.3 for sea surface and seabed, respectively, and use an ocean depth of 300 m. I use the method of images to compute the pressure signal in the water.

In a landmark paper, Southall et al. (2007) reviewed the current state of knowledge of hearing damage in marine mammals. They defined damage thresholds for various species groups that they categorized by hearing abilities. Of interest here are low-frequency cetaceans, midfrequency cetaceans, and high-frequency cetaceans. They defined two damage thresholds, one based on the peak pressure (PP) received by the animal and the other based on the sound exposure level (SEL). SEL is the integral of the acoustic power received by the animal, and it is weighted by the standardized hearing curve for its group. The SEL integral is defined to be over 24 h, and I have taken this to mean a complete transit of a long seismic line. The damage thresholds for an impulsive source are 230 dB (re μPa^2) for PP and 198 dB (re μPa^2-s) for SEL. I assume that the animal remains stationary in the water and does not flee.

R. Laws (✉)
Schlumberger Cambridge Research Ltd., Cambridge, Cambridgeshire CB3 0EL, UK
e-mail: rmlaws@slb.com

A.N. Popper and A. Hawkins (eds.), *The Effects of Noise on Aquatic Life*,
Advances in Experimental Medicine and Biology 730, DOI 10.1007/978-1-4419-7311-5_107,

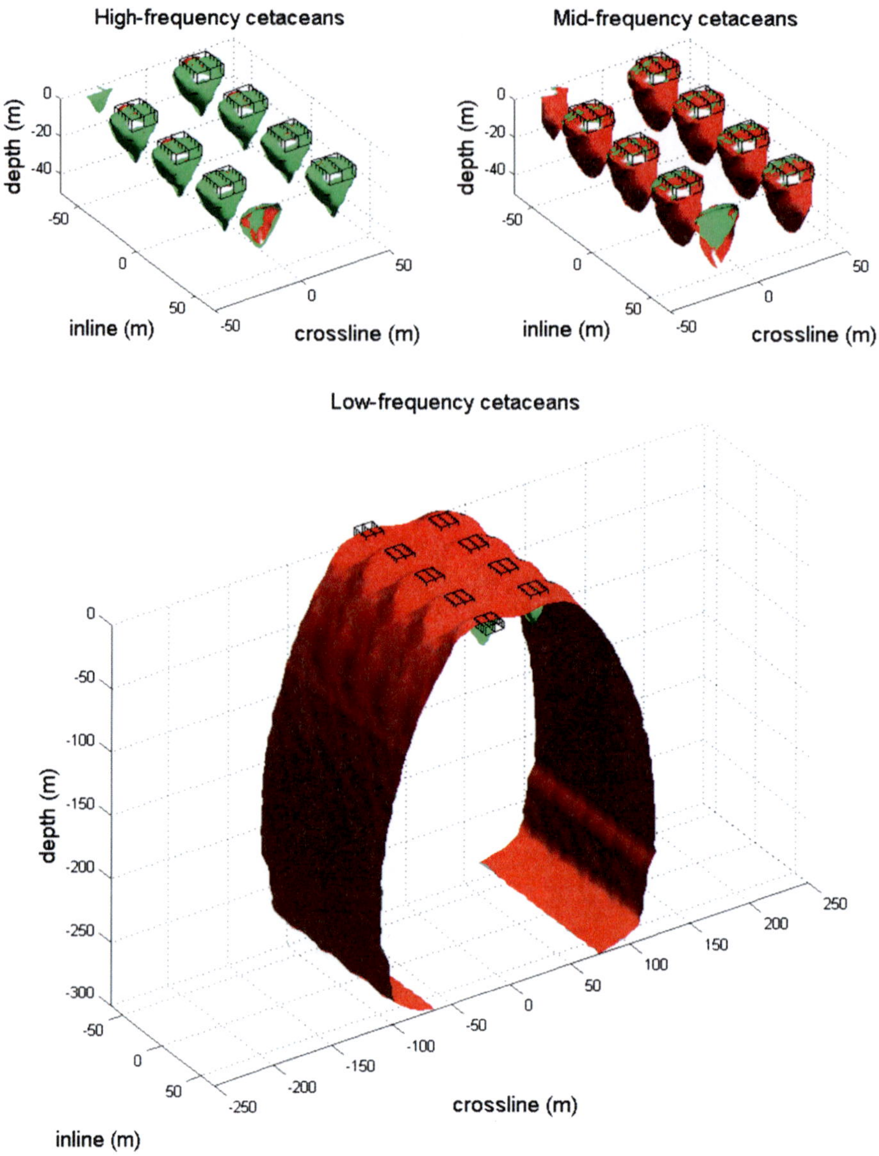

Fig. 1 Hearing damage zones for high-frequency cetaceans (top left), midfrequency cetaceans (top right), and low-frequency cetaceans (bottom). The peak pressure (PP) criterion is shown in green and the sound exposure level (SEL) criterion is shown in red. The figures show a slice across the seismic line, with the seismic source marked as a grid

Using the above, I calculated the zones around the seismic line where the damage thresholds defined by Southall et al. (2007) are exceeded. Figure 1 shows the damage zones for the three groups around a 3-D seismic line. The source array positions are marked with grids of black lines, and the vessel is sailing out of the page. Only a short "slice" across the line is shown, but the whole line was used in the calculation.

2 Damage Zones

The PP damage zone (Fig. 1, green) is confined to a roughly conical shape (point downward) centered on the source and extending down to a depth of ~30 m. It is the same for all three groups. There is such a PP damage zone around each shot of the survey. The SEL damage zones for midfrequency and high-frequency cetaceans are very similar to the PP zone. However, for low-frequency cetaceans, the SEL zone is much larger. It is tubular in shape, with a roughly circular cross-section radius of 150 m almost touching the surface along the sail line. This zone is large because the low-frequency cetacean hearing sensitivity overlaps more of the seismic band.

3 Exclusion Zone

Although it is not directly relevant to the calculations done here, it is instructive to bear in mind the size of the current :exclusion zone" around the seismic source. If a marine mammal is seen inside the zone, the source is switched off. The exclusion zone is 1,000 m wide. The calculations show that even the largest damage zone is much smaller than the exclusion zone.

4 How Many Shots Contribute to the SEL?

It is instructive to look at the whole of a short seismic line to see how many shots significantly contribute to the SEL for a low-frequency cetacean. Figure 2 shows that the significant part of the SEL is accumulated while the source is within about 500 m of the animal, that is, about 50 shots.

5 Conclusions

By combining the hearing damage criteria defined by Southall et al. (2007) with the measurements of the output of seismic source arrays and an assumption of no avoidance action by the animal, I have shown the extent of the hearing damage zone around a deep marine seismic line.

The midfrequency and high-frequency cetaceans have damage zones that are confined to a roughly conical region centered on the shot position, with a lateral extent of ~20 m and extending down to ~30 m.

The low-frequency cetaceans have a much larger damage zone and it results from the SEL criterion. This is because the hearing of this group overlaps more with the frequency range emitted by the seismic source. The low-frequency cetacean damage zone forms a continuous tube below the seismic line ~300 m wide and extending to a depth of ~300 m or more.

The seismic source "exclusion zone" is much larger than the largest damage zone.

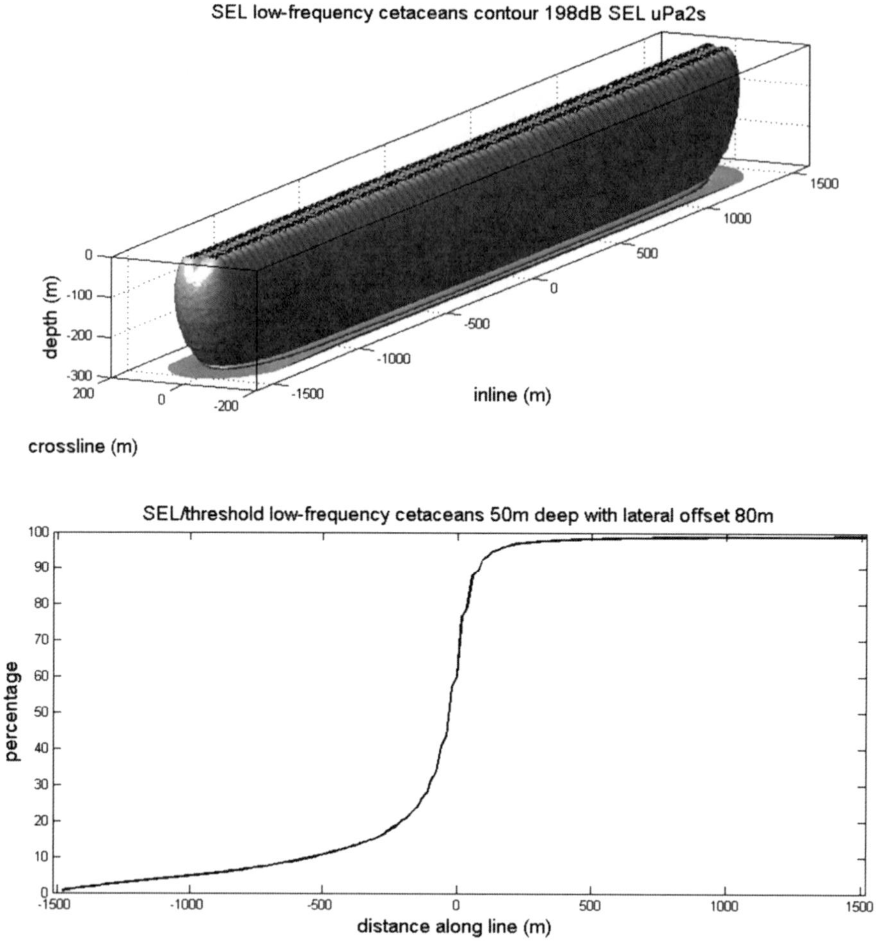

Fig. 2 Top: Whole of the SEL damage zone for low-frequency cetaceans around a 3-km seismic line. Bottom: How the SEL accumulates as a function of vessel position for an animal that is located at a depth of 50 m and at an offset 80 m from the center of the sail line (just outside the damage zone). The slight asymmetry results from the source directivity

Acknowledgments I thank Schlumberger Cambridge Research Ltd., Cambridge, UK, for allowing the use of their library of seismic source measurements for this study.

References

Laws RM, Hatton L, Haartsen M (1990) Computer modelling of clustered airguns. First Break 8:331–338.

Southall BL, Bowles AE, Ellison WT, Finneran JJ, Gentry RL, Greene CR Jr, Kastak D, Ketten DR, Miller JF, Nachtigall PE, Richardson WJ, Thomas JA, Tyack PL (2007) Marine mammal noise exposure criteria: Initial scientific recommendations. Aquat Mamm 33:411–522.

Ziolkowski AW, Parkes GE, Hatton L, Haugland T (1982) The signature of an air gun array: Computation from near-field measurements including interaction. Geophysics 47:1413–1421.

Modeling Cumulative Sound Exposure Over Large Areas, Multiple Sources, and Long Durations

Christine Erbe

1 Introduction

Guidelines and regulations for underwater noise exposure tend to focus on single events. For example, the noise field from a single strike on a pile during pile driving or from a single discharge of a seismic air gun array is mapped through either modeling or field measurements, and mitigation zones are defined based on ranges to thresholds commonly expressed as sound pressure level (SPL) or sound exposure level (SEL). In many situations, however, animals will experience ongoing or repeated exposure. The effects of cumulative exposure, the way in which effects from single exposures accumulate, and the potential for recovery between repeated exposures are poorly understood. Another (computational) challenge is to predict, through modeling, cumulative levels received by animals from multiple sources, from moving sources, and from multiple exposures over long durations and large areas. The following sections outline the steps to solve the computational problem for selected example cases.

2 Stationary Source, Moving Receiver

This example is based on pile-driving measurements taken in a dolphin habitat (Erbe 2009). Taking a bird's-eye view, the pile is driven at 0 m east and 0 m north and a dolphin travels in a straight line at a speed of 5 m/s, with a closest point of approach (CPA) at 200 m (Fig. 1, top). In situ transmission loss measurements showed that the received SEL as a function of the range (R) followed the equation SEL = $207 - 6 \log_{10}(R) - 0.04R$ for a large steel pile (150-cm outer diameter, 25-mm wall thickness, 30-m length) driven with a hydraulic hammer (14-t weight, 280-kJ maximum energy). The hammering interval was 1.8 s. Modeling the dolphin path from ~700 m west to 700 m east, this animal received 160 pulses, 80 pulses on either side of the CPA (shot numbers −80 to +80; Fig. 1, bottom). The SEL per shot is the received level for each shot accounting for transmission loss between the source and receiver and is maximum at the CPA (shot number 0; Fig. 1, bottom). The cumulative SEL is the sum of all levels and asymptotically approaches a value of 202 dB re 1 μPa^2-s in this scenario.

C. Erbe (✉)
Centre for Marine Science & Technology, Curtin University, Bentley, WA 6102, Australia; *formerly* JASCO Applied Sciences, Australia
e-mail: C.Erbe@curtin.edu.au

A.N. Popper and A. Hawkins (eds.), *The Effects of Noise on Aquatic Life*, Advances in Experimental Medicine and Biology 730, DOI 10.1007/978-1-4419-7311-5_108,

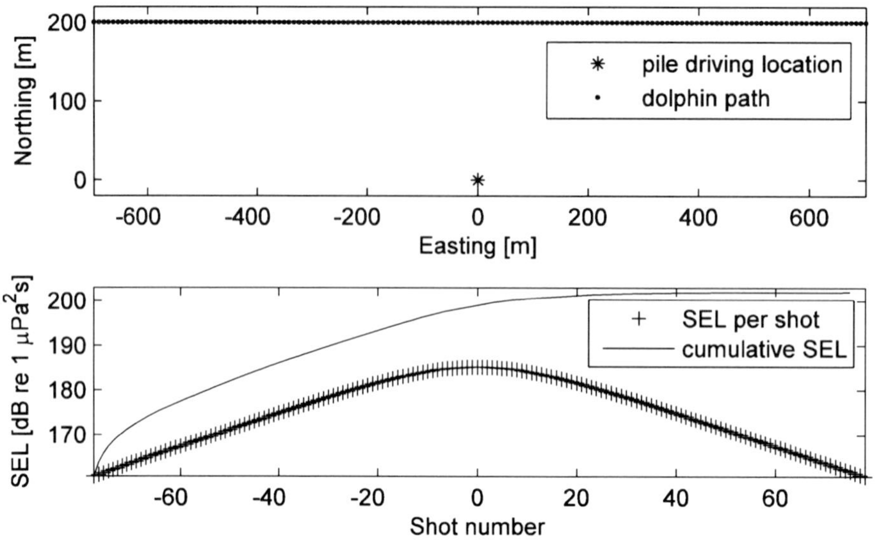

Fig. 1 Top: Dolphin swimming past pile-driving location. Bottom: Per shot and cumulative received sound exposure level (SEL)

3 Multiple Stationary Receivers, Moving Source

The reversed situation of a stationary receiver and a moving source can be modeled in the same way as the previous example. Sometimes, however, the modeling of multiple stationary receivers is desired, e.g., if the target species are benthic organisms underneath a seismic survey transect. Modeling a 1,500-in.^{3} generator/injector (GI) gun array with JASCO's air gun array source model (AASM; MacGillivray 2006) and using the ray-tracing sound-propagation model described in Erbe and Farmer (2000), cumulative received SEL from 9 shots spaced 25 m apart along a west-east transect was computed on a receiver grid spanning 120 m in Northing and 200 m in Easting (Fig. 2). Cumulative received levels are plotted at the seafloor, which was 10 m deep. Given the short source-receiver ranges, in particular right beneath the array, each air gun in the array was modeled separately and waveforms were superposed at each receiver location (Erbe and King 2009).

4 Multiple Stationary Receivers, Multiple Sources

A more complex situation arose when multiple intersecting transects were shot over a coral reef and where fish were expected not to flee the reef but to hide among the coral for the duration of the survey, thus receiving pulses for a few weeks (Erbe and King 2009). Thousands of shots were fired over a few hundred square kilometers, and the regulator was asked to compute maps of cumulative SELs that were to be overlaid with habitat maps to identify regions of high risk. It was computationally impractical to model every single sound-propagation path from every shot to every receiver. The approach taken followed these steps. 1) Place an evenly spaced receiver grid over the survey area. 2) Extract bathymetry profiles for all shot-receiver pairs. 3) Cluster bathymetry profiles with a self-organizing neural network. 4) Model transmission loss along all cluster centroids. 5) Extrapolate transmission loss for all other shot-receiver pairs. 6) Integrate energy at all receivers over all shots (= integration over time and area). The error introduced by the neural net was -1 ± 3 dB re 1 μPa^2-s. The error was largest where bathymetry varied "steeply" over short ranges (Erbe and King 2009).

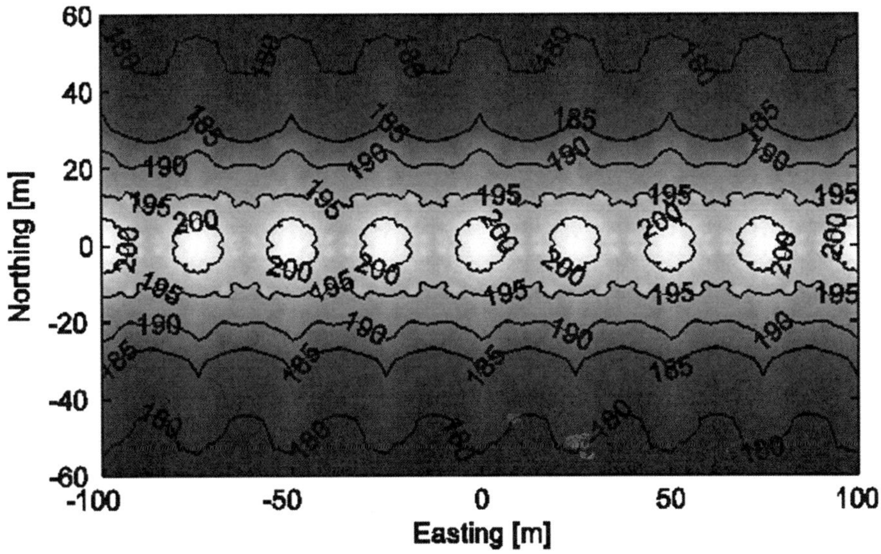

Fig. 2 Cumulative SEL on a receiver grid from 9 seismic air gun array discharges spaced 25 m apart along a west-east transect. Lighter shades correspond to higher SELs. Contours are drawn for SEL in dB re 1 μPa^2-s

5 Discussion

In the above example, bathymetry was the single most important factor affecting sound propagation. Bathymetry varied from being very steep on the outside of the reef to very flat in the reef center. Large coral outcrops existed all over the reef, sometimes reaching the water surface and stripping energy at high frequencies. Geology (geoacoustic parameters of the seafloor) and sound-speed profiles of the water column did not vary substantially over the reef. In other environments, where the geology or water properties are not homogeneous, environmental provinces should be defined and the model run for each province separately.

The tool is useful for moving sources or for very large numbers of sources where an integration in area (over all source locations) is desired. If only a few stationary sources exist, it will be easier to model sound propagation once for each source and to integrate over time. The tool is useful to assess the impact on marine species that are confined to the area modeled (i.e., they don't flee the area), e.g., fish at a coral reef or dugong confined inside a bay. The tool produces a sound-exposure map, which can be overlaid with habitat maps to estimate the percentage of habitat that receives certain threshold levels. Although it is feasible to model cumulative exposure over large areas and multiple sources, the biological effects of cumulative exposure remain largely unknown.

References

Erbe C (2009) Underwater noise from pile driving in Moreton Bay, QLD. Acoust Aust 37:87–92.

Erbe C, Farmer DM (2000) A software model to estimate zones of impact on marine mammals around anthropogenic noise. J Acoust Soc Am 108:1327–1331.

Erbe C, King AR (2009) Modeling cumulative sound exposure around marine seismic surveys. J Acoust Soc Am 125:2443–2451.

MacGillivray AO (2006) An acoustic modeling study of seismic airgun noise in Queen Charlotte Basin. MSc Thesis, University of Victoria, BC, Canada.

Sound-Recording Systems for Measuring Sound Levels During Seismic Surveys

Jan Tore Øvredal and Bjorn Totland

1 Introduction

Two new sound-recording systems were developed as part of a study on the effects of sound from seismic air guns on fish behavior. The systems were used to record sound pressure levels (SPLs) at several depths and distances from the seismic vessel during a geophysical survey carried out on a Norwegian fishing ground during the summer of 2009. The recordings were used to calculate sound exposure levels (SELs) and as input to models describing sound propagation of acoustic noise from seismic air gun arrays.

2 Sound-Recording Systems

To be able to record sound at several depths relative to both the surface and bottom without sound pollution from the support vessel, two separate hydrophone sound-recording systems were developed. The systems were designed to be able to function as stand-alone self-recording data loggers.

2.1 Drifting Antiheave Surface Buoy With Vertical Hydrophone String

A surface buoy system was developed to enable SELs to be recorded at three different depths (Fig. 1). To minimize vertical movements of the hydrophones due to wave actions, which create noise, the buoy was designed with a long slim shape that allows the waves to climb on the buoy instead of moving it up and down. The buoy was kept upright by lead weights and batteries placed in its base and by a sea anchor at the end of the hydrophone string. The buoy contained an UNO-2170 embedded computer with an internal hard drive for data logging and instrumentation control. A GPS receiver enabled the buoy to be tracked and a radio Ethernet link allowed remote

J.T. Øvredal (✉) • B. Totland
Institute of Marine Research, 5817 Bergen, Norway
e-mail: jan.tore.ovredal@imr.no; bjorn.totland@imr.no

A.N. Popper and A. Hawkins (eds.), *The Effects of Noise on Aquatic Life*,
Advances in Experimental Medicine and Biology 730, DOI 10.1007/978-1-4419-7311-5_109,

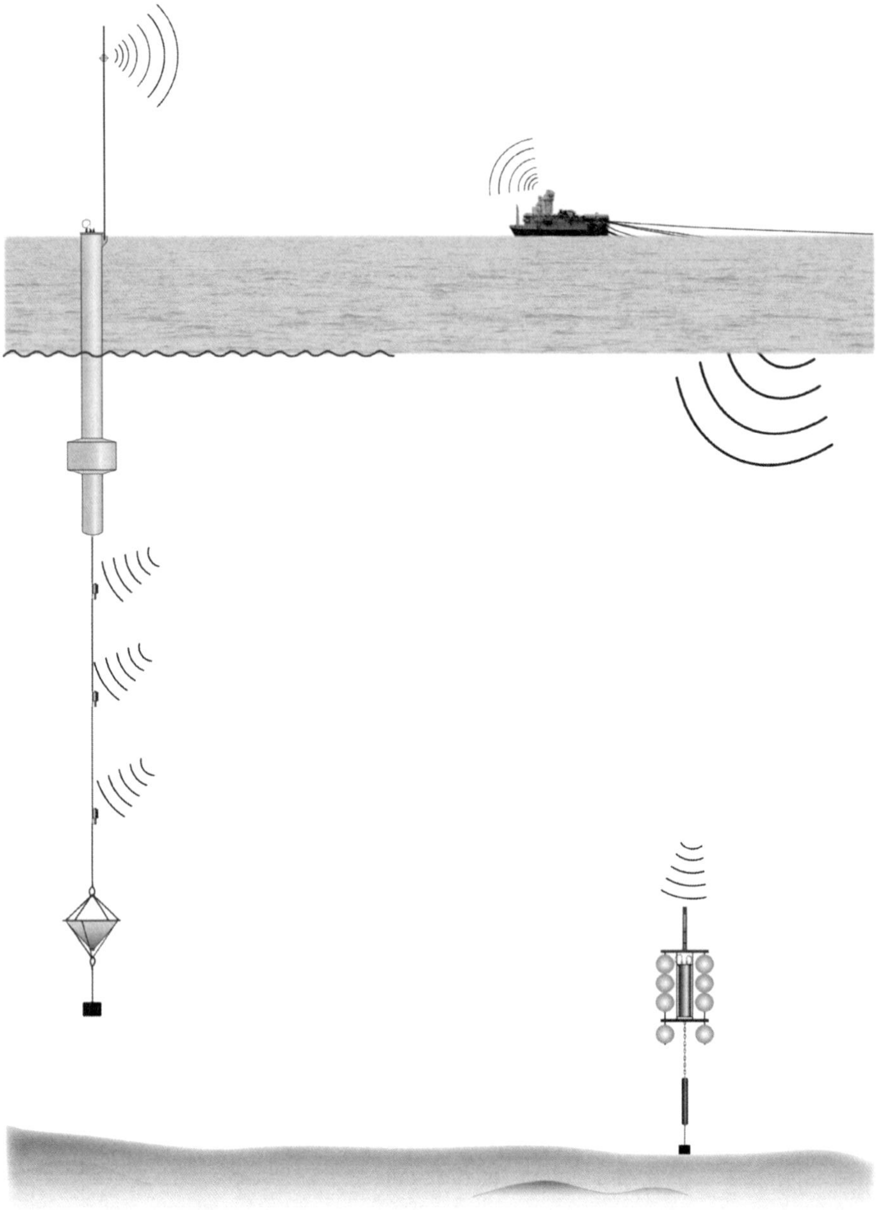

Fig. 1 The buoy systems

control and monitoring of the system from a distance of several nautical miles. Three Naxys 02345 Ethernet hydrophones were attached to the buoy via cables along a string at depths of 8, 32, and 64 m. The buoy also contained an Argos satellite transmitter for satellite tracking in case the buoy was lost. The system can be operated either as a stand-alone unit or from the research vessel via a radio Ethernet link. Sound data from the hydrophones can be recorded either continuously or at fixed times and intervals. GPS position data are logged continuously. The Naxys hydrophones are omni-directional, with a frequency range of 5 Hz to 300 kHz, a sensitivity of −211 dB re V/μPa and a configurable sampling frequency of 6–768 kHz. The hydrophone has an amplifier with an adjustable gain from 0 to 40 dB. Remote control software enables the operator on the research vessel to have a remote desktop view of the buoy computer via the Ethernet radio link, permitting real-time tracking of the buoy's position, monitoring of hydrophone signal, and adjustment of gain settings and recording intervals. Before deployment, the hydrophones were calibrated with a Brüel & Kjær 4229 piston calibrator. Star-Oddi DST depth tags attached to each hydrophone recorded their depths and vertical movements.

2.2 *Self-Recording Submersible Hydrophone Platform*

Four submersible acoustic hydrophone platforms were used to record sound on the seabed at a number of locations during seismic shooting (Fig. 1). An underwater housing of anodized aluminum connected to a Naxys 02345 Ethernet hydrophone via an Ethernet cable was mounted in a steel frame. The frame was made buoyant with floats and was attached to weights (100 kg) by a 3-m rope. An acoustic release unit enabled the unit to be retrieved when the recordings had been completed. The underwater housing contained an Advantech PCM-3370F-JOA1 single-board computer for data logging and system control. The computer used a flash drive for data storage to avoid noise from the disk during data logging. The electronics were powered by rechargeable A123 lithium-ion batteries with automatic low-battery capacity shut-down circuitry, enabling the system to operate for ~48 h. Logging interval, start time, gain, and sampling frequency were configured via remote control software with a serial connection before deployment. Recorded sound data were downloaded to an external PC by disconnecting the hydrophone and using the Ethernet connection.

3 Discussion

The sound-recording systems had to meet a number of criteria. A primary requirement was that they had to be able to operate in the same area and under the same weather conditions as the seismic vessel. The study area (Vesterålen, northern Norway) has a variety of bottom types and depths ranging from 50 to 900 m and can have rough weather and high seas. Because the research vessel was also engaged in acoustic mapping of fish distribution during the seismic survey, the systems had to be able to operate independently, with continuous logging of sound recordings for at least 48 h without the need for a cable connection to the vessel. A minimum signal-to-noise ratio (i.e., high sensitivity) was important to enable sound recordings to be made at distances up to 50 km from the seismic vessel. The surface buoy was therefore designed to record sound with minimum interference from sea noise. To enable recordings to be made over a wide range of distances (i.e., wide range of sound levels), the hydrophones had an amplifier with an adjustable gain. The depths at the fishing grounds were up to several hundred meters and the bottom hydrophone systems were constructed to operate at depths up to 500 m. The sound data were saved in a standard WAV file format for ease of postprocessing.

The sound recordings from the hydrophones were synchronized in time with the position of the sound source (i.e., the air gun array) during seismic shooting. The WAV files were processed to calculate SPL, SEL, peak pressure level (PPL), and impulse (I_{imp}) (Carey 2006; Madsen et al. 2006). These metrics provide various ways of describing the type and level of sound that affected the fish in the study area. The sound data recordings will also be used as input during the development of a sound-propagation model (Hovem 2007), which describes how sound from a seismic air gun array propagates.

References

Carey WM (2006) Sound sources and levels in the ocean. IEEE J Ocean Eng 31:61–75.

Hovem JM (2007) PlaneRay: An acoustic underwater propagation model based on ray tracing and plane wave reflection coefficients. In: Taroudakis M, Papadakis P (eds) Theoretical and computational acoustics 2007. Proceedings of the 8th International Conference on Theoretical and Computational Acoustics, 2–5 July 2007, Heraklion, Crete, Greece, pp 273–289.

Madsen PT, Johnson M, Miller PJO, Soto NA, Lynch J, Tyack P (2006) Quantitative measures of air-gun pulses recorded on sperm whales using acoustic tags during controlled exposure experiment. J Acoust Soc Am 120: 2366–2379.

First Description of the Sound Pressure and Particle Velocity Components of the Ambient Noise and Boat Noise Recorded at the WWF-Miramare Natural Marine Reserve, Trieste, Italy

Angelo Farina, Adriano Farina, Enrico Armelloni, Linda Sebastianutto, Carlo Franzosini, and Marta Picciulin

1 Introduction

Boat noise represents a chronic source of harassment for fish species, whose communication for inter- and intrasexual selection is based mainly on low-frequency sound signals (Amorim 2006). Investigating the impact of boat noise on target fish species is particularly relevant for coastal marine protected areas (MPAs), which are biologically rich locations deserving protection from anthropogenic pollutants. Although many fish species are primarily sensitive to the kinematic components of the sound field (Popper and Fay 1999), namely, to particle acceleration, boat noises have been characterized so far mostly by means of sound pressure measurements. In this work, the underwater acoustic background noise and the noise produced by a small outboard-engine boat moving at 6 knots were recorded inside the WWF-Natural Marine Reserve of Miramare (Trieste, Italy) by using a novel hydrophonic probe ("Soundfish") placed on the sea bottom (8 m depth). This allowed for characterization of the sound field not just in terms of sound pressure but also of the three Cartesian components of particle velocity.

A. Farina • E. Armelloni
Industrial Engineering Department, University of Parma, 43100 Parma, Italy

A. Farina
Department of Electronics, University of York, Heslington, York, YO10 5DD UK

L. Sebastianutto
Cognitive Neuroscience Sector, International School for Advanced Studies (SISSA), 34135 Trieste, Italy

C. Franzosini
WWF-Natural Marine Reserve of Miramare, 34014 Trieste, Italy

M. Picciulin (✉)
WWF-Natural Marine Reserve of Miramare, 34014 Trieste, Italy

CSEE-Department of Life Sciences,University of Trieste, 34127 Trieste, Italy
e-mail: marta.picciulin@gmail.com

A.N. Popper and A. Hawkins (eds.), *The Effects of Noise on Aquatic Life*,
Advances in Experimental Medicine and Biology 730, DOI 10.1007/978-1-4419-7311-5_110,

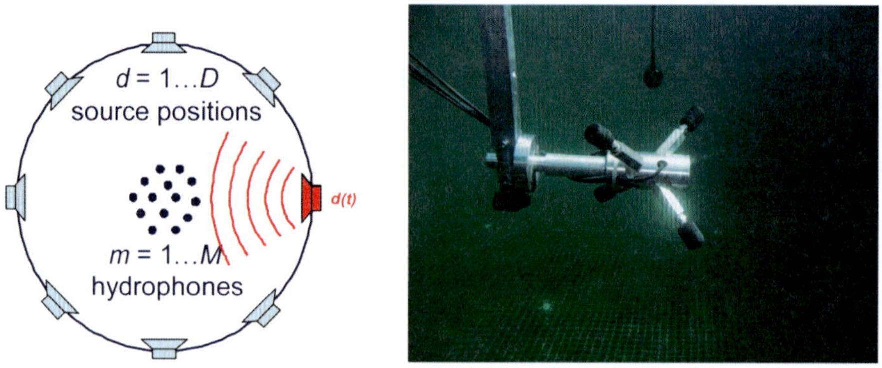

Fig. 1 Impulse response measurements performed inside a pool

2 Description of the "Soundfish" Probe and the Digital Signal Processing

The system is based on a modified ZOOM H2 digital sound recorder, renamed Brahma, capable of recording the signals coming from a probe consisting of four hydrophones placed at the vertices of a tetrahedron; this is the underwater equivalent of a Soundfield microphone. The recorder operates at 48 kHz and 24 bits and records standard uncompressed WAV files over a 16-GB secure digital (SD) card, which can be easily processed later on a PC. A software tool, named Brahmavolver, was developed for converting the raw signals coming from the four hydrophones to output signals representing, respectively, the sound pressure and the three Cartesian components of particle velocity. The processing is based on the use of a matrix of 4 × 4 finite impulse response (FIR) filters, currently 2,048 points long. In our approach (Farina et al. 2007), the filter coefficients are computed numerically, inverting a matrix (*M*) of measured impulse responses obtained with the sound source placed at a large number (*D*) of positions all around the probe, as shown in Figure 1.

3 Preliminary Tests in a Pool

A suitable number of impulse-response measurements were performed on the Soundfish probe inside the test pool kindly made available by WASS (Livorno, Italy), as shown in Figure 1. A turntable, controlled by our Aurora software, was employed for automatically rotating the probe in steps of 30° along both azimuth and elevation, yielding a set of 6 × 12 impulse responses. Figure 2 shows some of the results of these preliminary tests: the polar patterns of the pressure and particle velocity in two octave bands.

4 Field Recordings

The probe and the Brahma recorder were placed on the sea bottom at a depth of 8 m in the center of the protected area of the Miramare Reserve. A 30-min-long recording of the sea ambient noise (SAN) was performed, followed by recordings of a boat passing near the probe. The analysis of these recordings allowed for the computation of one-third octave band spectra of both

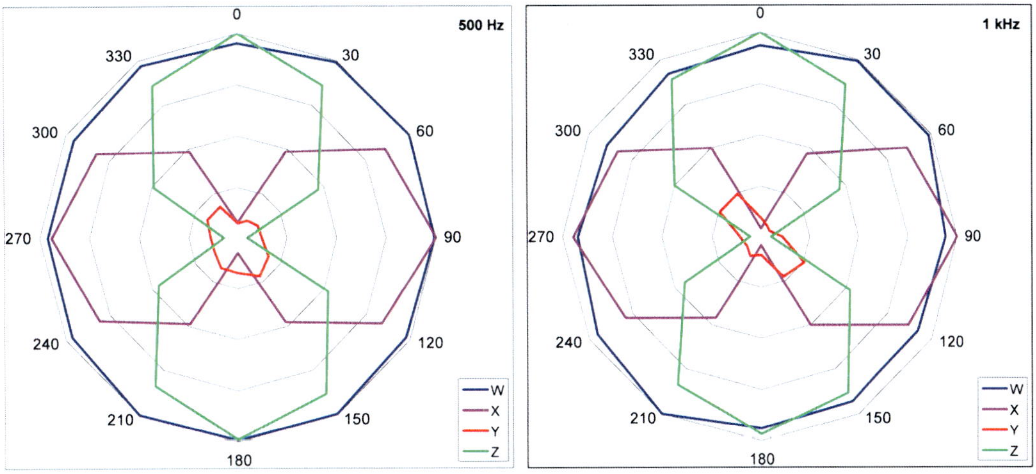

Fig. 2 Polar patterns of the Soundfish probe at 500 and 1,000 Hz

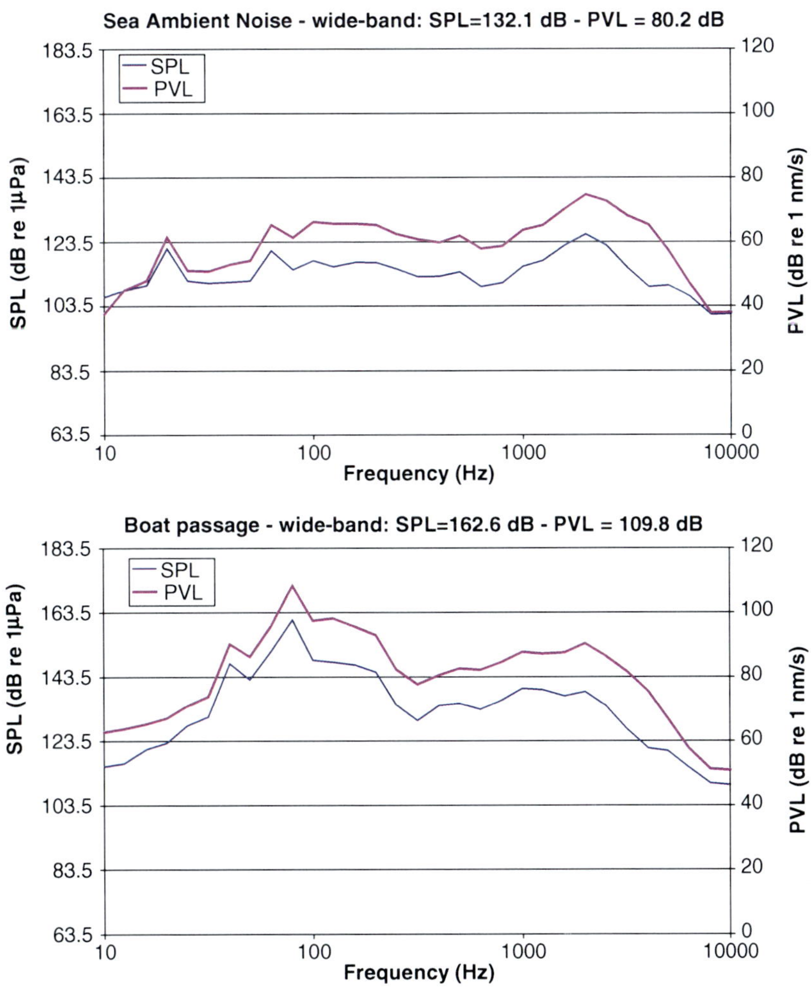

Fig. 3 Sound pressure level (SPL) and particle velocity level (PVL) spectra of sea ambient noise (left) and boat passage above the probe (right)

sound pressure level (SPL) and particle velocity level (PVL), computed with reference to the standard quantities for underwater acoustics (1 μPa and 1 nm/s, respectively). Furthermore, the particle acceleration levels (PALs) can be easily derived from the PVL values, following the procedure described in Picciulin et al. (2010).

Figure 3 presents the analysis of the recordings, showing the one-third octave band spectra in terms of SPL and PVL of the SAN and the boat passage.

Although only the overall magnitude of the particle velocity vector is reported here, the data obtained allow for computation of the direction of the vector, making it possible to know, at any instant, the position of the sound source, which resulted in good agreement with the real trajectory of the boat.

5 Conclusions

The new Soundfish probe can be employed for an analysis of the cause-effect relationship because at every instant, the position of the source relative to the receiver is known, along with the quantities relevant for assessing the impact of human-produced noise over marine species sensitive to either sound pressure or particle motion. The reliability of the new measurement system must now be assessed by employing it in a number of surveys, under different sea conditions, at different depths, and with various kinds of noise sources. It could also be advisable to repeat the calibration in the pool, employing narrower angular steps, for ensuring computation of even better digital filters.

References

Amorim MCP (2006) Diversity of sound production in fish. In: Ladich F, Collin SP, Moller P, Kapoor BG (eds) Communication in fishes. Science Publishers, Enfield, NH, pp 71–104.

Farina A, Capra L, Conti L, Martignon P, Fazi F (2007) Measuring spatial impulse responses in concert halls and opera houses employing a spherical microphone array. Proceedings of the 19th International Congress on Acoustics (ICA), 2–7 September 2007, Madrid, Spain.

Picciulin M, Sebastianutto L, Codarin A, Farina A, Spoto M, Ferrero EA (2010) In situ behavioural responses to boat noise exposure of *Gobius cruentatus* (Gmelin, 1789; fam. Gobiidae) and *Chromis chromis* (Linnaeus, 1758; fam. Pomacentridae) living in a Marine Protected Area. J Exp Mar Biol Ecol 386:125–132.

Popper AN, Fay RR (1999) The auditory periphery in fishes. In: Fay RR, Popper AN (eds) Comparative hearing: Fish and amphibians. Springer-Verlag, New York, pp 43–100.

Underwater Particle Acceleration Induced by a Wind Turbine in the Baltic Sea

Peter Sigray and Mathias H. Andersson

1 Introduction

During the last decade, there has been a major interest in wind energy production. The number of offshore wind farms is increasing rapidly. Several nations are planning to switch to "green" power, increasing the numbers even more. The public awareness and the general focus on the environment require that wind-based energy should be ecologically harmless. Lately, it has been realized that the aquatic environment is also a soundscape, not at least due to the fact that many species make use of sound for communication, mating, finding prey, and avoiding predators. Furthermore, most studies have been concerned with the effect of the sound pressure component of sound, even though many species are sensitive to particle motion. For this reason, a particle motion sensor was built and deployed in close vicinity to a wind turbine where the generated particle motion in the ocean was characterized. The results are presented here.

As part of a Swedish research program (led by Vindval, Stockholm, Sweden, and financed by the Swedish Energy Agency, Eskilstuna, Sweden) dealing with the environmental effects of wind power, the acoustical sound in close vicinity to an offshore wind turbine was investigated. The aim was to study the particle acceleration induced by a single wind turbine and relate the observed levels to fish hearing. The field trials were carried out in August and September 2007 in the Baltic Sea at the Utgrunden wind farm near the town of Kalmar, Sweden. During this period, the wind varied in both direction and speed. It was thus possible to make observations during different operational conditions.

2 Field Trials

A specially designed particle sensor was built to enable the measurement of particle motion. It consisted of a near neutrally buoyant sphere that co-oscillated with the acoustical sound. The operational frequency range was 0.1–300 Hz, limited by the built-in accelerometers. Before the trials, the sensor response was compared with hydrophones connected in gradient mode.

P. Sigray (✉)
Department of Underwater Research, Swedish Defence Research Institute, S-164 90 Stockholm, Sweden
e-mail: peter.sigray@foi.se

M.H. Andersson
Department of Zoology, Stockholm University, S-106 91 Stockholm, Sweden

A.N. Popper and A. Hawkins (eds.), *The Effects of Noise on Aquatic Life*,
Advances in Experimental Medicine and Biology 730, DOI 10.1007/978-1-4419-7311-5_111,

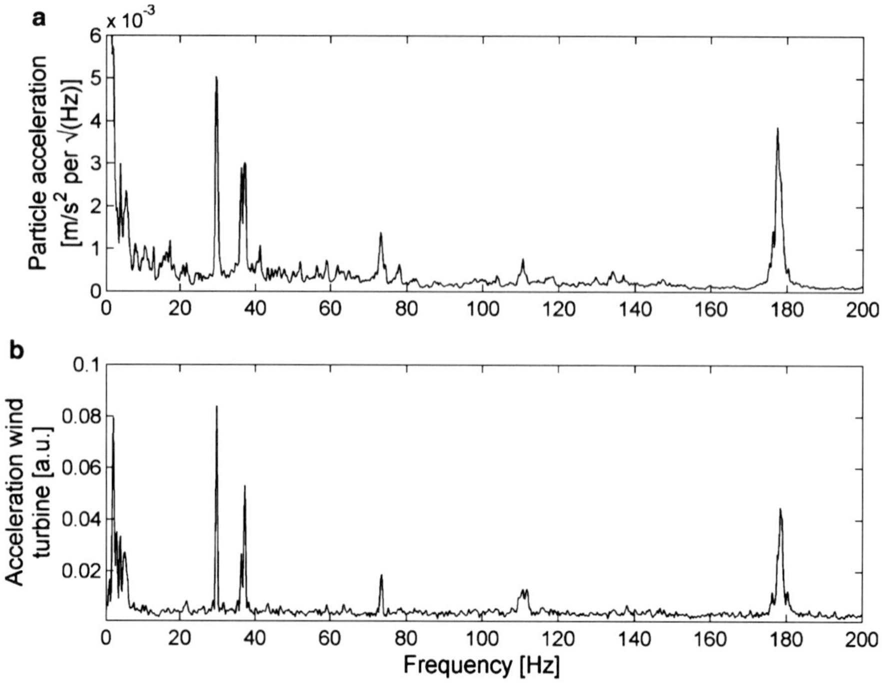

Fig. 1 Spectra showing generated accelerations in the ocean and on the wind turbine. (**a**) Particle acceleration measured by the particle motion sensor at a 1-m distance from the outer rim of the wind turbine foundation. (**b**) Vibration measured on the pillar of the wind turbine

A reference accelerometer was attached to the pillar of the wind turbine to measure the vibrations of the foundation. The data were compared with those from the particle motion sensor. It is clear that the observed particle motion in the sea was induced by vibrations generated by the wind turbine (cf. Fig. 1). The same frequencies as well as the amplitude relationships were observed, which strongly suggested that the sensor is doing justice to particle motion. Measured particle motion in the frequency range of 0.1–200 Hz is shown in Figure 2 for strong wind and for when the wind turbine was out of operation. As expected, the background level is the lowest, showing that the observed levels in the frequency interval of 0.1–200 Hz is affected by the wind turbine. Two interesting effects were observed. First, the blade-rotation frequencies (between 1 and 6 Hz) were not observed in the sea, indicating that low-frequency wave propagation was not effective in shallow environments. Second, below 2 Hz, the soundscape was dominated by wave-induced motion, and thus the influence of the wind turbine was most probably negligible.

3 Results

The analysis showed that the wind turbine generates a number of tones as well as a broadband background (cf. Fig. 1). At full power (wind speed greater than 11 m/s), 3 tones dominate at 29.5, 36.5, and 178 Hz, where the third tone was the strongest. At lower wind speed, the dominating tones were found at 29, 34, and 141 Hz, where the first tone was the strongest. During strong wind conditions, a dedicated test was performed where the wind turbine was started and allowed to reach full power production. Nonstationary tones were observed by the reference accelerometer as

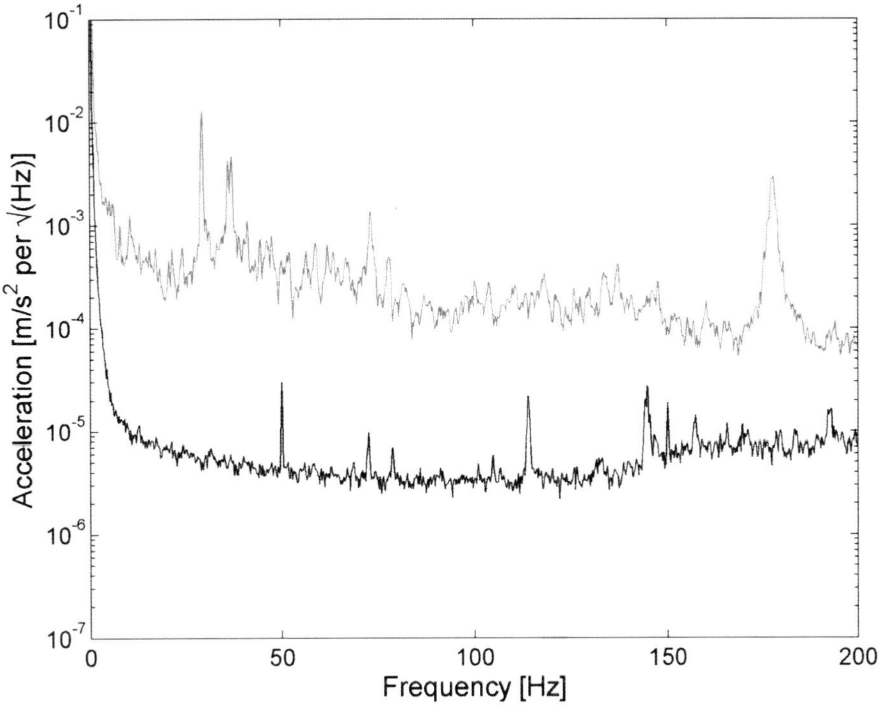

Fig. 2 Sound levels measured in the ocean 1 m from the wind turbine foundation. Black line, background level measured with a stopped wind turbine. Gray line, wind turbine generating power at 12 m/s wind speed

well as the particle motion sensor. Somewhat surprisingly, the maximum sound levels were generated at low wind speeds. A possible reason is that the gear box of the wind turbine was optimized for higher wind speeds and thus was noisier at lower speeds. It was possible to perform a test in which the sound in water was observed while the wind turbine was spinning up. It was clear that the tones are nonstationary; therefore, a better description is to characterize the tones as sliding. The sliding stops at ~10 m/s wind speed when full power production is reached. It was also observed that with weak winds, the wind turbine was frequently adjusting its pitch, resulting in the aforementioned sliding phenomena.

4 Effects on Fish

Particle motion levels were measured between 0.1- and 10-m distance from the outer rim of the foundation by moving the particle motion sensor. The sound levels were compared with known audiograms of *Gadus morhua* (Atlantic cod), *Perca fluviatilis* (European perch), *Salmo salar* (Atlantic salmon), and *Pleuronectes platessa* (European plaice). It can be concluded that at a 10-m distance, the levels are comparable in their hearing ability; thus the area where the wind turbine is affecting fish is small. Taking into account that fish disturbance requires even higher levels leads to the conclusion that this area is even smaller than 10 m. However, this conclusion is only valid for particle motion and hence is not applicable to sound pressure, which is known to be sensed by fish at larger distances.

5 Outlook

The awareness of the influence of sound on the aquatic environment has grown. New regulations are being prepared to put restrictions on aquatic sound levels (e.g., European Union Marine Strategy Framework Directive). The aim is to determine baseline levels, which are not to be exceeded. To be applicable, both components of sound have to be determined, i.e., pressure and particle motion. The latter can be studied employing the particle motion sensor as demonstrated here. Our present knowledge of specific sources and their effect is still scarce. Examples of important studies to be conducted are sound generated by ships, piling (Mueller-Blenkle et al., Chapter 89), and air guns, all of which are known to generate high levels of sound but also to map the general soundscape in anthropogenic choke points such as the sound between Denmark and Sweden that is known to be of great importance for some species such as *Anguilla anguilla* (European eel) (Andersson et al., Chapter 90).

Is the Ocean Really Getting Louder?

Michael Stocker and Tom Reuterdahl

In 1993, Donald Ross indicated a long-term trend of a low-frequency anthropogenic noise increase of 0.55 dB/yr between 1958 and 1975. This trend in ocean ambient-noise levels due to an expansion in global shipping has yielded an increase in the ambient-noise floor of the ocean that is anywhere from 6 to 12 dB higher than what it was in 1958 (depending on location). What became known as the "Ross Prediction" did not incorporate other anthropogenic sources of noise such as navigation and communication signals, noise from offshore fossil fuel exploration and extraction, and the noises from other marine industrial enterprises. There is a concern that the increase in ambient noise is masking biologically significant sounds, although the evidence for this is still scarce and somewhat speculative. Meanwhile, perhaps 90% of the biomass of complex vertebrates has been removed from the ocean since 1850 due to industrialized whaling and fishing operations (Meyers and Worm 2003; Roberts 2007).

Definitive population counts of marine mammals from prewhaling times are difficult to confirm; many models have been assembled by evaluating catch-and-kill records from commercial whaling logs. Townsend (1935) is one of the more commonly cited, although it only evaluates records from "primitive" whaling ships from 1751 through 1905, mostly under sail and not using explosive harpoon heads found from the early 20th century on.

Townsend (1935) estimates that only 5,114 bowhead whales were taken over the period of his study. Meanwhile, Bockstoce and Botkin (1977) indicated that prewhaling stocks of northern Pacific bowheads were ~30,000 and that some 18,650 were killed by 1914 when the industry collapsed. By 1978, Evans and Underwood estimated that Arctic populations of bowheads were between 2,000 and 3,000 animals, less than 10% of their 1847 Arctic population. Similarly, Kemf and Phillips (1994) indicated a prewhaling blue whale population of 275,000 whales, which is currently estimated to be "probably under 5,000."

Along with the uncertainties in population data, other uncertainties include unknown aggregation behavior that various whales would exhibit among larger populations. Although it is becoming apparent than most mysticetes have seasonal migrations driven by feeding and breeding opportunities, which would have some bearing on aggregation behavior, there is no density-dependent habitat selection analysis for mysticetes. Do animals increase or decrease their vocalizations in the presence of larger groups? Do they disperse or congregate? How do the contemporaneous variations in food supply impact their vocalization and aggregation behavior?

M. Stocker (✉) • T. Reuterdahl
Ocean Conservation Research, Lagunitas, CA 94938, USA
e-mail: mstocker@OCR.org

A.N. Popper and A. Hawkins (eds.), *The Effects of Noise on Aquatic Life*,
Advances in Experimental Medicine and Biology 730, DOI 10.1007/978-1-4419-7311-5_112,

Although the population counts vary widely, the operating assumption for the purpose of this inquiry is that "commercially viable" populations of whales had been so depleted by the mid-1940s that management of the fishery was largely prioritized toward rebuilding stocks and that perhaps as much as 90% of the great whales had been harvested since the beginning of commercial whaling in the mid-18th century.

Despite the disparities in population estimates and unknown behavioral responses to variability in population and food supplies, it is highly likely that the amplitude of biological noise from mysticete vocalizations was significantly lower in 1964 than in prewhaling times as a consequence of population depletions from industrial whaling. This paper attempts to model various biological noise scenarios in the North Pacific in 1800 and again in the 1958 baseline year of the "Ross Prediction."

Given all of the uncertainties, prewhaling marine biological noise levels can not be definitively determined. Nonetheless, this inquiry serves as a springboard for deeper discussions about the impacts of biological and mechanical ambient-noise levels on the marine bioacoustic habitat, whether the common masking assumptions are appropriate for marine mammals, whether mysticete filters for communication and other biologically significant sound sources might include a temporal component, and whether ambient-noise amplitude may be less of a biological aggravator than the characteristics of a particular noise source.

References

Bockstoce JR, Botkin DB (1977) The historical status and reduction of the Western Arctic bowhead whale (*Balena mysticetes*) population by the pelagic whaling industry 1848–1914. Report to the International Whaling Commission (Special Issue) 5:107–142.

Evans CD, Underwood LS (1978) How many bowheads. Oceanus 21:12–23.

Kemf E, Phillips C (1994) Whales in the wild. World Wildlife Fund, Gland, Switzerland.

Meyers R, Worm B (2003) Rapid worldwide depletion of predatory fish communities. Nature 423:280–283.

Roberts C (2007) Unnatural history of the sea. Island Press, Washington, DC.

Ross D (1993) On ocean underwater ambient noise. Acoust Bull 18:5–8.

Townsend CH (1935) The distribution of certain whales as shown by logbook records of American whaleships. Zoologica 19:1–50.

Part VIII
Science, Regulation, and Sound Exposure Criteria

The Complexity of Noise Impact Assessments: From Birdsong to Fish Behavior

Hans Slabbekoorn

Anthropogenic noise is on the rise. Sounds generated by human activities make the world more noisy in terms of sound levels as well as through expansion in time and space. The artificial noise penetrates all media (air, water, soil, vegetation) where it changes habitat acoustics for animals that are able to hear and for which hearing sounds may play a critical role in survival and reproduction (Slabbekoorn 2010). Awareness is also on the rise. Policy makers, industrial parties, and scientists are all increasingly aware of the potentially detrimental impact of noise pollution (Barber et al. 2009; Popper and Hastings 2009; Slabbekoorn and Ripmeester 2008; Southall et al. 2007). Dramatic physical consequences for animals in close proximity to sounds of loud intensities often draw the most attention. However, it becomes clear that more moderate noise levels, which are often widespread and long term, can also negatively affect many animals (Slabbekoorn et al. 2010). The increase in both anthropogenic noise levels and the awareness of the potential impact of high- and low-intensity sounds on animals leads to a need for adequate impact assessment methods.

The development of noise impact assessment methods is difficult, and there are many reasons why any future standardized procedure will likely be complex. First, any impact will depend on the transmission properties of the medium and the species-specific sensitivity to sound. Attenuation rates in water and air are very different and vary with locality and weather conditions, whereas species-specific hearing ranges vary considerably and sometimes do not even overlap. Furthermore, being aware of an anthropogenic sound does not necessarily mean being affected by it (Knudsen et al. 1992), and, similarly, behavioral changes associated with sound exposure can indicate, but are no proof of, negative consequences. There are many impact factors that can occur at the same time, that are not mutually exclusive, and that are often interrelated but not necessarily leading to additive effects (Table 1, Fig. 1).

The six main impact factors of anthropogenic noise include 1) physical damage, such as temporary or permanent hearing loss; 2) physiological stress, e.g., reflected by a rise in heart beat or cortisol level; 3) auditory masking, meaning a reduced detectability or recognizability of environmental or echolocation sounds or communicative signals; 4) spatial deterrence, by which animals move away from potentially favored feeding or breeding areas; 5) behavioral interruption, which can involve a breakdown of typical signal-response chains or interruption of activities such as schooling or spawning; and 6) signal modification, which refers to any temporal or spectral alteration of communicative signals. These impact factors do not stand alone and are interrelated in a complex network

H. Slabbekoorn (✉)
Behavioural Biology, Institute of Biology, Leiden University, 2333, BE Leiden, The Netherlands
e-mail: H.W.Slabbekoorn@biology.leidenuniv.nl

A.N. Popper and A. Hawkins (eds.), *The Effects of Noise on Aquatic Life*,
Advances in Experimental Medicine and Biology 730, DOI 10.1007/978-1-4419-7311-5_113,

Table. 1 An overview of all pairwise relationships among the six main impact factors of anthropogenic noise on animals

	Physical Damage	Physiological Stress	Auditory Masking	Spatial Deterrence	Behavioral Interruption	Signal Modification
Physical damage		+	+	+	+	+
Physiological stress	+		+	+	+	+
Auditory masking	n	n		+	+	+
Spatial deterrence	–	–	–		+	n
Behavioral interruption	n	n	n	+		n
Signal modification	n	+	–	n	+	

The impact factors are represented as cause in the first column and as consequence in the first row. One impact factor affects the other positively (+; making it worse) or negatively (–; leading to some release) or has no logical effect (n; neutral)

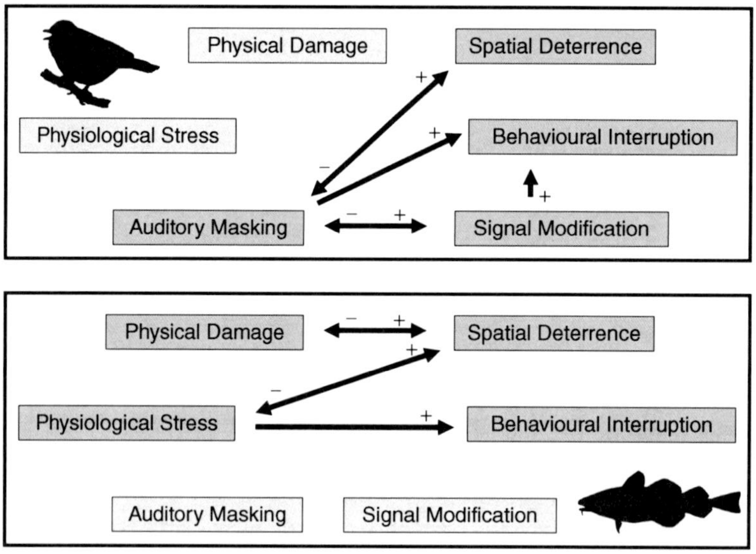

Fig. 1 Complexity of noise impact assessments and the parts that have received the most attention in studies in birds and fish. Relationships between two impact factors can be positively (+) or negatively (–) correlated, in which case one factor may be detrimental in itself but may reduce the impact of another at the same time

in which all pairwise combinations may influence each other in the sense that one is likely to make the other worse or likely to reduce the impact of the other (Table 1).

For example, any physical damage or physiological stress is likely to further increase the direct impact of anthropogenic noise on all other factors, whereas spatial deterrence may be detrimental for various reasons, but it could also lead to lower and shorter exposure levels that can release the negative impact through physical damage, physiological stress, or auditory masking. Auditory masking is in turn likely to contribute to the probability of behavioral interruption, spatial deterrence, and signal modification, of which the latter two yield two other examples of negative feedback loops through masking release (Brumm and Slabbekoorn 2005). Not all possible impact factors are always relevant and not all relationships are always critical to explore, but it is important to realize the full complexity of a proper noise impact assessment on individual fitness. Short-term consequences in one or two factors can only provide limited insight, whereas incorporation of the possible relationships among factors elucidates that some factors are inherent properties of noise exposure that animals just have to undergo while other factors may not need to be fully detrimental and can concern more or less adaptive response patterns.

Physiological and behavioral responses are often based on mechanisms evolved in the context of more natural noise patterns that could signal an increase or decrease in the probability of danger. Such responses can still benefit the animal in the context of artificial noise depending on species-specific response patterns and the impact factor. The impact of a predator escape response for animals nearby pile-driving sounds depends, e.g., on whether they flee or freeze. The impact of a habituation response can reduce spatial deterrence but at the same time leave auditory masking unaffected.

The impact factors are represented as cause in the first column and as consequence in the first row. One impact factor affects the other positively (+; making it worse) or negatively (–; leading to some release) or has no logical effect (n; neutral).

Studies on noise impact assessments in birds and fishes can provide a complementary picture because they have been biased to different factors (Fig. 1). On the one hand, many studies in birds have explored auditory masking and the consequences for spatial deterrence of breeding birds away from otherwise suitable habitat alongside noisy highways or industrial sites with noisy compressors (e.g., Francis et al. 2009). Also, the noise-dependent consequences for signal perception and signal production have been well studied in birds, which may both cause behavioral interruption in terms of a drop in response-eliciting capacity (Slabbekoorn and Ripmeester 2008). On the other hand, many studies in fish have explored the physical damage in hearing loss and physiological changes in cortisol secretion after sound exposure. Furthermore, understandably in the context of fisheries, several studies have addressed the spatial deterrence of fish, e.g., away from areas of seismic shooting or behavioral interruption of schooling behavior caused by vessel noise (Popper and Hastings 2009; Slabbekoorn et al. 2010). The impact factors that are not highlighted reflect relative data gaps (but see Ryals et al. 1999 for physical damage in birds; Codarin et al. 2009 and Hawkins and Chapman 1975 for auditory masking in fish).

In conclusion, conceptual similarities and complementary findings could mean a fruitful scientific integration when insights from bird and fish studies are combined in future research efforts to get a better understanding of the complexity of noise impact assessments.

References

Barber JR, Crooks JR, Fristup KM (2009) The costs of chronic noise exposure for terrestrial organisms. Trends Ecol Evol 25:180–189.

Brumm H, Slabbekoorn H (2005) Acoustic communication in noise. Adv Study Behav 35:151–209.

Codarin A, Wysocki LE, Ladich F, Picciulin M (2009) Effects of ambient and boat noise on hearing and communication in three fish species living in a marine protected area (Miramare, Italy). Mar Pollut Bull 58:1880–1887.

Francis CD, Ortega CP, Cruz A (2009) Noise pollution changes avian communities and species interactions. Curr Biol 19:1415–1419.

Hawkins AD, Chapman CJ (1975) Masked auditory thresholds in the cod, *Gadus morhua* L. J Comp Physiol 103:209–226.

Knudsen FR, Enger PS, Sand O (1992) Awareness reactions and avoidance responses to sound in juvenile Atlantic salmon, *Salmo salar* L. J Fish Biol 40:523–534.

Popper AN, Hastings MC (2009) The effects of human-generated sound on fish. Integr Zool 4:43–52.

Ryals BM, Dooling RJ, Westbrook E, Dent ML, MacKenzie A, Larsen ON (1999) Avian species differences in susceptibility to noise exposure. Hear Res 131:71–88.

Slabbekoorn H (2010) The impact of anthropogenic noise on animals. In: Breed M, Moore J (eds) Elsevier encyclopedia of animal behavior. Available via http://www.elsevier.com/wps/find/bookdescription.cws_home/723311/description#description.

Slabbekoorn, H, Bouton N, van Opzeeland I, Coers A, ten Cate C, Popper AN (2010) A noisy spring: The impact of globally rising underwater sound levels on fish. Trends Ecol Evol 25:419–427.
Slabbekoorn H, Ripmeester EAP (2008) Birdsong and anthropogenic noise: Implications and applications for conservation. Mol Ecol 17:72–83.
Southall BL, Bowles AE, Ellison WT, Finneran JJ, Gentry RL, Greene CR Jr, Kastak D, Ketten DR, Miller JH, Nachtigall PE, Richardson WJ, Thomas JA, Tyack PL (2007) Marine mammal noise exposure criteria: Initial scientific recommendations. Aquat Mamm 33:411–522.

Extrapolating Beyond Chinchillas: Behavioral Response Ambiguity Through the Lens of Variable Human Responses to Moderate Wind Farm Noise

Jim Cummings

1 Introduction

A fundamental challenge in assessing behavioral changes made in response to anthropogenic ocean noise is that we cannot talk to the animals involved; we can only observe (often subtle) behavioral changes and then make inferences or guesses about what they may be experiencing and what the motivations behind any changes, or lack of changes, may be. Of particular concern in recent years is the possibility that a subset of populations may be more sensitive to noise and thus be disproportionately affected by repeated exposures to chronic noise sources such as shipping, construction, or seismic surveys; some have even suggested that the most sensitive individuals may leave the area so quickly or permanently that studies of behavioral responses may be biased by the inclusion of predominantly more tolerant animals (Bejder et al. 2009). Also, marine mammals often move a moderate distance from a noise source, out of dangerous or "very annoying" range but not necessarily out of audible range; this has raised questions as to whether they may experience longer term elevations in stress levels even as they continue to engage in normal activities after being slightly displaced (Abdulla and London 2008; Bejder et al. 2009; Wright 2009).

It is clear that there is variability in behavioral responses to noise among ocean species and between individuals in a population. Similar individual variability in sensitivity to noise has been studied for decades among humans. As we try to ascertain the effects of anthropogenic noise on ocean creatures, there are two questions on the table. First, what can each species hear? The lack of direct measurements of hearing in wild populations of most aquatic species has led to the practice of extrapolating from results in easier-to-study species, perhaps most notably, in terms of stretching the extrapolation gap, chinchillas. The second question, essentially impossible to answer, is how are these sounds experienced by animals? And especially, how might individual variation in the experience relate to the observed variability in behavioral responses? Perhaps another leap of extrapolation is in order.

Recent surveys of residents living with moderate noise from nearby wind farms as well as studies of animal responses to wind farms may offer some insights that are relevant to ocean noise and animals. Similar to the results from studies of behavioral responses to ocean noise (Southall et al. 2007), there is no clear dose-response function between received noise levels at homes near wind farms and population-wide or individual responses, annoyance responses, sleep disruption, or relocation.

J. Cummings (✉)
Acoustic Ecology Institute, Santa Fe, NM 87508, USA
e-mail: cummings@acousticecology.org

A.N. Popper and A. Hawkins (eds.), *The Effects of Noise on Aquatic Life*, Advances in Experimental Medicine and Biology 730, DOI 10.1007/978-1-4419-7311-5_114,

Of particular note is a repeated pattern in which majorities of people who can hear wind farms are not particularly bothered by the noise, whereas a significant minority (10–40%, increasing in closer proximity to turbines) is more dramatically affected. This paper examines some of the experiential reports from neighbors, both those affected and not, as well as studies of annoyance responses near wind farms. In conclusion, I mention some of the potentially biologically significant effects of ocean noise exposure that may be informed by considering the likelihood that a subset of the population is more affected.

2 Individual Variability in Sensitivity to Sound

Distinct from individual, age-related, or extreme exposure-related differences in auditory sensitivity between individuals, there is a well-documented and long-studied spectrum of behavioral and psychological sensitivity to noise among humans. Some people simply are more bothered by any given type of noise source or decibel level of noise than are others. These variations have been studied since at least the 1970s; although they have been partially correlated with some external factors (such as attitudes toward the noise source), there is also a fundamental variability in coping with noise. A recent literature review (Minnesota Department of Health 2009) captures the essence of the conventional wisdom:

> "Sounds…can evoke different responses from individuals… Some people can dismiss and ignore the signal, while for others, the signal will grow and become more apparent and unpleasant over time (Moreira and Bryan, 1972; Bryan and Tempest, 1973). These reactions may have little relationship to will or intent, and more to do with previous exposure history and personality" (p. 15).

This individual variability in tolerance is also reflected in the subjective experiences reported by wind farm neighbors who can hear turbines at their homes as reported in the local press. In Vinalhaven, ME, one resident notes that for him, "It doesn't sound any different than when you've got the dishwasher running in your house. I have a brook that runs by my house, and I hear that more than I hear the turbines." Yet other neighbors a bit further away report the opposite reaction to the near parity of the brook and turbines: "As I watched the first rotation of the giant blades from our deck, my sense of wonder was replaced by disbelief and utter shock as the turbine noise revved up and up, past the sound of our babbling brook" (both quotes from Working Waterfront 2009).

Similarly, in a Wisconsin farmland wind farm that has generated an unusual amount of controversy over noise impacts, neighbors reacted very differently (Green Bay Press Gazette 2010). An apple orchard operator says turbines 300 ft from his property line make a slight "woof, woof, woof" sound but that it doesn't bother him or his customers, and a neighbor agrees: "You get a little whooshing sound once in a while. That doesn't bother me." Yet, another neighbor who has a turbine 500 ft from his property line says its sound, which he compares to a jet that never arrives, "is not for me. It's an invasion." He and his wife plan to sell their house and move elsewhere, a "permanent displacement" effect.

3 Quantifying the Variability in Sensitivity to Wind Farm Noise

Although such direct reports are illuminating, a series of research papers from Scandinavia offer the best big picture look at the patterns of response near wind farms (Pederson 2009; Pederson and Waye 2007; van den Berg et al. 2008; Waye 2009). Extensive surveys of neighbors of three large wind farms, totaling nearly 1,800 responses, show that although annoyance levels clearly rise with increasing received noise levels at all sound levels, even more people report not being bothered by the sounds. At 30–35 dB, over half of the rural residents reported hearing the turbines, whereas only 8–12% were annoyed; at 35–40 dB (within most global regulatory limits), 85% heard them, whereas just under 20% were annoyed; and at 40–45 dB (within common US regulatory limits), 95% heard them, whereas 45% were annoyed. The latter approaches the common 50% behavioral response rate sometimes used as a threshold in ocean noise assessment.

4 Variability in Response Among Wildlife Near Wind Farms

Likewise, studies of the effects of wind farms on wildlife (as well as reports of residents) show marked species and individual variability among birds, small mammals, and domestic and wild grazing species. One recent study of note (Pearce-Higgins et al. 2009) found that five local nesting species were relatively unaffected by living within or near wind farms, whereas 7 species were less likely to nest within 500–800 m of the turbines; of those displaced, the decrease ranged from 15 to 52%, again confirming a general pattern that many individuals tolerated the turbines, whereas a significant minority was more affected.

5 Discussion: Considerations for Marine Species

As we come to grips with this pattern, also seen in the ocean, that a significant minority of many species' populations appears to be more sensitive to disruption by anthropogenic noise, the implications will be particularly relevant in situations where a negative impact on a minority of the population may be especially problematic to population health and viability. This may include particularly stressed populations (such as the North Atlantic right whale), particularly sensitive times of life (especially newborns and perhaps old age), and situations in which synergistic effects with other factors (e.g., habitat degradation, toxins) may be triggered by noise related stress.

Of particular note are indications (Williams et al. 2006) that the lost energy intake resulting from reduced time spent foraging is likely to be the largest impact on the energy budget of disturbed individuals, far more substantial than energy expended in moving away from noise sources or vocalizing more or louder. Many recent studies suggest foraging is reduced by 15–35% in animals behaviorally disrupted by boat traffic, seismic surveys, and midfrequency active sonar (Cummings 2009).

A central feature of humans reporting noise-related annoyance is that for many of them, their experience with the noise triggers a variety of stress-related effects, including headaches, sleep disruption, irritability, and a decreased capacity to focus at work or school. Although these impacts have created much brouhaha about whether the noise itself is causing health effects, the more likely pathway of impact in most cases is that among the subset of the population that is more sensitive to noise, noise-related stress, and sleep disruption is the causal factor for various health effects. The many clearly articulated reports from humans who are more affected by living with a nearby noise source may be especially relevant to our appreciation for the experiential effects of chronic noise-related stress among ocean creatures.

In both wind farm planning and ocean management, it is often tempting to let the majority or the mean response become the ground for determination of risk and of regulatory protections, with minority responses considered to represent negligible impacts. However, just as communities near wind farms are coming to grips with the sometimes extreme effects on a minority of their neighbors, thanks to their ability to speak loudly and clearly about what they are experiencing, so too should ocean policy makers take into consideration the implications of individual variability in sensitivity to noise. If and when a significant minority of a population is more significantly or repeatedly affected by noise intrusions, the long-term impacts on populations are likely to be far from negligible.

References

Abdulla A, London O (eds) (2008) Maritime traffic effects on biodiversity in the Mediterranean Sea: Review of impacts, priority areas and mitigation measures. International Union for the Conservation of Nature Centre for Mediterranean Cooperation, Malaga, Spain.

Bejder L, Samuels A, Whitehead H, Finn H, Allen S (2009) Impact assessment research: Use and misuse of habituation, sensitization and tolerance in describing wildlife responses to anthropogenic stimuli. Mar Ecol Prog Ser 395:177–185.

Cummings J (2009) Does moderate anthropogenic noise disrupt foraging activity in whales and dolphins? Proceedings of the National Workshop to Examine the Effectiveness of Measures Used to Mitigate Potential Impacts of Seismic Sound on Marine Mammals, Canadian Science Advisory Secretariat/Fisheries and Oceans Canada Workshop, 12–13 May 2009, Ottawa, ON, Canada. Available via http://www.AcousticEcology.org/docs/Cummings_DFO2009_ForagingDisruptions.pdf.

Green Bay Press Gazette (2010) Wind farms cause controversy among community, January 24.

Minnesota Department of Health Environmental Health Division (2009) Public health impacts of wind turbines. Agency Report for the Minnesota Department of Commerce Office of Energy Security. Available via http://energyfacilities.puc.state.mn.us/resource.html?Id=24519.

Pearce-Higgins JW, Stephen L, Langston RHW, Bainbridge IP, Bullman R (2009) The distribution of breeding birds around upland wind farms. J Appl Ecol 46:1323–1331.

Pederson E (2009) Effects of wind turbine noise on humans. Proceedings of the Third International Meeting on Wind Turbine Noise, 17–19 June 2009, Aalborg, Denmark.

Pederson E, Waye KP (2007) Wind turbine noise, annoyance and self-reported health and well-being in different living environments. Occup Environ Med 64:480–486.

Southall BL, Bowles AE, Ellison WT, Finneran JJ, Gentry RL, Greene CR Jr, Kastak D, Ketten DR, Miller JH, Nachtigall PE, Richardson WJ, Thomas JA, Tyack PL (2007) Marine mammal noise exposure criteria: Initial scientific recommendations. Aquat Mamm 33:411–522.

van den Berg, F, Pederson E, Bouma J, Baaker R (2008) Project WINDFARMperception: Visual and acoustic impact of wind turbine farms on residents. Final Report, FP6-2005-Science-and-Society-20, Specific Support Action Project no. 044628.

Waye KP (2009) Perception and environmental impact of wind turbine noise. Proceedings of Inter-Noise 2009, 38th International Congress and Exposition on Noise Control Engineering, 23–26 August 2009, Ottawa, ON, Canada.

Williams R, Lusseau D, Hammond PS (2006) Estimating relative energetic costs of human disturbance to killer whales (*Orcinus orca*). Biol Conserv 133:301–311.

Working Waterfront (2010) Hard lessons from the Fox Islands Wind Project, December 2009. As electric co-op conducts sound experiment, Vinalhaven residents debate solution to turbine noise issue, February-March 2010.

Wright AJ (ed) (2009) Report of the workshop on assessing the cumulative impacts of underwater noise with other anthropogenic stressors on marine mammals: From ideas to action. Okeanos–Foundation for the Sea, 26–29 August 2009, Monterey, CA. Available via http://www.sound-in-the-sea.org/download/CIA2009_en.pdf.

Underwater Noise Effects From Wave Energy Devices on Marine Mammals: A Possible Approach

Sofia Patricio

1 Introduction

The past decades have seen a growing concern in the scientific community regarding the effects of underwater noise on marine life in general and particularly on marine mammals. Some marine species use sound as their primary sense, using it to communicate, socialize, feed, and echolocate. As a result, any change in underwater noise has the potential to disturb these species through physical, behavioral, perceptual, chronic, and indirect effects. Consequently, the increase in anthropogenic activities in the ocean is a growing concern that needs to be addressed.

Future large-scale implementation of wave energy converters (WECs) in the ocean may prove to be one additional source of underwater noise. Although it is not expected that each individual device will produce a high level of noise, the deployment of several devices on the same farm operating day and night may have an effect on the fauna.

In 2006, the Wave Energy Centre (WavEC) in collaboration with the Technological Research Centre of the University of Algarve (CINTAL-UAlg), Faro, Portugal, started research in this field through the project entitled Wave Energy Acoustic Monitoring (WEAM; http://www.wavec.org/index.php/31/weam/), which began in November 2007.

2 Wave Energy as a Possible New Noisy Activity

The increase in sea deployments of WECs raised concerns regarding possible environmental impacts. In a general overview, the knowledge about the environmental impacts of offshore renewable energy is growing; nonetheless, significant knowledge gaps remain. Developers and promoters have demonstrated particular concern regarding the possible effects of underwater noise emitted by WECs (Patrício et al. 2009a). There is little information on the characteristics of the noise produced during the different project phases of wave energy farms. Acoustical data on the sound produced during normal operations of wave energy technologies are not available yet.

S. Patricio (✉)
Wave Energy Centre, 1000 Lisbon, Portugal
e-mail: sofia@wave-energy-centre.org

A.N. Popper and A. Hawkins (eds.), *The Effects of Noise on Aquatic Life*,
Advances in Experimental Medicine and Biology 730, DOI 10.1007/978-1-4419-7311-5_115,

The research done in offshore wind energy regarding noise emissions can help guide the work needed to be done in wave energy, such as baseline studies, procedures, simulations, and monitoring, but it will not address the fundamental differences in noise between wind and wave energy devices or farms.

In offshore wind energy devices, noise is emitted into the water by vibration of the structure down to the bottom caused by blades or turbine movement (directly to water), noise transmission from air to water from blades or turbine, and cavitation around the structure. In contrast, WECs will produce the most noise under water and the project phases differ from wind farm projects mainly in the construction phase. The noise generated will be directly transmitted into the water column and, depending on the mooring system, to bottom sediments.

Just after characterization of the noise emitted by the WECs and potential coincidence with the hearing sensitivity range of marine animals, a first assessment of possible impacts on the animals can be attempted.

The noise emitted by each WEC, the acoustic signature, is expected to be produced from a variety of different components (mechanical or other moving parts) related to the device itself and/or by its interaction with the environment. There may be some similarity to vessel machinery noise that has been well studied by the underwater sound community. Machinery noise has been identified as being originated by 1) rotating parts (shafts and motor armatures); 2) repetitive discontinuities (gear teeth, armature slots, and turbine blades) like, for example, in Wells and Pelton turbines; 3) explosion in cylinders in internal combustion motors; 4) cavitation and fluid flow (pumps, pipes, cylinders, and valves); and 5) mechanical friction (Patrício et al, 2009b; Urick 1983).

3 Assessing the Impact

Performing in situ measurements in the area of the WEC may be sufficient for a minimal impact assessment. However, in some cases, this may not be sufficient to completely characterize the noise impact on the marine life (Patrício et al. 2009b). The observed noise may vary with time and space due to operating device conditions varying according to the sea state and the sound pressure level (SPL) depending on range, depth, and bearing. A complete spatial and time coverage via in situ measurements may become time consuming and expensive. Fortunately, there are very accurate acoustic propagation models that can be used as a complement to in situ measurements (Jensen et al. 1994).

A model can be a helpful tool to analyze how the sound will propagate from a single device or an array of devices. The deployment of an array of WECs can have a cumulative effect, increasing the sound generated or, in some special cases, resulting in the phenomenon of sound annulations in some frequencies. To be able to predict this, the oceanographic conditions (temperature, salinity, bathymetry, and sediments) as well as the acoustic signature of the WEC in different operational conditions need to be characterized in detail. Acoustic modeling could aid in setting up the layout of wave energy farms (geometry, number of devices, and distance apart) and the development of individual devices such that the environmental performance related to underwater acoustic noise is maximized (Patrício et al. 2009b).

Because the final aim is to find out to what extent an animal may be affected by a given noise source, it is necessary to proceed with a criterion. There are several studies that attempt to characterize the hearing sensitivity of marine species, which results in an audiogram. The most common effects that these studies can assess are behavioral changes or temporary damages in the hearing system, which are usually described as a temporary loss of hearing sensitivity, known as temporary threshold shift (TTS) (Richardson et al. 1995). Such studies can be used to evaluate the impact in a real scenario by using in situ measurements and application of available criteria. The result of this interpretation may be an evaluation in terms of influence zones where the experimenter roughly produces a set of plots as a function of space, indicating whether a certain level of impact is taking place or not. These influence zones usually range from audibility to temporary injury, permanent injury, or even death in very severe cases (Richardson et al. 1995).

Assessing the acoustic impact involves several steps. A possible approach to evaluate the impact of underwater noise of WECs on marine life could focus on measuring the frequency and duration of noise generated by WECs under different sea states and ocean conditions; monitoring and modeling the sound field from a single device and an array, including several farm configurations (device number, density, and geometry), to predict noise propagation; characterizing and monitoring the marine species; andcalculating the influence zones and assessing the possible effects (Patrício et al. 2009b).

To summarize, the three main steps are to characterize the acoustic receptor (the marine species), determining its acoustic sensitivity; characterize the acoustic channel through the physical properties of the ocean; and characterize the acoustic signature of the WECs.

References

Jensen F, Kuperman W, Porter M, Schmidt H (eds) (1994) Computational ocean acoustics. Springer-Verlag, New York.

Patrício S, Moura A, Simas T (2009a) Wave energy and underwater noise: State of art and uncertainties. OCEANS '09 IEEE Bremen, 11–14 May 2009, Bremen, Germany.

Patrício S, Soares C, Sarmento AJNA (2009b) Underwater noise modelling of wave energy devices. Proceedings of the 8th European Wave and Tidal Energy Conference, 7–10 September 2009, Uppsala, Sweden.

Richardson WJ, Greene CR Jr, Malme CI, Thomson DH (1995) Marine mammals and noise. Academic Press, San Diego, CA.

Urick R (1983) Principles of underwater sound. McGraw-Hill, New York.

Physical Biologists and Biological Physicists: Combining Biology and Physics in Research on the Effects of Noise on Aquatic Life

Douglas H. Cato

1 Introduction

Research into the effects of noise on aquatic life is interdisciplinary, requiring expertise from the physical as well as the biological scientists. This paper examines how physicists and biologists differ in their approach to research and how these differences need to be reconciled if interdisciplinary research is to be effective and avoid major pitfalls. Physicists (in classical physics) are used to dealing with relatively simple systems, and it is often possible to find deterministic ways of describing the systems (e.g., with mathematical models tested by measurement). Biologists deal with more complex systems that include a greater degree of randomness and usually have to treat their data statistically. There is no hard and fast division, however, and physicists can learn from the biologists and vice versa, and this is most effective when they work together.

Scientists are generally trained within a particular discipline, e.g., biology, physics, or chemistry, and in their research tend to specialize in one area of their discipline, e.g., animal behavior, physiology, underwater acoustics, or fluid dynamics. The substantial growth in scientific knowledge and the number of scientists have driven a general trend to increased specialization and compartmentalization in research, and scientists generally have limited knowledge of and little insight into other discipline areas. Practices and procedures of one discipline may be quite different from those of another. This includes methods of designing and conducting experiments and analyzing the resulting data. Even the approach to reporting may be different. These differences are well demonstrated when biology research is compared with physics research. The extent that scientists can work outside their discipline areas is therefore quite limited.

Although science has become more compartmentalized, there has been an increasing demand for interdisciplinary research, and the effects of noise on aquatic life is a good example. Interdisciplinary research presents a challenge that is most effectively met by forming teams of experts from the component discipline areas. For example, research into the behavioral effects of noise on aquatic life would require at least an expert on animal behavior and one on acoustic propagation and ambient noise. The better each understand the differences between the disciplines and the way that they work, the better the collaboration will be.

D.H. Cato (✉)
Defence Science and Technology Organisation, Pyrmont, NSW 2009, Australia

University of Sydney Institute of Marine Science, University of Sydney, Sydney, NSW 2006, Australia
e-mail: doug.cato@dsto.defence.gov.au; doug.cato@sydney.edu.au

A.N. Popper and A. Hawkins (eds.), *The Effects of Noise on Aquatic Life*, Advances in Experimental Medicine and Biology 730, DOI 10.1007/978-1-4419-7311-5_116,

In this paper, I discuss differences in the approach of biology and physics as applied to aquatic bioacoustics and the effects of noise and how these might affect research approaches, experimental design, and interpretation of the results. They are based on my experience of decades of collaborative work. Differences that are often apparently subtle can have significant effects. I hope that this will help early- and midcareer scientists in forming collaborations and getting the most out of interdisciplinary work.

Some scientists tend to be somewhat cavalier in their approach and chauvinistic in their attitude toward other disciplines and do not see the need to have expertise in the other disciplines involved when they embark on interdisciplinary research. I give examples of the problems this has caused. The history of the study of the effects of noise on aquatic life has many examples of misconceptions and errors that have arisen and continue to be widespread as a result of the failure of some scientists to obtain advice and participation from experts in fields other than their own.

2 Physics and Biology in Aquatic Bioacoustics and the Effects of Noise

Any research into the effects of noise on aquatic life involves both biology and physics. The physics includes underwater acoustics, which itself covers a number of fields, as well as other areas such as the acoustics of hearing. The biology includes animal behavior and communication, auditory perception, and ecology. It also includes those areas that are needed to investigate how the immediate effects of noise relate to longer term effects such as population dynamics and energy budgets. In this paper, I compare the areas of physics and biology that are usually involved in bioacoustics and the effects of noise on aquatic life rather than physics and biology in general.

Physics and biology are different in their approach to research and understanding of their subject matter. Physics, or at least classical physics, generally deals with simpler systems than biology. A common approach in physics is to develop an analogue of the real system of interest in the form of a theoretical model, and this is usually expressed mathematically. The idea is to obtain an insight into the physical process involved by developing a model that is as simple as can reasonably be achieved while retaining essence of the physical process. More complexity to bring the model closer to the real system can be added later. As in any area of science, the predictions of the theoretical models must, of course, be tested by measurements, and in the end, as with any discipline, the success of a theory depends on how well it is supported by measurements and observations. This is not to say that theoretical models are not used in biology but rather that physics is more amenable to mathematical modeling.

Biology deals with living systems, and in bioacoustics, this is typically the level of the whole animal or substantial components such as the senses, especially hearing. Animals are both very complex and very variable from one individual to the next, even between individuals in family groups. There may be many factors that affect this variation, and some may not be known. This imparts a degree of randomness that needs to be addressed statistically. As a result, any study of animals needs to include a large number of individuals to obtain a representative sample of the population. The observations and measurements are expressed as statistical distributions.

Mathematical models of physical systems may be deterministic in that the variables are directly related without the randomness of stochastic models as in biological systems, even though there are physical processes that need to be treated stochastically. Many physicists may not understand the importance of the stochastic process in biology, especially the need to obtain an adequate sample size in biological experiments.

3 Mathematical Models of Physics and Their Limitations

The development of mathematical models requires two important steps: 1) the assumptions: the development of a simplified analogue of the real system requires assumptions about the way the real system works, and 2) the approximations: in evaluating the mathematical expressions of the model, it is usually necessary to make approximations to allow the result to be calculated.

The assumptions and approximations used limit the application of the model to conditions in which these have negligible effect. The model will not be effective in other sets of conditions, but it may be possible to develop other models using different assumptions and approximations, and each may be effective within their own range of conditions.

The art in modeling is to choose the most effective assumptions and approximations. These should, on one hand, allow the model to be simplified enough to obtain a mathematically tractable result but, on the other hand, minimize the differences between the model and the real system. The way the chosen assumptions and approximations affect the model and its relationship to the real system is not always evident in advance or even in retrospect. An assumption may even void an essential component of the system.

Problems arise when the assumptions and approximations of a model are not considered. There are many examples where the physicist's model shows that an animal cannot perform some function and where the animal shows by its behavior that it is perfectly capable of doing just what the model "proved" that it cannot. An example is the claim by an underwater acoustician that although we can determine the angular direction of a sound source relative to our heads, we cannot tell whether the source is in front of us or behind us. Another (by a different acoustician) is the statement that turtles cannot localize sound because their ears are too close together for the frequencies of interest (this same reasoning would also say that many birds cannot localize). Both were using a simple model in which the ears are point receivers and the only information used by the animal is the time-of-arrival difference, a direct analogy with their own work with localization using hydrophone arrays. In both examples, the mistakes could have been avoided either by consulting the literature on hearing or by a simple experiment (it should only take a few moments for a person to realize that they can tell whether a source is in front or behind them). Hence biologists need to be skeptical of any mathematical model and physicists need to recognize the complexity of biological systems. Both need to be aware of what is already available in the literature.

4 Textbook Models May Be More a Means of Understanding a Phenomenon Than a Method of Predicting the Real World

Like scientists in other areas, physicists are trying to understand the phenomena of interest, and they will generally use mathematical models as the means of doing this. There are some important points to realize in this. A model intended to provide understanding of a process will tend to be simpler and will focus on the essence of the process. This is particularly true of ocean acoustics, where the results will depend on the environmental conditions. The textbook explanation may not be so useful if you are looking for an equation that is going to provide a reasonably accurate prediction of some effect of interest. It is up to the physicist to develop the theory further to predict results for particular environmental conditions.

Sound propagation in water provides examples to illustrate some of these effects. In an acoustically simple medium, i.e., one without any variation in sound speed or density and no absorption, sound radiates out from a point source equally in all directions. The sound emitted at any time radiates out as a spherical wave, i.e., as though on the surface of a sphere centered on the source. The power radiating out from the source is spread over the surface of a sphere and so is spread more thinly as the sphere expands. The sound intensity, which is the power per unit area, therefore decreases as the distance from the source, the radius (r) of the sphere, increases. Because the area is proportional to $1/r^2$, the intensity decreases in proportion to r^2. This is the familiar spherical or square law spreading.

This simple model gives an insight into the way sound radiates from a source (in fact, the way in which other forms of radiant energy travel from a source), but the model has a number of the assumptions described above. Under certain conditions, the assumptions do not have a large effect and spherical spreading will give a reasonable measure of the real propagation loss. Such conditions include distances relatively close to the source so that refraction does not have a large effect, distances well short of those required for the wave to reach the sea surface and bottom boundaries, and frequencies low enough for absorption to be negligible. Just how close to the source, how far from the boundaries, and how low the frequency must be for the assumptions to have an acceptably small effect is something that you have to determine. This is part of the art of developing and applying mathematical models of physical processes, and why it is important to have someone who is expert in this involved.

Shallow-water propagation is sometimes said to be by cylindrical rather than spherical spreading. This is based on the idea that the sea surface and bottom reflect energy back into the water column so that the sound can only spread horizontally as though on the surface of a cylinder. The power is spread over an area proportional to the distance r so that the intensity, or power per unit area, decreases as $1/r$ rather than $1/r^2$ as for spherical spreading. The propagation loss in decibels by this simple model would be $10 \log r$ for cylindrical spreading compared with $20 \log r^2$ for spherical spreading.

Although some books and papers give this simple model without discussions of limitations and some have used this simple model of $10 \log r$ to calculate propagation loss from a source in shallow water, this underestimates the loss by a substantial amount, usually some tens of decibels. The problem is that the sound is not constrained by the sea surface and bottom until it has travelled some distance; it will spread spherically initially. There will be some transition between cylindrical and spherical spreading. A simple calculation will show that if this transition occurs at around 100 m, cylindrical spreading would underestimate the loss by 20 dB. The real ocean is far more complicated. This model assumes that the bottom is a perfect reflector of sound and real sea floors rarely approach this ideal. Some sound travels into the bottom and some of the energy is absorbed, depending on the properties of the bottom. Some may be so poorly reflective that little is reflected back into the water column. A wide variation in loss is evident in shallow-water propagation.

There are many sophisticated theoretical models of propagation loss that can be used, but they are difficult to use and need an understanding of the theory to run them and this requires experts. Different models apply to different conditions and environments because of the assumptions involved. All require information about the ocean and the bottom, some of which may be difficult to obtain, and so this limits the application of the models. Direct measurement of propagation loss avoids modeling but is logistically difficult and again requires experts. It is also limited to the particular location and water column conditions of the measurement. Even so, the measurements can be used to ground truth the model and to improve its application to that location.

5 Problems of Inadequate Knowledge of Disciplines Involved

Problems may arise through lack of knowledge of one or more of the areas involved in the research. In some cases, this lack may not be readily apparent. Underwater acousticians will have little understanding of the complexity of animal behavior or the protocols required for adequate experimental design in behavioral studies. Everyone is familiar with animals and this may give the false impression that study of behavior is straightforward. The attitude may be that when the animal is exposed to a noise, we will observe what happens. That approach may then lead to inferences or interpretations that are not justified because of the failure to understand the difficulties of objective observation and analysis of animal behavior. There also may be inadequate experimental design because of lack of understanding of the need to obtain a representative sample. Part of the problem may be a lack of understanding that animal behavior is a significant field of study in its own right.

Behavioral biologists may be so used to applying statistics to experimental data that they apply this to the physical aspects without knowledge of what is already known and without considering if it could be more effectively studied deterministically. An example is an investigation of potential factors influencing an observed difference in vocalizations of dolphins resident in two geographically separated bays. The researcher tested to see whether this might be an effect of differences in the ambient noise between the two bays by taking a series of samples of the noise. Analysis showed that the samples were different by an amount that was statistically significant. There is a large amount of literature on ambient noise that shows that ambient noise shows wide temporal variation, but the researcher was not familiar with this. What had been determined was not a difference between the ambient noise in the bays, i.e., a geographic variation, but a temporal variation and a smaller difference than would commonly occur in each bay. It is well established that ambient noise varies widely as wind speed varies or as biological contribution varies. These need to be held constant to compare geographical variation.

6 Summary

Research in aquatic bioacoustics and the effects of noise is interdisciplinary and to be effective requires a collaboration of experts from all the fields involved. The full range of expertise is needed for adequate understanding of the processes involved, adequate experimental design, analysis and interpretation, and adequate knowledge of the research already published. The biologists need to understand how physicists work and make allowance, and vice versa. Both need to understand that the other will not be familiar with their practices and approach and that there will be a certain amount of negotiation and education on both sides.

However, the best reason to develop collaborations with other experts in interdisciplinary research is that it is such a rewarding experience from the insights it provides into other disciplines and from the opportunity to do really effective and very significant research, well beyond what the individuals might have achieved on their own.

"So, Am I Correct in My Understanding That a Decibel Is the Same as a Hertz?": The Quest for Informed, Objective Environmental Impact Analysis of Marine Anthropogenic Noise

John F. Polglaze

1 Introduction

Environmental impact assessment and evaluation, no matter how rigorous the underlying scientific research and analysis, is fundamentally and unavoidably a question of values judged within prevailing social, economic, cultural, and legal frameworks. Whatever researchers and professional environmental scientists may discern, even if done in a totally dispassionate, definitive, and objective manner, decisions will be based on other values and imperatives (hopefully) using the scientific data and analysis as illumination. In the arcane, often speculative arena of marine anthropogenic noise and its ecological effects, the need for decision makers to have a clear and effective understanding of the technical aspects of the subject is pivotal to the making of sensible, risk-based assessments and conclusions. Alas, this ideal is often not achieved, with less than optimal consequences for rational assessments and application of pragmatic management measures. This situation invariably results in the inefficient allocation of finite resources.

2 Background

Environmental impact evaluation of marine anthropogenic noise is a challenging arena. A range of anthropogenic noises need to be considered by regulators and industry when dealing with activities in marine and coastal areas, encompassing sources such as pile driving, seismic survey, dredging, drilling, pipe laying, underwater explosives, and sonar. Many represent minimal source levels or are of limited duration, barely warranting further assessment, yet environmental impact assessment, screening, and scoping processes have yet to attain the level of maturity where rational, informed judgments occur.

Given the irreconcilable complexities and uncertainties involved, there is wide variation in the summation and assessment of likely ecological impact from proposed noise-generating activities. Some regulators and proponents seek to undertake or require quantitative analyses, even if based on artificially derived or highly speculative input data, so that nominally "empirical" assessments are made.

J.F. Polglaze (✉)
PGM Environment, Perth, WA 6004, Australia
e-mail: john.polglaze@pgmenviro.com.au

A.N. Popper and A. Hawkins (eds.), *The Effects of Noise on Aquatic Life*,
Advances in Experimental Medicine and Biology 730, DOI 10.1007/978-1-4419-7311-5_117,

A lack of definitive data does not mean, however, that rational risk-based assessments cannot be arrived at, with concomitant, precautionary risk-mitigation measures adopted by proponents. The prevailing condition in some jurisdictions, however, is that overly restrictive, arguably unnecessary controls are imposed in response to populist causes rather than rational risk-based evaluations. This results in an inordinate effort and finite resources focused on areas of minimal risk at the expense of more pressing environmental needs.

The science of marine anthropogenic noise and its impact on marine fauna is inexact and characterized by extensive data deficiency. The august work of Southall et al. (2007) is rightly considered a watershed in the derivation of consistent metrics for the estimation of potential impacts of anthropogenic noise on marine mammals, but this work relies to a great extent on anecdotal observations and extrapolation and interpolation of data between species and across functional groups. Similarly, the population consequences of acoustic disturbance (PCAD) model (National Research Council 2005) is a useful conceptual tool for exploring any potential nexus between marine anthropogenic noise effects on individuals of a species and their translation into population level effects. Application of the PCAD model, however, is also hampered by a paucity of data.

The challenge for regulators and industry is achieving a common, coherent understanding of the physics of marine anthropogenic noise and acoustic propagation and the resultant ecological effects. This critical foundation should underpin environmental assessment of marine anthropogenic noise and the associated imposition of mitigation controls. Alas, regulatory processes frequently fall short of the ideal, with decisions made on the basis of preconceived notions and minimal understanding. In some jurisdictions, the prevalent impasse between proponents and opponents is often "resolved" through recourse to legal procedures. These adversarial processes are arguably unsuited and ill equipped to effectively deal with the uncertainties and technical complexities involved with marine anthropogenic noise, resulting in the perpetuation of questionable analyses and conclusions about marine anthropogenic noise and its likely ecological effects.

3 Examples of Marine Anthropogenic Noise Assessments

Recent examples are presented to illustrate where proponents and regulators have demonstrated an enlightened approach to these complex issues and are contrasted with cases where this has not been evident.

3.1 Examples of Iterative, Risk-Based Approaches

This section is intentionally left blank!

3.2 Examples of Noniterative, Non-Risk-Based Approaches

One example is a resources export terminal slated for an Australian port. It is fair to assume that an existing, long-established multipurpose harbor, hosting more than 800 annual ship visits while sustaining a small resident dolphin population, would likely be able to accommodate an additional 40 or so ship visits per year without a discernible, tangible effect on the dolphins. This view was not that of the regulators, who mandated that the environmental assessment was to include recordings and acoustic analysis of ship movements within the harbor and the associated noise modeling. This costly exercise produced results essentially no different from those derived from a literature review and the basic principles of acoustic propagation.

Ironically, within this same harbor, the regulator concerned employs outboard-powered aluminum dinghies to seek out the same dolphins for population studies, seemingly unaware that the noise from these boats will be as, if not more, audible to the dolphins than will the noise generated by large ships. Furthermore, there are no controls on either the number or activities of the multitude of small outboard fishing and recreational boats used by members of the general public in the same harbor.

Another omnibus example is that of United States Navy (USN) sonar and habitual recourse by environmental groups to litigation and subsequent deferral to US courts for arbitration. Since 2003 at least, the USN has been involved in ongoing court-mediated disputes. This has no doubt incurred a diversion of available technical assets and skewed research and assessment resources to countering claims made in litigation rather than on developing a more effective means of marine fauna protection. It is open to debate if emotive, adversarial processes, where the validity of assessment and cogitation hinges as much on the nuance of language and legal tactic as on the scientific analysis, are best suited to arrive at robust, risk-based conclusions for such scientifically complex questions. The inherently "daring" nature of relying on courts to determine the scientific merit of arguments of this ilk is starkly revealed in the case of one (who shall remain nameless) judge, who, when summing up several days of hearings before retiring to make his determination, uttered, "So, am I correct in my understanding that a decibel is the same as a hertz?"

4 Conclusions

The quest for informed, objective environmental impact analysis of marine anthropogenic noise continues. More than a decade since marine anthropogenic noise assessments became topical and commonplace and despite improvement in the quality and volume of data and the understanding of system complexities, there is scant evidence of any meaningful parallel improvement or maturation of the processes within which the significance of potential effects from marine anthropogenic noise is assessed. This situation could be improved with wider adoption of standard guidelines and impact avoidance and mitigation measures. Given the complexities and uncertainties involved, there is a pressing need for such development so that finite resources can be directed to where most ecological benefit can be realized.

References

Southall BL, Bowles AE, Ellison WT, Finneran JJ, Gentry RL, Greene CR Jr, Kastak D, Ketten DR, Miller JH, Nachtingall PE, Richardson WJ, Thomas JA, Tyack PL (2007) Marine mammal noise exposure criteria: Initial scientific recommendations. Aquat Mamm 33:411–522.

National Research Council (2005) Marine mammal populations and ocean noise: Determining when noise causes biologically significant effects. National Academies Press, Washington, DC.

Environmental Assessment of Offshore Wind Power Generation: Effect on a Noise Budget

James H. Miller, Gopu R. Potty, Kathleen J. Vigness-Raposa, David Casagrande, Lisa A. Miller, Jeffrey A. Nystuen, Peter M. Scheifele, and John G. Clark

1 Introduction

The construction, operation, and removal of offshore wind power facilities have the potential to increase the levels of noise in the surrounding waters. Increased shipping noise can be associated with all phases of the life cycle of wind farms. Pile driving during platform construction can be short-lived but can introduce very high intensity pulses into the ocean. Operational noise, while at a much lower level, can affect the surrounding area almost continuously for the life of the facility. The removal of wind platforms at the end of their operational life may use explosives, again short-lived but very intense.

In August 2008, the state of Rhode Island began planning for offshore renewable energy development with the establishment of a Special Area Management Plan (SAMP) for the ocean just south of Rhode Island. SAMPs are federally recognized management and regulatory tools that promote ecosystem-based management as well as reasonable coastal-dependent economic activity. The Ocean SAMP is an adaptive planning tool that promotes a balanced and comprehensive ecosystem-based management approach to the development and protection of Rhode Island's ocean-based resources, specifically offshore renewable energy. Part of the Ocean SAMP involves research into the potential effects of noise from construction, operation, and removal of offshore wind power facilities on marine animals. This paper concentrates on the low-level but almost continuous ocean noise from the operation of offshore wind turbines. The effects of this low-level operational noise on the marine environment may be subtle and difficult to observe without detailed measurements.

European investigators have measured underwater noise from offshore wind turbines for different wind speeds and power production levels. Betke et al. (2004) found noise levels averaging 112 dB re 1 μPa in a one-third octave band centered near 200 Hz at a distance of 110 m from a wind turbine

J.H. Miller (✉) • G.R. Potty • K.J. Vigness-Raposa • D. Casagrande • L.A. Miller
University of Rhode Island, Narragansett, RI 02881, USA
e-mail: miller@uri.edu

J.A. Nystuen
University of Washington, Seattle, WA 98105, USA
e-mail: nystuen@apl.washington.edu

P.M. Scheifele • J.G. Clark
University of Cincinnati, Cincinnati, OH 45221, USA
e-mail: scheifpr@ucmail.uc.edu

A.N. Popper and A. Hawkins (eds.), *The Effects of Noise on Aquatic Life*,
Advances in Experimental Medicine and Biology 730, DOI 10.1007/978-1-4419-7311-5_118,

at full power in the Utgruden wind farm in Sweden. Betke (2006) also measured noise levels averaging 118 dB re 1 μPa in a one-third octave band centered near 160 Hz at a distance of 100 m from a wind turbine at the Horns Rev wind farm in Denmark. The turbines in both measurements were mounted on monopiles. In Rhode Island, it is likely that the turbines will be mounted on lattice jacket structures (Schneider and Senders 2010).

In this paper, we apply the principles of noise budgets (Miller et al. 2008; Nystuen and Howe 2005) for estimating the potential biological effects from an eight-turbine farm planned for the waters south or southeast of Block Island, RI. We report on measurements of the ambient noise field using passive aquatic listener (PAL) systems (Ma et al. 2005). With the European measurements of wind turbine noise underwater, our measurements of the ambient noise, and transmission loss and estimates of sediment properties, we predict the noise budget for a site 10 km from the 8-turbine farm.

2 Noise Budgets

As outlined in Miller et al. (2008), in an ocean with constant sound speed and density, the instantaneous intensity of a wave far from a small source (n) is given by $I_n(f,t) = \mathrm{Re}\left[p_n(f,t)u_n^*(f,t)\right] = \dfrac{|p_n(f,t)|^2}{\rho c}$ where $p_n(f,t)$ is the acoustic pressure from the nth source in a band of frequencies (typically one-third octave) centered at frequency (f) and time (t) and $u_n(f,t)$ is the radial component of acoustic particle velocity. The average intensity in the frequency band is $\langle I_n(f)\rangle = \dfrac{1}{T\rho c}\int_0^T |p_n(f,t)|^2\, dt$ where T is the averaging time. If one is able to classify the nth source of sound for all times between 0 and T, a noise budget can be estimated using the average intensity from each source in the frequency band.

2.1 *Measured Noise Budget*

Two PAL systems were deployed in the waters south of Block Island (PAL labeled Eider at 41°7′ north latitude, 71°39′ west longitude and PAL labeled Puffin at 41°7.65′ north latitude, 71°33.6′ west longitude). Descriptions of the PAL instrument and its data-processing schema can be found in Nystuen et al. (2008). The instruments were deployed on 6 October 2008 and recovered on 14 November 2008. Figure 1 shows the computed average intensity noise budget for the one-third octave band centered at 500 Hz based on the data collected by the Eider PAL. The ambient noise field is dominated by shipping in this band (and practically all the frequency bands examined), with an average intensity of 5,086 pW/m^2, corresponding to 99 dB re 1 μPa in the one-third octave band. Wind-generated noise was the next most important source, with 3,869 pW/m^2, corresponding to 98 dB re 1 μPa in the one-third octave band. Biological sources including marine mammals and fish contributed 690 pW/m^2, corresponding to 90 dB re 1 μPa in the one-third octave band. Rain contributed 274 pW/m^2, corresponding to 86 dB re 1 μPa in the one-third octave band.

2.2 *The Effect of a Wind Farm in the Noise Budget*

Eight turbines are initially planned for the waters south of Block Island. Using transmission loss measurements from the area, the effect of the additional noise from the 8 turbines on the ambient

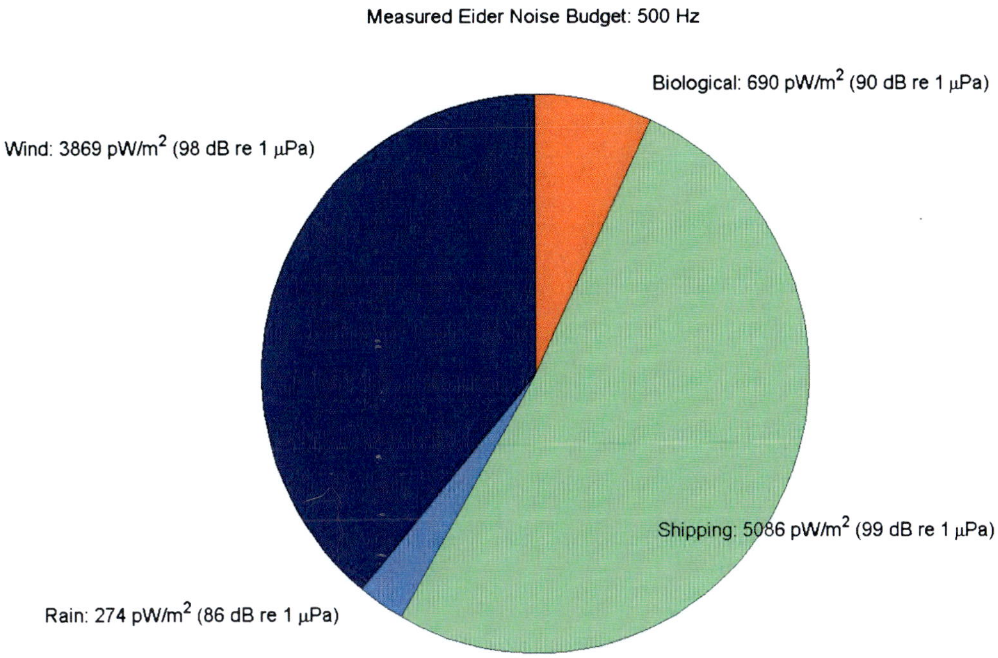

Fig. 1 The measured average intensity noise budget for the one-third octave band centered at 500 Hz based on the data collected by the Eider passive aquatic listener (PAL)

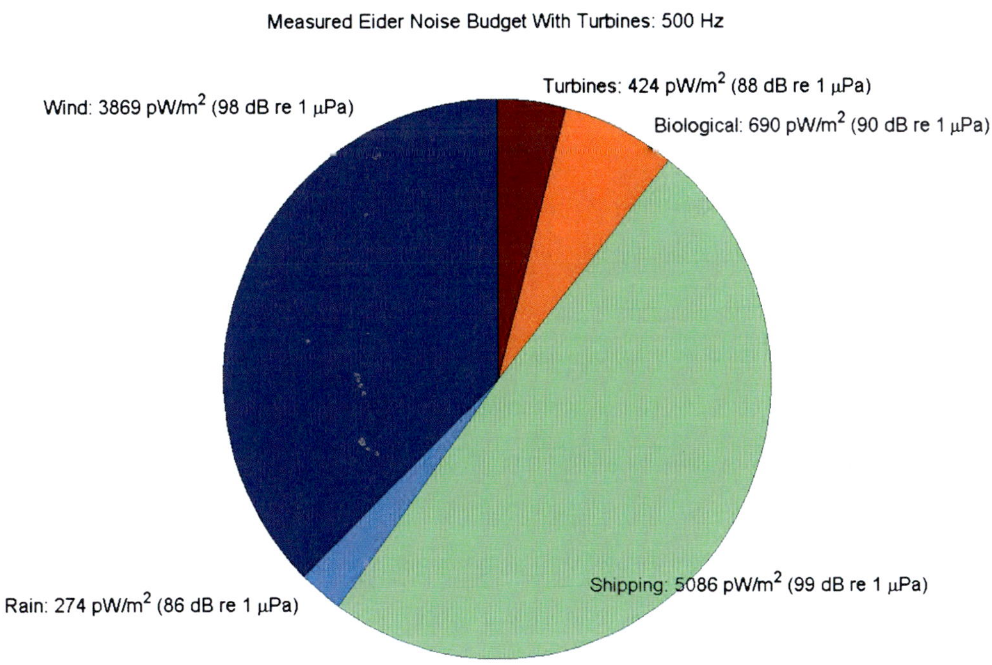

Fig. 2 The computed average intensity noise budget for the one-third octave band centered at 500 Hz based on the data collected by the Eider PAL with noise from 8 wind turbines at a range of 10 km

noise 10 km to the south of the wind farm was calculated. The modified budget is shown in Figure 2. The noise from the wind turbines are estimated to be 424 pW/m^2, corresponding to 88 dB re 1 μPa in the one-third octave band based on the data collected by Betke et al. (2004).

3 Conclusions

This paper has dealt with the low-level but almost continuous ocean noise from the operation of the offshore wind turbines and its potential effects on the marine environment. The effects of this low-level operational noise may be subtle and difficult to observe. At a range of 10 km from a planned eight-turbine wind farm in the waters south of Rhode Island, it is estimated that the effect on the ambient noise budget will be smaller than the average intensity of shipping, wind, and biological sources, adding 424 pW/m^2, corresponding to 88 dB re 1 μPa in the one-third octave band centered at 500 Hz.

References

Betke K (2006) Measurement of underwater noise emitted by an offshore wind turbine at Horns Rev. Report for Institut für technische und angewandte Physik GmbH (ITAP), Oldenburg, Germany.

Betke K, Schultz-von Glahn M, Matuschek R (2004) Underwater noise emissions from offshore wind turbines. Proceedings of the Joint Congress CFA/DAGA'04, 22–25 March 2004, Strasbourg, France, pp 591–592.

Ma BB, Nystuen JA, Lien RC (2005) Prediction of underwater sound levels from rain and wind. J Acoust Soc Am 117:3555–3565.

Miller JH, Bradley DL, Nystuen, JA (2008) Ocean noise budgets. Bioacoustics 17:133–136.

Nystuen JA, Amitai E, Anagnostou EN, Anagnostou MN (2008) Spatial averaging of oceanic rainfall variability using underwater sound: Ionian sea rainfall experiment 2004. J Acoust Soc Am 123:1952–1962.

Nystuen JA, Howe BM (2005) Ambient sound budgets. Proceedings of the Underwater Acoustics Measurements Conference, 28 June - 1 July, Heraklion, Crete.

Schneider JA, Senders M (2010) Foundation design: A comparison of oil and gas platforms with offshore wind turbines. MTSJ 44:32–51.

Development and Implementation of Criteria for Exposure of Western Gray Whales to Oil and Gas Industry Noise

Douglas P. Nowacek, Alexander Vedenev, Brandon L. Southall, and Roberto Racca

1 Introduction

Whether and how noise from human activities adversely affects marine life and what to do to mitigate negative impacts of industrial activities have been subjects of scientific research and regulatory interest for several decades, beginning with the observations of Payne and Webb (1971). In 1995, the US National Marine Fisheries Service (NMFS) began using underwater impulse noise criteria for estimating the physical injury to marine mammals [190 dB re 1 μPa root mean square (RMS) sound pressure levels for pinnipeds and most odontocete cetaceans and 180 dB re 1 μPa for mysticetes and sperm whales]. Subsequently, the High Energy Seismic Survey team (1999) concluded that exposure to air gun pulses with pulse-averaged received levels (RLs) above 180 dB re 1 μPa would likely result in significant behavioral, physiological, and/or hearing impacts. The NMFS has continued to use the 180-dB RL criterion for predicting injury from acoustic exposure for cetaceans and 190-dB RL for pinnipeds as well as a behavioral impact level of 160-dB RL; based primarily on observations of mysticete cetaceans reacting to air gun pulses (e.g., Malme et al. 1984), a 120-dB RL criterion has been applied by the NMFS in some conditions for some nonimpulsive "continuous" industrial noises.

Recently, Southall et al. (2007) reviewed and applied all available scientific literature in proposing noise-exposure criteria for marine mammals. Although specific (and quite different from the above) threshold values were proposed for predicting physical injury for different marine mammal hearing groups, Southall et al. concluded that the available data did not converge on a simple single-exposure level for predicting behavioral impacts. Rather, they suggested that contextual site- and

D.P. Nowacek (✉)
Duke University Marine Laboratory, Beaufort, NC 28516, USA
e-mail: dpn3@duke.edu

A. Vedenev
Shirshov Institute of Oceanology, Russian Academy of Sciences, Moscow, Russia 117997
e-mail: vedenev@ocean.ru

B.L. Southall
Southall Environmental Associates, Santa Cruz, CA 95060

University of California, Santa Cruz, Santa Cruz, CA 95064, USA
e-mail: Brandon.Southall@sea-inc.net

R. Racca
JASCO Applied Sciences, Victoria, BC V8Z 7X8, Canada
e-mail: Roberto.Racca@jasco.com

A.N. Popper and A. Hawkins (eds.), *The Effects of Noise on Aquatic Life*,
Advances in Experimental Medicine and Biology 730, DOI 10.1007/978-1-4419-7311-5_119,

species-specific factors are of critical importance in predicting behavioral responses and that an application of the most similar response data from other conditions to a situation of interest, taking into account location and species-specific factors, would be the most appropriate means of predicting behavioral impacts; they also suggested that various acoustic metrics were needed to most comprehensively predict possible impacts. Here, we provide a specific case study of such an approach in establishing and applying behavioral noise-exposure criteria using what empirical data are available and taking into account relevant biological factors and the conservation status of marine mammals that will be exposed to industrial noise in the course of an upcoming seismic survey in Russia.

The western gray whale (*Eschrichtius robustus*) is one of the most endangered cetaceans (Clapham et al. 1999). The overall range of the western gray whale population is not well understood. One area of the *E. robustus* habitat that has been well studied is their primary summer feeding ground off the northeast coast of Sakhalin Island, Russia. Since 1994, Russian and US research teams have been studying the behavior and distribution of the whales as well as documenting individual whales using photo identification methods (e.g., Weller et al. 2002). The collected data have established the importance of this area for *E. robustus* summer feeding and, particularly, the significance of the Sakhalin shelf for mother-calf pairs. The "core" area used by the whales is relatively small, reaching from shore ~5–10 km, approximately the 20-m isobath, offshore of Sakhalin. The Sakhalin shelf has also been identified by the oil and gas industry to be vitally important for energy development because they have discovered massive oil and gas reserves under the shelf. So, this brings about a nexus of a critically endangered population of whales and human activity that overlap geographically and temporally in very small spatial and temporal windows.

In 2002, the International Union for the Conservation of Nature (IUCN) convened the Independent Scientific Review Panel (ISRP) to evaluate the risks to *E. robustus* of planned exploration, construction, and oil/gas production by the Sakhalin Energy Investment Company (the Company). The ISRP completed a thorough review of the Company's plans (Reeves et al. 1995),and the success of the ISRP led to several subsequent reviews and the formation of the Western Gray Whale Advisory Panel (WGWAP) in 2006, also convened by IUCN (see http://www.iucn.org/wgwap/ for details and the history of engagement).

One area of significant concern has been the production of underwater noise by Company activities, both continuous noise from offshore construction and pulsed noise from seismic surveys. As a result of the ISRP review and the WGWAP process, we identified noise-exposure criteria and the methods for implementing them. Our overall goals were 1) to extend as much protection to the whales as possible given their conservation status and the importance of this habitat; 2) to use the best available and most relevant scientific information to set the criteria; 3) to advance the field of noise monitoring and mitigation for marine mammals exposed to industrial noise [e.g., incorporate emerging metrics such as sound exposure level (SEL)]; and 4) to ensure that the data collected, both acoustic and whale focused, could be analyzed with the intent of filling some of the data gaps we encountered.

2 Criteria

2.1 *Continuous Noise*

Continuous noise has been generated by various Company construction activities: the installation of a pipeline on the sea floor, the construction of a drilling platform, and ongoing activities such as the transit of crew change vessels. Much of the noise originated from the numerous vessels needed inter alia to move assets, to station keep while other activities occurred, and to perform the dredging and pipe laying along the offshore pipeline path.

Table 1 Criteria for daily continuous noise exposure for western gray whales on the Sakhalin shelf

Time of Exposure, min	Level B: RL for Start of Diagnostics, dB_{RMS} re 1 μPa	Level A: RL for Start of Mitigation, dB_{RMS} re 1 μPa
15	140	143
30	135	138
60	130	133
120	125	128
240	120	123
480	115	118

RL, received level; RMS, root mean square

Exposure criteria were developed for continuous noise based on the best available scientific information on the potential physiological consequences and behavioral response(s) of gray whales to underwater sounds. Generally, the harmful effects of noise on humans and other animals can be examined by reference to two consequences of exposure: 1) noise-induced hearing loss as a result of mechanical damage to or metabolic/chemical effects on the auditory system and 2) adverse effects on health resulting from psychological or behavioral reactions to aversive noise (see Kryter 1970). For this population of *E. robustus*, we had a third concern: the potential for a compromised nutritional state in whales that avoid prime feeding areas due to the proximity of those areas to noise-producing activities. Taking into account the Malme et al. (1984) 50% response at 120 dB for continuous noise, we infer that the indirect impact of extended exposure to noise at such a level could be substantial.

The WGWAP made a recommendation for acceptable daily noise-exposure levels (Table 1) measured at the perimeter of the feeding area, roughly the 20-m isobath offshore of Sakhalin, e.g., continuous sounds received at 140 dB $_{RMS}$ re 1 μPa within the feeding area have a daily allowable maximum of 15 min. Note that the relationships summarized in Table 1 can be extrapolated in both directions (i.e., higher or lower levels for more or less time, respectively). However, the extrapolation extends only to 146 dB for high levels, and if this level is ever recorded, mitigation should occur immediately. For low levels, the extrapolation continues down to 100 dB or ambient, whichever is higher. In practical te RMS, the capabilities of the measurement equipment had also to be taken into account because the scale extrapolation could not extend below the self-noise pedestal of the monitoring telemetry system.

2.2 *Pulsed Noise*

With respect to hearing damage, we adopted the Southall et al. (2007) criteria, specifically that permanent threshold shift onset is expected at 198 dB re 1 μPa^2-s cumulative SEL. Based on extensive modeling of the proposed air gun source, the propagation conditions, and the pulse characteristics downrange, we estimated the cumulative exposure obtained by considering the two shot lines nearest the feeding area and one line sufficiently offshore to allow for a "racetrack" turn, i.e., the maximum dose that could be expected in a 24-h period (Fig. 1). We discovered that the 198 dB_{SEL} was less conservative than the single-shot 180 dB_{RMS} re 1 μPa radius as estimated by modeling, so we adopted the 180 dB_{RMS} re 1 μPa criteria for the safety radius.

Determining acceptable levels for preventing or even minimizing behavioral disruption is less straightforward. No direct measurements of the effects on feeding *E. robustus* have been made, but the data from Malme et al. (1986) have been used previously to predict the sensitivity of feeding eastern gray whales and *E. robustus* (Reeves et al. 2005). The Malme et al. (1986) data led to estimated 10, 50, and 90% probabilities of gray whale avoidance reactions at 164, 170, and 180 dB $_{RMS}$

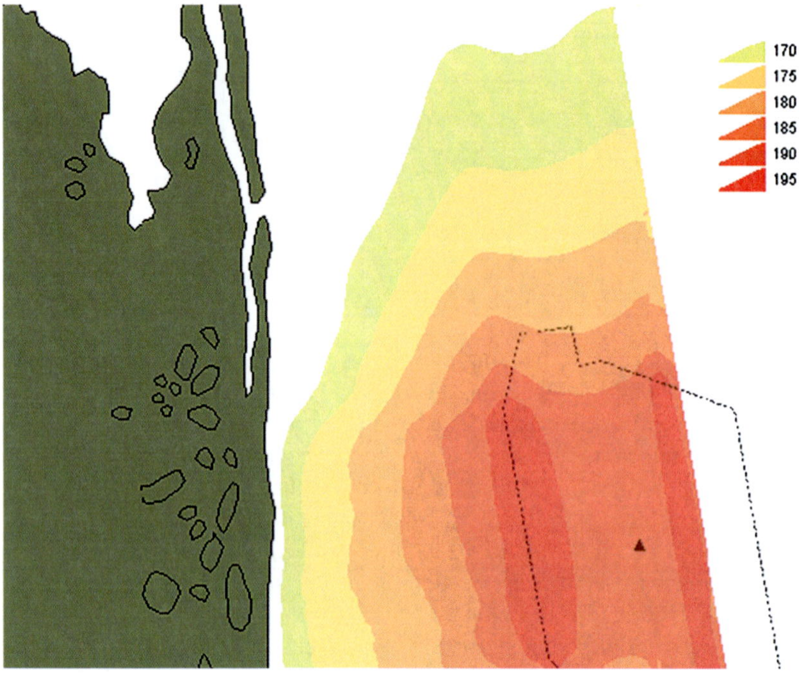

Fig. 1 Sound energy accumulated during shooting of the two most shoreward lines and one through the middle of the area. Inset: Characteristics of a seismic pulse as it spreads downrange

re 1 μPa, respectively. The RMS measurements reported by Malme et al. (1986) could not be compared directly with predicted values for the Sakhalin Energy four-dimensional survey because inter alia the seismic sources, pulse propagation, and bathymetry, the spectral characteristics differed substantially between the two scenarios. In particular, information on the rate and number of pulses to which the whales in the Malme et al. (1986) study were exposed would be required in order to determine their total sound exposure in units of dB_{SEL} and, despite information kindly provided by C. I. Malme (personal communication), the vessel approaches that resulted in the documented responses occurred on different days, i.e., the distance between a given whale and the vessel at the beginning of the approach was unknown. Given this, we realized that we could not estimate the dose by this method. The second potential method was based on an approach used by the US Navy to assess the risk to animals exposed to multiple sonar pings in an environmental impact statement (Jenkins 2005). However, the US Navy's method is unable to account for the changing amount of energy contained in the pulses as the seismic vessel approached and passed an animal or monitoring station rendered it inappropriate. Given the limitations in these possible approaches for the specific assessment of acoustic impacts for the critically endangered western grey whales, we decided 1) to use the widely accepted standard of 163 dB_{RMS} as a behavioral disturbance threshold; modeling showed this to correspond to 156 dB_{SEL} on a per-shot basis; and 2) that given the need to assess and improve a dose-based approach, measures of exposure level and duration should be included in the monitoring effort of future surveys as well as the subsequent data analysis.

Although in practice current authorizations by NMFS have called for a behavioral disturbance threshold for seismic surveys conducted near gray whales to be 160 dB_{RMS}, we believe the 163 dB_{RMS} level was appropriate because there is no scientific rationale for 160 dB_{RMS} other than perhaps it is below the 0.1 probability of disturbance. The first response observed by Malme et al. (1986) was at 149 dB_{RMS}, and thus the 160 dB_{RMS} level will not necessarily prevent all disturbances. In addition, significantly more mitigation measures will be in place (see WGWAP reports) for upcoming surveys conducted under these guidelines than are commonly employed in standard industry practice, providing additional assurance that as few whales as possible will be exposed even to the 163-dB_{RMS} level. Finally, the Company has endeavored to facilitate the postsurvey assessment of dose levels through measurement and analysis of received acoustic pulse data to yield SEL metrics.

3 Implementation of Criteria

For both the continuous and pulsed-noise cases, the WGWAP recommended to the Company a set of monitoring and mitigation strategies, and these strategies are explained in detail in the WGWAP documents available via http://www.iucn.org/wgwap/. One of the overarching recommendations was to complete industrial activities at times when the whales were present in the lowest numbers, e.g., the most effective means for minimizing the effects of seismic surveys is to complete them as early in the season as possible before many whales even arrive. In addition to the exposure criteria, the recommendations included equipment specifications, methods for monitoring, and even considerable work on defining where the monitoring should take place relative to the density of the whales. The program as described here may not fully address all aspects of the noise mitigation issues and may not be applicable to all situations, but it represents a more thorough and thoughtful approach to minimizing the effects of noise on these whales than previous attempts (Johnson et al. 2007) and has steps in place to generate valuable new data that will expand the limited current knowledge on responses of *E. robustus* to seismic surveys (Gailey et al. 2007).

References

Clapham PJ, Young SB, Brownell RL Jr (1999) Baleen whales: Conservation issues and the status of the most endangered populations. Mamm Rev 29:37–62.

Gailey G, Würsig B, McDonald TL (2007) Abundance, behavior, and movement patterns of western gray whales in relation to a 3-D seismic survey, Northeast Sakhalin Island, Russia. Environ Monit Assess 134:75–91.

High Energy Seismic Survey (1999) High Energy Seismic Survey review process and interim operational guidelines for marine surveys offshore Southern California. Report from High Energy Seismic Survey Team for California State Lands Commission and US Bureau of Ocean Energy, Management, Regulation and Enforcement, Camarillo, CA. Available via http://www.mms.gov/omm/pacific/lease/fullhessrept.pdf.

Jenkins K (2005) Undersea Warfare Training Range. Draft Overseas Environmental Impact Statement/Environmental Impact Statement pursuant to National Environmental Policy Act Section 102(2)(C) and Executive Order 12114.

Johnson S, Richardson WJ, Yazvenko SB, Blokhin SA, Gailey C, Jenkerson MR, Meier SK, Melton HR, Newcomer MW, Perlov AS, Rutenko SA, Würsig B, Martin CR, Egging DE (2007) A western gray whale mitigation and monitoring program for a 3-D seismic survey, Sakhalin Island, Russia. Environ Monit Assess 134:1–19.

Kryter KD (1970) Effects of noise on man. Academic Press, New York.

Malme CI, Miles PR, Clark CW, Tyack P, Bird JE (1984). Investigations of the potential effects of underwater noise from petroleum industry activities on migrating gray whale behavior. Phase II: January 1984 migration. Report 5586 by Bolt Beranek and Newman, Inc., Cambridge, MA, for US Bureau of Ocean Energy Management, Regulation, and Enforcement, US Department of the Interior, Anchorage, AK.

Malme C, Würsig B, Bird JE, Tyack (1986) Behavioral responses of gray whales to industrial noise: feeding observations and predictive modeling. Final Report of Principal Investigators, Outer Continental Shelf Environmental Assessment Program, National Oceanic and Atmospheric Administration, US Department of Commerce, 56:393–600.

National Marine Fisheries Service (NMFS) (1995) Small takes of marine mammals incidental to specified activities; offshore seismic activities in southern California. Notice of issuance of an incidental harassment authorization. Federal Register, October 17, 1995, vol. 60, no. 200, pp 53753–53760.

Payne R, Webb D (1971) Orientation by means of long range acoustic signaling in baleen whales. Ann NY Acad Sci 188:110–141.

Reeves RR, Brownell RL Jr, Burdin A, Cooke JG, Darling JD, Donovan GP, Gulland FMD, Moore SE, Nowacek DP, Ragen TJ, Steiner RG, VanBlaricom GR, Vedenev A, Yablokov AV (2005) Report of the Independent Scientific Review Panel on the impacts of Sakhalin II phase 2 on western North Pacific gray whales and related biodiversity: Report of the Independent Scientific Review Panel, International Union for Conservation of Nature and Natural Resources (IUCN), Gland, Switzerland, and Cambridge, UK.

Southall BL, Bowles AE, Ellison WT, Finneran JJ, Gentry RL, Greene CR Jr, Kastak D, Ketten DR, Miller JH, Nachtigall PE, Richardson WJ, Thomas JA, Tyack PL (2007) Marine mammal noise exposure criteria: Initial scientific recommendations. Aquat Mamm 33:411–522.

Weller DW, Burdin AM, Würsig B, Taylor BL, Brownell RL Jr (2002) The western gray whale: A review of past exploitation, current status and potential threats. J Cetacean Res Manage 4:7–12.

Strategic Assessment of the Risk Posed to Marine Mammals by the Use of Air Guns in the Antarctic: Concepts, Methods, Results, and Controversies

Olaf Boebel, Monika Breitzke, Elke Burkhardt, and Horst Bornemann

1 Introduction

During the past two years, the Alfred Wegener Institute (AWI) prepared a comprehensive, strategic assessment of the risk posed to marine mammals by the use of air guns for scientific, geophysical research in the Southern Ocean around Antarctica (Boebel et al. 2009). This strategic risk assessment focuses not only on a single activity (e.g., a specific expedition) as a risk assessment would but more generally considers the use of research air guns in this region's typical operational and environmental contexts, which show little interannual variation. The study attempts distinguishing between aspects of analysis (based on scientific knowledge and numerical calculations) and evaluation (based on a set of risk criteria and associated thresholds). The term assessment is used to describe the overall process, involving both analysis and evaluation.

2 Concepts and Methods

The analytical part commences with a synopsis of environmental (oceanographic and bathymetric) and operational characteristics from all seismic expeditions conducted in the Antarctic by the AWI during the past 22 years, obtaining the most typical scenarios for the geophysical research carried out there by our institution. This resulted in a set of 4 basic environmental scenarios that, combined with a set of 6 air gun configurations, were used to calculate single-shot acoustic fields (sound pressure level [SPL] and sound exposure level [SEL]) by numerical (finite difference 2.5-dimensional full waveform) modeling for a realistic ocean of 10 × 10 km dimension, resulting in a total of 24 different acoustic scenarios (Breitzke and Bohlen 2010).

The current state of ecological knowledge was compiled for the 14 cetacean and 6 pinniped species for which the Antarctic represents an important habitat. Ecological and physiological information such as dive cycles and hearing curves are subsequently used to guide estimations of cumulative exposure levels and to develop mitigation measures. Species status (in terms of endangerment) and migratory behavior enters the study in its final stage when population level risks are evaluated.

O. Boebel (✉) • M. Breitzke • E. Burkhardt • H. Bornemann
Alfred Wegener Institute, D-27568 Bremerhaven, Germany
e-mail: Olaf.Boebel@awi.de

A.N. Popper and A. Hawkins (eds.), *The Effects of Noise on Aquatic Life*,
Advances in Experimental Medicine and Biology 730, DOI 10.1007/978-1-4419-7311-5_120,

The identification of hazards, including associated evaluation criteria and threshold levels, provides the critical link between the sound propagation analysis and the risk evaluation. Because no legally binding set of numerical threshold levels exists for this ocean region at this time, we conducted a review of the pertinent literature. Three different risk categories were identified, for which a set of evaluation criteria was extracted from primarily three recent overview articles: 1) Southall et al. (2007), which provides numerical thresholds for the risk of "direct, immediate injury"; 2) Cox et al. (2006), from which a list of "abetting factors" was extracted for the risk of "indirect, immediate damage," i.e., the so-called beaked whale scenario; and 3) National Research Council (2005), which suggests a set of mostly qualitative evaluation criteria for the risk of "biologically significant acoustic disturbance."

By applying these criteria to the modeled acoustic fields (under the assumption of the ship following a straight course), critical exposure radii were calculated for single and multiple exposures (see Breitzke and Bohlen, Chapter 138). Finally, the resulting risk for individual animals and ensuing risks at the population level were evaluated, considering operational scenarios both with and without proposed mitigation measures in place.

3 Results

The analysis reveals that the risk for a marine mammal to incur "direct, immediate injury" from multiple exposures cannot be excluded in the immediate vicinity of the air gun clusters (<200 m). "Indirect, immediate damage" of an individual, however, appears rather unlikely because several important "abetting factors" as extracted from Cox et al. (2006) are not fulfilled in this study's context. A risk of "biologically significant acoustic disturbance," although it appears negligible for juveniles and adults, cannot be excluded for the (merely hypothetical, hitherto unobserved) possibility of individual mother-calf pair separations in the vicinity of the ship. The manifestation of any of these risks, however, depends on a whale actually being within the respective range of the ship. Hence, these risks are conditional on whale-ship encounters and need to be weighted with the probability thereof. This implies that the likelihood for a specific whale to be placed at any such risk is significantly reduced.

Our study estimates further that these findings result in a negligible risk at the population level as a result from the individual risk of "direct, immediate injury." Because the risk of "indirect, immediate damage" is already unlikely at the individual level, a transfer to consequences for the population can readily be excluded. It is only with regard to "biologically significant acoustic disturbance" that the "not to be excluded" possibility of an individual mother-calf separation may transfer to a small but "not to be excluded" possibility of population level consequences for the Antarctic blue whale due to their low population numbers. The probability of such an impact is, however, estimated to be smaller than estimates of the natural mortality rate or of the possible biological removal (PBR) as used in other contexts.

4 Controversial Issues

Noting the large existing gaps in the knowledge relevant to this issue and the fact that marine mammal behavior will never be fully predictable for an individual animal, it is unavoidable to base parts of this and other risk assessments on extrapolations of the current best knowledge, of statistical descriptions of typical behavior, and even on educated guessing. When such steps had to be taken here, we attempted to adhere to a conservative approach in our calculation and evaluation of contingent risks. The term conservative, therefore, stands for a selection of parameters or proxies, which chooses those that overestimate the risk while providing increased protection for the marine mammal. The phrase "a risk is not to be excluded" as used above therefore by no means implies that the risk is likely to occur. The phrase rather describes that our current knowledge is insufficient to be certain that a risk will not occur once the respective risk threshold is exceeded while, to the best of our knowledge, the risk will not occur at exposure levels below that threshold. Nevertheless, with the risk evaluation being critically dependent on the thresholds used, it was no surprise that some selections made in this study were met with disagreement when discussed with various stakeholders. Additionally, some rather fundamental issues also emerged as a possible matter of dissent. Most prominently, the following topics spurred the discussions.

- What is the subject of protection: an individual animal (commensurate with an animal welfare approach) or a local population or breeding group (the species/population protection approach)?
- Does the derivation of the "dual criteria" by Southall et al. (2007), which includes several (partially substantiated) extrapolations, provide conservative thresholds for injury levels or must this evaluation be challenged on the basis of (new) scientifically supported facts?
- What is a suitable acoustic metric to describe acoustic exposure in the context of "biologically significant acoustic disturbance," i.e., behavioral response?
- Last but not least, to what extent should the precautionary principle be applied, e.g., should a mere hypothetical scenario resulting in a residual risk provide sufficient reason to ban a proposed activity?

It is with these issues that the scientific community's expertise and guidance, preferably in form of peer-reviewed publications, would be most helpful to further develop balanced and objective risk assessments acceptable to the majority of stakeholders.

References

Boebel O, Breitzke M, Burkhardt E, Bornemann H (2009) Strategic assessment of the risk posed to marine mammals by the use of airguns in the Antarctic Treaty area. Information Paper IP 51, Agenda Item: CEP 8c, Antarctic Treaty Consultative Meeting XXXII, 6–17 April 2009, Baltimore, MD. Available via http://epic.awi.de/Publications/Boe2009c.pdf.

Breitzke M, Bohlen T (2010) Modelling sound propagation in the Southern Ocean to estimate the acoustic impact of seismic research surveys on marine mammals. Geophys J Int 181:818–846. Available via http://epic.awi.de/Publications/Bre2009f.pdf.

Cox TM, Ragen TJ, Read AJ, Vos E, Baird RW, Balcomb K, Barlow J, Caldwell J, Cranford T, Crum L, D'Amico A, D'Spain G, Fernández A, Finneran A, Gentry R, Gerth W, Gulland F, Hildebrand J, Houser D, Hullar T, Jepson RD, Ketten D, MacLeod CD, Miller P, Moore S, Mountain DC, Palka D, Ponganis P, Rommel S, Rowles T, Taylor B, Tyack P, Wartzok D, Gisiner R, Meads J, Benner L (2006) Understanding the impacts of anthropogenic sound on beaked whales. J Cetacean Res Manage 7:177–187.

National Research Council (2005) Marine mammal populations and ocean noise. National Academies Press, Washington, DC.

Southall BL, Bowles AE, Ellison WT, Finneran JJ, Gentry RL, Greene CR Jr, Kastak D, Ketten DR, Miller JH, Nachtigall PE, Richardson WJ, Thomas JA, Tyack PL (2007) Marine mammal noise exposure criteria: Initial scientific recommendations. Aquat Mamm 33:411–522.

Establishing a Safety Zone for Marine Mammals Due to Underwater Blasting

Norm Broner and Mike Huber

1 Introduction

Blasting operations to prepare for new infrastructure in port areas will result in underwater noise that has the potential to impact on marine animals in the vicinity. It is therefore necessary to establish a safety zone to protect marine species such as *Megaptera novaeangliae* (humpback whales), *Dugong dugon* (dugongs), *Tursiops aduncus* (bottlenose dolphins; also known as the Indo-Pacific bottlenose dolphins), and *Natator depressus* (flatback turtles) from potentially harmful underwater noise levels. This paper describes the derivation of such a zone as part of a project being declared a controlled action under the Australian Commonwealth Environment Protection and Biodiversity Conservation Act 1999.

2 Prediction of Underwater Blasting Noise Levels

Two sound metrics, the maximum instantaneous pressure (P_{max}) and sound exposure level (SEL), were to be predicted. The SEL integrates the received sound energy over all detonations in a shot and therefore directly addresses the potential cumulative impacts of successive detonations of the individual charges over short time intervals. In this instance, the modeling of 24 detonations of a 50-kg maximum instantaneous charge (MIC) represented the maximum potential sound emission from a shot. Acoustic modeling was conducted by Huson (2009) using the parabolic equation (PE) model and was based on

- a blasting design of 24 blastholes with a 50-kg MIC each and a 60-ms delay;
- confined blasts, using an unconfined charge size equivalence factor of 0.014, meaning that a confined 50-kg MIC blast is equivalent to a 0.7-kg MIC unconfined blast;
- bathymetry digitized from charts; water depth was up to 20 m; and
- assumed acoustic properties of the seabed and seawater.

N. Broner (✉) • M. Huber
Sinclair Knight Merz, Melbourne, VIC 8009, Australia
e-mail: nbroner@skm.com.au; mhuber@skm.com.au

A.N. Popper and A. Hawkins (eds.), *The Effects of Noise on Aquatic Life*,
Advances in Experimental Medicine and Biology 730, DOI 10.1007/978-1-4419-7311-5_121,

Sound emissions from blasting will consist of a series of 24 detonations at 60-ms intervals. At close range, these will be perceived as discrete pulses, but at a longer range, the sound will be perceived as a single event. The sound is dominated by low frequencies.

3 Blasting Noise Criteria

Blasting and other anthropogenic sound sources can have a hierarchy of effects on marine animals depending on their level of exposure and sensitivity, including severe organ trauma and mortality, permanent hearing loss (permanent threshold shift [PTS]), temporary hearing loss (temporary threshold shift [TTS]), and behavioral disturbance. Vulnerability to organ trauma depends on size, with small animals generally more sensitive than larger ones (Yelverton et al. 1975). Fishes are therefore more vulnerable to organ trauma than marine mammals and turtles.

Based on their low-frequency hearing, humpback whales are expected to be the fauna potentially occurring in the area that are most susceptible to blasting-induced TTS. Turtles and fishes also hear at low frequencies but have less-sensitive hearing and are therefore expected to be less likely to experience TTS than humpback whales. Dolphins are expected to be less vulnerable to blasting-induced TTS (and by extension PTS) than humpback whales, sea turtles, and fishes because dolphin hearing is poor at the low frequencies that dominate blasting noise. Dugongs are probably similar to dolphins in their vulnerability to TTS from blasting noise, but there is large uncertainty regarding dugong hearing.

The impact criteria used in this assessment are defined on the basis of criteria established in other studies and jurisdictions. The criteria adopted in this assessment are shown in Table 1.

Table 1 Summary of sound exposure criteria used in this assessment

Criterion	Effect/Application	Source	Comments
Peak pressure criteria			
224 dB re 1 μPa	Onset of TTS and behavioral disturbance in cetaceans. Also applied here to dugongs and turtles.	Southall et al. 2007	Application to turtles is conservative.
230 dB re 1 μPa	Onset of PTS and organ trauma in cetaceans. Also applied here to dugongs and turtles.	Southall et al. 2007	Application to organ trauma is conservative. Application to turtles is conservative.
SEL criteria			
183 dB re 1 μPa^2-s	TTS and behavioral disturbance in cetaceans. Also applied here to dugongs and turtles.	Southall et al. 2007	Application to turtles is conservative due to their poor hearing relative to cetaceans.
195 dB re 1 μPa^2-s	No-injury level for 0.1-g fish.	Yelverton et al. 1975; Hastings and Popper 2005	
198 dB re 1 μPa^2-s	PTS and organ trauma in cetaceans. Also applied here to dugongs and turtles.	Southall et al. 2007	Application to organ trauma is conservative. Application to turtles is conservative due to poor hearing relative to cetaceans.
200 dB re 1 μPa^2-s	No-injury level in 1-kg fish.	Yelverton et al. 1975; Hastings and Popper 2005	

The sound level exposure (SEL) criteria are applied without frequency weighting to baleen whales, turtles, and fishes. M-weighting for midfrequency cetaceans is applied to the temporary threshold shift (TTS) and behavioral disturbance in dolphins and dugongs. PTS, permanent threshold shift

4 Results

The predicted ranges to meet the adopted acoustic criteria are shown in Table 2.

Based on the results, it is recommended that the exclusion zone should be 1,150 m during initial preliminary blasting and subject to review on the basis of acoustic monitoring (and revised modeling if necessary). Humpback whales are more likely to be affected than the other animals considered because of their relatively sensitive low-frequency hearing. The 1,150-m range is still conservative for humpback whales because the acoustic criteria are based on the even more sensitive hearing of dolphins. For turtles, a 1,150-m exclusion zone to avoid TTS and behavioral disturbance is conservative because of the low hearing sensitivity of turtles relative to that of cetaceans. The 1,150-m range is also very conservative for dolphins and dugongs because of their relatively poor hearing at the low frequencies that dominate blast noise. This is demonstrated by the reduced range of 793 m to meet the SEL criteria for TTS and behavioral disturbance when M-weighting is applied.

5 Conclusions

The potential noise impact of underwater blasting on marine mammals needs to be considered. Given a good blasting design of 24 blastholes with a 50-kg MIC each and a 60-ms delay, a 1,150-m safety zone is recommended for protection of humpback whales, dolphins, dugongs, and turtles.

Table 2 Predicted ranges to meet the selected noise exposure criteria for confined shots of 20- and 50-kg MICs with a constant bottom depth of 13 m

				Range, m	
Criterion	Metric	Type of Impact	Applies To	20-kg MIC	50-kg MIC
P_{max} criteria, no-frequency weighting					
224 dB re 1 μPa	P_{max}	TTS/behavioral disturbance	Humpback whales, dolphins, dugongs, and turtles	165	230
230 dB re 1 μPa	P_{max}	PTS/organ trauma	Humpback whales, dolphins, dugongs, and turtles	80	115
SEL criteria, no-frequency weighting					
183 dB re 1 μPa^2-s	SEL	TTS/behavioral disturbance	Humpback whales and turtles	840	1,150
195 dB re 1 μPa^2-s	SEL	No injury	0.1-g fish	205	280
198 dB re 1 μPa^2-s	SEL	PTS/organ trauma	Humpback whales, dolphins, dugongs, and turtles	145	200
200 dB re 1 μPa^2-s	SEL	No injury	1 kg fish	115	155
SEL criteria, using M-weighting for midfrequency cetaceans					
183 dB re 1 μPa^2-s	SEL	TTS/Behavioral disturbance	Dolphins, dugongs	585	793

MIC, maximum instantaneous charge; P_{max}, maximum instantaneous pressure

References

Hastings MC, Popper AN (2005) Effects of sound on fish. Report prepared for the California Department of Transportation, Task Order 1, Contract No. 43A0139, by Jones & Stokes, Sacramento, CA.

Huson L (2009) Underwater sound propagation study. Blasting for Hay Point coal terminal berth 3. Job No. LHA273 July 2009, prepared by L. Huson & Associates Pty. Ltd., Woodend, VIC.

Southall BL, Bowles AE, Ellison WT, Finneran JJ, Gentry RL, Greene CR Jr, Kastak D, Ketten DR, Miller JH, Nachtigall PE, Richardson WJ, Thomas JA, Tyack PL (2007) Criteria for behavioral disturbance. Aquat Mamm 33:446–473.

Yelverton JT, Richmond DR, Hicks W, Saunders H, Fletcher ER (1975) The relationship between fish size and their response to underwater blast. Topical Report DNA 3677T prepared for the Defense Nuclear Agency by Lovelace Foundation for Medical Education and Research, Albuquerque, NM.

The Marine Mammal Protection Act: A Regulatory Approach to Identifying and Minimizing Acoustic-Related Impacts on Marine Mammals

Jaclyn N. Daly and Jolie Harrison

1 Introduction

In 1972, the US Congress passed the Marine Mammal Protection Act (MMPA), effectively giving the National Marine Fisheries Service (NMFS) of the National Oceanic and Atmospheric Administration (NOAA) responsibility for conserving and managing all species of cetaceans and pinnipeds (except walrus). The MMPA prohibits, with certain exceptions, the "take" (to harass, hunt, capture, collect, or kill or attempt to do so) of marine mammals in US waters and by US citizens on the high seas. Since 1982, the MMPA has allowed for the incidental taking of marine mammals by US citizens who engage in a specified activity (other than commercial fishing) provided an "incidental take authorization" (ITA) is issued under section 101(a)(5) of the MMPA. The NMFS's Office of Protected Resources is responsible for issuing these authorizations. Most ITAs cover sound-generating activities, such as naval training (e.g., utilizing sonar or explosives), seismic surveys, or marine construction, because they have the potential to result in marine mammal harassment.

To issue an ITA, the NMFS must determine that the taking will have a negligible impact on the affected species or stock and will not have an unmitigable adverse impact on their availability for taking for subsistence uses (where relevant). NMFS must also set forth the permissible methods of taking and the requirements pertaining to the practicable mitigation, monitoring, and reporting of such takings. An applicant must answer 14 questions (http://www.nmfs.noaa.gov/pr/permits/incidental.htm) designed to inform the NMFS of the nature and scope of the project, the number of marine mammals by species or stock anticipated to be taken and by what manner (i.e., serious injury or mortality, level A [injurious] harassment, or level B [behavioral] harassment), the impact of those takes on the species or stock, and proposed mitigation and monitoring measures, all of which assist the NMFS in making its determinations.

J.N. Daly (✉) • J. Harrison
National Marine Fisheries Service, National Oceanic and Atmospheric Administration,
Silver Spring, MD 20910, USA
e-mail: jaclyn.daly@noaa.gov; jolie.harrison@noaa.gov

A.N. Popper and A. Hawkins (eds.), *The Effects of Noise on Aquatic Life*,
Advances in Experimental Medicine and Biology 730, DOI 10.1007/978-1-4419-7311-5_122,

2 NMFS Approach to Acoustic Impacts Assessment

Marine mammals hear, produce, and use sound for various biological functions, including communication, foraging, navigation, and predator detection (Richardson et al. 1995; Southall et al. 2007). Interference with producing or receiving these sounds may result in adverse impacts. To analyze acoustic-related impacts from a specified activity, the NMFS reviews the application and the best scientific data available and considers the likely effectiveness of mitigation measures in minimizing any of the identified effects. If, from this analysis, the NMFS can make the necessary determinations described above, then an ITA may be issued.

The characteristics of a sound source play a large role in determining the potential impacts to marine mammals. For example, the NMFS considers whether the source generates impulsive or continuous noise, source frequencies (including if tonal or broadband) and sound pressure levels, sound-propagation rates, whether the source is moving or stationary, received levels, and duration of exposure. These factors must be considered in conjunction with one another. For example, a source may have a high source level. but the frequencies may be outside a species' functional hearing range and therefore may not pose a risk. Alternatively, a source with a lower source level, which emits chronic detectable noise, may pose greater threats than a higher level source operating in the short term and intermittently.

The context in which animals are exposed to noise also plays a significant role in the NMFS's impact analyses. Important factors include, among other things, age and reproductive status, use of the habitat (e.g., foraging or reproductive areas), and previous exposure to the source or other anthropogenic disturbance (i.e., is the animal tolerant or habituated) (Richardson et al. 1995; Southall et al. 2007). For example, anthropogenic noise may potentially disrupt mother-calf bonds. The behavior of a mother-calf pair to reestablish that bond, such as increasing call duration (e.g., Van Parijs and Corkeron 2001), and the associated fitness consequences due to bond disruption should be carefully considered.

Typical acoustic-related impacts that the NMFS considers in an analysis are auditory fatigue, behavioral reactions (e.g., avoidance, changes in travel, dive, reproduction, and foraging patterns), masking, and stress. These impacts are often difficult to quantify but the science is advancing. For example, scientists are measuring glucocorticoid levels to determine degrees of stress (e.g., Rolland et al. 2006) and developing algorithms to calculate communication masking space (e.g., Clark et al. 2009). The MMPA mandates that take be quantified. From a regulatory perspective, the most practical way to do this is to establish and apply acoustic harassment thresholds. Thresholds provide an objective method to estimate take (amount and severity) and help identify appropriate mitigation. Examples of some acoustic harassment thresholds that the NMFS uses are outlined in Table 1; however, note that these do not apply to all sound types (e.g., explosives and sonar), and the NMFS is in the process of revising these thresholds to better reflect current science. A challenge for many applicants without access to complex sound propagation models and marine mammal abundance information for discrete locations is determining the distances to these thresholds from the sound source (especially in coastal, heterogeneous environments) and calculating how many individuals may be taken from exposure to sound generated by their activity.

Table 1 Examples of NMFS's current in-water acoustic harassment threshold criteria

Sound Source	Level A Harassment	Level B Harassment
Impulsive (e.g., impact pile driving, air guns)	190 dB (pinnipeds) 180 dB (cetaceans)	160 dB
Continuous (e.g., vibratory pile driving, drilling)	190 dB (pinnipeds) 180 dB (cetaceans)	120 dB

NMFS, National Marine Fisheries Service. Thresholds are in root mean square (RMS) values re 1 μPa

3 Mitigation and Monitoring Measures

One of the most effective methods to reduce noise exposure-related impacts to marine mammals is spatial and/or temporal limitation of the activity where practicable. Other mitigation measures include delay of source start-up or shutdown if a marine mammal enters into a designated zone of exposure, source ramp-up to allow individuals time to leave an area before full power is reached, and use of sound attenuation devices (e.g., bubble curtains and pile caps) to reduce noise levels. Passive acoustic monitoring (PAM) may also be required in an ITA to augment visual observations through acoustical detection of marine mammal presence. The use of PAM to trigger mitigation measures (e.g., shutdowns) is less common because of the apparent nonavailability of PAM systems that can localize on marine mammals in real time and the practicality of deploying listening devices (e.g., hydrophone array and sonobuoy) during activities.

The NMFS typically requires protected species observers (PSOs) be on watch during activities that could result in take. PSOs are responsible for recording sighting, behavioral, and activity data and notifying a source operator if mitigation is necessary. These data are reported to the NMFS and aid in the analysis of similar activities where species and environmental conditions are comparable.

4 The Path Forward

The science of acoustics and noise-related impacts on marine mammals has become a hot topic in recent years and much progress is being made. However, there is still a plethora of information needed that would better inform the NMFS's impact analyses and regulatory decision-making processes. We suggest further development of, among others, the following research/technologies to better support policy decisions aimed at protecting and conserving marine mammals: data on behavioral responses to impulsive and continuous noise sources for which ITAs are issued (e.g., air guns and vibratory pile drivers); approach to linking individual responses with population-level effects; user friendly and cost-effective acoustic propagation-modeling programs; PAM systems with species identification and localization capabilities; and research techniques designed to assess the effectiveness of current mitigation measures.

References

Clark CW, Ellison W, Southall BL, Hatch L, van Parijs SM, Frankel A, Ponirakis D (2009) Acoustic masking in marine ecosystems: Intuitions, analysis, and implication. Mar Ecol Prog Ser 395:201–222.

Richardson WJ, Greene CR Jr, Malme CI, Thomson DH (1995) Marine mammals and noise. Academic Press, San Diego, CA.

Rolland R, Hamilton P, Kraus S, Davenport B, Gillett R, Wasser S (2006) Fecal sampling using detection dogs to study reproduction and health in North Atlantic right whales (*Eubalaena glacialis*). J Cetacean Res Manage 8:121–125.

Southall B, Bowles AE, Ellison WT, Finneran JJ, Gentry RL, Greene CR Jr, Kastak D, Ketten DR, Miller JH, Nachtigall PE, Richardson WJ, Thomas JA, Tyack PL (2007) Marine mammal noise exposure criteria: Initial scientific recommendations. Aquat Mamm 33:411–522.

Van Parijs S, Corkeron P (2001) Boat traffic affects the acoustic behavior of Pacific humpback dolphins *Sousa chinensis*. J Mar Biol Assoc UK 81:533–538.

Noise-Related Stress and Cumulative Impact Assessment

Andrew J. Wright

1 Introduction

Sublethal impacts of noise may suppress reproduction or accumulate to invisibly increase mortality rates. Chronic stress in humans has been linked with coronary disease, immune suppression, anxiety and depression, cognitive difficulties, and infertility (see Wright et al. 2007a and references therein). It is reasonable to assume that prolonged or repeated exposure to one or more sources of noise can induce chronic stress in marine species, either alone or in combination with exposure to other perceived or real threats (see Wright et al. 2007a,b and references therein). In turn, the consequences of chronic stress could make a population more susceptible to exposure to any additional threats (hereafter referred to as "drivers") or slow the recovery of a population in situations where one of a suite of drivers is removed.

Accordingly, successfully determining when cumulative exposures may result in population-level impacts, especially in endangered species or small populations, is a pressing management issue. Despite legislative mandates for full cumulative impact assessments (CIAs), current project-centric processes can actually hinder CIA efforts. Furthermore, consideration of cumulative impacts to date (e.g., Halpern et al. 2008, 2009) have not generally included noise, partly because there is much discussion within the bioacoustics community around how to assess total exposure from various noise sources alone. Finally, much of the data required for thorough CIAs are not available in marine mammals and other marine species. Accordingly, Okeanos – Foundation for the Sea convened an expert panel to consider the options and propose a way forward. The resulting discussions are summarized here. The original report (Wright 2009) should be considered the definitive version of this material in case of any conflict.

A.J. Wright (✉)
National Environmental Research Institute, Department of Arctic Environment,
Aarhus University, 4000 Roskilde, Denmark

Department of Environmental Science and Policy, George Mason University, Fairfax, VA 22030, USA
e-mail: awr@dmu.dk; marinebrit@gmail.com

A.N. Popper and A. Hawkins (eds.), *The Effects of Noise on Aquatic Life*,
Advances in Experimental Medicine and Biology 730, DOI 10.1007/978-1-4419-7311-5_123,

2 Assessing Cumulative Impacts

It was quickly realized that noise could not be given special attention in a CIA. Likewise, any CIA that did not include noise would be incomplete. However, because data availability is a key constraint for managers attempting CIAs, it was decided that various stages of CIA would be needed. Each step could build on previous ones as more data become available.

First, the distribution of physical habitats or ecosystems should be mapped out and then overlaid with the distribution of each (class of) driver present. Estimates of relative ecosystem vulnerability and/or driver importance can then be included to generate a cumulative exposure map (e.g., Halpern et al. 2008, 2009). These can be used to identify relatively pristine areas, important habitats already under strain, or "sacrifice" areas with already high impact for management decisions. Although the Halpern et al. (2008, 2009) examples required much data, simply overlaying maps of marine activities in a more conceptual way can effectively indentify areas of high and low human use. For model simplicity, it would probably be sufficient to include noise in two aggregated layers of different classes of noise with similar characteristics and/or impacts: 1) chronic or continuous noise (as from shipping) that can result in the masking of sounds of interest and the reduction in habitat value indefinitely and 2) acute noise sources (such as transient impulses) generally have higher peak noise levels, which can result in injury or even death under certain circumstances but tend to be more localized and present for shorter durations.

Population distributions can simply be overlaid on the total exposure maps to explore the average total exposure to a population. Here, the ecosystem vulnerability estimates used by Halpern et al. (2008, 2009) should be replaced by vulnerability estimates for the population of concern. If data are available, stage-based population models can estimate the population-level consequences of exposure. However, full consideration of cumulative impacts may well require the use of spatially explicit individual-based models (IBMs). In these, individual animals can move through the various exposure layers (see Clark et al. 2009 for an example of how this might be done for noise) rather than the total exposure index, accumulating a history of exposure to each driver. The aggregation of impacts would then occur on an individual basis instead of a population basis. These would then accumulate at the population level, producing a more accurate representation of the way that population-level effects are actually generated.

If data on the specific impacts of exposure to each driver on an animal and the subsequent physiological consequences are available, an additional modeling layer could be incorporated, representing the physiological processes within an individual. Within this layer, both the immediate effects of the various drivers and, most importantly, their subsequent consequences would be able to interact in various ways to ultimately influence the likelihood of mortality or the reproductive potential of each animal. Synergistic and antagonistic effects may be produced by the physiological network in this model layer, which could then be validated against published data on overall impacts. Unfortunately, it is still not clear how all the consequences of chronic stress could be incorporated into the model.

3 Present and Future Applications

The extent to which the above can be achieved for any given species is highly dependent on the data available. Cumulative exposure mapping will almost always be possible, at least to a certain extent. However, regulatory agencies may need to make efforts to aggregate information on the distribution of activities under their jurisdiction. In many cases, it is also possible to integrate population distributions into CIAs, allowing some level of population modeling. However, here the availability of reliable and appropriate datasets may become a limiting factor. IBMs incorporating the multiple drivers would require larger amounts of data, making them an option in fewer situations. However, for cetaceans, it is expected that full cumulative impact IBMs would be possible for at least two populations where relatively large amounts of data exist: *Eubalaena glacialis* (North Atlantic right whale) and *Orcinus orca* (southern resident killer whale). Once these models have been built, they could be used as proxies for other cetacean species in that they should identify combinations of drivers that could be particularly damaging or provide indications of thresholds where different consequences are likely to emerge. Despite the legal mandates, much of the above can only be incorporated into management given suitable political will and a certain amount of revision in management processes. For example, it would be useful for marine mammal Stock Assessment Reports in the United States to include a full report on the drivers to which each species is exposed. This would be a substantial step toward cumulative exposure mapping. Likewise, a shift away from project-based management, perhaps to ecosystem-based management, will inherently facilitate better CIAs as the focus will be placed more on the animals and their ecosystems.

References

Clark CW, Ellison WT, Southall BL, Hatch L, Van Parijs S, Fakel A, Ponirakis D (2009) Acoustic masking in marine ecosystems: Intuitions, analysis, and implications. Mar Ecol Prog Ser 395:201–222.

Halpern BS, Kappel CV, Selkoe KA, Micheli F, Ebert CM, Kontgis C, Crain CM, Martone RG, Shearer C, Teck SJ.(2009) Mapping cumulative human impacts to California Current marine ecosystems. Conserv Lett 2:138–148.

Halpern BS, Walbridge S, Selkoe KA, Kappel CV, Micheli F, D'Agrosa C, Bruno JF, Casey KS, Ebert C, Fox HE, Fujita R, Heinemann D, Lenihan HS, Madin EMP, Perry MT, Selig ER, Spalding M, Steneck R, Watson R (2008) A global map of human impact on marine ecosystems. Science 319:948–952.

Wright AJ (ed) (2009) Report of the workshop on assessing the cumulative impacts of underwater noise with other anthropogenic stressors on marine mammals: From ideas to action. Okeanos - Foundation for the Sea Symposium, 25-29 August 2009, Monterey, CA. Available via http://www.sound-in-the-sea.org/okeanos/sound.php?id=41. Accessed 1 May 2010.

Wright AJ, Aguilar Soto N, Baldwin AL, Bateson M, Beale CM, Clark C, Deak T, Edwards EF, Fernández A, Godinho A, Hatch LT, Kakuschke A, Lusseau D, Martineau D, Weilgart LS, Wintle BA, Notarbartolo-di-Sciara G, Martin V (2007a) Anthropogenic noise as a stressor in animals: A multidisciplinary perspective. Int J Comp Psychol 20:250–273.

Wright AJ, Aguilar Soto N, Baldwin AL, Bateson M, Beale C, Clark C, Deak T, Edwards EF, Fernández A, Godinho A, Hatch L, Kakuschke A, Lusseau D, Martineau D, Romero LM, Weilgart L, Wintle B, Notarbartolo di Sciara G, Martin V (2007b) Do marine mammals experience stress related to anthropogenic noise? Int J Comp Psychol 20:274–316.

Interim Criteria for Injury to Fish From Pile-Driving Activities: Recent Experiences

Richard Rodkin, Keith Pommerenck, and James Reyff

1 Introduction

In 2004, the California, Oregon, and Washington Departments of Transportation, US Federal Highway Administration, and state and federal resource agencies established the Fisheries Hydroacoustic Working Group (FHWG) to improve and coordinate information on fishery impacts from underwater sound caused by pile driving. The FHWG was supported by a panel of hydroacoustic and fisheries experts who prepared several papers that led to an *Agreement in Principle for Interim Criteria for Injury to Fish from Pile Driving Activities* in 2008. The key reports and documents that led to the *Interim Criteria* are briefly summarized. A recent monitoring effort completed in response to orders from various regulatory agencies is discussed. A recent preproject assessment and its implications for a planned construction project are presented. Recent research supporting revisions to the criteria are summarized.

2 Background on the Interim Criteria

Several key reports and documents were prepared for the FHWG: *The Effects of Sounds on Fish* (Hastings and Popper 2005), *Interim Criteria for Injury to Fish Exposed to Pile Driving Operations: A White Paper* (Popper et al. 2006), and *Update on Recommendations for Revised Interim Criteria for Pile Driving* (Carlson et al. 2007).

The Carlson et al. (2007) memo forms the basis for the *Agreement in Principle*. Criteria were recommended for three different effects on fish: 1) hearing loss due to temporary threshold shift (TTS); 2) damage to auditory tissues (generally sensory hair cells of the ear); and 3) damage to nonauditory tissues. Carlson et al. (2007) provided the basis for their interpretation and application of the new recommendations for *Interim Criteria*. The criteria for nonauditory tissue damage were based on several studies where fish were exposed to either relatively high-amplitude blasts (peak pressures of approximately 20 psi or 223 dB re 1 μPa) or sonar signals where maximum pressures often exceeded 197 dB re 1 μPa, with the accumulated sound exposure level (SEL) based on a relatively

R. Rodkin (✉) • K. Pommerenck • J. Reyff
Illingworth & Rodkin, Inc, Petaluma, CA 94952, USA
e-mail: rrodkin@illingworthrodkin.com; kpommerenck@illingworthrodkin.com; jreyff@illingworthrodkin.com

A.N. Popper and A. Hawkins (eds.), *The Effects of Noise on Aquatic Life*,
Advances in Experimental Medicine and Biology 730, DOI 10.1007/978-1-4419-7311-5_124,

short-term exposure to relatively high-level signal. The most stringent peak sound pressure level (SPL) criterion for all hearing generalist fish for all types of effects is 206 dB.

The interim cumulative SEL criteria for hearing generalists are nonauditory tissue damage: 183–213 dB, corresponding to a fish mass between 0.5 and 200 gm; auditory tissue damage: 189- to 213-dB SEL; and temporary threshold shift: 185-dB SEL.

Carlson et al. (2007) states the following: "Note, when there was variation in data, the values given are the minimum level resulting in *any effect* (emphasis added). Also, an 18-h period of rest was identified as the interval between exposures necessary to restart the calculation of cumulative SEL."

The Agreement in Principle for Interim Criteria for Injury to Fish from Pile Driving Activities (FHWG 2008), utilizing information in Carlson et al. (2007), identified SPLs of 206 dB peak and 187-dB cumulative SEL for all listed fish except those that are less than 2 gm. In that case, the criterion for the cumulative SEL is 183 dB. The *Agreement in Principle* does not specify the period over which the SEL is to be accumulated or a lower limit on single-strike sound levels to be included in the cumulative SEL. Stadler and Woodbury (2009) address these issues from the perspective of National Oceanic and Atmospheric Administration (NOAA) Fisheries in a paper presented at Inter-Noise 2009, Ottawa, ON. Stadler and Woodbury (2009) state, "NOAA Fisheries recognizes that a single-strike SEL below a certain level will not contribute to the overall cumulative SEL because there is virtually no effect on a fish. The single-strike SEL that has no effect is referred to as 'effective quiet,' but there are no data for estimating the SEL of effective quiet. Based on this uncertainty, NOAA Fisheries has adopted a conservative SEL for effective quiet of 150 dB." The 150-dB lower limit level is 37 dB below the interim criterion level for a cumulative SEL of 187 dB and requires exposure to 5,000 pile strikes to reach a cumulative SEL of 17 dB, a conservative threshold indeed. Regarding the rest period for recovery, Stadler and Woodbury (2009) state the following: "Although fishes are expected to recover from sub-injurious exposure to these sounds, the time required for recovery is unknown. It is unlikely that a fish will fully recover during the time between pile strikes (a few seconds), so it is crucial that a longer recovery period be chosen. NOAA Fisheries is currently using a 12-h recovery period and accumulates the SEL from all pile strikes that occur prior to a 12-h break in pile driving." In the absence of anything better, this guidance from NOAA has been followed.

3 Applying the Interim Criteria

3.1 A Field-Monitoring Experience

The California Department of Transportation is currently replacing bridges over rivers in northern California to meet certain seismic, scour, and bridge design standards. One project involves the impact driving of thirteen 2.2-m-diameter piles outside of the wetted channel of the river (with four of these piles at a distance of ~43 m from the wetted channel) and four additional 1.2-m test piles at the 43-m position. The piles are being driven inside cofferdams excavated to a depth of at least 20 ft below the surface. The pile driving would be divided between two separate construction seasons.

Using the best available data, the Biological Assessment for the project concluded that underwater sound levels would be substantially below the 206-dB peak single-strike threshold but that the accumulated SEL threshold during a maximum day of impact driving could be exceeded out to a distance of ~150 m from the driven piles.

The California Coastal Commission had ultimate permit authority, and their staff established rigorous acoustic and biological monitoring protocols, required a caged-fish study to expose salmonids to pile-driving noise at various distances from the pile-driving activity, and required a fish-exclusion zone with a fish passage 300 m long. This for a project with 17 piles, 9 of which were driven more

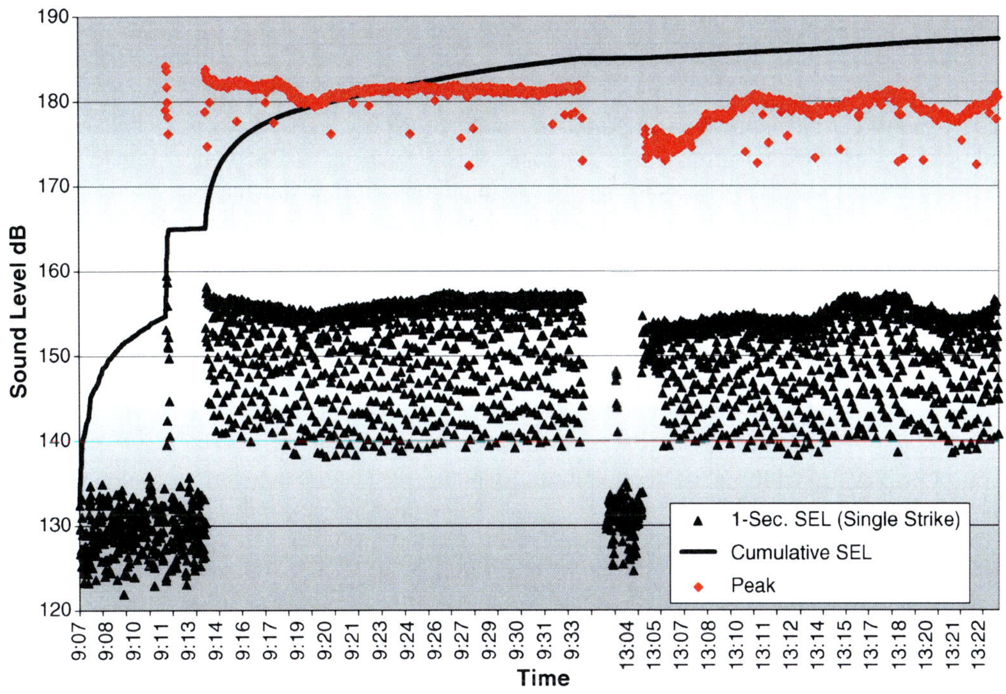

Fig. 1 Pile-driving sound levels measured in a northern California river. SEL, sound exposure level

than 40 m from the river, and no piles were actually driven in the water. The estimated cost for the monitoring and fish-cage study is $600,000.

Orders from the Coastal Commission required that the 187 dB cumulative SEL threshold not be reached during a day of pile driving (interpreted as 187.0 dB). Hydroacoustic monitoring in real time of the accumulating SEL to the nearest 0.1 dB therefore became the determining factor if pile driving should be stopped. The first monitoring occurred in early March 2009. There was swiftly flowing water in the river channel. Conditions in the river made it unsafe to enter so the hydrophones had to be deployed from the shore without the ability to investigate the subsurface conditions. It was evident that measured levels were elevated as a result of the strong current in the river. It was not possible to isolate the small contribution of low-level pile-driving noise from ambient noise during the direct measurement of the cumulative SEL.

Monitoring results for 11 March 2009 are shown in Figure 1. It shows the peak level and SEL for each 1-s interval as well as the cumulative SEL. These piles were located ~43 m from the edge of the wetted channel and 65 m from the hydrophone, the closest point in the river where conditions were suitable for monitoring. Two piles were driven during the day. The pile driven in the morning was driven to the required depth, but driving of the pile during the afternoon was stopped when the daily cumulative SEL at the hydrophone reached 187 dB. One wonders whether any fish received this cumulative dose, considering that one pile was driven in the morning and a second pile was driven in the afternoon. The peak pressure from a single strike never exceeded 195 dB and the SEL from a single strike never exceeded 160 dB. The cumulative SEL resulted from ~1,500 low-energy pulses. Nonetheless, the work was stopped pursuant to the permit requirements. Fish in the area the next morning were then exposed to elevated sound levels during the completion of the driving of the pile. Monitoring continued through the summer. The initial fish-cage study was completed. From the authors' perspective, this entire effort was poorly conceived but well executed.

3.2 A Preproject Assessment

A current project undergoing preproject assessment illustrates the effect of the cumulative SEL standard and its current application on proposed construction projects. Construction of a new bridge would require installation of four 2.6-m-diameter cast-in-steel-shell (CISS) piles. In addition, the construction of a trestle and falsework would use numerous 20-in.-diameter steel-shell piles. The large-diameter piles would be driven in a dewatered cofferdam. The small piles would be driven in the water, and with attenuation such as a bubble curtain, the assessment concluded that the peak pressure would be substantially below 206 dB and the single-strike SEL would be ~165 dB at 10 m. Based on the frequency of the hammer strikes, the analysis concluded that after 4 min of pile driving and 160 pulses, with attenuation in place, the 187-dB accumulated SEL threshold would be reached at a distance of 10 m from the pile. For the permanent CISS piles drive in a dewatered cofferdam, analysis from previous projects indicated an expected single-strike SEL of 165–175 dB just outside the cofferdam. Peak sound pressures were estimated to range from 185 to 195 dB, substantially below the single-strike threshold. The number of hammer strikes predicted to reach a cumulative SEL of 187 dB ranged from 16 to 160 strikes. Data from other projects indicated that it could take several thousand pile strikes to drive a pile over the period of ~1 h.

The engineers concluded that construction of the equipment trestles and falsework could not be completed within the work window deadlines established by resource agencies. Driving of each permanent pile would require several weeks even if the sound levels were at the lower limit of the range, making the construction of the piles impractical. Furthermore, partial driving of the pile could result in pile setup, a phenomenon that increases the capacity of the pile with time and makes the pile more resistant to driving to the specified tip elevation on subsequent days. The highest sound levels typically occur when the pile resistance is the greatest. For this project, the engineers were able to develop alternative engineering plans that would result in lower sound levels. This is a best case outcome but very unusual given the constraints normally associated with engineering options for the design of new bridge foundations.

4 Recent Caged-Fish Studies

Several recent studies that exposed caged fish to pile-driving sounds were summarized by Reyff (2010) in a presentation at the 2010 Transportation Research Board Annual Meeting, Washington, DC. These studies were limited to observable injuries that may have resulted during or shortly after exposure to pile-driving sounds. Although fish were examined for physical injury, the effects on hearing that include temporary threshold shift, permanent threshold shift, or hair cell damage were not evaluated. The studies included concrete piles driven at the Port of Oakland, CA, steel-sheet piles installed at the Port of Anchorage, AK (results similar but not yet reportable), steel piles driven in northern California previously discussed, and steel piles driven in Lake Washington in the Pacific Northwest. When compared with control groups of fish, physical injuries or adverse behavioral responses from exposed fish were not observed in any of the experiments.

Reyff (2010) summarized the results. In the Port of Oakland Wharf Reconstruction 2004, fishes exposed to single-strike SELs of 165 dB and a cumulative SEL of up to 191 dB experienced no physical trauma that could be related to exposure to underwater noise from pile driving. In the Port of Seattle Fishermen's Terminal Study 2006–07, juvenile Coho salmon exposed to maximum peak SPLs of up to 208 dB, an average single-strike SEL of 175 dB, and a cumulative SEL of 207 dB in 1 workday, resulting from 1,627 pile strikes, survived for the 10-day holding period, revealed no external or internal injuries related to pile-driving sound exposure, and readily consumed hatchery food during the first and subsequent feeding trials. Subtle behavioral changes of fish were noted in response to pile strikes. In the Mad River Replacement Project 2009, fish were exposed to the highest levels in the river where the cumulative SELs reached 194 dB. Again, fish showed no physical trauma that could be related to exposure to underwater noise from pile driving.

5 Conclusions

The cumulative SEL threshold contained in the *Agreement in Principle for Interim Criteria for Injury to Fish From Pile Driving Activities* (FHWG 2008) is having a significant effect on the cost and constructability of bridges in the western United States. The cumulative SEL threshold level cannot be met even with attenuation systems for piles driven in or near the water without limiting the driving during a workday and thereby increasing the construction period and potentially increasing the number of fish exposed to elevated sound levels. Recent studies of fish in cages exposed to pile driving show no physical trauma for fish exposed to levels significantly above a cumulative SEL of 187 dB. Extensive acoustical data from unattenuated and attenuated pile-driving sources are available to researchers interested in determining possible auditory effects resulting from actual pile-driving sounds. It is time to reexamine the cumulative SEL threshold in light of recent studies and conduct additional studies, if necessary, to evaluate auditory and nonauditory effects utilizing actual pile-driving pulses either in the laboratory or through additional field studies. Unless we want to continue to find only "no-effect" results, field fish-cage studies associated with construction projects should only be undertaken after independent experts conclude that useful information could result from the study.

References

Carlson T, Hastings M, Popper AN (2007) Update on recommendations for revised interim criteria for fish during pile driving activities. Memorandum to California and Washington Departments of Transportation. Available via http://www.dot.ca.gov/hq/env/bio/fisheries_bioacoustics.htm. Accessed 28 April 2010.

Fisheries Hydroacoustic Working Group (FHWG) (2008) Agreement in principle for interim criteria for injury to fish from pile driving activities. Available via http://www.dot.ca.gov/hq/env/bio/fisheries_bioacoustics.htm. Accessed 28 April 2010.

Hastings M, Popper AN (2005) Effects of sound on fish. Report to California Department of Transportation. Available via http://www.dot.ca.gov/hq/env/bio/fisheries_bioacoustics.htm. Accessed 28 April 2010.

Popper A, Carlton T, Hawkins A, Southall B, Gentry R (2006) Interim criteria for injury of fish exposed to pile driving operations: A white paper. Prepared for Fisheries Hydroacoustic Work Group. Available via http://www.dot.ca.gov/hq/env/bio/fisheries_bioacoustics.htm. Accessed 28 April 2010.

Reyff JA (2010) A summary of recent studies that exposed caged fish to pile driving sounds. Presented at the 2010 Transportation Research Board Annual Meeting, 10–14 January 2010, Washington, DC. Available via http://www.itre.ncsu.edu/ADC30/10WinterWrkshop_presentation_links.htm. Accessed 28 April 2010.

Stadler J, Woodbury D (2009) Assessing the effects to fishes from pile driving: Application of new hydroacoustic criteria. Report Presented at Inter-Noise 2009, 38th International Congress and Exposition on Noise Control Engineering, 23–26 August 2009, Ottawa, ON, Canada.

Marine Mammal Acoustics Exposure Analysis Models Used in US Navy Environmental Impact Statements

Douglas Wartzok, Christine Erbe, Wayne M. Getz, and Jeanette Thomas

1 Introduction

In compliance with the Marine Mammal Protection Act (MMPA), the Navy must prepare Environmental Impact Statements (EISs) and obtain Letters of Authorization (LOAs) allowing, as defined in the MMPA, incidental "takes" of marine mammals for its training activities. The Navy has prepared these documents for each of its training ranges addressing the possible effects of noise on marine mammals. The authors were asked to determine whether two Navy models and a model developed by Science Applications International Corporation (SAIC) were designed and implemented to obtain the best unbiased estimates for the level A and level B (types I and II) takes for midfrequency active sonar (MFAS).

All three models used National Marine Fisheries Service criteria for received sound exposure level (SEL) and sound pressure level (SPL) that would result in level A, level B type I, and level B type II harassment (Federal Register 2009). Level A harassment was considered to be the SEL that might cause a permanent threshold shift (PTS). For nonexplosive underwater acoustic signals, cetaceans were expected to experience PTS when the M-weighted SEL exceeded 215 dB re 1 μPa^2-s and the most sensitive pinnipeds (harbor seals and similar species) when the M-weighted SEL exceeded 203 dB re 1 μPa^2-s (see Southall et al. 2007 for details and references) Fig. 1. Level B harassment was subdivided into two types: type I where the SEL might cause a temporary threshold shift (TTS) and type II where the SPL might cause an observable behavioral change. For cetaceans, TTS was assumed to occur at a SEL greater than 195 dB re 1 μPa^2-s and for the most sensitive

D. Wartzok (✉)
Florida International University, Miami, FL 33199, USA
e-mail: wartzok@fiu.edu

C. Erbe
Centre for Marine Science & Technology, Curtin University, Bentley, WA 6102, Australia; *formerly* JASCO Applied Sciences, Australia
e-mail: C.Erbe@curtin.edu.au

W.M. Getz
University of California at Berkeley, Berkeley, CA 94720, USA
e-mail: wgetz@berkeley.edu

J. Thomas
Western Illinois University, Macomb, IL 61455, USA
e-mail: j-thomas@wiu.edu

A.N. Popper and A. Hawkins (eds.), *The Effects of Noise on Aquatic Life*,
Advances in Experimental Medicine and Biology 730, DOI 10.1007/978-1-4419-7311-5_125,

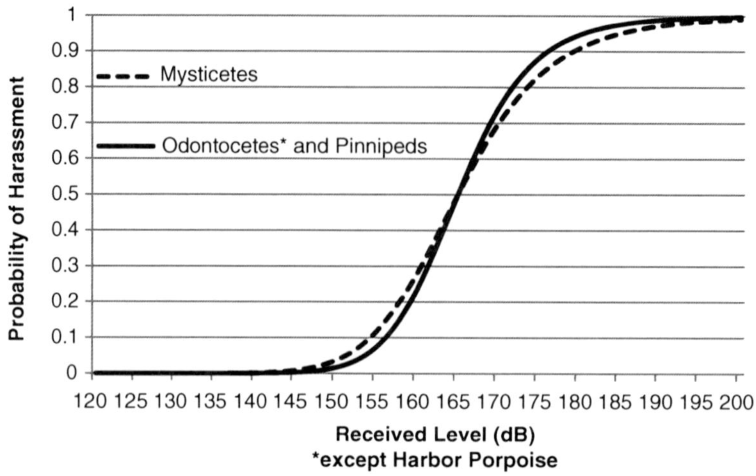

Fig. 1 Risk function for level B type II harassment

pinnipeds in water at a SEL greater than 183 dB re 1 μPa²-s (see Southall et al. 2007 for details and references). Level B type II harassment was based on a logistic-like risk function where the essential parameters are the basement value for a behavioral response (taken as 120-dB SPL for all marine mammals); a parameter *K* representing 50% risk (taken as 45-dB SPL for all marine mammals); and a risk transition sharpness parameter (i.e., an indication of how closely the risk function approaches a step function) that was set at 10 for odontocetes and pinnipeds and at 8 for mysticetes, which resulted in an increase in the proportion of the mysticete population harassed at lower SPL compared with the curve for odontocetes and pinnipeds. For harbor porpoises, a "particularly sensitive" species (Southall et al. 2007), a step function at 120 dB was used for the probability of harassment.

As an example of the type of situations for which these models need to be able to provide take estimates, the review specifically considered the use of MFAS for the Southern California (SOCAL) Range Complex EIS. The SOCAL Range Complex covers an area of 120,000 nautical square miles. The incidental-take authorization would cover 5 years (January 2009 to January 2014). The modelers do not know with any precision when or where MFAS activities may take place within the range during these 5 years. Nonetheless, they need to estimate the number of takes by harassment in each category for marine mammal species found in this area. The SOCAL Range Complex EIS noted that in 2007 postaction reports were filed for 797 h of MFAS in the SOCAL Range Complex.

2 The Models

The Naval Undersea Warfare Center (NUWC) in Newport, RI, has developed two marine mammal exposure-analysis models: the area-density model and the NUWC exposure model (NEMO). SAIC in Arlington, VA, has developed a third marine mammal exposure-analysis model. In all three models, sound propagation is modeled using the comprehensive acoustic simulation system based on Gaussian ray bundle propagation (CASS/GRAB). The acoustic field is modeled to a distance where the SPL drops below 120 dB re 1 μPa.

2.1 NUWC Area-Density Model

The model computes the three-dimensional (3-D) acoustic field of one or more simultaneously operating sources in a geographical region of operation and for the time of year (season) of operation based on operational specifications such as duty cycles and acoustic characteristics (such as frequency and beam pattern). NUWC handles bathymetry via three analysis points (shallow, sloping, and deep seafloor) and typically eight radials around each analysis point to calculate footprints that are then used throughout the region. To reduce computational memory requirements, the model then reduces the 3-D sound field to a two-dimensional (2-D) footprint by taking the maximum SEL and the maximum SPL over all depths.

The model then moves the sources through the operational area. The footprints are overlapped from one sound emission to the next. On a 2-D receiver grid that evenly spans the operational area, SELs are summed. For every new emission of sound, the model compares the maximum SPL received at all 2-D receiver grid points to the stored maximum SPL. If the new SPL exceeds the stored SPL, then it is replaced; otherwise not.

The NUWC model compares the cumulative SEL to assumed thresholds for PTS and TTS. The maximum SPL modeled is passed through the risk function to determine the probability of inducing a behavioral change.

Databases on marine mammal population density are accessed to estimate how many animals of which species are likely encountered in the geographic region at the time of operation. These animals are then evenly distributed over the 2-D receiver grid but can be distributed with more spatial structure if sufficient information is available on the distributions of particular marine mammal species in the zone of operation. Animal densities are then multiplied by the total area receiving SEL above threshold to yield the total number of exposures likely producing TTS or PTS. Animal densities are also multiplied by the probability of behavioral change and then summed over the 2-D grid to yield the total number of exposures resulting in a behavioral change, i.e., the number of takes.

2.2 NUWC NEMO Model

The NEMO model is a true 3-D model. NEMO uses the same sound propagation model as the area density model. However, in the NEMO model, SEL and SPL values are not maximized over all depths and reduced to a 2-D grid; rather a 3-D footprint of the source is kept. Marine mammals are no longer uniformly distributed in space but instead are modeled individually as "animats" using the marine mammal movement and behavior simulator (3MB) (Houser 2006). The animats' movements are stochastically determined by sampling from distributions of rates of movements in three dimensions, surface time, and time at depth. These distributions are dependent on the behavioral state of the animal. Transitions between behavioral states are also stochastically determined. NEMO keeps track of SEL and SPL for every individual animal.

2.3 SAIC Model

Environmental parameters are grouped into 5-20 provinces in the areas of interest (viz., SOCAL Range Complex or Atlantic Fleet active sonar training [AFAST] area), each having a constant sound speed profile, seafloor type, and water depth. Bathymetry is ignored except for water depth at the source location. Each province has a flat bottom. The entire operational area is split into these provinces. Sound propagation and bioacoustic impact are modeled once for each province. Harassment numbers are then combined into a weighted average, weighted according to how much of the operational area is represented by each province.

For each province, impact water volumes are computed for PTS, TTS, and behavioral risk. For PTS and TTS, the volume around the source in each depth bin where the SEL exceeds threshold value is calculated. For behavioral risk, the maximum SPLs are grouped into 0.5-dB bins and then a volume histogram is computed showing the volume around the source in each depth bin that falls into each SPL bin. The histogram is multiplied by the behavioral risk function to give the percentage of animals likely to experience behavioral harassment.

Animals are distributed evenly within their geographic areas of occurrence and with depth according to published dive profiles. The animals remain stationary relative to the ships.

3 Model Evaluation

A number of features are similar across the three models. They handle animal distribution at the level to which it is known. Lacking any specific knowledge of areas of operation, an average density over the range is used. They collapse annual variation in environmental conditions as related to acoustic propagation to two seasons, and when seasonal data are available on species occurrence within the range, these data are incorporated in the models.

Two of the models, NEMO and SAIC, distribute the animals in the *z*-axis as well as in the *x*-*y* plane. This approach is superior to the NUWC model that overestimates the animal exposures because the highest values of SEL and SPL in the water column are selected during the collapse of the 3-D calculations to a 2-D footprint. The SAIC model in which the animals are fixed in location is probably superior at this time to the NEMO model in which movement of the animats is drawn from various statistical distributions that themselves are not well known. The NEMO model is better suited for behavioral research at this stage of knowledge about the physical and biological parameters that influence an animal's behavior. In our opinion, the elegance of this model might lead to a false sense that we understand the movements of real animals much better than we do.

The models are conservative in that they overestimate the number of level B type I takes because they do not consider partial, or complete, recovery of the TTS during the 24-h accumulation period. A new version of NEMO should be able to record the time of arrival of the pings at each animat, and when more is known about TTS recovery, these data can lead to refined predictions of level B type I takes.

The behavioral risk function is used in all models by relating the probability of a response to the maximum level received over a 24-h period. However, an animal is not going to sit in the water being pinged at waiting for a higher SPL to come within the next 24 h. At any time that it receives a ping, it decides whether to react or not. With the next ping, it faces the same decision. The probability of response should be a function of the actual level received at any one time. The probability of an animal reacting at least once every 24 h is then equal to the cumulative probability for all pings received.

Neither the NUWC nor the SAIC models consider ambient noise. Although this will not change the propagation models, it can change the animal impact assessments. The SEL for TTS will likely be different under masked and nonmasked situations and the behavioral risk curve may be displaced under conditions of high ambient noise. Ambient noise may also affect the distribution of animals within the operation area.

The computational resources required to model out to a 120-dB SPL range could be much better used with a finer grain analysis at closer ranges (e.g., more provinces for the SAIC model; more radials and analysis points for NUWC models). Calculated level B type II takes can be increased based on the area of the curve that lies below the minimum decibel level modeled. For example, if the level B type II takes were modeled out to 157 dB, then the risk function suggests that increasing this number by 10% would yield approximately the same number of takes as a complete modeling out to the 120-dB range.

In the absence of any particular knowledge about future naval operations, NUWC models move the ships on random tracks. This will provide an unbiased estimate of takes as will the SAIC model in which ships move in straight lines when considered within the complete SAIC modeling paradigm.

The NEMO approach has the greatest potential to model the system but is well in advance of the available data and, by seeming to take into consideration more than is really known, leads to a false level of confidence about its predictions. If NEMO assigned animats to depth levels in proportion to the amount of time tracked individuals spent at various depths and left them in a stationary *x-y* position, as does the SAIC model, then NEMO would be the model that could best simulate the system.

The SAIC model is interesting in that, on the one hand, it is the farthest from simulating the system, but on the other hand, it is the best for providing the one thing required of a model in the EIS/LOA framework: the expected number of takes by species and by harassment level over large ranges with unspecified operations and over a 5-yr time horizon. SAIC has done some sensitivity analyses and found that over reasonable parameter ranges, the model is relatively robust.

4 Conclusions

The models need to determine the number of takes at three different harassment levels for a variety of species of marine mammals over the extended ranges of naval operations during the 5 years for which LOAs are sought based on the EIS. If the questions were more specific, the models would need to take into consideration more specific biological and physical variables. Because the models provide only the expected number of takes and provide no data on stochasticity as reflected in the variance in the number of takes, conduct no sensitivity analyses on how uncertainties in the data and processes propagate through the model, and require no scenario analyses, the model outputs are not very useful for management and mitigation.

Given the minimal knowledge available on animal responses to noise, modeling the animals as stationary and distributing them evenly throughout a habitat or province will neither consistently underestimate nor overestimate takes. Obviously, if we knew more about the response of animals to noise and other factors such as prey availability that affect their distribution, the take estimates could be improved, but lacking that knowledge and instead guessing the animals' behavior, e.g., avoidance, can lead to biased estimates.

The lack of information on the distribution of animals is a significant impediment to obtaining valid estimates of take. Obtaining better data on distribution, particularly with respect to environmental features defining habitat, is more important than fine tuning the movements of individual animals in response to noise.

The models are being inappropriately used to generate "take" rates as though these rates are both accurate and precise and have meaning in their own right. To the contrary, there is no evidence that the rates are accurate; they are likely biased, although probably on the conservative side; and because they have no stochasticity, they ignore issues relating to likely ranges of takes under variable conditions and are unable to predict the potential occurrence of catastrophic events. Furthermore, although the models have potential to be used as tools for designing less invasive exercises or operations, they are not used in this regard. They also need to be more closely tied to the design of monitoring and reporting procedures that can lead to better estimates of model parameters in the future.

References

Federal Register (2009) Taking and importing marine mammals; US Navy training in the Southern California Range Complex; Final rule. Fed Reg 74:3881–3918.

Houser DS (2006) A method for modeling marine mammal movement and behavior for environmental impact assessment. IEEE J Ocean Eng 31:76-81.

Southall BL, Bowles AE, Ellison WT, Finneran JJ, Gentry RL, Greene CR Jr, Kastak D, Ketten DR, Miller JH, Nachtigall PE, Richardson WJ, Thomas JA, Tyack PL. (2007) Marine mammal noise exposure criteria: Initial scientific recommendations. Aquat Mamm 33:411–522.

Status of the National Oceanic and Atmospheric Administration's Guidelines for Assessing Impacts of Anthropogenic Sound on Marine Mammals

Amy R. Scholik-Schlomer

1 Introduction

The National Oceanic and Atmospheric Administration (NOAA) is in the process of developing acoustic guidelines for assessing the effects of anthropogenic sound on marine mammal species under NOAA's jurisdiction. The guidelines will provide a summary of past policies and updated basic science-based guidance. The guidelines are intended for use by NOAA analysts/managers and relevant stakeholders, including federal agencies, and will increase the transparency and consistency of how acoustic effects on marine mammals are assessed in the context of our authorizations, permits, consultations, and exemptions under the various environmental statutes NOAA administers (e.g., Marine Mammal Protection Act, Endangered Species Act, and National Marine Sanctuaries Act).

The guidelines are expected to be organized in a manner that reflects the evolution of the underlying science and management decisions related to the effects of anthropogenic sound on marine mammals. Our past acoustic thresholds will be provided as well as updated procedures for assessing acoustic effects based on recent advances in science. NOAA is working toward numerical thresholds where appropriate and possible and general analytical paradigms in other cases (i.e., for instances where context- or environmental-specific factors reduce or eliminate the relevance of broadly applicable quantitative thresholds). Thus the acoustic guidelines will reflect qualitative considerations (e.g., masking, stress, cumulative impacts, and population consequences of sound exposure) as well as numerical thresholds for temporary (TTS) and permanent (PTS) threshold shift onset. An approach for updating acoustic criteria/thresholds and policy guidance is also presented. This abstract highlights a few topics that will be in the guidelines.

A.R. Scholik-Schlomer (✉)
National Marine Fisheries Service, National Oceanic and Atmospheric Administration, Silver Spring, MD 20910, USA
e-mail: amy.scholik@noaa.gov

A.N. Popper and A. Hawkins (eds.), *The Effects of Noise on Aquatic Life*, Advances in Experimental Medicine and Biology 730, DOI 10.1007/978-1-4419-7311-5_126,

2 Updated NOAA Acoustic Guidance

2.1 *Sound Source Categories*

NOAA recognizes that different sound source types pose different risks to marine mammals, especially in terms of effects on auditory structures, based on the physical features of the resulting waveforms. For these assessments, sound is divided into two broad categories (defined at the source). Impulsive sound sources are typically transient, brief, and broadband and consist of a rapid rise time (American National Standards Institute 1986). They can occur as a single event or be repetitive. Nonimpulsive sound sources can be broadband or tonal, brief or prolonged, and continuous or intermittent and typically do not have the rapid rise time that impulsive signals do (American National Standards Institute 1995). Although similar categories have been used in past assessments, they have not been specifically defined.

2.2 *Quantitative Paradigms*

2.2.1 Updated TTS and PTS Onset Thresholds

The goal is for NOAA's acoustic thresholds to reflect the best available science, which has increased dramatically over the last 10 years. The process for developing the guidelines began with reevaluating and updating all the current acoustic thresholds, starting with underwater thresholds for the onset of TTS (defined as a 6-dB threshold shift) and PTS (defined as a 40-dB shift). NOAA internally evaluated the scientific recommendations of the Southall et al. (2007) review as well as of relevant peer-reviewed studies published since Southall et al. (e.g., Lucke et al. 2009; Mooney et al. 2009a,b) in the context of its various environmental statutes.

Recently, the dual metrics of accumulated sound exposure level (SEL) and peak sound pressure level have been recommended as most appropriate for establishing TTS and PTS onset thresholds for marine mammals (Southall et al. 2007). NOAA is adopting these dual metrics for its updated thresholds. These thresholds are still being evaluated internally. The most updated information on these thresholds can be found at http://www.nmfs.noaa.gov/pr/acoustics/.

2.2.2 Behavioral Thresholds

NOAA's past approaches for assessing behavioral effects have primarily focused on broad and very simplistic numerical assessments based on the received sound level. However, significant developments in scientific knowledge are revealing that such simplistic approaches may underestimate or overestimate negative effects in that they fail to account for certain contextual factors that are likely critical in determining whether and how animals are impacted by a particular sound. Although the current state of science does not allow NOAA to update its previously established thresholds for behavior, it will be proposed that additional factors be considered qualitatively.

2.2.3 Marine Mammal Functional Hearing Groups and M-Weighting Functions

Current data indicate that not all marine mammal species have equal hearing capabilities (e.g., Richardson et al. 1995). To reflect this, Southall et al. (2007) recommended that marine

Table 1 Functional hearing groups

Functional Hearing Group	Estimated Hearing Range
Low-frequency cetaceans	7 Hz to 25 kHz
Mid-frequency cetaceans	150 Hz to 160 kHz
High-frequency cetaceans	200 Hz to 180 kHz
Pinnipeds in water	75 Hz to 75 kHz
Pinnipeds in air	75 Hz to 30 kHz

Adapted from the recommendations in Southall et al. (2007)

mammals be divided into five basic functional hearing groups based on estimated hearing ranges. NOAA concurs with dividing marine mammals into functional hearing groups but has preliminarily decided to extend slightly the estimated hearing range for low-frequency cetaceans, from 22 to 25 kHz (Table 1). This decision is based on data from Au et al. (2006) for humpback whales and from Frankel (2005) on gray whales, indicating that mysticetes may be able to hear beyond 22 kHz.

Based on the functional hearing groups and their estimated hearing ranges, Southall et al. (2007) established marine mammal-specific weighting functions (i.e., M-weighting functions) to better assess how a sound source of a particular frequency spectrum could affect members of a particular hearing group. NOAA considers M-weighting functions (based on the estimated hearing ranges presented in Table 1) appropriate when assessing the auditory effects of anthropogenic sound exposure to the inner ear and/or hearing ability, such as via PTS and TTS onset expressed in the accumulated SEL metric (not peak sound pressure metric, where flat-weighting is recommended).

2.3 *Qualitative Paradigms*

Although qualitative paradigms do not present means of directly quantifying sound source or receiver characteristics (often due to limited data or complexity/variability, as with behavioral responses to anthropogenic sound), we advise that they be considered within an assessment to the extent possible. As more data become available, there may be better means of quantifying these qualitative considerations (e.g., stress, masking, cumulative and synergistic effects) in the future.

2.3.1 Behavioral Disruption (Additional Considerations)

Assessing the severity of behavioral effects of anthropogenic sound exposure on marine mammals is challenging due in large part to the inherent complexity of behavioral responses. Behavioral reactions can vary not only among individuals but also within an individual depending on previous experience with a sound source, hearing sensitivity, sex, age, reproductive status, geographic location, season, health, social behavior, or other contextual factors (National Research Council 2003). It is recommended that these additional factors be considered qualitatively to the extent possible (Table 2).

Furthermore, certain species (e.g., harbor porpoises, beaked whales) or individuals (e.g., mother-calf pairs) may be particularly sensitive to anthropogenic sound. It is acknowledged that assessing behavioral effects for beaked whales and other particularly sensitive species/individuals will require a categorically different approach than many other cetacean species.

Table 2 Additional factors for consideration associated with behavioral disruption

Source Characteristics	Receiver Characteristics
Stationary vs. mobile sources (e.g., movement/speed of source: regular vs. irregular, toward or away from receiver)	Demographic factors (e.g., age, sex, offspring present)
Distance (horizontal/vertical) of sound source from receiver	Location of receiver and how long in area (transient vs. resident)
Onset of sound (gradual vs. sudden; rate of change of sound level)	Activity during exposure (e.g., resting, feeding, breeding, socializing)
Continuous vs. transient source (temporal property rather than waveform)	Previous exposure history with source (habituation, tolerance, sensitization)
Source directionality, location, and number	Hearing sensitivity/sensitive receptor

3 Updating Future NOAA Acoustic Guidance

NOAA's initial guidance for marine mammals will be revised, as the science allows, to update behavioral thresholds and various impact thresholds for explosives and airborne anthropogenic sounds. As more data become available, acoustic guidance also may be established for other protected species such as sea turtles and marine fishes. As with this guidance document, public review and external peer review will be integral to the creation of acoustic guidance.

NOAA will establish an intra-agency team consisting of staff from its various offices, regions, and science centers to reevaluate and update acoustic thresholds with new data every three to five years or as deemed necessary. In addition to evaluating new, applicable scientific studies, NOAA will also examine basic definitions, appropriate metrics, data standards, and methods to account for uncertainty, temporal and spatial considerations, and other relevant topics.

4 Where Are We in the Process?

NOAA's draft Acoustic Guidelines are undergoing an intra-agency review, which will be followed by an external peer-review process. The focus of the peer review is on the scientific and technical studies that have been applied and the manner that they have been applied in this guidance document. The external peer reviewers will not be asked to focus on policy decisions made within the document (e.g., the amount of uncertainty that is acceptable or the amount of precaution that should be embedded in the analysis). After the peer review, the public will be invited to provide comments on the document. Once the peer review and public comments are addressed, the acoustic guidelines will be released. Since our guidance document is in the process of being reviewed and evaluated, information provided in this summary paper is subject to change. Thus, for the most recent information on our acoustic guidance, please consult http://www.nmfs.noaa.gov/pr/acoustics/.

References

American National Standards Institute (1986) American national standard methods of measurement for impulse noise. ANSI S12.7–1986, Acoustical Society of America, New York.

American National Standards Institute (1995) American national standard of bioacoustical terminology. ANSI S3.20–1995, Acoustical Society of America, New York.

Au WWL, Pack AA, Lammers MO, Herman LM, Deakos MH, Andrews K (2006) Acoustic properties of humpback whale songs. J Acoust Soc of Am 120:1103–1110.

Frankel AS (2005). Gray whales hear and respond to a 21-25 kHz high-frequency whale-finding sonar. 16th Biennial Conference on the Biology of Marine Mammals, 12–16 December 2005, San Diego, CA, p 97.

Lucke K, Siebert U, Lepper PA, Blanchet MA (2009) Temporary shift in masked hearing thresholds in a harbor porpoise (*Phocoena phocoena*) after exposure to seismic airgun stimuli. J Acoust Soc Am 125:4060–4070.

Mooney TA, Nachtigall PE, Breese M, Vlachos S, Au WW (2009b) Predicting temporary threshold shifts in a bottlenose dolphin (*Tursiops truncatus*): The effects of noise level and duration. J Acoust Soc Am 125:1816–1826.

Mooney TA, Nachtigall PE, Vlachos S (2009a) Sonar-induced temporary hearing loss in dolphins. Biol Lett 5:565–567.

National Research Council (2003) Ocean noise and marine mammals. National Academies Press, Washington, DC.

Richardson WJ, Greene CR Jr, Malme CI, Thomson DH (1995) Marine mammals and noise. Academic Press, San Diego, CA.

Southall BL, Bowles AE, Ellison WT, Finneran JJ, Gentry RL, Greene CR Jr, Kastak D, Ketten DR, Miller JH, Nachtigall PE, Richardson WJ, Thomas JA, Tyack PL (2007) Marine mammal noise exposure criteria: Initial scientific recommendations. Aquat Mamm 33:411–522.

Regulatory Assessments of the Effects of Noise: Moving From Threshold Shift and Injury to Behavior

Craig E. Johnson

1 Introduction

The United States Endangered Species Act of 1973 (ESA) and Marine Mammal Protection Act of 1972 (MMPA) give the National Oceanic and Atmospheric Administration (NOAA) a mandate and the authority to protect numerous species of marine mammals, sea turtles, marine and anadromous fish, mollusks, and coral from activities conducted by government agencies, corporations, academic institutions, and private individuals, including activities that introduce anthropogenic noise into freshwater, estuarine, and marine ecosystems. To satisfy its obligations under these two laws, NOAA assesses the effects of sound-producing activities using an exposure-response risk-assessment methodology that is designed to satisfy the protective mandates of these laws.

This methodology begins with exposure analyses, which are designed to identify the number, age (or life stage), and gender of individuals that are likely to co-occur in space and time with any acoustic phenomena-produced physical, chemical, or biotic aspects of proposed actions (or "potential stressors") that are likely to have direct, indirect, interactive, or cumulative direct and indirect effects on individuals that might be exposed. Exposure analyses are followed by response analyses, which are designed to determine whether and how those individuals are likely to respond given their exposure. The assessments conclude with risk analyses, which begin with assessments of the probable risks any responses might mean for the individuals that are likely to be exposed (here, we measure risks to listed individuals using the individual's current or expected future reproductive success); changes in individual fitness are integrated to estimate probable changes in the viability of the population(s) those individuals represent; and changes in population viability are integrated to estimate probable changes in the viability of the species that comprise those populations.

These assessments are strongly influenced by our understanding of relationships between potential stressors, relevant exposure pathways, responses given those exposure pathways, and risks given those responses. For more than a decade, our assessments have emphasized direct acoustic exposure pathways centered on animal hearing mechanisms and the response of those hearing mechanisms to acoustic exposure (Southall et al. 2007). More recently, we have reconstructed the conceptual models underlying our assessments to recognize and, in some cases, emphasize the importance of animal behavior.

C.E. Johnson (✉)
National Marine Fisheries Service, National Oceanic and Atmospheric Administration, Silver Spring, MD 20910, USA
e-mail: craig.johnson@noaa.gov

A.N. Popper and A. Hawkins (eds.), *The Effects of Noise on Aquatic Life*, Advances in Experimental Medicine and Biology 730, DOI 10.1007/978-1-4419-7311-5_127,

2 Assessments Centered on Threshold Shift and Their Limitations

For more than a decade, assessments of the effects of anthropogenic sound on aquatic fauna have focused on the direct physical effects of exposing aquatic animals to the energy produced by a sound source. The exposure analyses for these assessments focused on identifying the number of individual animals that would be exposed (that is, co-occur in space and time) to particular sound sources and the received levels associated with the exposure in either sound exposure level (SEL; in dB re 1 μPa^2-s), peak sound pressure level for pile driving (in dB_{RMS} re 1 µPa at 1 m; RMS, root mean square), or both.

These assessments focused almost exclusively on determining whether an animal's auditory tissues were likely to be injured or damaged by exposure to sound pressure, with permanent threshold shift (permanent noise-induced losses in hearing sensitivity) being treated as equivalent to an injury and temporary threshold shift (temporary noise-induced losses in hearing sensitivity) being treated as equivalent to behavioral harassment. In most instances, substantial effort was dedicated to estimating received levels that would be treated as indicative of permanent or temporary threshold shift or, in the case of underwater detonations, tympanic membrane damage.

The outputs of these assessments generally consisted of estimates of the number of animals in one of five categories: 1) onset of massive lung injury or mortality (based on pounds per square inch), 2) tympanic rupture or slight lung injury, 3) permanent threshold shift, 4) temporary threshold shift, and 5) behavioral responses. In most cases, the behavioral responses in the last category have been treated as monolithic, with no distinction made between responses like long-distance avoidance of a sound field and abrupt, evasive dives.

These assessments have had numerous limitations. First, they treat "hearing" as a mechanical process that only involves structures in the ear that transduce sound pressure waves into vibrations and then to electrochemical impulses (rather than as a mechanical-cognitive-perceptual process). Second, they have emphasized the intensity of the sound [its received level (in decibels) or the integration of received energy] as the primary assessment metric while ignoring other attributes of an acoustic exposure (such as distance between an acoustic source and the receiving animal). Third, with the exception of lung injury, the outputs produced by these assessments have no explicit or implicit relationship to the fitness of the individuals that are expected to be exposed; that is, if we accept that an animal has experienced a threshold shift, we do not have sufficient information to make an inference about changes in the animal's fitness. To make any inference about the potential consequences of threshold shifts, we would need to know how much of a shift has occurred, the frequency ranges affected, and in the case of temporary threshold shifts, how long it might take for the hearing sensitivity to recover; this information is rarely provided in published and unpublished studies.

Finally, by treating behavioral responses as monolithic, assessments that have focused on threshold shifts and acoustic injury have not distinguished between behavioral responses that have direct fitness consequences (e.g., cows that abandon their calves), those that have probable fitness consequences (e.g., evasive dives in beaked whales), and those that are not known to have fitness consequences (e.g., auditory masking and vocal adjustments).

3 Moving Beyond Threshold Shift to Assessments Centered on Behavior

Because of the limitations in the assessment models discussed in the preceding section, we reconstructed our assessments on a model of animal behavior and behavioral decision making, which incorporates the cognitive processes involved in behavioral decisions. This revised model assumes that exposing aquatic fauna to anthropogenic sounds will primarily have an effect by changing the animal's behavior, although we retained physical trauma and noise-induced losses in hearing sensitivity (threshold shift) as components of the model. In addition, the model is based on an expanded conception of "hearing" that includes the cognitive processes an animal employs when it analyzes acoustic impulses (Blumstein and Bouskila 1996; Hudspeth 1997; Yost 2007), which includes the processes animals employ to integrate and segregate sounds and auditory streams and the circumstances under which they are likely to devote attentional resources to an acoustic stimulus.

Animals would then combine their perception of the acoustic stimulus with their assessment of the auditory scene (which includes other acoustic stimuli), their awareness of their behavioral state, physiological state (including whether they have noise-induced losses in hearing sensitivity), reproductive condition, and social circumstances to classify an acoustic stimulus and determine which specific behavior it will select from the set of behaviors that are appropriate to the auditory scene given its physiological and behavioral state when exposed and its experience.

This revised conceptual model produced several improvements in our assessments of the effects of anthropogenic noise on aquatic fauna. First, this revised model helped us recognize a broader set of variables associated with acoustic exposures, in addition to received level, that we had not explicitly considered in earlier assessments (e.g., the number of exposure events, duration of exposure events, frequency of exposure events, time interval between sequential exposure events, distance between an acoustic source and the receiving animal, the spectral characteristics of the waveform during the exposure, and the behavioral state of the animal when it is exposed). Second, the assessments produced by this model consider and are based on a broader array of behavioral and physical responses, and those responses explicitly connect to the fitness of animals that we expect to be exposed to anthropogenic noise (e.g., changing an animal's energy or time budgets, forcing animals to make life history trade-offs, changing social interactions among groups of animals). Because this approach explicitly connects acoustic exposure to the fitness of individual animals, the approach also allows us to assess potential population-level and species-level consequences of acoustic exposures.

Although this approach is more suitable for the assessments NOAA must conduct to satisfy the requirements of the ESA and MMPA, it would benefit from further study of the physical and behavioral responses of free-ranging animals that have been exposed to a variety of acoustic stimuli. Some exemplary studies have been conducted recently (e.g., Kvadsheim et al. 2007; Popper 2008) or are underway, but studies that are designed as either behavioral studies or whole animal studies will be especially important for future efforts to regulate and mitigate anthropogenic sound.

References

Blumstein DT, Bouskila A (1996) Assessment and decision making in animals: A mechanistic model underlying behavioral flexibility can prevent ambiguity. Oikos 77:569–576.

Hudspeth AJ (1997) How hearing happens. Neuron 19:947–950.

Kvadsheim PH, Benders F, Miller P, Doksæter L, Knudsen F, Tyack P, Nordlund N, Lam FP, Samarra F, Kleivane L, Godø OR (2007) Herring (sild), killer whales (spekkhogger) and sonar - The 3S-2006 cruise report with preliminary results. FFI Report No. 2007/011189, Forsvarets Forskningsinstitutt/Norwegian Defence Research Establishment, Kjellen, Norway.

Popper AN (2008) Effects of mid- and high-frequency sonars on fish. Report prepared for the US Department of the Navy Naval Undersea Warfare Center by Environmental BioAcoustics LLC, Rockville, MD, and Newport, RI.

Southall BL, Johnson CE, Scholik A, Adams T, Harrison J, Hollingshead K (2008) U.S. regulation of the effects of sound on marine life: NOAA's mandates and use of scientific information. Bioacoustics 17:275–278.

Yost WA (2007) Fundamentals of hearing. An introduction. Academic Press, New York.

United States Bureau of Ocean Energy Management, Regulation and Enforcement: Filling Data Gaps to Better Understand the Effects of Anthropogenic Noise on Marine Life

Jill Lewandowski, Elizabeth Burkhard, Kimberly Skrupky, and Deborah Epperson

1 Introduction

The offshore energy industries (both oil/gas and renewables) introduce anthropogenic noise into the marine environment through exploration (seismic), development and production (pile driving and drilling), decommissioning (explosives), and activities associated with these actions (i.e., icebreaking, support vessel traffic, and aircraft overflights). There are scientific uncertainties regarding how and what different marine animals hear, the behavioral and physiological effects of sound-producing activities on individual animals, and the significance of any effects to individuals on the populations. The uncertainties ultimately lead to more conservative protective measures, additional monitoring requirements, public criticism of environmental analyses and decision making, and, ultimately, litigation, additional costs, and delays in Bureau programs. The Bureau of Ocean Energy Management, Regulation and Enforcement (BOEMRE; formerly the Minerals Management Service), the US federal bureau responsible for regulation of these offshore industries, has developed intricate, comprehensive, and effective research programs aimed at filling key information and data gaps so as to better inform regulatory decision making and developing technologies to minimize the amount of noise put into the water.

2 BOEMRE Environmental Studies Program

The BOEMRE Environmental Studies Program (ESP) was established in 1973 to provide the information needed to predict, assess, and manage impacts from offshore energy and marine mineral exploration, development, and production activities on human, marine, and coastal environments. The research is geographically diverse, ranging from the Atlantic to the subtropical Gulf of Mexico

J. Lewandowski (✉) • E. Burkhard • K. Skrupky
United States Bureau of Ocean Energy Management, Regulation and Enforcement, Herndon, VA 20170, USA
e-mail: jill.lewandowski@boemre.gov; Elizabeth.Burkhard@boemre.gov; kimberly.skrupky@boemre.gov

D. Epperson
United States Bureau of Ocean Energy Management, Regulation and Enforcement,
New Orleans, LA 70123, USA
e-mail: Deborah.epperson@boemre.gov

A.N. Popper and A. Hawkins (eds.), *The Effects of Noise on Aquatic Life*,
Advances in Experimental Medicine and Biology 730, DOI 10.1007/978-1-4419-7311-5_128,

and the California offshore to the Alaska arctic offshore. It is also topically diverse and includes biology, physical oceanography, air and water quality, socioeconomic, and cultural resources research. Today, the ESP has evolved into a $30 million (USD) annual program that has significantly advanced science in many areas, such as the discovery of deep-sea chemosynthetic communities and acoustic impacts on marine life. The ESP works cooperatively with other federal agencies, academia, and industry (both domestic and international) to provide high-quality, peer-reviewed science to the public, stakeholders, and decision makers (see http://www.boemre.gov/eppd/sciences/esp/index.htm). BOEMRE environmental research consistently has been recognized for excellence with US Department of Interior Cooperative Conservation Awards and National Oceanographic Partnership Program Excellence Awards.

3 BOEMRE Technology Assessment and Research Program

The Technology Assessment and Research (TAR) Program was established in the 1970s to support research associated with operational safety and pollution prevention as well as oil spill response and cleanup capabilities. This program also is dedicated to investigating and assessing industry applications of technological innovations and ensuring that governing regulations, rules, and operational guidelines of the BOEMRE encompass the use of the best available and safest technologies. With this in mind, a portion of the TAR Program has focused on reducing the noise of industry operations through engineering adjustments (see http://www.boemre.gov/tarphome/). In view of the overlap of issues and challenges, participation in jointly funded projects with industry, other federal and state agencies, and international regulatory organizations has become the primary funding mechanism as well as a broader recognition that participation in these joint projects is the most effective and efficient means to leverage available funds.

4 Highlights of Key Acoustic-Related Research

For close to 40 years, the ESP and the TAR Program of the BOEMRE have continually produced high-quality scientific information and technological advancements that have addressed many data and information gaps related to acoustic issues. Some highlights of these efforts include

- a 7-yr, $11 million (USD) study on sperm whales in the Gulf of Mexico, their potential response to seismic exploration, and sperm whale prey studies;
- funding of National Research Council reviews (e.g., *Marine Mammals and Noise* [1995]);
- a 25+ yr bowhead whale research program assessing migration, feeding, and effects of industry activities on bowheads and other Arctic species;
- sound source characterization and reduction/mitigation of noise during production activities, dredging, seismic surveys, and pile driving;
- domestic and international workshops on acoustic-monitoring technologies and effects of seismic and pile-driving sounds on fish;
- $40 million (USD) over the life of the program in developing baseline information on protected species; and
- a summary and synthesis of seismic survey marine mammal observer reports for 2003–2008.

5 Identification of Existing Data Gaps

Despite the progress made to date, the BOEMRE and others recognize that some scientific uncertainty still exists. The BOEMRE also understands that the importance of collaboration and the leveraging of resources is critical in bringing our collective understanding of the effects of anthropogenic noise to the next level. For this reason, the BOEMRE participated in the US Joint Subcommittee on Ocean Science and Technology's Interagency Task Force on Anthropogenic Sound and the Marine Environment. The work of the Task Force produced a report detailing a US federal agency roadmap to focus and prioritize federal research efforts addressing marine anthropogenic sound over the next decade (see the full report at http://www.nmfs.noaa.gov/pr/pdfs/acoustics/jsost2009.pdf). The key data and information gaps from this report that are most relevant to the statutory mandates and responsibilities (and therefore are focused areas of acoustic impact research) of the BOEMRE are:

- to characterize industry sound sources and propagation characteristics and develop/refine models;
- to improve the ability to identify and understand biologically significant effects of sound exposure;
- to further develop and validate mitigation measures;
- to improve monitoring technologies in order to increase the effectiveness of mitigation;
- to reduce the footprint of existing sound sources and develop alternate, less adverse technologies;
- to support online databases of marine mammal research results and acoustic data;
- to standardize data collection/reporting/training of marine mammal observer programs;
- to develop methods to measure/model the cumulative effects of noise exposure on sensitive species; and
- to expand/improve baseline data for marine species particularly susceptible to anthropogenic sound.

6 Feeding Results Into Regulations

Importantly, all of the results from the ESP and the TAR Program feed directly into the BOEMRE Environmental Assessment Program (EAP; see http://www.boemre.gov/eppd/assessment/index.htm). The goal of the EAP is to prepare environmental analyses of project impacts, ensure compliance with over 15 environmental laws, and provide environmental policy guidance within the BOEMRE. The result of this process is an ongoing feedback loop where environmental assessments identify needed information, the BOEMRE research addresses these data gaps, and research results are then used to improve future environmental assessments. Stakeholder input and peer review are incorporated at every step in the process. Information garnered through BOEMRE-funded research is also integral in the BOEMRE development of mitigation aimed at lessening the potential for impacts to occur (including technology to reduce adverse sound levels), to monitor and test the effectiveness of implemented mitigation, and to adjust mitigation based on these monitoring results.

7 Summary

Protecting the environment while ensuring the safe development of our Nation's offshore energy (from both renewable and traditional sources) and marine mineral resources is a critical part of the mission of the BOEMRE. The BOEMRE, as with all federal agencies, must consider the potential environmental impacts for every decision made. This includes understanding the potential for and degree of adverse effects that may result from the introduction of anthropogenic noise into the marine environment from BOEMRE-regulated industry sources. The ESP and the TAR Program are integral in helping the BOEMRE achieve this mission because the strength and quality of the environmental decision making can only be as good as the science supporting it. Cumulatively, these research programs help the BOEMRE pursue an adaptive and ecosystem-based approach to its stewardship responsibilities.

Regulation of Sound in the Ocean: Recent and Future Possible Changes

Mark L. Tasker

1 Introduction

The regulation of sound in the ocean has proved difficult to achieve globally. Broadly, attempts have been made to ensure that sound does not harm marine life, but this has proved difficult to achieve primarily because there is very limited knowledge of which sounds or which features of sound can cause adverse effects. In some jurisdictions, the focus has been to avoid effects on animals that are particularly protected (e.g., cetaceans or some fish). Although this may be useful in regulating some sounds, there are limits to regulatory effectiveness because not all biota are considered and, additionally, key aspects of the biology of some protected biota (e.g., special and temporal distribution in the oceans) are not well known. In addition, regulation tends to be based around peak sound pressures or, more recently, around a total sound level over a period of time (sound exposure level), whereas adverse effects may derive from some other aspect of noise such as rise time or cumulative behavioral effects over a long period of time.

2 Recent Developments

To partially address some of the shortcomings of existing regulation, new legislation has been introduced in Europe and interpretation of the legislation has been clarified. In the United States, there have been some statements of intent but no actual changes yet.

2.1 European Union Marine Strategy Framework Directive

In the European Union (EU), the Marine Strategy Framework Directive (2008/56/EC) aims to improve the condition of all of Europe's seas and ensure that human usage of these seas is sustainable. The Directive requires EU member states to set a series of objectives for 11 Descriptors of

M.L. Tasker (✉)
Joint Nature Conservation Committee (JNCC), Inverdee House, Aberdeen AB11 9QA, UK
e-mail: mark.tasker@jncc.gov.uk

A.N. Popper and A. Hawkins (eds.), *The Effects of Noise on Aquatic Life*,
Advances in Experimental Medicine and Biology 730, DOI 10.1007/978-1-4419-7311-5_129,

Environmental Status. One of these descriptors, *Introduction of Energy, Including Underwater Noise, Is at Levels That Do Not Adversely Affect the Marine Environment*, concerns underwater noise. A scientific group, established by the European Commission to recommend ways forward, focused on sounds that affect relatively broad areas of the marine environment and developed three possible indicators of underwater sound.

The first of these indicators concerned low- and midfrequency impulsive sounds. The production of these sounds is at present regulated under environmental impact legislation. The amplitude of some of these sounds is such that a behavioral disturbance will occur to cetaceans and fish (and should an individual be very close to the source, physiological damage is likely also). Some of these high-amplitude sounds seem unavoidable for certain activities (e.g., geological survey using sound sources) and, therefore, some behavioral disturbance seems inevitable. Although little is known about the effect of many changes in behavior on the vital life functions of these animals (and therefore even less about possible population level effects), it seems a sensible approach to aim to reduce the amplitude of peak sound pressures or reduce sound exposure levels for the highest level sounds as well as attempt to limit possible cumulative effects through time. The group therefore suggested an indicator text: "The proportion of days within a calendar year, over areas of 15′ N × 15′ E/W in which anthropogenic sound sources exceed either of two levels, 183 dB re 1 μPa^2-s (i.e., measured as sound exposure level [SEL]) or 224 dB re 1 μPa peak (i.e., measured as peak sound pressure level) when extrapolated to 1 m, measured over the frequency band 10 Hz to 10 kHz." The two sound level thresholds were based on the work of Southall et al. (2007) and are the received levels likely to cause adverse physiological effects in cetaceans. To manage such an indicator, a "sound register" would need to be created that would enable the mapping of high-amplitude impulsive sounds in advance of them occurring; this would be based on the environmental impact assessments (or equivalent) that are mandatory in Europe for virtually all activities that might generate such sounds. Management of such sounds could then be through reducing (perhaps through mitigation) the sound level below the threshold or by setting an upper proportion of days in a year that such sounds could occur in an area of 15′ N × 15′ E/W.

After publication of the group's suggestion, a process involving the European Commission and EU member states led to the Commission to remove many of the suggested levels and suggest a draft indicator that at the time of writing (early May 2010) reads: "Proportion of days and their distribution within a calendar year, over areas of a determined surface and their spatial distribution, in which anthropogenic sound sources after mitigation exceed high levels measured as SEL or as peak sound pressure level (in dB re 1 μPa^2 at 1 m), measured over the frequency band 10 Hz to 10 kHz." There are some obvious faults in this (not least in the units!), and these are likely to be corrected before the proposal is voted on in June 2010. Should this (amended) proposal be accepted, it is not clear how "high levels" and "area of a determined surface" might be defined, but it seems likely that this will be left to EU member states working together, perhaps with some European Commission guidance.

A second indicator suggested by the group addressed high-frequency impulsive sounds. These sounds have been increasing in prevalence and derive particularly from certain navigational sonar often used on recreational vessels. It appears that in some cases, where such sonar is fitted to a vessel, it is not possible to turn it off, even if an operator wished to. The wording of the suggested indicator was "The total number of vessels that are equipped with sonar systems generating sonar pulses below 200 kHz should decrease by at least *x*% per year starting in [2012]." The group understood that pulses above 200 kHz would still be useable for navigation (several sonar systems already work at these frequencies) and these high frequencies would be above the hearing limits of marine mammals, in particular, harbor porpoise. Following the European Commission/EU member state process described above, this indicator has been dropped entirely.

The third indicator suggested by the group addressed low-frequency ambient noise levels. Low-frequency sound is increasing, at least in some parts of the oceans (McDonald et al. 2006).

These sounds are believed to mask low-frequency sounds used by whales (Clark et al. 2009). The group suggested that a suitable indicator might be, "The ambient noise level measured by statistical representative sets of observation stations in Regional Seas where noise within the 1/3 octave bands 63 and 125 Hz (centre frequency) should not exceed the baseline values of year [2012] or 100 dB (re 1 μPa root mean square [RMS]; average noise level in these octave bands over a year)." This indicator would be based on direct independent measurements. The choice of representative sets of observation stations would be left to EU member states working together and should benefit from existing networks of underwater observatories (e.g., European Seas Observatory Network [ESONET]). The choice of these octave bands is for signatures of anthropogenic noise that avoid most naturally generated sources (see Wenz 1962). The proposed threshold (100 dB re 1 μPa in the band) is based on being 10 log B above the current known maximum to take into account the fact that the measurement is integrated on one octave. The European Commission/EU member state process has led to the (early May) draft reading, "Trends in the ambient noise level within the 1/3 octave bands 63 and 125 Hz (centre frequency) (re 1 μPa RMS; average noise level in these octave bands over a year) measured by a statistically representative set of observation stations and with the use of models if appropriate."

2.2 *Interpretation of Existing European Directives*

Article 12 of the European Directive 92/43/EEC (the Habitats Directive) requires that certain wild animals including cetaceans, turtles, and Atlantic sturgeon are "strictly protected." Strict protection has recently been more certainly defined to include deliberate disturbance and injuring/killing as offenses. Following recent court cases, deliberate actions in this context include those where a person "consciously accepts the foreseeable results of his action." In other words, if a disturbance or injury is foreseeable, then there is a risk of an offense.

This redefinition of the offense has led the United Kingdom's Joint Nature Conservation Committee to draw up guidance that should be published in the near future to help developers, regulators, and courts assess 1) the likelihood of an offense being committed; 2) how this can be avoided; and 3) if it can't be avoided, the conditions under which the activity could go ahead under license. The likelihood of an activity resulting in injury or disturbance to one of the protected species will very much depend on the characteristics of the activity and the environment and the species concerned. Pursuing mitigation measures, alternative methods, locations, and/or times for carrying out proposed activities might, in some cases, be sufficient to reduce the risk of causing offense to negligible levels.

In relation to injury, the guidance proposes that a permanent shift in the hearing thresholds (PTS) of one of the protected species would constitute an injury offense and suggests the use of the Southall et al.'s (2007) precautionary criteria for injury. These criteria are based on quantitative sound level and exposure thresholds over which PTS onset could occur for different groups of species. If it is likely that one of the protected species could become exposed to sound at or above the levels proposed by Southall et al. (2007), then there is a risk that an injury offense could occur. The risk of an injury offense will be higher in areas where these species occur frequently and/or in high densities.

That the disturbance offense catches disturbance is significant because it is likely to be detrimental to the animals of one of the protected species or significantly affect their local abundance or distribution. Sporadic disturbances without any likely negative impact on the animals, i.e., trivial disturbances such as those resulting in short-term behavioral reactions, are not likely to result in an offense being committed. It is difficult to prescribe quantitative sound level criteria for the onset of disturbance because the level of sound received by the animal is not the only issue in determining its response and its significance. A disturbance offense is more likely when an activity causes persistent noise in an area for long periods of time. This guidance proposes that a disturbance offense is more likely to occur when there is a risk of 1) animals incurring a sustained or chronic disruption of behavior scoring 5 or more in the Southall et al. (2007) behavioral response severity scale or 2) animals being displaced from the area, with redistribution significantly different from natural variation. The risk of a disturbance offense being committed will therefore exist if there is a sustained noise in an area and/or a chronic noise exposure as a result of an activity. This risk is likely to be higher in regions where there are semiresident populations or where animals occur frequently and in high densities.

References

Clark CW, Ellison WT, Southall BL, Hatch L, van Parijs SM, Frankel A, Ponirakis D (2009) Acoustic masking in marine ecosystems: Intuitions, analyses, and implication. Mar Ecol Prog Ser 395:201–222.

McDonald MA, Hildebrand JA, Wiggins SM (2006) Increases in deep ocean ambient noise in the northeast Pacific west of San Nicolas Island, California. J Acoust Soc Am 120:711–718.

Southall BL, Bowles AE, Ellison WT, Finneran JJ, Gentry RL, Greene CR Jr, Kastak D, Ketten DR, Miller JH, Nachtigall PE, Richardson WJ, Thomas JA, Tyack PL (2007) Marine mammal noise exposure criteria: Initial scientific recommendations. Aquat Mamm 33:411–522.

Wenz GM 1962. Acoustic ambient noise in the ocean: spectra and sources. J Acoust Soc Am 34:1936–1956.

Marine Bioacoustics and the Regulation of Fisheries

Yury A. Kuznetsov, Lev N. Bocharov, Valeriy N. Akulin, and Michail Y. Kuznetsov

According to the UN General Assembly resolution on sustainable fisheries (2007) "... fish stocks in many parts of the world are overfished or subject to sparsely regulated and heavy fishing effort as a result of, among other things, illegal, unreported and unregulated fisheries, inadequate monitoring, and enforcement actions...." To deal with these problems, a set of principles for ecosystem management has been recommended by the Food and Agriculture Organization (FAO) of the United Nations in the form of a Code of Conduct for Responsible Fisheries. In terms of their impact on biota, fisheries can change the structure and function of marine ecosystems significantly. However, the theoretical rules for fisheries management have been formulated for the exploitation of homogeneous, ideal populations. They are not suitable for ensuring the sustainable development of the oceans and conservation of aquatic ecosystems in a wider sense.

Trawling accounts for 80% of the total world's fish catches. Assessing the condition and state of the fish stocks is done mainly by trawl and acoustic-trawl surveys. Because fish stocks show complex behavior and are subject to environmental change, simple trawl surveys cannot provide a complete picture of stock status. As a consequence, stochastic models describing the dynamics of exploited fish stocks are used. They often involve a mechanistic representation of the catch parameters (fishing effort, catch zone, coefficient of catchability). Fishing effort may be defined in terms of the volume of water sieved, the catch zone by the area swept by the trawl, and the coefficient of catchability, i.e., the proportion of fish retained from the volume of water sieved. Simple algorithms are used to estimate fish density on the ground, but they are often subject to great uncertainty. Canadian researcher Peter Larkin (1977), during the development of fishery theory, proclaimed an epitaph for the concept of maximum sustained yield as it props up the economy of a fishery at the expense of unreported discharge of catches and misrepresentation of statistics. The concept is ineffective in counteracting the disastrous situation in modern fisheries, i.e., illegal, unreported, and unregulated fisheries.

Fish themselves show species- and size-specific differences in behavior. Fishing nets have their own technical features and hydrodynamic behavior. Together these factors make an objective assessment of the fish stocks almost impossible. In considering the formation of fish aggregations and their behavior in the vicinity of the vessel and trawl, the acoustic field of the ship is considered

Y.A. Kuznetsov (✉)
Far East State Technical Institute of Fishery and Fish Industry, 690650 Vladivostok, Russia
e-mail: imf.dalrybvtuz@mail.ru

L.N. Bocharov • V.N. Akulin • M.Y. Kuznetsov
Pacific Scientific Research Fisheries Center, 690950 Vladivostok, Russia
e-mail: tinro@tinro.ru

A.N. Popper and A. Hawkins (eds.), *The Effects of Noise on Aquatic Life*,
Advances in Experimental Medicine and Biology 730, DOI 10.1007/978-1-4419-7311-5_130,

Fig. 1 Distance of reaction of *Clupea pallasii* to underwater noise of vessels

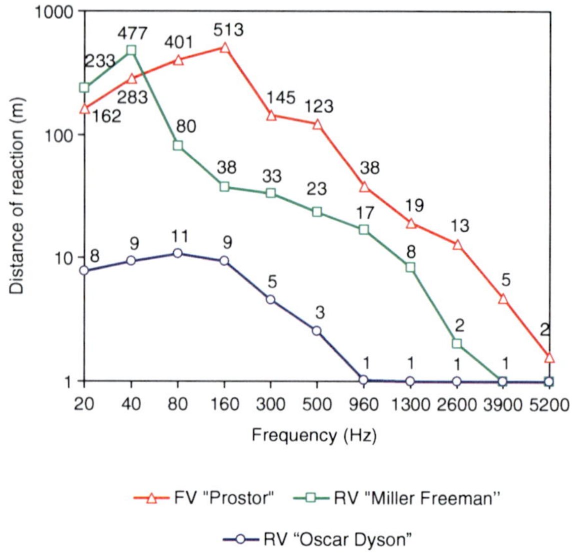

to be an important factor. Vabø et al. (2002), Mitson and Knudsen (2003), and more recently De Robertis and Wilson (2010), working with *Clupea harengus* and *Theragra chalcogramma*, have shown that fish schools dive and disperse at distances of 270-500 m from approaching vessels. The more usual shape and density of the fish schools can recover shortly after the vessel has passed. The characteristics of hearing in the fish and their swimming abilities affect the response to passage of the ship and trawl significantly. However, these features are not used in assessing the fisheries.

We carried out research work on hearing and locomotion in some Pacific species. Our research has shown that the reactions of fish to various stimuli with different spectral-energy and temporal characteristics vary significantly in terms of the direction moved and the duration and the latency of the response (Kuznetsov and Kuznetsov 2007). The frequency band in which fishing vessels create noise fits with the hearing sensitivity curve of the majority of species. We cannot yet predict the distance of the response of different species to the noise field generated by an approaching ship. Figure 1 demonstrates the estimated distance of reaction by *Clupea pallasii* (pacific herring) to a noisy vessel. Discrepancies in the assessment of the distance arises from the difference in noise levels emitted by different ships. For example, the noise-reduced research vessel *Oscar Dyson* (USA) results in reactions at much closer distances than the research vessel *Miller Freeman* (De Robertis and Wilson 2010). The noise of a large-capacity fishing vessel *Prostor* (Russian vessel) has a great effect on the behavior of herring.

As an alternative to the traditional approach, we consider new methods for examining marine ecosystems. Hydroacoustic methods and the tools of marine bioacoustics allow us to lift the "veil of secrecy" and enable us to discover the ways in which aquatic organisms interact with the natural world and also with the new stimuli provided by ships and fishing gears (Kuznetsov and Kuznetsov 2007).

To manage fisheries properly, we need to know how many fish there are. Our data allow us to predict the activity of fish close to vessels from a knowledge of the acoustic field and the swimming abilities of fish. It allows us to reduce any errors in the processes of echo-integration and estimation of the volume and composition of the catch through acoustic surveys and trawl surveys. We are able to compensate for the effects of noise. By measuring the characteristics of different vessels in a calibrated test area, we are able to derive information on the likely reactions of the fish. The data can be compiled together with information on vessel activities derived from satellite monitoring systems (vessel monitoring system [VMS]) and then delivered to users by means of a user friendly interface. Existing VMSs (e.g., in Russia) provide information on the positioning of ships, daily reports on catches, and the pattern of operating activities (Koshkareva et al. 2005).

With modern VMSs, users will be more demanding in terms of the quality of information. Shipowners, fishing operators onboard the vessels, fisheries institutes, control authorities, and fishery managers are interested in obtaining more detailed information on catches. Their need for information requires additional analysis of the real mechanisms controlling catches and bycatches. Techniques will need to be standardized in accordance with updated technical regulations. The final result should be stable instruments for the control of fisheries. These methods should enable us to uphold the principles set out by the FAO.

References

De Robertis A, Wilson CD (2010) Silent ships sometimes do encounter more fish. 2. Concurrent echosounder observations from a free-drifting buoy and vessels. ICES J Mar Sci 67:996-1003. Doi:10.1093/icesjms/fsp301.

Koshkareva LA, Obraztsov FA, et al. (2005) Monitoring of fishery (in Russian). Petropavlovsk-Kamchatski, Russian Federation.

Kuznetsov YA, Kuznetsov MY (2007) Substantiation and development of methods and means of fishery bioacoustics (in Russian). Dalrybvtuz, Vladivostok, Russian Federation.

Larkin PA (1977) An epitaph for the concept of maximum sustained yield. Trans Am Fish Soc 106:1–11.

Mitson RB, Knudsen HP (2003) Causes and effects of underwater noise on fish abundance estimation. Aquat Living Resour 16:255–263.

Vabø, R, Olsen K, Huse I (2002) The effect of vessel avoidance of wintering Norwegian spring-spawning herring. Fish Res 58:59–77.

Update From the OEER Association on Research Findings Associated With the Assessment of the Impacts of Seismic Exploration on Marine Invertebrates

Jennifer M. Matthews

1 Introduction

If asked what you know about Nova Scotia, you might mention a diverse landscape that is largely resource based, surrounded by the Atlantic Ocean. You most likely would not refer to the province's emerging offshore oil and gas industry. You almost certainly would not mention Nova Scotia as an incubator of scientific studies on the effects of seismic exploration on marine life; however, this is increasingly becoming the case. Nova Scotia's offshore has a rich, diverse marine life, potential as a major petroleum producer, and a commitment to addressing the scientific gaps on what is known and unknown about the possible impacts on marine species during offshore exploration. Nova Scotia's Offshore Energy Environmental Research (OEER) Association is leading this effort. The OEER Association was established by the government of Nova Scotia in 2007 as an independent, not-for-profit research body, strengthened by the province's first-rate universities and research communities and committed to collecting the very best research to benefit government policy makers and industry stakeholders. The association has a particular interest in seismic-invertebrate research.

This paper discusses the evolution of seismic-invertebrate research in Nova Scotia, where the OEER Association currently stands in terms of an integrated, science-based approach to seismic-invertebrate research, and where the Association is headed.

2 Evolution of OEER Association's Seismic-Invertebrate Research

Nova Scotia has a rich fishing tradition and a more recent petroleum industry that has become one of the province's leading economic drivers. The existence of both of these industries in this small region makes it critical to use scientific study, rigor, and evidence to determine potential impacts of seismic energy on marine invertebrates. In September 2007, the OEER Association held a workshop

J.M. Matthews (✉)
Offshore Energy Environmental Research Association, Halifax, NS B3J 3P7, Canada
e-mail: j.matthews@offshoreenergyresearch.ca

A.N. Popper and A. Hawkins (eds.), *The Effects of Noise on Aquatic Life*,
Advances in Experimental Medicine and Biology 730, DOI 10.1007/978-1-4419-7311-5_131,

with more than 40 stakeholders from academia, government, fishing, exploration, research, and consulting. Together, this group considered existing research and set key priorities for future offshore research. The group debated three major questions. 1) What are the current issues of concern to stakeholders regarding seismic exploration and its impacts on invertebrates? 2) What research and capacity development activities should be supported to meet the knowledge needs of stakeholders? 3) What is the best operational framework to support relevant research? At the close of the workshop, five research priorities emerged: noise sources, behavioral issues, biological functions, population impacts, and socioeconomic and environmental outcomes.

3 Seismic-Invertebrate Research in 2010

A request for proposals (RFP) entitled *Assessing the Impacts of Seismic Exploration on Marine Invertebrates* was released in July 2009. The RFP builds on previous experience and will help develop qualitative and/or quantitative proof to confirm or disprove potential negative impacts of seismic exploration on invertebrates. The OEER Association and external peer reviewers examined submissions and in February 2010 awarded funding to three research groups. These studies will be completed in 2011.

3.1 Dr. Geoffrey Lee-Dadswell, Cape Breton University

Title: *Physics of the Interaction Between a Crab and a Seismic Test Pulse – Stage 3: Continued Development of a Mathematical Model and Testing of Model via Simulation*

Dr. Lee-Dadswell's project builds on earlier phases of his original research that has been funded by the OEER Association since 2007. This is a multiphased study examining how seismic energy interacts with *Chinocetes opilio* (snow crab). Dr. Lee-Dadswell is currently completing the third phase of his research project, building on his previous work by improving the mathematical model previously developed and producing a modeling software package for use by other researchers. The main goal is to remove several assumptions and approximations in previous mathematical modeling to more accurately model the motion of *Chinocetes opilio* tissue when exposed to seismic energy from an air gun pulse.

3.2 Dr. Chris Purcell, Defence Research and Development Canada (DRDC)

Title: *Feasibility of a Marine Vibroseis System to Minimize Potential Impacts of Seismic Surveying on Commercial Marine Invertebrates*

Dr. Purcell and his team from DRDC as well as consultants from Hurley Environment Ltd. and GeoSpectrum Technologies Inc. are conducting a study to investigate the feasibility of using a marine vibroseis system to conduct seismic testing. The study focuses on forming a hypothesis and designing experiments to determine if the impact of seismic energy is reduced by using a marine vibroseis system with a lower peak intensity and longer pulse duration. The team will develop specifications for a system that could replace conventional air guns and investigate the feasibility of using the modular projector system (MPS) as a marine vibroseis source.

3.3 Dr. Mikio Moriyasu et al., Fisheries and Oceans Canada (DFO)

Title: *Establishment of Baseline Biological Data on Snow Crab (*Chinocetes opilio*) Offshore Cape Breton for Future Assessment of Potential Impacts of Seismic Noise on Snow Crab*

Dr. Moriyasu leads a team from the DFO along with researchers from Threshold Associates, Atlantic Veterinary College (University of Prince Edward Island), University of New Brunswick, Dalhousie University, and Gulf Aquarium and Marine Station Cooperative. This team is conducting a study that aims to improve understanding of the fundamental biological characteristics of *Chinocetes opilio* in their natural habitat and the physiological effects of handling. This is a multi-phased research project.

4 OEER Association and Seismic-Invertebrate Research Beyond 2010

In the years ahead, the OEER Association will continue to focus on its commitment to strong science and research, with the goal of filling knowledge gaps. With a solid foundation and growing expertise, Nova Scotia's offshore seismic-invertebrate research will continue to provide significant knowledge and insight for coastal stakeholders in Nova Scotia and abroad.

Managing Underwater Noise in European Waters: Implementing the Marine Strategy Framework Directive

Mark Tasker, Mats Amundin, Michel Andre, Anthony D. Hawkins, William Lang, Thomas Merck, Amy Scholik-Schlomer, Jonas Teilmann, Frank Thomsen, Stefanie Werner, and Manell Zakharia

1 Introduction

The European Union's (EU) Marine Strategy Framework Directive (2008/56/EC) aims to improve the condition of all Europe's seas and ensure that human usage of these seas is sustainable. The directive will work by requiring EU member states to set a series of objectives for 11 Descriptors of Environmental Status. One of these descriptors, *Introduction of Energy, Including Underwater Noise, Is at Levels That Do Not Adversely Affect the Marine Environment*, concerns underwater noise.

M. Tasker (✉)
Joint Nature Conservation Committee, Inverdee House, Aberdeen AB11 9QA, UK
e-mail: mark.tasker@jncc.gov.uk

M. Amundin
Kolmårdens Djur and Naturpark, 618 92 Kolmården, Sweden

M. Andre
Laboratori d'Aplicacions Bioacustiques, Universitat Politecnica de Catalunya, 08800 Vilanova i la Geltrú, Barcelona, Spain

A.D. Hawkins
Loughine Limited, Kincraig, Blairs, Aberdeen AB12 5YT, UK

W. Lang
21 Kingfisher Drive, Mandeville, LA 70448, USA

T. Merck
Federal Agency for Nature Conservation, Isle of Vilm, 18581 Putbus, Germany

A. Scholik-Schlomer
Office of Protected Resources, National Marine Fisheries Service, National Oceanic and Atmospheric Administration, Silver Spring, MD 20910, USA

J. Teilmann
National Environmental Research Institute, University of Aarhus, DK-4000 Roskilde, Denmark

F. Thomsen
Cefas, Lowestoft, Suffolk NR33 0HT, UK

S. Werner
Federal Environment Agency, 06844 Dessau-Rosslau, Germany

M. Zakharia
North Atlantic Treaty Organization Undersea Research Centre, 19126 La Spezia, Italy

A.N. Popper and A. Hawkins (eds.), *The Effects of Noise on Aquatic Life*, Advances in Experimental Medicine and Biology 730, DOI 10.1007/978-1-4419-7311-5_132,

A task group consisting of the authors was established to recommend to the European Commission indicators that might be used by member states to define good environmental status (GES) under this descriptor. The task group focused on sounds that affect relatively broad areas of the marine environment and developed three possible indicators of underwater sound.

2 Recommendations to the European Commission

2.1 Indicator for Low- and Midfrequency Impulsive Sounds

High-amplitude low- and midfrequency impulsive anthropogenic sounds (e.g., impact pile driving, seismic surveys, explosions, and some sonar systems) are those that have caused the most public concern, particularly in relation to perceived negative effects on marine mammals and fish. Laboratory studies have found both physiological and behavioral effects in a variety of marine organisms, whereas field studies have shown behavioral disturbance and in some cases death (physiological effects are difficult to study in the field). In principle, sound input is likely to have greater adverse effects at higher sound amplitudes and with a greater number of inputs (persistence). Lower frequency sounds will affect a wider area but may not be detectable by all organisms and therefore be less likely to have an adverse effect. The following indicator was proposed as a way of geographically quantifying the occurrence of these sounds: the proportion of days within a calendar year over areas 15′ N × 15′ E/W in which anthropogenic sound sources exceeded either of 2 levels, 183 dB re 1 μPa^2-s (i.e., measured as sound exposure level [SEL]) or 224 dB re 1 μPa peak (i.e., measured as peak sound pressure level) when extrapolated to 1 m, measured over the frequency band of 10 Hz to 10 kHz.

Recording would be on the basis of subdivisions of European seas. The indicator is focused on those impulsive noise sources that are most likely to have adverse effects (as listed above). Most sources would be quantifiable from either relevant statutory impact assessments or reports from activities rather than on direct independent measurement. The proportion of days would be set by member states and could be based on a review of relevant activities in the immediate past and on their view on sustainable impact. It is up to member states to evaluate noise distribution and occurrence and whether or not the seas are in GES. The size of the grid rectangle was chosen as a compromise. An index sensitive to small changes in activity would have small rectangles while large rectangles are likely to be administratively easier to use. The choice of frequency bandwidth (10 Hz to 10 kHz) is based on the observation that sounds at higher frequencies do not propagate as far as sounds within this frequency band. This choice of bandwidth also excludes most depth-finding and fishery sonar systems. The task group recommended that all values be reviewed in the future in the light of any new scientific publications.

2.2 Indicator for High-Frequency Impulsive sounds

Depth-sounding sonar systems on small vessels typically use frequencies between 50 and 200 kHz. Sonar usage, particularly on leisure boats, is increasing and is unregulated. Some marine mammals use frequencies up to ~180 kHz for communication, and thus there is an overlap in frequency usage. There has been little research on the effects of these sonar systems and the scientific evidence for adverse effects is limited. However, the sounds are similar to those used in acoustic alarms (pingers) that are designed to scare small cetaceans from gill and tangle nets and can therefore have a potential to cause adverse effects for those animals using high-frequency sound. A precautionary approach would be to reduce the usage of sonar systems working at frequencies below 200 kHz. There are potential navigational concerns, but because some existing sonar systems use these high frequencies, these concerns must be minor. A possible initial indicator would be that the total number of vessels that are equipped with sonar systems generating sonar pulses below 200 kHz should decrease by at least $x\%$ per year starting in 2012. This indicator does not include a measure of the use of small vessels or the use of sonar on them because these would be virtually impossible to monitor; the number of vessels that have such sonar systems should be a sufficient proxy for the use of these sonar systems. The target percentage decrease (x) in usage would be set by member states depending on how rapidly a reduction was deemed necessary.

2.3 Indicator for Low-Frequency Continuous Sound

Ambient noise is defined as background noise without distinguishable sound sources. It includes natural (biological and physical processes) and anthropogenic sounds. Research has shown increases in ambient noise levels in the past 50 years, mostly due to shipping activity. This increase might result in the masking of biological relevant signals (e.g., communication calls in marine mammals and fish), considerably reducing the range over which individuals are able to exchange information. It also has been demonstrated that marine mammals can alter their communication signals in noisy environments. This response may suffice in moderate situations, but at some higher threshold, adverse consequences seem possible. It is further likely that prolonged and/or intense exposure to increased ambient noise leads to physiological and behavioral stress. Thus chronic exposure to noise can impair important biological functions and may lead to consequences that are as severe as those induced by acute exposure. A possible initial indicator would be that the ambient noise level measured by statistically representative sets of observation stations in regional seas where noise within the one-third octave bands of 63 and 125 Hz (center frequency) should not exceed the baseline values of year (2012) or 100 dB (re 1 μPa root mean square [RMS]; the average noise level in these octave bands over a year).

This indicator would be based on direct independent measurements. The choice of representative sets of observation stations is left to the member states working together and should benefit from existing networks of underwater observatories. The choice of these octave bands is on the basis of scientifically justifiable signatures of anthropogenic noise that avoids most naturally generated sources. The baseline year would be set when the observatory system for a regional sea is established while the suggested cap on ambient noise is suggested to avoid ambient noise levels that are likely to be harmful. Although shipping activities are forecast to increase, the ability to reduce noise in new ship designs can make this possible.

3 What Happened Next

The European Commission has used our work to suggest that indicators 1 and 3 be used by member states to assess GES. It is not known why indicator 2 was not adopted. The commission decided not to be prescriptive of the areas to be used in assessing indicator 1 and also to open a debate on the decibel values suggested in that indicator. Further developments and decisions will occur over the next years.

Part IX
Monitoring, Management, and Mitigation

Acoustic-Based Cetacean Detection in Irish Waters

Eugene McKeown

1 Introduction

The south and west coasts of Ireland are one of the best whale-watching areas in Europe. Twenty-four different cetacean species have been recorded in Irish waters (Wilson and Berrow 2006); arguably, Ireland has more marine mammal than native land mammal species. We had an active whaling industry in the early 1900s, with an average of 100 large whales killed during each year of operation. In 1991, Ireland became the first country in Europe to declare its waters a marine mammal sanctuary, protecting pinnipeds and cetacea up to 320 km off the Irish Coast.

Acoustic detection of cetaceans has been carried out in Ireland since the late 1980s, and in recent years, there has been an increased emphasis on the acoustic detection of cetaceans for mitigation purposes because of an increase in seismic exploration, investment in offshore renewable energy, and significant port and infrastructure development being carried out. The European Union (EU) Marine Strategy Framework Directive (2008/56/EC) has defined anthropogenic noise as "pollution" and requires objective measurement and reporting of the environmental state of Irish waters.

The National Parks and Wildlife Service (NPWS) has, so far, designated three special areas of conservation under the EU Habitats Directive (92/43/EEC) for cetacean protection. Projects located in these areas require environmental impact studies and appropriate assessments to be carried out at the planning stage. These surveys, in addition to research funded by the Marine Institute and the NPWS, are significant sources of acoustic data. Biospheric Engineering Ltd. has carried out acoustic surveys in inshore waters since 2004, including recording some species for the first time in Irish inshore waters.

2 Irish Waters

The Republic of Ireland extends to over 70,000 km^2 of land. Our waters, however, are much larger in extent. Ireland's offshore territory is currently defined by the 200-nautical mile limit (426,872 km^2), with the additional successful United Nations Convention on the Law of the Sea (UNCLOS) submission claim of porcupine abyssal plain adding a further 39,495 km^2. There are currently Irish

E. McKeown (✉)
Biospheric Engineering Limited, Bearna, County Galway, Ireland
e-mail: eugene@biospheric.ie

A.N. Popper and A. Hawkins (eds.), *The Effects of Noise on Aquatic Life*,
Advances in Experimental Medicine and Biology 730, DOI 10.1007/978-1-4419-7311-5_133,

claims submitted to UNCLOS for the Hatton-Rockall area and Celtic Sea and Bay of Biscay area for a further 457,977 km^2, a combined total of almost 13 times the land area!

This extensive portion of the Northeast Atlantic features both deep-ocean and coastal habitats, including abyssal plains over 4,000 m deep, deep-water canyon systems, and extensive continental shelf waters less than 200 m deep. The most productive areas appear to be along the continental shelf edge and the inshore waters. The continental shelf edge is located ~100 km from the coast at the southwest and northwest corners of the island. The North Atlantic drift and prevailing winds make the island a "lee" shore as well as being a waypoint on north-south migration in the Northeast Atlantic. This unique set of circumstances provides almost year-round cetacean activity, weather permitting!

3 Irish Cetacean Species

Irish cetacean species include both *Balaenoptera musculus* (blue whale) and *Phocoena phocoena.* Our waters include all the usual inshore species in addition to deep-diving ocean species such as *Physter macrocephalus* (sperm whale), *Globicephala melas* (pilot whale), and beaked whales such as *Ziphius cavirostris* (Cuvier's beaked whale) and *Mesoplodon mirus* (True's beaked whale).

The Irish Whale and Dolphin Group (IWDG) maintains a database of sightings averaging ~2,000 sightings/yr, the vast majority of which are recorded inshore or from headlands. The most common species sighted are *Delphinus delphis* (common dolphin) followed by *Phocoena phocoena*, *Tursiops truncatus* (bottlenose dolphin), *Balaenoptera acutorostrata* (minke whale), *Balaenoptera physalus*, *Globicephala melas*, *Grampus griseus* (Risso's dolphin), *Megaptera novaeangliae* (humpback whale), and *Orcinus orca* (killer whale) (Berrow 2001). It is unusual for a whale-watching boat trip to return without a sighting, and an average trip would result in three species being sighted.

4 Engineering Challenges in Irish Waters

A well-known Irish ballad has the line "All God's creatures have a place in the choir, some sing low and some sing higher." The line is particularly apt in the context of acoustic monitoring of cetaceans when fin whale calls at 20 Hz and harbor porpoise echolocation clicks at 130,000 Hz plus have to be considered.

According to the Nyquist theorem, the sampling frequency must be at least twice the highest frequency present in the original signal to avoid aliasing. For *Phocoena phocoena*, therefore, a sampling rate of 300,000 Hz is required. This can only be obtained using specialized analog-to-digital hardware. Although PC sound cards have expanded in frequency range, none has yet reached this capability. A number of devices and hydrophones that can provide this sampling frequency are now commercially available.

It is not unusual in Ireland to encounter *Balaenoptera physalus* in the company of both *Delphinus delphis* and *Phocoena phocoena* feeding on Clupea harengus (herring) and *Sprattus sprattus* (sprat) off the south coast in winter. Similarly, *Phocoena phocoena* can be found in the company of *Balaenoptera acutorostrata* and occasionally *Delphinus delphis* or indeed *Grampus griseus* near the Blasket Islands in summer.

When two species with widely different frequency ranges are encountered, it is desirable but difficult to record both with a single acoustic setup. Such a recording can confirm the presence of each species and in some cases indicate the number of animals present.

The combination of *Delphinus delphis* and *Phocoena phocoena* in one area leads to a second engineering challenge, that of dynamic range. Shy animals such as *Phocoena phocoena* echolocating in directional beam patterns can be difficult to detect, whereas *Delphinus delphis* or *Tursiops truncatus* when close to the hydrophone can easily saturate the hydrophone. The use of 24-bit depth-sampling devices and increasing the preamplifier supply voltage can help to overcome these difficulties.

Irish weather conditions can also present a significant engineering challenge. We have one of the world's greatest wave resources off our west coast, with average wave heights in summer of 2 m. Wave heights of 3-4 m are regularly recorded in winter, with occasional 10-m waves recorded inshore. Annual average wind speeds vary from 3 to 8 m/s (Beaufort force 3 to 5) depending on where you are, the highest values being recorded on the west and north coasts. Ireland is a leader in offshore renewable energy research, and although these are ideal conditions for energy generation, they present a challenge for acoustic recordings due to platform noise.

Close inshore work can be carried out most weeks, but even in the summertime, working more than 1 km from shore requires careful planning and attention to both wind and wave forecasts. Recording from a boat in rough sea states practically prohibits the detection of low-frequency vocalizations. Due to the rocky nature of the coast, techniques such as dropping a hydrophone to the bottom are not practical options. To overcome some of these difficulties, we have been developing low-noise recording techniques, such as separating the hydrophone from the boat, and autonomous recording buoys.

5 Overcoming the Difficulties

Due to significant mixing forces, shallow coastal water does not generally have any significant thermocline. For this reason, hydrophones located 3-5 m below the surface are as sensitive to biological sounds as those located at greater depth. To reduce platform noise, we have developed a technique employing a 150-m string of buoyant rope and buoys along which we run a cable to 2 or 3 hydrophones. The hydrophones are located in an array configuration, with two close together and a third located some distance away. The close pair are used to localize a vocalization, with left/right ambiguity, and the third hydrophone is used to confirm the direction and estimate a range for the signal. The system has been tested in a number of estuaries using calibrated sources and has been found to work well. A separation of 150 m is not sufficient, however, to eliminate all platform noise, and a reliable operation is only possible in good sea states due to uncertainty regarding the precise location of the hydrophones.

In 2008, we began to develop a self-contained recording buoy for short-term deployment. This device was used to measure noise levels from onshore blasting that could be detected in the marine environment and to provide data on attenuation due to distance. The device comprises a hydrophone attached to a small notebook computer inside a waterproof housing on a buoy. The device can be deployed in more severe weather conditions than boat-based measurements. The system works well for short-term deployment during mitigation for blasting or pile driving but does not provide real-time data.

To obtain real-time data, e.g., the presence or absence of cetaceans, we have recently been working with the Marine Institute and IBM to develop a system for deployment as part of the SMARTBAY Galway project. This system uses WiMax (a long-range wireless broadband) connection to relay the signal back to shore. Due to bandwidth limitations on the signal over water, it is necessary to employ "smart" monitoring techniques to limit the data to be transmitted. We are using a buffering and envelope detection algorithm to determine when species are present. When detected, data are then transmitted to shore for further analysis. Further developments of the system will include dynamic sampling rate changes and miniaturization in addition to reducing power consumption. The ultimate goal is a network of autonomous buoys capable of relaying real-time acoustic data over a wide area.

6 Conclusions

The variety of species and the challenging marine environment in Irish coastal waters present unique but interesting challenges to monitoring cetacean vocalizations. Developments over the past 20 years have been significant, and new wireless technologies provide a wonderful opportunity to gain real-time data on the marine acoustic environment.

References

Berrow S (2001) Biological diversity of cetaceans (whales, dolphins and porpoises) in Irish waters. In: Nunn J (ed) Marine biodiversity in Ireland and adjacent waters. Estuarine & Coastal Sciences Association Conference, Ulster Museum, 26–27 April 2001, Belfast, Ireland, pp 115–120.

Wilson J, Berrow S (2006) A guide to the identification of the whales and dolphins of Ireland. Irish Whale and Dolphin Group, Kilrush, Ireland.

Real-Time Monitoring of Noise in Cetacean Acoustic Niches

Michel André, Mike van der Schaar, Serge Zaugg, Ludwig Houégnigan, Antonio Sánchez, Alex Mas, and Joan V. Castell

1 Source of Noise Pollution

Sources of sound produced by human activities induce physical, physiological, and behavioral effects on marine fauna (mammals, reptiles, fish, and invertebrates), effects that can be diverse depending on the proximity to the signal source. These impacts include a reduction in the abundance of fish species of up to 50% in zones under exploration, changes in cetacean behavior and migration routes, and a distinct range of physical injuries in both marine vertebrates and invertebrates. There may be further long-term consequences due to chronic exposure, and sound can indirectly affect animals due to changes in the accessibility of prey, which may also suffer the adverse effects of acoustic pollution (Richardson et al. 1995). These damages could significantly impair the conservation of already endangered species that use acoustically contaminated areas for migratory routes, reproduction, and feeding.

For many reasons, evaluating the acoustic impact of artificial sound sources in the marine environment is a complex and expensive proposition. First, we face the relative lack of information on the sound-processing and analysis mechanisms in marine organisms. Although we are capable of cataloging and recording the majority of these signals, we still do not know enough about the important role they play in the balance and development of populations. Second, the possible impact of sound emissions may not only concern auditory reception systems but might also interfere on other sensorial and systemic levels, possibly lethal for the affected animal. Complicating the situation even more is the fact that a prolonged or punctual exposure to a determined noise can have negative short-, medium-, and long-term consequences not immediately observed. The lack of provision and research resources contributes to the greatest difficulty in obtaining objective data that will allow the efficient control of anthropogenic noise in the ocean.

In addition, we find ourselves with a most pressing problem that relates to the homogenization of measurements. At the moment, there is no well-defined protocol for measuring marine acoustic pollution or any agreement on the enunciation of these measurements. Although the effects of noise on the marine environment are increasing, the variability of the available parameters to measure these effects leads to heterogeneous or fragmented results that appear of little use in orientating preventive and precise management actions (André et al. 2010).

M. André (✉) M. van der Schaar • S. Zaugg • L. Houégnigan • A. Sánchez • A. Mas • J.V. Castell
Laboratory of Applied Bioacoustics, Technical University of Catalonia, 08034 Barcelona, Spain
e-mail: michel.andre@upc.edu

A.N. Popper and A. Hawkins (eds.), *The Effects of Noise on Aquatic Life*,
Advances in Experimental Medicine and Biology 730, DOI 10.1007/978-1-4419-7311-5_134,

Finally, most studies lack information on the long-term effects of noise sources on specific populations. There are very few data on current ambient noise levels in most regions and even less historical data. Information on trends is not available for any European waters. According to the Marine Mammal Commission (2007), underwater ambient sound levels will increase over time with more human activity (shipping, offshore construction) in the marine environment. It should be further noted that the potential increase in ambient sound levels will not affect all areas equally but specific regions where offshore activity is high, e.g., some of the exclusive economic zones around northwest Europe (see OSPAR Commission 2009). Potential effects might not be proportionate to pollution levels due to variation in sound propagation and, most importantly, the distribution of marine life that is sensitive to sound.

2 Cetaceans and Acoustic Niches

To some extent, all marine animals, most notably cetaceans, depend on acoustic exchange for a great number of activities and vital behaviors such as communication, geographical orientation, habitat relationships, feeding, and a wide range of endeavors within the broader social group (cohesive action, warnings, and maternal relationships).

The cetacean acoustic capabilities are the result of an evolution of the marine fauna to adapt to the sea habitat, the acoustic aspects of which are of vital importance. This acoustic habitat is nowadays distorted by anthropogenic noise contributions that have direct and indirect effects at individual and population levels.

In particular, with regard to the exchange of information, individuals and, accordingly, populations rely on an acoustic habitat for establishing and maintaining normal interactions. When such a habitat is degraded, acoustic communication may be degraded, with dysfunctional consequences in the social system.

In other words, each species has its specific acoustic niche within a larger acoustic habitat that probably requires its own level of acoustic comfort to behave according to its evolution path. This then leads to the concept of an "acoustic ecology" and the concern for its conservation.

A consideration of the acoustic ecology implies that there are costs associated with the modifications in acoustic habitat (e.g., in the reduction of feeding efficiency, mating success, predator avoidance), and these costs can affect primarily individuals and then populations. Unlike other natural modifications in the habitat, these anthropogenic impacts are both quite recent and relatively fast in an evolutionary time frame relevant to cetacean species adaptations.

3 Monitoring Cetacean Sounds and Anthropogenic Noise

Monitoring cetacean distribution from their sounds in their habitats was initiated in the 1970s. The technique was rapidly applied to tracking whales over large distances. Advances in electronics, computers, and numerical analysis now make passive acoustic monitoring (PAM) technology more accessible and affordable to relatively small research budgets. Various systems have been used, including shore-cabled and radio-linked systems, drifting buoys, and arrays of autonomous recorders for versatile and long-term deployments. The goal of such PAM systems is the continuous mapping of the presence and distribution of whales over ocean basins and assessing their densities. Recently, this analysis was performed in quasi real time. Nevertheless, their performance in effectively accomplishing these tasks depends on the characteristics of the targeted whale calls, the environment, the type of equipment used, and its deployment and configuration (André et al. 2010; Houégnigan et al. 2010; Zaugg et al. 2010). This performance may significantly vary from case to case. The success of PAM first depends on the capacity to isolate the targeted calls from the rest of the acoustic signal in which they are imbedded or mixed, especially for distant sources and low signal-to-noise ratios (SNRs).

Ocean noise level also exhibits considerable variability in space and time in response to fluctuating natural sources such as wind, ice, rain, sounds produced by various organisms, and anthropogenic sources (National Research Council 2003). Anthropogenic sound sources have a broad range of characteristics, including source level, frequency content, duty cycle/pattern of occurrence, and movement (i.e., stationary or mobile). Sound sources can also vary between coastal and open ocean regions. For example, shipping activity as a whole adds a component to ocean basin noise levels, whereas an individual ship can create a dominant but time-limited noise source within a local area. However, as mentioned before, there are almost no data on current ambient noise levels in most regions and even less accessible historical data.

The combination of both requirements (the monitoring of cetacean sounds and the assessment of noise effects) requires finding a protocol that takes into account the acoustic niches of the species and the variety of anthropogenic noise sources.

4 Automated Real-Time Monitoring of Acoustic Events and Noise

PAM has the potential to be implemented continuously and over long time periods, resulting in large and representative datasets. However, this inevitably leads to a high rate of audio data acquisition that could be problematic when the data need to be transmitted, stored, or analyzed. For observatories with a limited power supply, e.g., radio-linked autonomous buoys, transmission, storage, or additional data processing (e.g., automated classification, data compression) have to be optimized, which may imply the loss of potentially interesting information. For cabled observatories where power and communication are not an issue, limitations arise with storage. In any case, the need on one hand for immediate mitigating actions when facing acoustic events that could be harmful to individuals or populations and, on the other hand, the necessity of long-term monitoring of noise call for the development of a robust technique able to provide both historical statistical data on noise and alarms on specific acoustic events, i.e., a fully automated real-time detection and classification system that would be able to provide this information while minimizing technical costs (storage, computation time). The approach proposed here divides the recording bandwidth in frequency bands that cover the acoustic niche of most species and secondly applies to these bands a series of detectors and classifiers (as well as localization and tracking algorithms) that also allow assessment of the short-, medium-, and long-term contributions of noise sources in these acoustic niches (André et al. 2010; Houégnigan et al. 2010; Zaugg et al. 2010).

Figure 1 gives an overview of the system that the Laboratory of Applied Bioacoustics has developed and is currently applying at several underwater observatories in Europe (European Seafloor Observatory Network of Excellence [ESONET]) and North-East Pacific Time-Series Undersea Networked Experiments (NEPTUNE) Canada as well as in radio-linked, stand-alone, and autonomous buoys. The live audio data stream as well as the output of the statistical analysis (noise measurements, acoustic presence of species over time) can be accessed online at http://listentothedeep.com. This contribution to the real-time assessment of noise interactions with cetaceans as well as to the long-term management of anthropogenic sound sources represents a step toward an improvement of the acoustic ecology status for marine organisms.

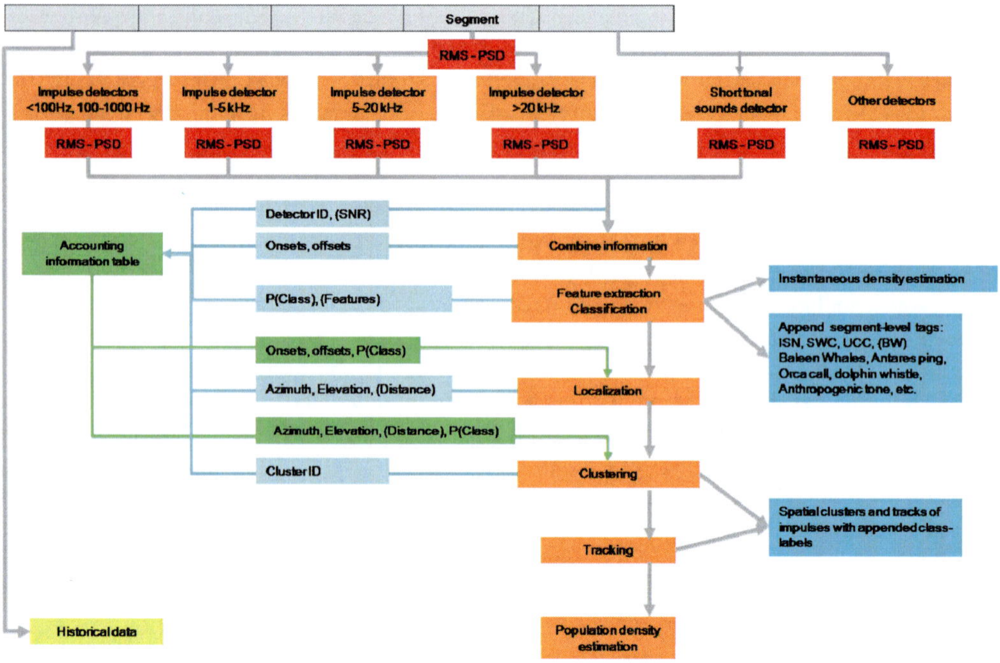

Fig. 1 Overview of the architecture of the acoustic data management. RMS, root mean square; PSD, power spectral density

Acknowledgments This study was funded by the European Commission (EC) under Contract 036851, EC FP6, and Listening to the Deep-Ocean Environment (LIDO) of the European Seafloor Observatory Network of Excellence (ESONET) and is included in the Canadian Advanced Network and Research for Industry and Education (CANARIE), Data from the Deep, Judgment from the Crowd (North-East Pacific Time-Series Undersea Networked Experiments [NEPTUNE] Canada), and Astronomy with a Neutrino Telescope and Abyss environmental RESearch (ANTARES).

References

André M, Morell M, Mas A, Solé M, van der Schaar M, Houégnigan L, Zaugg S, Castell JV, Baquerizo C, Rodríguez Roch L (2010) Best practices in management, assessment and control of underwater noise pollution. CONAT150113NS2008029, Laboratory of Applied Bioacoustics, Technical University of Catalonia, Barcelona, Spain. Available via http://www.lab.upc.es.

Houégnigan L, van der Schaar M, Zaugg S, André M (2010) Space-time and hybrid algorithms for the passive acoustic localisation of sperm whales and vessels. Appl Acoust 71:1000–1010.

Marine Mammal Commission (2007) Marine mammals and noise: A sound approach to research and management. A Report to Congress from the Marine Mammal Commission, Bethesda, MD..

National Research Council (2003) Ocean noise and marine mammals. National Academies Press, Washington, DC.

OSPAR Commission (2009) Overview of the impacts of anthropogenic underwater sound in the marine environment. OSPAR Convention for the Protection of the Marine Environment of the North-East Atlantic, Biodiversity and Ecosystems Series, Publication No. 441/2009.

Richardson WJ, Greene CR Jr, Malme CI, Thomson DH (1995) Marine mammals and noise. Academic Press, San Diego, CA.

Zaugg S, van der Schaar M, Houégnigan L, Gervaise C, André M (2010) Real-time acoustic classification of sperm whale clicks and shipping impulses from deep sea observatories. Appl Acoust 71:1011–1019.

What Does Ecosystem Acoustics Reveal About Marine Mammals in the Bering Sea?

Jennifer L. Miksis-Olds, Jeffrey A. Nystuen, and Susan E. Parks

1 Introduction

The Bering Sea is a highly productive marine ecosystem that supports valuable commercial fisheries, numerous marine mammal populations, and local communities. The ecological dynamics of the eastern Bering Sea shelf are a product of complex oceanographic processes in which physical processes are ultimately linked to biological production. Knowledge of how the ecosystem responds to variability in climatic patterns and events (i.e., local storms) that influence ice cover and other physical-biological interactions will be critical in the future understanding of this important ecosystem. From a bottom-up perspective, the timing, magnitude, and dominant species of phytoplankton and zooplankton may shift in response to reductions in the ice cover. From a top-down perspective, longer periods of open-water availability allows for an increase in marine mammal and human utilization of the habitat area. The indirect impacts of these shifts, particularly by humans, are largely unknown.

Sound is capable of propagating over great distances in the ocean, and loud, low-frequency sources can propagate halfway around the globe given the right conditions (Munk et al. 1994). For this reason, acoustic technology is useful in detecting a variety of environmental signals in the marine environment, including natural physical phenomena, biological response to changing physical conditions, and anthropogenic activities. Natural sources of sound that contribute to ambient noise levels in the Bering Sea include wind, waves, sea ice, precipitation, and marine organisms. Variations in the ambient noise level as a function of frequency at a specific location can change by as much as 10–20 dB from day to day, minute to minute, and even second to second based on variations on the noise sources (Richardson et al. 1995). Noise generated by human activities is both explicitly introduced into the environment for a specific purpose, as in the case of navigational sonar or seismic exploration, or as an indirect by-product of activities such as shipping, construction, and dredging.

Passive acoustic recordings in the marine environment are used to detect, localize, identify, and track sources of sound. By examining the spectral characteristics of ocean sound, it is possible to identify local sound sources, the percentage of time that a particular source is present, and the loudness

J.L. Miksis-Olds (✉) • S.E. Parks
Applied Research Laboratory, The Pennsylvania State University, State College, PA 16801, USA
e-mail: jlm91@psu.edu

J.A. Nystuen
Applied Physics Laboratory, University of Washington, Seattle, WA 98105, USA

A.N. Popper and A. Hawkins (eds.), *The Effects of Noise on Aquatic Life*, Advances in Experimental Medicine and Biology 730, DOI 10.1007/978-1-4419-7311-5_135,

of that source. Sound pressure levels at a given frequency have previously been related to physical environmental parameters affecting biological productivity and include wind speed, sea state, sea ice cracking due to thermal tension, and precipitation (Farmer and Xie 1989; Makris and Dyer 1986; Nystuen et al. 2008; Wenz 1962).

Knowledge about physical and biological parameters in the aquatic environment can also be gained with active acoustics. Active acoustics involves the emission of sound and corresponding reception of echo intensities traveling back to the transducer from targets in the environment. Analysis of received echoes provides information related to ocean depth, surface conditions, ocean temperature, currents, and presence of marine organisms ranging in size from plankton to marine mammals (Medwin and Clay 1998).

The combination of passive and active acoustics provides information about ecosystem dynamics underlying the interactions between physical ocean processes, biological response, and human activities in the Bering Sea. Passive acoustics is used to monitor physical processes and weather as well as the sound produced by marine mammals and humans (i.e., vessels), whereas active acoustics provides a time series of zooplankton and fish density and distribution in the water column. Synoptic measurements such as these allow for the interpretation of acoustic measurements in the context of the larger ecosystem, including multiple trophic levels and the physical environment. Understanding the interactions between different components of the ecosystem is critical in ultimately understanding and predicting the cumulative and synergistic effects of sound on marine mammals as the Arctic continues to change.

2 Acoustic Ecology of the Bering Sea

A combination of active and passive acoustic technology was deployed on National Oceanic and Atmospheric Administration (NOAA) Fisheries Oceanography Coordinated Investigations (FOCI) moorings at two locations (M2 in the southeast and M5 in the central region) along the 70-m isobath of the eastern Bering Sea shelf (Stabeno et al. 2008). Passive aquatic listeners (PALs) were used to record environmental sound levels and transient signals produced by marine mammals, humans, and geophysical processes (Nystuen et al. 2008). Acoustic water column profilers (AWCP; ASL Environmental Sciences, Sidney, BC, Canada) operating at 125, 200, and 460 kHz were deployed on each mooring to record acoustic backscatter associated with fish and zooplankton.

Examples of interannual, seasonal, and spatial variability between and within the two different sites are illustrated in Figure 1. The acoustic soundscape is dominated by geophysical sources during the fall (wind, precipitation) and a combination of marine mammals and ice in the spring (Fig. 1). Sound levels at 1 kHz also dramatically increased in the spring with the presence of vocalizing ice-breeding or associated marine mammals at M5 (Fig. 1c, and d). Ice was not present at M2 in 2009, and no ice-breeding or associated marine mammals were detected at this location. Environmental sound levels in the southeastern Bering Sea are louder during the nonwinter months and contain a greater number of vessel detections than those in the central Bering Sea.

Acoustic backscatter at M5 peaked in the fall and rapidly decreased with the onset of sea ice in winter. A spring bloom after the retreat of ice was observed, but it was less intense and shorter in duration than the fall bloom. Intense vertical migration patterns and greater intensity acoustic backscatter measured in the fall also coincided with the detection of humpback whale and gray whale vocalizations. These species were not detected at this location at other times of the year, and acoustic detections of these species ended in December before the arrival of seasonal ice.

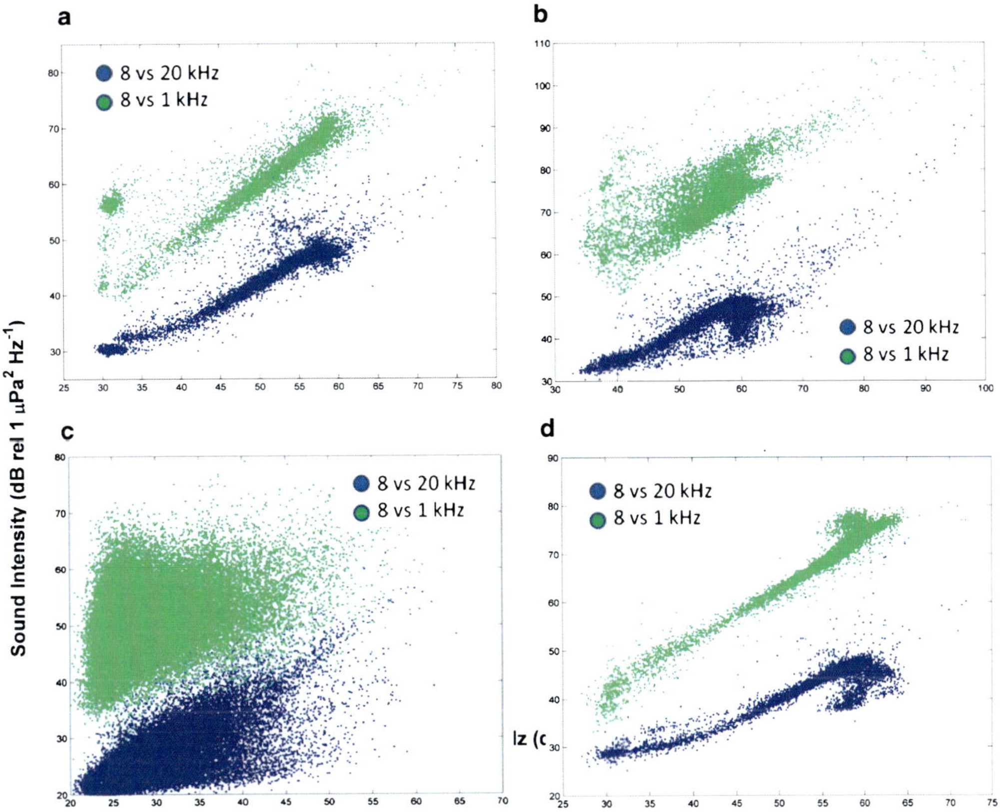

Fig. 1 Soundscapes from M2 in Fall 2004 (**a**), M2 in Fall 2008 (**b**), M5 in Spring 2008 (**c**), and M5 in Fall 2008 (**d**). The *x* axis is the sound level at 8 kHz. The *y*-axis is the sound level at either 20 kHz (blue) or 1 kHz (green). The soundscapes in a, b, and d show a linear pattern indicating an environment dominated by wind. Sound levels increase linearly as wind speed increases. The variability at M2 in Fall 2008 (**b**) is due to the presence of vocalizing marine mammals and ships. The variability at M5 in Spring 2008 (**c**) is due to sounds produced by sea ice and ice-breeding or associated marine mammals. Interannual variability is demonstrated by comparing a and b. Seasonal variability is illustrated by comparing c and d, and spatial variability is observed by comparing b and d

3 Conclusions

Acoustic sampling is playing a vital role in the investigation of how multiple ecological factors influence marine mammal habitat use in the Arctic because acoustic sensors can operate year-round to record environmental signals at times of the year when the area is typically inaccessible to traditional, ship-based sampling methodologies. It is during the ice-covered and transition periods that critical overturning events between the shelf and adjacent waters are likely to occur and impact marine mammals throughout the year; it is also during this period that there is historically a void in observations. Synoptic measurements from active and passive acoustic instruments provide the capability to detect and interpret biophysical interactions associated with the presence of different marine mammal species.

Acoustic time series reveal that upper and lower trophic-level dynamics are tightly coupled to sea ice in the Bering Sea. The integration of active and passive acoustic time series indicates that migrating humpbacks and gray whales are present in the central Bering Sea during the fall when zooplankton levels are elevated but not at other times of the year. Differences in the acoustic soundscapes within and between the two Bering Sea locations highlight the relative absence of human activity and presence of ice-breeding pinnipeds and ice-associated bowhead whales in the central region compared with the southeastern region. The increase in variability as well as elevated sound levels at this location during the winter/spring suggests that the acoustic communication systems of ice-dependent or associated animals evolved in a loud and complex acoustic environment.

Environmental sound levels recorded during the spring at mooring M5 in 2009 may be considered a baseline of "normal" seasonal patterns where sea ice is present for a portion of the year and human activity is minimal. It is highly likely that the acoustic environment of the Bering Sea will be altered as the area experiences warming due to climate change. The Bering Sea has already experienced significant warming (~3°C) over the last several decades that has been closely associated with a marked decrease in sea ice concentration, duration, and maximum extent over the area (Stabeno et al. 2007; Wang and Overland 2009). Direct climate effects will be linked to ice coverage, and indirect acoustic effects will occur as humans begin to use areas previously inaccessible due to ice and the timing of vocalizations produced by ice-dependent marine mammals shifts in response to changes in ice coverage. How this will impact the diverse sub-Arctic marine mammal species is unknown, but extreme care should be taken in interpreting the effects of sound on animals in this area because their entire ecosystem will be in a state of flux.

References

Farmer DM, Xie Y (1989) The sound generated by propagating cracks in sea ice. J Acoust Soc Am 85:1489–1500.

Makris NC, Dyer I (1986) Environmental correlates of pack ice noise. J Acoust Soc Am 79:1434–1440.

Medwin H, Clay CS (1998) Fundamentals of acoustics. Academic Press, San Diego, CA.

Munk WH, Spindel RC, Baggeroer A, Birdsall TG (1994). The Heard Island feasibility test. J Acoust Soc Am 96:2330–2342.

Nystuen JA, Amitai E, Anagnostou EN, Anagnostou MN (2008) Spatial averaging of oceanic rainfall variability using underwater sound: Ionian Sea rainfall experiment 2004. J Acoust Soc Am 123:1952-1962.

Richardson WJ, Greene CR Jr, Malme CI, Thomson DH (1995) Marine mammals and noise. Academic Press, San Diego, CA.

Stabeno PJ, Bond NA, Salo SA (2007) On the recent warming of the southeastern Bering Sea shelf. Deep-Sea Res Part II Top Stud Oceanogr 54:23–26.

Stabeno PJ, Napp JM, Whitledge TE (2008) Sentinels for Bering Sea ecosystem change: 2006-2007. Project 602 Final Report, North Pacific Research Board, Anchorage, AK.

Wang M, Overland JE (2009) A sea ice free Arctic within 30 years? Geophys Res Lett 36:L07502.

Wenz GM (1962) Underwater ambient noise in the ocean: Spectra and sources. J Acoust Soc Am 34:1936–1956.

Ship Traffic Noise Distribution in the Strait of Gibraltar: An Exemplary Case for Monitoring Global Ocean Noise Using Real-Time Technology Now Available for Understanding the Effects of Noise on Marine Life

Thomas Folegot

1 Introduction

The management and conservation of the world's oceans require synthesis of spatial data on the distribution and intensity of human activities and their impact on marine ecosystems. There are many detrimental influences at sea; among them, man-made noise is recognized as being a serious threat to aquatic life. To assess the impact, one must first determine the sound levels received by the marine life.

Ocean noise is difficult to estimate for the following reasons. 1) There is a diversity of man-made sound sources, such as underwater explosions, ships, seismic exploration, offshore construction (e.g., offshore wind farms and hydrocarbon production), industrial activities, sonar of various types, and acoustic devices. 2) Underwater noise propagates well in the ocean as a function of local bathymetry, temperature, and salinity. Depending on location, season, and local climate conditions, sound in the ocean can propagate along very long distances and concentrate at various depths ranging from tens to hundreds of kilometers from the sound source.

Quiet-Oceans has developed a global acoustic prediction tool that combines real-time environmental data with human-generated noise sources, including ship noise. This is needed to synthesize the acoustic data that represents the three-dimensional noise levels and distribution. This tool is optimized for real-time calculations using innovative parallelization technologies. Easily implementable, this technology brings new capabilities to assess the evolution of underwater sound levels and distribution in the world's oceans. Its application supports scientific studies, which quantify and prioritize direct and indirect anthropogenic pressures on aquatic life.

This paper illustrates the potential of the technology by imaging the sound distribution in the Strait of Gibraltar during June 2008 at the rate of one image per hour. The Strait of Gibraltar provides examples of 1) intense ship traffic, 2) complex Atlantic and Mediterranean oceanography, and 3) marine mammal habitats and migration routes. Data from the Strait of Gibraltar are analyzed, and the potential of the technology is demonstrated and illustrated.

T. Folegot (✉)
Quiet-Oceans, 29280 Plouzane, France
e-mail: thomas.folegot@quiet-oceans.com

A.N. Popper and A. Hawkins (eds.), *The Effects of Noise on Aquatic Life*, 601
Advances in Experimental Medicine and Biology 730, DOI 10.1007/978-1-4419-7311-5_136,

2 An Innovative Imaging Technology

Recent regulation combined with emerging technologies now makes it possible to estimate the sound levels received by marine animals at any time and almost any place. Meteorological and oceanographic prediction systems provide reliable wind field, water temperature, and salinity data for present and forecasting modes. The automated identification system (AIS) imposed by the International Maritime Organization since 2004 provides real-time information about the location, type, and behavior of any vessel larger than 300 tons. When coupled with measured and modeled vessel noise databases and with ocean acoustic-modeling tools, a reliable three-dimensional acoustic field is produced in quasi real time.

This deterministic tool assesses the acoustic, spatial, and temporal variability produced by the rapidly changing distribution of sound sources on the surface of the ocean and the spatial and temporal variability of climate and ocean. This technology was implemented in the Strait of Gibraltar.

3 Distribution of Ocean Noise in the Area of Gibraltar

The fusion of oceanographic, bathymetric, and real-time ship traffic data has been implemented to dynamically monitor the ocean noise pollution during June 2008 in a 400- × 200-km area centered on the Strait of Gibraltar. The real position of the vessels cruising in the area was obtained from the AIS network, and a template from 3 June is shown in Figure 1. Each dot represents a broadband noise source located near the surface. The sound field generated by each source is estimated according to the oceanography and bathymetry in the area around the vessel. Each individual three-dimensional

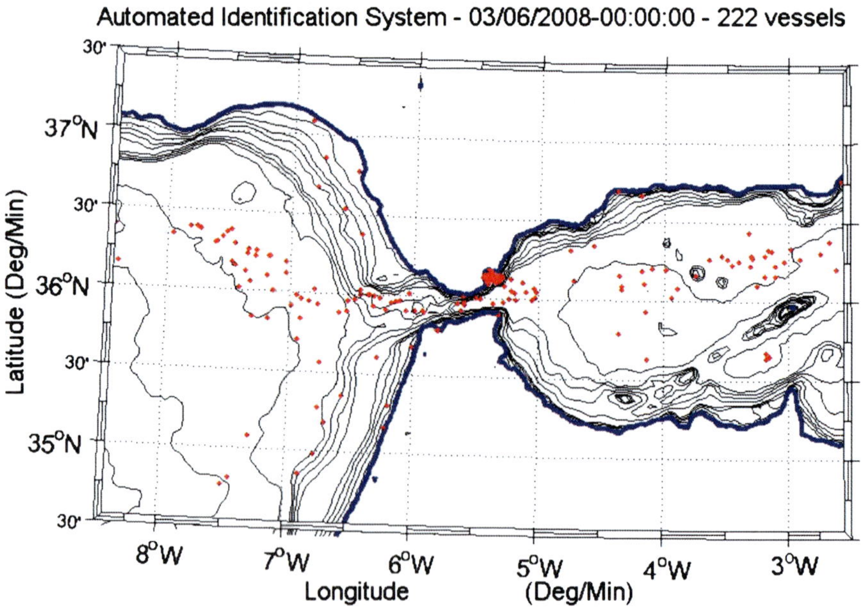

Fig. 1 The maritime traffic of ships (dots) larger than 300 tons (roughly a 50-m-long vessel) as of 3 June at 00:00 Coordinated Universal Time (UTC) is mainly organized along an east-west axis. In the Mediterranean Sea, the traffic flows in a very organized manner, whereas in the Atlantic Ocean, the traffic spreads toward the northwest and southwest. Full video at a one-image-per-hour rate is available at http://www.quiet-oceans.com

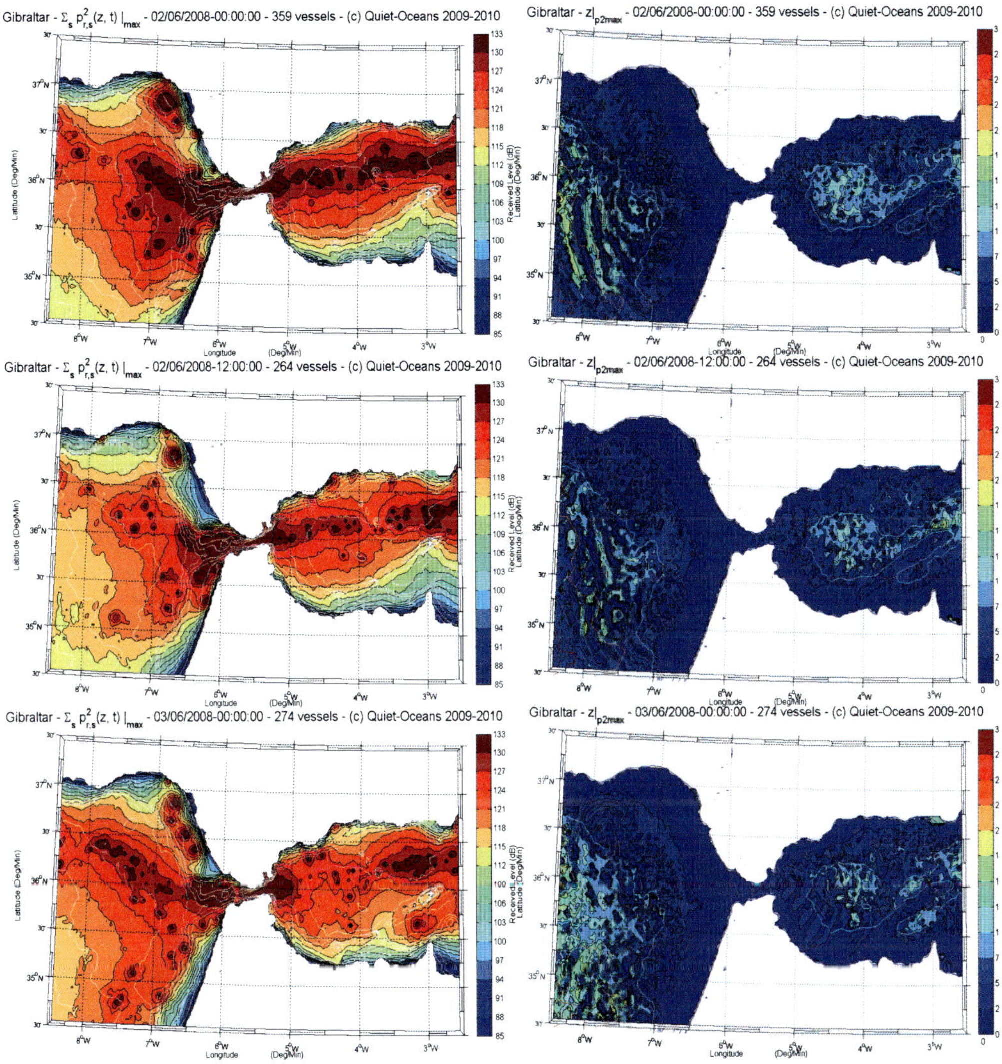

Fig. 2 Left: The sound generated by ship traffic has been simulated in 3 dimensions and the distribution of maximum energy is represented in decibels every 12 h for the first days of June 2008. Right: Depth (in meters) of maximum energy every 12 h for the first days of June 2008. Full video at a one-image-per-hour rate is available at http://www.quiet-oceans.com

energy distribution is summed to produce the total sound that the animals receive. The repeated calculation at regular times leads to a survey of the noise distribution over time. An extract of snapshots is shown in Figure 2, right, arbitrarily represented every 12 h at 250 Hz. The noise is not evenly distributed with depth and the corresponding maximum energy is displayed in Figure 2, left.

A one-image-per-hour ocean noise monitoring has been implemented and has led to three significant results. Those results are not dependent on frequency but are based on the geometric properties of the acoustic propagation. 1) The most intense noise levels are concentrated inside the narrowest and shallowest part of the Strait but expand on both sides in very large areas in both the Atlantic and Mediterranean basins. 2) Rapid and large changes in the maximum noise level distribution occur on both the Atlantic and Mediterranean sides. 3) On both sides of the Strait in June 2008, the most intense noise varied between 700 and 1,300 m in depth faster than the hourly timescale (Fig. 2, left, cyan and green). Noise pollution is concentrated in the deepest layers of the ocean but not uniformly in either space or time as vessels move on the surface. Foraging activities have been observed at those depths by local cetaceans such as pilot whales and Cuvier beaked whales, resident in the Strait of Gibraltar and in the Alboran Sea. This result suggests that they probably experience increasing difficulties operating their biosonar to detect prey and provide for their feeding needs.

4 Conclusions

Quiet-Oceans has developed an essential technology that aims to contribute to the understanding of noise impact on aquatic life. As demonstrated in the Gibraltar area, the complexity of the ocean noise is assessed. Deterministic spatial ocean noise prediction is produced "continuously" over large areas, providing a rich and fine description. This contributes to the ability to understand the mechanisms of noise impacts on marine life and to assess and possibly demonstrate disturbance and masking effects. Habitat models usually focus on environmental parameters. Quiet-Oceans provides a man-made noise layer to habitat models and correlates aquatic life behavior and maritime human activities.

Future developments of the technology will assimilate continuous in situ measurements and improve the quality of the assessment of absolute noise levels in real time.

Are There Technological Alternatives to Air Guns for Oil and Gas Exploration to Reduce Potential Noise Impacts on Cetaceans?

Linda Weilgart

1 Introduction

Seismic air gun surveys, used to find oil and gas deposits beneath the ocean floor, produce loud sharp impulses that can raise noise levels substantially over large areas. These surveys can last months, and the noise they produce is virtually ubiquitous in some areas of the world's oceans. Although noise impacts from seismic surveys on marine life (fish, marine mammals, and even invertebrates) are well documented, the biological relevance of these impacts on wild populations remains controversial among the various stakeholders. Rather than address the controversy or evaluate the evidence for or against an impact, the purpose of a workshop held in Monterey, CA, in 2009 was to examine quieter, potentially less harmful technologies that might be able to, at least partially, replace air guns.

Supported by Okeanos – Foundation for the Sea, a multidisciplinary group of geophysical scientists, seismologists, biologists, and regulators met to find ways to make marine seismic surveys quieter. The participants agreed that marine life would benefit from a quieter ocean. This is their consensus report.

2 Findings and Recommendations

The most effective acoustic mitigation remains not exposing marine life (i.e., through avoidance) to additional anthropogenic noise.

- Regulators together with the hydrocarbon and seismic survey industry should reduce sound levels to the lowest practicable and/or use alternative technologies to reduce noise exposure.
- Impulsive sources like air guns have the potential to physically impact marine life because of the sharp rise times and high peak pressures of air guns. Behavioral effects are also possible even at large distances from the air guns.

L. Weilgart (✉)
Department of Biology, Dalhousie University, Halifax, NS B3H 4J1, Canada

Okeanos - Foundation for the Sea, D-64297 Darmstadt, Germany
e-mail: lweilgar@dal.ca

A.N. Popper and A. Hawkins (eds.), *The Effects of Noise on Aquatic Life*,
Advances in Experimental Medicine and Biology 730, DOI 10.1007/978-1-4419-7311-5_137,

2.1 Air Guns

- Air gun design can be optimized to reduce unwanted energy.
- Imaging deep geological targets requires a low-frequency (<200 Hz) acoustic source. Currently, seismic air guns produce broadband acoustic energy (>200 Hz) and in directions (both inline and horizontal to the plane of interest) that are not useful. Thus unnecessary acoustic energy (noise) should be reduced through array, source, and receiver design optimization.
- Lower source levels could be achieved through better system optimization, i.e., a better pairing of source and receiver characteristics, and better system gain(s). For example, new receiver technologies, such as fiber-optic receivers, may allow the use of quieter sources through a higher receiver density and/or a lower system noise floor.

2.2 Use of Alternative Technologies With Air Guns and/or Instead of Air Guns

Controlled sources generally put the same level of geophysically useable energy into the water as impulsive sources like air guns but over a longer period of time and at a resulting lower peak sound level, i.e., they are quieter. For example, for a rough calculation in the near field, a 1-s oscillatory/vibrator/projector pulse puts the same level of geophysically useful energy into the water as a 10-ms pulse from an air gun but is 100 times quieter, resulting in a 10,000-fold reduction in the area of ensonification. These sources include technologies such as the electromechanical modern marine vibrator, low-frequency acoustic projector (driving cylinder, e.g., LISA, a low-frequency electromagnetic transducer system), the solid-state piezoceramic Helmholz resonator (e.g., The Naval Research Laboratory's DTAGS), and other nonimpulsive, oscillating sound sources. Furthermore, controlled sources can produce sound over the frequency range desired, generating signals that can be specifically designed to minimize the impact on marine mammals and maximize geological interpretability (e.g., pseudorandom sequences).

Controlled sources, by using a sweep rather than an impulse, can reduce the amplitude (peak levels) by 30 dB by spreading out the energy over time. The use of pseudonoise (PN) sequences could reduce the acoustic footprint further (perhaps by an additional 20 dB/Hz by spreading out frequencies over time), but more research is needed to fully understand how to implement these sequences in an effective and optimized way.

- In certain situations and with certain non-air gun source types, placing the sources and/or receivers near or on the seafloor can reduce the required source level as well as the amount of sound that needs to travel through the water column. For example, marine vibrators can operate close to the seabed and accomplish increased penetration relative to shallow towing.
- A controlled source offers improved receiver optimization possibilities compared with air guns. For instance, a combination of fiber-optic sensors with a reduced bandwidth seismic source, such as a marine vibrator, may make the most optimal use of these technologies.
- Front-loading the exploration workflow with the use of silent technologies (e.g., controlled source electromagnetics [CSEM]/3-dimensional [3-D] electromagnetics, gravity, and gravity gradiometry) could optimize the exploration process and require less sound. For instance, if 2-D air gun surveys followed by quieter technologies (e.g., 3-D CSEM) do not show promising targets, proceeding with 3-D seismic surveys may not be worthwhile. Conversely, one may optimize 3-D seismic activities based on the results from 2-D seismic and 3-D CSEM.
- Technologies such as marine vibrators, microseismic monitoring (passive seismic), and fiber-optics could reduce the need for 4-D air gun surveys used to monitor the movement of oil or gas in an exploited reservoir over time.

- Regulators and/or the seismic and hydrocarbon industries should fund or undertake research into impacts on marine animals of alternative technologies.
- Although some air gun alternative technologies are available now or will be in the next 1–5 years, an increase in research and development (R&D) funding for alternative exploration technologies will accelerate development and expand the application window. Governments should encourage the development and use of alternative technologies in an environmentally sensitive manner through both regulatory changes and additional funding to regulatory bodies, scientists, and engineers.

3 Coordination/Incentives

- Governments should discontinue programs that discourage the utilization of non-air gun technologies. Rather, they should develop incentives for any environmentally beneficial alternative technologies.
- The academic geophysical community should be encouraged to research quieter alternatives to air guns, with the aid of government and/or industry funding.
- Regulators should encourage and help fund research and development of quieter, alternative sources and their impact assessments.

Acknowledgments This is based on the report of the Workshop on Alternative Technologies to Seismic Airgun Surveys for Oil and Gas Exploration and Their Potential for Reducing Impacts on Marine Mammals, Monterey, CA, 31 August to 1 September 2009, held by Okeanos – Foundation for the Sea, Darmstadt, Germany. Available via http://www.sound-in-the-sea.org/download/AirgunAlt2010.pdf.

The participants were Ron Brinkman, Chris Clark, John Diebold, Peter Duncan, Rob Habiger, Leila Hatch (Chair), John Hildebrand, Phil Nash, Jeremy Nedwell, Dave Ridyard, Rune Tenghamn, Peter van der Sman, Lindy Weilgart, Warren Wood, and John Young. I thank Andrew Wright (Coorganizer), Sarah Dolman, Jo Hastie, Michael Jasny, Tanja Winkler, and, especially, Dieter Paulmann.

Modeling Cumulative Sound Exposure Along a Seismic Line to Assess the Risk of Seismic Research Surveys on Marine Mammals in the Antarctic Treaty Area

Monika Breitzke and Thomas Bohlen

1 Introduction

Based on the Protocol for the Protection of the Antarctic Environment established in 1991, all activities south of 60° S are subject to an environmental impact assessment (EIA). This also applies to seismic research surveys, which have come into public and regulatory bodies' focus due to their potential impact on marine mammals. To assess the potential risk of air gun shots on marine mammals, numerical modeling of sound propagation is an essential tool. In advance of a seismic survey, computed sound fields are used to derive critical exposure-zone radii within which certain hearing thresholds are exceeded. Up to now, only ray-tracing approaches, which take the geometry of an air gun array and the signatures of its individual air guns into account but neglect the sound-velocity profile of the water column and the seafloor, are used in EIAs for sound-propagation modeling of single shots. However, increasing demands to consider the effects resulting from a stratified water column, the interactions with the seafloor, and the cumulative effects resulting from multiple shots fired along a seismic line accentuate the need to use full waveform methods. Therefore, here we summarize the results of a detailed 2.5-dimensional (2.5-D) finite-difference (FD) modeling study (Breitzke and Bohlen 2010), which fulfills these requirements, approximates compact air gun clusters deployed by the *R/V Polarstern* in polar regions by "point source equivalents," and simulates marine mammals as static receivers. It is a contribution to a strategic risk-assessment study on the impact of seismic research surveys on marine mammals in the Antarctic Treaty Area (Boebel et al. 2009).

2 Study Area and Model Parameters

The modeling study focuses on the Amundsen/Bellingshausen and Weddell Seas, where most of the multichannel seismic (MCS) research activities were conducted with the *R/V Polarstern* since it was put into service in 1982. Most of these MCS lines were collected during the austral summer months

M. Breitzke (✉)
Alfred-Wegener-Institute for Polar and Marine Research (AWI), 27570 Bremerhaven, Germany

University of Bremen, 28359 Bremen, Germany
e-mail: Monika.Breitzke@uni-bremen.de

T. Bohlen
Karlsruhe Institute of Technology (KIT), 76187 Karlsruhe, Germany

A.N. Popper and A. Hawkins (eds.), *The Effects of Noise on Aquatic Life*,
Advances in Experimental Medicine and Biology 730, DOI 10.1007/978-1-4419-7311-5_138,

and covered water depths between ~200 and 1,000 m and ~3,000 and 4,500 m. Typical sound-velocity profiles show an ~150- to 250–m-thick, cold, low-velocity sound channel close to the sea surface, which is overlaid by a thin (~10- to 30-m) layer of warmer water masses with higher sound velocity. Further below, a positive sound-velocity gradient occurs. Typical deep- and shallow-water models with 3,000- and 400-m water depth, respectively, are derived from these data. For the seafloor, a normal incidence reflection coefficient of 0.2 is assumed. The model size and seismic profile length amount to 10 km.

3 Methods

The modeling approach consists of three steps: 1) modeling the seismic source, 2) modeling sound propagation in the ocean generated by a single shot, and 3) modeling cumulative sound exposure levels (SELs) due to multiple shots fired along a seismic line and exposure histories received by static marine mammals.

The seismic source signatures are computed by the NUCLEUS source modeling package. Totally, four compact air gun configurations are considered: a G gun (8.5 l), a 3 GI-gun cluster (2.2 l/5.2 l), an 8 G-gun cluster (68.2 l), and an 8 G-gun cluster + 1 Bolt 1500 LL (100.9 l). These configurations are approximated by "point source equivalents." That is, in the case of the single G gun, the notional signature [$n(t)$] computed with NUCLEUS is used as a source signal and in the case of the compact air gun clusters, the time-integrated far-field signature [$f(t)$] computed with NUCLEUS is used as a source signature [$\tilde{n}(t) = \int f(t)dt$].

A 2.5-D FD code, which implies cylindrical symmetry, is used for modeling sound propagation due to a single shot (Bohlen 2002). From the resulting grid of synthetic seismograms, distributed equidistantly over the model, 2-D sound pressure and SEL fields of a single shot are derived.

The cumulative acoustic impact of multiple shots is determined by superposing 3-D SEL fields of single shots moving along the seismic line. This implies that, first, the 2-D SEL field of the single shot is extrapolated to a 3-D SEL field by rotating it around the cylinder axis of the 2.5-D FD model. Then, the 3-D SEL field is shifted along the seismic line according to the ship speed and shot interval to simulate the 2nd, 3rd, 4th, and so forth, shot. Finally, the SELs of the shifted fields are added to get the 3-D cumulative SEL field. Additionally, time-dependent exposure histories received by static marine mammals are extracted.

4 Results

The modeling approach is applied to 13 different model configurations ranging from a semi-infinite isovelocity model to deep- and shallow-water isovelocity and Amundsen/Bellingshausen and Weddell Sea models. The results show the typical dipolelike directivity in the case of single shots. In the case of multiple shots, a tubular cumulative SEL field develops along the seismic line, with maximum levels vertically beneath the line. Compared with a semi-infinite model, the seafloor reflections in the deep- and shallow-water models enhance the acoustic impact of single shots close to the sea surface and contribute to the seismically induced noise levels, particularly at far offsets. Cumulative SELs decrease more slowly in shallow waters than in deep waters due to the stronger reverberations. This leads to seismically induced noise levels at far offsets and close to the sea surface of ~140–145 dB re 1 μPa^2-s for the 8 G-gun cluster in the deep-water isovelocity model and to ~10-dB higher SELs in shallow waters. The near-surface sound channel and the sound-velocity gradient in the Southern Ocean models cause distortions of the SEL field contour lines, particularly in the

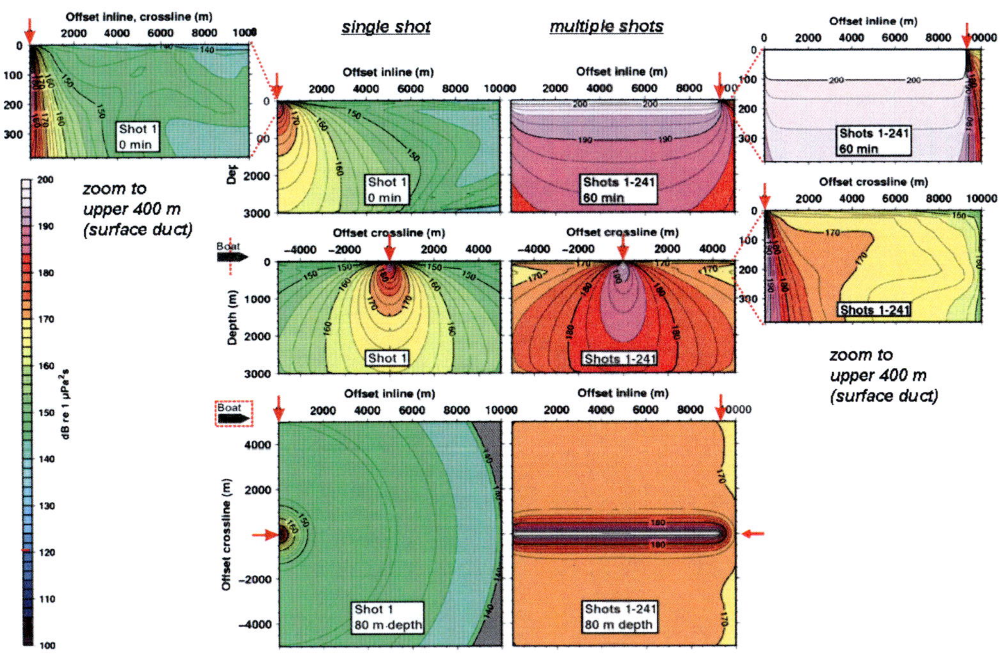

Fig. 1 Deep-water model for the Amundsen/Bellingshausen Sea. Center columns show sound exposure level (SEL) inline- and crossline-depth sections and the horizontal section in the sound channel axis 80 m below the sea surface generated by a single shot (center left) and 241 superposed shots (center right) of the 8 G-gun cluster. Left and right columns display a zoom to the upper 400 m. Red arrows mark the position of the ship after 0 and 60 min of survey time at 0- and 9,26-m offset inline. Modified from Breitzke and Bohlen (2010)

deep-water models (Fig. 1). However, only low-to-moderate levels not relevant for critical exposure-zone radii are affected. Time-dependent exposure histories received close to the sea surface depend strongly on the marine mammal's offset crossline. At greater depths, the crossline position becomes less important and exposure histories mainly depend on the animal's diving depth. Exposure-zone radii derived for different hearing thresholds depend mainly on the air gun configuration and are almost independent of the water depth, the sound-velocity profile, and the physical properties of the seafloor. In the case of single shots, the radii can roughly be predicted from the source level using a spherical-spreading law. In the case of multiple shots, the radii increase with lower thresholds according to a cylindrical-spreading law, and the source level and shot interval contribute to the intercept.

References

Boebel O, Breitzke M, Burkhard E, Bornemann H (2009) Strategic assessment of the risk posed on marine mammals by use of airguns in the Antarctic Treaty area. Information Paper IP 51, Agenda Item CEP 8c, Antarctic Treaty Consultative Meeting XXXII, 6–17 April 2009, Baltimore, MD.

Bohlen T (2002) Parallel 3-D viscoelastic finite difference modelling. Comput Geosci 28:887–899.

Breitzke M, Bohlen T (2010) Modelling sound propagation in the Southern Ocean to estimate the acoustic impact of seismic research surveys on marine mammals. Geophys J Int 181:818–846.

North Atlantic Treaty Organization Marine Mammal Risk Mitigation Principles and Guidelines

Kendra L. Ryan and Stephane Jespers

1 Introduction

Atypical mass strandings of beaked whales in spatial and temporal proximity to active sonar experiments and military exercises increased concern about the possible effects of tactical active sonar on marine mammals. North Atlantic Treaty Organization (NATO) nations using tactical active sonar during maritime activities were linked to several of these strandings, most notably in 1996 in Kyparissiakos Bay, Greece, and in 2000 in the Bahamas (D'Amico 2009). Several hypotheses have been suggested for the cause of these strandings; however, the precise behavioral and/or physiological cause is still unknown. In addition, numerous tactical active-sonar exercises and experiments have occurred without associated atypical mass-stranding events. Therefore, a combination of factors such as the presence of multiple ships, the continuous transmission of sound, constricted geographical area, and strong surface duct have been suggested for mass-stranding events to occur as a result of the transmission of tactical active sonar.

Past research efforts to understand the causes of these mass-stranding events related to tactical active sonar have included studies to understand the auditory systems of marine mammals. Numerous studies documented that intense or prolonged sounds impact the physiology of marine mammals in various ways; for example, sounds interfere with mammals' ability to hear biologically significant signals or communications, a condition called auditory masking. Mammals may also undergo a threshold shift of their hearing capabilities, either temporarily or permanently, as a result of auditory fatigue or damage to auditory components such as sensory hair cells or ear membranes.

2 Mitigation Actions by NATO

In May 1996, an atypical mass-stranding event in Kyparissiakos Bay, Greece, was associated with a Supreme Allied Commander Atlantic (SACLANT) Undersea Research Centre (now known as the NATO Undersea Research Centre [NURC]) active-sonar experiment (Frantzis 1998).

K.L. Ryan (✉)
North Atlantic Treaty Organization Undersea Research Centre, 19126 La Spezia, Italy
e-mail: ryan@nurc.nato.int

S. Jespers
Directorate General of Armaments, 75005 Paris, France
e-mail: stephane.jespers@dga.defense.gouv.fr

A.N. Popper and A. Hawkins (eds.), *The Effects of Noise on Aquatic Life*,
Advances in Experimental Medicine and Biology 730, DOI 10.1007/978-1-4419-7311-5_139,

The NURC subsequently hosted a bioacoustics panel composed of subject matter experts to investigate the relationship between the sonar experiment and the stranding event. The panel concluded that although an acoustic link was neither clearly established nor eliminated as a direct or indirect cause of the May 1996 stranding, additional research was needed to determine biological and behavioral characteristics of the species involved (SACLANT Undersea Research Centre 1998). The recommendations of this panel prompted the creation of NATO's first research project on this topic, the sound ocean living marine resources project, now known as the marine mammal risk mitigation (MMRM) project.

During the next several years, this project supported NATO's environmental protection policy. An international network of scientists with knowledge of sonar and marine mammal distribution, behavior, and bioacoustics was established. Numerous sea trials were conducted to understand the natural history of the species involved in the stranding events. A database was generated containing information related to the distribution, behavior, and bioacoustics of Mediterranean Sea marine mammals. From these data, an internal staff instruction was written to ensure that active acoustic research was planned and performed in an environmentally sensitive manner (Ryan 2009).

The MMRM project and mitigation instructions were only applicable to the NURC's activities and not to NATO's military maritime activities. Therefore, the Maritime Capability Group (MCG) 2, under the NATO Naval Armaments Group (NNAG), convened an ad hoc working group for MMRM. This working group, composed of scientific and military experts from 10 NATO member nations, is developing risk-mitigation principles and guidelines applicable to NATO military maritime activities.

3 Principles

The purpose of the principles document is to list the main environmental protection principles to be observed, the organization to be set up, and the implementation to be executed before and during the conduct of active-sonar activities. These principles apply to the use of active sonar during the preparation for and execution of NATO-led maritime activities. Maritime activities are those performed for the purpose of improving the ability or effectiveness of naval forces, e.g., sea trials, system or concept experimentations, and training exercises. These principles apply to all NATO and non-NATO participants taking part in NATO-led maritime activities. The document outlines four fundamental principles. 1) Under all conditions, NATO-led forces must strive to respect European Parliament principles and policies. 2) NATO and participating commanders must consider the potential effects of active-sonar operations on marine mammals at the earliest opportunity in their planning and throughout the execution and conduct of NATO-led maritime activities. 3) The best practicable and feasible mitigation actions must be applied. 4) Observed adverse effects to marine mammals must be immediately reported to appropriate NATO authorities.

4 Guidelines

The objective of the planning guidelines document is to provide planning guidance before and suggested mitigation measures during the execution of NATO-led active-sonar maritime activities to mitigate the risk to marine mammals.

The rationale of the guidelines is that marine mammals may experience physiological effects or behavioral changes that could increase their risk of injury when in proximity to active-sonar use. Specifically, these guidelines are established to enable NATO-led maritime activities to be conducted while minimizing the risk 1) to a marine mammal of a temporary threshold shift that may result from the use of intensive sound sources; 2) to porpoises of a temporary threshold shift and significant behavioral disturbance that may result from the use of intensive sound sources; and 3) of beaked whale stranding as a result of a behavioral reaction to active sonar.

MMRM is primarily achieved in the planning phase before the accomplishment of active-sonar maritime activities. Planning includes the consideration of marine mammal distribution information and habitats when deciding on the area of maritime activity, the establishment of visual and passive acoustic-monitoring plans, the establishment of ranges where active-source transmissions will be stopped when in the presence of marine mammals, the consideration of active-source transmission characteristics, and the planning of vessel movements. During the planning process, before the execution of NATO-led maritime activities, a planning document stating the agreed MMRM measures shall be completed by and exchanged between activity planners. The MMRM planning document shall include the agreed guidelines to ensure sonar operations are in accordance with NATO environmental protection policies.

5 Conclusions

Atypical mass strandings of marine mammals are temporally and spatially related to NATO active-sonar maritime activities. In addition to strandings, marine mammals may experience auditory injury from high received levels of anthropogenic noise such as active sonar. Recognizing that its use of sound in the ocean may pose a risk to marine mammals, NATO developed a proactive research program to gather information on sensitive species and to develop tools and techniques to minimize the risk to marine mammals. Results from this research and other programs in the field were used by the ad hoc working group on MMRM to develop common NATO principles and guideline documents to assist NATO maritime forces in the execution of tactical active-sonar maritime activities while mitigating the risk to marine mammals. NATO will continue its commitment to facilitate the integration of environmental protection into all NATO-led military activities.

References

D'Amico A, Gisiner, RC, Ketten, DR, Hammock JA, Johnson C, Tyack PL, Mead J (2009) Beaked whale strandings and naval exercises. Aquat Mamm 35:452–472.

Frantzis A (1998) Does acoustic testing strand whales? Nature 392:29.

Ryan K (2009) NATO Undersea Research Centre marine mammal risk mitigation rules and procedures. Special Publication NURC-SP-2009-002, NATO Undersea Research Centre, La Spezia, Italy. Available via http://www.nurc.nato.int/publications/pubs/2009/NURC-SP-2009-002.pdf. Accessed 2 May 2010.

SACLANT Undersea Research Centre (1998) Bioacoustics panel and marine mammal environmental and mitigation procedures panel. SACLANTCEN M-133, NATO Undersea Research Centre, La Spezia, Italy.

Mitigating Impacts of Underwater Noise From Dredging on Beluga Whales in Cook Inlet, Alaska

Christopher A. Hoffman

1 Introduction

The Port of Anchorage, near the head of Cook Inlet in south central Alaska, receives ~80% of all the goods that enter Alaska. Several glacial rivers deposit their large silt load into Cook Inlet, and each year the US Army Corps of Engineers (Corps) dredges ~1,650,000 m^3 of silt from in front of the port. Without annual dredging from May through October each year, the water depth at the port would be inadequate for cargo vessels.

It is common to observe *Delphinapterus leucas* (beluga whales) near the Port of Anchorage, especially in the summer when dredging is underway. In October 2008, the Cook Inlet distinct population segment of *Delphinapterus leucas* was listed as threatened under the Endangered Species Act. Although it was clear that dredging would need to continue at the Port of Anchorage, it was also clear that existing protection measures for *Delphinapterus leucas* would need to be reevaluated.

2 Project Approval Process

As part of the Endangered Species Act consultation process, the Corps prepared a biological assessment to examine the potential effects of underwater noise from dredging on *Delphinapterus leucas*. The Corps' analysis focused on underwater noise sources and potential impacts on *Delphinapterus leucas* and offered several mitigation measures. Based on this analysis, the National Marine Fisheries Service (NMFS) concurred with the Corps' assessment that dredging and disposal may affect but would not adversely affect *Delphinapterus leucas* in Cook Inlet. Although the regulatory framework for determining different levels of harassment for cetaceans and pinnipeds is based on sound pressure levels (SPLs) without regard for species hearing sensitivities, it is clear that both marine life and construction processes would benefit if greater emphasis was placed on frequencies and species auditory thresholds.

C.A. Hoffman (✉)
United States Army Corps of Engineers, Alaska District, Elmendorf Air Force Base, AK 99506, USA
e-mail: Christopher.A.Hoffman@usace.army.mil

A.N. Popper and A. Hawkins (eds.), *The Effects of Noise on Aquatic Life*,
Advances in Experimental Medicine and Biology 730, DOI 10.1007/978-1-4419-7311-5_140,

2.1 Impact Analysis for *Delphinapterus leucas*

Delphinapterus leucas have often been observed within a few meters of cargo ships and dredges at the Port of Anchorage, but it remains unclear if *Delphinapterus leucas* simply tolerate the underwater noise or have become habituated. It is also possible that during certain operational phases the ship or dredge noise is at a frequency that does not cross the auditory threshold of *Delphinapterus leucas*. Although *Delphinapterus leucas* can hear over a wide range of frequencies, their hearing is best in the 10- to 100-kHz range. This range is above most noise associated with dredging, and at low frequencies (<100 Hz), the hearing threshold of *Delphinapterus leucas* may only approximate or exceed the 1-Hz band levels recorded for vessel noise at the Port of Anchorage (Blackwell and Greene 2002). The work of Blackwell and Greene, funded in large part by NMFS, was critical for the assessment at the Port of Anchorage. It was the centerpiece of the Corps' biological assessment because it went beyond a simple analysis of SPLs and frequencies of various underwater noise sources. Instead, it analyzed underwater noise relative to *Delphinapterus leucas* hearing sensitivity and thus provided greater protection to *Delphinapterus leucas* and more realistic mitigation measures for dredging and disposal. Although this report did not include noise from dredging, the characterization of tugboat noise in the Cook Inlet operating environment was critical. Although underwater noise data on dredging are very limited, Scientific Fishery Systems, Inc. (2009) measured underwater noise from a clamshell dredge in Cook Inlet. These measurements confirmed the assertion by Richardson et al. (1995) that tug noise was typically stronger than the dredge noise. Like the tug noise reported by Blackwell and Greene (2002), underwater noise data from dredging could be evaluated relative to *Delphinapterus leucas* hearing sensitivities.

3 Benefits for *Delphinapterus leucas*

It can be beneficial to consider potential underwater impacts to *Delphinapterus leucas* or any aquatic animal in the context of how that species perceives sound. Specifically, a decibel measurement of sound such as a species-specific hearing threshold (ht), commonly written as dB_{ht}(Species) level, as recently detailed by Nedwell et al. (2007), holds promise for improved analyses of underwater noise impacts in the future. At a minimum, a dB_{ht}(Species) level analysis would present a more accurate picture of where anthropogenic underwater noise approaches or crosses the auditory threshold for *Delphinapterus leucas* based on available audiograms. Although there are variations in *Delphinapterus leucas* audiograms, the limited data, when viewed graphically, put potential impacts in far greater perspective than a generalized narrative for all toothed whales. Whether behavioral or electrophysiological, audiograms are likely to be refined and improved over time, although challenges are likely to persist for large cetaceans.

4 Benefits for Construction

Mitigation measures can be improved with a dB_{ht}(Species) level analysis because they can be targeted to either reduce underwater noise where it crosses the *Delphinapterus leucas* hearing threshold or allow for more effective shutdown radii based on the relation of sound signatures of various construction activities to *Delphinapterus leucas* audiograms. Obviously, targeted and effective mitigation measures benefit *Delphinapterus leucas* with better protection. An added benefit may be achieved when development proponents or contractors understand how underwater noise from their project might impact aquatic animal life. Rather than simply being told they cannot exceed a certain

SPL, it can be graphically demonstrated that certain frequencies of the project underwater noise cross an aquatic animal's auditory threshold, perhaps to such a degree that the underwater noise will be injurious. This conceptual "buy in" by development proponents and contractors may ease the negotiations with regulatory agencies during the permit process and reduce some of the litigation that often accompanies underwater activities such as construction, seismic surveys, and oil and gas exploration and production. Additionally, if certain frequencies from anthropogenic noise exceed the auditory threshold for an aquatic species to a potentially harmful degree, it may be possible to focus research and development on modifying machinery or targeting mitigation methods so that SPLs in a specific frequency range can be greatly reduced.

5 Conclusions

Although audiograms still need to be refined for many species and have not even been developed for large cetaceans, using a dB_{ht}(Species) level analysis can be beneficial. While a regulatory framework focused on dB_{ht}(Species) level analysis may be years away, analyses in environmental documents should move beyond simply reporting SPLs and focus on frequencies and species audiograms to put project-derived noise into context. In this manner, it is possible to better understand potential impacts to aquatic life and develop more effective mitigation measures.

References

Blackwell SB, Greene CR Jr (2002) Acoustic measurements in Cook Inlet, Alaska, during August 2001. Greeneridge Report 271–1 prepared by Greeneridge Sciences, Inc., Santa Barbara, CA, for National Marine Fisheries Service, National Oceanic and Atmospheric Administration, Anchorage, AK. Available via http://www.fakr.noaa.gov/protectedresources/whales/beluga/CI_Acoustics_Final.pdf. Accessed 15 April 2010.

Nedwell JR, Turnpenny AWH, Lovell J, Parvin SJ, Workman R, Spinks JAL, Howell D (2007) A validation of the dB_{ht} as a measure of the behavioural and auditory effects of underwater noise. Subacoustech Report Reference 534R1231 prepared by Subacoustech Ltd., Southampton, UK, for the UK Department of Business, Enterprise and Regulatory Reform, London, UK. Available via http://www.subacoustech.com/information/downloads/reports/534R1231.pdf. Accessed 22 April 2010.

Richardson WJ, Green CR Jr, Malme CI, Thomson, DH (1995) Marine mammals and noise. Academic Press, San Diego, CA.

Scientific Fishery Systems, Inc. (2009) 2008 Underwater noise survey during construction pile driving, Port of Anchorage Marine Terminal Development Project. Report 08–06 prepared by Scientific Fishery Systems, Inc,, Anchorage, AK, for the US Department of Transportation Maritime Administration, Washington, DC, and Port of Anchorage, Anchorage, AK. Available via http://www.portofanchorage.org/Library/2008%20Underwater%20Noise%20Survey%20Rept%201-7-09.pdf. Accessed 17 April 2010.

Exploration and Production Sound and Marine Life Joint Industry Programme: Research Progress and Applications

Jennifer L. Michael, John Campbell, David Hedgeland, Michael Jenkerson, H. Rodger Melton, Lori Notor, Russell D. Tait, Sarah L. Tsoflias, and Gary Wolinsky

1 Introduction

A group of 14 oil and gas companies and the International Association of Geophysical Contractors (IAGC) through the International Association of Oil and Gas Producers (OGP) have been funding the Exploration and Production (E&P) Sound and Marine Life Joint Industry Programme (JIP). The JIP funds research to reduce the uncertainty around the risk of negatively impacting marine animal populations during E&P activities. Although there is little to no scientific evidence demonstrating significant negative impacts to marine animal populations, there are some gaps in the scientific knowledge that can contribute to uncertainty in industry risk assessments. In the absence of complete

J.L. Michael (✉) • G. Wolinsky
Chevron Energy Technology Company, Richmond, CA 94801, USA
e-mail: Jennifer@chevron.com; gawo@chevron.com

J. Campbell
International Association of Oil and Gas Producers, London, SE1 8NL, UK
e-mail: john.campbell@ogp.org.uk

D. Hedgeland
Petroleum Geo-Services (PGS) Geophysical, Weybridge, Surrey KT13 0NY, UK
e-mail: david.hedgeland@pgs.com

M. Jenkerson
ExxonMobil Exploration Company, Houston, TX 77002, USA
e-mail: Mike.Jenkerson@exxonmobil.com

H.R. Melton
ExxonMobil Upstream Research Company, Houston, TX 77098, USA
e-mail: Rodger.melton@exxonmobil.com

L. Notor
ConocoPhillips Company, Houston, TX 77252, USA
e-mail: Lori.r.notor@conocophilliips.com

R.D. Tait
ExxonMobil Production Company, Houston, TX 77002, USA
e-mail: russell.d.tait@exxonmobil.com

S.L. Tsoflias
International Association of Geophysical Contractors, Houston, TX 77008, USA
e-mail: sarah.tsoflias@iagc.org

A.N. Popper and A. Hawkins (eds.), *The Effects of Noise on Aquatic Life*,
Advances in Experimental Medicine and Biology 730, DOI 10.1007/978-1-4419-7311-5_141,

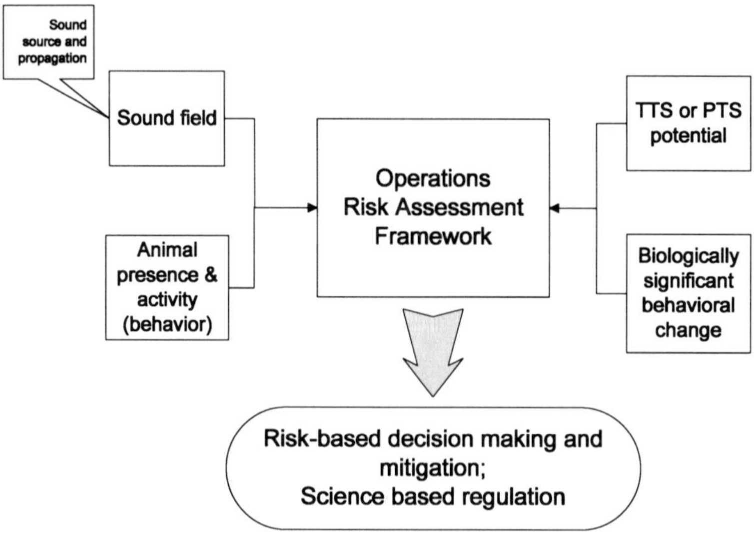

Fig. 1 Summary of the major areas of Joint Industry Programme (JIP)-funded research, all of which supports developing a risk assessment framework for industry operations based on sound science. TTS, temporary threshold shift; PTS, permanent threshold shift

data, environmental regulators may, and do, impose conservative restrictions on E&P activities intended to prevent or mitigate the possibility of significant impacts (Tsoflias and Gill 2008).

The JIP-funded research by independent researchers and contractors will advance understanding of the characteristics of industry sound sources, how sound emitted from these sources propagates in the marine environment, and what the effects of these sounds are on marine life. The results of this research will assist in identifying appropriate mitigation and monitoring strategies. JIP research is informing the scientific community and regulators about the potential environmental effects of operations, ultimately resulting in improved science-based decisions by regulators and increased permitting efficiency for oil and gas activities.

The JIP supports a research program to test scientific hypotheses and build a more comprehensive understanding of the potential environmental risk(s) from oil and gas operations (Fig. 1). The goals for this information are to inform and update policy decision makers and regulatory development processes affecting offshore oil and gas exploration; determine the basis for mitigation measures that are protective of marine life, cost effective, and credible with stakeholders; and feed into planning for efficient E&P project development that is environmentally protective.

2 Results

JIP-funded research is divided into five broad categories.

- *Sound Source Characterization and Propagation*: Research to better define the characteristics of E&P industry sound and how it propagates.
- *Physical, Physiological, and Hearing Effects*: Research to measure the physical and physiological effects of sound on targeted marine life. To determine if and how industry sound may impact marine life, it is critical to understand what and how animals hear and how sound exposure affects hearing ability.

- *Behavioral Reactions of Marine Life and Biological Significance*: Research to understand behavioral responses of marine life to industry sound and what, if any, biological significance this exposure might have.
- *Mitigation, Monitoring, and Analysis/Management of Monitoring Data*: Research to reduce exposure of animals to sound. Such research allows industry to operate with reduced risk in areas that are subject to restrictions and low light and visibility conditions.
- *Research Tools*: Development of improved tools for collecting data in the other research categories.

JIP-funded research has made substantial progress toward describing industry sound sources and propagation. Conducted in the Gulf of Mexico in 2007, the 3D Seismic Source Characterization Project recorded seismic source array output in three dimensions. Twenty paired sensitive and desensitized hydrophones were deployed at a range of depths on three separate moorings. Special positioning equipment was used to locate these moorings continuously in three dimensions. Environmental acoustic recording system (EARS) buoys recorded the wide bandwidth data from the air gun array for a full range of takeoff angles and azimuths. The data were acquired with sufficient shot records to establish a statistically valid sample of sound pressure level and spectral characteristics in over 1,000 azimuth/takeoff angle bins at frequencies up to 25 kHz. The data are currently being analyzed and will be published starting in mid-2010.

A calibrated range for air gun measurements operated in a Norwegian fjord from 2007 to 2010, making high-fidelity, broadband measurements up to 50 kHz of single and cluster air guns using up to 20 hydrophones and 6 velocity sensors. Near- and far-field source signature measurements were made for a subset of single air guns and two-gun clusters over a wide range of operating conditions. These data will be used to improve existing air gun modeling codes, thereby allowing modeled estimations of air gun sound emissions as an alternative to field measurements.

Monitoring technologies have been developed that will detect marine animals in the water column. PAMGuard is a versatile, real-time, open-source suite of software that includes and extends the key capabilities of older open-source PAM software such as the International Fund for Animal Welfare suite and Ishmael while maintaining the same look and feel. After several years of development and a series of research-based field trials, a core version of the software is now available for industry use.

Environmental risk assessments need to include sound from all E&P activities in addition to air guns. Acoustic-source signature data for nonseismic E&P sound sources such as drilling, dredging, pile driving, and vessels have been collected from a number of different geographic areas and oceanographic conditions by industry, consultants, and academia and are available at http://www.soundandmarinelife.org.

Additional progress has been made toward understanding the sound effects on animal tissue, behavior, and the biological significance of any such effects. The JIP has supported several research projects to advance potential use of the population consequences of acoustic disturbance (PCAD) framework to identify situations where behavioral effects are more likely to be significant. Currently, PCAD provides a useful conceptual framework to guide development of mitigation and monitoring strategies.

In 2009, the JIP extended funding for an additional three years (approximately $11 million program total). Initially, a greater percentage of this funding will be spent on a behavioral response study (BRS) on migrating humpback whales. The BRS will measure the behavioral effects of a full seismic array on humpback whales and will test the reaction to ramp up and components of ramp up. Conducted from 2010 to 2013 with migrating humpback whales off Australia's east and west coasts, the study will build stepwise assessment from one air gun to a full commercial array.

3 Conclusions

JIP-funded studies are providing credible scientific information that is informing decision makers and regulators of viable risk assessments and is assisting in developing enhanced mitigation and monitoring technologies and methods. Research is assessing, developing, and adapting the technology needed to enable offshore E&P operations (particularly seismic surveys) to continue under limited surface visibility conditions while minimizing the risk of adverse impacts on marine life populations.

Scientific knowledge about the consequences associated with anthropogenic sound in the marine environment will help oil and gas project planners undertake risk assessments and develop mitigation plans to address potentially significant risks. In addition, this information and data will inform the scientific community and regulators about potential environmental risks of operations. These data will allow regulators to make decisions based on scientific data, with the potential to reduce permitting delays for industry operations. Improved scientific understanding of the potential environmental risks to marine life from oil and gas operations is critical to developing effective mitigation strategies.

References

Tsoflias SL, Gill GC (2008) E&P industry's challenges with managing mitigation guidelines for the protection of marine life during marine seismic operations. Paper 111950-MS, SPE International Conference on Health, Safety, and Environment in Oil and Gas Exploration and Production, 15–17 April 2008, Nice, France.

Managing Anthropogenic Underwater Noise in the Northwest Territories, Canada

Bruce W. Hanna, Peter A. Cott, Amanda A. Joynt, and Lois Harwood

Potential impacts on aquatic life associated with anthropogenic underwater noise are an emerging concern worldwide (Popper et al. 2005). In the Northwest Territories (NWT), Canada, there has been an increased interest in natural resources development and the potential for impacts to both marine and freshwater areas. Fish (including marine mammals and other aquatic life) and their habitats are managed by Canada's Department of Fisheries and Oceans (DFO) under the federal Fisheries Act and the Oceans Act. The DFO reviews project proposals, assesses potential impacts, provides scientific and regulatory advice based on the current state of knowledge, identifies existing data gaps, and initiates steps to address them. If residual impacts are expected with mitigative measures in place, the DFO is able to account for and authorize impacts under the Fisheries Act. Here we identify the types of noise-related impacts that are pertinent to the species and industry activities in the NWT and how these concerns are being addressed by the DFO.

As exploration and development increases in the NWT, concerns have been raised regarding anthropogenic sound impacts in both freshwater and marine ecosystems. For example, known hydrocarbon reserves beneath the Beaufort Sea are being further delineated using air gun-based two-dimensional and three-dimensional seismic surveys. There is concern that intense seismic sounds and the large spatial extent of the seismic coverage may lead to injury, disturbance, displacement, or masking of communication of marine mammals, in particular *Balaena mysticetus* (bowhead whale) during feeding or migrations. Recent research has identified both physical and behavioral effects on *Balaena mysticetus* resulting from the use of air guns in marine seismic surveys, although the extent of such effects and their biological significance is difficult to assess (Harwood et al. 2009, 2010). Impacts will vary depending on a number of factors including the ambient noise level; the activity of the animals at the time that they are affected; the availability of alternative habitats elsewhere; and the type, duration, and frequency of the underwater noise. In extreme or repeated cases, deflection or displacement of migrating or feeding *Balaena mysticetus* could compromise the survival or reproduction of affected individuals.

Exposure to underwater noise of 160 dB is thought to cause behavioral changes in baleen whales, with noise >180 dB potentially causing physical injury such as hearing loss. The DFO recommends an air gun array shutdown if a marine mammal enters a specified shutdown zone relating to exposure

B.W. Hanna (✉) • P.A. Cott • L. Harwood
Department of Fisheries and Oceans, Yellowknife, NWT X1A 1E2, Canada
e-mail: Bruce.Hanna@dfo-mpo.gc.ca

A.A. Joynt
Department of Fisheries and Oceans, Inuvik, NWT X0E 0T0, Canada

A.N. Popper and A. Hawkins (eds.), *The Effects of Noise on Aquatic Life*,
Advances in Experimental Medicine and Biology 730, DOI 10.1007/978-1-4419-7311-5_142,

to 180 dB (the extent of the zone varies with water depth and substrate). Other mitigative measures include "ramping up" of firing air gun arrays to encourage and allow time for marine mammals to vacate the area, temporal and spatial restrictions in certain critical habitats, and onboard marine mammal observers. On a large scale, the creation of marine protected areas (MPAs) can allow for specific regulations to be established in a given area. For example, in the nearshore Beaufort Sea, the proposed MPA will protect *Delphinapterus leucas* (beluga whales) and their estuarine habitats and ensures subsistence-harvesting opportunities for the local Inuvialuit people (Government of Canada 2010). In addition, as part of an Alaskan initiative, the DFO is involved with Alaskan regulators and harvesters in the deployment of satellite transmitters on *Balaena mysticetus*. One study objective is to monitor the behavior of individual whales in relation to active seismic surveys (Alaska Department of Fish and Game 2009).

Recent exploratory drilling activity in the Beaufort Sea has been conducted during periods of ice cover. This coincides with the breeding, pupping, lactation, and basking periods of the *Phoca hispida* (ringed seals) an abundant and widespread species in the Beaufort Sea (Smith 1987). A recent study was undertaken to identify and evaluate potential impacts from exploratory drilling in the ice-covered season on the local *Phoca hispida* population. The study provided important baseline information on the use of the nearshore Beaufort Sea by *Phoca hispida* during the spring and is a benchmark for any future studies involving multiple or longer term drilling operations. The results suggest that one season of drilling had no detectable effect on *Phoca hispida* in the study area compared with winters when drilling did not take place (Harwood et al. 2007).

In freshwater systems, energy sources, such as air guns and explosives used for seismic surveys, generate noise that has the potential to cause impacts (Cott et al. 2003). The use of air guns in riverine systems was thought to cause physical or behavioral impacts to fishes such as hearing damage and the interruption of migrations. To address concerns related to the use of air guns in riverine environments, DFO coordinated a study examining the impacts of air gun-generated noise on the physiology and behavior of fishes in the Mackenzie River. Although temporary hearing loss was documented in some species (Popper et al. 2005), no evidence of physical damage to the inner ear structure was found (Song et al. 2008) and no significant behavioral effects were identified (Jorgensen and Gyselman 2009).

To protect fish from the effects of explosives, DFO required companies conducting seismic programs to monitor instantaneous pressure changes (IPCs) in the water column when detonating charges in water bodies not frozen to the bottom. When broad-based monitoring results indicated that there were IPCs detrimental to fish, the DFO decided to no longer authorize (i.e., allow the killing of fish) seismic programs using dynamite in water bodies (Cott et al. 2003). To conduct these programs in the NWT, the onus is now on the company to demonstrate that a conservative IPC level of 50 kPa or less can be achieved consistently during a preoperation testing phase. If this cannot be accomplished, the water body in question is removed from the seismic program because impacts to fish may occur (Cott and Hanna 2005). A study was conducted in collaboration with industry to investigate the effects of explosives on fish eggs and fry and to identify an IPC threshold that is adequately protective. Tissue damage in fry occurred at an IPC < 70 kPa; however, no impact was noted to the eggs (Godard et al. 2008). More research is being conducted to determine the IPC thresholds for adult fish (D. Godard, unpublished data).

Similarly, noise generated from other aspects of northern development such as pile driving or winter road construction traffic can potentially impact fish (Mann et al, 2009). Baseline sound levels generated from a variety of sources associated with exploration activity, from drilling to aircraft landing on the ice, were documented (Mann et al. 2009). These data can be correlated with information on the hearing capability of fish (see Mann et al. 2007) that may be in a program area to identify and implement effective mitigation measures. Under-ice fish deterrents have also been tested as a means to encourage fish to avoid areas but with limited success, likely due to the low metabolism of fish in the winter (Racca et al. 2004). Research is currently under way to determine if the winter spawning *Lota lota* (burbot) vocalize. If they do, they may be vulnerable to anthropogenic sounds interfering with their vocalizations, hindering reproductive success (P. Cott, unpublished data).

Although tangible progress has been made, there are still challenges facing the DFO relating to the management of impacts of industrial underwater noise. Significant gaps in baseline ecological information for both the marine and freshwater systems of the NWT remain, making it difficult for the government to keep pace with industry. Other unknowns include the nature, type, and extent of cumulative effects; thresholds that might cause detrimental effects to different species and life stages; and how impacts may be exacerbated by overarching factors such as climate change and pollution. The use of current research to adaptively manage sound-related impacts to aquatic life is paramount in an area with such large gaps in baseline data. As we move forward, it will be essential to identify areas of common interest and resources and to continue to expand on collaboration opportunities with our regulatory and scientific colleagues within and outside the NWT.

References

Alaska Department of Fish and Game (2009) Satellite tracking of Western Arctic bowhead whales. Alaska Department of Fish and Game, Juneau, AK. Available via www.wildlife.alaska.gov./index.cfm?adfg=marinemammals.bowhead. Accessed 6 May 2010.

Cott P, Hanna B (2005) Monitoring explosive-based winter seismic exploration in waterbodies, NWT 2000–2002. In: Armsworthy SL, Cranford PJ, Lee K (eds) Proceedings of the Offshore Oil and Gas Environmental Effects Monitoring Workshop: Approaches and Technologies, Bedford Institute of Oceanography, 26–30 May 2003, Dartmouth, NS, Canada, pp 473–601.

Cott PA, Hanna BW, Dahl JA (2003) Discussion on seismic exploration in the Northwest Territories 2000–2003. Can Manuscr Rep Fish Aquat Sci 2648: vi +36.

Godard DR, Peters L, Evans R, Wautier K, Cott PA, Hanna B, Palace V (2008) Development of histopathology tools to assess instantaneous pressure change-induced effects in rainbow trout (*Oncorhynchus mykiss*) early life stages. Environmental Studies Research Funds Report 164, Winnipeg, MB, Canada.

Government of Canada (2010) Tarium Niryutait marine protected areas regulations. Canada Gazette, 10 April 2010, Ottawa, ON, Canada. Available via http://www.gazette.gc.ca/rp-pr/p1/2010/2010-04-10/html/reg2-eng.html#reg. Accessed 6 May 2010.

Harwood L, Smith TG, Melling H (2007) Assessing the potential effects of near shore hydrocarbon exploration on ringed seals in the Beaufort Sea Region 2003–2006. *Environmental Studies Research Funds Report* 162.

Harwood L, Joynt A, Kennedy D, Pitt R, Moore S (2009) Spatial restrictions and temporal planning as measures to mitigate potential effects of seismic noise on cetaceans: A working example from the Canadian Beaufort Sea, 2007–2008. Department of Fisheries and Oceans Canadian Science Advisory Secretariat Research Document 2009/040, iv + 14 p.

Harwood LA, Auld J, Joynt A, Moore SE (2010) Distribution of bowhead whales in the SE Beaufort Sea during late summer, 2007–2009. Department of Fisheries and Oceans Canadian Science Advisory Secretariat Research Document 2009/111, v + 22 p.

Jorgensen JK, Gyselman EC (2009) Hydroacoustic measurements of the behavioral response of arctic riverine fishes to seismic airguns. J Acoust Soc Am 126:1598–1609.

Mann D, Cott P, Horne B (2009) Under-ice noise generated from diamond exploration in a Canadian sub-arctic lake and potential impacts on fishes. J Acoust Soc Am 126:2215–2222.

Mann DA, Cott PA Hanna BW, Popper AN (2007) Hearing in eight species of northern Canadian freshwater fishes. J Fish Biol 69:1–12.

Popper AN, Smith ME, Cott PA, MacGillivray AO, Austin ME, Mann DA (2005) Effects of exposure to seismic airgun use on hearing of three fish species. J Acoust Soc Am 117:3958–3971.

Racca RG, Hannay DE, Murray RB, Griffiths WB, Muller M (2004) Testing fish deterrents for use under-ice in the Mackenzie Delta area. Environmental Studies Research Funds Report 145, Calgary, AB, Canada.

Smith TG (1987) The ringed seal, *Phoca hispida*, of the Canadian western arctic. Can Bull Fish Aquat Sci 216:1–81.

Song J, Mann DA, Cott PA, Hanna BW, Popper AN (2008) Ear structure in seven species of northern Canadian freshwater fishes, including effects of seismic exposure on ear tissues in select species. J Acoust Soc Am 124:1360–1366.

SIRENA '10: Beyond the Mediterranean

Jeffery Haun, Kendra Ryan, and Arnold B-Nagy

1 Introduction

Atypical mass strandings of marine mammals in close spatial and temporal proximity to experiments and naval exercises involving tactical active sonar systems have increased concern about the possible impact of this type of anthropogenic sound. The North Atlantic Treaty Organization (NATO) Undersea Research Centre (NURC) has been at the forefront of research regarding the interaction of marine mammals and active sonar. In 1999, NURC established one of the first research programs in the world to address this topic, the Sound, Ocean and Living Marine Reso0urces (SOLMAR) project, now known as the Marine Mammal Risk Mitigation (MMRM) project.

One of the objectives of the project was to develop tools and procedures with which an experimenter could determine the presence of marine mammals, specifically, the species of interest, Cuvier's beaked whale (*Ziphius cavirostris*), near a sonar source before and during its use (D'Amico et al. 2003). Historically, visual observations have been the primary method of determining marine mammal presence; however, this method is limited due to the high percentage of time that beaked whales are submerged (Tyack et al. 2006), difficulty seeing the low profile of beaked whales when they are on the surface (Barlow and Gisiner 2006), and adverse weather conditions. As a result of these limitations, the project is now focusing on passive acoustic detection technologies and associated algorithms to detect, classify, and localize (DCL) marine mammals. Several of these technologies were tested during the SIRENA '10 sea trial.

2 Application to Risk Mitigation

Current methods of MMRM include visual monitoring, ramp-up (i.e., commencing active transmissions at low sound pressure levels and increasing the strength of the transmission over time until the required level is achieved), and temporal or spatial restrictions of active sonar use. The effectiveness

J. Haun (✉) • K. Ryan • A. B-Nagy
North Atlantic Treaty Organization Undersea Research Centre, 19126 La Spezia, Italy
e-mail: haun@nurc.nato.int; ryan@nurc.nato.int; b-nagy@nurc.nato.int

A.N. Popper and A. Hawkins (eds.), *The Effects of Noise on Aquatic Life*,
Advances in Experimental Medicine and Biology 730, DOI 10.1007/978-1-4419-7311-5_143,

of these methods may be limited due to environmental conditions, technological limitations, animal behavior, and operating restrictions. Passive acoustic monitoring complements these methods by allowing additional detections during adverse environmental conditions or during the mammal's diving behaviors that were previously not addressed.

Passive acoustic monitors should not be relied on as the sole tool for the determination of the location of an animal for which subsequent mitigation techniques are implemented. This method has its limitations. Due to the high frequency of the beaked whale signals, the area of detection by passive acoustic monitors is limited to 3–4 km (Pavan et al. 2009). Yet, depending on the environmental conditions, the received sound pressure levels of the active sonar may still be significant at distances over tens of kilometers from the source. In addition, depending on the active sonar activity, the area that requires monitoring may be defined in the hundreds of kilometers. Therefore, various methods of deploying passive acoustic monitors such as seeding an area with buoys, deploying persistent autonomous oceanographic gliders, or using research vessels with towed bodies should be explored.

3 Description

The SIRENA '10 sea trial was conducted from 1 May to 14 June 2010 off the southwest coast of Portugal in the North Atlantic Ocean. The main reason why the cruise was located in this area was that it is a NATO exercise operating area and the data collected will provide a baseline as to the types and density of animals that occur in this area. It also provides invaluable information for the habitat-modeling work that we have undertaken and continue to refine for use in mitigating the risk to marine mammals from any NATO exercises in the ocean. This effort involved the cooperation of 42 scientists and engineers representing 17 organizations from 7 countries. A total of five passive acoustic systems, three detection and classification algorithms, one identification and data-collection system, and one visual data-collection system were tested. As well as the visual and acoustic data that were collected, there were both remote sensing and in situ measurements of oceanographic parameters collected for inclusion in the habitat-modeling project at NURC. All operations were conducted onboard the *NRV Alliance*.

The operations on the *NRV Alliance* were divided into three phases. During all phases, line-transect surveys were performed while acoustic and visual methods were used to determine the presence and absence of cetaceans. In addition, during phase 1, the bottom-mounted passive acoustic detection buoys were placed in beaked whale "hot spots." The buoys were then recovered at the end of phase 3.

4 Conclusions

Stranding of beaked whales in temporal and spatial relation to the use of tactical active sonar brought attention to a possible relationship between marine mammal and sonar. The continued use of active sonar for military operations, oil and gas exploration, and scientific research necessitates implementation of MMRM techniques and technologies. Several passive acoustic technologies including arrays, a towed body, and a buoy were evaluated during the SIRENA '10 sea trial. These technologies will be included in the suite of tools used for MMRM.

References

Barlow J, Gisiner R (2006) Mitigating, monitoring and assessing the effects of anthropogenic sound on beaked whales. J Cetacean Res Manage 7:239–249.

D'Amico A, Bergamasco A, Zanasca P, Carniel S, Nacini E, Portunato N, Teloni V, Mori C, Barbanti R (2003) Qualitative correlation of marine mammals with physical and biological parameters in the Ligurian Sea. IEEE J Ocean Eng 28:29–43.

Pavan G, Fossati C, Priano M, Manghi M (2009) Acoustic detection of Cuvier's beaked whales (*Ziphius cavirostris*). In: Dolman S, McLeod C, Evans PGH (eds) Proceedings of the Workshop "Beaked Whale Research," ECS Special Publication Series No. 51, 21st Annual Meeting of the European Cetacean Society, 26 April 2007, San Sebastian, Spain, pp 31–35.

Tyack PL, Johnson M, Soto NA, Sturlese A, Madsen PT (2006) Extreme diving of beaked whales. J Exp Biol 209:4238–4253.

Planning Is Critical to Ensure Effective Mitigation of Naval Activities

Sarah J. Dolman

1 Introduction

Navies self-regulate their impacts and set their own mitigation strategies. Current onboard mitigation efforts are generally untested but are probably insufficient for many species. Furthermore, increasing scientific evidence demonstrates that the ranges required for successful mitigation based on safety zones are usually larger than is feasible to monitor with current real-time onboard practices. Additionally, the potential exists for detrimental cumulative impacts arising from multiple exposures to sonar in conjunction with other military activities that include exercises incorporating a range of vessels such as warships, carriers, aircraft (including helicopters), and submarines and with a variety of events such as missile and ordnance testing and "sinking" exercises as well as with civilian activities. The adoption by all navies of effective, long-term, and meaningful management measures in the planning stage is an urgent priority. Fortunately, a number of navies have already undertaken considerable work to protect marine wildlife, demonstrating that environmental duty of care does not need to come at the expense of navy training.

2 Case Study: North Atlantic Treaty Organization Joint Warrior Exercise

The North Atlantic Treaty Organization (NATO) conducted the Joint Maritime Course (JMC) military training exercise in coastal waters and in deeper waters to the north and west of Scotland between 1946 and 2006. The JMC occurred three times a year, in March, June, and November. Military jets, submarines, warships (including minesweepers and submarine hunters), landing craft, powerboats, and sonobuoys were utilized during these exercises. Many of these activities and their combination over a 60-yr time frame can potentially impact marine mammals (Parsons et al. 2000). The JMC has recently been replaced by the Joint Warrior (JW) exercise, a twice yearly NATO event usually

S.J. Dolman (✉)
Whale and Dolphin Conservation Society (WDCS) Wildlife Centre, Spey Bay, Moray IV32 7PJ, UK

School of Biological Sciences, University of Aberdeen, Aberdeen AB24 2TZ, UK
e-mail: sarah.dolman@wdcs.org

A.N. Popper and A. Hawkins (eds.), *The Effects of Noise on Aquatic Life*, Advances in Experimental Medicine and Biology 730, DOI 10.1007/978-1-4419-7311-5_144,

occurring for two or three weeks in spring (around April) and autumn (September-October). In all, up to 30 ships, 5 submarines, and 85 aircraft take part, with the majority of the assets coming from the UK forces, with up to 12 other NATO and allied nations taking part.

The west coast of Scotland is home to a diversity of marine species, including 24 species of cetaceans, vulnerable basking sharks, and internationally significant species of seabirds. Yet, the Ministry of Defence (MOD) has not conducted an environmental impact assessment (EIA) to fully assess the potential impacts, including cumulative and in-combination effects, from ongoing naval activities in this region (Dolman et al. 2009).

With this specific case study in mind, some points toward planning to ensure effective management and mitigation in all military activities are considered.

3 Planning in Naval Exercises to Ensure Protection of Marine Species

An ongoing commitment on all, but not limited to, exercise areas should include conducting full EIAs based on peer-reviewed science. To enable the identification and evaluation of potential impacts from activities on the marine environment and the assessment of reasonable alternatives, an EIA is a legislated requirement for all marine users, except navies, under the EIA directive. Such a process is undertaken so that environmental uncertainties and consequences can be fully understood early in the decision-making process, with public transparency. Although European navies may be exempt from this form of assessment, the most effective way to ensure that all possible impacts are considered is to conduct a detailed and thorough environmental assessment. EIAs have been undertaken for the global operation of the UK's Sonar 2087 (QinetiQ 2003), and the US Navy (USN) is currently undertaking EIAs/environmental impact statements (EISs) for all ranges and operating areas off the US coast (and some overseas ranges including the Mariana Islands range complex) but has yet to conduct such an assessment for any of its activities in European waters. EIAs can clearly be achieved without compromising navy training.

Current exposure standards in many countries are flawed as detailed in Parsons et al. (2008). Assessments to predict the potential impact of military sonar based on physical damage to cetaceans are erroneous (Parsons et al. 2008). The MOD's current use of physical injury as a criterion for harm is not appropriate because in many sonar exposures, it is more likely to be behavioral change rather than physiological impacts that results in a negative impact. The current level of "sonar risk assessment" that is undertaken by the MOD does not provide the level of detail that is required by a recognized EIA (as under the EIA directive) and the criterion on which the current level of sonar risk assessment to assess physical injury due to active sonar is based is not appropriate for considering disturbance. Consideration should include all activities involved in exercises, including but not limited to active sonar activities.

Additional commitments should include 1) funding for the ongoing (long-term) and independent collection of field survey data (marine mammal distribution and abundance and sound-propagation studies, including modeling verification); 2) explicitly identifying and avoiding sensitive areas (including bathymetric features of possible importance to cetaceans) and employing other spatiotemporal restrictions in known sensitive (feeding, breeding, calving) and legislated protected sites; 3) identifying a limited number of exercise locations where cetaceans are found in low densities; 4) reporting back to enable a better understanding of how effective the guidelines are and how well implemented, what any observed impacts to cetaceans are, and to help in understanding the effectiveness of any mitigation measures undertaken as well as employing adaptive management procedures to update mitigation practices accordingly; 5) transparency and enforcement in legislative compliance; 6) acknowledging the limitations and conducting in situ real-time passive acoustic monitoring (PAM) and aerial surveys to ensure a more effective detection of cetaceans than by using visual surveys alone; 7) consideration of all navies operating in European waters; and 8) transparency and cooperation with stakeholders throughout (including nature conservation agencies within governments, conservation groups, scientists, and those involved with a stranding/incident response).

4 Conclusions

The MOD manages its own environmental responsibilities, allowing for limited accountability. A number of elements need to be undertaken before the extent of the potential effects of naval activities can begin to be understood. The adoption of effective, long-term, and meaningful management measures in the planning stage, including EIA as a starting point, is clearly important.

References

Dolman SJ, Weir CR, Jasny M (2009) Comparative review of naval marine mammal guidance implemented during naval exercises. Mar Pollut Bull 58:465–477.

Parsons ECM, Birks I, Evans PGH, Gordon JCD, Shrimpton JH, Pooley S (2000) The possible impacts of military activities on cetaceans in West Scotland. Eur Res Cetaceans 14:185–191.

Parsons ECM, Dolman SJ, Wright AJ, Rose NA, Burns WCG (2008) Navy sonar and cetaceans: Just how much does the gun need to smoke before we act? Mar Pollut Bull 56:1248–1257.

Qineti Q (2003) Sonar 2087 and the environment. Report prepared for the Defence Procurement Agency, Ministry of Defence, UK.

Standardizing Protected Species Observer Requirements in the United States

Kyle Baker, Deborah Epperson, Howard Goldstein, Kimberly Skrupky, Brad Smith, Gregg Gitschlag, Jill Lewandowski, and Teresa Turk

1 Introduction

Seismic surveys remain at consistent levels in US waters, and new marine areas are now available for leasing of oil, gas, and renewable ocean energy projects. There is an increasing need to design a protected species observer (PSO) program for marine mammals protected under the US Marine Mammal Protection Act and species listed as endangered or threatened under the US Endangered Species Act to address expanding offshore energy monitoring and data collection needs in US waters. The National Marine Fisheries Service (NMFS; the lead agency for protected species conservation) and the Bureau of Ocean Energy Management, Regulation and Enforcement (BOEMRE;

K. Baker (✉)
National Marine Fisheries Service, National Oceanic and Atmospheric Administration,
St. Petersburg, FL 33701, USA
e-mail: kyle.baker@noaa.gov

D. Epperson
Bureau of Ocean Energy Management, Regulation and Enforcement, New Orleans, LA 70123, USA
e-mail: deborah.epperson@boemre.gov

H. Goldstein • T. Turk
National Marine Fisheries Service, National Oceanic and Atmospheric Administration,
Silver Spring, MD 20910, USA
e-mail: howard.goldstein@noaa.gov; teresa.turk@nooa.gov

K. Skrupky • J. Lewandowski
Bureau of Ocean Energy Management, Regulation and Enforcement, Herndon, VA 20170, USA
e-mail: kimberly.skrupky@boemre.gov; jill.lewandowski@boemre.gov

B. Smith
National Marine Fisheries Service, National Oceanic and Atmospheric Administration,
Anchorage AK 99513, USA
e-mail: brad.smith@noaa.gov

G. Gitschlag
National Marine Fisheries Service, National Oceanic and Atmospheric Administration,
Galveston, TX 77551, USA
e-mail: gregg.gitschlag@noaa.gov

A.N. Popper and A. Hawkins (eds.), *The Effects of Noise on Aquatic Life*,
Advances in Experimental Medicine and Biology 730, DOI 10.1007/978-1-4419-7311-5_145,

the lead agency with oversight of offshore energy [oil, gas, minerals extraction, and renewable energy] in US federal waters) conducted a national review (K. Baker, D. Epperson, T. Turk, H. Goldstein, K. Skrupky, J. Lewandowski, B. Smith, and G. Gitschlag, NOAA Technical Memorandum, in preparation) of the seismic survey PSO program. This review resulted in recommended actions to implement standards that would benefit protected species management. Core components of the program include standards for PSO qualifications, PSO training, data collection, and reporting.

2 PSO Qualifications

Under the proposed standards, all new PSO training candidates would be required to have 1) a bachelor's degree from an accredited college or university, with a major in one of the natural sciences and a minimum of 30 semester hours or equivalent in the biological sciences; 2) at least one undergraduate course in math or statistics; and 3) experience with data entry on computers. Successful qualification to become a PSO would require completion of required coursework, e.g., field exercises, homework, and tests, in an NMFS-approved training course with a passing grade of 80% or greater. In addition to educational and training requirements, PSOs would need to assert that he or she has no financial conflict of interest with the seismic survey or operator, meet minimum physical/medical conditions to perform assigned tasks, be able to clearly and concisely communicate verbally and in writing in English, and have the ability to work legally in the United States.

3 PSO Training

PSOs play a vital role in collecting data for the NMFS and BOEMRE on species presence, number of animals exposed in the noise zones of influence, distance of animals relative to seismic surveys, behavioral reactions, and other important data. The consistency and quality of PSO training and performance are critical to the successful management of a PSO program. The skill set required to be an effective PSO ranges from protected species survey techniques, data-recording protocols, species biology and behavior; expertise in passive acoustic-monitoring equipment and computer software, and familiarity with offshore industry operations. NMFS has identified criteria for PSOs that may be used to assess the qualifications of PSOs to be approved to perform mitigation, monitoring, and data collection during offshore activities. Currently, there are a number of training programs in the United States and elsewhere that provide training of varying content and quality. Standardized training-program requirements will result in PSOs being consistently trained with the scientific knowledge, observer techniques, data-collection methods, and a quality assurance/quality control process. To meet consistent training needs, the NMFS is considering agreements and partnerships to establish regional PSO training centers.

4 Data Collection and Reporting

Data reporting is routinely required through federal permits and licenses to monitor the effectiveness of mitigation measures and the take of protected species. Standards are expected to improve data quality, analysis of datasets between different regions, and evaluations of mitigation effectiveness and inform adaptive management strategies. Standardized forms and software will minimize discrepancies among datasets and will allow the synthesis and comparison of datasets whose data collection and reporting are otherwise required and collected independently of one another. The importance of accurate and complete reporting of the results of the mitigation measures and their effectiveness at reducing or avoiding take are critically important components of any adaptive management strategy. The NMFS and BOERME are developing strategies necessary to migrate to electronic data collection and submission of reports to ensure greater consistency in reporting and ease in importing the data into a database for subsequent analysis.

5 Summary

Currently, there is considerable geographic variation in the data-collection requirements and procedures for US-permitted seismic surveys conducted for energy and research purposes. The development of standards to train PSOs, standardization of data collection, submission standards, and development of quality assurance and quality control standards will make data analysis more efficient and robust. The implementation of national standards for the Seismic Survey PSO Programs will increase the integrity of data collected and reported for more effective protected species management.

Using Passive Acoustics to Monitor the Presence of Marine Mammals During Naval Exercises

Anurag Kumar, Jene Nissen, Joel Bell, and Mandy Shoemaker

1 Introduction

The US Navy has been collecting data on the presence of marine mammals in conjunction with Navy exercises involving active sonar activity. During July 2008, the US Navy performed antisubmarine warfare training exercises in Onslow Bay, North Carolina, using midfrequency active sonar (1–10 kHz). The exercises were conducted in one of the potential sites of the proposed east coast Undersea Warfare Training Range (USWTR). As part of the monitoring effort for these exercises, five bottom-mounted passive acoustic recorders were deployed in Onslow Bay. The recordings contain hundreds of odontocete vocalizations, including *Globicephala* sp. (pilot whales) and *Physeter macrocephalus* (sperm whales), occurring before, during, and after the sonar events.

2 Methods

During 6–27 July 2008, five bottom-mounted marine acoustic recording units (MARUs) from Cornell University, Ithaca, NY, were deployed ~7 km apart near the shelf break in Onslow Bay. This configuration was chosen to maximize the spatial coverage area because the exact location of the exercise can vary. Although the spacing between the MARU locations was greater than the average detection range of most odontocetes, recordings and subsequent vocal activity identified on each MARU were considered to be independent from each other.

Each recorder sampled continuously at 32 kHz during the duration of the deployment. The MARUs were deployed ~1 wk before a planned US Navy exercise involving active sonar activity was to occur at the site. This allowed us to start monitoring for marine mammal vocal activity before

A. Kumar (✉) • J. Bell • M. Shoemaker
Naval Facilities Engineering Command Atlantic, Norfolk, VA 23508, USA
e-mail: anurag.kumar@navy.mil; joel.t.bell@navy.mil; mandy.shoemaker@navy.mil

J. Nissen
United States Fleet Forces Command, Norfolk, VA 23551, USA
e-mail: richard.j.nissen@navy.mil

A.N. Popper and A. Hawkins (eds.), *The Effects of Noise on Aquatic Life*, Advances in Experimental Medicine and Biology 730, DOI 10.1007/978-1-4419-7311-5_146,

the Navy exercise. We analyzed these acoustic recordings for odontocete vocalizations (clicks, whistles, and burst-pulse sounds) using a combination of long-term spectral averages (LTSAs) and visual review of spectrograms and for sonar activity using a spectrogram correlation detector.

2 Results

Sonar activity was detected primarily from 16 to 18 July, with some activity also detected from 26 to 27 July, resulting in a total of 5 days with recorded sonar activity and 17 days without sonar present. From these records, we found odontocete vocal events, most of which were classified as unidentified odontocetes (i.e., odontocetes that could not yet be classified as to species). Of the vocal events identified, ~200 vocal events were made by *Physeter macrocephalus* and 13 were likely made by *Globicephala* sp. Because of a significant amount of hard drive noise, we were unable to examine the data for low-frequency calls of large whales below 200 Hz.

Unidentified odontocete vocal events were detected both day and night on each day. The *Physeter macrocephalus* clicks were detected on two MARUs in deeper water (>270 m) throughout the entire recording period mainly at night (from 2000h to 0600h).

The *Globicephala* sp. vocalizations were also detected on only the two deeper MARUs, but unlike *Physeter macrocephalus*, they were detected sporadically throughout the day and night and only from 14 to 18 July.

On a few occasions, we also found unidentified odontocetes that appeared to be mimicking sonar signals with frequency-modulated whistles of similar frequencies immediately after the sonar signal. The mimicry events would last for several sequential pings before ceasing to be heard.

The duration of the vocal events ranged from 1 min to >12 h, with an average duration of 41 ± 1.23 min. The total duration of vocal events by day for each classification group were compared with the total duration of sonar activity by day during the recording period. No statistical correlation was found between the number of vocal events heard and the duration of sonar activity.

4 Conclusions

The recorded acoustic data indicate that marine mammals were present in Onslow Bay when sonar was used. The majority of the vocalizations detected were whistles and clicks from unidentified odontocetes, with a number of identified clicks from *Physeter macrocephalus* and whistles and clicks from *Globicephala* sp. On several occasions, the *Physeter macrocephalus* clicks appeared to have been produced by a single individual; however, the clicks were not localized so we cannot rule out the possibility that there were multiple whales. It is possible that the same individual was consistently foraging near the shelf break at night during July 2008. The probable pilot whale vocalizations recorded on 17 July 2008, were recorded shortly after a sighting of *Globicephala* sp. was made by the University of North Carolina, Wilmington (UNCW) aerial survey team flying in the area on the same day.

The unidentified odontocete vocal events are most likely from either offshore bottlenose dolphins (*Tursiops truncatus*) or Atlantic spotted dolphins (*Stenella frontalis*) because these are the most commonly sighted species of odontocetes in the area based on 2 yr of periodic areal and shipboard surveys.

There was no correlation between the use of sonar and the daily duration of odontocete vocal activity. We noted several instances when odontocete vocalizations overlapped with midfrequency sonar and a few instances where odontocetes apparently mimicked the sonar signals. From this project, we can determine that marine mammals are being exposed to sonar during this naval exercise, but we do not know at what levels and for what lengths of time they are exposed. Due to the extent of the spacing between the MARU locations, we were not able to determine location for any of the vocal events.

Overall, these recorders yield important information about the presence of odontocetes during these naval sonar exercises in Onslow Bay. Since this recording event, the Navy was able to arrange two more deployments in 2009 in conjunction with planned naval exercises occurring at the proposed USWTR site off Jacksonville, FL.

Development of a National Database and Standards for Protected Species Observer Data in the United States

Howard Goldstein, Deborah Epperson, Kyle Baker, Kimberly Skrupky, Brad Smith, Gregg Gitschlag, Jill Lewandowski, and Teresa Turk

1 Background

Many noise-producing activities such as naval sonar, seismic surveys, and coastal and offshore development have the potential to be harmful to marine species protected under the US Marine Mammal Protection Act (MMPA) and Endangered Species Act (ESA). Under MMPA, ESA, and Outer Continental Shelf Lands Act (OCLSA) requirements, protected species observers (PSOs) are commonly required to monitor protected species and the offshore activities, follow mitigation protocols, collect data, and report to government agencies responsible for managing natural resources and regulating many noise-producing activities.

H. Goldstein (✉) • T. Turk
National Marine Fisheries Service, National Oceanic and Atmospheric Administration,
Silver Spring, MD 20910, USA
e-mail: Howard.Goldstein@noaa.gov; Teresa.Turk@noaa.gov

D. Epperson
Bureau of Ocean Energy Management, Regulation and Enforcement, New Orleans, LA 70123, USA
e-mail: Deborah.Epperson@boemre.gov

K. Baker
National Marine Fisheries Service, National Oceanic and Atmospheric Administration,
St. Petersburg, FL 33701, USA
e-mail: Kyle.Baker@noaa.gov

K. Skrupky • J. Lewandowski
Bureau of Ocean Energy Management, Regulation and Enforcement, Herndon, VA 20170, USA
e-mail: Kimberly.Skrupky@boemre.gov; Jill.Lewankowski@boemre.gov

B. Smith
National Marine Fisheries Service, National Oceanic and Atmospheric Administration,
Anchorage, AK 99513, USA
e-mail: Brad.Smith@noaa.gov

G. Gitschlag
National Marine Fisheries Service, National Oceanic and Atmospheric Administration,
Galveston, TX 77551, USA
e-mail: Gregg.Gitschlag@noaa.gov

A.N. Popper and A. Hawkins (eds.), *The Effects of Noise on Aquatic Life*,
Advances in Experimental Medicine and Biology 730, DOI 10.1007/978-1-4419-7311-5_147,

The National Marine Fisheries Service (NMFS) and the Bureau of Ocean Energy Management, Regulation and Enforcement (BOEMRE) formed a PSO working group (PSOWG) in 2008 to review offshore energy PSO programs. The PSOWG recommended standardized training, data collection, and reporting procedures and the establishment and maintenance of a national database to house data for routine analysis. To increase the quality and utility of PSO data collected during offshore energy projects, improving the collection and reporting of quality data is important to monitor the impacts of offshore activities and to inform future adaptive management decisions. Using seismic surveys as a model for the national PSO program, standardized training in data-collection methods and reporting procedures, development of standardized forms, and establishment of a centralized database will allow the natural resource agencies to more effectively manage offshore resources and potential impacts on protected species.

2 Development of National Standards for PSO Data

Data reporting is routinely required through Federal authorizations to monitor the effectiveness of monitoring and mitigation measures and the potential impacts of noise-producing activities on protected species. Data standards are expected to improve data quality and analysis of datasets between different regions, improve evaluations of monitoring and mitigation effectiveness, and inform adaptive management strategies. Standardized forms and software used by PSOs will minimize discrepancies among datasets and will allow the synthesis and comparison of datasets in which data collection and reporting are otherwise required and collected independently of one another. The importance of accurate and complete reporting of the results of the monitoring and mitigation measures and their effectiveness at reducing or avoiding potential environmental impacts incidental to noise-producing activities are important components of any successful adaptive management strategy.

Currently, there is considerable variation in the data-collection requirements and procedures between the authorized activities occurring in US waters off the coasts of the Gulf of Mexico, Atlantic Ocean, Pacific Ocean, Pacific Islands, and Alaska and on the high seas (international waters) for seismic surveys conducted for purposes of oil and gas exploration and production and research. Past challenges in attempts to analyze these data include no standardized collection and reporting procedures, formatting issues, and discrepancies in the metrics reported. Requirements for data collection and reporting should include information on PSO effort, survey details, and species sightings. Reports should be submitted in a standardized format and in standard metric units. The development of quality assurance and quality control standards and the development of data quality, format, electronic submission, and analysis standards will make data analysis more efficient and robust. The implementation of national standards for these aspects of seismic survey PSO programs would increase the integrity of data collected and reported for more effective protected species conservation management.

3 Development of a National Database

The NMFS and BOEMRE are proposing development of a national database for collecting data submitted in monitoring reports required under MMPA, ESA, and OCSLA authorizations. Importing and organizing electronic data into a national database would ensure greater consistency and timeliness in reporting as well as for subsequent analysis. Other potential benefits of a national database include monitoring an applicant's compliance with authorizations; analyzing cumulative effects on the human environment under the National Environmental Policy Act (NEPA); presence or absence, distribution, and abundance of protected species; and disseminating information to the public who are interested in protected species issues, such as sharing or contributing to the Ocean Biogeographic Information Systems Spatial Ecological Analysis of Megavertebrate Populations (OBIS-SEAMAP) database.

Coordinated Management of Anthropogenic Noise From Offshore Construction Off Sakhalin Island for Protection of the Western Gray Whale

Roberto Racca

1 Introduction

The main population of a critically endangered cetacean species, *Eschrichtius robustus* (western gray whale), spends the summer months foraging on benthos-rich feeding grounds in the near-shore waters of the Piltun-Astokh region off northeastern Sakhalin Island in the Russian Far East. Significant oil and gas interests also make this region a very active offshore development area, with activities ranging from subsea pipeline deployment to installation of hydrocarbon production platforms.

Starting in 2004, a multiyear construction program for the Sakhalin II project was planned and conducted according to the principles of anthropogenic noise mitigation tailored to minimize exposure of the whales to levels considered disruptive to foraging and calf rearing. This effort involved advanced noise-propagation modeling of each season's activities to assess their potential impact, iterative adjustment of operational schedules to minimize noise exposure within the whale feeding area, and subsequent real-time monitoring of sound levels during construction at various locations to ensure that limits were not exceeded. To date, this operation is likely the most complex offshore development yet performed using such an extensive and integrated paradigm of noise forecasting, planning, monitoring, and mitigation response.

2 Modeling and Planning

Each construction season from 2005 to 2007 was characterized by a main focus. In 2005, the floating into place and lowering to the seafloor of the concrete gravity-based structure (CGBS) for a new hydrocarbon production platform. In 2006, the dredging of a pipeline route connecting two offshore platforms to Sakhalin island (the pipeline had to be buried in the sediment at water depths up to 30 m to avoid winter ice scour) and the laying of said pipeline, In 2007, the floating into place and installation atop the CGBS of the entirely prebuilt topsides (including everything from drilling machinery to living quarters) of the new platform. For each of these years, the aggregate underwater noise levels from the vessels to be involved in the operations were preassessed through modeling

R. Racca (✉)
JASCO Applied Sciences, Victoria, BC V8Z 7X8, Canada
e-mail: Roberto.Racca@jasco.com

A.N. Popper and A. Hawkins (eds.), *The Effects of Noise on Aquatic Life*,
Advances in Experimental Medicine and Biology 730, DOI 10.1007/978-1-4419-7311-5_148,

to ensure that they would not exceed acceptable limits for the whale population feeding along shore. Before the complex 2006 pipeline installation, different staging options were evaluated to minimize estimated noise footprints, and a full comparative environmental assessment (CEA) was carried out in 2004 to select the pipeline routing that would least impact the whale population during construction. The planning of every year of activity required determination of acoustic source levels for every major vessel involved, definition of the spatial-temporal scenarios of the operation, and modeling of the acoustic footprints; each of these requirements is discussed below

2.1 Source-Level Measurements

Any source of underwater sound can be characterized as having an intrinsic acoustic strength at each frequency that is independent of the environment in which the sound propagates. This source level (SL) is conceptualized as being measured at 1 m from a pointlike emitter. Clearly, it is impossible to register such a quantity directly for a large marine structure such as a dredger 100 m or more in length; it is possible, however, to measure acoustic levels at a sufficient distance from the source and, with due consideration to the environment in which the measurement is taken, "back propagate" the sound levels to the notional 1-m reference. Knowing the SL for a particular vessel at a given regimen of operation enables estimation of its acoustic footprint in any environment through propagation modeling.

For certain underwater sound sources, such as geophysical survey air gun arrays, it is possible and often preferable to model the SL from physical principles rather than attempting to measure it. This is not at all easy for a complex mechanical system such as a vessel, where resonances and couplings among different components shape the overall acoustic emission. All of the SL values used in the acoustic modeling of Sakhalin II construction scenarios were therefore obtained in advance through SL measurements of the vessels to take part in a given operation or of closely similar proxies. Extensive SL measurements were carried out in 2004 on many of the vessels to be involved in the Piltun-Astokh construction as they performed similar activities at a development site a few tens of kilometers to the south, outside the region of potential acoustic impact on the western gray whale population. In some cases, however, vessels due to participate in the construction could be measured in advance only while operating in vastly different geographic locations; indeed, the SL of a large pipe-laying barge was measured by members of the project team off the north coast of Australia.

2.2 Scenario Definitions

For operations such as the 2005 CGBS installation or the 2007 topsides float-in, it was relatively simple to "storyboard" the activities in terms of a few notional arrangements of towing vessels and support craft. Because of the concentration of vessels and sustained engine power levels required to position and hold a very large structure on a precise station, the most critical phase from a noise emission standpoint involved a precisely planned arrangement of tugs and could therefore be defined precisely for the model scenario. Much more complex was the modeling of the 2006 dredging and pipe laying in which dozens of vessels would operate in large "spreads" (clusters) at constantly shifting locations along the pipeline route. The condensing of months of planned activity into a finite number of cases for aggregate acoustic footprint modeling was simplified by defining, along the pipeline route, several notional stations at which to place predetermined arrangements of

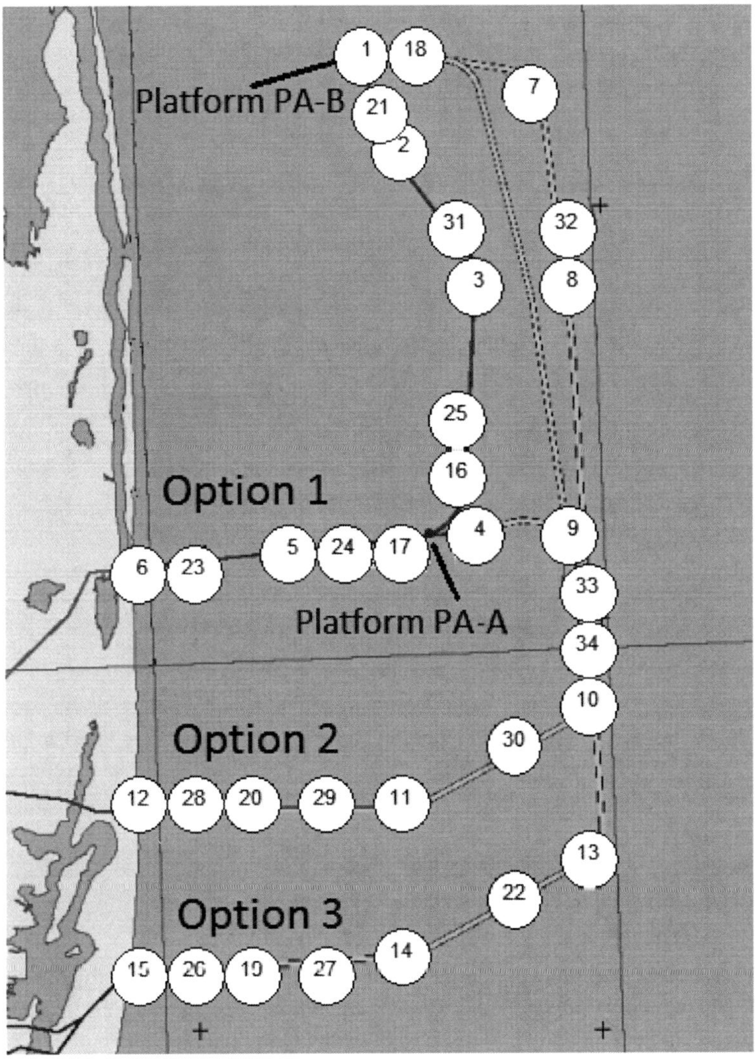

Fig. 1 Notional sound modeling stations for construction spreads along three possible pipeline routes

vessels defining the construction spreads. The construction season was then temporally subdivided into broad activity phases corresponding to a static distribution of spreads at given stations based on the temporal-geographical progression of pipeline construction. This yielded 10 consecutive model scenarios, each with up to 10 noise-source epicenters (clusters of vessel sources) at the notional stations. The sound propagating from each individual source in a scenario was modeled to yield an aggregate footprint for that phase of construction. This process was performed in the context of the CEA for three pipeline routing options, which resulted in the eventual selection of the longest route to landfall as providing optimally mitigated construction noise conditions. Figure 1 shows the full set of operation modeling stations along the three pipeline routing options, giving a feel for the extensive level of effort involved in the footprint forecasting.

2.3 *Modeling and Footprint Rendering*

An advanced acoustic propagation modeling software, JASCO's Marine Noise Propagation Model (MONM), was used to estimate the noise footprints of individual vessels in each construction scenario, which were then added to produce aggregate noise-level maps. MONM is a proprietary implementation of the widely used parabolic equation code RAM (Collins 1993), modified to account for shear-wave losses at the seabed, an important consideration in the shallow-water, absorptive bottom environment on the Sakhalin shelf. MONM uses a complex density method (Zhang and Tindle 1995) to implement the shear-wave energy conversion in a significantly faster computational manner than other approaches. MONM has been used for a variety of environments and extensively validated against measurements; for the Sakhalin II application, it was specifically tuned over the relevant frequency range (from a few tens of hertz to a few kilohertz) by adaptively refining the propagation parameters for the best match of the transmission loss (TL) estimates of the model with results of dedicated TL studies conducted over numerous transects in the relevant geographic area.

For each individual noise source in a scenario, MONM estimated frequency-specific TL in one-third octave bands along a fan of radials from the source location (also accounting for depth at the site of principal sound emission, usually the propellers or thrusters for a vessel). Combined with the measured SL in the same bands, these yielded radial footprints of frequency-specific received levels that were then summed to give broadband levels. Finally, the radial footprints of all sources in a scenario were recast onto a common geo-referenced grid and summed into an aggregate noise-level map for a given phase of an operation. Figure 2 shows a sample modeled noise-level map of concurrent dredging and pipe-laying operations in the southern part of the pipeline on the landfall segment. Only the principal vessels are labeled; additional vessels (such as barge support tugs) were in fact modeled to generate this map.

3 Monitoring and Mitigation

During each construction season, a real-time acoustic-monitoring infrastructure was deployed to enable constant verification of anthropogenic noise levels propagating into the western gray whale feeding area, whose expected notional boundary in the 2006 season, estimated from population surveys in prior years, is shown as a shaded outline in Figure 2. Noise levels were monitored with moored underwater acoustic sensors with radio buoys that broadcast telemetric data to an onshore station. Four such moorings were deployed in construction years 2005 and 2007 and five in 2006 at locations following approximately the notional outline of the feeding area shown in Figure 2. The telemetric stations broadcast (over unused channels of the marine VHF band) a modulated signal that carried the full acoustic waveform information, thus enabling its analysis onshore to yield not only broadband SLs but also spectral distributions and, if necessary, direct audio playback to help identify sources. A team of acousticians took turns around the clock during active phases of construction to monitor the signals for any trend toward thresholds of potential behavioral influence on the whales, an eventuality that, thanks to the extensive preseason model-based planning, could reasonably be expected to be exceptional in nature.

Mitigation criteria were based on a multitier fixed-threshold paradigm aimed at responding both to mildly impacting but sustained exposure and to more significantly impacting transient exposure. Although a proportional dose-based criterion was not adopted for mitigation-response triggering, a time-windowed cumulative index was used by the acoustician on watch to evaluate trends. Throughout construction, there was never an outright exceedance of the criteria that would require a halt in activities, and in a few instances, the monitoring team was able to advise operations coordinators of unexpected rising trends that were then reversed through localized corrective actions.

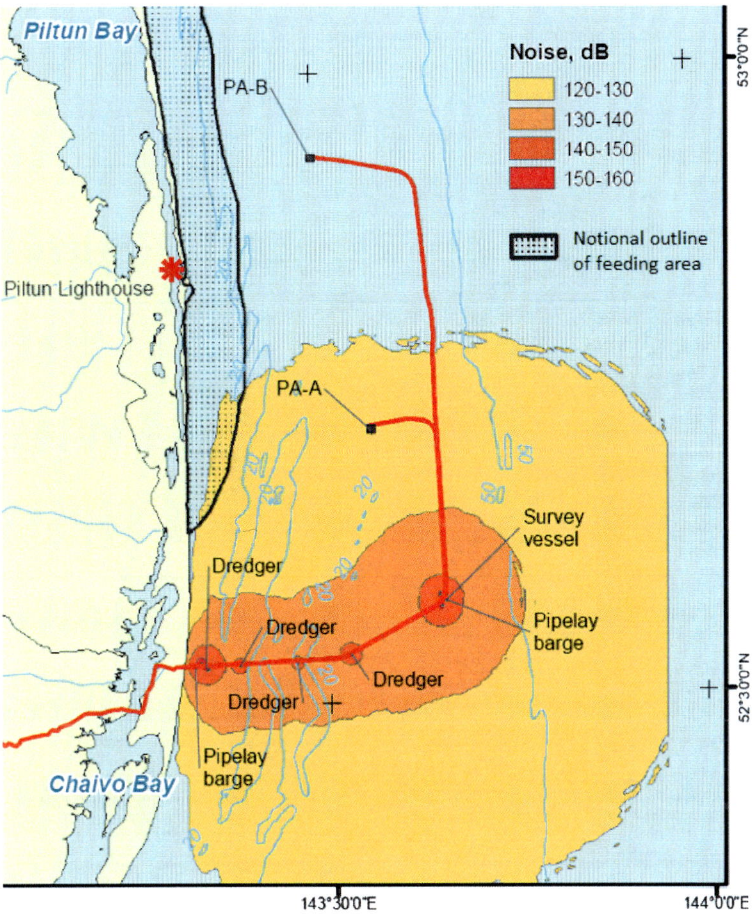

Fig. 2 Modeled acoustic footprint of a representative pipeline construction phase

4 Conclusions

The coordinated noise-management program described here is commensurate in its complexity to the level of endangerment of the whale population it is designed to protect. It must be recognized that such extensive measures, in both planning and executing an operation, may prove unfeasible or, at any rate, unwarranted for other ecosystems, activities, or geographic regions. This approach, however, sets a quality standard for comprehensive, end-to-end noise-impact prevention and could potentially serve well in a variety of circumstances.

References

Collins MD (1993) The split-step Padé solution for the parabolic equation method. J Acoust Soc Am 93:1736–1742.

Zhang ZY, Tindle CT (1995) Improved equivalent fluid approximations for a low shear speed ocean bottom. J Acoust Soc Am 98:3391–3396.

Part X
Workshops and Concluding Remarks

Workshop One: Risk Analysis

Thomas J. Carlson, Christ de Jong, and Rene P.A. Dekeling

The workshop looked at the assessment of risk to aquatic animals exposed to anthropogenic sound. The discussion focused on marine mammals given the worldwide attention being paid to them at the present time, particularly in relationship to oil and gas exploration, ocean power, and increases in ship traffic.

It was noted that risk assessment for fish exposed to sound was of most concern when sound exposure affected the catch of fishers, especially as a result of changes in fish behavior. Impacts on fish from physical injury to individuals or through population effects, such as a temporary interruption in spawning behavior, were currently of secondary concern.

It was clear to all that the nature of risk assessment may vary greatly with the goals of the assessment. An example was given of the business risk assessments conducted by oil and gas companies when an activity such as exploration is being considered. Here the risk assessment is focused on the complexity and costs of obtaining access to an area of interest given regulatory concern for the well-being of marine mammals and fish. Typically, ordinal-scaled data are used to rank alternative outcomes where the distance between outcomes is uncertain, but their rank in terms of risk of either a positive or negative outcome is clear. Sums of ranks are used to assess the merits of alternative strategies to pursue access to the area of interest, where both the risk of exposure to animals present in the target area and the business risk of investment in the activity being considered are components of the analysis.

The group also discussed risk assessment where both exposure and response are quantifiable and measured using interval and ratio scales. Examples were numerous and included the risk analysis possible for individuals and populations given assessment of exposure and dose-response functions. The discussion of the range of application of risk assessment led to identification of a number of needs that could lead to an improved assessment of the risk to aquatic animals of underwater sound and the communication of the results of those assessments to others.

T.J. Carlson (✉)
Pacific Northwest National Laboratory, Battelle, Portland, OR 97204, USA
e-mail: Thomas.Carlson@pnl.gov

C. de Jong
MON/Acoustics, TNO Science and Industry, 2600 AD Delft, The Netherlands
e-mail: christ.dejong@tno.nl

R.P.A. Dekeling
DG Water, 2500 EX Den Haag, The Netherlands
e-mail: rene.dekeling@minvenw.nl

A.N. Popper and A. Hawkins (eds.), *The Effects of Noise on Aquatic Life*,
Advances in Experimental Medicine and Biology 730, DOI 10.1007/978-1-4419-7311-5_149,

It was agreed by all that the primary reason for the assessment of risk, particularly as related to marine mammals and commercially important fish stocks, is communication between those with expertise in business, scientific, and engineering disciplines and those relying on such expertise to make decisions. A need for clarity in the definition and communication of the scope, approach, and purpose of risk assessment between subject matter experts and decision makers before the assessment was initiated was identified. A need to keep risk scenarios distinct was also evident. It was agreed that the more distinct the purpose and scope of a risk analysis, the greater the probability of successful completion and communication to others.

Workshop participants also discussed the difficulty and complexity of dealing with uncertainties and unknowns in risk assessments. Reference was made to newsmakers (e.g., often politicians) making statements about "not knowing what we don't know" when assessing the risk of alternative decisions given a difficult situation. It was agreed that paralysis was not the appropriate response when faced with such circumstances. Instead, a cautious approach should be taken, with provision for including alternative actions in any risk analysis. Strategic factors should be considered in assessing risk from exposure to anthropogenic sound. Advice from both experts and decision makers was needed.

The group observed that a recently completed assessment of available data had resulted in thresholds for the exposure of marine mammals to most categories of anthropogenic sound and that such an assessment was underway for fish and sea turtles. It was also observed that such assessments were based on sparse data in many cases and were generally based on measures of either the onset of physical injury or a surrogate for injury onset such as a temporary threshold shift in hearing. It was noted that the effects on behavior were becoming more important considerations in the analysis of the risk on the impact on aquatic animals from noise exposure. It was also noted that little information was available to permit identification of acceptable outcomes for behavioral responses to sound. At the moment, we just have thresholds for exposure leading to behavioral responses. The key idea here is that the behavioral response to sound exposure does not necessarily mean that the well-being of the reacting animals has been affected adversely.

Regulators in the session identified a need for improvements in the analysis and reporting of events such as the stranding of marine mammals. It was noted that other industries, transportation being the example cited, have strategies and standards for the investigation of events such as airplane crashes. Such "root cause analysis" has the objective of identifying, without ambiguity, the factors that led to the event and communicating that analysis to industry and government so that the probability of their reoccurrence can be reduced. Workshop participants noted that although the evaluation of such events, particularly for marine mammals, has been improved, the assessments still lack the standardization and rigor that would make them more informative and useful. "Root cause analysis" should be undertaken for significant events of consequence to marine mammals and other aquatic animals. All agreed that the resulting data would have a high value in future risk assessments and could identify exposures to be avoided.

Workshop participants agreed on the need for standardization in underwater exposure metrics and that such standardization would help both the conduct and presentation of risk analyses. Participants noted that not only was the mathematical form of metrics important but also clarification of when either particular metrics or the preferred measure and standards for acquisition and processing of underwater noise signals were appropriate. All agreed that the science of underwater acoustics is complicated and that standardization of the basics from the mathematical forms for metrics to their acquisition and processing is needed. Those in the group who were members of standards committees or were otherwise familiar with the development and application of standards noted that actions in some of the identified areas were underway and agreed to carry the workshop findings into their standards development activities.

A concluding discussion by the workshop participants focused on the means for providing not only data but also information of the importance for risk assessments. Workshop participants agreed that not all of the information or elements of data that would be of use in risk assessments were available in the peer-reviewed literature. The idea of a Web site that could serve as a "clearing house" or other venue for peer-reviewed literature, reports, and other information of value for risk assessment was explored.

Workshop Two: Long-Term and Cumulative Effects of Acoustic Exposure Along With Other Stressors

Robert Gisiner and George Frisk

1 Introduction

The working group considered long-term and cumulative effects of acoustic exposure, including the effects of acoustic exposure in combination with other stressors. The discussion was initiated by three short informal presentations invited by the organizers. This paper summarizes the main points that emerged from the workshop. In general, we have avoided specific attribution of comments because the points might not always be original to the speaker or the speaker might not always have been identified. We are very grateful to the workshop participants who generously shared their expertise and insights with us and apologize for any errors, omissions, or misinterpretations on our part. That said, we hope these notes will be useful to others exploring this complex and challenging topic.

2 Introductory Presentations

Andrew Wright (Chapter 123) provided an overview of a 2009 workshop organized by the Okeanos Foundation on the cumulative effects of underwater noise (Wright 2009). Dr. Wright noted the difficulty of managing and assessing the effects of any sound source without understanding the contributions of other sources of noise in an organism's environment. He also stressed the shortcomings of a project-by-project approach to the assessment and management of effects, acknowledging that he and his colleagues were not alone in their appreciation of this weakness in current legislative and regulatory mechanisms for managing the effects of sound. Increasing emphasis on ecosystem-based management by the conservation and resource management community was consistent with this shifting emphasis from single-species, single-stressor management to consideration of the community of organisms within an ecosystem and all the stressors affecting that community. The Okeanos workshop also focused on the challenges of trying to anticipate and manage future noise regimens

R. Gisiner (✉)
Chief of Naval Operations, Energy, Installations and Environment Division
and Oceanography Department, Naval Postgraduate School, Monterey, CA 93943, USA
e-mail: bob.gisiner@navy.mil

G. Frisk
Department of Ocean and Mechanical Engineering, Florida Atlantic University,
Dania Beach, FL 33004, USA
e-mail: gfrisk@.seatech.fau.edu

A.N. Popper and A. Hawkins (eds.), *The Effects of Noise on Aquatic Life*,
Advances in Experimental Medicine and Biology 730, DOI 10.1007/978-1-4419-7311-5_150,

and their effects. Toward that end, the Okeanos Foundation, like others, is currently exploring the merits of individual-based models and sound-field mapping as tools for future underwater sound assessment and management.

Doug Wartzok (Wartzok et al., Chapter 125), Chair of a National Research Council (NRC) panel that examined this topic (NRC 2005), reviewed the population consequences of acoustic disturbance (PCAD) model that had emerged from the panel's deliberations. Like the Okeanos model, the PCAD model employed an individual-based approach to conceptualizing and modeling effects. The relationship between the effects of stressors on individuals and the consequent effect on population dynamics was formulated within the context of existing population risk-tolerance models like the potential biological removal (PBR) model developed by the US National Marine Fisheries Service, with the intent of preventing the decline of healthy populations or delaying recovery for depleted populations.

Mike Weise (Office of Naval Research) reported on a newly convened workshop process supported by the Office of Naval Research, energy industry, and other partners. The workshops follow an independent expert working group format developed by the National Center for Ecological Analysis and Synthesis (NCEAS). The workshop team has been tasked with examining a number of marine mammal populations known to present very rich datasets for the type of analysis set forth in the PCAD model. Dr. Weise reported that although the group's work is not yet published, they have discovered that the available data for well-studied species like elephant seals and bottlenose dolphins support a surprisingly complete implementation of the PCAD model. Subsequent efforts to populate PCAD models with existing data will look at the available data for right whales and beaked whales. The initial unpublished results suggest that we may be better able to construct complex, realistic models of long-term, cumulative effects than had initially been anticipated when the PCAD model was formulated in 2007.

3 Terminology and Definitions

One of the first emphases to emerge from our discussions was the need for consistency and clarity in terminology use to describe stress and cumulative effects. A definition of stress itself was recognized as being difficult at this time, although the medical significance of stress assessment is driving a considerable research effort on the topic, including the relationship between behavioral or psychological stress and its physiological manifestations. Stress was noted to be a normal part of life, integral to stimulating and maintaining healthy neuroendocrine responses and immune system activity. Predicting when stress becomes excessive or undesirable remains difficult. Individual differences in response to stressors are clearly subject, like all other traits, to natural selection and will play a role in the observed population effects; we can expect the response within a population to be nonuniform even under ideal circumstances as organisms balance stress-response capabilities with other biological processes that also contribute to survival and reproductive success. The group also noted that energy budget models have also been used as a surrogate for direct measurement of the physiological effects of stressors, but these models were not discussed in detail.

For cumulative effects, we concurred that it was important to use two different terms to discriminate the effects that occur over time (cumulative effects) from those that co-occur or overlap in time (aggregate effects). The interactions between stressors were recognized to be potentially linear (additive) or nonlinear (multiplicative, synergistic, exacerbating, or logarithmic). In describing the effects of multiple recurrences of the same stressor or the effects of combinations of stressors, development of an interaction factor will be needed in considering nonlinear cumulative or aggregate effects. Examples of chemical and biological effects that illustrate these interactions are well documented, but such interactive effects between multiple acoustic stressors or between acoustic stressors and other stressors are still not well documented.

4 Relevant Enabling Concepts

The group discussed the relationship between the short-term, immediately observable consequences of underwater noise and the long-term consequences. The most frequently emphasized and readily observed effect is behavioral change, but the degree to which observed behavioral responses predict significant deleterious long-term consequences for the individual or population are not easy to predict. This was the particular focus of the NRC PCAD workshop (NRC 2005) and a recognized problem in current environmental protection legislation, particularly the Marine Mammal Protection Act, in which all observable behavioral responses to disturbance are scored equally in permits and risk analyses.

The concept of homeostasis versus allostasis was discussed as a heuristic mechanism for discriminating normal or tolerable variations in environmental stress from variation that presents meaningful survival and reproductive consequences to the individual and population, often characterized as "biologically significant" effects of a stressor or cumulative stress.

5 Modeling Cumulative Effects

The modeling efforts described by Wright, Wartzok, and Weise at the start of our discussion were not designed specifically or exclusively for the effects of sound on marine mammals; they are general models, capable of modeling effects of any and all stressors on any species. All concurred that although these models represent a considerable leap forward in the quantitative expression of the effects of sound on the marine environment, alternative modeling approaches should continue to be considered. For example, we noted that the review by Wartzok et al. earlier in this conference (Chapter 125) of three US Navy models of acoustic effects found that under some circumstances, a simple model with simplifying assumptions may be a more effective decision-making tool than a model that attempts to capture the detailed biology and dynamic acoustic environment of a given sound-exposure scenario.

6 Suggestions

Although considerable work remains to be done in defining stress, in assessing the contributions from multiple stressors, and in deriving a measure of cumulative biological effects, the workshop participants did concur on some practical suggestions for dealing with the environmental risk from man-made sound. First among these was a recommendation to identify priority species, stressors, and locations for greatest attention.

Although almost every marine ecosystem contains multiple natural and man-made stressors, the group considered that the most effective action would be to focus on controllable sources of stress. Although animals may experience considerable stress from decadal climate variation and associated movement or loss of prey, that is an uncontrollable source of stress, whereas management of controllable sources of stress, usually man-made, would be more likely to produce a positive effect, e.g., relocating or curtailing competing fisheries during a decadal climatic oscillation, reducing chemical contamination in a noisy harbor area, or managing the use of different sound sources over time to reduce synergistic effects from multiple sound sources operating concurrently.

The group also recommended efforts to assess the relative importance of different stressors. In some cases, it may not be necessary or useful to identify and monitor the full suite of stressors, recognizing that in many cases, a single dominant stressor can overshadow aggregate effects. This had been the implicit prevailing reasoning behind the focus on the effects of military midfrequency sonar systems. Although the sonar systems are not common or persistent sources of underwater noise, the unique combination of loudness, frequency structure, and mode of use for midfrequency antisubmarine sonar systems appears to produce effects that are qualitatively different from the known effects of other more common sources of anthropogenic noise such as ship noise. However, the persistence and continuing increase in sound sources associated with commercial shipping and offshore energy exploration and development (both fossil fuel and alternative energy) could potentially have an overall cumulative effect that is greater than that from louder but less ubiquitous sources such as sonar systems. As stated earlier in this report, efforts to manage and reduce the effects of man-made noise must occur within the context of a fuller understanding of all contributors to ocean noise along with their interactions with each other and with other nonacoustic stressors.

References

National Research Council (NRC) (2005) Marine mammal populations and ocean noise: Determining when noise causes biologically significant effects. National Academies Press, Washington, DC.

Wright AJ (ed) (2009) Report of the workshop on assessing the cumulative impacts of underwater noise with other anthropogenic stressors on marine mammals: From ideas to action. Available via http://okeanos-foundation.com/topics-and-projects/symposia/completed-symposia/.

Workshop Three: Approaches to Behavioral Analysis of Responses to High-Energy Acoustic Sources

Svein Løkkeborg and Frank Thomsen

1 Background and Objectives

The behavioral effects of high-energy acoustic sources such as pile driving, seismic surveys, explosions, and sonar on marine mammals and fish have received increasing attention by scientists, policy makers, and stakeholders. In both mammals and fish, behavioral responses can potentially occur at relatively low levels of noise exposure and, therefore, impact zones can be quite large. These responses may prevent marine mammals and fish from reaching breeding or spawning sites, finding food, and acoustically locating mates. Any of these effects could lead to potential long-term effects on reproduction and population parameters. In the case of fish, avoidance reactions can also result in displacement away from potential fishing grounds and might result in reduced catches. Studies have been undertaken using a variety of methodologies ranging from purely qualitative observations to controlled laboratory experiments.

The idea of the workshop was to bring together some of the concepts and ideas on behavioral analysis of responses to high-energy acoustic sources that have already been discussed, both in the literature and during the Cork Conference. We wished to promote a wider discussion of some pressing issues such as general issues (objectives, field vs. laboratory); the design of behavior experiments; acoustical problems of presentation and measurement; behavior metrics: how do we assess behavioral changes?; and so what? How do we assess the biological significance of behavioral changes?

The workshop was divided into three discussion slots covering these issues. Each slot was opened with short introductory remarks by Rebecca Dunlop (University of Queensland, Brisbane, Australia; experimental design and behavioral-response analysis), Ronald Kastelein (Sea Mammal Research Company [SEAMARCO], Harderwijk, The Netherlands; acoustical problems), and Jacob Tougaard (Aarhus University, Roskilde, Denmark; quantifying behavior/behavior metrics) followed by an open discussion.

S. Løkkeborg (✉)
Institute of Marine Research, 5005 Bergen, Norway
e-mail: svein.lokkeborg@imr.no

F. Thomsen
Cefas, Lowestoft, Suffolk NR33 0HT, UK
e-mail: frank.thomsen@cefas.co.uk

A.N. Popper and A. Hawkins (eds.), *The Effects of Noise on Aquatic Life*,
Advances in Experimental Medicine and Biology 730, DOI 10.1007/978-1-4419-7311-5_151,

2 Discussion

2.1 Experimental Design and Behavioral-Response Analysis

This part of the workshop addressed the important aspects to consider when planning behavioral-response studies. Behavioral studies may be time consuming and expensive to conduct, and the sample size per experiment is often low. With some idea of the "normal variance," a power analysis is a useful tool when setting up the number of groups (i.e., the sample size) for a particular experiment. Variations within a group may be reduced by using uniform groups (e.g., male vs. female), but it is important to note that extrapolation with respect to sex, age, and species must be made with great caution. Pilot studies can provide valuable information about sample size and variance when designing experimental procedures. The importance of avoiding pseudoreplications when carrying out behavior experiments was discussed. This problem can be overcome by ensuring that enough and appropriate replicates are chosen. In deciding what type of behavioral-response variables are to be observed and measured, it is essential to consider the biological significance of changes in behavior.

The inferences that can be drawn from behavioral studies was a central topic in the discussion. It was emphasized that we have to be cautious. Conclusions should only be made with respect to the type of source employed and the stimulus sounds to which the experimental animals were exposed. It is essential to use control groups to attribute observed responses and effects to the sound source studied. Studies that do not have a control may be confounded by explanatory variables other than treatment effect. Thus to be able to attribute unambiguously behavioral changes to a particular sound source (i.e., to demonstrate treatment effects), the experiment needs to include replicate controls and a before/during/after design.

2.2 Acoustical Problems

The importance of involving experts in acoustics in the experiments was discussed and emphasized. Because of the complexity in sound presentation and measurement in both laboratory and field studies, these issues should be handled by persons sufficiently trained in acoustics. In principle, using a real sound source in the experiment is recommended. However, this might not always be possible or feasible. For example, some types of sound have very high source levels (pile driving, seismic air guns) and such sources may injure animals. It is also possible that the exposed individuals will move out of range of the observer. Playback studies often investigate responses to particular sound pressure levels or frequencies, which is only possible using speakers systems where the sound can be manipulated. Thus, although it is desirable to use real sound sources, the exact design of the study depends on the questions asked.

The question whether to perform studies in the wild or under laboratory conditions is a complex one. It was noted that laboratory studies usually provide greater control over experimental conditions and might be more cost effective. It was considered important, however, that the response being measured should be biologically relevant, e.g., it should fall within the range of behavior shown in the wild. Studies in the field could yield very powerful results because animals would show "natural" behaviors. But field studies can be costly and gaining full control over experimental conditions is challenging. In essence, the methodology must depend on the specific objectives of the study. Much can be learned from controlled experiments in the laboratory that is not possible in field studies and vice versa. We therefore need both laboratory and field studies to better understand how aquatic animals react to anthropogenic sounds. It is important to test predictions from the laboratory in the field. In particular, the consequences of behavioral changes for fitness and survival have to be tested in field studies.

2.3 *Quantifying Behavior: Behavior Metrics*

Because the discussion on the former two topics was very time consuming, only a relatively small amount of time could be dedicated to the issue of how to quantify behavior. It was noted that there is a hierarchy of goals related to the quantification of behavioral reactions: determination of the zone (range) of reaction, quantification of impact, and establishment of exposure criteria. Several examples of relevant behavior metrics were mentioned during the workshop (movement, swimming speed, startle response, dive profile, respiration rate and force, vocalization). The behavior metrics measured could be related to physiological changes. Measuring group behavior (individuals pooled into a group) is often the easiest but fails to capture the most sensitive and susceptible animals. Habituation versus adaptation to sounds was discussed in relation to the quantification of behavioral responses. Habituation may occur when the benefit of staying is higher than the cost of avoidance (moving away).

Workshop Four: International Regulatory Issues

Amy R. Scholik-Schlomer and Camille Mageau

1 Introduction and Context

The goal of the two-hour workshop was to discuss the issue of underwater noise in the context of international regulations.

To place the topic in context, Camille Mageau, Workshop Chair, listed the diversity of participants (e.g., industry, government, advisory agencies, and consulting firms), commenting that they illustrate the breadth of interests involved in domestic and international discussions on impacts and mitigation/management of underwater noise in the marine environment. The Chair also reviewed the different types of expertise and parties most often needed in the development of regulations and the usual conditions that need to be met before an issue is subject to regulation.

The expertise most often needed to support a regulatory initiative includes generators of knowledge: scientists from academia, government, and nongovernmental organizations conducting studies to inform decisions; providers of advice: those who synthesize and translate knowledge for use by policy makers and regulators; recipients of advice: those who translate advice into policy, regulatory, and/or nonregulatory tools and decisions; and stakeholders: those who have the strongest stake in the quality of the advice given, including those who are required to comply with policies or regulations.

The overarching conditions leading to international regulation of an issue are international recognition of the problem; governance: international or, at the very least, a regional management body to regulate and enforce regulations; political will to address the issue on an international level; and tractability of the issue: whether regulation is the best approach (workshop participants decided to focus the discussion on this topic).

It was acknowledged by the workshop participants that the issue of underwater noise is inherently complex. There are multiple sources of concern, evolving technologies, various techniques to mitigate impacts, constantly evolving knowledge, and many areas where uncertainty still exists. It was also noted that funding to understand the potential impacts of "new sources" (e.g., alternative energy),

A.R. Scholik-Schlomer (✉)
National Marine Fisheries Service, National Oceanic and Atmospheric Administration, Silver Spring, MD 20910, USA
e-mail: amy.scholik@noaa.gov

C. Mageau
Fisheries and Oceans Canada, Ottawa, ON K1A 0E6, Canada
e-mail: camille.mageau@dfo-mpo.gc.ca

A.N. Popper and A. Hawkins (eds.), *The Effects of Noise on Aquatic Life*, Advances in Experimental Medicine and Biology 730, DOI 10.1007/978-1-4419-7311-5_152,

about which very little is known, will be particularly challenging and that the international science community is still struggling to better understand "traditional sources" (e.g., seismic, sonar).

All of these factors often make it difficult to effectively translate rapidly evolving science into regulations. Additionally, it was recognized that many of the conditions needed for the regulation of noise at the international level are not in place and that many of the challenges associated with ocean noise cannot be addressed all at once. To be most effective, the issue needs to be broken into smaller and more manageable parts. For example, the regulation of underwater noise first needs to be resolved on a domestic level before it can effectively be managed internationally. This is challenging because most policies and statutes were not created to deal directly with noise, leading most countries to address management of ocean noise through nonregulatory or indirect means such as environmental impact assessment processes.

With these acknowledgements, the workshop participants decided to focus their discussions on the tractability of the issue, identifying major knowledge gaps and challenges to addressing these gaps and common concerns and ways to collectively improve the international community's ability to move forward toward possible solutions. Everyone agreed that sound science is needed to create sound policies and, at this point, getting agreement on the science is probably easier than getting agreement on the policy. Thus the discussions focused on issues where potential progress could be made within the brief time available for this workshop.

2 Perspectives

Following the Chair's introduction, brief presentations were made by Mark Tasker (Joint Nature Conservation Committee [JNCC]), Jill Lewandowski (Bureau of Ocean Energy Management, Regulation and Enforcement [BOEMRE]), and John Young (ExxonMobil Exploration Company). These presenters provided three perspectives on the international regulation of underwater noise that helped stimulate further discussion.

Mark Tasker: There are no global regulation standards for industry to follow, but some regulation standards have been developed nationally and are being developed on a European scale. These may include explicit standards in relation to the treatment of noise in environmental assessments. There has been a distinct change in emphasis toward behavioral disturbance and masking and away from direct injury as undesirable effects of noise.

Jill Lewandowski: There needs to be a mechanism/network established to share information on the latest science and regulation of the noise issue. A general awareness of the current state of knowledge would likely help produce more scientifically sound policies.

John Young: There needs to be a balance between the human interests/needs and the environment. Industry often has its own standards ("best practices") that are applied globally. Many navies also have established "best practices" for operating beyond international borders

3 General Discussion and Conclusions

The group discussion focused on several topics ranging from information sharing and education to legal, societal, and funding issues. Recognizing that the intent of the workshop was to discuss international regulatory issues but that a number of prerequisites needed for the "regulation" of ocean noise at the international level were absent, the discussion focused on how some of these "prerequisites" had been addressed domestically through the use of streamlined environmental assessments processes, the development of government-endorsed codes of practice, and shared government/industry research projects. The discussion focused on positive ways to move forward. The group was successful in identifying the key issues as well as some potential means of working together better on this topic. A summary is presented in Section 3.4.

3.1 *Information Sharing*

One of the issues identified during the discussion was the need to better and more regularly facilitate exchanges among the science, policy, funding, and regulatory communities (i.e., beyond the *Effects of Noise on Aquatic Life* meetings). One idea put forward was the establishment of a Web-based information sharing network, with the contents uploaded and organized on a country-by-country basis (i.e., each country would have its own Web page within the sharing network). This Web-based correspondence system would 1) help those countries with fewer resources and capacity rapidly access and track research, policy, and regulatory and nonregulatory developments in other countries; 2) allow for better consistency in how assessments are completed and decisions are made (e.g., risk assessment protocols); and 3) promote an accelerated dissemination of scientific knowledge. It could also aid in identifying the best practices for assessments and ways to mitigate and monitor effects as well as generating a list of common priorities for research. Most importantly, the Web-based system would allow for the establishment of a worldwide professional network of contacts (i.e., those working on regulatory issues).

Regulators at the workshop presented identified data gaps and uncertainty as some of the greatest challenges. This information sharing network might help to more effectively identify common global priorities and needs. It was also discussed that it may be best for the information sharing network to be structured according to noise sources on an activity-specific basis rather than dealing with all sources/activities together. This strategy might allow for quicker and more productive progress in synthesizing known information, identifying gaps, and translating knowledge into policy and guidance (e.g., in the United States, the Fisheries Hydroacoustic Working Group [FHWG], which focused on pile driving associated with transportation-related activities on the US West coast, was able to develop interim criteria for fish injury by forming a small group of regulators, scientists, and stakeholders specifically focusing on this activity and its effects on fishes). Many agreed that this was a worthwhile idea but recognized that a country and/or an agency would need to take the initiative and establish an information sharing network into which other countries could "feed."

3.2 *Education, Communication, and Information Needs*

Public perception and understanding of underwater noise and its potential impact is fraught with misunderstanding and misinformation resulting in delays and/or cancellation of research programs and industry-related programs. As a whole, all groups mentioned above need to do a better job of educating the public, those in the legal system (e.g., judges dealing with legal issues associated with noise), and governments in a way that accurately and fairly communicates the level of knowledge/uncertainty and mitigation associated with the impact of noise in the aquatic environment. All of these considerations are particularly important given the competition for research funding. Additionally, both scientific data and policy need to be translated into something the public can easily understand.

Industry and regulators also have certain needs. From an industry perspective, there needs to be better communication and understanding of what should be mitigated (e.g., injury, behavioral disturbance). Industry also wants to know what to expect when operating across jurisdictional boundaries and would prefer as much consistency as possible. Regulators have a need for peer-reviewed scientific papers as well as summary papers, specifically written with regulators as the target audience, as standards to support global decision making. A set of regulatory-specific summary papers would identify what is known and what is unknown as well as ensure that everyone is considering and using the same data. These papers could be also be distributed by the information sharing network mentioned above. The group realistically raised the question as to who could take the lead in writing these background synthesis papers and who would fund and conduct the peer reviews.

3.3 Additional Factors for Consideration Among Regulators

Although most of the conference focused on the science of studying underwater noise, regulators face additional considerations beyond science in the application of policy in the "real world." Often forgotten by those who do not regularly deal with policy issues, the societal/socioeconomic consequences of an action are important considerations. One of the challenges is that societal values can vary locally and are often difficult to quantify. Furthermore, regulators must simultaneously interface with multiple user groups using the same space (e.g., industry, fisheries). Risk assessments also need to include the risk of not undertaking the activity being assessed (e.g., societal, economic, ecological), and regulators need to effectively communicate these considerations to the public.

3.4 Potential Ways Forward

During the conference, behavioral and cumulative impacts were identified as the primary issues where more knowledge is needed. To address these data gaps, longer term field studies are required. Challenges associated with conducting these studies range from social acceptability to funding supportbecause these types of studies are expensive and there are limited funders willing to invest in these more complicated, longer term studies. Recognizing that industry has developed research funding partnerships, participants suggested that a funding mechanism is needed where governments and industry can work together to more efficiently and effectively apply their funds to address issues of common concern. Again, an information sharing network might be one potential mechanism for accomplishing this.

Establishing regional noise budgets that include spatial, temporal, and spectral aspects would help set priorities, identify sources that need to be further addressed, and offer a means to better assess cumulative effects from multiple sources. Once noise budgets are established, they could be integrated with other forms of stressors within the environment (e.g., Halpern et al. 2008) and help regulators and managers focus on those activities most problematic from the "noise" perspective. These noise budgets would be useful from both an ecosystem perspective and an environmental compliance perspective.

In summary, the workshop discussion focused on gaining enough knowledge to better address concerns from a science and policy perspective. It was concluded that starting with small, manageable steps and moving to address consistency across boundaries was the most realistic approach. The need to develop common standards and codes of practice, which might support streamlining environmental assessment processes, and to educate the public, the judiciary, and funders would be a wise investment in time and effort. The establishment of an information sharing network and, ultimately, better communication among the various groups involved in this issue was seen as a huge step in the right direction.

References

Halpern, BS, Walbridge S, Selkoe KA, Kappel CA, Micheli F, D'Agrosa C, Bruno JF, Casey KS, Ebert C, Fox HE, Fujita R, Heinemann D, Lenihan HS, Madin EMP, Perry MT, Selig ER, Spalding M, Steneck R, Watson R (2008) A global map of human impact on marine ecosystems. Science 319:948–952.

Summing Up

Roger L. Gentry

1 Introduction

This summary covers the 56 oral presentations that were given in the regular sessions of the Second Conference on the Effects of Noise on Aquatic Life that took place in Cork, Ireland, in August 2010. The summary presents my subjective view, based on experience in this field, of findings that I found new or noteworthy. Of course, I recognize that any member of the audience would likely have different views and interpretations on all that was heard at the meeting, just as readers of this volume will have their own views.

2 General Trends Since the 2007 Nyborg Conference

The Cork Conference emphasized several general subjects that were only touched on at the First Conference on the Effects of Noise on Aquatic Life that took place in Nyborg, Denmark, in 2007. This change in emphasis shows that our thinking in this field is evolving.

2.1 Masking

The most commonly discussed new topic, featured in seven talks, was auditory masking, This subject seems to be much on people's minds these days, possibly because of the paper by Clark et al. (2009) about masking from shipping noise. Southall (Chapter 1) mentioned masking in his keynote address, and Dooling and Therrien (Chapter 17) and Fay (Chapter 29) discussed the extensive masking data that are available for birds and fish, respectively, in their presentations. Au, Richlen, and Lammers (Chapter 28) discussed the masking of fish calls by snapping shrimp and boats and the masking of humpback whale calls by boats. All these talks focused on the negative aspects of masking. Nachtigall, Supin, and Breese (Chapter 10), on the other hand, discussed the positive benefits of masking, at least forward masking that he sees as an essential element of the automatic gain control

R.L. Gentry (✉)
ProScience Consulting LLC, Dickerson, MD 20842, USA
e-mail: roger.gentry@comcast.net

A.N. Popper and A. Hawkins (eds.), *The Effects of Noise on Aquatic Life*,
Advances in Experimental Medicine and Biology 730, DOI 10.1007/978-1-4419-7311-5_153,

system he has been studying in echolocating cetaceans. Johnson (Chapter 127) listed masking as a noise effect that regulations should but do not presently include.

To me, a critical paper to consider was that of Reichmuth (Chapter 4) who discussed the various ways in which animals are released from masking. Clark et al. (2009) mentioned none of these release processes and thereby may have left the impression that masking is always total. Masking is a very complex phenomenon that should be included in the regulatory process and in the next version of the marine mammal noise-exposure criteria (Southall et al. 2007). However, much remains to be learned if these inclusions are to be science based. Fortunately, the US Office of Naval Research has started a new initiative on masking that may eventually produce the needed data, just as their former initiative produced needed data on temporary threshold shift (TTS).

2.2 *Context*

Another much-discussed topic in this conference was that exposure context may be a better index of behavioral response than the dose-response relationship. A dose-response relationship has been sought for decades but always unsuccessfully. It was found (US Department of the Navy 2007) that when migrating gray whales were exposed to a sound source placed in their migratory path, they began to deviate around it at a received level of ~120 dB re 1 μPa. But when the source was moved a mile outside the corridor, the animals showed no response to that level. The initial response of the migrating whales was caused by the context of the exposure (source placement), not by the exposure magnitude. In this conference, Dunlop, Noad, and Cato (Chapter 65) showed how context has become a central feature of the Australian humpback whale study that is about to start. Context will be woven into their multivariate analysis and modeling effort to measure the responses of humpback whales to seismic arrays. Southall's study (Chapter 1) of beaked whales and other odontocetes in Southern California also focuses on context through multivariate analysis. Both studies involve a very large number of possibly explanatory variables. Since Nyborg, it seems we have undergone a paradigm shift in our thinking about and measuring the behavioral effects of sound exposure.

2.3 *High-Frequency Exposures*

Finneran (Chapter 44) presented new data on TTS onset and growth rates when animals are exposed to frequencies higher than in his initial study (~3 kHz). He showed that at frequencies of 20 kHz and higher, TTS onset occurs at a lower sound exposure level and that it grows at a faster rate toward permanent threshold shift (PTS). These findings imply that changes are needed in the weighting functions and criteria for injury onset in the Southall et al. (2007) noise-exposure criteria. In the mid-1990s, the Acoustic Thermometry of Ocean Climate (ATOC) program raised concerns about the effects of low-frequency sound (~75 Hz) on whales. After 2000, the concern shifted to midfrequencies (3–10 kHz) with the sonar-related stranding of beaked whales in the Bahamas. Finneran's new data shift our emphasis to the higher frequencies (>20 kHz) at which most marine mammals hear best.

Ellison and Frankel (Chapter 98) mentioned this high-frequency problem in their talk, and Nachtigall and Supin (Chapter 10) expressed concern in their talk about the effect of exposure to high frequencies in cetaceans that echolocate. It now seems necessary to determine whether any high-frequency sources were being used whenever any new marine mammal strandings occur. We may have been on the wrong track by focusing on low and midfrequencies because those frequencies seem to be the ones to which the marine mammal ear is most robust.

2.4 *Long-Term Monitoring*

In his keynote address, Southall (Chapter 1) called for the long-term passive acoustic monitoring (PAM) of sites where offshore operations were to occur. He referred to this as a "cradle-to-grave" approach. The oil and gas industry has also called for this kind of monitoring. They refer to it as "life-of-the-field," meaning that PAM is used before exploration begins, continues through the operational phase, and ends with structure removal decades later. In their talk, Miller et al. (Chapter 118) appeared to make a start in this direction. He reported on acoustic measurements and modeling of noise budgets at a presumptive wind turbine site. He then modified the model to simulate a 12-dB decrease in the source level (SL) and multiplied the output of a single turbine by the intended number of future turbines to predict the future sound field. This kind of planning is highly desirable.

2.5 *Metrics*

The search for a metric that provides a good index for acoustic effects continues. Sound exposure level (SEL) now seems not to be the silver bullet we were hoping it would be. In his talk and in print (Finneran et al. 2010), Finneran (Chapter 44) showed that the relationship between SEL and TTS breaks down as duration of the exposure increases. Exposures with longer durations will produce a larger TTS. Le Prell (Chapter 43) mentioned kurtosis as a possibly useful metric. Kurtosis was considered by Southall et al. (2007) and is being considered by the Acoustical Society of America (ASA) working group that is writing noise-exposure criteria for fish and turtles. Peak pressure seems to have fallen out of favor as a key metric because it is not a good index to acoustic effects. Also, during a question and answer period, Tony Hawkins reminded the audience that some fish do not sense pressure at all and that exposure criteria in units of particle motion are needed. Casper, Halvorsen, and Popper (Chapter 20) reported that some shark audiograms have been obtained using a shaker to create particle acceleration. But criteria in those units have not yet been attempted. This lack of measurement of particle motion seriously impedes our knowledge of fish hearing and response.

2.6 *New Taxa*

At this conference, four papers were given on hearing and sound production in taxa that were not discussed at Nyborg. Mooney et al. (Chapter 28) presented some interesting and important results on hearing in squid. Staaterman et al. (Chapter 37) discussed sound production in mantis shrimp, which are a component of the bottom infauna. And Piniak et al. (Chapter 18) and Lavender, Bartol, and Bartol (Chapter 19) reported some very welcome data on hearing in turtles. These new directions must be continued in the future if we are to have a comprehensive understanding of the effects of underwater noise.

3 Individual Papers on Noteworthy Subjects

In addition to new themes that emerged since the Nyborg meeting, several individual papers reported new techniques, directions, or findings that seemed noteworthy. This list does not imply importance or merit relative to other papers. It is a subjective list. The papers are discussed in the order in which they were presented.

3.1 Le Prell, Chapter 43

This paper reviewed the subject of oxidative stress and the possible role of vitamins and antioxidants as mitigation for TTS. This subject has not been discussed in any investigation of TTS in marine mammals that the author is aware of. Le Prell cautioned the authors of Southall et al. (2007) not to dismiss TTS as noninjurious because her work showed that robust TTS can progress to PTS. However, "robust" in this context meant ~40 dB of shift. The Southall et al. criteria defined 40 dB of shift as the onset of PTS, not robust TTS, and defined TTS onset as 6 dB of shift. Apparently, there are no data showing whether such a small amount of shift can lead to PTS.

3.2 Erbe, Chapter 108

This paper discussed why it is erroneous to estimate the source level of a distributed source by back-calculation from the far field. This error occurs quite often and greatly overestimates the actual source levels that animals may experience. Also, it called for studies on chronic sound exposure that are badly needed but rarely suggested these days. Several opportunities exist to measure chronic exposure, especially in fish. The same studies could look for a stress response to exposure.

3.3 Carlson, Chapter 51

This talk was an excellent tutorial on how to conduct an injury study in the context of a risk assessment. Carlson first established a panel of tissue injury types, reduced this to numbers, and used them in a rigorous way to assess the risk in animals and to mitigate the cause of the risk, in this case, the blades of hydroturbines. His data led to a redesign of the turbine blades. This did away with the injuries he had studied and coincidentally increased turbine output by ~8%.

3.4 Halvorsen et al., Chapter 52

Halvorsen et al. reported a that uses a wave tube to simulate pile-driving strikes. Their goal is to set science-based injury thresholds for pile driving. The wave tube has much potential for future work. It can produce exposures to sound pressure, particle motion, or any combination of the two. These data will be very helpful to users like Rodkin, Pommerenck, and Reyff (Chapter 124) who called for science-based mitigation measures in pile-driving operations.

3.5 Houser et al., Chapter 59

An experiment is underway that measures the behavioral and physiological responses of naïve dolphins and sea lions to intense sound. None of the animals were conditioned to accept intense exposure as part of a test so their responses will suggest how wild animals might react to unexpected anthropogenic noise. Regulators have been asking for such data for some time, yet no one has produced them.

3.6 *Cranford and Krysl, Chapter 15*

Whole head scans of beaked whales and other toothed whales were used to identify a possible new sound transmission pathway from the environment to the cochlea. The visualization methods included animation of the movements of anatomical features. This approach greatly simplified the understanding of how complex anatomical systems function.

3.7 *Laws, Chapter 107*

This paper presented three-dimensional graphics that portrayed the size and shape of safety zones around an air gun array that would result from applying the Southall et al. (2007) tissue injury criterion. It showed clearly that the criterion expressed in SEL is much more protective of marine mammals than the criterion expressed in peak pressure. It also showed how bottom characteristics can affect the shape of the safety zone. This paper combined modern visualization methods with current exposure criteria to produce images that will aid regulators and users alike.

3.8 *Boyd*

A behavioral-response study on beaked whales has shown that it is very difficult to intentionally deliver an intense acoustic exposure to a submerged marine mammal. The common misconception is that because intense sources are used at sea, marine mammals will receive extreme exposures. But propagation losses reduce extreme exposures to occurring within a fairly small range, meaning that intense exposures are probably less common than might be expected. Also, this study showed that beaked whales are sensitive to many forms of noise, not just sonar, and that their response is to move away slowly underwater for a long distance. Just how this escape response is related to sonar-related whale stranding is still a matter of conjecture.

3.9 *Tougaard et al., Chapter 61*

This paper quantified the distance that harbor porpoises were displaced by a sound source and the length of time they stayed away. The extent and duration of displacement are the simplest, yet perhaps the most important, behavioral responses we can measure at this point given the current state of our knowledge. This study seemed to me exceptional in the simplicity of its methods and the clarity of its results.

3.10 *Løkkeborg et al., Chapter 95*

Løkkeborg reported on an important repeat in 2009 of a study by Engås et al. (1996) on the effects of air guns on commercial fisheries. This has been a contentious issue with fishers for over a decade. The present results showed that fish did not move 18 km away from the source as the earlier study had shown. Instead, they moved around only locally and stopped feeding. Løkkeborg attributed this difference in findings to the greater number of shots that the Engås et al. study produced in a much smaller space than in the present work.

3.11 *Costa, Chapter 96*

This paper reported on field research that attempts to obtain data needed to create numeric transfer functions between cells of the population consequences of acoustic disturbance (PCAD) model (National Research Council 2005). The data describe the flexibility that two different species of pinnipeds (seals) have in a foraging effort, specifically elephant seals that store fat for a year and deliver it to their young as a large bolus in a short period of time versus fur seals that deliver small amounts of fat from local foraging over a long period. Hopefully, numeric transfer functions will give the PCAD model some predictive power. The model is a promising tool in assessing the effects of noise exposure on animals.

3.12 *Regulatory Issues*

Three different regulators (Lewandowski et al. [Bureau of Ocean Energy Management, Regulation and Enforcement], Chapter 128; Tasker [Joint Nature Conservation Committee], Chapter 129; and Sholik-Schlomer [National Oceanic and Atmospheric Administration], Chapter 126) described the process each agency uses to bring science into policy and criteria decisions. Rodkin et al. (Chapter 124), reported how the full regulatory process can adversely affect users and the animals to be protected. The examples he gave suggested that after the policy and criteria decisions were made, regulations were applied that were excessive and misdirected. These four talks followed a paper by Cato (Chapter 116; presented by Robert McCauley) that concluded that physicists and biologists see the world differently and need to work together. Apparently, regulators and users as well as environmental advocates see the world differently. It seems possible that perceptual differences among all parties involved in underwater noise issues may be as difficult to resolve as the science issues that are presently at issue.

4 Future Directions

4.1 *High-Frequency Sources*

Results reported in this conference call for more research on the effects of high-frequency sound sources on marine mammal ears. 1) More data are needed on TTS onset and growth rate within the band of best hearing for selected marine mammal species. 2) New data are needed on the behavioral responses of marine mammals to high-frequency sources. 3) The Southall et al. (2007) noise-exposure criteria and frequency-weighting curves need to be modified in accordance with these new data. 4) In marine mammal mass-stranding events, the output characteristics of all sources being operated should be examined, not just the sources of traditional concern (e.g., sonar, air guns). 5) Depending on the results of the above actions, new regulatory controls on high-frequency sources may be required.

4.2 *Behavioral Response*

Behavioral response seems to have replaced death or direct tissue injury as the major concern about acoustic exposure. However, our community remains undecided about where to begin research because of the following unanswered questions. 1) Which behavioral responses deserve our attention first? Clearly, not all responses are equally important. 2) On which species should we invest our time and resources? Behavioral responses likely do not generalize across species, so choices must be made. 3) How do we observe and measure behavioral response in aquatic animals, especially fish? Until researchers and funding groups together confront these and other questions, our progress will likely remain unsatisfactory.

4.3 Masking

More research is needed on auditory masking, especially using realistic maskers and realistic sounds being masked. More research is needed on the various ways and extents to which animals derive release from masking. Only through this kind of research can we rationally evaluate the importance of masking as an effect of anthropogenic noise and design-appropriate exposure criteria, mitigation measures, and protective regulations.

4.4 Particle Motion

Research is increasing on hearing and the effects of noise on aquatic species that rely on particle motion instead of sound pressure. Funding groups should encourage this trend because without such data, the effects of anthropogenic noise on some fish and invertebrates will remain unknown and exposure criteria will not be possible.

4.5 Long-Term Passive Acoustic Monitoring

The technological means to detect marine mammal species and estimate their population densities using passive acoustic monitoring (Marques et al. 2009) are largely available. It is time to apply them in evaluating the long-term effects of human activities on species composition and abundance. Areas of intense human activity are of greatest concern, as the recent massive oil spill in the Gulf of Mexico has shown.

4.6 Applied Versus Basic Research

Realistically, we can expect to continue receiving new information about the effects of noise on aquatic animals on a piecemeal basis. Most of the research in this field is funded by groups that are concerned about a specific problem or small set of problems. Research funded by navies will continue to focus on sonar issues, industry research will focus on air guns and extraction operations, and research funded by hunting or fishing groups will focus on any source that jeopardizes harvests. These are the realities of modern funding. Ideally, we would have a source of funds for basic research that is not constrained by such boundaries. However, the few agencies that fund true basic research have shown little interest in this field.

References

Clark CW, Ellison WT, Southall BL, Hatch L, Van Parijs SM, Frankel A, Ponirakis D (2009) Acoustic masking in marine ecosystems: Intuitions, analysis, and implication. Mar Ecol Prog Series 395:201–222.

Engås A, Løkkeborg S, Ona E, Soldal AV (1996) Effects of seismic shooting on local abundance and catch rates of cod (*Gadus morhua*) and haddock (*Melanogrammus aeglefinus*). Can J Fish Aquat Sci 53:2238–2249.

Finneran JJ, Carder DA, Schlundt CE, Dear RL (2010) Growth and recovery of temporary threshold shift at 3 kHz in bottlenose dolphins: Experimental data and mathematical models. J Acoust Soc Am 127:3256–3266.

Marques, TA, Thomas L, Ward J, DiMarzio N, Tyack PL (2009) Estimating cetacean population density using fixed passive acoustic sensors: An example with Blainville's beaked whales. J Acoust Soc Am 125:1982–1994.

National Research Council (2005) Marine mammal populations and ocean noise: Determining when noise causes biologically significant effects. National Academies Press, Washington, DC.

Southall BL, Bowles AE, Ellison WT, Finneran JJ, Gentry RL, Greene CR Jr, Kastak D, Ketten DR, Miller JH, Nachtigall PE, Richardson WJ, Thomas JA, Tyack PL (2007) Marine mammal noise exposure criteria: Initial scientific recommendations. Aquat Mamm 33:411–522.

US Department of the Navy (2007) Surveillance towed array sensor system-low frequency active (SURTASS-LFA) phase II. Available via http://www.surtass-lfa-eis.com/Research/index.htm.

ERRATUM

Whistles of Bottlenose Dolphins: Group Repertoires and Geographic Variations in Brazilian Waters

Lilian S. Hoffmann, Elton Ferlin, Pedro F. Fruet, Rodrigo C. Genovês, Fernanda P. Valdez, Juliana Di Tullio, Glauco Caon, and Thales R. Freitas

A.N. Popper and A. Hawkins (eds.), *The Effects of Noise on Aquatic Life*,
Advances in Experimental Medicine and Biology 730, DOI 10.1007/978-1-4419-7311-5, pp. 141–144,

---

DOI 10.1007/978-1-4419-7311-5_154

In the chapter "Whistles of Bottlenose Dolphins: Group Repertoires and Geographic Variations in Brazilian Waters" (DOI 10.1007/978-1-4419-7311-5_31), in *The Effects of Noise on Aquatic Life*, the name of the corresponding author Lilian S. Hoffmann was incorrectly represented as Lisa S. Hoffmann. The author's name is correct as it appears below.

Lilian S. Hoffmann
Universidade Federal do Rio Grande do Sul
Porto Alegre, RS
CEP 90050-170 Brazil
e-mail: liliansander@gmail.com

The online version of the original chapter can be found at
http://dx.doi.org/10.1007/978-1-4419-7311-5_31

Index

A.N. Popper and A. Hawkins (eds.), *The Effects of Noise on Aquatic Life*,
Advances in Experimental Medicine and Biology 730, DOI 10.1007/978-1-4419-7311-5,

C

D

Printed by Publishers' Graphics LLC USA
SRO20120410-01
2012